New Computational Paradigms

❖❖ New ❖❖❖ Computational Paradigms

Changing Conceptions of What is Computable

S. Barry Cooper

Benedikt Löwe

Andrea Sorbi

Editors

 Springer

S. Barry Cooper
Department of Pure Mathematics
University of Leeds
Leeds LS2 9JT
United Kingdom
pmt6sbc@leeds.ac.uk

Benedikt Löwe
Institute for Logic, Language and
    Computation
University of Amsterdam
Plantage Muidergracht 24
1018 TV Amsterdam
The Netherlands
bloewe@science.uva.nl

Andrea Sorbi
Department of Mathematics and
    Computer Science "Roberto Magari"
Università di Siena
Pian dei Mantellini 44
53100 Siena
Italy
sorbi@unisi.it

ISBN-13: 978-1-4419-2263-2   e-ISBN-13: 978-0-387-68546-5
DOI: 10.1007/978-0-387-68546-5

Mathematics Subject Classification (2000): 68Q01, 68Q05

Cover illustration: Created by Sven Geier of the California Institute of Technology. The image, an example of fractal art, is entitled "Deep Dive."

Printed on acid-free paper

9 8 7 6 5 4 3 2 1

springer.com

Contents

Preface

June 2005 saw the coming together, in Amsterdam, of the first meeting of a new research community, which sought to renew, in the new century, the ground-breaking legacy of Alan Turing. *Computability in Europe* (CiE) originated with a 2003 proposal for EU funding but rapidly developed into a Europe-wide network of over 400 researchers from 17 countries, around 70 institutions, and a number of different research disciplines (mathematics, computer science, physics, biology, philosophy, and logic). This book of invited (and rigorously refereed) articles showcases the diversity, excitement, and scientific innovation of that first meeting, and the powerful multidisciplinarity that it injected into computational research.

Many of the contributions to be found here reflect the necessity to deal with computability in the real world—computing on continuous data, biological computing, physical computing, etc.—which has focused attention on new paradigms of computation, based on biological and physical models. This book looks at new developments in the theory and practice of computation from a mathematical and predominantly logical perspective, with topics ranging from classical computability to complexity, biocomputing, and quantum computing. Traditional topics in computability theory are also covered as well as relationships among proof theory, computability and complexity theory, and new paradigms of computation arising from biology and quantum physics and issues related to computability with/on the real numbers. The book is addressed to researchers and graduate students in mathematics, philosophy, and computer science with a special interest in foundational issues. Logicians and theoretical physicists will also benefit from this book.

Since that first conference, CiE has become more than the sum of its parts, reasserting an older tradition of scientific research. This more thoughtful approach is what this 1944 quotation from Einstein[1] seems to refer to:

---

[1] A. Einstein to R. A. Thornton, unpublished letter dated Dec. 7, 1944; in Einstein Archive, Hebrew University, Jerusalem.

So many people today—and even professional scientists—seem to me like someone who has seen thousands of trees but has never seen a forest. A knowledge of the historical and philosophical background gives that kind of independence from prejudices of his generation from which most scientists are suffering. This independence created by philosophical insight is— in my opinion—the mark of distinction between a mere artisan or specialist and a real seeker after truth.

There is a parallel between the competitive hyperactive specialism of parts of computer science (and logic) and that of the string theory community that Lee Smolin[2] focuses on in his recent book. He pinpoints:

... a more brash, aggressive, and competitive atmosphere, in which theorists vie to respond quickly to new developments ... and are distrustful of philosophical issues. This style supplanted the more reflective, philosophical style that characterized Einstein and the inventors of quantum theory, and it triumphed as the center of science moved to America and the intellectual focus moved from the exploration of fundamental new theories to their application.

This book embodies what is special about what CiE is trying to do in taking computational research beyond the constraints of "normal science," while building a cohesive research community around fundamental issues of computability.

Leeds   *S. Barry Cooper*
Amsterdam   *Benedikt Löwe*
Siena   *Andrea Sorbi*
July 2007

---

[2] L. Smolin, *The Trouble With Physics: The Rise of String Theory, the Fall of a Science and What Comes Next*, Houghton Mifflin, 2006.

List of Contributors

Serikzhan Badaev
Kazakh National University, Almaty
050038, Kazakhstan
badaev@kazsu.kz

Johan van Benthem
Institute for Logic, Language &
Computation (ILLC), University of
Amsterdam, Amsterdam, 1018 TV, the
Netherlands
johan@science.uva.nl
and
Department of Philosophy, Stanford
University, Stanford, CA 94305,
U.S.A.
johan@csli.stanford.edu

Olivier Bournez
INRIA Lorraine and LORIA (UMR
7503 CNRS-INPL-INRIA-Nancy2-
UHP), BP239 54506 Vandœuvre-Lès-
Nancy, France
Olivier.Bournez@loria.fr

Vasco Brattka
Laboratory of Foundational Aspects
of Computer Science, Department of
Mathematics and Applied Mathematics,
University of Cape Town, Rondebosch
7701, South Africa
Vasco.Brattka@uct.ac.za

Samuel R. Buss
Department of Mathematics, University
of California, San Diego, La Jolla, CA
92093-0112, U.S.A.
sbuss@math.ucsd.edu

Manuel L. Campagnolo
DM/ISA, Technical University of
Lisbon, Tapada da Ajuda, 1349-017
Lisboa, Portugal
and
SQIG/IT Lisboa
mlc@math.isa.utl.pt

Abbas Edalat
Department of Computing, Imperial
College London, London SW7 2AZ,
United Kingdom
ae@doc.ic.ac.uk

Sergey Goncharov
Institute of Mathematics of Siberian
Branch of Russian Academy of
Sciences, Novosibirsk 6300090,
Russia
gonchar@math.nsc.ru

Joel David Hamkins
The College of Staten Island of The City
University of New York, Mathematics,
Staten Island, NY 10314, U.S.A.
and

The Graduate Center of The City
University of New York, Ph.D. Program
in Mathematics, New York, NY 10016,
U.S.A.
jhamkins@gc.cuny.edu
http://jdh.hamkins.org

Peter Hertling
Institut für Theoretische Informatik und
Mathematik, Fakultät für Informatik,
Universität der Bundeswehr München,
85577 Neubiberg, Germany
Peter.Hertling@unibw.de

Andrew Hodges
Wadham College, University of
Oxford, Oxford OX1 3PN, United
Kingdom
andrew.hodges@wadh.ox.ac.uk
http://www.turing.org.uk

Ulrich Kohlenbach
Department of Mathematics, Technische Universität Darmstad, D-64289
Darmstadt, Germany
kohlenbach@mathematik.tu
-darmstadt.de
http://www.mathematik.
tu-darmstadt.de/
~kohlenbach

Yuri Matiyasevich
Steklov Institute of Mathematics,
St. Petersburg 191023, Russia
yumat@pdmi.ras.ru
http://logic.pdmi.ras.ru/
~yumat

Elvira Mayordomo
Departamento de Informática e
Ingeniería de Sistemas, Universi-
dad de Zaragoza, 50018 Zaragoza,
Spain
elvira@unizar.es

Russell Miller
Queens College of The City University
of New York, Mathematics, Flushing,
New York 11367, U.S.A.
and
The Graduate Center of The City
University of New York, Ph.D. Program
in Computer Science, New York, NY
10016, U.S.A.
Russell.Miller@qc.cuny.edu

Yiannis N. Moschovakis
Department of Mathematics, University of California, Los Angeles, CA
90095-1555, U.S.A.
and
Graduate Program in Logic, Algorithms
and Computation ($M\Pi\Lambda A$), University
of Athens, Panepistimioupolis, 15784
Zografou, Athens, Greece
ynm@math.ucla.edu

Dag Normann
Department of Mathematics, The
University of Oslo, Blindern N-0316,
Oslo, Norway
dnormann@math.uio.no

Dana Pardubská
Department of Computer Science,
Comenius University, 842 48 Bratislava,
Slovakia
pardubska@fmph.uniba.sk

Vasilis Paschalis
Graduate Program in Logic, Algorithms
and Computation ($M\Pi\Lambda A$), University
of Athens, Panepistimioupolis, 15784
Zografou, Athens, Greece
pasva@yahoo.com

Gheorghe Păun
Institute of Mathematics of the Ro-
manian Academy, 014700 Bucureşti,
Romania

and
Research Group on Natural Computing,
Department of Computer Science
and Artificial Intelligence, University of Sevilla Avda, 41012 Sevilla,
Spain
george.paun@imar.ro
gpaun@us.es

Michael Rathjen
Department of Pure Mathematics,
University of Leeds, Leeds LS2 9JT,
United Kingdom
and
Department of Mathematics, Ohio
State University, Columbus, OH 43210,
U.S.A.
rathjen@math.ohio-state.edu

Grzegorz Rozenberg
Department of Computer Science,
University of Colorado at Boulder
Boulder, CO 80309, U.S.A.
and
Leiden Institute of Advanced Com-
puter Science (LIACS), Leiden
University, 2300 RA Leiden, the
Netherlands
rozenber@liacs.nl

Helmut Schwichtenberg
Mathematisches Institut der Universität München, D-80333 München,
Germany
schwicht@math.lmu.de

Daniel Seabold
Department of Mathematics, Hofstra University, Hempstead, NY
11549-1030, U.S.A.
matdes@hofstra.edu

Wilfried Sieg
Department of Philosophy, Carnegie
Mellon University, Pittsburgh, PA
15213, U.S.A.
sieg@cmu.edu

Viggo Stoltenberg-Hansen
Department of Mathematics, Uppsala University, S-75106 Uppsala,
Sweden
viggo@math.uu.se

John V. Tucker
Department of Computer Science,
University of Wales Swansea, SA2 8PP,
Wales
J.V.Tucker@swansea.ac.uk

Steve Warner
Department of Mathematics, Hofstra University, Hempstead, NY
11549-1030, U.S.A.
matsjw@hofstra.edu

Klaus Wehirauch
Faculty of Mathematics and Computer
Science, University of Hagen, 58084
Hagen, Germany
Klaus.Weihrauch@fernuni-hagen.de

Jiří Wiedermann
Institute of Computer Science, Academy
of Sciences of the Czech Republic, 182
07 Prague 8, Czech Republic
jiri.wiedermann@cs.cas.cz

Alan Turing's Legacy
and New Computational Paradigms

Alan Turing, Logical and Physical

Andrew Hodges

Wadham College, University of Oxford, Oxford OX1 3PN, United Kingdom
andrew.hodges@wadh.ox.ac.uk
http://www.turing.org.uk

Summary. Alan M. Turing (1912–1954), the founder of computability theory, is generally considered a pure logician. But his ideas involved the practical and physical implementation of logical structure, particularly concerned with the relationship between discrete and continuous, and his scientific work both began and ended in theoretical physics.

1 Delilah day

We shall begin in the middle, and with an Alan Turing who may not be familiar to readers of this volume. 'It is thought,' he wrote, 'that ... a very high degree of security indeed can be obtained. There is certainly no comparison in security with any other scrambler of less than ten times the weight. For tank-to-tank and plane-to-plane work, a rather less ambitious form of key will probably be adequate. Such a key unit might be of about the same size as the combining unit.' The day was no ordinary day: it was 6 June 1944, and the system was no ordinary system; it was the 'Delilah' real-time speech and fax scrambler, devised and developed by Turing with the assistance of one engineer, Donald Bayley. According to this report (Turing 1944), the work had begun in early May 1943. It was conducted at the British Secret Service station at Hanslope Park, Buckinghamshire.

Turing's speech cryptosystem illustrates many features typical of his work but not widely known. (1) Speech secrecy characterized not only the content, but the fate, of this work. This 'top secret' report emerged only in 2004. Further work by Turing may yet be found to shed fresh light on his computational proposals. (2) This particular project was marked by Turing's ambition to combine both the *logical* demand of cryptographic security with the *physical* constraint of a small mobile unit for operational use. (3) The system combined both *discrete* and *continuous* elements, for the sampling and the key was discrete, whilst the modular addition of signal amplitudes was achieved by analogue electronic components. (4) The comparison with 'any other scrambler' reflects the outcome of his top-level mission to the United States at the end of 1942, which gave him uniquely privileged access to advanced American systems—so secret that he was not allowed to speak about it to British colleagues. It was typical for him, on his return, to seek to outdo Bell Telephone Laboratories with his single brain, and to build a better system with his own hands. This ambition also served as the learning curve in electronics he needed in 1945 for his computer design. (5) Turing's wording indicates authoritative judgment, and not the submitting of a proposal for the approval of superiors. In another report (Turing 1942), he commented of his U.S. Navy counterparts that: 'I am persuaded that one cannot very well trust these people where a matter of judgement in cryptography is concerned.' Those judgements ranged extensively over questions of what would now be called software and hardware: 'I think we can make quite a lot of use of their machinery.'

I have introduced Turing in this unconventional form in order to shed a different light on the man whom many readers of this volume will think of as a pure logician, who founded computability theory and modern computer science through attacking Hilbert's Entscheidungsproblem in 1935–1936. As Newman (1955) put it, Turing was 'at heart more an applied than a pure mathematician'. It might be more true to say that Turing had resisted this Cambridge classification from the outset. He attacked every kind of problem—from arguing with Wittgenstein, to the characteristics of electronic components, to the petals of a daisy. He did so on the basis of immense confidence in the power of mathematical analysis, in whatever field he chose. But Newman's comment gives a correct impression that Turing began (and ended) with the physical world.

Turing's first known serious study, at the age of 16, was Einstein's own semi-popular text of *The Theory of Relativity*, a present from his grandfather. His next encounter was with Eddington's *The Nature of the Physical World,* from which he learnt about quantum mechanics, then fresh and new. He was already research-minded, somehow understanding more of general relativity than Einstein's book explained, and seeing the problem posed by Schrödinger's configuration space. His first reading at the level of contemporary research was of von Neumann's *Mathematische Grundlagen den Quantenmechanik* in 1933. His comment, 'very interesting, and not at all difficult reading, although the applied mathematicians seem to find it rather strong,' conveys not merely that he thought the applied mathematicians a little weak. It indicates his zest for serious new mathematics with which to study the nature of the physical world. Such was the start, and it was only two years later that Turing had the idea of the Turing machine as a definition of effective calculability.

2 The shock of the new

Turing was stimulated by Newman's 1935 lectures, which culminated in the question of the Entscheidungsproblem. By Spring 1936, he had a fully worked out theory of Turing machines, with which to give a fully satisfactory and negative answer to Hilbert. His own famous opening (Turing 1936), of which Newman was perhaps the first reader, was: 'a number is computable if its decimal can be written down by a machine.' Thus Turing introduced physical ideas into logic. But Turing had worked in ignorance of Church's definition of effective calculability in terms of the lambda-calculus. Turing had to add an appendix demonstrating the mathematical equivalence of his definition, and this delayed publication until the end of 1936.

Newman (1955) wrote that 'it is difficult today to realize how bold an innovation it was to introduce talk about paper tapes and patterns punched in them, into discussions of the foundations of mathematics', and the image was indeed a dose of Modern Times. Yet Church's review of *On computable numbers* (Church 1937) did not express anything other than welcome, even though that reviewer had the reputation of being cautious to the point of pedantry. It was generous in spirit despite the fact that it must have been disconcerting for Church that a young unknown, a complete outsider, had given a more satisfactorily direct and 'intuitive' account of effective calculation than the lambda-calculus. Moreover, Church was actually bolder in his physical imagery than Turing was:

> The author [Turing] proposes as a criterion that an infinite sequence of digits 0 and 1 be "computable" that it shall be possible to devise a computing machine, occupying a finite space and with working parts of finite size, which will write down the sequence to any number of terms if allowed to run for a sufficiently long time. As a matter of convenience, certain further restrictions are imposed in the character of the machine, but these are of such a nature as obviously to cause no loss of generality — in particular, a human calculator, provided with pencil and paper and explicit instructions, can be regarded as a kind of Turing machine.

In a later sentence (in the review of Post's work, immediately following) Church referred to Turing's concept as computability by an *arbitrary machine,* subject only to such finiteness conditions.

It is an interesting question as to what Turing, who had started work in mathematical physics, thought were the physical connotations of his 1936 work. Firstly, there is the question of the building of working Turing machines, and the universal machine in particular. Newman, and Turing's friend David Champernowne, later attested to discussions of it even at that time. But no written material has reached us. It is certainly hard to see how Turing could have failed to see that the atomic machine operations could be implemented with the sort of technology used in automatic telephone exchanges and teleprinters. The second, more difficult, question is what Turing thought the structure and limitations of computability had to say about the physical world.

Turing certainly brought an idea of physical action into the picture of computation, which was different from Church's lambda-calculus. But his thorough analysis was of the human calculator, with arguments for finiteness based not on physical space and size, but on human memory and states of mind. It is strange that Church did not simply quote this model in his review, but instead portrayed the human calculator as a particular case of an apparently more general finite 'machine' with 'working parts'. Nowadays his assertion about what can be computed by an 'arbitrary machine', emphasing its generality, and characterizing it in terms of space and size, reads more like the 'physical Church-Turing thesis'. If Church was trying to dispel the idea that computability had something to do with the scope of physical machines, his words were singularly ill-chosen for that purpose. But it seems that no one in the early era, including Turing, thought out and analysed the distinction between the rule-based human worker and a more general physical mechanism.

In his post-war writing, Turing made free use of the word 'machine' for describing mechanical processes and made no attempt to alert his readers to any distinction between human worker-to-rule and physical system—a distinction that, nowadays, would be considered important. Thus Turing (1948) referred to written-out programs, for a human to follow, as 'paper machines'. The imagery is of a human acting out the part of a machine. Indeed, he stated that any calculating machine could be imitated by a human computer, again the reverse of the 1936 image. He referred often to the rote-working human calculator as a model for the way a computer worked and a guide as to what it could be made to do in practice. But he also referred to the advantage of the universal machine being that it could replace the 'engineering' of special-purpose machines. Most importantly, he appealed to the idea of simulating the brain as a physical system. So in later years Turing readily appealed to general ideas of physical mechanisms when discussing the scope of computability. But as for what he had in mind in 1936, we cannot know.

A salient factor in this blank is that Turing left no notes or precursor papers to explain the development of his ideas. His 1934 work on the Central Limit Theorem (stimulated by a lecture of Eddington), though original, gave no indication of the dramatic work he would do a year later. But there is one small point that suggests that he probably had a pre-existing technical interest in constructive analysis of the real numbers, and for a curious reason.

Turing's fellow mathematics scholar at King's College, and probably his closest friend, was David Champernowne, later a distinguished economist. In 1933, Champernowne published a short paper while still a second-year student—surely a competitive spur to Turing. He made a contribution to the theory of normal numbers, a subject springing from measure theory. A normal number is one whose digits are uniformly distributed in any number base. It was known that almost all numbers are normal, and yet no one had constructed a specific example. Champernowne (1933) observed and proved that $0.123456789101112\ldots$ is normal in the decimal scale. Turing took up his friend's subject and worked on generalizing this construction so as to satisfy the full criterion of normality. We know this because his notes were handwritten on the reverse side of six pages of the typescript of *On Computable Numbers,* as illustrated in (Hodges 2006). This shows the date of Turing's work (which was never published, and according to the commentary of Britton (1992) was flawed) to be 1936 or later. But given that the impetus came in 1933, it seems very possible that his ideas about constructive procedures for defining real numbers germinated before the stimulus from Newman's lectures in 1935.

Turing's note on normal numbers (Turing 1936?) uses the terms 'mechanical process' and 'constructive enumeration', which show the closeness to the ideas of *On Computable Numbers*. Indeed these notes were written on the reverse sides of pages of the typescript including those on 'computable convergence'. This is the section he must have hoped to expand into a new constructive treatment of analysis, judging by the hostage to fortune he left in the introduction saying that he would 'soon' give a further paper on the computable definition of real functions. The difficulty posed by the non-unique decimal representation of the reals seems to have stopped this project in its tracks. Although Turing's method in his correction note (Turing 1937) was the first step in modern computable analysis, Turing himself never followed up this lead. Nevertheless, he was always mindful of continuous mathematics, which entered into much of his mathematical work.

Passing to Turing's work in logic after 1936, readers of this volume will be familiar with his ordinal logics (Turing 1939), with new ideas now known as relative computability and Turing degrees. The question I will address, however, is that of what connection the logical had with the physical. Feferman (2001) sees the work, which was Turing's 1938 Ph.D. thesis at Princeton, as being a diversion into purely mathematical logic. But Newman (1955) emphasised the interpretation of Turing's work in terms of the mental 'intuition' required to see the truth of a formally unprovable Gödel sentence. I would therefore see Turing as to some extent trying to follow his 1936 analysis of a mind carrying out a mechanical process, with an analysis of a mind when not so restricted.

But if Turing was seriously thinking about mental operations, we may ask what Turing thought the brain was doing when it 'saw the truth' in a manner that could not be modelled computably. In 1936–1939 Turing wrote nothing concerning the physical brain, and this question may seem too fanciful to consider. But it is not, for we know he brought to the 1930s an older fascination with that fundamental problem of science, the conflict of free will and physical determinism. His juvenile but highly serious and deeply felt essay 'Nature of Spirit' (Turing c. 1932), influenced by Eddington's views, held that the brain was governed by quantum-mechanical physics, and that one could thereby rescue human free will from Laplacian predictability. Did these youthful thoughts influence his ideas of intuition in 1938, or had he long since dropped them? We can only conjecture.

In contrast, we do know that in 1937–1939 Turing's yen for physical engineering was strong, even though he was very busy and prolific with algebra and analysis. His zeta-function machine involved cutting cogwheels to effect an approximation to a Fourier series; it involved both discrete and continuous elements. His relay multiplier was closer to the Turing machine, being a discrete system to effect binary multiplication. And this eccentricity, designed with a cipher system in mind, certainly paid off.

On 1 November 1939, the Bletchley Park cryptanalysts (Knox et al. 1939) reported on the attack on naval Enigma messages to be made by 'the machine now being made at Letchworth, resembling, but far larger than the Bombe of the Poles (superbombe machine).' It would be hard to overstate the significance of this machine to the course of the Second World War, or the miracle of Turing, a logician with a penchant for practical implementation, being in exactly the right place at the right time—at least from the British point of view. In 1948 Turing wrote of how machinery was generally considered as limited to straightforward, repetitive tasks until 'about 1940', and it can hardly be doubted that by citing that date he was alluding to the impact of the 'superbombe'—the Turing-Welchman Bombe with its stunning parallel logic for solving the plugboard complication of the Enigma enciphering machine used for German military communications. From that time onwards, Turing led the way at Bletchley Park in showing the power of algorithms. Some of these were implemented on physical machines, others through people carrying out processes such as the 'Banburismus' that exploited his new Bayesian inference methods (Good 1992, Good 1993, Good 2001). Turing's mechanization of judgment through quantified 'weight of evidence' (essentially the same as Shannon's measure) was perhaps a key step in persuading him of the scope for the mechanization of human intelligence. In 1943, he discussed such ideas with Shannon, and in 1944 he was speaking more concretely of 'building a brain' to Donald Bayley his electronic assistant.

It is striking to see those words 'purely mechanical', as used by Turing in 1938 in his formal account of computability, now being used to describe operations that seemed of magical power, as in this description of the electronic Colossus—and, of course, the statistical theory it used:

> It is regretted that it is not possible to give an adequate idea of the fascination of a Colossus at work: its sheer bulk and apparent complexity; the fantastic speed of thin paper tape round the glittering pulleys... the wizardry of purely mechanical decoding letter by letter (one novice thought she was being hoaxed); the uncanny action of the typewriter in printing the correct scores without and beyond human aid . . . (Anon 1945)

As with his contemporaries in the whole 'cybernetic' movement, wartime importance and success imbued Turing in the late 1940s with a sense of the power of science and engineering. There is an obvious difference between his pre-war and post-war discussion of computing. Whilst in 1936 he had limited the discussion to that of a human being carrying out a mechanical process, from 1945 onwards he took the view that all mental processes fell within the scope of computable operations, famously getting to sonnet-appreciation in (Turing 1950). Turing's use of the word 'brain' after 1943 indicates that he had by then decided that the scope of the computable could be broadened. The ground had already been laid in 1936, by the bold argument that a process need not actually be written down in 'instruction notes', but could be embodied in 'states of mind'. But Turing's post-war views still marked a definite shift. For they implied a definite rejection of the view that uncomputable 'intuitive' steps are taken by the human mind when it recognises true but formally unprovable Gödel statements.

One sentence in Turing's proposal for the ACE computer (Turing 1946) is evidence of this shift: it is a claim that a computer might play 'very good chess' if 'occasional serious mistakes' were allowed. This cryptic reference to mistakes was amplified and clarified by the argument of (Turing 1947) that once the possibility of mistakes is allowed, Gödel's theorem becomes irrelevant. According to this argument, true 'seeing of truth' is not actually expected of people, but only a good attempt. As Turing put it, on the principle of 'Fair Play for Machines', the same standard should be applied to a computer and its mistakes likewise forgiven. Many people see this 'mistakes' argument as vital to Artificial Intelligence now—e.g., from different sides, Davis (Davis 2000) and Penrose (Penrose 1994). It was essential for Turing's developing machine-intelligence ideas to find a reason to reject the Gödelian view that truth-seeing intuition is an actual uncomputable step taken by the mind. (It is no surprise that Gödel later objected to Turing's theory of the mind, although he had originally endorsed Turing's definition of computability.) Turing's use of the word 'brain'—as opposed to the 'states of mind' in 1936—also indicates his renewed interest in its *physical* nature.

In 1945–1946, however, Turing was busy on the physical side of the new computer, the practical universal computing machine as he called it. The Delilah gave him hands-on practice in electronics that he extended into studying the physics of delay lines. His ACE report gave detailed accounts of possible physical storage mechanisms. His vision for future software—the vision of coding all known processes, and allowing the machine itself to take over all routine aspects of the coding—was more important than hardware. But this too showed interest from the start in applications to continuous mathematics, to which we now turn.

3 Everything is really continuous

Turing is best known to the public for the Turing Test: its drama and wit, rules, and role-play have fascinated a wide audience. But behind this lay the very much less glamorous idea of the discrete machine model, to which Turing devoted at least as much attention. The theory was prepared in (Turing 1948) with his definition of 'controlling' as opposed to 'active' machinery. Turing pointed out that it is the information-theoretic or logical structure that we are interested in, not the particular form of matter and physical action that implements it. In his radio broadcast (Turing 1952) he gave the famous line, 'We are not interested in the fact that the brain has the consistency of cold porridge. We don't want to say "This machine's quite hard, so it isn't a brain, so it can't think."' No one has ever claimed that a universal Turing machine could take over the functions of an 'active' machine, of which Turing's example was 'a bulldozer'.

Turing then classified the machines he was interested in as either discrete or continuous. It should be noted that Turing's examples of 'machines' were physical objects. It is also a non-trivial point that he made no mention of discrete machines calling on uncomputable 'oracles', although if such theoretical constructs had been on his mind, he had every opportunity to define them and classify them as a type of machine in this paper. On the contrary, Turing dismissed what he called the Mathematical Argument against machine intelligence, i.e., the argument from the existence of algorithmically unsolvable problems. Those who claim to detect references to uncomputable functions in his major 1948 and 1950 papers have to explain why Turing so perversely threw readers off the scent by giving short shrift to the significance of uncomputability, whilst emphasising the computable operations of a digital computer. It is now a popular idea to see an infinite random sequence as a source of uncomputable data. When Turing discussed random input, however, he gave the most cursory description of what this implied, saying that the (pseudo-random, computable) digits of π would do, or perhaps a 'roulette wheel'—i.e., a physical system with sensitive dependence on initial data.

Similarly, having sketched his various networks of logical gates, Turing stated the strategy of trying them out on a computer. (Oddly, it appears that he never did so when the Manchester computer became available to him, but Teuscher (2000) has performed a thorough investigation.) Turing stressed that the point of these new ideas was to gain insight into the apparently non-mechanical aspects of mind. Had he wished to associate these new ideas with the uncomputable 'intuition' of 1938, he could have said so; instead, he placed them firmly within the realm of what could be done with computers, by the modification of programs through 'learning' processes. He emphasised the surprising power of the computable, and indeed of the entirely finite. The infinitude that comes into Turing's discussion in 1950 is the theoretical infinitude of a Turing machine tape, i.e., all within the scope of computability.[1]

When Turing defined *continuous* machines, was he trying to escape from this prison of computability? Not at all, for his central interest was a thesis about the brain, claiming it as a continuous machine that can effectively be regarded as discrete, and then simulated by a computer.

---

[1] The description of digital computers in (Turing 1950) seems to have confused readers such as Copeland (2000), who places great weight on an apparent allusion to machines with infinitely many states. Turing focusses on a practical, finite, computer and considers all its storage as internal to the machine. It has no external 'tape'. This simplified semi-popular account makes it awkward for Turing to indicate the full scope of computability, in which an unlimited external tape is essential. Turing does it by discussing a 'theoretical' extension to 'unlimited store', carefully explaining that only a finite amount can be used at any time. As this store clearly includes the unlimited *tape* of the standard Turing machine description, his reference to a 'theoretical' computer with 'infinite' store is all within the scope of computability. Turing's reference to machines with 'finitely many states' is a reference to those Turing machines that only use a finite portion of tape, i.e., totally finite machines. Indeed, his emphasis was on the capacity of such finite machines, confirmed by his (low) numerical estimate of the number of bits of storage required. See also Hodges (2006).

It is worth noting that Turing's several references to everything being 'really' continuous were based on considerable experience with continuous analysis and mathematical physics. His example of a continuous machine was 'a telephone', but few of his readers would have guessed at the sophistication of the relationship between discrete and continuous that had gone into his secret Delilah work. His mathematical and practical work was full of such relationships. In (Turing 1947) he made a start on a more abstract account of the relationship by describing discrete states as disjoint topological regions in the continuous configuration space of a physical system. This 1947 talk is also important in that Turing opened by explaining the new digital computer as *superior* to the analogue differential analyser, on the principle that greater accuracy, without limit, could always be achieved by more software. This was a major strategic question, physicists such as Hartree having devoted much effort to the construction of differential analysers, and Turing had to know what he was talking about. Turing also chose the example of solving an ordinary differential equation to illustrate to mathematicians what a digital computer could do. In this as in so many other examples, the application of computable processes to the modelling of the physical world was uppermost in his thought.

But the physical brain was his real centre of interest. In 1948 he classified it as probably 'Continuous Controlling, but is very similar to much discrete machinery', and that 'there seems every reason to believe' that brains could effectively be treated as discrete. One such reason was stated in (Turing 1950) in a section on 'the Argument from Continuity in the Nervous System'. It is worth noting that this is an argument against the thesis that the brain, as a physical system, can be modelled by a digital computer. Thus Turing's answer to this argument gives his reasons for saying that the brain *could* in principle be so simulated. The answer he gave is perhaps not what readers would have expected. For it was not an argument from the discrete neurons and their gate-like behaviour. Instead, Turing drew attention to a general feature of non-trivial continuous systems:

> The displacement of a single electron by a billionth of a centimetre at one moment might make the difference between a man being killed by an avalanche a year later, or escaping. It is an essential property of the mechanical systems which we have called 'discrete state machines' that this phenomenon does not occur.

Although Poincaré had long before pointed out what is now called the 'butterfly effect' of sensitive dependence on initial conditions, this breakdown of effective predictability was not then as well known as it is now. In this discussion of brain simulation Turing has apparently assumed as obvious the general idea of discrete approximation through finite-difference methods, as he had explained in 1947, and jumped immediately to discuss this more advanced problem that very few in 1950 would have thought of posing—that for the brain 'a small error in the information about the size of a nervous impulse impinging on a neuron, may make a large difference to the size of the outgoing impulse.' (It was, perhaps, really *his own* objection that he was answering.) Turing's argument was that the details of such chaotic effects are of no functional significance, so that one could simulate their effect within a discrete state machine just by random choices.[2]

Nowadays one would see Turing's treatment of continuous dynamical systems and randomness as inadequate; just a shadow of future investigations in computable analysis, analogue computing, and the intricate analysis of chaotic phenomena. But it does clearly illustrate the way he related the concept of computability to very general physical mechanisms.

4 Back to the nature of the physical world

Turing's work in morphogenesis, which got going after 1950 with what was one of the first examples of computer use in serious scientific research, might perhaps have led him to study chaotic phenomena, if he had wished to follow that route. It seems, however, that his interest remained entirely on the special stable solutions relating to biological structures that emerged by numerical methods from his partial differential equations. But the germ of another physical idea was developing. It was enunciated in a BBC radio talk on 'Can a digital computer think?' (Turing 1951). This was mainly a précis of his 1950 paper, though this time with a completely explicit programme of simulating the physical brain on a computer. Turing explained the principle that a universal machine could simulate any machine but added a new note: this would only be possible for machines:

> of the sort whose behaviour is in principle predictable by calculation. We certainly do not know how any such calculation should be done, and it has even been argued by Sir Arthur Eddington that on account of the Indeterminacy Principle in Quantum Mechanics no such prediction is even theoretically possible.

It was Copeland (Copeland 1999) who drew attention to the seriousness of this suggestion of Turing that quantum-mechanical uncertainty creates a problem for Turing machine simulation. His sentence certainly marks a change of view from 1950, because then Turing had answered the 'continuity of the nervous system' argument without mentioning this deeper problem. It indicates a new quest for a more fundamental analysis. Turing only wrote this one sentence on the relationship of computation to quantum mechanics, so we know no more of when or why he had decided that it now had to be taken more seriously. It should be noted that Turing does not actually assert a view of his own but attributes a view to Eddington; indeed the words 'it has even been argued' might suggest that he saw it as a rather fanciful suggestion of Eddington's. Nevertheless it seems that he did take it seriously as a further argument against machine intelligence.

---

[2] Penrose (1989, p. 173) has a similar conclusion that chaotic effects are of no 'use' to the brain as regards its function for thought or intelligence.

The reason for thinking this is that in 1953—regardless of all his problems with criminal trial, punishment, and surveillance—Turing tried to formulate 'a new quantum mechanics', suggesting there was something wrong with the standard axioms he had learned from von Neumann in 1933. Moreover, he focussed on the principle of wave-function 'reduction' that underlies the indeterminacy or uncertainty principle. In his last months he came up with the 'Turing Paradox' (Gandy 1954)—that the standard principles of quantum mechanics imply that in the limit of continuous observation a quantum system cannot evolve. The few remarks reported by Gandy suggest that he was trying to make his quantum mechanics more finitistic, so it seems very probable that he was hoping to defeat this 'Eddington argument', but we cannot tell where his ideas might have led if they had not been cut off by death.

It is striking, given Eddington's influence on his youthful ideas, that Turing returned to this source. It is a puzzle as to why he had not done so before, for instance in the 1948 discussion where he examined thermodynamic and other physical constraints on computation. This is yet another unanswerable question. It is striking also that the argument Turing attributed to Eddington has become, since the 1980s, the Penrose argument. Penrose has located in wave-function reduction a place where there is room for some unknown physical law that cannot be simulated by a computable process and says that this could underlie the physical function of the brain. This is rather like a serious development of the view Turing took in his youthful essay of 1932, which perhaps Turing considered afresh when pondering his 'paradox'.

Turing's 'paradox', later known as the 'watched pot problem' in quantum measurement theory, is now called the Quantum Zeno Effect and is relevant to the new technology of quantum 'interaction-free measurement'. This points to the evolving question of what computation means, when twentieth-century physics is taken seriously. Turing was galloping ahead in 1953–1954: learning about spinors and tensors for representing particles and gauge theory for forces. He was as well prepared in 1954 as he had been in 1935 for a new and unpredictable burst of advanced mathematics put to 'applied' use. Some people reject the idea of Turing ever shifting his views, but change and development is typical of the creative life of science. Discovery and experience oblige it. One paradigm shift after another, not a static 'thesis', to be held and defended as a dogmatic 'position', has characterised the development of twentieth-century physics. In Turing's spirit one should expect and welcome new computational paradigms, if soundly based on physical reality.

References

[Anon 1945] Anon (1945), General Report on Tunny, p. 327. This internal GCHQ report of 1945 was unsigned but is attributed to I. J. Good, D. Michie and G. Timms. Available in the National Archives, HW 25

[Britton 1992] J. L. Britton (1992), Introduction and notes to Turing's note on normal numbers. In J. L. Britton (ed.) *The Collected Works of A. M. Turing: Pure Mathematics*. North-Holland, 1992

[Champernowne 1933] D. G. Champernowne (1933), The construction of decimals normal in the scale of ten. J. Lond. Math. Soc. 8:254–260

[Church 1937] A. Church (1937), Review of Turing (1936). J. Symbolic Logic 2:42–3

[Copeland 1999] B. J. Copeland (1999), A lecture and two radio broadcasts on machine intelligence by Alan Turing. In K. Furukawa, D. Michie, and S. Muggleton (eds.), *Machine Intelligence 15*. Oxford University Press, 1999

[Copeland 2000] B. J. Copeland (2000), Narrow versus wide mechanisms: including a re-examination of Turing's views on the mind-machine issue. J. Philos. 96:5–32

[Davis 2000] M. Davis (2000), *The Universal Computer*. Norton, 2000

[Feferman 2001] S. Feferman (2001), Preface to 'Systems of logic based on ordinals'. In R. O. Gandy and C. E. M. Yates (eds.) *The Collected Works of A. M. Turing: Mathematical Logic*. North-Holland, 2001

[Gandy 1954] R. O. Gandy (1954), letter to M. H. A. Newman, available at www.turingarchive.org, item D/4. Text in R. O. Gandy and C. E. M. Yates (eds.) *The Collected Works of A. M. Turing: Mathematical Logic*. North-Holland, 2001

[Good 1992] I. J. Good (1992), Introductory remarks for the article in *Biometrika* **66** (1979), 'A. M. Turing's statistical work in World War II'. In J. L. Britton (ed.) *The Collected Works of A. M. Turing: Pure Mathematics*. North-Holland, 1992

[Good 1993] I. J. Good (1993), Enigma and Fish. In F. H. Hinsley and A. Stripp (eds.) *Codebreakers*. Oxford University Press, 1993

[Good 2001] I. J. Good (2001), Commentary on Turing's manuscript 'Minimum cost sequential analysis'. In R. O. Gandy and C. E. M. Yates (eds.) *The Collected Works of A. M. Turing: Mathematical Logic*. North-Holland, 2001

[Hodges 1983] A. Hodges (1983), *Alan Turing: the Enigma*. Burnett, London; Simon & Schuster; new edition Vintage, 1992

[Hodges 2006] A. Hodges (2006), The essential Turing, book review. Notices Amer. Math. Soc. 53:1190–1199

[Knox et al. 1939] A. D. Knox, P. F. G. Twinn, W. G. Welchman, A. M. Turing and J. R. Jeffreys (1939), Report dated 1 November 1939. In National Archives, HW 14/2

[Newman 1955] M. H. A. Newman (1955), Alan Mathison Turing. Biographical memoirs of Fellows of the Royal Society 1:253–263

[Penrose 1989] R. Penrose (1989), *The Emperor's New Mind*. Oxford University Press, 1989

[Penrose 1994] R. Penrose (1994), *Shadows of the Mind*. Oxford University Press, 1994

[Teuscher 2000] C. Teuscher (2000), *Turing's Connectionism, an Investigation of Neural Network Architectures*. Springer, 2002. See also C. Teuscher, Turing's connectionism, in C. Teuscher (ed.) *Alan Turing: Life and Legacy of a Great Thinker*. Springer, 2004

[Turing c. 1932] A. M. Turing (c. 1932), Handwritten essay: Nature of Spirit. Photocopy available in www.turingarchive.org, item C/29. Text in (Hodges 1983, p. 63)

[Turing 1936] A. M. Turing (1936), On computable numbers, with an application to the Entscheidungsproblem. Proc. London Math. Soc. (2) 42:230–265

[Turing 1936?] A. M. Turing (1936?), A note on normal numbers, manuscript and typescript available at www.turingarchive.org, item C/15. Text in J. L. Britton (ed.) *The Collected Works of A. M. Turing: Pure Mathematics*. North-Holland, 1992

[Turing 1937] A. M. Turing (1937), On computable numbers, with an application to the Entscheidungsproblem. A correction. Proc. London Math. Soc. (2) 43:544–546

[Turing 1939] A. M. Turing (1939) Systems of logic based on ordinals. Proc. London Math. Soc. (2) 45:161–228

[Turing 1942] A. M. Turing (1942), typescript 'Report on cryptographic machinery available at Navy Department, Washington', dated 28 November 1942, in the National Archives, HW 57/10

[Turing 1944] A. M. Turing (1944), Speech System 'Delilah' — Report on Progress, typescript dated 6 June 1944. I am indebted to Ralph Erskine for locating this document in the National Archives, HW 62/2. A description of the Delilah, with photographs, was given in (Hodges 1983)

[Turing 1946] A. M. Turing (1946), Proposed electronic calculator, copy of typescript available at www.turingarchive.org, item C/32. Text published in various forms, e.g. in D. C. Ince (ed.) *The Collected Works of A. M. Turing: Mechanical Intelligence*. North-Holland, 1992

[Turing 1947] A. M. Turing (1947), Lecture to the London Mathematical Society, 20 February 1947, typescript available at www.turingarchive.org, item B/1. Text published in various forms, e.g. in in D. C. Ince (ed.) *The Collected Works of A. M. Turing: Mechanical Intelligence*. North-Holland, 1992

[Turing 1948] A. M. Turing (1948), Intelligent machinery, National Physical Laboratory report, typescript available at www.turingarchive.org, item C/11. Text published in various forms, e.g. in B. J. Copeland (ed.) *The Essential Turing*. Oxford University Press, 2004

[Turing 1950] A. M. Turing (1950), Computing machinery and intelligence. Mind 59:433–460

[Turing 1951] A. M. Turing (1951), Can digital computers think? BBC talk, typescript available at www.turingarchive.org, item B/5. Text published in B. J. Copeland (ed.) *The Essential Turing*. Oxford University Press, 2004

[Turing 1952] A. M. Turing (1952), Can automatic calculating machines be said to think?, Radio discussion, typescript available at www.turingarchive.org, item B/6. Text published in B. J. Copeland (ed.) *The Essential Turing*. Oxford University Press, 2004

The Turing Model of Computation and its
Applications to Logic, Mathematics, Philosophy, and
Computer Science

Computability and Numberings

Serikzhan Badaev[1] and Sergey Goncharov[2]

[1] Kazakh National University, Almaty 050038, Kazakhstan badaev@kazsu.kz
[2] Institute of Mathematics of Siberian Branch of Russian Academy of Sciences,
Novosibirsk 6300090, Russia gonchar@math.nsc.ru

Introduction

The theory of computable numberings is one of the main parts of the theory of numberings. The papers of H. Rogers [36] and R. Friedberg [21] are the starting points in the systematical investigation of computable numberings. The general notion of a computable numbering was proposed in 1954 by A.N. Kolmogorov and V.A. Uspensky (see [40, p. 398]), and the monograph of Uspensky [41] was the first textbook that contained several basic results of the theory of computable numberings. The theory was developed further by many authors, and the most important contribution to it and its applications was made by A.I. Malt'sev, Yu.L. Ershov, A. Lachlan, S.S. Goncharov, S.A. Badaev, A.B. Khutoretskii, V.L. Selivanov, M. Kummer, M.B. Pouer-El, I.A. Lavrov, S.D. Denisov, and many other authors.

S.S. Goncharov and A.T. Nurtazin found applications of the theory of computable numberings to the theory of computable models, more precisely, to the problem of decidability of prime and saturated models [30]. Later S.S. Goncharov applied computable numberings of families of partial computable functions to the problem of characterizing autostability on the base of Scott's families [22] and established the existence of models with finite algorithmic dimension (on the base of the duality, founded by him, between the problem of the possible number of computable Friedberg numberings for families of c.e. sets and the problem of the existence of models with finite algorithmic dimension) by constructing families of c.e. sets with any finite number of Friedberg numberings up to equivalence [23, 24]. The problems arising in

- estimate of the complexity of isomorphisms between different representations of computable models,

- description of the autostable models,

- classification of the definable relations in computable models, etc.

led to investigation of computable numberings not only for families of partial computable functions and c.e. sets but also for families of constructive objects of a more general nature. In addition, in computability theory, one meets uniform computations for families of a special kind of relations and functions having high algorithmic complexity, and in the theory of computable models, very often we have to deal with computable classes of computable models.

All this was a strong motivation for S.S. Goncharov and A. Sorbi to propose in 1997 a new approach to the notion of computable numbering for general families of objects, which admit a constructive description in formal languages with a Gödel numbering for formulas [31].

Since then a lot of problems have been considered in the study of computable numberings for families of sets in the arithmetical hierarchy (see [7]–[15], and [33]–[35]) and in the hierarchy of Ershov (see [6], [16], [29], and [39]). Applications of generalized computable numberings were also pursued. The paper of S.S. Goncharov and J. Knight [27] offered an approach to the classification problem based on computable Friedberg numberings. And, in [26], and [28], computable numberings in all levels of the hyperarithmetical hierarchy, including the infinite ones, already have been applied to study problems in the theory of computable models.

In this paper, we study some problems relative to computable numberings in the sense of Goncharov–Sorbi for families of sets in the hyperarithmetical hierarchy. In section 1, we introduce a notion of computable numbering of a family of hyperarithmetical sets. In section 2, we continue to go along the line of research devoted to the problem of the isomorphism types of Rogers semilattices for families of arithmetical sets, which was initiated in the papers [7], [11]–[14].

We refer to the handbooks [37], [38], and [5] for the notions and standard notations on computability theory and computable infinite formulas. Undefined notions of the theory of numberings can be found in [18], and [19]. For more background on generalized computable numberings, see the articles [7], and [9].

1 Computable numberings in the hyperarithmetical hierarchy

A surjective mapping ν of the set ω of natural numbers onto a nonempty set $\mathcal{A}$ is called a *numbering* of $\mathcal{A}$. Suppose that $\mathcal{A}$ is a family of objects that admit constructive descriptions. By this we mean that one can define a language $\mathcal{L}$ (henceforth identified with a corresponding set of "well-formed formulas") and an interpretation of (fragments of) this language via an onto partial mapping $i \colon \mathcal{L} \longrightarrow \mathcal{A}$. For any object $a \in \mathcal{A}$, each formula Φ of $\mathcal{L}$ such that $i(\Phi) = a$ is interpreted as a "description" of a. Suppose further that $G \colon \omega \longrightarrow \mathcal{L}$ is a Gödel numbering.

Following [31], we propose:

Definition 1.1. A numbering ν of $\mathcal{A}$ is called *computable in $\mathcal{L}$ with respect to an interpretation i* if there exists a computable mapping f such that $\nu = i \circ G \circ f$.

It is immediate to see that Definition 1.1 does not depend on the choice of the Gödel numbering G. Hence, via identification of $\mathcal{L}$ with ω through some fixed Gödel numbering, the above definition states that ν is computable if there is some computable function f from ω to $\mathcal{L}$ such that $\nu = i \circ f$.

Definition 1.1 has a wide scope of applications, based on suitable choices of $\mathcal{L}$ and i. Let's, at first, consider the families of arithmetical sets. As language $\mathcal{L}$ we take in this case the collection of arithmetical first-order formulas in the signature $\langle +, \cdot, 0, s, \leqslant \rangle$, and i will be a mapping associating each formula with the corresponding set defined by that formula in the standard model $\mathfrak{N}$ of Peano arithmetic. For $v \in N$, denote by $\mathbf{v}$ an arithmetic term defining v, that is, the term $s(s(\ldots s(0) \ldots))$, in which the symbol s occurs v times. Then $A \in \Sigma^0_{n+1}$ if and only if there exists an arithmetic Σ_{n+1} formula $\Phi(v)$ such that

$$v \in A \Leftrightarrow \mathfrak{N} \models \Phi(\mathbf{v}).$$

If $\mathcal{A}$ is a family of Σ^0_{n+1} sets, then, by [31, Proposition 2.1], a numbering $\nu \colon N \to \mathcal{A}$ is computable with respect to interpretation i if and only if

$$\{\langle m, v \rangle \mid v \in \nu(m)\} \in \Sigma^0_{n+1}.$$

Thus, despite the strong hierarchy theorem, [37, §14.5], a computable numbering ν of a family $\mathcal{A} \subseteq \Sigma^0_{n+1}$ may be thought of as an enumeration procedure for the sequence $\nu(0), \nu(1), \ldots$ of Σ^0_{n+1} sets, which is uniformly computable with respect to the oracle $\emptyset^{(n)}$.

It seems promising to generalize in a straightforward manner this description of a computable numbering ν in terms of relative computability of the relation $\{\langle m, v \rangle \mid v \in \nu(m)\}$ for families of sets from any level of the arithmetical hierarchy, to families from any arbitrary level of the hyperarithmetical hierarchy.

Classes of the hyperarithmetical hierarchy

We need some notions from the textbook [37] to make our paper self-contained. First we remember Kleene's system of notations for computable ordinals. This system consists of a set $\mathcal{O}$ of notations, together with a partial ordering $<_\mathcal{O}$.

The ordinal 0 gets notation 1.

If a is a notation for α, then 2^a is a notation for $\alpha + 1$. Then $a <_\mathcal{O} 2^a$, and also, if $b <_\mathcal{O} a$, then $b <_\mathcal{O} 2^a$.

Suppose α is a limit ordinal. If φ_e is a total function, giving notations for an increasing sequence of ordinals with limit α, then $3 \cdot 5^e$ is a notation for α. For all n, $\varphi_e(n) <_{\mathcal{O}} 3 \cdot 5^e$, and if $b <_{\mathcal{O}} \varphi_e(n)$, then $b <_{\mathcal{O}} 3 \cdot 5^e$.

The sequence of oracles $\{\emptyset^{(n)}\}_{n \in \omega}$ is extended with the family of sets $H(a), a \in \mathcal{O}$, by transfinite induction on the ordinals $|a|_{\mathcal{O}}$ as follows.

(1) $H(1) = \emptyset$,

(2) $H(2^a) = H(a)'$,

(3) $H(3 \cdot 5^e) = \{\langle u, v \rangle \mid u <_{\mathcal{O}} 3 \cdot 5^e \ \& \ v \in H(u)\}$.

Now, following Kleene, we define the classes $\Sigma_\alpha^0, \Pi_\alpha^0, \Delta_\alpha^0$ of the hyperarithmetical hierarchy for all computable ordinals $\alpha \geq \omega$. For infinite α, a relation is said to be $\Sigma_\alpha^0, \Pi_\alpha^0$, or Δ_α^0 if it is, respectively, c.e., co-c.e., or computable relative to $H(a)$, *for some* $a \in \mathcal{O}$ with $|a|_{\mathcal{O}} = \alpha$. By a theorem of Spector, such a relation will be c.e., co-c.e., or computable relative to $H(a)$ *for every* $a \in \mathcal{O}$ with $|a|_{\mathcal{O}} = \alpha$. It is important for us to recall the well-known lack of uniformity in the definition of the classes $\Sigma_\alpha^0, \Pi_\alpha^0, \Delta_\alpha^0$ when we pass from finite to infinite computable ordinals. For finite α, say $\alpha = n$, the Σ_n^0 relations are the ones that are c.e. relative to $H(a)$ where $|a|$ is $n - 1$. There is the same lack of uniformity for Π_α^0 and Δ_α^0 relations.

If α is a computable limit ordinal and $a \in \mathcal{O}$ is a notation for α, then for every $n \in \omega$, the classes of the hyperarithmetical hierarchy may be also defined by $H(a)$–forms.

- $\Sigma_{\alpha+n}^0 = \Sigma_{n+1}^{H(a)}$,

- $\Pi_{\alpha+n}^0 = \Pi_{n+1}^{H(a)}$,

- $\Delta_{\alpha+n}^0 = \Delta_{n+1}^{H(a)}$.

We often use relativized forms of the sets $H(a)$. Let X be any set of natural numbers. Then

(1) $H^X(1) = X$,

(2) $H(2^a) = H^X(a)'$,

(3) $H(3 \cdot 5^e) = \{\langle u, v \rangle \mid u <_{\mathcal{O}} 3 \cdot 5^e \ \& \ v \in H^X(u)\}$.

We will need some details of Kleene's notion of partial recursive function relative to an oracle X, as is done in [37, § 9.2.]:

$$\varphi_z^X = \{\langle x, y \rangle \mid \exists u \exists v \left(\langle x, y, u, v \rangle \in W_{\rho(z)} \ \& \ D_u \subseteq X \ \& \ D_v \subseteq \overline{X} \right)\}; \quad (1)$$

here $\rho(z)$ is a computable function with some special properties.

Ash–Knight's classification of the infinitary computable formulas

Let α be any constructive ordinal, and let $\mathcal{A}$ be a family of Σ_α^0 sets. Our aim is to show that a numbering $\nu\colon N \to \mathcal{A}$ is computable in the sense of Goncharov–Sorbi if and only if

$$\{\langle m, v \rangle \mid v \in \nu(m)\} \in \Sigma_\alpha^0.$$

But we still have not defined the notion of a computable numbering for a family $\mathcal{A} \subseteq \Sigma_\alpha^0$, for constructive ordinals $\alpha \geq \omega$. To do this, we need to specify a language suitable for descriptions of Σ_α^0 sets, as well as an interpretation of these descriptions.

Evidently, we cannot restrict our descriptions to finite first-order formulas of arithmetics. We need a language with more expressive opportunities, namely, the language $L_{\omega_1\omega}$ with countable disjunctions and conjunctions. Indeed, to keep valid very productive tools like the compactness theorem, one is forced to consider some admissible fragments of the language $L_{\omega_1\omega}$. We will give an inductive definition of a language of computable infinitary formulas that has been used to characterize isomorphism types of computable models in terms of so-called Scott's rank (see [5]).

We follow [5] to give a classification of the family of computable infinitary formulas for any computable signature. Let $\{x_i : i \in \omega\}$ be a countable set of variables of a language L. Computable infinitary formulas are classified as *computable* Σ_α, or *computable* Π_α, for various computable ordinals α. Roughly speaking, they are infinitary formulas in which the disjunctions and conjunctions are over c.e. sets. In predicate formulas, only finitely many free variables are allowed, and for both predicate and propositional languages, only formulas in normal form are considered.

To each formula Φ, we associate a tuple of variables $\bar{x}$, including the free variables of Φ. We define the class of computable infinitary formulas by induction on the complexity, which is a computable ordinal. The computable Σ_0- and Π_0-formulas are the finitary open formulas.

For a computable ordinal $\alpha > 0$, a computable Σ_α formula $\Phi(\bar{x})$ is the disjunction of a c.e. set of formulas of the form $(\exists \bar{y})\psi$, where ψ is a computable Π_β formula for some $\beta < \alpha$ and $\bar{y}$ includes the variables of ψ that are not in $\bar{x}$ ($\bar{y}$ may also include some variables from $\bar{x}$).

Similarly, a computable Π_α formula $\Phi(\bar{x})$ is the conjunction of a c.e. set of formulas of the form $(\forall \bar{y})\psi$, where ψ is a computable Σ_β formula for some $\beta < \alpha$ and $\bar{y}$ includes the variables of ψ not in $\bar{x}$.

The informal notions given above are sufficient for us, but, for $\alpha \geq 2$, they are not precise. We refer to the textbook of C. Ash and J. Knight [5] for formal definitions as well as for their original Gödel numbering of the family of computable infinitary formulas.

Infinitary computable formulas have remarkable properties and have many applications in computable model theory (see [1]–[5], and [26]). For instance, we can illustrate this by the following two statements.

Proposition 1.2 ([5],[26]). *If $\mathcal{A}$, $\mathcal{B}$ are computable structures satisfying the same computable infinitary sentences, then $\mathcal{A} \cong \mathcal{B}$.*

Proposition 1.3 ([5],[26]). *Suppose $\bar{a}$, $\bar{b}$ are tuples satisfying the same computable infinitary formulas in a computable structure $\mathcal{A}$. Then there is an automorphism of $\mathcal{A}$ taking $\bar{a}$ to $\bar{b}$.*

To study Scott's ranks and the problems of auto-stability and algorithmic dimension as well as definability problems for computable structures, one has to extend the notion of a computable numbering for families of sets from finite levels of the arithmetical hierarchy to a notion of computable numbering for families of sets from infinite levels of the hyperarithmetical hierarchy [28].

Relations definable by computable infinitary formulas

Theorem 1.4 ([5, Theorem 7.5 (a)]). *For any computable structure $\mathcal{A}$, if Φ is a computable Σ_α formula, then $\Phi^{\mathcal{A}}$ is in the class Σ_α^0, and if Φ is a computable Π_α formula, then $\Phi^{\mathcal{A}}$ is in the class Π_α^0 of the hyperarithmetical hierarchy. Moreover, uniformity holds.*

Therefore, every relation definable in the standard model $\mathfrak{N}$ of arithmetic by an infinitary computable Σ_α or a Π_α formula is, respectively, a Σ_α^0 or a Π_α^0 set. We will show that there are no new definable relations in $\mathfrak{N}$.

Lemma 1.5. *For every computable ordinal α, if $a \in \mathcal{O}$ is a notation for α, then $H(2^a)$ is definable in $\mathfrak{N}$ by a Σ_α formula, which is computable uniformly in a.*

Proof. We prove this lemma by transfinite induction on α. The statement of the lemma holds for any finite ordinal α (see [37]).

Let $a = 3 \cdot 5^e$ for some e. Then

$$x \in H(2^a) \Leftrightarrow \exists y \exists u \exists v \left(\langle x, y, u, v \rangle \in W_{\rho(x)} \,\&\, D_u \subseteq H(a) \,\&\, D_v \subseteq \overline{H(a)} \right).$$

Since $H(a) = \{ \langle b, m \rangle \mid b <_{\mathcal{O}} a \,\&\, m \in H(b) \}$, it follows that

$$x \in H(a) \Leftrightarrow (\exists b <_{\mathcal{O}} a) \exists m \left(x = \langle b, m \rangle \,\&\, m \in H(b) \right).$$

The relation $x = \langle b, m \rangle \,\&\, m \in H(b)$ is definable in a $\mathfrak{N}$ by Σ_β formula $\psi_b(x, m)$ for some ordinal $\beta = |b|_{\mathcal{O}}$, which is less than the limit ordinal α. We can consider

$\psi_b(x, m)$ as a $\Pi_{\beta+1}$ formula, which is computable uniformly in b. Therefore the relation $x \in H(a)$ is definable in $\mathfrak{N}$ by an infinite disjunction of formulas $\exists m \psi_b(x, m)$ over the c.e. set $\{b \mid b <_\mathcal{O} a\}$. So the relation $D_u \subseteq H(a)$ is definable by a Σ_α formula computable uniformly in a and u.

It is also true that

$$x \notin H(a) \Leftrightarrow (\forall b <_\mathcal{O} a)\forall m(x \neq \langle b, m \rangle)\vee$$

$$(\exists b <_\mathcal{O} a)\exists m \, (x = \langle b, m \rangle \, \& \, m \notin H(b)).$$

The relation $x = \langle b, m \rangle \, \& \, m \notin H(b)$ is definable in $\mathfrak{N}$ by a Π_β formula $\theta_b(x, m)$ for some ordinal $\beta = |b|_\mathcal{O}$ that is less than the ordinal α. By induction, $\theta_b(x, m)$ is computable uniformly in b and therefore the relation $x \notin H(a)$ is definable in $\mathfrak{N}$ by an infinite disjunction of formulas $\exists m \theta_b(x, m)$ over the c.e. set $\{b \mid b <_\mathcal{O} a\}$ and one Π_1 formula. So the relation $D_v \subseteq \overline{H(a)}$ is definable by a Σ_α formula computable uniformly in a and v.

Gathering all the facts proved above we obtain that the relation $x \in H(2^a)$ is definable in $\mathfrak{N}$ by a Σ_α formula, which is computable uniformly on a.

Finally suppose that α is infinite and $a = 2^b$ for some b. By induction we have that the relation $x \in H(a)$ is definable in $\mathfrak{N}$ by a Σ_β formula where $\beta = |b|_\mathcal{O}$. As in the previous case we have

$$x \in H(2^a) \Leftrightarrow \exists y \exists u \exists v \left(\langle x, y, u, v \rangle \in W_{\rho(x)} \, \& \, D_u \subseteq H(a) \, \& \, D_v \subseteq \overline{H(a)} \right).$$

By the definitions we obtain that the relation $D_u \subseteq H(a)$ is definable in $\mathfrak{N}$ by a finite conjunction of Σ_β formulas, and therefore, it is definable by a Σ_α formula computable uniformly in u and a. Similarly the relation $D_v \subseteq \overline{H(a)}$ is presented by a finite conjunction of Π_β formulas, and therefore, it is definable in $\mathfrak{N}$ by a Σ_α formula that is computable uniformly in v and a. Taking infinite disjunction over all u, v, and y such that $\langle x, y, u, v \rangle \in W_{\rho(x)}$, we conclude that in this case the relation $x \in H(2^a)$ is also definable in $\mathfrak{N}$ by a Σ_α formula that is computable uniformly on a. □

Theorem 1.6. *For every computable ordinal α and every set X from the class Σ_α^0 of the hyperarithmetical hierarchy, X is definable in $\mathfrak{N}$ by an infinitary computable Σ_α formula.*

Proof. The claim of the theorem is true for any finite ordinal. Suppose that α is a computable infinite ordinal, and let a be a notation for α. Let $X \in \Sigma_\alpha^0$. Then X is c.e. relative to $H(a)$, and hence, X is 1-reducible to $(H(a))' = H(2^a)$. If f is a computable function that does the reduction of X to $H(a)$, then

$$\forall x(x \in X \Leftrightarrow \exists y(y = f(x) \, \& \, y \in H(2^a))).$$

Now Lemma 1.5 implies that X is definable in $\mathfrak{N}$ by a computable an infinitary Σ_α formula. □

Corollary 1.7. *For every computable ordinal α and every set $X \in \omega$, $X \in \Sigma_\alpha^0$ if and only if X is definable in $\mathfrak{N}$ by an infinitary computable Σ_α formula.*

Hyperarithmetical numberings

Let $\mathcal{L}$ be the family of infinitary computable formulas. We will denote by Φ the Gödel numbering of $\mathcal{L}$ given by C. Ash and J. Knight in [5].

Definition 1.8. Let α be a computable ordinal. A numbering ν of a family $\mathcal{A} \subseteq \Sigma_\alpha^0$ is called Σ_α^0– *computable* if there exists a computable function f such that $\{\Phi_{f(i)} | i \in \omega\}$ is a set of Σ_α formulas of Peano arithmetic and

$$\nu(m) = \{x \in \omega \mid \mathfrak{N} \models \Phi_{f(m)}(\mathbf{x})\};$$

here $\mathbf{x}$ stands for the numeral for x. The set of Σ_α^0–computable numberings of $\mathcal{A}$ will be denoted by $\mathrm{Com}_\alpha^0(\mathcal{A})$.

In other words, a Σ_α^0–computable numbering is just a computable numbering in the sense of Definition 1.1.

Theorem 1.9. *A numbering ν of a family $\mathcal{A} \subseteq \Sigma_\alpha^0$ is Σ_α^0– computable if and only if $\{\langle m, x \rangle \mid x \in \nu(m)\}$ is Σ_α^0.*

Proof. $\Rightarrow$. Let f be a computable function such that $\{\Phi_{f(i)} | i \in \omega\}$ is a set of Σ_α formulas and for all x, m

$$x \in \nu(m) \Leftrightarrow \mathfrak{N} \models \Phi_{f(m)}(\mathbf{x}).$$

Let $\theta(m, x)$ be the infinite disjunction

$$\bigvee_{n \in \omega} (m = f(n) \;\&\; \Phi_{f(n)}(x)).$$

Every $\Phi_{f(n)}$ is a Σ_α formula; i.e., $\Phi_{f(n)}$ is a disjunction of formulas of form $\exists \overline{y} \Psi_{g(n,i)}$ over some c.e. set $W_{h(n)}$:

$$\Phi_{f(n)} = \bigvee_{i \in W_{h(n)}} \exists \overline{y} \Psi_{g(n,i)}.$$

The functions g, h are computable (see [5]), and for every n and every i, $\Psi_{g(n,i)}$ is a Π_β formula with $\beta < \alpha$. Therefore, $\theta(m, x)$ is the disjunction of the formulas $\exists \overline{y}(m = f(n) \;\&\; \Psi_{g(n,i)})$ over a c.e. set, and hence, θ is a Σ_α formula. By Theorem 1.4, the set $\{\langle m, x \rangle \mid x \in \nu(m)\}$ is Σ_α^0.

$\Leftarrow$. Let $\{\langle m, x \rangle \mid x \in \nu(m)\}$ be Σ_α^0. By Theorem 1.6, there exists a Σ_α formula $\eta(m, x)$ such that

$$x \in \nu(m) \Leftrightarrow \mathfrak{N} \models \eta(\mathbf{m}, \mathbf{x}).$$

It is easy to check by transfinite induction on the ordinal notations that, for every $m \in \omega$, the formula $\eta(\mathbf{m}, x)$, of one free variable x, is Σ_α. Obviously, an index of this formula can be effectively found, uniformly from m. □

Corollary 1.10. *A numbering ν of a family $\mathcal{A} \subseteq \Sigma_\alpha^0$ is Σ_α^0–computable if and only if $\{\langle m, x \rangle \mid x \in \nu(m)\}$ is definable in $\mathfrak{N}$ by some Σ_α formula.*

Numbering $\nu : \omega \mapsto \mathcal{A}$ of a family $\mathcal{A}$ of hyperarithmetical sets is also called *hyperarithmetical* if $\{\langle m, x \rangle \mid x \in \nu(m)\}$ is definable in $\mathfrak{N}$ by some computable (infinitary) formula of Peano arithmetic. A family $\mathcal{A}$ for which $\mathrm{Com}_\alpha^0(\mathcal{A}) \neq \emptyset$ is called Σ_α^0–*computable*. If the ordinal α is finite, then we usually use use the term *arithmetical numbering*.

The function f in Definition 1.8 can be chosen Σ_α^0 computable because in the definition of Σ_α formulas, one can replace disjunctions over c.e. sets with disjunctions over hyperarithmetical sets (see [5, Proposition 7.11]).

We now revise some of the basic definitions of the theory of numberings. Two numberings ν, μ of $\mathcal{A}$ can be compared by defining $\nu \leq \mu$ (ν is *reducible to* μ) if there is a computable function f such that $\nu = \mu \circ f$. Two numberings ν and μ are *equivalent* (written $\nu \equiv \mu$) if $\nu \leq \mu$ and $\mu \leq \nu$.

The equivalence $\equiv$ partitions the set $\mathrm{Com}_\alpha^0(\mathcal{A})$ into the equivalence classes $\deg(\mu)$ of all Σ_α^0–computable numberings μ of $\mathcal{A}$, thus originating a quotient structure, denoted by $\mathcal{R}_\alpha^0(\mathcal{A})$. The latter forms an upper semilattice under the partial ordering induced by $\leq$, where the join of two numberings ν and μ is defined by $(\nu \oplus \mu)(2n) = \nu(n)$ and $(\nu \oplus \mu)(2n + 1) = \mu(n)$ induces the least upper bound of $\deg(\mu)$ and $\deg(\nu)$. $\mathcal{R}_\alpha^0(\mathcal{A})$ is called the *Rogers semilattice* of $\mathcal{A}$.

2 Isomorphism types of Rogers semilattices

One of the global aims of the theory of computable numberings is to investigate the isomorphism types of the Rogers semilattices. Furthermore, we will consider nontrivial families of sets only, i.e., families that contain at least two sets. As the first stage of this research we study the differences in the isomorphism types of Rogers semilattices of computable numberings for the families of sets lying in different levels of the arithmetical hierarchy (see [11]–[14]). The strongest result that has been obtained in this direction is as follows. For every two nontrivial families of sets taken from two different finite levels of the arithmetical hierarchy, if the gap in the levels is not less than 3, then the corresponding Rogers semilattices of computable numberings are not isomorphic [14]. Roughly speaking, we try to extend this statement to the families of sets taken from infinite levels of the hyperarithmetical hierarchy. We should note that this is mainly a straightforward relativization of the proofs from [14].

Theorem 2.1. *For any computable ordinals $\alpha > 0$ and β and for every Σ^0_α– computable family $\mathcal{A}$ and every nontrivial Σ^0_β– computable family $\mathcal{B}$, if $\alpha + 3 \leq \beta$, then the Rogers semilattices $\mathcal{R}^0_\alpha(\mathcal{A})$ and $\mathcal{R}^0_\beta(\mathcal{B})$ are not isomorphic.*

Proof. Let $\alpha > 0$ and β be any computable ordinals such that $\beta \geq \alpha + 3$. Let $a, b \in \mathcal{O}$ stand for some notations of ordinals α and $\alpha + 3$, respectively. Let $\mathcal{A}$ be any Σ^0_α– computable family, and let $\mathcal{B}$ be a Σ^0_β– computable family, which contains at least two sets.

We will construct in the Rogers semilattice $\mathcal{R}^0_\beta(\mathcal{B})$ an interval that forms a Boolean algebra not isomorphic to any interval in the Rogers semilattice $\mathcal{R}^0_\alpha(\mathcal{A})$.

For the ease of the reader we recall only necessary notions and statements that allow us to formulate the requirements for constructing the desired interval.

Definition 2.2. If ρ is a numbering of a family $\mathcal{A}$, and C is a nonempty c.e. set, with f a computable function such that $\mathrm{range}(f) = C$, then we define $\rho_C \leftrightharpoons \rho \circ f$.

The definition does not depend on f: If we define ρ_C starting from any other computable function g such that $\mathrm{range}(g) = C$, then we get a numbering that is equivalent to the one given by f. The assignment $C \mapsto \rho_C$ from c.e. sets to numberings (up to equivalence of numberings) is called *Lachlan operator*.

Lemma 2.3. *For every pair A, B of c.e. sets and for every pair of numberings τ, ρ, we have:*

(1) The following are equivalent:

    (a) $\rho_A \leqslant \rho_B$;

    (b) there is a partial computable function φ satisfying $\mathrm{dom}(\varphi) \supseteq A$, $\varphi[A] \subseteq B$ and for all $x \in A$, $\rho(x) = \rho(\varphi(x))$;

(2) if $A \subseteq B$, then $\rho_A \leqslant \rho_B$;

(3) if $\rho_A \leqslant \rho_B$, then $\rho_B \equiv \rho_{A \cup B}$;

(4) if $\tau \leqslant \rho$, then $\tau \equiv \rho_C$ for some c.e. set C;

(5) if $\tau \leqslant \rho$, and $\tau \equiv \rho_C$, for some c.e. set C, then for every γ such that $\tau \leqslant \gamma \leqslant \rho$ there exists a c.e. set D with $C \subseteq D$ and $\gamma \equiv \rho_D$;

(6) $\rho_{A \cup B} \equiv \rho_A \oplus \rho_B$.

Proof. See Lemma 2.2 in [10]. □

In what follows, the symbol $[\eta, \theta]$ denotes the following interval of degrees in $\mathcal{R}^0_\alpha(\mathcal{A})$:

$$[\eta, \theta] \leftrightharpoons \{\deg(\mu) \mid \eta \leq \mu \leq \theta\}.$$

Now we estimate the complexity of any interval of Rogers semilattice $\mathcal{R}^0_\alpha(\mathcal{A})$ if it is Boolean algebra. The notion of an **X**– computable Boolean algebra plays a key role in establishing our claim. Recall (see [25]) that a Boolean algebra $\mathfrak{A}$ is called **X**–*computable* if its universe, operations, and relations are **X**– computable.

Lemma 2.4. *Let* $\eta, \theta \in \mathrm{Com}^0_\alpha(\mathcal{A})$. *If* $[\eta, \theta]$ *is a Boolean algebra, then it is* $H(b)$– *computable.*

Proof. Given η and θ as in the hypothesis of the lemma, we first observe that by (4) and (5) of Lemma 2.3 there exists a c.e. set C such $\eta \equiv \theta_C$ and

$$[\eta, \theta] = \{\deg(\theta_X) \mid X \text{ is c.e. and } X \supseteq C\}.$$

For every i, let $U_i \leftrightharpoons C \cup W_i$. This gives an effective listing of all c.e. supersets of C. By Lemma 2.3 (1b), for every i, j, we have $\theta_{U_i} \leqslant \theta_{U_j}$ if and only if

$$\exists p[\forall x(x \in U_i \Rightarrow \exists y(\varphi_p(x) = y \,\&\, y \in U_j))$$

$$\&\, \forall x \forall y(x \in U_i \,\&\, \varphi_p(x) = y \Rightarrow \theta(x) = \theta(y))].$$

Since $\theta \in \mathrm{Com}^0_\alpha(\mathcal{A})$, this implies, by Theorem 1.9, that the binary relation $z \in \theta(x)$ is c.e. relative to the oracle $H(a)$. Therefore, the binary relation $\theta(x) = \theta(y)$ is a $\forall\exists$–predicate relative to the oracle $H(a)$.

Simple calculations show now that $\theta_{U_i} \leqslant \theta_{U_j}$ is a $\Sigma^0_{\alpha+2}$–relation in i, j.

Let us consider the equivalence relation ε on ω defined by

$$(i, j) \in \varepsilon \Leftrightarrow \theta_{U_i} \leqslant \theta_{U_j} \,\&\, \theta_{U_j} \leqslant \theta_{U_i}.$$

Let $B \leftrightharpoons \{x \mid \forall y(y < x \Rightarrow (x, y) \notin \varepsilon)\}$. Define a bijection $\psi_1 : B \to [\eta, \theta]$, by letting $\psi_1(i) = \deg(\theta_{U_i})$, for all $i \in B$. It is evident that ψ_1 induces on B a partially ordered set $\mathfrak{B}$, which is a Boolean algebra isomorphic to $[\eta, \theta]$. The interval $\mathfrak{B}$ is an $H(b)$– computable partially ordered set since $(H(a))''' = H(b)$. It follows from [17] (see also [25, Theorem 3.3.4]) that $\mathfrak{B}$ with respect to the corresponding Boolean operations is $H(b)$-computable too. $\qquad\square$

Lemma 2.5 (L. Feiner). *Let* $\mathfrak{F}$ *be a computable atomless Boolean algebra. Then for every* **X** *there is an ideal* J *such that* J *is* **X**–*c.e. and the quotient* $\mathfrak{F}/J$ *is not isomorphic to any* **X**– *computable Boolean algebra.*

Proof. See [20]. $\qquad\square$

Below, we will use the following notations. For a given c.e. set H, $\{V_i \mid i \in \omega\}$ denotes an effective listing of all c.e. supersets of the set H defined, for instance, by $V_i \leftrightharpoons H \cup W_i$, for all i. We will assume for convenience that $V_0 = H$. Let ε_H stand for the distributive lattice of the c.e. supersets of H. For a given c.e. set $V \supseteq H$, let V^* denote the image of V under the canonical homomorphism of ε_H onto ε_H^* (i.e., ε_H modulo the finite sets), and let $\subseteq^*$ denote the partial ordering relation of ε_H^*. Obviously, if J is an ideal in ε_H, then $J^* \leftrightharpoons \{V^* \mid V \in J\}$ is an ideal in ε_H^*.

As is known (see, for instance, [25]), if $\mathfrak{A}$ is a Boolean algebra and J is an ideal of $\mathfrak{A}$, then the universe of the quotient Boolean algebra $\mathfrak{A}/J$ is given by the set of equivalence classes $\{[a]_J \mid a \in \mathfrak{A}\}$ under the equivalence relation $\equiv_J$ given by

$$a \equiv_J b \Leftrightarrow \exists c_1, c_2 \in J(a \vee c_1 = b \vee c_2),$$

and the partial ordering relation is given by

$$[a]_J \leq_J [b]_J \Leftrightarrow a - b \in J,$$

where $a - b$ stands for $a \wedge \neg b$.

Lemma 2.6. *Let $\mathcal{B}$ be a Σ_β^0–computable family, $\mu \in \mathrm{Com}_\beta^0(\mathcal{B})$, and let H be any c.e. set such that $\mu(H) = \mathcal{B}$ and ε_H^* is a Boolean algebra. Let $\psi_2 : \varepsilon_H \longrightarrow [\mu_H, \mu]$ be the mapping given by $\psi_2(V_i) = \deg(\mu_{V_i})$ for all i, and let I be any ideal of ε_H. Then ψ_2 induces an isomorphism of ε_H^*/I^* onto $[\mu_H, \mu]$ if and only if for every i, j*

(1) $V_i \in I \Rightarrow \mu_{V_i} \leqslant \mu_H$;

(2) $V_i - V_j \notin I \Rightarrow \mu_{V_i} \nleqslant \mu_{V_j}$ (where $V_i - V_j \leftrightharpoons (V_i \setminus V_j) \cup H$).

Proof. See Lemma 4 in [14]. □

By Lemma 2.4, all Boolean intervals of $\mathcal{R}_\alpha^0(\mathcal{A})$ are $H(b)$–computable Boolean algebras. Therefore, to show the theorem, it is sufficient:

(i) to consider a computable atomless Boolean algebra $\mathfrak{F}$ and an ideal J of $\mathfrak{F}$ as in Feiner's Lemma such that J is c.e. in $H(b)$ and $\mathfrak{F}/J$ is not isomorphic to any $H(b)$–computable Boolean algebra,

(ii) to find Σ_β^0–computable numberings ν and μ of $\mathcal{B}$ such that the interval $[\nu, \mu]$ of $\mathcal{R}_\beta^0(\mathcal{B})$ is a Boolean algebra isomorphic to $\mathfrak{F}/J$.

First, we consider item (i) above. Let $\mathfrak{F}$ be a computable atomless Boolean algebra. According to a famous result of Lachlan [32], there exists a hyperhypersimple set H such that ε_H^* is isomorphic to $\mathfrak{F}$. We fix such a set H.

We refer to the textbook of Soare [38] for the details of a suitable isomorphism χ of ε_H^* onto $\mathfrak{F}$. We only notice that starting from a computable listing $\{b_0, b_1, \ldots\}$ of the elements of $\mathfrak{F}$, one can find a Σ_3^0–computable Friedberg numbering $\{B_0, B_1, \ldots\}$ of a subfamily of the family ε_H such that $\varepsilon_H^* = \{B_0^*, B_1^*, \ldots\}$ and $\chi(B_i^*) = b_i$.

We will use the techniques for embedding posets into intervals of Rogers semilattices, which have been developed in [10]. Let J be any $H(b)$ –c.e. ideal of $\mathfrak{F}$ satisfying the conclusions of Lemma 2.5, and let $\hat{J} = \{j \in \omega \mid b_j \in J\}$. Then $\hat{J}$ is an

$H(b)$ –c.e. set, $I^* \rightleftharpoons \{B_j^* \mid j \in \hat{J}\}$ is an ideal of ε_H^*, and $\mathfrak{F}/J$ is isomorphic to ε_H^*/I^*. So, instead of the Boolean algebra $\mathfrak{F}/J$ in item (ii) above, we can consider ε_H^*/I^*.

Let $I \rightleftharpoons \{V \mid V \in \varepsilon_H \ \& \ V^* \in I^*\}$, and let $\hat{I} = \{i \in \omega \mid V_i^* \in I^*\}$. Obviously, I is an ideal of ε_H.

Lemma 2.7. *The relations "$V_i \in I$" (equivalently: "$i \in \hat{I}$"), in i, and "$V_i - V_j \in I$", in i, j, are both $H(b)$ –c.e.*

Proof. Straightforward relativization of the proof of Lemma 5 in [14]. $\square$

Since $\beta \geq \alpha + 3$ we can use the oracle $H(b)$ and apply Lemma 2.6 to construct a suitable numbering μ of $\mathcal{B}$ and consider the corresponding mapping ψ_2 that will give us an isomorphism of ε_H^*/I^* onto the interval $[\mu_H, \mu]$.

The requirements

First of all, we need the numbering μ to satisfy the requirement:

$$\mathbf{B} : \mu[H] = \mathcal{B}$$

to guarantee that μ_H is a numbering of the whole family $\mathcal{B}$. Then in view of Lemma 2.6, we must satisfy, for every i, j, p, the requirements:

$$\mathbf{P}_i : V_i \in I \Rightarrow \mu_{V_i} \leqslant \mu_H,$$
$$\mathbf{R}_{i,j,p} : V_i - V_j \notin I \Rightarrow \mu_{V_i} \not\leqslant \mu_{V_j} \text{ via } \varphi_p,$$

where by "$\mu_{V_i} \not\leqslant \mu_{V_j}$ via φ_p" we mean that φ_p does not reduce μ_{V_i} to μ_{V_j} in the sense of Lemma 2.3(1b).

We take any numbering $\nu \in \mathrm{Com}_\beta^0(\mathcal{B})$ and try to construct a numbering μ that meets the above requirements $\mathbf{B}, \mathbf{P}_i, \mathbf{R}_{i,j,p}$. Evidently, to get $\mu \in \mathrm{Com}_\beta^0(\mathcal{B})$ starting from any uniform enumeration of the numbering ν, we have to avoid using oracles that are stronger than the oracles used in ν–computations. Since $\beta \geq \alpha + 3$, the oracles relative to which we can make uniform computations in the numbering ν have complexity at least $H(b)$. This lower complexity boundary for the oracles that we intend to use in our strategies to construct μ is essential. And this forces us to partition the rest of the proof into two cases.

CASE 1 $\beta > \alpha + 3$. In this case we can even use oracles of complexity equal to or greater than $H(2^b)$. And Lemma 2.7 implies that the conditions $V_i \in I$ or $V_i - V_j \notin I$ in the requirements $\mathbf{P}_i$ and $\mathbf{R}_{i,j,p}$ are effectively recognizable by any oracle of complexity equal or higher than $H(2^b)$. The strategies and the construction for building the numbering μ in this case are relativized versions of the corresponding strategies and construction from [14, Theorem 1].

CASE 2 $\beta = \alpha + 3$. In this case we use only oracle $H(b)$. Lemma 2.7 implies that a condition $V_i \in I$ is eventually recognizable by this oracle. As to the negative condition $V_i - V_j \notin I$ in the requirement $\mathbf{R}_{i,j,p}$, we constantly try to destroy the reducibility $\mu_{V_i} \not\leq \mu_{V_j}$ via φ_p until (if ever) the condition $V_i - V_j \in I$ is eventually recognized by the oracle $H(b)$.

We refer to [14, Theorem 2] for the corresponding versions of the strategies and the construction in the setting of the arithmetical hierarchy. $\square$

Acknowledgments

The first author was partially supported by the State grant of Kazakhstan "The Best Teacher of Higher Education, 2005." The second author was partially supported by the RFBR grant 05-01-00819 and a grant from the scientific schools of Russia 4413.2006.1

References

1. Ash, C. J.: Categoricity in hyperarithmetical degrees. Annals of Pure and Applied Logic, **34**, 1–14 (1987)
2. Ash, C. J., Knight, J. F.: Pairs of computable structures. Annals of Pure and Applied Logic, **46**, 211–234 (1990)
3. Ash, C. J., Knight, J. F.: Possible degrees in computable copies. Annals of Pure and Applied Logic, **75**, 215–221 (1995)
4. Ash, C. J., Knight, J. F.: Possible degrees in computable copies II. Annals of Pure and Applied Logic, **87**, 151–165 (1997)
5. Ash, C. J., Knight, J. F.: Computable Structures and the Hyperarithmetical Hierarchy. Elsevier Science, Amsterdam (2000)
6. Badaev, S. A.: On Rogers semilattices. Lecture Notes in Computer Science, **3959**, 704–706 (2006)
7. Badaev, S. A., Goncharov, S. S.: Theory of numberings: open problems. In: Cholak, P., Lempp, S., Lerman, M., Shore, R. (eds) Computability Theory and its Applications. Contemporary Mathematics, 257. American Mathematical Society, Providence (2000)
8. Badaev, S. A., Goncharov, S. S.: Rogers semilattices of families of arithmetic sets. Algebra and Logic, **40**, 283–291 (2001)
9. Badaev, S. A., Goncharov, S. S., Sorbi, A.: Completeness and universality of arithmetical numberings. In: Cooper, S. B., Goncharov, S. S. (eds) Computability and Models. Kluwer/Plenum Publishers, New York (2003).
10. Badaev, S. A., Goncharov, S. S., Podzorov, S.Yu., Sorbi, A.: Algebraic properties of Rogers semilattices of arithmetical numberings. In: Cooper, S. B., Goncharov, S. S. (eds) Computability and Models. Kluwer/Plenum Publishers, New York (2003)

11. Badaev, S. A., Goncharov, S. S., Sorbi, A.: Isomorphism types and theories of Rogers semilattices of arithmetical numberings. In: Cooper, S. B., Goncharov, S. S. (eds) Computability and Models. Kluwer/Plenum Publishers, New York (2003)

12. Badaev, S. A., Goncharov, S. S., Sorbi, A.: Elementary properties of Rogers semilattices of arithmetical numberings. In: Downey, R., Ding, D., Tung, S. H., Qiu, Y. H., Yasugi, M., Wu, G. (eds) Proceedings of the 7-th and 8-th Asian Logic Conferences. World Scientific, Singapore (2003)

13. Badaev, S. A., Goncharov, S. S., Sorbi, A.: On elementary theories of Rogers semilattices. Algebra and Logic, **44**, 143–147 (2005)

14. Badaev, S. A., Goncharov, S. S., Sorbi, A.: Isomorphis types of Rogers semilattices for the families from different levels of arithmetical hierarchy. Algebra and Logic, **45**, 361–370 (2006)

15. Badaev, S. A., Podzorov, S.Yu.: Minimal coverings in the Rogers semilattices of Σ_n^0-computable numberings. Siberian Mathematical Journal, **43**, 616–622 (2002)

16. Badaev, S. A., Talasbaeva Zh. T.: Computable numberings in the Hierarchy of Ershov. In: Goncharov, S.S., Ono, H., Downey, R. (eds) Proceedings of 9-th Asian Logic Conference. World Scientific Publishers, Singapore (2006)

17. Dzgoev, V. D.: Constructive enumerations of Boolean lattices. Algebra and Logic, **27**, 395–400 (1988)

18. Ershov, Yu.L.: Theory of Numberings. Nauka, Moscow (1977).

19. Ershov, Yu.L.: Theory of numberings. In: Griffor, E. R. (ed) Handbook of Computability Theory. North-Holland, Amsterdam (1999)

20. Feiner, L.: Hierarchies of Boolean algebras. Journal of Symbolic Logic, **35**, 365–374 (1970)

21. Friedberg, R. M.: Three theorems on recursive enumeration. Journal of Symbolic Logic, **23**, 309–316 (1958)

22. Goncharov, S. S.: The quantity of nonautoequivalent constructivizations. Algebra and Logic, **16**, 169–185 (1977)

23. Goncharov, S. S.: Computable single-valued numerations. Algebra and Logic, **19**, 325–356 (1980)

24. Goncharov, S. S.: On the problem of number of non-self-equivalent constructivizations. Algebra and Logic, **19**, 401–414 (1980)

25. Goncharov, S. S.: Countable Boolean Algebras and Decidability. Plenum, Consultants Bureau, New York (1997).

26. Goncharov, S. S.: Computability and computable models. In: Mathematical Problems in Applied Logic II. Springer, Heidelberg (2007)

27. Goncharov, S. S., Knight, J. F.: Computable structure/non-structure theorems. Algebra and Logic, **41**, 351–373 (2002)

28. Goncharov, S. S., Harizanov, V. S., Knight, J. F., McCoy, C., Miller, R. G., Solomon, R.: Enumerations in computable structure theory. Annals of Pure and Applied Logic, **136**, 219–246 (2005)

29. Goncharov, S. S., Lempp, S., Solomon, D. R.: Friedberg numberings of families of n-computably enumerable sets. Algebra and Logic, **41**, 81–86 (2002)

30. Goncharov, S. S., Nurtazin, A. T.: Constructive models of complete solvable theories. Algebra and Logic, **12**, 67–77 (1973)

31. Goncharov, S. S., Sorbi, A.: Generalized computable numerations and non-trivial Rogers semilattices. Algebra and Logic, **36**, 359–369 (1997)

32. Lachlan, A. H.: On the lattice of recursively enumerable sets. Transactions of the American Mathematical Society, **130**, 1–37 (1968)

33. Podzorov, S.Yu.: Initial segments in Rogers semilattices of Σ_n^0–computable numberings. Algebra and Logic, **42**, 121–129 (2003)
34. Podzorov, S.Yu.: Local structure of Rogers semilattices of Σ_n^0–computable numberings. Algebra and Logic, **44**, 82–94 (2005)
35. Podzorov, S.Yu.: On the definition of a Lachlan semilattice. Siberian Mathematical Journal, **47**, 315–323 (2006)
36. Rogers, H.: Gödel numberings of partial computable functions. Journal of Symbolic Logic, **23**, 49–57 (1958)
37. Rogers, H.: Theory of Recursive Functions and Effective Computability. McGraw-Hill, New York (1967)
38. Soare, R. I.: Recursively Enumerable Sets and Degrees. Springer-Verlag, Berlin Heidelberg (1987)
39. Talasbaeva, Zh.T.: Positive numberings of families of sets in the Ershov hierarchy. Algebra and Logic, **42**, 413–418 (2003)
40. Uspensky, V. A.: Kolmogorov and mathematical logic. Journal of Symbolic Logic, **57**, 385–412 (1992)
41. Uspensky, V. A.: Lectures on Computable Functions. Fiz-MatGiz, Moscow (1960)

Computation as Conversation

Johan van Benthem

Institute for Logic, Language & Computation (ILLC), University of Amsterdam,
Amsterdam, the Netherlands
johan@science.uva.nl
and
Department of Philosophy, Stanford University, Stanford, CA, U.S.A.
johan@csli.stanford.edu

Summary. Against the backdrop of current research into 'logical dynamics' of information, we discuss two-way connections between conversation and computation, leading to a broader perspective on both.

1 Information flow for children and logical dynamics

The Amsterdam Science Museum *NEMO* organizes regular Kids' Lectures on Science.[1] Imagine 60 children aged around 8 sitting in a small amphitheatre—with parents present in the wings, but not allowed to speak. Last February, it was my pleasure to give one on Logic. While preparing for the event, I got more and more worried. How does one talk logic to such an audience, without boring or upsetting them? Was there *anything* in common between children that age and the abstractions that drive one's university career? How to even start? My first question was this:

The Restaurant In a restaurant, your father has ordered Fish, your mother ordered Vegetarian, and you have Meat. Out of the kitchen comes some new person with the three plates. What will happen? The children got excited, many little hands were raised, and one said: "He asks who has the Meat". "Sure enough", I said: "He asks, hears the answer, and puts the plate. What happens next?" Children said, "He asks who has the Fish!" Then I asked once more what happens next? And now one could see the Light of Reason start shining in those little eyes. One girl shouted: "He does not ask!" Now, *that* is logic ....

---

[1] See http://www.nemo-amsterdam.nl/.

After that, we played a long string of scenarios, including card games, Master Mind, and Sudoku, and we discussed what best questions to ask and conclusions to draw.[2] In my view, the Restaurant is about the simplest realistic logical scenario. Several basic informational actions take place intertwined: questions, answers, and inferences, and the setting crucially involves more than one agent. Moreover, successive speech acts can be analyzed for their informational content once they have taken place, but they can also be planned beforehand: what best to ask, and how best to answer? The program of 'Logical Dynamics' (van Benthem 1996) is about identifying and analyzing such scenarios, moving, in particular, the information-carrying events into the logical systems themselves. And once we take that view, we need a congenial account of *computation*. What happens during a conversation is that information states of children—singly and in groups—change over time, in a systematic way triggered by various communicative events. In this universe of states and possible transitions between them, the long experience of computer scientists in modeling computation becomes relevant, from Turing's first 'single-minded' computers to dealing with the multi-agent Internet. Please note that this is not a matter of computational 'implementation', the subservient stance some computer scientists assume vis-a-vis other academic disciplines. We care rather about fundamental ideas and the general cultural contribution of Informatics.

This paper is largely a discussion of known results and what they mean or suggest in a broader setting. Proofs and further details are found in the cited literature.

2 Multi-agent information models and epistemic logic

The first step in modeling conversation is a good notion of state and, hence, the 'static component' of the total enterprise. For simple scenarios like the above, a logical apparatus exists, viz. *epistemic logic* (Hintikka 1962, Fagin et al. 1995). In the Restaurant scenario, the initial information state for the waiter from the kitchen had six possible arrangements for the three dishes over the three of us. As far as the new waiter is concerned, all are options, and he only 'knows' what is true in *all* of them. The new information that I have the Meat reduces this uncertainty to only two: 'Fish-Vegetarian' or 'Vegetarian-Fish' for my father and mother. Either way, the waiter now knows that I have the Meat. Then hearing that my father has the Fish reduces this to one single option: the waiter has complete information about the correct placement of the dishes and does not need to ask any further question, even though he may still have to perform an inference to make this vivid to himself:

---

[2] The program included a Magic session with a card trick that failed to defy Logic in the end—plus a nonscheduled case of *crying*, a less common speech act in Academia. But that is another story.

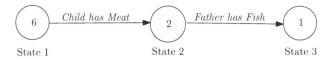

State 1   State 2   State 3

Epistemic logic: language and models   Here is some basic epistemic logic, as far as needed here. The syntax has a classical propositional base with added modal operators $K_i\phi$ ('i *knows that* ϕ') and $C_G\phi$ ('ϕ is *common knowledge* in group G'):

$$p \mid \neg\phi \mid \phi \vee \psi \mid K_i\phi \mid C_G\phi.$$

The states of our informational processes are *models* for this language, i.e., triples $M = (W, \{\sim_i \mid i \in G\}, V)$ where W is a set of worlds, the $\sim_i$ are binary accessibility relations between worlds that agent i cannot distinguish as viable candidates for the real situation[3], and V is a propositional valuation. The fundamental epistemic truth condition for knowledge of an agent is then as follows:

$$M, s \models K_i\phi \text{ iff for all } t \text{ with } s \sim_i t: M, t \models \phi.$$

This language can define an existential dual of knowledge $\neg K_j \neg\phi$ (or $\langle j\rangle\phi$): agent j considers it *possible that* ϕ, plus other useful expressions such as $K_j\phi \vee K_j\neg\phi$: agent j knows *whether* ϕ. In particular, *multi-agent interaction* is a crucial feature. For example, in asking a 'normal' question, a questioner Q conveys he does not know if ϕ: $\neg K_Q\phi \wedge \neg K_Q\neg\phi$. Moreover, usually he also thinks that the addressee A might know, which can be stated as an iterated two-agent assertion $\langle Q\rangle(K_A\phi \vee K_A\neg\phi)$.

State transitions: information flow and model update   Levels of knowledge about others occurred in the second scenario that was played with the children in *NEMO*:

> ***The Cards*** Three cards were given to three volunteers who stepped up: *1* got Red, *2* White, and *3* Blue. Each child could see his or her own card but not those of the others (I was circling my little volunteers to make sure). Child *2* was allowed one question, and she asked *1*: "Do you have the blue card?" *1* answered truthfully: "No". Which child figured out what in this process?

I asked beforehand, and all said they knew nothing. I asked again right after the question, and now Child *1* said he knew the cards. His reasoning, as whispered to me: "She would not have asked if she had the blue card herself. So, *3* has it." After the answer was given, children *1* and *2* said they knew the cards and *3* still did not. But (with a little help) *3* did understand why the others knew the cards.

All this can be analyzed in words, but here is how things would look in an epistemic state transition framework. The initial situation again has six options, and the uncertainty lines indicate what players hold possible from where they are:

---

[3] One often takes these relations to be equivalence relations, but this is optional.

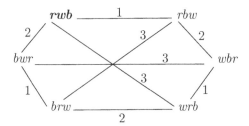

2's question, seen as informative[4], *eliminates* all worlds with second position '*b*':

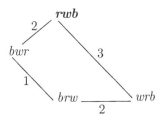

We see at once that, in the real world ***rwb***, *1* has no uncertainty line going out, and hence he knows the cards there. (We also see that *3* knows this, as it happens at both *rwb* and *wrb*.) Next, *1*'s answer eliminates all worlds with first position '*b*':

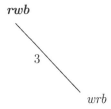

This reflects the final situation of the children.

Group knowledge   Once again, multi-agent interaction is crucial. Indeed, the children even achieve a new level of knowledge that is sui generis, viz. *common knowledge*: in addition to what they know about the facts of the situation, they also know that the others know, and so on, up to any iteration. Common knowledge occurs in philosophy, linguistics, and economics as a prerequisite for coordinated action. Technically, this new notion is defined as follows over our models:

$$M, s \models C_G \phi \text{ iff for all } t \text{ that are reachable from } s \text{ by some}$$
$$\text{finite sequence of } \sim_i \text{ steps } (i \in G) : M, t \models \phi.$$

This multi-agent view may seem far from standard logic and computation where single agents draw inferences or make calculation steps. But real argumentation is an interactive process, and even in the heartland of computation, very early on, Turing emphasized the crucial social character of using computers and learning.[5]

---

[4] Taking questions in this innocent way need not be sensible in the setting of *real games*!

[5] cf. (Turing 1950). Wilfried Sieg explained to me how Turing emphasized social learning.

Belief and other attitudes of agents   Knowledge is just one informational attitude of agents. One can also model *beliefs*, *probabilities*, and so on, using a broader variety of accessibility relations. A simple epistemic structure suffices for our aims, but we will mention less simplistic versions with agents' beliefs occasionally.

Summarizing then, our initial *NEMO* example is not 'child's play'. Conversational scenarios are a basic human ability involving sophisticated interactive knowledge that needs to be understood in depth. And thus, they provide a rich subject of study for Informatics, where logical and computational notions make good sense.

3 Conversation as computation: update actions

Communicative events range from simple public statements to complex private ones: recall my whispered conversation with child *1*. And much more subtle scenarios exist in our lives. To move this inside our logic, we need an explicit account of relevant actions and their effects. Here a powerful metaphor comes into play:

Conversation is Computation!

Conversation is really an interactive form of computation, much as present-day computational systems have many agents engaged in a wide variety of tasks. Technically, then, conversational processes, and communication in general, may be modeled using existing systems from the computational tradition. In this paper, we will focus mainly on *dynamic logic*, originally developed as a logical account of programs and their effects (Pratt 1976), which has gradually evolved into a general theory of action. We start with the simplest mechanism of information flow.

Public announcement as world elimination   Public announcements of true propositions P change the current situation as follows. For any model M, world s, and formula P true at s, $(M \mid P, s)$ (*M relativized to P at s*) is the submodel of M whose domain is the set $\{t \in M \mid M, t \models P\}$. In a picture, one goes

Crucially, truth values of formulas may change in such an update step: most notably, because agents who did not know that P now do after the announcement. This truth value change can be quite subtle over time, including even cases where statements make themselves false.[6] One needs logics to keep this all straight.

Product update with event models   Whispering is a public announcement in a subgroup of a larger group, but it is only partially observable to the others. Hiding, secrets, and limited observation are ubiquitous in everyday communication. Consider

---

[6] Truly announce P = "You do not know that p, but it is really true" – and P becomes false.

your *email*. The epistemic-dynamic role of *cc* is a public announcement. But the more sophisticated button *bcc* achieves a *partial* announcement that can even mislead other participants. More complex scenarios arise in computer security and in the arena of *games*, which are often designed to manipulate information flow. Partial observation of events may be analyzed as the following construction for changing models (Baltag et al. 1998). Scenarios where information flows in different ways for different agents can be represented in

$$\text{Event models} \quad A = (E, \{\sim_i \mid i \in G\}, \{PRE_e \mid e \in E\}).$$

Here E collects all relevant events. The uncertainty relations $\sim_i$ encode which events agents cannot distinguish. For example, when the children checked their cards, the girl with the white card could not tell '*1*'s seeing red' from '*1*'s seeing blue'. Now, information flow occurs because events e have *preconditions* PRE_e for their occurrence (say, my having a red card, not knowing the answer to my question, etc.). When you observe an event, you learn that something must have been the case for this to happen.

The following *Update Rule* encodes the resulting mechanism of information flow:

For any epistemic model (M, s) and event model (A, e), the *product model* $(M \times A, (s, e))$ has a distinguished new world (s, e), and then

(a) a domain $\{(s, e) \mid s$ a world in M, e an event in A, $(M, s) \models PRE_e\}$,

(b) accessibility relations $(s, e) \sim_i (t, f)$ iff *both* $s \sim_i t$ *and* $e \sim_i f$,

(c) the valuation for atomic formulas p at (s, e) is that for s in M.[7]

Product update models a wide variety of information scenarios. And the universe of models with product update $M \times A$ has a rich logical and computational structure.[8]

Belief and other dynamic phenomena   Knowledge was just one feature in information flow. If we also model agents' beliefs and expectations, product update can describe events affecting belief, including *misleading* actions, leading to false beliefs. Moreover, we need not just record *information update*. We can also model *belief revision*, a more agent-dependent phenomenon, which can depend on very different 'policies' for different types of agent, more conservative or more radical.[9]

---

[7] This stipulation of 'inertia' basically says that physical facts do not change under communication. This constraint can be lifted easily to let the system deal with genuine nonepistemic world change.

[8] Unlike with world elimination, epistemic product models can now get *larger* under update. But there is a counteracting force to this growth in complexity, as later models may be *bisimilar* with earlier ones, making the iterated epistemic long-term process cycle (van Benthem 2006C).

[9] Different policies even multiply when we define updates for further relevant phenomena in communication and interaction, such as changes in preferences or goals (van Benthem & Liu 2005).

4 Dynamic-epistemic logics of informative events

Given all these interesting actions that transform epistemic models, we want to study them explicitly. Now, keeping track of truth value changes for epistemic assertions can be as tricky as finding out what a particular program achieves over time. Thus, it is useful to keep track of both the statics and the dynamics in one logical calculus. Relevant frameworks from the computational literature include temporal logic, process algebra, or linear logic. Here, we choose *dynamic logic* (Kozen et al. 2000) with its two levels of expressions π for programs and propositions ϕ describing the successive states produced by these. The main operator of the language is

$[\pi]\phi$:   "after any successful execution of π, ϕ holds in the resulting state".

This language stays close to that of modal logic, the lingua franca of much of computational logic, and it treats dynamic processes as being equivalent up to *bisimulation*, probably the most widely used notion of process equivalence. Still, this section is not meant as propaganda for any approach but as a demonstration of how computational logic of conversation and much more is entirely feasible.

Dynamic epistemic logic of public announcement   The language of *public announcement logic PAL* is the epistemic language with added action expressions:

$$\begin{array}{lll} \text{Formulas} & P: & p \mid \neg\phi \mid \phi \vee \psi \mid K_i\phi \mid C_G\phi \mid [A]\phi \\ \text{Action expressions} & A: & P! \end{array}$$

Here, treating announcements as actions and having them explicitly inside modalities of the language comes from dynamic logic. The semantics is this:

$$M, s \models [P!]\phi \quad \text{iff} \quad \textit{if } M, s \models P, \textit{ then } M \mid P, s \models \phi.$$

There is a complete calculus of information flow under public announcement—i.e., a complete logic of basic communication (Plaza 1989, Gerbrandy 1999):

Theorem. *PAL* without common knowledge is axiomatized completely by the usual laws of epistemic logic plus the following *reduction axioms*:

$$\begin{array}{ll} [P!]q \leftrightarrow (P \rightarrow q) & \text{for atomic facts } q \\ [P!]\neg\phi \leftrightarrow (P \rightarrow \neg[P!]\phi) \\ [P!](\phi \wedge \psi) \leftrightarrow ([P!]\phi \wedge [P!]\psi) \\ [P!]K_i\phi \leftrightarrow (P \rightarrow K_i[P!]\phi). \end{array}$$

Methodology   These axioms describe conversation in an elegant style, analyzing effects of assertions in a compositional way by recursion on the 'postconditions' behind the dynamic modalities. Thus, they reduce every formula of our dynamic-epistemic language eventually to a formula in the static epistemic language (cf. the 'regression procedure' of (Reiter 2001)). In terms of the logic, the reduction procedure shows that *PAL* is *decidable*, since the static epistemic base logic is decidable.[10]

This method of 'dynamification' applies to a wide range of informational events. First, choose a static language with models that represent information states for groups of agents. Next analyze the relevant informational events as update models changing the static ones. These updates are then described explicitly in a dynamic extension of the language, which can also state the effects of events using propositions that hold after their occurrence. The resulting logics have a two-tier setup:

$$\boxed{static\ basic\ logic} \text{————} \boxed{dynamic\ extension}$$

At the static level, one gets a complete axiom system for one's chosen models. The computational analysis then adds a set of dynamic *reduction axioms* for effects of events. Thus every formula is equivalent to a static one—and hence, if the static base logic is decidable, so is its dynamic extension. In principle, this modular dynamic epistemic design is independent from specific properties of the static models. For example, the *PAL* axioms do not depend on assumptions about epistemic accessibility relations. Its completeness theorem holds just as well if the static models are arbitrary, validating the minimal modal logic K as some minimal logic of *belief*.

Technical issues   Sometimes, treating conversation as computation changes our ideas about an underlying static system. For example, the completeness theorem for *PAL* omits *common knowledge* after announcements. To get a reduction axiom for formulas $[P!]C_G\phi$, one must enrich epistemic logic beyond its standard version, cf. (van Benthem et al. 2005). *Conditional common knowledge* $C_G(P, \phi)$ says that ϕ is true in all worlds reachable via some finite path of accessibilities running entirely through worlds with P. Then we get the valid reduction law: $[P!]C_G\phi \leftrightarrow C_G(P, [P!]\phi)$. Conditional common knowledge is not definable in the basic epistemic language, but it is bisimulation-invariant, and completeness proofs are easily generalized.[11] There is an analogy here with *conditional* assertions $\phi \Rightarrow \psi$ in belief revision, which state what we would believe were the antecedent to be considered (van Benthem 2006A). *PAL* has a modal bisimulation-based model theory, with many interesting issues of expressive power and computational complexity.[12]

General dynamic epistemic logic   A more general product update for communicative and observational scenarios can also be dealt with in this dynamic logic format. The language of *dynamic-epistemic logic* (*DEL*) has the following syntax:

---

[10] This reduction does not settle computational *complexity*: basic epistemic logic is **Pspace**-complete, but translation via the axioms may increase the length of formulas exponentially. cf. Section 6.

[11] Indeed, *PAL* with conditional common knowledge is axiomatized completely by adding just one more valid reduction law $[P!]C_G(\phi, \psi) \leftrightarrow C_G(P \wedge [P!]\phi, [P!]\psi)$.

[12] cf. (van Benthem 2006D) for a survey of many open problems in this area.

$$p \mid \neg\phi \mid \phi \vee \psi \mid K_i\phi \mid C_G\phi \mid [A, e]\phi :$$

with (A, e) any event model with actual event e. The semantic key clause is

$$M, s \models [A, e]\phi \qquad \text{iff} \qquad M \times A, (s, e) \models \phi.$$

(Baltag et al. 1998) then showed completeness in this wider setting:

Theorem. *DEL* is effectively axiomatizable and decidable.

The key reduction axiom is the one extending that for public announcement:

$$[A, e]K_i\phi \leftrightarrow PRE_e \rightarrow \wedge\{K_i[A, f]\phi)) \mid f \sim_i e \text{ in } A\}.$$

Further challenges   Again consider *common knowledge* or *belief*. Just try to figure out what common beliefs hold in the following email scenario. Agent *1* sent message e, but in such a way that the other agent *2* believes that message *f* was sent:

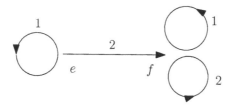

(van Benthem et al. 2005) extends *DEL* to a logic *LCC* using ideas from dynamic logic and μ-calculus to get complete sets of axioms for such scenarios.[13]

Dynamic logics for belief revision and preference change   The above format also provides complete logics for events of *belief revision* and even more general *preference change*. These involve conditional beliefs and compositional axioms for changes in them after 'hard facts' such as public announcements $P!$ or 'soft facts': weaker triggers for belief revision $\Uparrow P$ that may be overridden later on.[14]

Model change and other dynamic frameworks   The general idea behind update mechanisms for knowledge or belief is *definable model change*. One selects or even creates new individual objects (the worlds) out of old ones and then redefines the relevant relations between them. There are other systems than dynamic logic in the

---

[13] Another relevant issue is the 'view of agents' in product update. They satisfy *Perfect Memory* and *No Miracles*: learning only occurs through observation of suitable events. (van Benthem & Liu 2004) shows that this is complete—(Liu 2006) looks at much greater *diversity* of epistemic agents.

[14] Just to show the format, here are two reduction axioms for new beliefs after hard and soft triggers: $[P!]B_i(\phi \mid \psi) \leftrightarrow P \rightarrow B_i([P!]\phi \mid P \wedge [P!]\psi)$, $[\Uparrow P]B(\phi \mid \psi) \leftrightarrow (E(P \wedge [\Uparrow P]\psi) \wedge B([\Uparrow P]\phi \mid P \wedge [\Uparrow P]\psi)) \vee B([\Uparrow P]\phi \mid [\Uparrow P]\psi)$. Here E is the existential modality "in at least one world". For details, cf. (van Benthem 2006A, Baltag & Smets 2006).

computational literature with a similar flavour. For example, *process algebra* is a family of calculi for constructing new processes out of given ones. Indeed, our product update $M \times A$ respects bisimulation in the standard process-algebraic sense. In our view, *DEL* is a nice calculus of model change intermediate between dynamic logic and process algebra, which combines an 'external language' for defining processes with an 'internal language' describing properties of states within these processes. Merging major computational process paradigms may be a good idea in general.

5 Program structures in conversation

Genuine computation involves control over long *sequences* of actions. Likewise, conversation involves many assertions governed by *program constructions*. When talking with our dean, we first praise the current state of the Faculty of Science and then ask for funding. And what we say depends on his current state. We commiserate when he looks troubled and joke when he looks happy. Finally, there is an iterative process of 'flattery'. We keep saying nice things until his brow clears and the right moment for our funding request has come. Thus, conversation involves all the basic operations from sequential programming: (a) sequential composition ;, (b) guarded choice *IF ... THEN ... ELSE ...*, and (c) guarded iterations *WHILE ... DO ....*

A much-quoted concrete example is the puzzle of the 'Muddy Children':

> *After playing outside, two of three children have mud on their foreheads. They all see the others, but not themselves, so they do not know their own status. Now their father comes and says: "At least one of you is dirty". He then asks: "Does anyone know if he is dirty?" The children answer truthfully. As this question–answer episode repeats, what will happen?*

Nobody knows in the first round. Next, the muddy children argue as follows. 'If I were clean, the one dirty child I see would have seen only clean kids, and so she would have known that she was dirty. But she did not. So I must be dirty, too!' Thus both know their status in the second round. The third child knows it is clean one round later. The puzzle easily extends to more clean and dirty children.[15]

Clearly, all three preceding program constructions occur here: sequential assertion, guarded action (children must respond differently depending on what they know), and iteration: the process repeats until common knowledge is achieved.

Adding full dynamic logic   To analyze complex conversations, *PAL* or *DEL* must be extended with *propositional dynamic logic PDL*, which has a test operation $?\phi$ on

---

[15] For a concrete update sequence describing this scenario, cf. (Fagin et al. 1995, van Benthem 2006C).

propositions, plus the three regular operations of sequential composition ;, choice $\cup$, and iteration $*$. We display the major valid axioms here[16]:

$(a) \quad [\phi?]\psi \leftrightarrow (\phi \rightarrow \psi),$

$(b) \quad [\pi_1; \pi_2]\phi \leftrightarrow [\pi_1][\pi_2]\phi,$

$(c) \quad [\pi_1 \cup \pi_2]\phi \leftrightarrow ([\pi_1]\phi \wedge [\pi_2]\phi),$

$(d) \quad [\pi^*]\phi \leftrightarrow (\phi \wedge [\pi][\pi^*]\phi),$

and

$(e) \quad (\phi \wedge [\pi^*](\phi \rightarrow [\pi]\phi)) \rightarrow [\pi*]\phi.$

These axioms work by recursion on the first argument of our modal statements $[\pi]\phi$, rather than the second. It is known that *PDL* as a system of arbitrary actions is completely axiomatized by these principles—and indeed, it is decidable.[17]

Further constructions   But conversation also involves other program operations. It is crucial to the Muddy Children puzzle that the children answer *simultaneously*. This is parallel composition of individual actions, as in distributed computing and process algebra. *PAL* treats simultaneous speech as announcing a conjunction, and thus $(\phi \wedge \psi)!$ is a simple analogue of a parallel composition $\phi! \parallel \psi!$.[18]

Temporal logic   All this eventually embeds dynamic epistemic logics into broader *epistemic temporal logics* over branching trees of events (Fagin et al. 1995) and (Parikh & Ramanujam 2003). The latter links up with another process view in computer science, viz. temporal logics in the style of Pnueli, Clarke, and others. cf. (van Benthem & Pacuit 2006) for connections with our current setting.

6 Complexity of logical tasks

Computation involves a balance between representation and processing of data, and so do logical systems. Although dynamic epistemic logics provide a rich account of effects of events that carry information, their expressive power has a price in terms of *computational complexity*. Indeed, any logical system can be used for a variety of core tasks that all involve computational complexity.

Model checking   We start with *model checking*, i.e., determining whether $M, s \models \phi$ for a given model M and formula ϕ. For basic epistemic logic, this task is ***P-time*** in the size of formulas and models (Vardi 1997). In our conversational setting, model checking *DEL*-formulas corresponds to computing the effects of informational events in a given informational situation. (van Benthem et al. 2005) shows

---

[16] The axioms for π^* say that a universal modality $[\pi^*]$ is a greatest fixed-point operator.

[17] Combining *PDL* with epistemic logic into a richer version of *DEL* will involve recursions on both actions and postconditions. The precise nature of this joint approach remains to be understood.

[18] No explicit axiomatization is known yet for this parallel operator $\parallel$ in *PAL* or *DEL*.

that model-checking complexity remains **P-time** for arbitrary formulas ϕ of *DEL*.[19] Thus, verifying the effects of a given conversational plan is an easy task.

Satisfiability But the more ambitious task is *conversation planning*: how do we set up a setting in order to achieve certain desired effects? This can still be cast as a model checking problem when the epistemic 'space' is given beforehand (see below), but in general one asks for the existence of some information model satisfying some specified properties. This is the problem of *satisfiability* (*SAT*): when does a given formula have a model? The *SAT* problem for basic epistemic logic is **Pspace**-complete. The axioms for *PAL* provide a *SAT* reduction to this system, but given the shape of the axioms, this might be exponential. (Lutz 2005) provides a better reduction that shows that *SAT* complexity for *PAL* remains **Pspace**-complete.[20] Although this might suggest that dynamifying a base logic does not affect complexity, further dynamic epistemic logics still have surprises in store. In particular, the above combination of *PAL* with the program operations of dynamic logic *PDL*, i.e., the combination of two systems, each of which are decidable, leads to a surprise:

Theorem. (Miller & Moss 2005) *PAL* with *PDL* operations is *undecidable*.[21]

More concretely, designing puzzles like Muddy Children and solving conversation planning problems in them can be extremely hard!

Complexity of further tasks? Besides standard model checking and satisfiability, there may be other natural complexity issues for dynamic-epistemic logics. For instance, a set of admissible assertions, or a more general conversational protocol over some given initial model, generates a model M with all possible trajectories for an informational process. In such a model, we can ask for conversational plans achieving intended effects, say in the form of *PDL* programs as above that are guaranteed to move from the initial state to some state satisfying some goal proposition ϕ. The resulting intermediate model-checking problems asks whether an executable *PDL* program π exists such that $[\pi]\phi$ holds at the current state s in the given model M. This is not quite ordinary model-checking, but it is not full-fledged *SAT* either. Here is another variant issue. What actions are worth counting in our update setting? For instance, is there an analogue of the computational notion of *communication complexity* defined in (Yao 1979)? Finally, on the more empirical side, once partial observation of events is considered as in *DEL*, one expects intuitive complexity jumps from public to private announcement, or from speaking the truth to lying. But so far, not all relevant intuitions and empirical folk wisdom about such thresholds have been turned into precise mathematics yet.

Danger zones Many authors have explained (Halpern & Vardi 1989, Marx 2006, van Benthem & Blackburn 2006, van Benthem & Pacuit 2006) how modal logics practice the art of 'living dangerously' at the edge of undecidability. With expressive

---

[19] This complexity is *EXPTIME* for the full language with all *PDL* operations.

[20] The *SAT*-complexity for *DEL* probably remains the same.

[21] The proof uses *infinite* epistemic models: it is not known whether it holds with just finite models.

power *tree*-oriented, they are decidable guarded-quantifier formalisms. But when dangerous patterns become definable, in particular two-dimensional *grids* with two confluent relations, they tend to become undecidable—and may even incur non-arithmetical complexity. In a dynamic epistemic setting, geometric confluence reflects commutativity laws for modalities (Halpern & Vardi 1989, van Benthem 2001) that may make logics undecidable—though the precise recipe for disaster is delicate. For knowledge and action, an equivalence between $K[e]\phi$ (knowing that an event e produces a certain result ϕ) and $[e]K\phi$ (an event e's producing knowledge that ϕ) amounts to the semantic condition that agents have *perfect memory*. Thus, writing logics for well-endowed idealized agents can drive up complexity!

7 Reversing the direction: computation as conversation

We have amply shown by now how conversation can be viewed as computation, leading to interesting issues that can be studied by combining techniques from philosophical and computational logic. But this link also suggests an *inversion in perspective*. In particular, lower bound results concerning complexity often establish that some other problem of known complexity can be reduced to the current one. And though these reductions may be technical, usually, they convey a lot more useful information, often of a semantic nature—and hence, they establish stronger analogies than mere 'equi-difficulty'. To see this more concretely, take our analysis of conversation as a form of computation. The simple point that we wish to make now is that complexity analysis, as available in known results, also allows us to view

Computation as Conversation!

Realizing computation as conversation   High-complexity results are often taken to be bad news, as they say that some logical task is hard to perform. But the good news here is that, by the very same token, an interesting transfer happens: the logic manages to encode significant problems with mathematical content. For instance, consider the famous result that *SAT* in propositional logic is *NP*-complete. Reversing the perspective, this result also means that solving just one basic logical task has *universal computational power* for a large class of problems encountered in practice. Moreover, the proof of *NP*-completeness for propositional *SAT* even gives a simple *translation* from arbitrary computational tasks to logical ones.[22] The same reversal applies to other complexity classes. For example, ***Pspace***-complete is the solution complexity for many natural games (Papadimitriou 1994, van Emde Boas 2002) and, hence, being able to solve *SAT* problems in our base logic, i.e., the ability to create consistent epistemic scenarios suffices for solving lots of games.

Now, in this same light, consider the above result from (Miller & Moss 2005). What they prove is essentially that each *tiling problem*—and hence also each significant

---

[22] A course in propositional logic is at the same time one in universal computation, *if* only you knew the key . . . .

problem about computability by Turing machines—can be reduced effectively to a *SAT* problem in *PAL* + *PDL*. I literally take this result to mean the following:

> *Conversation has Universal Computing Power*   Any significant computational problem can be realized as one of conversation planning.

Even so, in this technical sense, 'computation as conversation' is mainly a metaphor. In what follows, I take one more step, which does not require us to 'take sides'.

8 Merging computation and conversation

The real benefit of bringing together computation and conversation is not reduction of one to the other. It is creating a broader theory with interesting new questions. In particular, a theory of computation that absorbs ideas from conversation must incorporate the dynamics of *information flow* and *social interaction*. We will mainly discuss one way of doing this here. It starts from known algorithms and then adds further structure. We proceed by a series of examples, as our aim is merely to show how many new questions can be asked at once in this setting, without established answers. At the end, we note a few more general trends.

Epistemizing algorithms   Consider the basic computational issue of Graph Reachability (*GR*). Given a graph G with distinguished points x, y, is there a chain of directed arrows in G leading from x to y? This task can be solved in ***Ptime*** in the size of the graph: there are fast quadratic-time algorithms finding a path (Papadimitriou 1994). The same analysis holds for the task of reachability of some point in G satisfying a general goal condition ϕ. *GR* models search problems in general, and the solution algorithm performs two closely related tasks: determining whether a route exists at all, and giving us an actual *plan* to get from x to y. We consider various ways of introducing knowledge and information.

Knowing you have made it   Suppose you are an agent trying to reach a goal region ϕ but with only limited observation of the graph in which you are moving. In particular, you need not know, at any point x, at which precise location you are. Thus, the graph G is now a model $(G, R, \sim)$ with accessibility arrows but also epistemic *uncertainty links* between nodes. A first epistemization of *GR* merely asks for the existence of some plan that will lead you to a point *that you know to be in the goal region ϕ*. (Brafman et al. 1993) analyzes a practical setting for this, with a robot whose sensors do not tell her exactly where she is standing. In this case, it seems reasonable to add a *test* to the task, inspecting current nodes to see whether we are definitely in the goal region: $K\phi$. Given the ***P***-time complexity of model checking for modal-epistemic languages, the new search task remains ***P***-time.

Having a reliable plan   In this setting, further issues arise. What about the plan itself? If we are to trust it, should not we require that we *know it to be successful*? Consider the following graph, with an agent at the root trying to reach a ϕ-point:

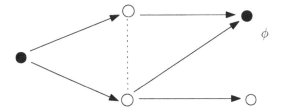

The dotted line indicates the agent cannot tell the two intermediate positions apart. A plan that achieves the goal is *Up ; Across*. But after following one part of this, the agent no longer knows where he is and, in particular, whether moving *Across* will reach the ϕ-point, or rather moving *Up*. Let us first formulate the requirement. Suppose for simplicity that a plan is just a finite sequence a of arrows. We may then require initial knowledge that this will work: $K[a]K\phi$. But this is just at the start: we may also want to be sure at all intermediate stages that the remainder of the plan will work. This would require truth of all formulas

$$[a_1]K[a_2]K\phi, \qquad \text{where } a = a_1;a_2{}^{23}$$

The existence of such a 'transparent' plan can still be checked in **Ptime**, since the number and size of the relevant assertions only increases polynomially. But this quickly gets more complex with plans defined by more complex *PDL* programs. It is not obvious how to even *define* the right notion of epistemic reliability, and we suspect that it may lead to new languages beyond *DEL* and *PDL*.[24]

Different types of agent   But there is more to the epistemic setting in the preceding example. Note that the agent in the graph has *forgotten* her first action: otherwise, she could not be uncertain between the two nodes in the middle. Our earlier *DEL*-style agents with *Perfect Recall* would not be in this mess, as they can only have uncertainties about what other agents did. And the earlier mentioned commutation law $K[a]\phi \rightarrow [a]K\phi$ which holds for them will automatically derive intermediate knowledge from initial knowledge $K[a]K\phi$. But there are many kinds of epistemic agent: with perfect recall, with finite memory bounds, etc.[25] Thus, epistemized algorithms naturally go together with questions about what sorts of agents are to be running them—and the complexity of these tasks-for-agents can vary accordingly.

Epistemic plans   But also, in an epistemic setting, the notion of a plan itself requires further thought. A plan is a sort of program that can react to circumstances, via conditional instructions such as *IF α THEN do a ELSE b*. The usual understanding of the test condition α is that one finds out if it holds and then chooses an action accordingly. But for this to work, the agent has to be able to *perform* that test! Say,

---

[23] If an agent has Perfect Recall, $K[a]\phi$ implies $[a]K\phi$, and the initial formula implies all the others. But for bounded agents, our distinction makes sense.

[24] As in (van Benthem 2001), some version of the epistemic μ-calculus may be needed, at least for reliable strategies of players in a game. These are related to the 'uniform strategies' of game theory.

[25] No standard taxonomy of this diversity exists yet: cf. (Liu 2006) for a first overview.

we ask a computer to check the current value of some variable or a burglar to check whether the safe has a Yale lock or some inferior brand. But in the above graph, the plan '*IF* you went *Up*, *THEN* move *Across ELSE* move *Up*', though correct as an instruction for reaching the goal, is no use, as the agent has no way of deciding which alternative holds. There are two ways of dealing with this. One is to include knowledge into programs (Fagin et al. 1995). We make actions dependent on conditions like 'the agent knows α' that can always be decided, provided agents have epistemic introspection.[26] Suitable epistemic programs are automatically transparent in the above sense (van Benthem 2001). The other option is to define a notion of '*executable plan*' in an epistemic model M, making sure that agents can find out whether a test condition holds at any stage where this is needed. But so far, I have not found a definition for epistemic executability that satisfies me.

Dynamifying static logics: update actions   Finding out whether a proposition holds involves actions of communication or observation, and hence, we move beyond epistemized static logics to dynamic ones. Then we could model the above test conditions α as explicit actions of *asking* whether α holds. This requires richer multi-agent models, though, where one can query other agents, or perhaps Nature, about certain things. We will not pursue this topic here, but the logic *DEL* in this paper is a showcase of 'dynamification'. Thus, it should be well suited for analyzing dynamified algorithms—and so are epistemic variants of *PDL* or the μ-calculus.[27]

Multi-agent scenarios and interactive games   Several epistemic scenarios in the preceding discussion suggest adding more than one agent, moving from traditional lonely algorithmic tasks to more social ones. For example, reaching a goal and knowing that you are there naturally comes with variants where *others* should *not* know where you are. Examples in the literature include the 'Moscow Puzzle' (van Ditmarsch 2002), where people have to tell each other the cards that they have without letting a third party present know the solution. Card games, or the earlier-mentioned use of email, provide many further examples. This social interactive perspective comes out even more in the setting of *games* and interaction between different players. Indeed, games have been proposed as a very general model of computation (Abramsky 2006), and new logical questions about them abound (van Benthem 2005B).

Reachability and sabotage   Turning algorithms into games involves the 'prying apart' of existing algorithms into games with different roles for different agents. Early examples are logic games in the style of Lorenzen, Ehrenfeucht, or Hintikka (cf. the survey in (van Benthem 1999)). A more algorithmic example is in (van Benthem 2005A). Consider again Graph Reachability. The following picture gives a travel network between two European capitals of logic and computation:

---

[26] Interestingly, some heuristic algorithms in (Gigerenzer & Todd 1999) have this flavour.

[27] An extreme case of this setting are pure information games, where all moves are actions of asking questions and giving answers, and players go for goals like 'being the first to know'.

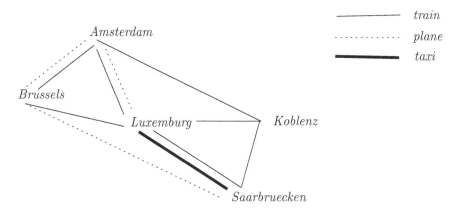

It is easy to plan trips either way. But what if the transportation system breaks down and a malevolent demon starts canceling connections, anywhere in the network? At every stage of our trip, let the demon first take out one connection. Now we have a two-player *sabotage game*, and the question is who can win it where. Some simple reasoning will show that, from Saarbruecken, a German colleague still has a winning strategy. But the Dutch situation is less rosy: Demon has the winning strategy.

This example suggests a general transformation for any algorithmic task. It becomes a *sabotaged* one when it is cast as a game with obstructing players. This raises several new questions, e.g., about logical languages describing these games, and players' plans (*strategies*) in them. In particular, how does the *computational complexity* of the original task change when we need to solve the new game? For sabotaged Graph Reachability, it has been shown in (Rohde 2005) that this complexity moves up from low **P**-time to **Pspace**-completeness. That is, the problem now takes a polynomial amount of memory space, which makes it of the complexity of Go or Chess.[28]

Catch Me If You Can   But there is no general rule predicting when a newly created game becomes more complex than its algorithmic ancestor. Again consider graphs, the setting par excellence for algorithmic tasks, but now with another game variant of GR. 'Obstruction' could also mean that some other player tries to catch me en route, making it impossible for me to continue. It is easy to cast this as a game, too:

> Starting from an initial position (G, x, y) with me located at x and you at y, I move first, then you, and so on. I win if I reach my goal region in some finite number of moves without meeting you. You win in all other cases.[29]

This game, too, is very natural, and it models a wide variety of realistic situations, such as warfare, or avoiding certain people at receptions.[30] But this time, the

---

[28] Link cutting games also have other interesting interpretations. (van Benthem 2006B) has a variant dual to the above where a Teacher tries to trap a Student into reaching a certain state of knowledge.

[29] Thus, you win: if you catch me before I am in the goal region, if I get stuck, or if the game continues indefinitely. Other natural ways of casting these conditions would allow draws.

[30] Fabius Maximus Cunctator tried to win a war by avoiding his enemy Hannibal throughout.

computational complexity stays lower.[31] Solving *Catch Me If You Can* still only takes **Ptime** in the size of the graph! This can be seen by the analysis of the analogous 'Cat & Mouse' game in (Greenlaw et al. 1991).[32, 33]

Adding knowledge and observation again   In actual warfare, catching games naturally involves limited observation and partial knowledge. In such *games of imperfect information*, players need not be able to see where the others are, and solution complexity may go up to **Pspace** and beyond. (Sevenster 2006) is an extensive study of various epistemized algorithms in this setting, using connections with the '*IF* logic' of (Hintikka & Sandu 1997) to clarify their properties. In particular, he shows that the situation is delicate. For example, consider that mild form of warfare called the game of 'Scotland Yard'. Here the invisible player who tries to avoid getting caught has to reveal her position after every k moves for some fixed k. But then the game can be turned into one of perfect information by re-encoding players' moves making k-sequences of old moves into single steps. (van Benthem 2001, van Benthem 2005B) study many other aspects of merging *DEL* with game theory.[34]

Rephrasing the issues in game theory?   From a genuine game-theoretic viewpoint, many other questions may become relevant, however. For example, Sevenster's major complexity results are in the *IF* tradition of asking whether some player has a winning strategy even when hampered by lack of knowledge. But the most crucial feature of finite games of imperfect information, both mathematically and in practice, is the existence of something more delicate: *Nash equilibria in mixed strategies*, letting players choose moves with certain probabilities. Maybe it is the resulting *game values* that we should be after for gamified algorithms. Thus, gamification as generalized computation should also make us pause and think about the most natural counterparts to the properties of algorithms when they were still pure.

This is just one of many issues when we take game structure seriously. Imperfect information games also invite explicit events of observation and communication (Osborne & Rubinstein 1994, van Benthem 1999, van Benthem 2001). Moreover, they fit naturally with the *parallel action* mentioned earlier, as much of game theory is about simultaneous choice of moves by players. And then: why two players, and not more? For example, even inside the heartland of logic games, it has been proposed that argumentation, often cast as a tennis match, really needs a 'Proponent', an 'Opponent', and a *Judge*. Thus, our view of algorithms in a social setting naturally

---

[31] The difference with the Sabotage game is that the graph remains fixed during the game.

[32] I owe this reference to Merlijn Sevenster, who also points out the finer complexity difference that Reachability is **NL**-complete, whereas Cat & Mouse is **Ptime**-complete.

[33] A direct argument is as follows. The game can be recast as a graph game over an extended graph with positions (G, x, y) counting players' moves as described while allowing you 'free moves' when I am caught or get stuck. Now we let you win if you can keep moving forever. It is known that graph games like this can be solved in **Ptime**. One can see this as a modal model-checking problem for formulas $<>^n T$ with n the graph size.

[34] (van Otterloo 2005, van Benthem 2007) study extensive games with explicit actions of announcing relevant facts or even players' intentions concerning their future moves.

merges computer science, logic, and game theory with new links and new research questions running all across.

9 Toward a general theory: transformations and merges

Our discussion in the preceding section has been just a bunch of examples, trying to convey the pleasure of exploring an interactive epistemic viewpoint on computation. But it also suggests several more systematic topics.

Epistemizing logics   One broad concern is the design of appropriate *logical languages* for these new structures. This might seem a simple matter of combining components like dynamic and epistemic logic, but it can be much more interesting.[35] Next, relevant tasks for these languages can fall into the cracks of the standard notions of complexity. For example, natural planning problems seem intermediate between model checking and satisfiability. They ask, in a given model M with state s, whether some epistemic plan exists that takes us from s to the set of goal states. Thus, *epistemizing logics* is a nontrivial exercise, when done with a heart.

Epistemizing and gamifying algorithms   Next, there is the issue of finding *general transformations* on algorithms behind the above examples. Instead of botany, one would want general results on what these do to the solution complexity of the original task. The dissertations (Rohde 2005, Sevenster 2006) were the first steps.

A bit quixotically, what we are doing here can be seen as *dynamification* once more, but now at a meta-level. We have been using dynamic viewpoints to transform given problems in their original guise, and now, we are trying to make that process itself into an object of logical study. This is one way of seeing more unity in the diverse examples and logics that arise when 'computation and conversation' are mixed together. But there are other ways. In particular, promising *convergences* can be observed between various systems for describing 'computation and conversation', witness the comparison among dynamic epistemic logic, epistemic temporal logic, modal product logics, and other paradigms in (van Benthem & Pacuit 2006). More generally, one broad aim of theory construction in this arena is as follows:

Epistemized process theory   Moving toward fundamental theories of computation and bringing in explicit considerations of observation and conversation suggest epistemic versions of existing process theories, such as Process Algebra. As the latter includes an explicit account of 'communication channels', making the connection seems appropriate.[36] The same points apply to interaction and game semantics for

---

[35] For example, (van Benthem 1999) shows how even the issue of finding 'the epistemic version' of propositional dynamic logic is not at all simple, as *DEL* suggests a two-level approach, providing both *states* and *arrows* with uncertainty relations, giving us a range of options for matching logical languages.

[36] (Dechesne & Wang 2007) compares renderings of communication scenarios in *DEL* and Process Algebra.

computation. For example, standard models for linear logic achieve nondeterminacy by moving to infinite games. But nondeterminacy reigns in simple finite games with imperfect information, suggesting epistemic versions of linear game semantics. Also, strategies in linear logic crucially involve switching across games, and using information about moves in one to make the best moves in the other (Abramsky 2006), which is again well within our circle of ideas.[37] Of course, as we noted, there is also the issue of how all this relates to existing *game theory*. Perhaps, the current contacts among logic, computer science, and game theory may be viewed as preliminaries to a new theory with aspects of all three.

10 Conclusions

This paper fits in a broad current trend. Bringing together computation and broader information-based activities of conversation and communication is in the air, and it has been there for at least two decades. It may be seen with the epistemic analyses of communication protocols in (Fagin et al. 1995), with calculi of distributed computing like Milner's *CSP*, and of course, with modern theories of agents and intelligent information systems. We have tried to show here that this trend is more than a metaphor by pointing at concrete logics that deal with it, and at a sequence of interesting new issues that arise when we merge the two agendas systematically. To some, the resulting theory may look strange at first, as it combines hard-core computational logic with epistemic logics from the 'softer' philosophical tradition—something that may look even more outrageous when we add, not just knowledge, but also agents more ephemeral beliefs, and who knows, even their intentions and most intimate desires. Still, we think computation plus information update and belief revision is a perfectly viable marriage. It is rich in theory, and also, it fits very well with modern computation in societies of interacting agents. Indeed, recent research programs like 'social software' (Parikh 2002) even take this into activist mode and propose not just analyzing existing social procedures in this style but even designing new better ones. In this, social software is like 'mechanism design' in game theory but pursued by sophisticated computational techniques.

As a counter-point to such 'soft' social settings, it needs to be said that the Dynamic Turn advocated in this paper is also observable in hard-core areas like physics. Recent interfaces between computer science and quantum mechanics emphasize the dynamic interactions of observing agents with physical systems in operator-based Hilbert spaces. Accordingly, systems of dynamic logic and game semantics for linear logics are crossing over from computation to the foundations of physics, as well as the practices of quantum computation. (Abramsky & Coecke 2004, Baltag & Smets 2004), and some entries in (Rahman et al. 2004) are samples of this trend. For what this might mean in a broader information theory, cf. (Abramsky 2006).

---

[37] In recursion theory, a precursor is (Condon 1988) on Turing machines run by agents with limited observation—though for specialized complexity-theoretic purposes.

Another way of stating the main point of this paper is that computation is a pervasive and fundamental category across the sciences and humanities, provided we cast it in its proper generality, linking up with epistemic logics broadly construed. In one direction, our dynamic epistemic systems show how this introduces significant computational models into the study of what used to be thought of as preserves for linguists and philosophers. In the opposite direction, we can 'epistemize' and 'dynamify' existing logics and algorithms, to get interesting broader theories. Returning to our Introduction, it should be clear that this is much more than 'implementation' in an auxiliary sense, but rather a way of letting fundamental ideas from computer science play the central academic role that they so clearly deserve.

Despite all these grand perspectives, this paper was written by a logician, as biased—Heaven knows—as the next person. This may be a good place for a disclaimer. Despite the amount of space devoted to *dynamic epistemic logic* in this paper, we have used it mainly as a 'search-light system' for interesting phenomena, not as the final word on the structure and flow of information. Indeed, even from the viewpoint of the *NEMO* Restaurant, we have still missed crucial aspects of the children's activities! The waiter or the card players do not just update with new information, but they also *infer* things already at their disposal. But valid conclusions from existing information do not change a current *DEL* information state. To describe this finer dynamics, another process structure is needed.[38] Likewise, our discussion of testing conditions for algorithms or games suggests that we have left out the dynamics of *evaluation* (van Benthem 1996). The logical core tasks of inference and model-checking have their own dynamics, which goes beyond our framework here. Thus, even the logical foundations of information and computation remain wide open.[39]

Finally, what about *Computing in Europe*? I would like to believe that a broad stance on any subject matter, reflecting a certain erudition beyond one's immediate specialty is a crucial aspect of European culture. The view of computation offered in this paper qualifies. Also, pursuing theoretical interests without immediate practical gain seems a well-established European value, even though I admit it may be one of the old leisure class, rather than one of the hectic yuppies of today. But in this summer season, it is another image that intrigues me, based on just one passage in this paper. The undecidability of dynamic epistemic logic with iteration shows how the most difficult computational problems can be solved in *successful conversation*. Thus, what is going on at, say, all those Parisian terraces as I write these lines, on the last day of the Tour de France, is one gigantic parallel computer. 'Computing' usually evokes an image of boring machines, or even more boring nerds. Wouldn't it be great if *Computing in Europe* were the *Art of Conversation in the Old World*?

---

[38] This can be done, but no consensus exists. (van Benthem 1996, Chapter 11) dynamifies Herbrand models for Prolog to model such inferential steps, whereas (Abramsky 2006) presents a more general universe of abstract information states, where inference or computation steps do increase information.

[39] cf. the forthcoming *Handbook of the Philosophy of Information*, P. Adriaans & J. van Benthem, eds., 2007.

Acknowledgement. I would like to thank various audiences that have been exposed to these ideas, at the ESF Workshop 'Games and Verification' in Cambridge 2006, the ESSLLI Workshop on 'Knowledge and Rationality' in Malaga 2006, the Project 'Games, Action & Social Software' at NIAS Wassenaar 2006, and two Indian Logic Conferences held in Kolkata and Mumbai 2007. In particular, I would like to thank Merlijn Sevenster for helpful comments on complexity and games and Andrea Sorbi for expert editorial help far beyond the call of duty.

References

[Abramsky 2006] Abramsky S. (2006) Information, Processes, and Games. To appear in Adriaans P., van Benthem J. (eds.), Handbook of the Philosophy of Information. Elsevier Science Publishers, Amsterdam

[Abramsky & Coecke 2004] Abramsky S., Coecke B. (2004) A Categorical Semantics of Quantum Protocols. In Proceedings of the 19th Annual IEEE Symposium on Logic in Computer Science (LiCS '04), IEEE Computer Science Press

[Baltag et al. 1998] Baltag A., Moss L., Solecki S. (1998) The Logic of Public Announcements, Common Knowledge and Private Suspicions. In Proceedings TARK 1998, Morgan Kaufmann Publishers, Los Altos, 43–56

[Baltag & Smets 2004] Baltag A., Smets, S. (2004) The Logic of Quantum Programs. In Proceedings of the 2nd International Workshop on Quantum Programming Languages, TUCS General Publication No 33, Turku Center for Computer Science. Extended version: LQP: The Dynamic Logic of Quantum Information. Oxford Computing Lab & Philosophy, Free University Brussels

[Baltag & Smets 2006] Baltag A., Smets S. (2006) Dynamic Belief Revision over Multi-Agent Plausibility Models. In Proceedings LOFT 2006, Department of Computing, University of Liverpool

[van Benthem 1996] van Benthem J. (1996) Exploring Logical Dynamics. CSLI Publications, Stanford

[van Benthem 1999] van Benthem J. (1999) Logic in Games. Lecture Notes, ILLC Amsterdam

[van Benthem 2001] van Benthem J. (2001) Games in Dynamic Epistemic Logic. In Bonanno G., van der Hoek W. (eds.) Bulletin of Economic Research, 53:4, 219–248

[van Benthem 2005A] van Benthem J. (2005) An Essay on Sabotage and Obstruction. In D. Hutter (ed.), Mechanizing Mathematical Reasoning, Essays in Honor of Jörg Siekmann on the Occasion of his 69th Birthday. Springer, LNCS, 2605:268–276

[van Benthem 2005B] van Benthem J. (2005) Open Problems in Logic and Games. In Artemov S., Barringer H., d'Avila Garcez A., Lamb L., Woods J. (eds.) Essays in Honour of Dov Gabbay. King's College Publications, London, 229–264

[van Benthem 2006A] van Benthem J. (2006) Dynamic Logic of Belief Revision. ILLC Tech Report, DARE electronic archive, University of Amsterdam. To appear in J. Applied Non-Classical Logics

[van Benthem 2006B] van Benthem J. (2006) Living With Rational Animals. Invited Lecture at Workshop on Knowledge and Rationality, 18th ESSLLI Summer School, Malaga

[van Benthem 2006C] van Benthem J. (2006) One is a Lonely Number: On the Logic of Communication. In: Chatzidakis Z., Koepke P., Pohlers W. (eds.) Logic Colloquium '02. ASL and A.K. Peters, Wellesley MA, 96–129 [van Benthem 2006D] van Benthem J. (2006) Open Problems in Update Logic. In Gabbay D., Goncharov S., Zakharyashev M. (eds.) Mathematical Problems from Applied Logic I. Springer, New York, Novosibirsk, 137–192

[van Benthem 2007] Rationalizations and Promises in Games. Bejing Philosphical Review, Chinese Academy of Social Sciences

[van Benthem et al. 2005] van Benthem J., van Eijck J., Kooi B. (2005) A Logic for Communication and Change. ILLC & CWI Amsterdam & philosophy department, Groningen. First version in van der Meijden R. et al. (eds.) Proceedings TARK 2005, Singapore. Extended version in: Inform. and Comput. 2006

[van Benthem & Blackburn 2006] van Benthem J., Blackburn P. (2006) Modal Logic: A Semantic Perspective. Tech Report, ILLC, Amsterdam. In: Blackburn P., van Benthem J., Wolter F. (eds.) Handbook of Modal Logic. Elsevier, Amsterdam, 2007

[van Benthem & Liu 2004] van Benthem J., Liu F. (2004) Diversity of Logical Agents in Games. Philosophia Scientiae 8:2, 163–178

[van Benthem & Liu 2005] van Benthem J., Liu F. (2005) Dynamic Logic of Preference Upgrade. ILLC Tech Report, DARE electronic archive, University of Amsterdam. To appear in: J. Applied Non-Classical Logics

[van Benthem & Pacuit 2006] van Benthem J., Pacuit E. (2006) The Tree of Knowledge in Action. Tech Report, ILLC Amsterdam. In: Proceedings AiML 2006, Melbourne

[Brafman et al. 1993] Brafman R., Latombe J-C, Shoham Y. (1993) Towards Knowledge-Level Analysis of Motion Planning. Proceedings AAAI 1993, 670–675

[Condon 1988] Condon A. (1988) Computational Models of Games. PhD Thesis. Computer Science Department, University of Washington

[Dechesne & Wang 2007] Dechesne F., Wang Y. (2007) Dynamic Epistemic Verification of Security Protocols. In van Benthem J., Ju S., Veltman F. (eds.) A Meeting of the Minds. Proceedings LORI, Bejing 2007. College Publications, London, 129–143

[van Ditmarsch 2002] van Ditmarsch H. (2002) Keeping Secrets with Public Communication. Department of Computer Science. University of Otago

[van Emde Boas 2002] van Emde Boas P. (2002) Models for Games and Complexity. Lecture Notes. ILLC, Amsterdam

[Fagin et al. 1995] Fagin R., Halpern J., Moses Y., Vardi M. (1995) Reasoning about Knowledge. MIT Press, Cambridge (Mass.)

[Gerbrandy 1999] Gerbrandy J. (1999) Bisimulations on Planet Kripke. Dissertation DS-1999-01. Institute for Logic, Language and Computation. University of Amsterdam

[Gigerenzer & Todd 1999] Gigerenzer G., Todd P. M., ABC Research Group (1999) Simple Heuristics That Make Us Smart. Oxford University Press

[Greenlaw et al. 1991] Greenlaw R., Hoover H., Ruzzo W. (1991) A Compendium of Problems Complete for P. University of Alberta, Computer Science Department, Technical Report 91–11

[Halpern & Vardi 1989] Halpern J., Vardi M. (1989) The Complexity of Reasoning about Knowledge and Time, I: lower bounds. J. Comput. System Sci., 38:1, 195–237

[Hintikka 1962] Hintikka J. (1962) Knowledge and Belief. Cornell University Press, Ithaca

[Hintikka & Sandu 1997] Hintikka J., Sandu G. (1997) Game-Theoretical Semantics, In van Benthem J., ter Meulen A. (eds.) Handbook of Logic and Language, Elsevier, Amsterdam, 361–410.

[Kozen et al. 2000] Kozen D., Harel D., Tiuryn J. (2000) Dynamic Logic. MIT Press, Cambridge (Mass.)

[Liu 2006] Liu F. (2006) Diversity of Logical Agents. ILLC Research Report, University of Amsterdam. Presented at Workshop on Bounded Agents, ESSLLI Malaga 2006

[Lutz 2005] Lutz C. (2005) Complexity and Succinctness of Public Announcement Logic. LTCS Report 05-09, Technical University Dresden

[Marx 2006] Marx M. (2006) Complexity of Modal Logics. To appear in Blackburn P., van Benthem J., Wolter F. (eds.) Handbook of Modal Logic. Elsevier, Amsterdam

[Miller & Moss 2005] Miller J., Moss L. (2005) The undecidability of iterated modal relativization. Studia Logica, 79:3, 373–407

[Osborne & Rubinstein 1994] Osborne M., Rubinstein A. (1994) A Course in Game Theory. MIT Press, Cambridge (Mass.)

[van Otterloo 2005] A Security Analysis of Multi-Agent Protocols. Dissertation. Department of Computing, University of Liverpool, & ILLC, University of Amsterdam, DS-2005-05

[Papadimitriou 1994] Papadimitriou C. (1994) Computational Complexity. Addison-Wesley

[Parikh 2002] Parikh R. (2002) Social Software. Synthese 132:187–211

[Parikh & Ramanujam 2003] Parikh R., Ramanujam R. (2003) A Knowledge Based Semantics of Messages. CUNY New York & Chennai, India. In van Benthem J., van Rooy R. (eds.) Special Issue on Information Theories, J. Logic Lang. Inform., 12:4, 453–467

[Plaza 1989] Plaza J. (1989) Logics of Public Announcements. In Proceedings 4th International Symposium on Methodologies for Intelligent Systems

[Pratt 1976] Pratt V. (1976) Semantical Considerations on Floyd-Hoare Logic. In Proceedings 17th Ann. IEEE Symposium on Foundations of Computer Science, 109–121

[Rahman et al. 2004] Rahman S., Gabbay D., Van Bendegem J-P , Symons J. (2004) Logic, Epistemology, and the Unity of Science, Vol. I. Kluwer, Dordrecht

[Reiter 2001] Reiter R. (2001) Knowledge in Action. MIT Press, Cambridge (Mass.)

[Rohde 2005] Rohde Ph. (2005) On Games and Logics over Dynamically Changing Structures. Dissertation. Rheinisch-Westfälische Technische Hochschule Aachen

[Sevenster 2006] Sevenster M. (2006) Branches of Imperfect Information: Logic, Games, and Computation. Dissertation DS-2006-06. ILLC Amsterdam

[Turing 1950] Turing A. M. (1950) Computing machinery and intelligence. Mind 59:433–460

[Vardi 1997] Vardi M. (1997) Why is Modal Logic So Robustly Decidable?. In Immerman N., Kolaitis Ph. (eds.) Descriptive Complexity and Finite Models. American Mathematical Society.

[Yao 1979] Yao A. C. (1979) Some Complexity Questions Related to Distributed Computing. In Proceedings of the 11th STOC, 209–213

Computation Paradigms in Light of Hilbert's Tenth Problem

Yuri Matiyasevich*

Steklov Institute of Mathematics, St. Petersburg 191023, Russia
yumat@pdmi.ras.ru
http://logic.pdmi.ras.ru/~yumat

Summary. This is a survey of a century-long history of interplay between Hilbert's tenth problem (about solvability of Diophantine equations) and different notions and ideas from Computability Theory. The present paper is an extended version of [83].

1 Statement Of The Problem: Intuitive Notion Of Algorithm

In the year 1900, the prominent German mathematician D. Hilbert delivered to the *Second International Congress of Mathematicians* (held in Paris) his famous lecture titled *Mathematische Probleme* [41]. There he put forth 23 (groups of) problems that were, in his opinion, the most important open problems in mathematics that the pending 20th century would inherit from the passing 19th century. Problem number 10 was stated as follows:

> **10. Entscheidung der Lösbarkeit einer diophantischen Gleichung.**
> Eine diophantische Gleichung mit irgendwelchen Unbekannten und mit ganzen rationalen Zahlkoeffizienten sei vorgelegt: *man soll ein Verfahren angeben, nach welchem sich mittels einer endlichen Anzahl von Operationen entscheiden läßt, ob die Gleichung in ganzen rationalen Zahlen lösbar ist.* [2]

---

* The author is very grateful to Martin Davis and Grant Olney Passmore for their comments that corrected English and improved the presentation. Support from the Council for Grants of the President of the Russian Federation under grant NSh-8464.2006.1 is also acknowledged.

[2] **10. Determination of the Solvability of a Diophantine Equation.** Given a Diophantine equation with any number of unknown quantities and with rational integral numerical coefficients: *Devise a process according to which it can be determined by a finite number of operations whether the equation is solvable in rational integers.*

A *Diophantine equation* is an equation of the form

$$P(x_1, \ldots, x_m) = 0 \tag{1}$$

where P is a polynomial with integer coefficients. Hilbert raised the question about solving Diophantine equations in *"rational integers"* that were nothing else but numbers $0, \pm 1, \pm 2, \ldots$, which we will call just *integers*. In the last section of the paper, we will investigate a more general version of the problem in which solutions are allowed to be arbitrary *"algebraic integers."*

A method demanded by Hilbert would allow us to recognize also solvability of Diophantine equations in *natural numbers* $0, 1, 2, \ldots$, namely, equation (1) has a solution in natural numbers if and only if equation

$$P(p_1^2 + q_1^2 + r_1^2 + s_1^2, \ldots, p_m^2 + q_m^2 + r_m^2 + s_m^2) = 0 \tag{2}$$

has a solution in arbitrary integers. Without lost of generality in this paper, we will deal with solving Diophantine equations in natural numbers, and respectively, all italic lowercase Latin letters will range over $0, 1, 2, \ldots$ (unless otherwise specified).

Since Diophantus's time (3rd century A.D.), number-theorists have found solutions for plenty of Diophantine equations and have proved the unsolvability of a large number of other equations. However, for different classes of equations, or even for different individual equations, one had to invent different specific methods. In the tenth problem, Hilbert asked for a *universal* method for recognizing the solvability of Diophantine equations; i.e., in modern terminology, the tenth problem is a *decision problem* (the only one among the 23 problems).

Note that Hilbert did not use the word "algorithm" in his statement of the tenth problem. Instead, he used the rather vague wording *"a process according to which it can be determined by a finite number of operations ..."*. Although he could have used the word "algorithm," it would not really have helped much to clarify the statement of the problem because, at that time, there was no rigorous definition of the general notion of an algorithm. What existed was a number of examples of particular mathematical algorithms (such as the celebrated Euclidean algorithm for finding the greatest common divisor of two integers) and an intuitive conception of an algorithm in general.

Does this imply that Hilbert's tenth problem was ill-posed? Not at all. The absence of a general definition of an algorithm was not in itself an obstacle to finding a positive solution of Hilbert's tenth problem. If somebody invented the required *"process,"* it would presumably be clear that in fact this process does the job, so an intuitive conception of an algorithm would be sufficient for positive solution of the tenth problem, which was, most likely, Hilbert's expectation.

It took 70 years before Hilbert's tenth problem was solved in the negative sense: *There exists no algorithm (i.e., no Turing Machine, no recursive function, and so on) that would tell for an arbitrary Diophantine equation whether it has a solution.* However, the following question naturally arises: *would Hilbert accept this technical result as a "solution" of his problem?* The following citation from Hilbert's address [41] suggests that he would be completely satisfied by the work done by logicians with respect to his tenth problem:

> Mitunter kommt es vor, daß wir die Beantwortung unter ungenügenden Voraussetzungen oder in unrichtigem Sinne erstreben und infolgedessen nicht zum Ziele gelangen. Es entsteht dann die Aufgabe, die Unmöglichkeit der Lösung des Problems unter den gegebenen Voraussetzungen und in dem verlangten Sinne nachzuweisen. Solche Unmöglichkeitsbeweise wurden schon von den Alten geführt, indem sie z. B. zeigten, daß die Hypotenuse eines gleichschenkligen rechtwinkligen Dreiecks zur Kathete in einem irrationalen Verhältnisse steht. In der neueren Mathematik spielt die Frage nach der Unmöglichkeit gewisser Lösungen eine hervorragende Rolle, und wir nehmen so gewahr, daß alte schwierige Probleme wie der Beweis des Parallelenaxioms, die Quadratur des Kreises oder die Auflösung der Gleichungen 5. Grades durch Wurzelziehen, wenn auch in anderem als dem ursprünglich gemeinten Sinne, dennoch eine völlig befriedigende und strenge Lösung gefunden haben.
>
> Diese merkwürdige Tatsache neben anderen philosophischen Gründen ist es wohl, welche in uns eine Überzeugung entstehen läßt, die jeder Mathematiker gewiß teilt, die aber bis jetzt wenigstens niemand durch Beweise gestützt hat–ich meine die Überzeugung, daß ein jedes bestimmte mathematische Problem einer strengen Erledigung notwendig fähig sein müusse, sei es, daß es gelingt, die Beantwortung der gestellten Frage zu geben, sei es, daß die Unmöglichkeit seiner Lösung und damit die Notwendigkeit des Mißlingens aller Versuche dargetan wird.[3]

---

[3] Occasionally it happens that we seek the solution under insufficient hypotheses or in an incorrect sense and, for this reason, do not succeed. The problem then arises: to show the impossibility of the solution under the given hypotheses or in the sense contemplated. Such proofs of impossibility were effected by the ancients, for instance, when they showed that the ratio of the hypotenuse to the side of an isosceles triangle is irrational. In later mathematics, the question as to the impossibility of certain solutions plays a preeminent part, and we perceive in this way that old and difficult problems, such as the proof of the axiom of parallels, the squaring of circle, or the solution of equations of the 5th degree by radicals have finally found fully satisfactory and rigorous solutions, although in another sense than that originally intended.

It is probably this important fact along with other philosophical reasons that gives rise to conviction (which every mathematician shares, but which no one has as yet supported by a proof) that every definite mathematical problem must necessarily be susceptible of an exact settlement, either in the form of an actual answer to the question asked or by the proof of the impossibility of its solution and therewith the necessary failure of all attempts.

2 Equations: From Words To Numbers

The research directed toward the negative solution started at the end of the 1940s. At that time A. A. Markov [72] and E. L. Post [102] established, independently, the undecidability of the so-called *word problem* for semigroups. This problem was stated by A. Thue [120] in 1914 and is known also as *Thue's problem*. This was the first decision problem that arose naturally in mathematics and was finally shown to be undecidable.

Both Post and Markov were interested in Hilbert's tenth problem. Post wrote already in [101] that Hilbert's tenth problem "begs for an unsolvability proof." Markov considered the following approach to Hilbert's tenth problem to be plausible. Among decision problems that deal, like the Thue problem, with words, there is one problem that looks as though it is very close to Hilbert's tenth problem, namely, the problem of decidability of *word equations*. Solving a word equation in the two-letter alphabet $\{\alpha_0, \alpha_1\}$ (which is essentially the general case) can be reduced easily to solving a particular Diophantine equation thanks to the following fact: every 2×2 matrix with natural number entries and determinant equal to zero can be represented in a unique way as the product of matrices $A_0 = \left(\begin{smallmatrix} 1 & 0 \\ 1 & 1 \end{smallmatrix}\right)$ and $A_1 = \left(\begin{smallmatrix} 1 & 1 \\ 0 & 1 \end{smallmatrix}\right)$, and hence, a word $\alpha_{i_1} \ldots \alpha_{i_n}$ can be represented by the four entries of the matrix $A_{i_1} \ldots A_{i_n}$.

However, this natural approach did not succeed. The reason for this failure was discovered much later: in 1977 G. S. Makanin [70] found a decision procedure for word equations. Nevertheless, one could try to revive Markov's idea by considering a wider class of word equations. In [74] (for further development, see [85]) I introduced a different way of coding words by numbers based on the Fibonacci numbers. This encoding allows us to express by Diophantine equations not only the equality of concatenations of several words but also the equality of the lengths of such concatenations. Thus, the undecidability of Hilbert's tenth problem would immediately follow from the undecidability of systems of equations in words and their lengths, but the latter remains an open problem.

3 Davis's Conjecture: From Algorithms To Sets

M. Davis began to tackle Hilbert's tenth problem at the end of the 1940s (he was inspired by the above-cited phrase from the paper of his teacher, Post). He considered *Diophantine sets* that are sets of natural numbers having *Diophantine representations*, i.e., definitions of the form

$$a \in \mathfrak{M} \iff \exists x_1 \ldots x_m [P(a, x_1, \ldots, x_m) = 0] \tag{3}$$

where P is again a polynomial with integer coefficients, but now one of its variables is selected as a *parameter*. Davis's aim was to give a characterization of the whole class of Diophantine sets. Computability theory immediately furnishes a condition that is necessary for a set to be Diophantine: every Diophantine set is, evidently, effectively enumerable. Davis conjectured ([17, 18]) that this necessary condition is also sufficient:

Davis's conjecture. *A set of natural numbers is Diophantine if and only if it is effectively enumerable.*

Effectively enumerable sets can be defined via the notion of an algorithm, but the concepts can be defined in reverse order: having given an independent definition of a effectively enumerable set, one can develop the whole theory of computability in terms of effectively enumerable sets instead of algorithms; examples of such an approach can be found in G.S. Tseitin's paper [122] and P. Martin-Löf's book [73]. Thus Davis's conjecture opened a way to base computability theory on the number-theoretical notion of a Diophantine set.

4 Davis's Conjecture: First Step To The Proof Via Arithmetization

Davis made the first step to proving his conjecture by showing in [18] that every effectively enumerable set $\mathfrak{M}$ has an almost Diophantine representation:

Theorem (M. Davis). *Every effectively enumerable set $\mathfrak{M}$ has a representation of the form*

$$a \in \mathfrak{M} \iff \exists z \forall y_{\leq z} \exists x_1 \dots x_m [P(a, x_1, \dots, x_m, y, z) = 0] \tag{4}$$

where P is a polynomial with integer coefficients and $\forall y_{\leq z}$ is the bounded universal quantifier "for all y not greater than z."

A representation of this type became known as the *Davis normal form*. To obtain it, Davis started in [18] with a representation of the set $\mathfrak{M}$ by an arbitrary arithmetical formula with any number of bounded universal quantifiers. The existence of such arithmetical formulas for every effectively enumerable set was demonstrated by K. Gödel in his classical paper [28]. Because of the bound on the universal quantifiers, each such formula defines an effectively enumerable set and, hence, these formulas could be used as a foundation for Computability Theory.

5 Original Proof Of Davis: Post's Normal Forms

According to a footnote in Davis's paper [18], the idea of obtaining the representation (4) by combining universal quantifiers from a general arithmetic representation was from the (anonymous) referee of the paper. The original proof of Davis (outlined in [17] and given with details in [20]) was quite different. Namely, Davis arithmetized *Post normal forms*. These forms are a special case of more general *canonical forms* introduced by Post in [100] as a possible foundation of computability theory (and the above-cited book [73] uses just Post canonical forms). In contrast to general canonical forms, normal forms have rules of the very simple form

$$\frac{Px}{xQ} \tag{5}$$

meaning that if a word has a prefix P, one can cut this prefix and add the suffix Q. Post proved that, despite the simplicity of such rules, normal forms are as powerful as general canonical forms; the simplicity of rules (5) allowed Davis to perform the arithmetization using only one universal quantifier.

6 Davis's Conjecture Proved: Effectively Enumerable Sets Are Diophantine

It took two decades before Davis's conjecture became a theorem (for historical details see, for example, [80]; for an extensive bibliography on Hilbert's tenth problem visit [129]). The following weaker result from M. Davis, H. Putnam, and J. Robinson [21] was a milestone on the way to the proof of Davis's conjecture:

DPR-Theorem. *For every effectively enumerable set $\mathfrak{M}$, there exists a representation of the form*

$$a \in \mathfrak{M} \iff \exists x_1 \ldots x_m [E(a, x_1, x_2, \ldots, x_m) = 0] \tag{6}$$

where E is an exponential polynomial, *i.e., an expression constructed by combining the variables and particular integers using the traditional rules of addition, multiplication, and exponentiation.*

The final step in the proof of Davis's conjecture was accomplished in [75], and nowadays the corresponding theorem is often called the

DPRM-Theorem. *The notion of Diophantine set and the notion of effectively enumerable set coincide.*

Thus a (seemingly narrow) notion from the Number Theory turned out to be equivalent to the very general notion from the Computability Theory.

7 Existential Arithmetization I: Turing Machines

Already the very first proof of the DPRM-theorem given in [75] was constructive in the sense that as soon as an effectively enumerable set $\mathfrak{M}$ is presented in any standard form, it is possible to find a corresponding Diophantine representation (3). According to this proof this should be done in the following four steps:

1. construction of an arithmetical formula with many bounded universal quantifiers;

2. transformation of this formula into Davis normal form (4);

3. elimination of the single bounded universal quantifier at the cost of passing to exponential Diophantine equations, thus getting an exponential Diophantine representation (6);

4. elimination of the exponentiation.

Now that we know that in fact no universal quantifier is necessary at all, it would be more natural to try to perform the whole arithmetization by using only purely existential formulas. From a technical point of view, the success of such an approach crucially depends on the selection of an appropriate device for representation of effectively enumerable sets.

For the first time such a purely existential arithmetization was achieved in [77] with the set $\mathfrak{M}$ being recognized by a Turing machine; a simplified way of constructing Diophantine representation by arithmetization of Turing machines is presented in [80]; yet another construction based on Turing machines is given in [9] (see [10] for a more easily accessible publication).

8 Existential Arithmetization II: Register Machines

When arithmetizing Turing machines, one first has to introduce a method to represent the content of the tape of the machine by numbers. In this respect another kind of abstract computing device, *register machines*, turned out to be more suitable as a starting point for constructing Diophantine representations. Register machines were introduced almost simultaneously by several authors: J. Lambek [65], Z. A. Melzak [90], M. L. Minsky [91, 92], and J. C. Shepherdson and H. E. Sturgis [109]. Like Turing machines, register machines have very primitive instructions, but in addition, they deal directly with numbers rather than with words. This led to a "visual proof" of the simulation of register machines by Diophantine equations (see [56, 57, 58, 82]).

9 Existential Arithmetization III: Partial Recursive Functions

Another classical tool for the foundations of Computability Theory is the notion of a *partial recursive function*. Existential arithmetization of these functions was done in

[81] where Diophantine representations are constructed inductively, alongside construction of a partial recursive function from the initial functions. To deal with primitive recursion and the minimization operator, it turned out to be useful to generalize the notion of partial recursive function: instead of dealing, say, with a one-argument function f, it was more convenient to work with a function F, defined on arbitrary n-tuples of natural numbers by

$$F(\langle a_1, \ldots, a_n \rangle) = \langle f(a_1), \ldots, f(a_n) \rangle. \tag{7}$$

10 Universality In Number Theory: Collapse Of Diophantine Hierarchy

The DPRM-theorem allows the transfer of several ideas from Computability Theory to Number Theory. One example of such a transfer is the existence of a *universal Diophantine equation*, i.e., an equation

$$U(a, k, y_1, \ldots, y_M) = 0 \tag{8}$$

with the following property: *for an arbitrary Diophantine equation*

$$P(a, x_1, \ldots, x_m) = 0 \tag{9}$$

there exists an (effectively computable) number k_P such that for arbitrary values of the parameter a, the equation (9) has a solution in $x_1, \ldots, x_m$ if and only if the equation

$$U(a, k_P, y_1, \ldots, y_M) = 0 \tag{10}$$

has a solution in $y_1, \ldots, y_M$. This implies that the traditional number-theoretical hierarchy of Diophantine equations of degree $1, 2, \ldots$ with $1, 2, \ldots$ unknowns collapses at some level. Of course, there is a trade-off between the degree and the number of unknowns. One can construct a universal Diophantine equation of degree **d** with **m** unknowns, where $\langle \mathbf{d}, \mathbf{m} \rangle$ is any of the following pairs:

$$\langle 4, 58 \rangle, \ \langle 8, 38 \rangle, \ \langle 12, 32 \rangle, \ \langle 16, 29 \rangle, \ \langle 20, 28 \rangle, \ \langle 24, 26 \rangle, \ \langle 28, 25 \rangle, \ \langle 36, 24 \rangle,$$

$$\langle 96, 21 \rangle, \ \langle 2668, 19 \rangle, \ \langle 2 \times 10^5, 14 \rangle, \ \langle 6.6 \times 10^{43}, 13 \rangle, \ \langle 1.3 \times 10^{44}, 12 \rangle, \quad (11)$$

$$\langle 4.6 \times 10^{44}, 11 \rangle, \ \langle 8.6 \times 10^{44}, 10 \rangle, \ \langle 1.6 \times 10^{45}, 9 \rangle$$

(the (easy) best bound **d = 4** of degree is from T. Skolem [117]; the best bound **m = 9** of the number of unknowns (non-negative) from the author of [78] is presented with details by Jones in [51] where also all the "intermediate" pairs (11) are given; the bound **m = 11** in the case of integer solutions is announced by Zh.-W. Sun in [119]).

As positive results we can mention only the (trivial) decidability of Diophantine equations in one unknown and the decidablity of equations of degree two

([33, 34, 114]), so the gap between decidable and undecidable is large. For exponential Diophantine equations, the situation is much better: already three unknowns are sufficient for undecidability ([79]); moreover, one can confine exponentiation to the special unary case 2^x ([55], see also [80, Sect. 8.2]). This restriction is of interest because H. Levitz [66] gave a decision procedure for such unary (base-2) exponential Diophantine equations in one unknown, and so only the case of two unknowns remains open.

Although the existence of a universal Diophantine equation immediately follows from the DPRM-theorem and the existence of, say, a universal Turing machine, the very idea of the existence of such a universal object in the theory of Diophantine equations seemed quite implausible not only to number-theorists, but to some logicians (see [62]).

The existence of a universal Diophantine equation is an example of a result that is number-theoretical in its statement, but that was originally proved by tools from Computability Theory; today such an equation (8) can be constructed by purely number-theoretical methods (see [80, Ch. 4]).

It is interesting to note the following difference between universal Diophantine equations and, say, universal Turing machines. The latter typically have a very peculiar construction, whereas a universal Diophantine equation can have a regular structure. As soon as we know the mere existence of a universal Diophantine equation of degree **d** with **m** unknowns (any pair from equation (11) would suit), we can treat the equation

$$\sum_{i_0+\cdots+i_m\leq d} (k_{i_0,\ldots,i_m} - k)a^{i_0}y_1^{i_1}\ldots y_m^{i_m} = 0$$

as universal by agreeing to index Diophantine equations not by a single number as in equation (10) but by the tuple consisting of all the k's. If indexing by a single number is desired, one can combine all these k's into a single index by using any polynomial assuming each of its values at most once; for example, the following equation is universal in the sense of equations (8)–(10):

$$\left(\sum_{i_0+\cdots+i_m\leq d} (z_{i_0,\ldots,i_m} - z)\,a^{i_0}y_1^{i_1}\ldots y_m^{i_m}\right)^2 +$$

$$\left(k - \sum_{i_0+\cdots+i_m\leq d} z_{i_0,\ldots,i_m}\left(z + \sum_{i_0+\cdots+i_m\leq d} z_{i_0,\ldots,i_m}\right)^{i_0(d+1)^0+\cdots+i_m(d+1)^m}\right)^2 = 0$$

The idea of such a transparent universal Diophantine equation was stated by N. K. Kossovskii in [61].

11 Growth Of Solutions: Speeding Up Diophantine Equations

Another example of a transfer of ideas from Computability Theory to Number Theory is as follows. M. Davis [19] used the DPRM-theorem to get for Diophantine
equations an analog of a speed-up theorem of M. Blum [8]. Namely, *for every total computable function $\alpha(a, x)$, one can construct two one-parameter Diophantine
equations*

$$P_1(a, x_1, \ldots, x_k) = 0, \qquad P_2(a, x_1, \ldots, x_k) = 0 \tag{12}$$

such that

*(i) for every value of the parameter a, exactly one of these two equations has a
solution;*

(ii) if Diophantine equations

$$Q_1(a, y_1, \ldots, y_l) = 0, \qquad Q_2(a, y_1, \ldots, y_l) = 0 \tag{13}$$

*are solvable for the same values of the parameter a as, respectively, equations
(12), then one can construct a third pair of Diophantine equations*

$$R_1(a, z_1, \ldots, z_m) = 0, \qquad R_2(a, z_1, \ldots, z_m) = 0 \tag{14}$$

such that

– *these equations are again solvable for the same values of the parameter a as,
respectively, equations (12);*

– *for all sufficiently large values of the parameter a for every solution $y_1, \ldots, y_l$
of one of the equations (13), there exists a solution $z_1, \ldots, z_m$ of the corresponding equation (14) such that*

$$y_1 + \cdots + y_l > \alpha(a, z_1 + \cdots + z_m) \tag{15}$$

This formulation of a Diophantine speed-up theorem was given in terms of an arbitrary total function for the sake of the greatest generality; by substituting for
α any particular (fast growing) total computable function, one would obtain a
purely number-theoretic result that, however, has never been imagined by number-
theorists.

12 Diophantine Machines: Capturing Nondeterminism

The DPRM-theorem allows one to treat Diophantine equations as computing devices.
This was done in a picturesque way by L. Adleman and K. Manders in [1]. Namely,
they introduced the notion of *Non-Deterministic Diophantine Machine*, NDDM for
short.

A NDDM is specified by a parametric Diophantine equation (9) and works as follows: on input a, it guesses the numbers $x_1, \ldots, x_m$ and then checks (9); if the equality holds, then a is accepted.

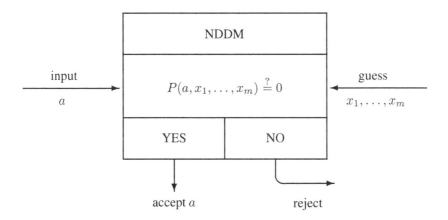

The DPRM-theorem is exactly the statement that NDDMs are as powerful as, say, Turing machines; i.e., every set acceptable by a Turing machine is accepted by some NDDM and, of course, *vice versa*.

The idea behind the introduction of a new computing device was as follows: in NDDM we have full separation of guessing and deterministic computation, and the latter is very simple—just the calculation of the value of a polynomial.

13 Unambiguity: Equations With Unique Solution

NDDMs are essentially nondeterministic computing devices. For such devices, non-determinism is sometimes fictitious in the sense that at most one path can lead to accepting; if this is so, one speaks about *unambiguous computations*. The corresponding property for (exponential) Diophantine representations was called *single-foldness*: a representation (3) or (6) is called a *single-fold representation* if for any given value of the parameter a, there exists at most one choice of the unknowns $x_1, \ldots, x_m$.

The existence of single-fold exponential Diophantine representations for every effectively enumerable set was established in [76] and later was improved to the existence of single-fold exponential Diophantine representations with only three existential variables (see [54, 80]).

The existence of single-fold (or even weaker *finite-fold*) Diophantine representations is a major open problem; the positive answer would shed light on some difficulties met in Number Theory in connection with effectivization of some results about Diophantine equations (for more details, see, for example, [76, 80]).

Single-fold exponential Diophantine representations have found applications in descriptive complexity (see Section 17 below).

14 Diophantine Complexity: D vs. NP

Although the DPRM-theorem implies that NDDMs are as powerful as any other abstract computational device, the intriguing crucial question remains open: *how efficient are the NDDMs?* Adleman and Manders conjectured that in fact NDDMs are as efficient as Turing machines.

For the latter there are two natural complexity measures: TIME and SPACE. For NDDMs there is only one natural complexity measure that plays the role of both TIME and SPACE. This measure is SIZE, which is the size (in bits) of the smallest solution of the equation (it is not essential whether we define this solution as the one with the smallest possible value of $\max\{x_1, \ldots, x_m\}$, or of $x_1 + \cdots + x_m$).

Adleman and Manders obtained in [1, 71] the first results comparing the efficiency of NDDMs and Turing machines by estimating the SIZE of a NDDM simulating a Turing machine with TIME in special ranges.

Imposing bounds on the SIZE, we can define a corresponding complexity class. It was shown by A. K. Vinogradov and N. K. Kossovskii [125] that in this way one can define all Grzegorczyk classes starting from $\mathcal{E}^3$. Of course, the lower classes are of greater interest, and what is typical, they turned out to be more difficult.

Adleman and Manders [1, 71] also introduced the class $\mathbf{D}$ consisting of all sets $\mathfrak{M}$ having representations of the form

$$a \in \mathfrak{M} \iff \exists x_1 \ldots x_m \left[P(a, x_1, \ldots, x_m) = 0 \,\&\, |x_1| + \cdots + |x_m| \leq |a|^k \right]$$

where $|a|$ denotes, as usual, the (binary) length of a. It is easy to see that $\mathbf{D} \subseteq \mathbf{NP}$ and the class $\mathbf{D}$ is known (see [71]) to contain $\mathbf{NP}$-complete problems, but otherwise the class $\mathbf{D}$ is little understood. Adleman and Manders asked whether in fact $\mathbf{D} = \mathbf{NP}$. Recently Ch. Pollett [98] showed that this is so provided that $\mathbf{D} \subseteq \mathbf{co\text{-}NLOGTIME}$ and indicated several other ways to tackle the $\mathbf{D} = \mathbf{NP}$ question.

R. Venkatesan and S. Rajagopalan [124] considered the *Randomized Diophantine Problem* and proved that it is average-case complete; unfortunately, their proof is conditional, and their assumption (on the existence of a Diophantine equation with a special property) is equivalent to $\mathbf{D} = \mathbf{NP}$.

H. Lipmaa [68] introduced $\mathbf{PD}$, the "deterministic part" of the class $\mathbf{D}$, and used Diophantine equations for secure information exchange protocols.

Arithmetical definitions of the class **NP** via an analog of the Davis normal form (4) were given by B. R. Hodgson and C. F. Kent [42, 60] and by S. Yukna [127, 128].

15 Algorithms For Algorithms: Undecidable Properties Of Programs

Algorithms are implemented in the form of programs. Recognizing the equivalence of two programs and simplification of programs are important for practice. Of course, in a general setting, these problems are undecidable, and Hilbert's tenth problem shows that they are so even if we restrict ourselves to programs of a very simple structure. In fact, the following two programs with input values $x1, \ldots, xm$ are equivalent (and the latter is the simplest program equivalent to the former) if and only if the corresponding Diophantine equation has no solution:

```
if  P(x1,...,xm)=0           print(0)
    then print(1)
    else print(0)
```

This easy fact was remarked by T. Kasami and N. Tokura [59] with the emphasis that these programs do not contain loops.

However, one of these two programs contains a control structure, `if ... then ... else`. O. H. Ibarra and B. S. Leininger [44, 45, 46] showed that as soon as we replace multiplication by integer division, it will be possible to get rid of control structures and they obtained numerous undecidability results about straight-line programs, even having small numbers of input and local variables. Some of these results were quantitively improved by D. Shiryaev [110]; namely, he reduced the number of variables required for undecidability. It is interesting to note that for this reduction he used not the mere undecidability of Diophantine equations with a small number of unknowns but an intermediate technical result from the paper [86] where the number of unknowns in undecidable Diophantine equations was reduced only to 13.

Other undecidable properties of programs were established with the use of the undecidability of Hilbert's tenth problem in [7, 35, 38, 40, 47, 104, 126].

16 Parallel Computations: Calculation Of A Polynomial On A Petri Net

Petri nets and *systems of vector addition* were introduced as tools for describing parallel computations. A system of vector addition S consists of a finite set of vectors

$V_1, \ldots, V_k$ of equal size; the components of these vectors are *integers*. From a given vector A of the same size with natural number components, one is allowed to go in one step to any vector $A + V_i$ provided that all components of this sum are also *natural numbers*. Repeating such transitions many times, we obtain the set $\mathcal{R}_S(A)$ of all vectors reachable from A in the system S.

The *containment problem* is to decide, given two systems of vector addition S_1 and S_2 and an initial vector A, whether $\mathcal{R}_{S_1}(A) \subseteq \mathcal{R}_{S_2}(A)$; similarly, in the *equivalence problem*, one is interested to know whether $\mathcal{R}_{S_1}(A) = \mathcal{R}_{S_2}(A)$.

M. Rabin (unpublished) used the undecidability of (exponential) Diophantine equations to prove that the containment problem for systems of vector addition (and hence also for Petri nets, because the latter easily simulate systems of vector addition) is undecidable (see papers of M. Hack [37] and T. Araki and T. Kasami [3] where also a stronger result, the undecidability of the equivalence problem for systems of vector addition, was obtained; for other presentations of these results, see [80, Section 10.2], [106]).

The crucial point in proving these results was a definition (introduced by Rabin) of a calculation of the values of a function $F(x_1, \ldots, x_m)$ by a system of vector addition. Namely, a system S is said to calculate this function if $F(x_1, \ldots, x_m)$ is the largest value of the $m + 1$st component of vectors reachable in this system from the vector $\langle x_1, \ldots, x_k, 0, \ldots, 0 \rangle$. Rabin showed that all polynomials with non-negative integer coefficients are computable in this sense by systems of vector addition. Given an equation (1), we can find polynomials $P_1(x_1, \ldots, x_m)$ and $P_2(x_1, \ldots, x_m)$ with non-negative integer coefficients such that

$$P^2(x_1, \ldots, x_m) = P_2(x_1, \ldots, x_m) - P_1(x_1, \ldots, x_m). \tag{16}$$

Now the equation (1) has no solution in natural numbers if and only if

$$P_1(x_1, \ldots, x_m) + 1 \leq P_2(x_1, \ldots, x_m)$$

for all values of $x_1, \ldots, x_m$. Using vector addition, calculating in Rabin's sense polynomials $P_1(x_1, \ldots, x_m) + 1$ and $P_2(x_1, \ldots, x_m)$, one can reduce Hilbert's tenth problem to the containment problem for systems of vector addition; the equivalence problem for such systems requires a more subtle construction.

E. W. Mayr and A. R. Meyer [87] showed that the *finite* containment and equivalence problems for Petri nets, being decidable, have complexity that cannot be bounded by any primitive recursive function. These results were improved by R. R. Howell, L. E. Rosier, D. T. Huynh, and H.-Ch. Yen in [43].

17 A Step Above Hilbert's Tenth Problem: Computational Chaos In Number Theory And Game Theory

Diophantine equations are undecidable. However, every Diophantine set is effectively enumerable, and hence, its *descriptive complexity* is the least possible: for every polynomial P, the initial segment of the set $\mathfrak{M}$ from (3), i.e., the intersection of the set $\mathfrak{M}$ with the set

$$\{a \mid a \leq N\} \tag{17}$$

can be coded by $O(\log(N))$ bits only (it is sufficient to know the cardinality of this intersection in order to effectively compute the intersection itself). However, we can reach the maximal descriptive complexity by considering questions that are only slightly more complicated than those from Hilbert's tenth problem. G. Chaitin [14] constructed a one-parameter exponential Diophantine equation such that the set

$$\{a \mid \exists^{\infty} x_1 \ldots x_m[E(a, x_1, x_2, \ldots, x_m) = 0]\} \tag{18}$$

requires N bits (up to an additive constant) for *prefix-free coding* of its intersection with the set (17); here $\exists^{\infty}$ means the existence of infinitely many solutions of the equation. Informally, one can say that the set (18) is completely chaotic.

More recently T. Ord and T. D. Kieu [93] constructed another exponential Diophantine equation that for every value of a has only finitely many solutions, but the parity of the number of solutions again has completely chaotic behavior in the sense of descriptive complexity.

The proofs by Chaitin and Ord and Kieu looked like clever but *ad hoc* tricks. I [84] made the following generalization: instead of asking whether the number of solutions is finite/infinite or even/odd, one can ask whether the number of solutions belongs to any fixed decidable infinite set with infinite complement (with respect to the set $\{0, 1, 2, \ldots, \infty\}$).

All these results were obtained for exponential Diophantine equations because they are based on the existence of single-fold exponential Diophantine representations; the existence of similar chaos among genuine Diophantine equations is a major open question.

K. Prasad [103] translated Chaitin's result from the question about the infinitude of the number of solutions of an exponential Diopohantine equation to the question of the infinitude of the number of Nash equilibria in multi-person, noncooperative games.

18 Unification: It Is Hard To Make Things Equal

Most systems of automatic reasoning are based on *algorithms of unification* (for details, see, for example, [4, 115]). Hilbert's tenth problem itself can be stated as a unification problem: *Given two polynomials $P(x_1, \ldots, x_m)$ and $Q(x_1, \ldots, x_m)$, can we make them equal by substituting particular natural numbers for the free variables $x_1, \ldots, x_m$?* Although in the most general settings the unification problem is undecidable, in applications one usually deals with restricted classes of pairs of terms to be unified. For example, J. A. Robinson [108] presented a unification algorithm for the first-order predicate calculus. Hilbert's tenth problem was first used in a new proof of the undecidability of third-order unification given by L. D. Baxter [6] and then for establishing the undecidability of second-order unification done by W. D. Golfarb [29].

The difficulty in using Hilbert's tenth problem for obtaining the undecidability of unification in logical calculi can be explained as follows: in such calculi we have to deal with variables (for individual objects, functions, and predicates) in a formal way, that is without giving any *interpretation* for them, in particular, we can say directly neither that individual variables range over integers nor that particular functional symbols denote addition and multiplication. A natural number n can be easily represented by the term

$$T_n = \underbrace{F(F(\ldots F(x) \ldots))}_{n \text{ times}} \tag{19}$$

addition $n + m$ corresponds now to substitution of T_m for x in T_n but defining multiplication requires some trick possible with second-order variables but is impossible in the first-order case.

For other undecidability results about unification obtained from the undecidability of Hilbert's tenth problem, see [5, 11, 12, 22, 26, 69, 121].

19 Simple Set: Diophantine Games Are Difficult

J. Jones [49], based on ideas of M. Rabin [105], introduced *Diophantine games*. Suppose that two players, *P*eter and *U*rsula, select the values, respectively, of the parameters $a_1, \ldots, a_m$ and of the *u*nknowns $x_1, \ldots, x_m$ in the Diophantine equation

$$P(a_1, \ldots, a_m, x_1, \ldots, x_m) = 0 \tag{20}$$

and they do this in turn in the following order: $a_1, x_1, \ldots, a_m, x_m$. Trivially, the undecidability of Hilbert's tenth problem implies the impossibility to determine, given a polynomial P, which of the players has a winning strategy. More interesting, even if we know that Ursula has a winning strategy, this strategy may turn out not to be effectively computable. For proving this, the mere undecidability of Hilbert's tenth problem is not sufficient, one needs the full power of the DPRM-theorem in order to construct a Diophantine representation for *simple sets*, the existence of which was established by Post [101].

It is somewhat surprising that despite the need to use such nontrivial objects as simple sets, a Diophantine game with uncomputable strategy for Ursula can be defined by a relatively compact polynomial. Jones [50] showed that this is so for the following game:

$$\left\{\{a_1 + a_6 + 1 - x_4\}^2 \cdot \left\{\left\langle (a_6 + a_7)^2 + 3a_7 + a_6 - 2x_4\right\rangle^2\right.\right.$$

$$+ \left\langle \left[(x_9 - a_7)^2 + (x_{10} - a_9)^2\right]\left[(x_9 - a_6)^2 + (x_{10} - a_8)^2((x_4 - a_1)^2\right.\right.$$

$$+ (x_{10} - a_9 - x_1)^2)\big]\left[(x_9 - 3x_4)^2 + (x_{10} - a_8 - a_9)^2\right]\left[(x_9 - 3x_4 - 1)^2\right.$$

$$\left. + (x_{10} - a_8 a_9)^2\right] - a_{12} - 1\right\rangle^2 + \left\langle [x_{10} + a_{12} + a_{12}x_9 a_4 - a_3]^2\right.$$

$$\left. + [x_5 + a_{13} - x_9 a_4]^2\right\rangle\right\} - x_{13} - 1\right\}\{a_1 + x_5 + 1 - a_5\}\left\{\left\langle (x_5 - x_6)^2\right.\right.$$

$$\left. + 3x_6 + x_5 - 2a_5\right\rangle^2 + \left\langle \left[(a_{10} - x_6)^2 + (a_{11} - x_8)^2\right]\left[(a_{10} - x_5)^2\right.\right.$$

$$+ (a_{11} - x_7)^2((a_5 - a_1)^2 + (a_{11} - x_8 - a_2)^2)\big]\left[(a_{10} - 3a_5)^2\right.$$

$$+ (a_{11} - x_7 - x_8)^2\big]\left[(a_{10} - 3a_5 - 1)^2 + (a_{11} - x_7 x_8)^2\right] - x_{11} - 1\right\rangle^2$$

$$\left. + \left\langle [a_{11} + x_{11} + x_{11}a_{10}x_3 - x_2]^2 + [a_{11} + x_{12} - a_{10}x_3]^2\right\rangle\right\} = 0$$

More results about undecidability and complexity of Diophantine games can be found in [52, 53, 123].

Another kind of game was introduced by A. H. Lachlan [64] as a possible tool to establish results about the lattice of effectively enumerable sets. He conjectured that for these games it can be decided which of the two players has the winning strategy. He obtained partial results in this direction, but recently M. Kummer [63] proved many results about the undecidability of Lachlan's games using the undecidability of Hilbert's tenth problem.

K. Prasad [103] proved for "traditional," multiperson, noncooperative games with polynomial *payoff functions* that there is no algorithm to decide whether a game has a *Nash equilibrium in pure strategies*; for a similar result for *mixed strategies*, one would need single-fold representations (for a definition, see Sect. 13), and thus currently, undecidability is established only for the case when the payoff functions are exponential polynomials (for a definition, see Sect. 6).

20 Continuous Variables: Limitations Of Computer Algebra

General systems of computer algebra, such as MATHEMATICA and MAPLE, are oriented for dealing with formulas with variables ranging over continuous domains like real or complex numbers. Nevertheless, the undecidability of Hilbert's tenth problem put a lot of restrictions on what computer algebra systems can do. Here is just one example from the work of J. Denef and L. Lipshitz [24].

Consider the partial differential equation

$$(1 - t_1) \cdots (1 - t_m) P \left(t_1 \frac{\partial}{\partial t_1}, \ldots, t_m \frac{\partial}{\partial t_m} \right) Y(t_1, \ldots, t_m) = 1 \qquad (21)$$

where P is a polynomial with integer coefficients. It is easy to verify that if it has a power series solution

$$Y(t_1, \ldots, t_m) = \sum_{x_1, \ldots, x_m} c_{x_1, \ldots, x_m} t_1^{x_1} \ldots t_m^{x_m} \qquad (22)$$

then

$$c_{x_1, \ldots, x_m} P(x_1, \ldots, x_m) = 1 \qquad (23)$$

Thus the differential equation (21) has a solution if and only if the Diophantine equation

$$P(x_1, \ldots, x_m) = 0 \qquad (24)$$

has no solution, and hence, no computer algebra system can tell, given a polynomial P, whether the differential equation (21) has a solution.

More examples of the use of (exponential) Diophantine equations for establishing undecidability results in continuous domains can be found in [2, 13, 16, 25, 32, 94, 107, 116, 118] (see also [80, Ch. 9]).

21 DNA Recombination And Metabolism: Models Of Computation Motivated By Biology

Many models of computations are nowadays suggested by biology. Although biologists are interested in knowing how adequate to nature such models are, computer scientists are interested in establishing first of all that a new model of computations is as powerful as are conventional models. The DPRM-theorem turned out to be a convenient tool for proving the universality of several biologically motivated models of computations.

The *splicing operation* was introduced by T. Head [39] as formalization of the *recombination* of DNA molecules. A *splicing rule* is defined by a quadruple of words $\langle U_1, U_2, U_3, U_4 \rangle$, and it allows transformation of a pair of words of the form $\langle x_1 U_1 U_2 x_2, y_1 U_3 U_4 y_2 \rangle$ into the pair $\langle x_1 U_1 U_4 x_2, y_1 U_3 U_2 y_2 \rangle$ (for a more motivated, detailed, and picturesque description of splicing, see Sect. 11 of Gh. Paun's paper [95] in this volume). Splicing rules by themselves can generate only regular languages; however, with some additional controls, they are as powerful as, say, Turing machines, and P. Frisco [27] used the power of Diophantine equations in a (new) proof of this fact.

Membrane computing, motivated by metabolism in living cells, was introduced by Paun in 1998 (see Sect. 14 in his paper [95] in this volume). Á. R. Jiménez and M. J. P. Jiménez [48] and C. Li., Z. Dang, O. H. Ibarra, and H.-Ch. Yen [67] used the DPRM-theorem in order to establish the computational power of different versions of membrane computing.

22 Other Kinds Of Impossibilities: Non-Algorithmical Corollaries Of Algorithmical Results

Most of the examples given above of applications of the DPRM-theorem were of an algorithmic nature: either one proves undecidability of some decision problem or one establishes the universality of some device. There are a number of cases where the undecidability of Hilbert's tenth problem was used to prove the impossibility of something essentially different from the existence of an algorithm. To obtain such corollaries one could contrast Hilbert's tenth problem with some problem known to be decidable, and analyze what is the obstacle for reducing Hilbert's tenth problem to this other problem.

For example, R. L. Goodstein and R. D. Lee proved in [31] the decidability of the existence of solutions of equations constructed in terms of the operations

$$x + y, \ x \times y, \ 1 \dot{-} x, \ x \dot{-} 1 \tag{25}$$

and in [30], Goodstein deduced from this that the binary subtraction $x \dot{-} y$ cannot be expressed as a composition of operations (25).

Similarly, E. M. Gurari and E. M. Ibarra [36] established the decidability of a special class of counter machines and deduced from this and the undecidability of Hilbert's tenth problem that the language $\{u_1^i u_2^j u_3^k | k = ij\}$ cannot be recognized by any machine from this class.

Having established the undecidability of the equivalence problem for straight-line programs with many inputs and the decidability of this problem for straight-line programs with a single input and a restricted set of operations, Ibarra and Leininger [44] concluded that no straight-line program with one input using only operations from this set computes a map from natural numbers onto the set of all pairs of natural numbers.

23 Future Research: Back To Diophantus

Most of Hilbert's 23 problems are in fact groups of related problems. The tenth problem is, however, stated as a single question. We saw in Section 1 that the method for recognizing solvability of Diophantine equations in *"rational integers"* asked by

Hilbert *explicitly* would give us also a method for recognizing solvability of Diophantine equations in natural numbers. Recognizing the solvability of Diophantine equations in different rings of *algebraic integers* can also be reduced to Hilbert's tenth problem, so *implicitly* Hilbert asked for many algorithms. But now we know that there is no algorithm in the case of integer solutions; however, this fact has no immediate corollaries for other rings because reductions in the other direction are not obvious.

J. Denef [23] (see also [80, Sect. 7.3]) showed that solving a Diophantine equation in integers can be reduced to solving another Diophantine equation in *Gaussian integers*, that is, numbers of the form $a + bi$ where $i = \sqrt{-1}$ and a and b are real integers. More precisely, he proved that rational integers have a Diophantine definition in the ring of Gaussian integers, namely, he constructed a particular Diophantine equation

$$P(\chi_0, \chi_1, \ldots, \chi_m) = 0 \tag{26}$$

such that

- in every solution of the equation (26) in Gaussian integers $\chi_0, \chi_1, \ldots, \chi_m$ the value of χ_0 is a rational integer;

- for every rational integer χ_0 there exist values of $\chi_1, \ldots, \chi_m$ that together with this value χ_0 produce a solution of the equation (26).

Of course, this reduction gave also a counterpart of the DPRM-theorem for the ring of Gaussian numbers. If one is interested only in the undecidability of Diophantine equations in some ring, it would be sufficient to construct a *Diophantine model of integers* in this ring.

The result of Denef was extended by different authors to other classes of rings of algebraic numbers (see surveys [96, 111] or the recent book [113] of A. Shlapentokh), and this did not require any new ideas or methods from Computability Theory. Surprisingly, the general case of a ring of algebraic numbers from finite extension of rational numbers still remains open, and probably there are deep number-theoretical obstacles for this. In particular, B. Mazur [88, 89] (for further development, see [15, 99, 112, 113]) put forward a number of conjectures that imply the impossibility of constructing a Diophantine model of integers in some rings, so solving the analog of Hilbert's tenth problem for such rings might require new algorithmical ideas.

One of these "bad" rings would be (according to Mazur's conjectures) the ring of all rational numbers. In fact, Diophantus himself was solving equations neither in integers nor in natural numbers but in (positive) rational numbers. It is easy to reduce solvability of Diophantine equations in rational numbers to solvabilty of Diophantine equations in rational integers so a positive solution of the tenth problem would also have given a method for solving Diophantine equations in rational numbers. There is a somewhat less evident way to reduce the solvability of Diophantine equations in rational numbers to the solvabilty of homogeneous Diophantine equations in rational integers, and vice versa (see, for example, [80, Sect. 7.4]). These equations have many nice properties, and perhaps for this particular subclass of Diophantine equations, there is a "process" such as Hilbert had asked for in the case of general Diophantine equations. This is one of the most important open problems closely related to Hilbert's tenth problem. Progress in this direction has been modest. The difficulty might be explained by the fact that conventional proofs of the undecidability of Hilbert's tenth problem heavily use divisibility properties of integers that are trivial for rational numbers. A possible way (incompatible with Mazur's conjectures) to overcome this obstacle is suggested by Th. Pheidas [97].

References

1. Adleman L., Manders K.: Diophantine complexity. In: 17th Annual Symposium on Foundations of Computer Science, 81–88 (1976)
2. Adler A.: Some recursively unsolvable problems in analysis. Proceedings of the American Mathematical Society, **22**(2):523–526 (1969)
3. Araki, T., Kasami, T.: Some undecidable problems for Petri nets. Systems-Computers-Controls, **7**(1):20–28 (1976); Japanese original: Denshi Tsushin Gakkai Ronbunshi, **59D**:25–32 (1976)
4. Baader, F., Siekmann, J. H.: Unification theory. In: Handbook of Logic in Artificial Intelligence and Logic Programming, Vol. 2, 41–125, Oxford Univ. Press, New York (1994)
5. Baaz, M.: Note on the existence of most general semi-unifier. Arithmetic, Proof Theory, and Computational Complexity (Prague, 1991), 20–29, Oxford Logic Guides, Vol. 23, Oxford Univ. Press, New York (1993)
6. Baxter, L. D.: The undecidability of the third order dyadic unification problem. Information and Control, **38**(2):170–178 (1978)
7. Bezem, M., Keuzenkamp, J., Undecidable goals for completed acyclic programs. New Generation Comp. **12**:209–213 (1994)
8. Blum M.: A machine-independent theory of the complexity of recursive functions. Journal of the ACM, **14**(2):322–336 (1967)
9. Boas, P. van E.: Dominos are forever. In: Priese, L. (ed) Report on the 1st GTI-workshop, Reihe Theoretische Informatik, Universität-Gesamthochschule Paderborn, 75–95 (1983)
10. Boas, P. van E.: The convenience of tillings. Lect. Notes Pure Appl. Math. **187**:331–363 (1997)
11. Bockmayr, A.: A note on a canonical theory with undecidable unification and matching problem. Journal of Automated Reasoning, **3**(4):379–381 (1987)
12. Burke, E. K.: The undecidability of the unification problem for nilpotent groups of class ⩾ 5. J. London Math. Soc. (2). **48**:52–58 (1993)
13. Caviness B. F.: On canonical forms and simplification. Journal of the ACM, **17**(2):385–396 (1970)
14. Chaitin G.: Algorithmic Information Theory. Cambridge University Press, Cambridge, England (1987)
15. Cornelissen, G., Zahidi, K.: Topology of Diophantine sets: remarks on Mazur's conjectures. Contemp. Math., **270**:253–260 (2000)

16. Da Costa, N. C. A., Doria, F. A.: Undecidability and incompleteness in classical mechanics. Int. J. Theor. Physics, **30**(8):1041–1073 (1991)
17. Davis M.: Arithmetical problems and recursively enumerable predicates (abstract). J. Symbolic Logic, **15**(1):77–78 (1950)
18. Davis M.: Arithmetical problems and recursively enumerable predicates. J. Symbolic Logic, **18**(1):33–41 (1953)
19. Davis M.: Speed-up theorems and Diophantine equations. In Rustin R. (ed.) Courant Computer Science Symposium 7: Computational Complexity, 87–95. Algorithmics Press, New York (1973)
20. Davis M.: Computability and Unsolvability. Dover Publications, New York (1982)
21. Davis, M., Putnam, H., Robinson, J.: The decision problem for exponential Diophantine equations. Ann. Math. (2), **74**:425–436 (1961). Reprinted in Feferman, S. (ed.) The collected works of Julia Robinson, Collected Works, **6**, American Mathematical Society, Providence, RI (1996)
22. Degtyarev, A., Voronkov, A.: Simultaneous rigid E-unification is undecidable. Lecture Notes in Computer Science, **1092**:178–190 (1996)
23. Denef, J.: Hilbert's Tenth Problem for quadratic rings. Proc. Amer. Math. Soc., **48**(1):214–220 (1975)
24. Denef, J., Lipshitz, L.: Power series solutions of algebraic differential equations. Mathematische Annalen, **267**(2):213–238 (1984)
25. Denef, J., Lipshitz, L.: Decision problems for differential equations. J. Symbolic Logic, **54**(3):941–950 (1989)
26. Farmer, W. M.: Simple second-order languages for which unification is undecidable. Theoretical Computer Sci., **87**:25–41 (1991)
27. Frisco, P.: Diophantine equations and splicing: a new demonstration of the generative capacity of H systems. Lect. Notes Computer Science, **2054**:43–52 (2001)
28. Gödel, K.: Über formal unentscheidbare Sätze der Principia Mathematica und verwandter Systeme. I. Monatsh. Math. und Phys. **38**(1):173–198 (1931)
29. Goldfarb, W., D.: The undecidability of the second-order unification problem. Theoretical Computer Science, **13**(2):225–230 (1981)
30. Goodstein, R. L.: Hilbert's tenth problem and the independence of recursive difference. J. London Math. Soc. (Second Series), **10**(2):175–176 (1975)
31. Goodstein, R. L., Lee, R. D.: A decidable class of equations in recursive arithmetic. Zeitschrift für Mathematische Logik und Grundlagen der Mathematik, **12**:235–239 (1966)
32. Grigor'ev, D. Yu., Singer, M. F.: Solving ordinary differential equations in terms of series with real exponents. Trans. Amer. Math. Soc., **327**(1):329–351 (1991)
33. Grunewald, F., Segal, D.: How to solve a quadratic equation in integers. Math. Proc. Cambridge Philos. Soc., **89**(1):1–5 (1981)
34. Grunewald, F., Segal, D.: On the integer solutions of quadratic equations. Journal of the Reine Angew. Math., **569**:13–45 (2004)
35. Gurari E. M.: Decidable problems for powerful programs. J. ACM, **32**(2):466–483, (1985)
36. Gurari, E. M., Ibarra, O. H., Two-way counter machines and Diophantine equations, J. ACM, **29**(3):863–873 (1982)
37. Hack, M.: The equality problem for vector addition systems is undecidable. Theoretical Computer Science, **2**(1):77–95 (1976)
38. Harel, D., Pnueli, A., Stavi, J.: Propositional dynamic logic of nonregular programs. Journal Computer and System Sciences, **26**(2):222–243 (1983)

39. Head, T.: Formal language theory and DNA: An analysis of the generative capacity of specific recombinant behavior. Bull. Math. Biology, **49** (1987)

40. Hickey, T., Mudambi, S.: Global compilation of Prolog. J. Logic Programming, **7**:193–230 (1989)

41. Hilbert, D.: Mathematische Probleme. Vortrag, gehalten auf dem internationalen Mathematiker Kongress zu Paris 1900. Nachr. K. Ges. Wiss., Göttingen, Math.-Phys.Kl. 253-297 (1900). See also Hilbert, D.: Gesammelte Abhandlungen, Springer, Berlin **3** (1935) (Reprinted: Chelsea, New York (1965)). English translation: Bull. Amer. Math. Soc., **8**:437–479 (1901-1902); reprinted in: Browder (ed.) Mathematical Developments arising from Hilbert Problems, Proceedings of Symposia in Pure Mathematics **28**, American Mathematical Society, 1–34 (1976)

42. Hodgson, B. R., Kent, C. F.: A normal form for arithmetical representation of NP-sets. J. Computer System Sci., **27**(3):378–388 (1983)

43. Howell, R. R., Rosier, L. E., Huynh, D. T., Yen H.-Ch.: Some complexity bounds for problems concerning finite and 2-dimensional vector addition systems with states. Theoretical Computer Science, **46**(2–3):107–140 (1986)

44. Ibarra, O. H., Leininger, B. S.: The complexity of the equivalence problem for straight-line programs. Conference Proceedings of the Twelfth Annual ACM Symposium on Theory of Computing, Los Angeles, California, 273–280 (1980)

45. Ibarra, O. H., Leininger, B. S.: Straight-line programs with one input variable. SIAM Journal on Computing, **11**(1):1–14 (1982)

46. Ibarra, O. H., Leininger, B. S.: On the simplification and equivalence problems for straight-line programs, J. ACM, **30**(3):641–656 (1983)

47. Ibarra, O. H., Rosier L. e.: The equivalence problem and correctness formulas for a simple class of programs. Lecture Notes Comp. Sci., **176**:330–338 (1984)

48. Jiménez, Á. R., Jiménez, M. J. P.: Generation of Diophantine sets by computing P systems with external output. Lect. Notes Comp. Sci., **2509**:176–190 (2002)

49. Jones, J. P.: Recursive undecidability—an exposition. Amer. Mathem. Monthly, **81**(7):724–738 (1974)

50. Jones, J. P.: Some undecidable determined games. International Journal of Game Theory, **11**(2):63–70 (1982)

51. Jones, J. P.: Universal Diophantine equation. J. Symbolic Logic **47**:549–571 (1982)

52. Jones, J. P.: Computational complexity of winning strategies in two players polynomial games (in Russian). Zapiski Nauchnykh Seminarov Leningradskogo Otdeleniya Matematicheskogo Instituta im. V. A. Steklova AN SSSR (LOMI), **192**:69–73 (1991)

53. Jones, J. P., Fraenkel, A.S.: Complexities of winning strategies in Diophantine games. J. Complexity, **11**:435–455 (1995)

54. Jones, J. P., Matijasevič, Ju. V.: Exponential Diophantine representation of recursively enumerable sets. In: Stern, J. (ed) Proceedings of the Herbrand Symposium: Logic Colloquium'81, Studies in Logic and the Foundations of Mathematics, **107**:159–177, North Holland, Amsterdam (1982)

55. Jones, J. P., Matijasevič, Ju. V.: A new representation for the symmetric binomial coefficient and its applications. Annales Sci. Mathém. du Québec, **6**(1):81–97 (1982)

56. Jones, J. P., Matijasevič, Ju. V.: Direct translation of register machines into exponential Diophantine equations. In: Priese, L. (ed) Report on the 1st GTI-workshop, Reihe Theoretische Informatik, Universität-Gesamthochschule Paderborn, 117–130 (1983)

57. Jones, J. P., Matijasevič, Ju. V.: Register machine proof of the theorem on exponential Diophantine representation of enumerable sets. J. Symbolic Logic, **49**(3):818–829 (1984)

58. Jones, J. P., Matijasevič, Ju. V.: Proof of recursive unsolvability of Hilbert's tenth problem. Amer. Math. Monthly, **98**(8):689–709 (1991)
59. Kasami, T., Nobuki, T.: Equivalence problem of programs without loops. Systems-Computers-Controls, **2**(4):83–84 (1971); Japanese original: Denshi Tsushin Gakkai Ronbunshi, **54-C**:657–658 (1971)
60. Kent, C. F., Hodgson, B. R.: An arithmetical characterization of NP. Theor. Computer Science, **21**(3):255–267 (1982)
61. Kosovskiĭ, N. K.: On Diophantine representations of the sequence of solutions of Pell equation.   Zapiski Nauchnykh Seminarov Leningradskogo Otdeleniya Matematicheskogo Instituta im. V. A. Steklova AN SSSR (LOMI), **20**:49–59 (1971)
62. Kreisel, G., Davis, M., Putnam, H., Robinson, J.: The decision problem for exponential Diophantine equations. Mathem. Reviews, **24**: #A3061, 573 (1962)
63. Kummer, M.:The complexity of recursion theoretic games. Trans. Amer. Math. Soc., **358**:1, 59–86 (electronic) (2006)
64. Lachlan, A. H.: On some games which are relevant to the theory of recursively enumerable sets. Ann. Math. (2), **91**:291–310 (1970)
65. Lambek, J.: How to program an infinite abacus. Canad. Math. Bull., **4**:295–302 (1961)
66. Levitz, H.: Decidability of some problem pertaining to base 2 exponential Diophantine equations, Zeitschrift Mathematische Logik Grundlagen Mathematik, **31**(2):109–115 (1985)
67. Li, C., Dang, Z., Ibarra, O. H., Yen, H.-Ch.: Signaling P systems and verifications problem. Lecture Notes Comput. Sci., **3580**:1462–1473 (2005)
68. Lipmaa, H.: On Diophantine complexity and statistical zero-knowledge arguments. Lecture Notes Computer Science, **2894**:398–415 (2003)
69. Livesey, M., Siekmann, J., Szabó, P., and Unvericht, E.: Unification problems for combinations of associativity, commutativity, distributivity and idempotence axioms. In: William H. J., Jr. (ed), Proceedings of the Fourth Workshop on Automated Deduction, 175–184, Austin, Texas, (1979)
70. Makanin, G. S.: The problem of solvability of equations in a free semigroup (in Russian). Math. Sbornik, **103**:147–236 (1977); English transl. in: Math. USSR Sbornik,   Math. USSR Sbornik, **32**(2):129–198 (1977)
71. Manders, K. L., Adleman, L.: NP-complete decision problems for binary quadratics. J. Comput. System Sci., **16**(2):168–184 (1978)
72. Markov, A. A.: Impossibility of certain algorithms in the theory of associative systems (in Russian), Dokl. Akad. Nauk SSSR, **55**(7):587–590 (1947). Translated in: Compte rendus de l'Académie des Sciences de l'U.R.S.S., **55**:583–586 (1947)
73. Martin-Löf, P. Notes on Constructive Mathematics. Almqvist & Wikseil, Stockholm (1970)
74. Matiyasevich, Yu. V.: The connection between Hilbert's tenth problem and systems of equations between words and lengths (in Russian), Zap. nauch. Seminar. Leningr. otd. Mat. in-ta AN SSSR, **8**:132–144 (1968). English translation: Seminars in Mathematics, V. A. Steklov Mathematical Institute, **8**:61–67 (1970)
75. Matiyasevich, Yu. V.: Enumerable sets are Diophantine (in Russian). Dokl. AN SSSR, **191**(2):278–282 (1970); Translated in: Soviet Math. Doklady, **11**(2):354–358
76. Matiyasevich, Yu. V.: Existence of noneffectivizable estimates in the theory of exponential Diophantine equations (in Russian). Zapiski Nauchnykh Seminarov Leningradskogo Otdeleniya Matematicheskogo Instituta im. V. A. Steklova AN SSSR (LOMI), **40**:77–93 (1974); Translated in: Journal of Soviet Mathematics, **8**(3):299–311 (1977)

77. Matiyasevich, Yu. V.: A new proof of the theorem on exponential Diophantine representation of enumerable sets (in Russian). Zapiski Nauchnykh Seminarov Leningradskogo Otdeleniya Matematicheskogo Instituta im. V. A. Steklova AN SSSR (LOMI), **60**:75–92 (1976); Translated in: Journal of Soviet Mathematics, **14**(5):1475–1486 (1980)

78. Matiyasevich, Yu. V.: Some purely mathematical results inspired by mathematical logic, In: Proceedings of Fifth International Congress on Logic, Methodology and Philosophy of science, London, Ontario, 1975, Reidel, Dordrecht, 121–127 (1977)

79. Matiyasevich, Yu. V.: Algorithmic unsolvability of exponential Diophantine equations in three unknowns (in Russian), In: Markov, A.A., Homich V.I. (eds), Studies in the Theory of Algorithms and Mathematical Logic, Computing Center Russian Academy Sci., Moscow, 69–78 (1979); Translated in Selecta Mathematica Sovietica, **3**:223–232 (1983/1984)

80. Matiyasevich, Yu. V.: Desyataya Problema Gilberta. Fizmatlit, Moscow, (1993). English translation: Hilbert's Tenth Problem. MIT Press, Cambridge, MA (1993). French translation: Le dixième Problème de Hilbert, Masson, Paris Milan Barselone (1995). URL: http://logic.pdmi.ras.ru/~yumat/H10Pbook,

81. Matiyasevich, Yu.: A direct method for simulating partial recursive functions by Diophantine equations. Annals Pure Appl. Logic, **67**:325–348 (1994)

82. Matiyasevich, Yu.: Hilbert's tenth problem: what was done and what is to be done Contemporary mathematics, **270**:1–47, (2000)

83. Matiyasevich, Yu.: Hilbert's tenth problem and paradigms of computation, Lecture Notes Computer Science, **3526**:310–321 (2005)

84. Matiyasevich, Yu.: Diophantine flavor of Kolmogorov complexity. Trans. Inst. Informatics and Automation Problems National Acad. Sciences of Armenia, **27**:111–122 (2006)

85. Matiyasevich, Yu.: Word Equations, Fibonacci Numbers, and Hilbert's Tenth Problem. URL: http://logic.pdmi.ras.ru/~yumat/Journal/jcontord.htm

86. Matijasevič, Yu., Robinson, J.: Reduction of an arbitrary Diophantine equation to one in 13 unknowns. Acta Arithmetica, **27**:521–553 (1975)

87. Mayr E. W., Meyer, A. R.: The complexity of the finite containment problem for Petri nets. Journal of the ACM, **28**(3):561–576 (1981)

88. Mazur, B.: The topology of rational points. Experimental Mathematics, **1**(1):35–45 (1992)

89. Mazur, B.: Questions of decidability and undesidability in Number Theory. J. Symbolic Logic, **59**(2):353–371 (1994)

90. Melzak, Z. A.: An informal arithmetical approach to computability and computation. Canad. Math. Bull., **4**:279–294 (1961)

91. Minsky, M. L.: Recursive unsolvability of Post's problem of "tag" and other topics in the theory of Turing machines. Ann. of Math. (2), **74**:437–455 (1961)

92. Minsky, M. L.: Computation: Finite and Infinite Machines. Prentice Hall, Englewood Cliffs, NJ (1967)

93. Ord, T., Kieu, T. D.: On the existence of a new family of Diophantine equations for Ω. Fundam. Inform. **56**(3):273–284 (2003)

94. Pappas, P.: A Diophantine problem for Laurent polynomial rings. Proceedings of the American Mathematical Society, **93**(4):713–718 (1985)

95. Paun, Gh.: From cells to (silicon) computers, and back. This volume, pages 343–371

96. Pheidas, Th., Zahidi, K.: Undecidability of existential theories of rings and fields: a survey. Contemp. Math., **270**:49–105 (2000)

97. Pheidas, Th.: An effort to prove that the existential theory of Q is undecidable. Contemp. Math., **270**:237–252 (2000)

98. Pollett, Ch.: On the bounded version of Hilbert's tenth problem. Arch. Math. Logic, **42**(5):469–488 (2003)
99. Poonen, B.: Hilbert's tenth problem and Mazur's conjecture for large subrings of $\mathbb{Q}$. J. Amer. Math. Soc., **16**(4):981–990 (2003)
100. Post, E. L.: Formal reductions of the general combinatorial decision problem. Amer. J. Math., **65**:197–215 (1943); reprinted in: The Collected Works of E. L. Post, Davis, M. (ed), Birkhäuser, Boston (1994).
101. Post, E. L.: Recursively enumerable sets of positive integers and their decision problems. Bull. Amer. Math. Soc., **50**:284–316 (1944); reprinted in: The Collected Works of E. L. Post, Davis, M. (ed), Birkhäuser, Boston (1994).
102. Post, E. L.: Recursive unsolvability of a problem of Thue. J. Symbolic Logic, **12**:1–11 (1947); reprinted in: The Collected Works of E. L. Post, Davis, M. (ed), Birkhäuser, Boston (1994).
103. Prasad, K.: Computability and randomness of Nash equilibrium in infinite games. J. Mathem. Economics, **20**(5):429–442 (1991).
104. Reif, J. H., Lewis, H. R.: Efficient symbolic analysis of programs, J. Computer System Sci., **32**(3):280–314 (1986)
105. Rabin M. O.: Effective computability of winning strategies. In: Dresher, M., Tucker, A. W., Wolff, P. (eds), Contributions to the Theory of Games. Volume III, Annals of Mathematics Studies, **39**:147–157 Princeton University Press, Princeton, NJ (1957)
106. Reutenauer, Ch., Aspect Math'ematiques des R'eseaux de Pétri. Masson, Paris Milan Barcelone Mexico (1989); Engl. transl: The Mathematics of Petri Nets, Prentice-Hall, Englewood Cliffs, NJ (1990)
107. D. Richardson, Some undecidable problems involving elementary functions of a real variable. J. Symbolic Logic, **33**(4):514–520 (1968)
108. Robinson, J. A.: A machine-oriented logic based on the resolution principle, J. Assoc. Comput. Mach. **12**:23–41 (1965)
109. Shepherdson, J. C., Sturgis, H. E.: Computability of recursive functions, J. ACM **10**(2):217–255 (1963)
110. Shirayev, D. V.: Undecidability of some decision problems for straight-line programs (in Russian), Kibernetika, **1**:63–66 (1989)
111. Shlapentokh, A.: Hilbert's tenth problem over number fields, a survey. Contemp. Math., **270**:107–143 (2000)
112. Shlapentokh, A.: A ring version of Mazur's conjecture on top[ology of rational points. Int. Math. Res. Notes, **2003**(7):411–423 (2003)
113. Shlapentokh, A.: Hilbert's Tenth Problem. Diophantine Classes and Extensions to Global Fields. Cambridge Univ. Press, Cambridge, England (2007)
114. Siegel, C. L.: Zur Theorie der quadratischen Formen. Nachrichten Akademie Wissenschaften in Göttingen. II. Mathematisch-Physikalische Klasse, **3**:21–46 (1972)
115. Siekmann, J. H., Unification theory. J. Symbolic Comp., **7**:207–274 (1989)
116. Singer, M. F.: The model theory of ordered differential fields. J. Symbolic Logic, **43**:1, 82–91 (1978)
117. Skolem, Th.: Über die Nicht-charakterisierbarkeit der Zahlenreihe mittels endlich oder abzählbar unendlich vieler Aussagen mit ausschliesslich Zahlenvariablen, Fundamenta Mathematicae, **23**:150–161 (1934)
118. Stallworth D. T., Roush, F. W.: An undecidable property of definite integrals. Proceedings of the American Mathematical Society, **125**(7):2147–2148 (1997)
119. Sun, Zh.-W.: Reduction of unknowns in Diophantine representations. Science in China (Scientia Sinica) Ser. A., **35**(3):257–269 (1992)

120. Thue, A.: Problem über Veränderungen von Zeichenreihen nach gegebenen Regeln. Vid. Skr. I. Mat.-natur. Kl., **10**:493–524 (1914). Reprinted in: Thue, A.: Selected Mathematical Papers, Oslo (1977)

121. Tiden, E., Arnborg, S.: Unification problems with one-sided distributivity, J. Symbolic Computation, **3**:183–202 (1987)

122. Tseitin, G.S.: A method of presenting the theory of algorithms and enumerable sets (in Russian). Trudy Matematicheskogo instituta im. V. A. Steklova 72 (1964) 69–99. English translation in: Am. Math. Soc. Translat., II. Ser. **99**:1–39 (1972)

123. Tung, Sh. P. The bound of Skolem functions and their applications. Information and Computation, **120**:149–154 (1995)

124. Sivaramakrishnan Rajagopalan, S.: Average case intractability of matrix and Diophantine problems. Proceedings Twenty-Fourth Annual ACM Symposium Theory Comput., Victoria, British Columbia, Canada, 632–642 (1992)

125. Vinogradov, A. K., Kosovskiĭ, N. K.: A hierarchy of Diophantine representations of primitive recursive predicates (in Russian). Vychislitel'naya tekhnika i voprosy kibernetiki, no. 12, 99–107. Lenigradskiĭ Gosudarstvennyĭ Universitet, Leningrad (1975)

126. Wolfson, O.: Parallel evaluation of Datalog programs by load sharing. J. Logic Programming, **12**:369–393 (1992)

127. Yukna, S.: Arithmetical representations of classes of computational complexity (in Russian). Matematicheskaya logika i eё primeneniya, no. 2, 92–107, Institut Matematiki i Kibernetiki Akademii Nauk Litovskoĭ SSR, Vil'nyus (1982)

128. Yukna, S.:. On arithmetization of computations (in Russian). Matematicheskaya logika i eё primeneniya, no. 3, 117–125, Institut Matematiki i Kibernetiki Akademii Nauk Litovskoĭ SSR, Vil'nyus (1983)

129. URL: `http://logic.pdmi.ras.ru/Hilbert10`

Elementary Algorithms and Their Implementations*

Yiannis N. Moschovakis[1] and Vasilis Paschalis[2]

[1] Department of Mathematics, University of California, Los Angeles, CA 90095-1555, USA, and Graduate Program in Logic, Algorithms and Computation ($M\Pi\Lambda A$), University of Athens, Panepistimioupolis, 15784 Zografou, Athens, Greece, ynm@math.ucla.edu

[2] Graduate Program in Logic, Algorithms and Computation ($M\Pi\Lambda A$), University of Athens, Panepistimioupolis, 15784 Zografou, Athens, Greece, pasva@yahoo.com

In the sequence of articles [3, 4, 5, 6, 7], Moschovakis has proposed a mathematical modeling of the notion of *algorithm*—a set-theoretic "definition" of algorithms, much like the "definition" of real numbers as Dedekind cuts on the rationals or that of random variables as measurable functions on a probability space. The aim is to provide a traditional foundation for the theory of algorithms, a development of it within axiomatic set theory in the same way as analysis and probability theory are rigorously developed within set theory on the basis of the set theoretic modeling of their basic notions. A characteristic feature of this approach is the adoption of a very abstract notion of algorithm that takes *recursion* as a primitive operation and is so wide as to admit "non-implementable" algorithms: *implementations* are special, restricted algorithms (which include the familiar *models of computation*, e.g., Turing and random access machines), and an algorithm is implementable if it is *reducible* to an implementation.

Our main aim here is to investigate the important relation between an (implementable) algorithm and its implementations, which was defined very briefly in [6], and, in particular, to provide some justification for it by showing that standard examples of "implementations of algorithms" naturally fall under it. We will do this in the context of (deterministic) *elementary algorithms*, i.e., algorithms that compute (possibly partial) functions $f : M^n \rightharpoonup M$ *from* (relative to) finitely many given partial functions on some set M; and so part of the paper is devoted to a fairly detailed exposition of the basic facts about these algorithms, which provide the most important examples of the general theory but have received very little attention in the

*The research for this article was co-funded by the European Social Fund and National Resources—(EPEAEK II) PYTHAGORAS in Greece.

general development. We will also include enough facts about the general theory, so that this paper can be read independently of the articles cited above—although many of our choices of notions and precise definitions will appear unmotivated without the discussion in those articles.[1]

A second aim is to fix some of the basic definitions in this theory, which have evolved—and in one case evolved back—since their introduction in [3].

We will use mostly standard notation, and we have collected in the Appendix the few, well-known results from the theory of posets to which we will appeal.

1 Recursive (McCarthy) programs

A (pointed, partial) *algebra* is a structure of the form

$$\boldsymbol{M} = (M, 0, 1, \boldsymbol{\Phi}) = (M, 0, 1, \{\phi^{\boldsymbol{M}}\}_{\phi \in \Phi}), \tag{1}$$

where 0, 1 are distinct points in the universe M, and for every *function constant* $\phi \in \Phi$,

$$\phi^{\boldsymbol{M}} : M^n \rightharpoonup M$$

is a partial function of some arity $n = n_\phi$ associated by the *vocabulary* Φ with the symbol ϕ. The objects 0 and 1 are used to code the truth values, so that we can include relations among the primitives of an algebra by identifying them with their characteristic functions,

$$\chi_R(\vec{x}) = R(\vec{x}) = \begin{cases} 1, & \text{if } R(\vec{x}), \\ 0, & \text{otherwise.} \end{cases}$$

Thus $R(\vec{x})$ is synonymous with $R(\vec{x}) = 1$.

Typical examples of algebras are the basic structures of arithmetic

$$\boldsymbol{N}_u = (\mathbb{N}, 0, 1, S, \mathrm{Pd}), \quad \boldsymbol{N}_b = (\mathbb{N}, 0, 1, \mathrm{parity}, \mathrm{iq}_2, \mathrm{em}_2, \mathrm{om}_2)$$

which represent the natural numbers "given" in unary and binary notation. Here $S(x) = x + 1, \mathrm{Pd}(x) = x \mathbin{\dot-} 1$ are the operations of successor and predecessor, $\mathrm{parity}(x)$ is 0 or 1 accordingly as x is even or odd, and

$$\mathrm{iq}_2(x) = \text{the integer quotient of } x \text{ by } 2, \quad \mathrm{em}_2(x) = 2x, \quad \mathrm{om}_2(x) = 2x + 1$$

are the less familiar operations that are the natural primitives on "binary numbers." (We read $\mathrm{em}_2(x)$ and $\mathrm{om}_2(x)$ as *even* and *odd* multiplication by 2.) These are *total*

---

[1] An extensive analysis of the foundational problem of defining algorithms and the motivation for our approach is given in [6], which is also posted on http://www.math.ucla.edu/∼ynm.

algebras, as is the standard (in model theory) structure on the natural numbers $N = (\mathbb{N}, 0, 1, =, +, \times)$ on $\mathbb{N}$. Genuinely partial algebras typically arise as *restrictions* of total algebras, often to finite sets: if $\{0, 1\} \subseteq L \subseteq M$, then

$$\boldsymbol{M} \restriction L = (L, 0, 1, \{\phi^M \restriction L\}_{\phi \in \Phi}),$$

where, for any $f : M^n \rightharpoonup M$,

$$f \restriction L(x_1, \ldots, x_n) = w \iff x_1, \ldots, x_n, w \in L \ \& \ f(x_1, \ldots, x_n) = w.$$

The (explicit) *terms* of the language $\mathsf{L}(\Phi)$ of programs in the vocabulary Φ are defined by the recursion

$$A :\equiv 0 \mid 1 \mid \mathsf{v}_i \mid \phi(A_1, \ldots, A_n) \mid \mathsf{p}_i^n(A_1, \ldots, A_n)$$
$$\mid \text{if } (A_0 = 0) \text{ then } A_1 \text{ else } A_2, \quad (\Phi\text{-terms}),$$

where v_i is one of a fixed list of *individual variables*; p_i^n is one of a fixed list of n-ary (partial) *function variables*; and ϕ is any n-ary symbol in Φ. In some cases we will need the operation of *substitution* of terms for individual variables: if $\vec{\mathsf{x}} \equiv \mathsf{x}_1, \ldots, \mathsf{x}_n$ is a sequence of variables and $\vec{B} \equiv B_1, \ldots, B_n$ is a sequence of terms, then

$$A\{\vec{\mathsf{x}} :\equiv \vec{B}\} \equiv \text{the result of replacing each occurrence of } \mathsf{x}_i \text{ by } B_i.$$

Terms are interpreted naturally in any Φ-algebra $\boldsymbol{M}$, and if $\vec{\mathsf{x}}, \vec{\mathsf{p}}$ is any list of variables that includes all the variables that occur in A, we let

$$[\![A]\!](\vec{x}, \vec{p}) = [\![A]\!]^M(\vec{x}, \vec{p})$$
$$= \text{the value (if defined) of } A \text{ in } \boldsymbol{M}, \text{ with } \vec{\mathsf{x}} := \vec{x}, \vec{\mathsf{p}} := \vec{p}. \quad (2)$$

Now each pair $(A, \vec{\mathsf{x}}, \vec{\mathsf{p}})$ of a term and a sufficiently inclusive list of variables defines a (partial) *functional*

$$F_A(\vec{x}, \vec{p}) = [\![A]\!](\vec{x}, \vec{p}) \quad (3)$$

on M, and the associated *operator*

$$F_A'(\vec{p}) = \lambda(\vec{x})[\![A]\!](\vec{x}, \vec{p}). \quad (4)$$

We view F_A' as a mapping

$$F_A' : (M^{k_1} \rightharpoonup M) \times \cdots \times (M^{k_n} \rightharpoonup M) \to (M^n \rightharpoonup M), \quad (5)$$

where k_i is the arity of the function variable p_i, and it is easily seen (by induction on the term A) that it is continuous.

A *recursive* (or McCarthy) *program* of $\mathsf{L}(\Phi)$ is any system of *recursive term equations*

$$A \quad : \quad \begin{cases} \mathsf{p}_A(\vec{\mathsf{x}}) = \mathsf{p}_0(\vec{\mathsf{x}}) = A_0 \\ \qquad \mathsf{p}_1(\vec{\mathsf{x}}_1) = A_1 \\ \qquad \vdots \\ \mathsf{p}_K(\vec{\mathsf{x}}_K) = A_K \end{cases} \quad (6)$$

such that $p_0 \equiv p_A, p_1, \ldots, p_K$ are distinct function variables; $p_1, \ldots, p_K$ are the only function variables that occur in $A_0, \ldots, A_K$; the only individual variables that occur in each A_i are in the list $\vec{x}_i$; and the arities of the function variables are such that these equations make sense. The term A_0 is the *head* of A, and it may be its only *part*, since we allow $K = 0$. If $K > 0$, then the remaining parts $A_1, \ldots, A_K$ comprise the *body* of A. The *arity* of A is the number of variables in the head term A_0. Sometimes it is convenient to think of programs as (extended) *program terms* and rewrite (6) in the form

$$A \equiv A_0 \text{ where } \{p_1 = \lambda(\vec{x}_1)A_1, \ldots, p_K = \lambda(\vec{x}_K)A_K\}, \tag{7}$$

in an expanded language with a symbol for the (abstraction) λ-operator and mutual recursion. This makes it clear that the function variables $p_1, \ldots, p_K$ and the occurrences of the indivudual variables $\vec{x}_i$ in A_i ($i > 0$) are *bound* in the program A, and that the putative "head variable" p_A does not actually occur in A. An individual variable z is free in A if it occurs in A_0.

To interpret a program A on a Φ-structure M, we consider the system of *recursive equations*

$$\begin{cases} p_1(\vec{x}_1) = [\![A_1]\!](\vec{x}_1, \vec{p}) \\ \quad \vdots \\ p_K(\vec{x}_K) = [\![A_K]\!](\vec{x}_K, \vec{p}). \end{cases} \tag{8}$$

By the basic Fixed Point Theorem 8.1, this system has a set of least solutions

$$\overline{p}_1, \ldots, \overline{p}_K,$$

the mutual fixed points of (the body of) A, and we set

$$[\![A]\!] = [\![A]\!]^M = [\![A_0]\!](\overline{p}_1, \ldots, \overline{p}_K) : M^n \rightharpoonup M.$$

Thus the *denotation* of A on M is the partial function defined by its head term from the mutual fixed points of A, and its arity is the arity of A.

A partial function $f : M^n \rightharpoonup M$ is M-*recursive* if it is computed by some recursive program.

Except for the notation, these programs are introduced by John McCarthy in [2]. McCarthy proved that *the N_u-recursive partial functions are exactly the Turing-computable ones*, and it is easy to verify that these are also the N_b-recursive partial functions.

In the foundational program we describe here, we associate with each Φ-program A of arity n and each Φ-algebra M, a *recursor*

$$\text{int}(A, M) : M^n \rightsquigarrow M,$$

a set-theoretic object that purports to model the algorithm expressed by A on M and determines the denotation $[\![A]\!]^M : M^n \rightharpoonup M$. These *referential intensions* of

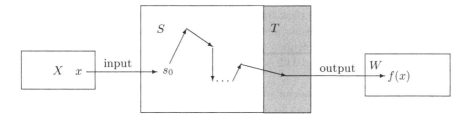

Fig. 1. Iterator computing $f : X \rightharpoonup W$.

Φ-programs on M model "the algorithms of M"; taken all together, they are the *first-order* or *elementary algorithms*, with the algebra M exhibiting the partial functions *from* (relative to) *which* any particular elementary algorithm is specified.

It naturally turns out that the N_u-algorithms are not the same as the N_b-algorithms on $\mathbb{N}$, because (in effect) the choices of primitives in these two algebras codify distinct ways of representing the natural numbers.

2 Recursive machines

Most of the known computation models for partial functions $f : X \rightarrow W$ on one set to another are captured faithfully by the following well-known, general notion:

Definition 2.1 (Iterators). For any two sets X and Y, an *iterator* (or *abstract machine*) $i : X \rightsquigarrow Y$ is a quintuple $(\text{input}, S, \sigma, T, \text{output})$, satisfying the following conditions:

(I1) S is an arbitrary (nonempty) set, the *set of states* of i;

(I2) input $: X \to S$ is the *input function* of i;

(I3) $\sigma : S \to S$ is the *transition function* of i;

(I4) $T \subseteq S$ is the set of *terminal states* of i, and $s \in T \implies \sigma(s) = s$;

(I5) output $: T \to Y$ is the *output function* of i.

A *partial computation* of i is any finite sequence $s_0, \ldots, s_n$ such that for all $i < n$, s_i is not terminal and $\sigma(s_i) = s_{i+1}$. We write

$$s \to_i^* s' \iff s = s', \text{ or there is a partial computation with } s_0 = s, s_n = s',$$

and we say that i *computes* the partial function $\bar{i} : X \rightarrow Y$ defined by

$$\bar{i}(x) = w \iff (\exists s \in T)[\text{input}(x) \to_i^* s \ \& \ \text{output}(s) = w].$$

For example, each Turing machine can be viewed as an iterator $i : \mathbb{N} \rightsquigarrow \mathbb{N}$, by taking for states the (so-called) "complete configurations" of i, i.e., the triples (σ, q, i) where σ is the tape, q is the internal state, and i is the location of the machine, along with the standard input and output functions. The same is true for random access machines, sequential machines equipped with one or more stacks, etc.

Definition 2.2 (Iterator isomorphism). An isomorphism between two iterators i_1, i_2: $X \rightsquigarrow Y$ with the same input and output sets is any bijection $\rho : S_1 \rightarrowtail\hspace{-0.6em}\twoheadrightarrow S_2$ of their sets of states such that the following conditions hold:

(MI1) $\rho(\text{input}_1(x)) = \text{input}_2(x) \quad (x \in X)$.

(MI2) $\rho[T_1] = T_2$.

(MI3) $\rho(\tau_1(s)) = \tau_2(\rho(s)), \quad (s \in S_1)$.

(MI4) If $s \in T_1$ and s is *input accessible*, i.e., $\text{input}_1(x) \rightarrow^* s$ for some $x \in X$, then $\text{output}_1(s) = \text{output}_2(\rho(s))$.

The restriction in (MI4) to input accessible, terminal states holds trivially in the natural case, when every terminal state is input accessible; but it is convenient to allow machines with "irrelevant" terminal states (much as we allow functions $f : X \rightarrow Y$ with $f[X] \subsetneq Y$), and for such machines it is natural to allow isomorphisms to disregard them.

For our purpose of illustrating the connection between the algorithm expressed by a recursive program and its implementations, we introduce *recursive machines*, perhaps the most direct and natural implementations of programs.

Definition 2.3 (Recursive machines). For each recursive program A and each Φ-algebra M, we define the *recursive machine* $i = i(A, M)$ that computes the partial function $[\![A]\!]^M : M^n \rightharpoonup M$ denoted by A as follows.

First we modify the class of Φ-terms by allowing all elements in M to occur in them (as constants) and by restricting the function variables that can occur to the function variables of A:

$$B :\equiv 0 \mid 1 \mid x \mid \mathsf{v}_i \mid \phi(B_1, \ldots, B_n) \mid \mathsf{p}_i(B_1, \ldots, B_{k_i})$$
$$\mid \text{if } (B_0 = 0) \text{ then } B_1 \text{ else } B_2 \quad (\Phi[A, M]\text{-terms}),$$

where x is any member of M, viewed as an *individual constant*, like 0 and 1. A $\Phi[A, M]$-term B is *closed* if it has no individual variables occurring in it, so that its value in M is fixed by M (and the values assigned to the function variables $\mathsf{p}_1, \ldots, \mathsf{p}_K$).

The states of i are all finite sequences s of the form

$$a_0 \ \ldots \ a_{m-1} : b_0 \ \ldots \ b_{n-1},$$

where the elements $a_0, \ldots, a_{m_1}, b_0, \ldots, b_{n-1}$ of s satisfy the following conditions:

- Each a_i is a function symbol in Φ, or one of $p_1, \ldots, p_K$, or a closed $\Phi[A, M]$-term, or the special symbol ?; and

- Each b_j is an individual constant, i.e., $b_i \in M$.

The special symbol ":" has exactly one occurrence in each state, and that the sequences $\vec{a}, \vec{b}$ are allowed to be empty, so that the following sequences are states (with $x \in M$):

$$x : \qquad : x \qquad :$$

The *terminal states* of i are the sequences of the form

$$: w;$$

i.e., those with no elements on the left of ":" and just one constant on the right; and the *output* function of i simply reads this constant w; i.e.,

$$\text{output}(: w) = w.$$

The states, the terminal states, and the output function of i depend only on M and the function variables that occur in A. The input function of i depends also on the head term of A,

$$\text{input}(\vec{x}) \equiv A_0\{\vec{x} :\equiv \vec{x}\} : ,$$

where $\vec{x}$ is the sequence of individual variables that occur in A_0.

The transition function of i is defined by the seven cases in the Transition Table 1, i.e.,

$$\sigma(s) = \begin{cases} s', & \text{if } s \to s' \text{ is a special case of some line in Table 1,} \\ s, & \text{otherwise,} \end{cases}$$

and it is a function, because for a given s (clearly) at most one transition $s \to s'$ is *activated* by s. Notice that only the external calls depend on the algebra M, and only the internal calls depend on the program A—and so, in particular, all programs with the same body share the same transition function.

Theorem 2.4. *Suppose A is a Φ-program with function variables $\vec{p}$, M is a Φ-algebra, $\overline{p}_1, \ldots, \overline{p}_K$ are the mutual fixed points of A, and B is a closed $\Phi[A, M]$-term. Then for every $w \in M$,*

$$[B]^M(\overline{p}_1, \ldots, \overline{p}_K) = w \iff B : \to^*_{i(A,M)} : w. \tag{9}$$

In particular, with $B \equiv A_0\{\vec{x} :\equiv \vec{x}\}$,

$$[A]^M(\vec{x}) = w \iff A_0\{\vec{x} :\equiv \vec{x}\} : \to^*_{i(A,M)} : w,$$

and so the iterator $i(A, M)$ computes the denotation $[A]^M$ of A.

Table 1. Transition table for the recursive machine $i(A, M)$.

(pass)	$\vec{a} \; \underline{x} : \; \vec{b} \; \rightarrow \; \vec{a} \; \underline{: x} \; \vec{b} \quad (x \in M)$
(e-call)	$\vec{a} \; \underline{\phi_i} : \vec{x} \; \vec{b} \; \rightarrow \; \vec{a} \; \underline{: \phi_i^M(\vec{x})} \; \vec{b}$
(i-call)	$\vec{a} \; \underline{p_i} : \vec{x} \; \vec{b} \; \rightarrow \; \vec{a} \; \underline{A_i\{\vec{\mathsf{x}}_i := \vec{x}\} :} \; \vec{b}$
(comp)	$\vec{a} \; \underline{h(A_1, \ldots, A_n)} : \; \vec{b} \; \rightarrow \; \vec{a} \; \underline{h \; A_1 \; \cdots \; A_n} : \; \vec{b}$
(br)	$\vec{a} \; \underline{\text{if} \; (A = 0) \; \text{then} \; B \; \text{else} \; C} : \; \vec{b} \; \rightarrow \; \vec{a} \; \underline{B \; C \, ? \, A} : \; \vec{b}$
(br0)	$\vec{a} \; \underline{B \; C \, ? : 0} \; \vec{b} \; \rightarrow \; \vec{a} \; \underline{B} : \; \vec{b}$
(br1)	$\vec{a} \; \underline{B \; C \, ? : y(\neq 0)} \; \vec{b} \; \rightarrow \; \vec{a} \; \underline{C} : \; \vec{b}$

- The underlined words are those that trigger a transition and change.
- $\vec{x} = x_1, \ldots, x_n$ is an n-tuple of individual constants.
- In the *external call* (e-call), $\phi_i \in \Phi$ and arity$(\phi_i) = n_i = n$.
- In the *internal call* (i-call), p_i is an n-ary function variable of A defined by the equation $p_i(\vec{\mathsf{x}}) = A_i$.
- In the *composition transition* (comp), h is a (constant or variable) function symbol with arity$(h) = n$.

Outline of proof. First we define the partial functions computed by $i(A, M)$ in the indicated way,

$$\widetilde{p}_i(\vec{x}_i) = w \iff \mathsf{p}_i(\vec{x}_i) : \rightarrow^* \; : w,$$

and show by an easy induction on the term B the version of (9) for these,

$$[\![B]\!]^M(\widetilde{p}_1, \ldots, \widetilde{p}_K) = w \iff B : \rightarrow^*_{i(A,M)} \; : w. \tag{10}$$

When we apply this to the terms $A_i\{\vec{\mathsf{x}}_i := \vec{x}_i\}$ and use the form of the internal call transition rule, we get

$$[\![A_i]\!]^M(\vec{x}_i, \widetilde{p}_1, \ldots, \widetilde{p}_K) = w \iff \widetilde{p}_i(\vec{x}_i) = w,$$

which means that the partial functions $\widetilde{p}_1, \ldots, \widetilde{p}_K$ satisfy the system (8), and in particular

$$\overline{p}_1 \leq \widetilde{p}_1, \ldots, \overline{p}_K \leq \widetilde{p}_K.$$

Next we show that for any closed term B as above and any system $p_1, \ldots, p_K$ of solutions of (8),

$$B : \rightarrow^* w \implies [\![B]\!]^M(p_1, \ldots, p_K) = w;$$

this is done by induction of the length of the computation, which establishes the hypothesis, and setting $B \equiv A_i\{\vec{\mathbf{x}}_i := \vec{x}_i\}$, it implies that

$$\widetilde{p}_1 \leq p_1, \ldots, \widetilde{p}_K \leq p_K.$$

It follows that $\widetilde{p}_1, \ldots, \widetilde{p}_K$ are the least solutions of (8); i.e., $\widetilde{p}_i = \overline{p}_i$, which together with (10) completes the proof.

Both arguments appeal repeatedly to the following trivial but basic property of recursive machines: *if* $s_0, s_1, \ldots, s_n$ *is a partial computation of* $\mathsf{i}(A, M)$ *and* $\vec{a}^*, \vec{b}^*$ *are such that the sequence* $\vec{a}^* s_0 \vec{b}^*$ *is a state, then the sequence*

$$\vec{a}^* s_0 \vec{b}^* , \vec{a}^* s_1 \vec{b}^* , \ldots, \vec{a}^* s_n \vec{b}^*$$

is also a partial computation of $\mathsf{i}(A, M)$. $\qquad\qquad\qquad\qquad\qquad\square$

3 Monotone recursors and recursor isomorphism

An algorithm is *expressed directly* by a definition by mutual recursion, in our approach, and so it can be modeled by the *semantic content* of a mutual recursion. This is most simply captured by a tuple of related mappings, as follows.

Definition 3.1 (Recursors). For any poset X and any complete poset W, a (*monotone*) *recursor* $\alpha : X \leadsto W$ is a tuple

$$\alpha = (\alpha_0, \alpha_1, \ldots, \alpha_K),$$

such that for suitable, complete posets $D_1, \ldots, D_K$:

(1) Each *part* $\alpha_i : X \times D_1 \times \cdots D_K \to D_i$, $(i = 1, \ldots, k)$ is a monotone mapping.

(2) The *output mapping* $\alpha_0 : X \times D_1 \times \cdots \times D_K \to W$ is also monotone.

The number K is the *dimension* of α; the product $D_\alpha = D_1 \times \cdots \times D_K$ is its *solution set*; its *transition mapping* is the function

$$\mu_\alpha(x, \vec{d}) = (\alpha_1(x, \vec{d}), \ldots, \alpha_K(x, \vec{d})),$$

on $X \times D_\alpha$ to D_α; and the function $\overline{\alpha} : X \to W$ *computed by* α is

$$\overline{\alpha}(x) = \alpha_0(x, \vec{d}(x)) \quad (x \in X),$$

where $\vec{d}(x)$ is the least fixed point of the system of equations

$$\vec{d} = \mu_\alpha(x, \vec{d}).$$

By the basic Fixed Point Theorem 8.1,

$$\overline{\alpha}(x) = \alpha_0(x, \vec{d}_\alpha^\kappa(x)), \text{ where } \vec{d}_\alpha^\xi(x) = \mu_\alpha(x, \sup\{\vec{d}_\alpha^\eta(x) \mid \eta < \xi\}), \qquad (11)$$

for every sufficiently large ordinal κ—and for $\kappa = \omega$ when α is continuous. We express all this succinctly by writing[2]

$$\alpha(x) = \alpha_0(x, \vec{d}) \text{ where } \{\vec{d} = \mu_\alpha(x, \vec{d})\}, \tag{12}$$

$$\overline{\alpha}(x) = \alpha_0(x, \vec{d}) \overline{\text{where}} \{\vec{d} = \mu_\alpha(x, \vec{d})\}. \tag{13}$$

The definition allows $K = \text{dimension}(\alpha) = 0$, in which case[3] $\alpha = (\alpha_0)$ for some monotone function $\alpha_0 : X \to W$, $\overline{\alpha} = \alpha_0$, and equation (12) takes the awkward (but still useful) form

$$\alpha(x) = \alpha_0(x) \text{ where } \{ \ \}.$$

A recursor α is *continuous* if the mappings α_i ($i \leq K$) are continuous and *discrete* if X is a set (partially ordered by $=$) and $W = Y \cup \{\bot\}$ is the bottom liftup of a set. A discrete recursor $\alpha : X \rightsquigarrow Y_\bot$ computes a partial function $\overline{\alpha} : X \to Y$.

Definition 3.2 (The recursor of a program). Every recursive Φ-program A of arity n determines naturally the following recursor $\mathfrak{r}(A, M) : M^n \rightsquigarrow M_\bot$ relative to a Φ-algebra M:

$$\mathfrak{r}(A, M) = [\![A_0]\!]^M(\vec{x}, \vec{p})$$
$$\text{where } \{p_1 = \lambda(\vec{x}_1)[\![A_1]\!]^M(\vec{x}_1, \vec{p}), \ldots, p_K = \lambda(\vec{x}_K)[\![A_K]\!]^M(\vec{x}_K, \vec{p})\}. \tag{14}$$

More explicitly (for once), $\mathfrak{r}(A, M) = (\alpha_0, \ldots, \alpha_K)$, where $D_i = (M^{k_i} \rightharpoonup M)$ (with $\text{arity}(\mathsf{p}_i) = k_i$) for $i = 1, \ldots, K$, $D_0 = (M^n \rightharpoonup M)$, and the mappings α_i are the continuous operators defined by the parts of the program term A by (4); so $\mathfrak{r}(A, M)$ is a continuous recursor. It is immediate from the semantics of recursive programs and (11) that the recursor of a program computes its denotation,

$$\overline{\mathfrak{r}(A, M)}(\vec{x}) = [\![A]\!]^M(\vec{x}) \quad (\vec{x} \in M^n). \tag{15}$$

Notice that the transition mapping of $\mathfrak{r}(A, M)$ is independent of the input $\vec{x}$, and so we can suppress it in the notation:

$$\mu_{\mathfrak{r}(A, M)}(\vec{p}) = \mu_{\mathfrak{r}(A, M)}(\vec{x}, \vec{p}) = (\alpha_1(\vec{p}), \ldots, \alpha_K(\vec{p})). \tag{16}$$

---

[2] Formally, "where" and "$\overline{\text{where}}$" denote operators that take (suitable) tuples of monotone mappings as arguments, so that where $(\alpha_0, \ldots, \alpha_K)$ is a recursor and $\overline{\text{where}}(\alpha_0, \ldots, \alpha_n)$ is a monotone mapping. The recursor-producing operator "where" is specific to this theory and not familiar, but the mapping producing $\overline{\text{where}}$ is one of many fairly common, recursive program constructs for which many notations are used, e.g.,

$$(\text{letrec}([d_1\ \alpha_1] \ldots [d_K\ \alpha_K])\ \alpha_0) \text{ or } (d_1 = \alpha_1, \ldots d_K = \alpha_K) \text{ in } \alpha_0.$$

[3] Here $D_\alpha = \{\bot\}$ (by convention or a literal reading of the definition of *product poset*), $\mu_\alpha(x, d) = d$, and (by convention again) $\alpha_0(x, \bot) = \alpha_0(x)$.

Caution: *The recursor* $\mathfrak{r}(A, M)$ *does not always model the algorithm expressed by* A *on* M, because it does not take into account the *explicit steps* that may be required for the computation of the denotation of A. In the extreme case, if $A \equiv A_0$ is an explicit term (a program with just a head and empty body), then

$$\mathfrak{r}(A, M) = [\![A_0]\!]^M(\vec{x}) \text{ where } \{ \ \}$$

is a trivial recursor of dimension 0—and it is the same for all explicit terms that define the same partial function, which is certainly not right. We will put off until Section 7 the correct and complete definition of $\mathrm{int}(A, M)$, which models more faithfully the algorithm expressed by a recursive program. As it turns out, however,

$$\mathrm{int}(A, M) = \mathfrak{r}(\mathrm{cf}(A), M)$$

for some program $\mathrm{cf}(A)$ that is associated canonically with each A, and so the algorithms of an algebra M will all be of the form (14) for suitable A's.

Definition 3.3 (Recursor isomorphism). Two recursors $\alpha, \beta : X \rightsquigarrow W$ (on the same domain and range) are *isomorphic*[4] if they have the same dimension K, and there is a permutation $(l_1, \ldots, l_K)$ of $(1, \ldots, K)$ and poset isomorphisms $\rho_i : D_{\alpha, l_i} \to D_{\beta, i}$, such that the induced isomorphism $\rho : D_\alpha \to D_\beta$ on the solution sets preserves the recursor structures; i.e., for all $x \in X, \vec{d} \in D_\alpha$,

$$\alpha_0(x, \vec{d}) = \beta_0(x, \rho(\vec{d})),$$

$$\rho(\mu_\alpha(x, \vec{d})) = \mu_\beta(x, \rho(\vec{d})).$$

In effect, we can reorder the system of equations in the body of a recursor and replace the components of the solution set by isomorphic copies without changing its isomorphism type.

It is easy to check, directly from the definitions, that isomorphic recursors $\alpha, \beta : X \rightsquigarrow W$ compute the same function $\overline{\alpha} = \overline{\beta} : X \to W$; the idea, of course, is that *isomorphic recursors model the same algorithm*, and so we will simply write $\alpha = \beta$ to indicate that α and β are isomorphic.

4 The representation of abstract machines by recursors

We show here that the algorithms expressed by abstract machines are represented faithfully by recursors.

---

[4] A somewhat coarser notion of recursor isomorphism was introduced in [7] to simplify the definitions and proofs of some of the basic facts about recursors, but it proved not to be a very good idea. We are reverting here to the original, finer notion introduced in [5].

Definition 4.1 (The recursor of an iterator). For each iterator

$$i : (\text{input}, S, \sigma, T, \text{output}) : X \rightsquigarrow Y,$$

as in Definition 2.1, we set

$$\mathfrak{r}(i) = p(\text{input}(x))$$
$$\text{where } \{p(s) = \text{if } (s \in T) \text{ then output}(s) \text{ else } p(\sigma(s))\}. \quad (17)$$

This is a continuous, discrete recursor of dimension 1, with solution space the function poset $(S \rightharpoonup Y)$ and output mapping $(x, p) \mapsto p(\text{input}(x))$, which models the *tail recursion* specified by i.

It is very easy to check directly that for each iterator $i : X \rightsquigarrow Y$,

$$\overline{\mathfrak{r}(i)}(x) = \overline{i}(x) \quad (x \in X), \quad (18)$$

but this will also follow from the next, main result of this section.

Theorem 4.2. *Two iterators* $i_1, i_2 : X \rightsquigarrow Y$ *are isomorphic if and only if the associated recursors* $\mathfrak{r}(i_1)$ *and* $\mathfrak{r}(i_2)$ *are isomorphic.*

Proof. By (17), for each of the two given iterators i_1, i_2,

$$\mathfrak{r}_i = \mathfrak{r}(i_i) = (\text{value}_i, \mu_i),$$

where for $p \in D_i = (S_i \rightharpoonup Y)$ and $x \in X$,

$$\text{value}_i(x, p) = p(\text{input}_i(x)),$$
$$\mu_i(x, p) = \mu_i(p) = \lambda(s)[\text{if } (s \in T_i) \text{ then output}_i(s) \text{ else } p(\tau_i(s))] : S_i \rightharpoonup Y.$$

Part 1. Suppose first that $\rho : S_1 \rightarrowtail\!\!\!\rightarrow S_2$ is an isomorphism of i_1 with i_2, and let $f_i : S_i \rightharpoonup Y$ be the least-fixed-points of the transition functions of the two iterators, so that

$$f_i(s) = \text{output}_i(\tau_i^{|s|}(s)) \text{ where } |s| = \text{the least } n \text{ such that } \tau_i^n(s) \in T_i.$$

In particular, this equation implies easily that

$$f_1(s) = f_2(\rho(s)).$$

The required isomorphism of $\mathfrak{r}_1$ with $\mathfrak{r}_2$ is determined by a poset isomorphism of the corresponding solution sets $(S_1 \rightharpoonup Y)$ and $(S_2 \rightharpoonup Y)$, which must be of the form

$$\pi(p)(\rho(s)) = \sigma_s(p(s)) \quad (s \in S_1), \quad (19)$$

by Proposition 8.4. We will use the given bijection $\rho : S_1 \rightarrowtail\!\!\!\rightarrow S_2$ and bijections $\sigma_s :$ $Y \rightarrowtail\!\!\!\rightarrow Y$ which are determined as follows.

We choose first for each $t \in T_1$ a bijection $\sigma_t^* : Y \rightarrowtail\!\!\!\rightarrow Y$ such that

$$\sigma_t^*(\text{output}_1(t)) = \text{output}_2(\rho(t)) \quad (t \in T_1),$$

and then we consider two cases:

(a) If $f_1(s)\uparrow$ or there is an input accessible s' such that $s \rightarrow^* s'$, let $\sigma_s(y) = y$.

(b) If $f_1(s)\downarrow$ and there is no input accessible s' such that $s \rightarrow^* s'$, let $\sigma_s = \sigma_t^*$, where $t = \tau_1^{|s|}(s) \in T_1$ is "the projection" of s to the set T_1.

Lemma. For all $s \in S_1$,

$$\sigma_{\tau_1(s)} = \sigma_s \quad (s \in S_1). \tag{20}$$

Proof. If $f_1(s)\uparrow$ or there is an input accessible s' such that $s \rightarrow^* s'$, then $\tau_1(s)$ has the same property, and so σ_s and $\sigma_{\tau_1(s)}$ are both the identity; and if $f_1(s)\downarrow$ and there is no input accessible s' such that $s \rightarrow^* s'$, then $\tau_1(s)$ has the same properties, and it "projects" to the same $t = \tau_1^n(s) \in T_1$, so that $\sigma_s = \sigma_{\tau_1(s)} = \sigma_t^*$.   (Lemma)$\square$

Now π in (19) is an isomorphism of $(S_1 \rightharpoonup Y)$ with $(S_2 \rightharpoonup Y)$ by Proposition 8.4, and it remains only to prove the following two claims:

(i) $\text{value}_1(x, p) = \text{value}(x, \pi(p))$, i.e., $p(\text{input}_1(x)) = \pi(p)(\text{input}_2(x))$. This holds because $s = \text{input}_1(x)$ is input accessible, and so σ_s is the identity and

$$\pi(p)(\text{input}_2(x)) = \pi(p)(\rho(\text{input}_1(x)) = \sigma_s(p(\text{input}_1(x))) = p(\text{input}_1(x)).$$

(ii) $\pi(\mu_1(p)) = \mu_2(\pi(p))$; i.e., for all $s \in S_1$,

$$\pi(\mu_1(p))(\rho(s)) = \mu_2(\pi(p))(\rho(s)). \tag{21}$$

For this we distinguish three cases:

(iia) $s \in T_1$, and it is input accessible. Now (a) applies in the definition of σ_s so that σ_s is the identity and we can compute the two sides of (21) as follows:

$$\pi(\mu_1(p))(\rho(s)) = \mu_1(p)(s) = \text{output}_1(s),$$
$$\mu_2(\pi(p))(\rho(s)) = \text{output}_2(\rho(s)),$$

and the two sides are equal by (MI4) whose hypothesis holds in this case.

(iib) $s \in T_1$, but it is not input accessible, so that (b) applies. Again

$$\pi(\mu_1(p))(\rho(s)) = \sigma_s(\mu_1(p)(s)) = \sigma_s(\text{output}_1(s)),$$
$$\mu_2(\pi(p))(\rho(s)) = \text{output}_2(\rho(s)),$$

and the two sides are now equal by the choice of $\sigma_s = \sigma_s^*$.

(iic) $s \notin T_1$. In this case we can use (20):

$$\pi(\mu_1(p))(\rho(s)) = \sigma_s(\mu_1(p)(s)) = \sigma_s(p(\tau_1(s))),$$
$$\mu_2(\pi(p))(\rho(s)) = \pi(p)(\tau_2(\rho(s))) = \pi(p)(\rho(\tau_1(s))) = \sigma_{\tau_1(s)}(p(\tau_1(s))),$$

and the two sides are equal by (20).

This completes the proof of **Part 1**.

Part 2. Suppose now that $\pi : (S_1 \rightharpoonup Y) \rightarrowtail\!\!\!\!\rightarrow (S_2 \rightharpoonup Y)$ is an isomorphism of $\mathfrak{r}_1$ with $\mathfrak{r}_2$, so that by Proposition 8.4,

$$\pi(p)(\rho(s)) = \sigma_s(p(s)) \quad (p : S_1 \rightharpoonup Y),$$

where $\rho : S_1 \rightarrowtail\!\!\!\!\rightarrow S_2$ and for each $s \in S_1$, $\sigma_s : Y \rightarrowtail\!\!\!\!\rightarrow Y$. Moreover, since π is a recursor isomorphism, we know that

$$p(\text{input}_1(x)) = \pi(p)(\text{input}_2(x)), \tag{22}$$
$$\pi(\mu_1(p))(\rho(s)) = \mu_2(\pi(p))(\rho(s)). \tag{23}$$

We will use these identities to show that ρ is an isomorphism of $\mathfrak{i}_1$ with $\mathfrak{i}_2$.

(a) $\rho[T_1] = T_2$, (MI2).

From the definition of the transition maps μ_i, it follows immediately that $\text{output}_i = \mu_i(\bot)$, and hence

$$\text{output}_2 = \mu_2(\bot) = \mu_2(\pi(\bot)) = \pi(\mu_1(\bot)) = \pi(\text{output}_1);$$

and then using the representation of π, for all $s \in S_1$,

$$\text{output}_2(\rho(s)) = \pi(\text{output}_1)(\rho(s)) = \sigma_s(\text{output}_1(s)),$$

so that $\text{output}_1(s)\!\downarrow \iff \text{output}_2(\rho(s))\!\downarrow$, which means precisely that $\rho[T_1] = T_2$.

(b) *If* $s = \text{input}_1(x)$, *then* σ_s *is the identity and* $\rho(s) = \rho(\text{input}_1(x)) = \text{input}_2(x)$, (MI1).

If t is such that $\rho(t) = \text{input}_2(x)$, then by (22), for all p,

$$p(s) = \pi(p)(\text{input}_2(x)) = \pi(p)(\rho(t)) = \sigma_t(p(t));$$

and if we apply this to

$$p(u) = \begin{cases} y, & \text{if } u = s, \\ \bot, & \text{otherwise,} \end{cases}$$

and consider the domains of convergence of the two sides, we get that $t = s$, and hence $y = \sigma_s(y)$ and $\rho(s) = \rho(t) = \text{input}_2(x)$.

(c) *If s is input accessible, then for every $t \in Y$, $\sigma_s(y) = y$.*

In view of (b), it is enough to show that if σ_s is the identity, then $\sigma_{\tau_1(s)}$ is also the identity. This is immediate when $s \in T_1$, since $\tau_1(s) = s$ in this case. So assume that s is not terminal, and that σ_s is the identity, which with (23) gives, for every p,

$$\mu_2(\pi(p))(\rho(s)) = \pi(p)(\mu_1(p))(\rho(s)) = \sigma_s(\mu_1(p)(s)) = \mu_1(p)(s).$$

If $s \notin T_1$, then $\rho(s) \notin T_2$ by (a), and so this equation becomes

$$\pi(p)(\tau_2(\rho(s))) = p(\tau_1(s)).$$

If t is such that $\rho(t) = \tau_2(\rho(s))$, then this identity together with the representation of π gives

$$\sigma_t(p(t)) = \pi(p)(\rho(t)) = \pi(p)(\tau_2(\rho(s))) = p(\tau_1(s));$$

and, as above, this yields first that $t = \tau_1(s)$ and then that $\sigma_{\tau_1(s)}$ is the identity.

(d) *If s is terminal and input accessible, then* $\mathrm{output}_2(\rho(s)) = \mathrm{output}_1(s)$, (MI4).

As in (b), and using (c) this time,

$$\mathrm{output}_2(\rho(s)) = \pi(\mathrm{output}_1)(\rho(s)) = \sigma_s(\mathrm{output}_1(s)) = \mathrm{output}_1(s).$$

Finally:

(e) *For all $s \in S_1$, $\rho(\tau_1(s)) = \tau_2(\rho(s))$.*

This identity holds trivially if $t \in T_1$, in view of (a), and so we may assume that $t \notin T_1$ and apply the basic (23), which with the representation of π yields for all p and s,

$$\sigma_s(\mu_1(p)(s)) = \mu_2(\pi(p))(\rho(s)) = \pi(p)(\tau_2(\rho(s))).$$

If, as above, we choose t so that $\rho(t) = \tau_2(\rho(s))$, this gives

$$\sigma_s(p(\tau_1(s))) = \pi(p)(\rho(t)) = \sigma_t(p(t));$$

and by applying this to some p, which converges only on $\tau_1(s)$, we get $t = \tau_1(s)$, so that $\tau_2(\rho(s)) = \rho(t) = \rho(\tau_1(s))$, as required. □

5 Recursor reducibility and implementations

The notion of *simulation* of one program (or machine) by another is notoriously slippery: in the relevant Section 1.2.2 of one of the standard (and most comprehensive) expositions of the subject, van Emde Boas [1] starts with

Intuitively, a simulation of [one class of computation models] M by [another] M' is some construction which shows that everything a machine $M_i \in M$ can do on inputs x can be performed by some machine $M'_i \in M'$ on the same inputs as well;

goes on to say that "it is difficult to provide a more specific formal definition of the notion"; and then discusses several examples which show "how hard it is to define simulation as a mathematical object and still remain sufficiently general."

The situation is somewhat better at the more abstract level of recursors, where a very natural relation of *reducibility* of one recursor to another seems to capture a robust notion of "algorithm simulation." At the same time, the problem is more important for recursors than it is for machines: because this modeling of algorithms puts great emphasis on understanding *the relation between an algorithm and its implementations*, and this is defined by a reducibility.

In this section we will reproduce the relevant definitions from [6] and we will establish that the recursor $\mathfrak{r}(A, M)$ determined by a program on an algebra M is implemented by the recursive machine $\mathfrak{i}(A, M)$.

Definition 5.1 (Recursor reducibility). Suppose $\alpha, \beta : X \rightsquigarrow W$ are recursors with the same input and output domains and respective solution sets

$$D_\alpha = D_1^\alpha \times \cdots \times D_K^\alpha, \quad D_\beta = D_1^\beta \times \cdots \times D_L^\beta,$$

and transition mappings

$$\mu_\alpha : X \times D_\alpha \to D_\alpha, \quad \mu_\beta : X \times D_\beta \to D_\beta.$$

A *reduction* of α to β is any monotone mapping

$$\pi : X \times D_\alpha \to D_\beta$$

such that the following three conditions hold, for every $x \in X$ and every $d \in D_\alpha$:

(R1) $\mu_\beta(x, \pi(x, d)) \leq \pi(x, \mu_\alpha(x, d))$.

(R2) $\beta_0(x, \pi(x, d)) \leq \alpha_0(x, d)$.

(R3) $\overline{\alpha}(x) = \overline{\beta}(x)$.

We say that α is *reducible* to β if a reduction exists,

$$\alpha \leq_r \beta \iff \text{there exists a reduction } \pi : X \times D_\alpha \to D_\beta.$$

Recursor reducibility is clearly reflexive and (easily) transitive.

If $K = L = 0$ so that the recursors are merely functions, this means that they are identical, $\alpha_0(x, \bot) = \beta_0(x, \bot)$; and if $K = 0$ while $L > 0$, then these conditions degenerate to the existence of some monotone $\pi : X \to D_\beta$, such that

$$\mu_\beta(x, \pi(x)) \le \pi(x), \quad \beta_0(x, \pi(x)) \le \alpha_0(x, \perp), \quad \overline{\alpha}(x) = \alpha_0(x, \perp) = \overline{\beta}(x),$$

which simply means that β computes $\overline{\alpha}$—i.e., they hold of any β such that $\overline{\beta} = \overline{\alpha}$ with $\pi(x) = d^\kappa_\beta(x)$ as in (11). In the general case, (R1) and (R2) imply (by a simple ordinal recursion) that for all x and ξ,

$$d^\xi_\beta(x) \le \pi(x, d^\xi_\alpha(x)), \quad \beta_0(x, d^\xi_\beta(x)) \le \alpha_0(x, d^\xi_\alpha(x)),$$

and then (R3) ensures that in the limit,

$$\overline{\beta}(x) = \beta_0(x, d^\kappa_\beta(x)) = \alpha_0(x, d^\kappa_\alpha(x)) = \overline{\alpha}(x);$$

thus β computes the same map as α but possibly "slower," in the sense that each iterate $d^\xi_\beta(x)$ gives us part (but possibly not all) of the information in the corresponding stage $d^\xi_\alpha(x)$—and this in a uniform way. It is not completely obvious, however, that the definition captures accurately the intuitive notion of *reduction* of the computations of one recursor to those of another, and the main result in this section aims to provide some justification for it.

Definition 5.2 (Implementations). A recursor $\alpha : X \rightsquigarrow Y \cup \{\perp\}$ into a flat poset is *implemented* by an iterator $\mathfrak{i} : X \rightsquigarrow Y$ if it is reducible to the recursor representation of $\mathfrak{i}$; i.e.,

$$\alpha \le_r \mathfrak{r}(\mathfrak{i}).$$

Theorem 5.3. *For each recursive program A and each algebra $\boldsymbol{M}$, the recursive machine $\mathfrak{i}(A, \boldsymbol{M})$ associated with A and $\boldsymbol{M}$ implements the recursor $\mathfrak{r}(A, \boldsymbol{M})$ defined by A on $\boldsymbol{M}$; i.e.,*

$$\mathfrak{r}(A, \boldsymbol{M}) \le_r \mathfrak{r}(\mathfrak{i}(A, \boldsymbol{M})).$$

To simplify the construction of the required reduction, we establish first a technical lemma about Scott domains, cf. Definition 8.2:

Lemma 5.4. *Suppose $\alpha : X \rightsquigarrow W$ and the solution set D_α has the Scott property, and set*

$$D^*_\alpha = \{d \in D_\alpha \mid (\forall x)[d \le \mu_\alpha(x, d)]\}. \tag{24}$$

Then D^*_α is a complete subposet of D_α, and for any recursor $\beta : X \rightsquigarrow W$, $\alpha \le_r \beta$ if and only if there exists a monotone map $\pi^* : X \times D^*_\alpha \to D_\beta$ that satisfies (R1) – (R3) of Definition 5.1 for all $x \in X, d \in D^*_\alpha$.

Proof. We assume the hypotheses on α and $\pi^* : X \times D^*_\alpha \to D_\beta$, and we need to extend π^* so that it has the same properties on D_α.

(1) *The subposet D^*_α is complete, and for each $d \in D$, the set*

$$X^*_d = \{d^* \in D^*_\alpha \mid d^* \le d\}$$

is directed, so that

$$\rho(d) = \sup X^*_d \in D^*_\alpha.$$

Proof. For the first claim, it is enough to show that if $X \subseteq D_\alpha^*$ is directed and $d = \sup X$, then $d \in D_\alpha^*$: this holds because for any $d^* \in X$, $d^* \leq d$, and so

$$d^* \leq \mu_\alpha(x, d^*) \leq \mu_\alpha(x, d),$$

using the characteristic property of D_α^* and the monotonicity of μ_α; and now taking suprema, we get the required $d \leq \mu_\alpha(x, d)$.

For the second claim, if $d_1, d_2 \in X_d^*$, then they are compatible, and so their supremum $d^* = \sup\{d_1, d_2\}$ exists, $d^* \leq d$, and it suffices to show that $d^* \in D_\alpha^*$; this holds because $d_i \leq \mu(x, d_i) \leq \mu_\alpha(x, d^*)$ by the characteristic property of D_α^* and the monotonicity of μ_α, and so $d^* \leq \mu_\alpha(x, d^*)$ by the definition of d^*.   **(1)** $\square$

We now extend π^* to all of D_α by

$$\pi(x, d) = \sup\{\pi^*(x, d^*) \mid d^* \in X_d^*\}.$$

(2) (R1): *For all x and d, $\mu_2(x, \pi(x, d)) \leq \pi(x, \mu_1(x, d))$.*

Proof. We must show that

$$\mu_2(x, \sup\{\pi^*(x, d^*) \mid d^* \in D_\alpha^* \ \& \ d^* \leq d\})$$
$$\leq \sup\{\pi^*(x, e^*) \mid e^* \in D_\alpha^* \ \& \ e^* \leq \mu_1(x, d)\}. \quad (25)$$

If $d^* \in D_\alpha^*$ and $d^* \leq d$, then $d^* \leq \rho(d)$; hence, $\pi^*(x, d^*) \leq \pi^*(x, \rho(d))$, and if we take suprema and apply μ_2, we get

$$\mu_2(x, \sup\{\pi^*(x, d^*) \mid d^* \in D_\alpha^* \ \& \ d^* \leq d\}) \leq \mu_2(x, \pi^*(x, \rho(d)))$$
$$\leq \pi^*(x, \mu_1(x, \rho(d))),$$

using the hypothesis, that (R1) holds for π^*; thus, to complete the proof of (25), it is enough to verify that
$$\mu_1(x, \rho(d)) \leq \mu_1(x, d),$$
which, however, is clearly true, since $\rho(d) \leq d$ and μ_1 is monotone.   **(2)** $\square$

(3) (R2) *For all x and d, $\beta_0(x, \pi(x, d)) \leq \alpha_0(x, d)$.*

Proof. Arguing as in the proof of **(2)**, we conclude that

$$\beta_0(x, \pi(x, d)) = \beta_0(x, \sup\{\pi^*(x, d^*) \mid d^* \in X_d^*\}) \leq \beta_0(x, \pi^*(x, \rho(d)))$$
$$\leq \alpha_0(x, \rho(d)) \leq \alpha_0(x, d),$$

where we appealed to (R2) for π^* in the next-to-the-last inequality.   **(3)** $\square$

Finally, (R3) holds because the computation of $\overline{\alpha}$ takes place entirely within D_α^* on which the given π^* coincides with π and is assumed to satisfy (R1)–(R3). $\square$

Proof of Theorem 5.3. We fix a recursive program A and an algebra M and set

$$\alpha = \mathfrak{r}(A, M) = (\alpha_0, \alpha_1, \ldots, \alpha_K), \quad \beta = (\beta_0, \beta_1) = \mathfrak{r}(\mathfrak{i}(A, M))$$

as these are defined in Definitions 3.2, 2.3, and 4.1, so that

$$D_0 = (M^n \rightharpoonup M), \quad D_\alpha = D_1 \times \cdots \times D_K \quad (\text{with } D_i = (M^{k_i} \rightharpoonup M));$$

the mappings α_i are the continuous operators defined by the parts A_i of the program A by (4); the transition mapping μ_α of α (defined in (16)) is independent of the input $\vec{x}$; $D_\beta = (S \rightharpoonup M)$ with S the set of states of the recursive machine $\mathfrak{i}(A, M)$; $\beta_0(d)(\vec{x}) = d(A_0\{\vec{x} :\equiv \vec{x}\} :)$; and

$$\mu_\beta(d) = \beta_1(d) = \lambda(s)[\text{if } (s =: w' \text{ for some } w') \text{ then } w' \text{ else } d(\sigma(s))],$$

where σ is the transition mapping σ of the recursive machine and (like μ_α) is independent of the input $\vec{x}$. To complete the proof, we need to define a monotone mapping $\pi : D_\alpha \rightarrow D_\beta$ so that (tracing the definitions) the following two conditions hold:

(R1) For all $\vec{p} \in D_\alpha$, $\beta_1(\pi(\vec{p})) \leq \pi(\mu_\alpha(\vec{p}))$; this means that for all $w \in M$,

$$\pi(\mu_\alpha(\vec{p}))(: w) = w, \tag{R1a}$$

and for every nonterminal state s of the recursive machine $\mathfrak{i}(A, M)$ and any $w \in M$,

$$\pi(\vec{p})(\sigma(s)) = w \implies \pi(\mu_\alpha(\vec{p}))(s) = w. \tag{R1b}$$

(R2) For all $\vec{x} \in M^n, \vec{p} \in D_\alpha$ and $\vec{x} \in M^n, w \in M$,

$$\pi(\vec{p})(A_0\{\vec{x} :\equiv \vec{x}\} :) = w \implies [\![A_0]\!]^M(\vec{x}, \vec{p}) = w.$$

The third condition (R3) is independent of the reduction π and holds by Theorem 2.4, (18), and (15).

For each tuple $\vec{p} \in D_\alpha$, let

$$M\{\vec{p} := \vec{p}\} = (M, 0, 1, \{\phi^M_{\phi \in \Phi}\} \cup \{p_1, \ldots, p_n\})$$

be the *expansion* of the algebra M by the partial functions $p_1, \ldots, p_K$ in the vocabulary $\Phi \cup \{p_1, \ldots, p_K\}$ of the given program A. The states of $\mathfrak{i}(A, M)$ are also states of $\mathfrak{i}(A, M\{\vec{p} := \vec{p}\})$—but differently interpreted, since $p_1, \ldots, p_K$ have now been fixed and so calls to them are *external* (independent of the program A) rather than *internal*. For any state $\vec{a} : \vec{b}$, we set

$$\pi(\vec{p})(\vec{a} : \vec{b}) = w \iff \vec{a} : \vec{b} \rightarrow^*_{M\{\vec{p} := \vec{p}\}} w$$

$$\iff \text{the computation of } \mathfrak{i}(A, M\{\vec{p} := \vec{p}\})$$

$$\text{which starts with } \vec{a} : \vec{b} \text{ terminates in the state } : w.$$

Since M and the $\Phi[A, M]$-terms are fixed, these computations depend only on the *assignment* $\{\vec{\mathsf{p}} := \vec{p}\}$, and we will call them $\{\vec{\mathsf{p}} := \vec{p}\}$-*computations*; they are exactly like those of $i(A, M)$, except when one of the p_i is called with the correct number of arguments; i.e., when p_i is n-ary, then for any $x_1, \ldots, x_n$,

$$\vec{a}\ \mathsf{p}_i : x_1 \cdots x_n\ \vec{b} \to_{\{\vec{\mathsf{p}} := \vec{p}\}} \vec{a} : p_i(x_1, \ldots, x_n)\ \vec{b}.$$

We establish that π has the required properties in a sequence of lemmas.

(1) *The mapping π is monotone.*

Proof. If $\vec{p} \le \vec{q}$ and $\pi(\vec{p})(s) = w$, then the finite $\{\vec{\mathsf{p}} := \vec{p}\}$-computation starting with s calls only a finite number of values of the partial functions $p_1, \ldots, p_K$ and terminates in the state $: w$; the $\{\vec{\mathsf{p}} := \vec{q}\}$-computation starting with s will then call the same values of $q_1, \ldots, q_K$, it will get the same answers, and so it will reach the same terminal state $: w$.   **(1)**$\square$

(2) *For each $\Phi[A, M]$-term B,*

$$\pi(\vec{p})(B :) = [\![B]\!]^M(\vec{p}).$$

Proof is easy, by induction on B.   **(2)**$\square$

By Lemma 5.4, it is enough to verify (R1) and (R2) for $\vec{p} \in D_\alpha^*$, and we will appeal to this in **(5)** below.

(3) (R1a): $\pi(\mu_\alpha(\vec{p}))(: w) = w$.

Proof. This is immediate from the definition of π, since the $\{\vec{\mathsf{p}} := \mu_\alpha(\vec{p})\}$-computation starting with $: w$ terminates immediately, without looking at the partial function assigned to any p_i.   **(3)**$\square$

(4) (R1b) *for the case where s is a state of the form*

$$\vec{a}\ \mathsf{p}_i : x_1 x_2 \cdots x_n\ \vec{b}$$

with $\mathrm{arity}(\mathsf{p}_i) = n$.

Proof. By the transition table of the recursive machine $i(A, M)$,

$$\vec{a}\ \mathsf{p}_i : x_1 x_2 \cdots x_n\ \vec{b} \to \vec{a}\ A_i\{\vec{\mathsf{x}}_i :\equiv \vec{x}\} : \vec{b},$$

and so the hypothesis of (R1b) in this case gives us that

$$\pi(\vec{p})(\vec{a}\ A_i\{\vec{\mathsf{x}}_i :\equiv \vec{x}\} : \vec{b}) = w,$$

which means that w is the output of the $\{\vec{\mathsf{p}} := \vec{p}\}$-computation

$$\left.\begin{array}{c} \vec{a}\, A_i\{\vec{\mathsf{x}}_i :\equiv \vec{x}\} : \vec{b} \\ \vdots \\ : w \end{array}\right]\ \{\vec{\mathsf{p}} := \vec{p}\}.$$

By **(2)** above,

$$\left.\begin{array}{c} \vec{a}\, A_i\{\vec{\mathsf{x}}_i :\equiv \vec{x}\} : \vec{b} \\ \vdots \\ \vec{a} : [\![A_i\{\vec{\mathsf{x}}_i :\equiv \vec{x}\}]\!](\vec{p})\ \vec{b} \end{array}\right]\ \{\vec{\mathsf{p}} := \vec{p}\},$$

and so, comparing outputs, we conclude that $\vec{a}$ and $\vec{b}$ are empty and that

$$w = [\![A_i\{\vec{\mathsf{x}}_i :\equiv \vec{x}\}]\!](\vec{p}).$$

On the other hand, by the definition of π and μ_α, there is a $\{\vec{\mathsf{p}} := \mu_\alpha(\vec{p})\}$-computation

$$\left.\begin{array}{c} \mathsf{p}_i : x_1 x_2 \cdots x_n \\ : [\![A_i\{\vec{\mathsf{x}}_i :\equiv \vec{x}\}]\!](\vec{p}) \end{array}\right]\ \{\vec{\mathsf{p}} := \mu_\alpha(\vec{p})\},$$

which gives us precisely the required conclusion of (R1b) in this case,

$$\pi(\mu_\alpha(\vec{p}))(\vec{a}\ \mathsf{p}_i : x_1 x_2 \cdots x_n\ \vec{b}) = w. \tag{4}\ \square$$

(5) (R1b) *otherwise, i.e., when s is not a state of the form $\vec{a}\ \mathsf{p}_i : x_1 x_2 \cdots x_n\ \vec{b}$.*

Proof. The transition table of $\mathrm{i}(A, \boldsymbol{M})$ now gives

$$s \ \to\ \vec{a}' : \vec{b}'$$

for some state $\vec{a}' : \vec{b}'$, and the hypothesis of (R1b) guarantees a computation

$$\left.\begin{array}{c} \vec{a}' : \vec{b}' \\ \vdots \\ : w \end{array}\right]\ \{\vec{\mathsf{p}} := \vec{p}\}.$$

Since the transition $s \ \to\ \vec{a}' : \vec{b}'$ does not refer to any function letter p_j, it is equally valid for $\mathrm{i}(A, \boldsymbol{M}\{\vec{\mathsf{p}} := \vec{p}\})$, and so we have a computation

$$\left.\begin{array}{c} s \\ \vec{a}' : \vec{b}' \\ \vdots \\ : w \end{array}\right]\ \{\vec{\mathsf{p}} := \vec{p}\}.$$

But $\vec{p} \leq \mu_\alpha(\vec{p})$ since $\vec{p} \in D_\alpha^*$, and so by **(1)**,

$$
\left. \begin{array}{c} s \\ \vec{a}' : \vec{b}' \\ \vdots \\ : w \end{array} \right] \{ \vec{\mathsf{p}} := \mu_\alpha(\vec{p}) \},
$$

which means that $\pi(\mu_\alpha(\vec{p}))(s) = w$ and completes the proof of (R1b).   **(5)**$\square$

(6) (R2) *holds; i.e.*,

$$
\pi(\vec{p})(A_0\{\vec{\mathsf{x}} :\equiv \vec{x}\} :) = w \implies [\![A_0]\!]^M(\vec{x}, \vec{p}) = w.
$$

Proof. This is an immediate consequence of **(2)**, which for $B \equiv A_0\{\vec{\mathsf{x}} :\equiv \vec{x}\}$ gives
$\pi(\vec{p})(A_0\{\vec{\mathsf{x}} :\equiv \vec{x}\} :) = [\![A_0\{\vec{\mathsf{x}} :\equiv \vec{x}\}]\!]^M(\vec{p}) = [\![A_0]\!]^M(\vec{x}, \vec{p})$.   **(6)**$\square$

This completes the proof of Theorem 5.3. $\square$

Although the technical details of this proof certainly depend on some specific features of recursive machines, the idea of the proof is general and robust: it can be used to show that any reasonable "simulation" of a McCarthy program A by an abstract machine implements the recursor $\mathfrak{r}(A, M)$ defined by A—and this covers, for example, the usual "implementations of recursion" by Turing machines or random access machines of various kinds.

6 Machine simulation and recursor reducibility

On the problem of defining formally the notion of one machine simulating another, basically we agree with the comments of van Emde Boas [1], quoted in the beginning of Section 5, that it is not worth doing. There is, however, one interesting result that relates a specific, formal notion of machine simulation to recursor reducibility.

Definition 6.1 (Machine simulation, take 1). Suppose

$$
i_i = (\text{input}_i, S_i, \sigma_i, T_i, \text{output}_i) : X \rightsquigarrow Y \quad (i = 1, 2)
$$

are two iterators on the same input and output sets. A **formal simulation** of i_1 by i_2 is any function $\rho : S_2 \to S_1$ such that the following conditions hold:

1. For any state $t \in S_2$, if $\rho(t) \to_1 s \in S_1$, then there is some $t' \in S_2$ such that $t \to_2^* t'$ and $\rho(t') = s$.

2. If $t_0 = \text{input}_2(x)$, then $\rho(t_0) = \text{input}_1(x)$.

3. If $t \in T_2$, then $\rho(t) \in T_1$ and $\text{output}_1(\rho(t)) = \text{output}_2(t)$.

4. If $t \in S_2 \setminus T_2$ and $\rho(t) \in T_1$, then there is path

$$t \rightarrow_2 t_0 \rightarrow_1 \cdots \rightarrow_2 t_k \in T_2$$

in $\mathfrak{i}_2$ such that $\text{output}_1(\rho(t)) = \text{output}_2(t_k)$ and $\rho(t_i) = \rho(t)$ for all $i \leq k$.

This is one sort of formal notion considered by van Emde Boas [1], who (rightly) does not adopt it as his only (or basic) formal definition.

Theorem 6.2 (Paschalis [8]). *If $\mathfrak{i}_2$ simulates $\mathfrak{i}_1$ formally by Definition 6.1, then $\mathfrak{r}(\mathfrak{i}_1) \leq_r \mathfrak{r}(\mathfrak{i}_2)$.*

We do not know whether the converse of this result holds or the relation of this notion of simulation with natural alternatives that involve maps $\rho : S_1 \rightarrow S_2$ (going the other way). The theorem, however, suggests that perhaps the robust notion $\mathfrak{r}(\mathfrak{i}_1) \leq_r \mathfrak{r}(\mathfrak{i}_2)$ may be, after all, the best we can do in the way of giving a broad, formal definition of what it means for one machine to simulate another.

7 Elementary (first-order) recursive algorithms

The term
$$E \equiv \text{if } (\phi_1(x) = 0) \text{ then } y \text{ else } \phi_2(\phi_1(y), x)$$

intuitively expresses an algorithm on each algebra M in which ϕ_1, ϕ_2 are interpreted, the algorithm that computes its value for any given values of x and y. We can view it as a program

$$E \quad : \quad \mathsf{p}_0(x, y) = \text{if } (\phi_1(x) = 0) \text{ then } y \text{ else } \phi_2(\phi_1(y), x) \qquad (26)$$

with empty body, but the recursor $\mathfrak{r}(E, M)$ of E constructed in Definition 3.2 does not capture this algorithm—it is basically nothing but the partial function defined by E. This is because, in general, the recursor $\mathfrak{r}(A, M)$ of a program A captures only "the recursion" expressed by A, and there is no recursion in E. The theory of *canonical forms* that we will briefly outline in this section makes it possible to capture such explicit algorithms (or parts of algorithms) by the general construction $(A, M) \mapsto \mathfrak{r}(A, M)$. In particular, it yields a robust notion of *the elementary algorithms* of any given algebra M.

To ease the work on the syntax that we need to do, we add to the language a function symbol cond of arity 3 and the abbreviation

$$\text{cond}(A, B, C) \equiv_{df} \text{if } (A = 0) \text{ then } B \text{ else } C;$$

this is not correct semantically, because cond is not a *strict partial function*, but it simplifies the definition of Φ-terms that now takes the form

$$A :\equiv 0 \mid 1 \mid \mathsf{v}_i \mid \mathsf{c}(A_1, \ldots, A_n), \qquad\qquad (\Phi\text{-terms})$$

where c is any constant (ϕ_i, cond) or variable function symbol (p_i^n) of arity n.

A term is *immediate* if it is an individual variable or the "value" of a function variable on individual variables,

$$X :\equiv \mathsf{v}_i \mid \mathsf{p}_i^n(\mathsf{u}_1, \ldots, \mathsf{u}_n) \qquad\qquad (\text{Immediate terms})$$

so that, for example, $\mathsf{u}, \mathsf{p}(u_1, u_2, u_1)$ are immediate when p has arity 3. Computationally, we think of immediate terms as "generalized variables" that can be accessed directly, like the entries $a[i], b[i, j, i]$ in an array (string) in some programming languages.

Definition 7.1 (Program reduction). Suppose A is a program as in (6),

$$\mathsf{p}_i(\vec{\mathsf{x}}_i) = \mathsf{c}(A_1, \ldots, A_{j-1}, A_j, A_{j+1}) \qquad\qquad (27)$$

is one of the equations in A, *and A_j is not immediate*. Let q be a function variable with arity(q) = arity(p) that does not occur in A. The *one-step reduction of A determined by* p_i, j, *and* q yields the program B constructed by replacing (27) in A by the following two equations:

$$\mathsf{p}_i(\vec{\mathsf{x}}_i) = \mathsf{c}(A_1, \ldots, A_{j-1}, \mathsf{q}(\vec{\mathsf{x}}_i), A_{j+1}),$$
$$\mathsf{q}(\vec{\mathsf{x}}_i) = A_j.$$

We write

$$A \Rightarrow_1 B \iff \text{there is a one-step reduction of } A \text{ to } B,$$
$$A \Rightarrow B \iff B \equiv A \text{ or } A \Rightarrow_1 A_1 \Rightarrow_1 \cdots \Rightarrow_1 A_k \equiv B,$$

so that the *reduction relation* on programs is the reflexive and transitive closure of one-step reduction.

Let size(A) be the number of nonimmediate terms that occur as arguments of function symbols in the parts of A. A program is *irreducible* if size(A) = 0, so that no one-step reduction can be executed on it.

Each one-step reduction lowers size by 1, and so, trivially:

Lemma 7.2. *If $A \Rightarrow_1 A_1 \Rightarrow_1 \cdots \Rightarrow_1 A_k$ is a sequence of one-step reductions starting with A, then $k \leq$ size(A); and A_k is irreducible if and only if $k =$ size(A).*

The reduction process clearly preserves the head function variable of A, and so it preserves its arity. It also (easily) preserves denotations,

$$A \Rightarrow B \implies [\![A]\!]^M = [\![B]\!]^M,$$

but this would be true even if we removed the all-important immediacy restriction in its definition. The idea is that much more is preserved: we will claim, in fact, that if $A \Rightarrow B$, then A *and* B *express the same algorithm in every algebra*.

Caution. This notion of reduction is a syntactic operation on programs, which models (very abstractly) *partial compilation*, bringing the mutual recursion expressed by the program to a useful form before the recursion is implemented *without committing to any particular method of implementation of recursion*. No real computation is done by it.

We illustrate the reduction process by constructing a reduction sequence starting with the explicit term E in (26) and showing on the right the parameters we use for each one-step reduction:

$$E \; : \; \mathsf{p}_0(x, y) = \text{if } (\phi_1(x) = 0) \text{ then } y \text{ else } \phi_2(\phi_1(y), x) \qquad (\mathsf{p}_0, 1, \mathsf{q}_1)$$

$$E_1 \; : \; \begin{aligned} \mathsf{p}_0(x, y) &= \text{if } (\mathsf{q}_1(x, y) = 0) \text{ then } y \text{ else } \phi_2(\phi_1(y), x) \\ \mathsf{q}_1(x, y) &= \phi_1(x) \end{aligned} \qquad (\mathsf{p}_0, 3, \mathsf{q}_2)$$

$$E_2 \; : \; \begin{aligned} \mathsf{p}_0(x, y) &= \text{if } (\mathsf{q}_1(x, y) = 0) \text{ then } y \text{ else } q_2(x, y) \\ \mathsf{q}_2(x, y) &= \phi_2(\phi_1(y), x) \\ \mathsf{q}_1(x, y) &= \phi_1(x) \end{aligned} \qquad (\mathsf{q}_2, 1, q_3)$$

$$E_3 \; : \; \begin{aligned} \mathsf{p}_0(x, y) &= \text{if } (\mathsf{q}_1(x, y) = 0) \text{ then } y \text{ else } q_2(x, y) \\ \mathsf{q}_2(x, y) &= \phi_2(\mathsf{q}_3(x, y), x) \\ \mathsf{q}_3(x, y) &= \phi_1(y) \\ \mathsf{q}_1(x, y) &= \phi_1(x) \end{aligned}$$

Now E_3 is irreducible, and so the reduction process stops.

We will not take the space here to argue that E_3 expresses *as a mutual recursion* the same *explicit* algorithm that is intuitively expressed by E. But note that the order of the equations in the body of E_3 is of no consequence: Definition 3.3 ensures that we can list these in any order without changing the isomorphism type of the recursor $\mathfrak{r}(E_3, M)$. This reflects our basic understanding that the intuitive, explicit algorithm expressed by E does not specify whether the evaluations that are required will be done in parallel, or in sequence, or in any particular order, except, of course, where the nesting of subterms forces a specific order for the calls to the primitives—and this is exactly what is captured by the structure of the irreducible program E_3.

To call $\mathfrak{r}(E_3, M)$ "the algorithm expressed by E," we must show that it is independent of any particular reduction of E to an irreducible term, and to do this we must abstract from the specific, fresh variables introduced by the reduction process and the order in which the new equations are added.

Definition 7.3 (Program congruence). Two programs A and B are *congruent* if B can be constructed from A by an alphabetic change (renaming) of the individual and functions variables and a permutation of the equations in the body of A. This is obviously an equivalence relation on programs that agrees with the familiar term-congruence on programs with empty body. We write:

$$A \equiv_c B \iff A \text{ and } B \text{ are congruent.}$$

It follows directly from this definition and Definition 3.3 that congruent programs have isomorphic recursors in every algebra $\boldsymbol{M}$,

$$A \equiv_c B \implies \mathfrak{r}(A, \boldsymbol{M}) = \mathfrak{r}(B, \boldsymbol{M}).$$

Theorem 7.4 (Canonical forms).

Every program A is reducible to a unique up to congruence irreducible term $\mathrm{cf}(A)$, *its canonical form.*

In detail, every program A has a canonical form $\mathrm{cf}(A)$ *with the following properties:*

(1) $\mathrm{cf}(A)$ *is an irreducible program.*

(2) $A \Rightarrow \mathrm{cf}(A)$.

(3) If $A \Rightarrow_1 A_1 \Rightarrow_1 \cdots \Rightarrow_1 A_k$ is any sequence of one-step reductions and $k = \mathrm{size}(A)$, then $A_k \equiv_c \mathrm{cf}(A)$.

Part (3) gives a method for computing $\mathrm{cf}(A)$ up to congruence and implies the first claim, that it is the unique up to congruence term that satisfies (1) and (2): because if $A \Rightarrow B$ and B is irreducible, then $A \Rightarrow_1 A_1 \Rightarrow_1 \cdots \Rightarrow_1 A_k \equiv B$ with $k = \mathrm{size}(A)$ by Lemma 7.2, and hence, $B \equiv_c \mathrm{cf}(A)$ by (3).

Outline of proof. To construct canonical forms, we fix once and for all some ordering on all the function variables, and starting with A, we execute a sequence of $\mathrm{size}(A)$ one-step reductions, selecting each time the lowest $\mathsf{p}, j, \mathsf{q}$ for which a reduction can be executed. The last program of this sequence satisfies (1) and (2), and so we only need verify (3), for which it suffices to check that

$$A \Rightarrow_1 B \implies \mathrm{cf}(A) \equiv_c \mathrm{cf}(B). \tag{28}$$

For this reason, the basic fact is that one-step reductions commute, in the following, simple sense: if we label them by showing their parameters,

$$A \xrightarrow{\mathsf{p}_i, j, \mathsf{q}} B$$

$$\iff B \text{ results by the one-step reduction on } A \text{ determined by } \mathsf{p}_i, j, \mathsf{q},$$

then,

$$A \xrightarrow{\mathsf{p}_i, j, \mathsf{q}} B \xrightarrow{\mathsf{p}_k, l, \mathsf{r}} C, \implies \text{ for some } B', A \xrightarrow{\mathsf{p}_k, l, \mathsf{r}} B' \xrightarrow{\mathsf{p}_i, j, \mathsf{q}} C. \tag{29}$$

The obvious conditions here are that $q \not\equiv r$ and that either $i \neq k$ or $j \neq l$, so that all four one-step reductions indicated can be executed, but, granting these, (29) follows immediately by the definitions. Finally, (28) can be verified using (29) by an easy "permutability" argument, which we will skip. $\square$

A more general version of the Canonical Form Theorem 7.4 for *functional structures* was established in [4], and there are natural versions of it for many richer languages, including a suitable formulation of the typed λ-calculus with recursion (PCF). This version for McCarthy programs is especially simple to state (and prove) because of the simplicity in this case of the definition of reduction in Definition 5.1.

Definition 7.5 (Referential intensions). The *referential intension* of a McCarthy program A in an algebra M is the recursor of its canonical form,

$$\mathrm{int}(A, M) = \mathfrak{r}(\mathrm{cf}(A), M);$$

it models the elementary algorithm expressed by A in M.

In this modeling of algorithms then, Theorem 5.3 establishes that *every elementary algorithm* $\mathrm{int}(A, M)$ *expressed by a term* A *in an algebra* M *is implemented by the recursive machine* $\mathfrak{i}(\mathrm{cf}(A), M)$ *of the canonical form of* A.

8 Appendix

A subset $X \subseteq D$ of a poset D is *directed* if

$$x, y \in X \implies (\exists z \in X)[x \leq_D z \ \& \ y \leq_D z],$$

and D is *complete* (a *dcpo*) if every directed $X \subseteq D$ has a (necessarily unique) *supremum* (least upper bound),

$$\sup X = \min\{z \in D \mid (\forall x \in X)[x \leq_D z]\}.$$

In particular, every complete poset (with this definition) has a least element,

$$\perp_D = \sup \emptyset,$$

and for each set Y, its *bottom liftup*

$$Y_\perp = Y \cup \{\perp\}$$

is the complete *flat* poset that has just one new element $\perp \notin Y$ put below all the members of X,

$$x \leq_{Y_\perp} y \iff x = \perp \lor x = y \quad (x, y \in Y_\perp).$$

It will also be convenient to view each set X as (trivially) partially ordered by the identity relation, so that X is a subposet of $X_\perp$.

The (Cartesian) *product* $D = D_1 \times \cdots \times D_n$ of n posets is ordered component-wise,

$$x \leq_D y \iff x_1 \leq_{D_1} y_1 \ \& \ \cdots \ \& \ x_n \leq_{D_n} y_n$$
$$(x = (x_1, \ldots, x_n), y = (y_1, \ldots, y_n)),$$

and it is complete, if $D_1, \ldots, D_n$ are all complete.

A mapping $\pi : D \to E$ from one poset to another is *monotone* if

$$x \leq_D y \implies \pi(x) \leq_E \pi(y) \qquad (x, y \in D),$$

and *continuous* if in addition, for every directed, nonempty subset of D and every $w \in D$,

$$\text{if } w = \sup X, \text{ then } \pi(w) = \sup \pi[X].$$

If D and E are complete posets, then this conditions takes the simpler form

$$\pi(\sup X) = \sup \pi[X] \quad (X \text{ directed, nonempty}),$$

but it is convenient to allow D to be arbitrary (for example, a product of complete posets and sets) in the definition.

A *poset isomorphism* is any bijection $\pi : D \rightarrowtail\!\!\!\rightarrow D_2$ that respects the partial ordering,

$$d_1 \leq_D d_2 \iff \pi(d_1) \leq_E \pi(d_2),$$

and it is automatically monotone and continuous.

Theorem 8.1 (Fixed Point Theorem). *Every monotone mapping $\pi : D \to D$ on a complete poset to itself has a least fixed point*

$$\overline{x} = (\mu x \in D)[x = \pi(x)],$$

characterized by the following two properties:

$$\overline{x} = \pi(\overline{x}), \quad (\forall y)[\pi(y) \leq_D y \implies \overline{x} \leq y];$$

in fact, $\overline{x} = \overline{x}^\kappa$ for every sufficiently large ordinal number κ, where the transfinite sequence $\{\overline{x}^\xi\}_\xi$ (of iterates of π) is defined by the recursion

$$\overline{x}^\xi = \pi(\sup\{\overline{x}^\eta \mid \eta < \xi\}), \tag{30}$$

and we may take $\kappa = \omega$ if π is continuous.

Moreover, if $\pi : D \times E \to D$ is monotone (respectively, continuous) and D is complete, then the mapping

$$\rho(y) = (\mu x \in D)[\pi(x, y) = x] \quad (y \in E)$$

is monotone (respectively, continuous), $\rho : E \to D$.

When $D = D_1 \times \cdots \times D_n$ is a product of complete posets, then the Least Fixed Point Theorem guarantees the existence of *canonical* (least) solutions

$$\bar{x}_1 : E \to D_1 \ldots, \bar{x}_n : E \to D_n$$

for each system of monotone, recursive equations with parameters

$$
\begin{aligned}
x_1(y) &= \pi_1(x_1, \ldots, x_n, y) \\
x_2(y) &= \pi_2(x_1, \ldots, x_n, y) \\
&\quad \cdots \\
x_n(y) &= \pi_n(x_1, \ldots, x_n, y);
\end{aligned}
\tag{31}
$$

the solutions are monotone, and if each π_i is continuous, then they are continuous.

For any sets $X_1, \ldots, X_n, Y$, a *partial function* $f : X_1 \times \cdots \times X_n \rightharpoonup Y$ is any mapping

$$f : X_1 \times \cdots \times X_n \to Y_\perp.$$

Sometimes we identify f with its obvious liftup

$$\hat{f} : X_{1,\perp} \times \cdots \times X_{n,\perp} \to Y_\perp,$$

which takes the value $\perp$ if any one of its arguments is $\perp$. The *composition* of partial functions is defined using these liftups: for $x \in X_1 \times \cdots \times X_n, w \in Y$,

$$
\begin{aligned}
&f(g_1(x), \ldots, g_n(x)) = w \\
&\quad \Longleftrightarrow (\exists u_1, \ldots, u_n)[g_1(x) = u_1 \ \& \ \cdots \ \& \ g_n(x) = u_n \ \& \ f(u_1, \ldots, u_n) = w].
\end{aligned}
$$

We will use the familiar notations for *convergence* and *divergence* of partial functions,

$$f(x)\!\downarrow \ \Longleftrightarrow \ f(x) \neq \perp, \quad f(x)\!\uparrow \ \Longleftrightarrow \ f(x) = \perp.$$

For any poset D and any complete poset E, the function spaces

$$
\begin{aligned}
\mathrm{Mon}(D \to E) &= \{\pi : D \to E \mid \pi \text{ is monotone}\}, \\
\mathrm{Cont}(D \to E) &= \{\pi : D \to E \mid \pi \text{ is continuous}\}
\end{aligned}
$$

are complete posets with the pointwise partial ordering,

$$\pi \leq \rho \ \Longleftrightarrow \ (\forall x \in D)[\pi(x) \leq_E \pi(y)];$$

this is also true of the partial function spaces

$$(X_1 \times \cdots X_n \rightharpoonup Y) = \{f : X_1 \times \cdots \times X_n \rightharpoonup Y\} = \mathrm{Mon}(X_1 \times \cdots X_n \to Y_\perp),$$

with which we are primarily concerned in this article. We list here two properties of them that we will need.

Definition 8.2 (Scott domains). Two points d_1, d_2 in a poset D are *compatible* if their doubleton $\{d_1, d_2\}$ has an upper bound in D; and D has the *Scott property*, if any two compatible points $d_1, d_2 \in D$ have a (necessarily unique) least upper bound:

$$(\exists e)[d_1 \leq e \ \& \ d_2 \leq e] \implies \sup\{d_1, d_2\} \text{ exists.}$$

Proposition 8.3. *Every partial function poset $(X \rightharpoonup Y)$ has the Scott property; and if $D_1, \ldots, D_n$ have the Scott property, then their product $D_1 \times \cdots \times D_n$ also has the Scott property.*

Proof. For compatible $d_1, d_2 : X \rightharpoonup Y$, clearly $\sup\{d_1, d_2\} = d_1 \cup d_2$, which is the least common extension of d_1 and d_2. The second statement follows by induction on n, using the obvious fact that $D_1 \times \cdots \times D_n \times D_{n+1}$ and $(D_1 \times \cdots \times D_n) \times D_{n+1}$ are isomorphic. □

The next proposition gives a normal form for all poset isomorphisms between partial function spaces, which we need in the proof of Theorem 4.2 (and it may be well known but we could not find it in the literature).

Proposition 8.4. *For any sets S_1, S_2, Y_1, Y_2, suppose $\rho : S_1 \rightarrowtail\!\!\!\twoheadrightarrow S_2$ and for each $s \in S_1$, $\sigma_s : Y_1 \rightarrowtail\!\!\!\twoheadrightarrow Y_2$ are given bijections, and set*

$$\pi(p)(\rho s) = \sigma_s(p(s)) \qquad (p : S_1 \rightharpoonup Y_1, \pi(p) : S_2 \rightharpoonup Y_2). \tag{32}$$

Then π is a poset isomorphism, and every poset isomorphism

$$\pi : (S_1 \rightharpoonup Y_1) \rightarrowtail\!\!\!\twoheadrightarrow (S_2 \rightharpoonup Y_2)$$

satisfies (32) with suitable ρ, $\{\sigma_s\}_{(s \in S_1)}$.

Proof. We skip the easy verification that the map defined by (32) is a poset isomorphism.

For the converse, fix an isomorphism $\pi : (S_1 \rightharpoonup Y_1) \rightarrowtail\!\!\!\twoheadrightarrow (S_2 \rightharpoonup Y_2)$, and for each $s \in S_1$ and each $y \in Y_1$, let

$$p_s^y(t) = \begin{cases} y, & \text{if } t = s, \\ \bot, & \text{otherwise,} \end{cases}$$

so that $p_s^y : S_1 \rightharpoonup Y_1$. Each p_s^y is minimal above $\bot$ (atomic) in $(S_1 \rightharpoonup Y_1)$ and, in fact, every minimal above $\bot$ point of $(S_1 \rightharpoonup Y_1)$ is p_s^y for some s and t. It follows that each π-image $\pi(p_s^y) : S_2 \rightharpoonup Y_2$ is a minimal (above $\bot$) partial function in $(S_2 \rightharpoonup Y_2)$ that converges on a single point in S_2, and so we have functions $\rho : S_1 \times Y_1 \rightarrow S_2$ and $\sigma : S_1 \times Y_1 \rightarrow Y_2$ so that

$$\pi(p_s^y)(t) = q_{\rho(s,y)}^{\sigma(s,y)}(t) = \begin{cases} \sigma(s,y), & \text{if } t = \rho(s,y), \\ \bot, & \text{otherwise.} \end{cases} \tag{33}$$

(1) *For all s, y_1, y_2, $\rho(s, y_1) = \rho(s, y_2)$.*

Proof. If $\rho(s, y_1) \neq \rho(s, y_2)$ for some $s, y_1 \neq y_2$, then the partial functions $q_{\rho(s,y_1)}^{\sigma(s,y_1)}$ and $q_{\rho(s,y_2)}^{\sigma(s,y_2)}$ are compatible (since they have disjoint domains of convergence) with least upper bound given by their union

$$q = q_{\rho(s,y_1)}^{\sigma(s,y_1)} \cup q_{\rho(s,y_2)}^{\sigma(s,y_2)};$$

but then the inverse image $\pi^{-1}(q)$ is above both $p_s^{y_1}$ and $p_s^{y_2}$, which are incompatible, which is absurd.   (1)□

We let $\rho(s) = \rho(s, y)$ for any (and all) $y \in Y_1$, and we set $\sigma_s(y) = \sigma(s, y)$, so that the basic definition (33) becomes

$$\pi(p_s^y) = q_{\rho(s)}^{\sigma_s(y)}, \qquad (\rho : S_1 \to S_2, s \in S_1, \sigma_s : Y_1 \to Y_2, y \in Y_1). \qquad (34)$$

(2) *The map* $\rho : S_1 \twoheadrightarrow S_2$ *is a bijection.*

Proof. To see that ρ is an injection, suppose $\rho(s_1) = \rho(s_2)$ and fix some $y \in Y_1$. If $\sigma_{s_1}(y) = \sigma_{s_2}(y)$, then $q_{\rho(s_1)}^{\sigma_{s_1}(y)} = q_{\rho(s_2)}^{\sigma_{s_2}(y)}$, and so $p_{s_1}^y = p_{s_2}^y$, which implies that $s_1 = s_2$, since these two equal partial functions have respective domains of convergence, the singletons $\{s_1\}$ and $\{s_2\}$; and if $\sigma_{s_1}(y) \neq \sigma_{s_2}(y)$, then the two partial functions $q_{\rho(s_1)}^{\sigma_{s_1}(y)}$ and $q_{\rho(s_2)}^{\sigma_{s_2}(y)}$ are incompatible, which means that their π-preimages $p_{s_1}^y$ and $p_{s_2}^y$ must also be incompatible—which can only happen if $s_1 = s_2$.

To see that ρ is surjective, suppose $t \in S_2$, fix some $w \in Y_2$, and set

$$r(u) = \begin{cases} w, & \text{if } u = t, \\ \bot, & \text{otherwise.} \end{cases} \qquad (u \in S_2).$$

This is a minimal point in $(S_2 \rightharpoonup Y_2)$, so there exist $s \in S_1, y \in Y_1$ such that $\pi(p_s^y) = r$—which means that $q_{\rho(s)}^{\sigma_s(y)} = r$, and hence $t = \rho(s)$, since the respective domains of convergence of these two partial functions are $\{\rho(s)\}$ and $\{t\}$.   (2)□

(3) *For each* $s \in S_1$, *the map* $\sigma_s : Y_1 \twoheadrightarrow Y_2$ *is a bijection.*

Proof. If $\sigma_s(y_1) = \sigma_s(y_2)$, then $q_{\rho(s)}^{\sigma_s(y_1)} = q_{\rho(s)}^{\sigma_s(y_2)}$, so that $p_s^{y_1} = p_s^{y_2}$, which implies $y_1 = y_2$; thus σ_s is an injection. And finally, for each $w \in Y_2$, let

$$r(t) = \begin{cases} w, & \text{if } t = \rho(s), \\ \bot, & \text{otherwise;} \end{cases}$$

this is a minimal point in $(S_2 \rightharpoonup Y_2)$, and so there is some s', y such that

$$\pi(p_{s'}^y) = q_{\rho(s')}^{\sigma_{s'}(y)} = r;$$

by considering the domains of convergence of these two points we conclude as above that $\rho(s') = \rho(s)$, so that $s' = s$ by **(2)**, and then considering their values, we get the required $t = \sigma_s(y)$.   (3)□

This concludes the proof of the Proposition.   □

References

1. P. van Emde Boas. Machine models and simulations. In Jan van Leeuwen, editor, *Handbook of Theoretical Computer Science, Vol. A, Algorithms and Complexity*, pages 1–66. Elsevier and MIT Press, 1994.
2. J. McCarthy. A basis for a mathematical theory of computation. In P. Braffort and D Herschberg, editors, *Computer Programming and Formal Systems*, pages 33–70. North-Holland, 1963.
3. Y. N. Moschovakis. Abstract recursion as a foundation of the theory of algorithms. In M. M. Richter et. al., editors, *Computation and Proof Theory*, Vol. 1104, pages 289–364. Springer-Verlag, Berlin, 1984. Lecture Notes in Mathematics.
4. Y. N. Moschovakis. The formal language of recursion. *The Journal of Symbolic Logic*, 54:1216–1252, 1989.
5. Y. N. Moschovakis. A mathematical modeling of pure, recursive algorithms. In A. R. Meyer and M. A. Taitslin, editors, *Logic at Botik '89*, Vol. 363, pages 208–229. Springer-Verlag, Berlin, 1989. Lecture Notes in Computer Science.
6. Y. N. Moschovakis. On founding the theory of algorithms. In H. G. Dales and G. Oliveri, editors, *Truth in Mathematics*, pages 71–104. Clarendon Press, Oxford, 1998.
7. Y. N. Moschovakis. What is an algorithm? In B. Engquist and W. Schmid, editors, *Mathematics unlimited – 2001 and Beyond*, pages 929–936. Springer, 2001.
8. V. Paschalis. Recursive algorithms and implementations, 2006. (M.Sc. Thesis, in Greek).

Applications of the Kleene–Kreisel Density Theorem to Theoretical Computer Science

Dag Normann

Department of Mathematics, The University of Oslo, Blindern N-0316 Oslo, Norway
dnormann@math.uio.no

Summary. The Kleene–Kreisel density theorem is one of the tools used to investigate the denotational semantics of programs involving higher types. We give a brief introduction to the classic density theorem, then show how this may be generalized to set theoretical models for algorithms accepting real numbers as inputs, and finally survey some recent applications of this generalization.

1 Introduction

Classical Computability Theory is the study of what may actually be computed, when the objects used for inputs are finite entities like integers or words in a finite alphabet. Of course, relativized computability, complexity issues, and other aspects of genuine computations will be considered to be in the realm of classical computability theory as well.

In *Generalized Computability Theory* we analyze mathematical structures that support alternative forms of computations or computation-like phenomena. Normally, it is the mathematical structures that are important, the concepts of computability are adjusted to these structures. We will use "CT" for "Computability Theory" and "GCT" for "Generalized Computability Theory."

In Computer Science the strategy is often different. There the actual programs and computations are what is important, and the mathematical models used for denotational semantics are of importance only to the extent they help us analyze the programming languages and programs. The split in attitude between CT and TCS (Theoretical Computer Science) is not absolute, and the same concepts of computability over the same mathematical structures will be studied occasionally both in CT and in TCS.

In this paper we will give a survey of the theory for continuous functionals of higher types, and we will focus on recent nontrivial applications of the Kleene–Kreisel density theorem and its generalizations. The paper will be semi-technical, with some formal definitions, but only indications of proofs. For a general overview of the interplay between the CT-approaches and the TCS-approaches to computations in higher types in general, see Normann [20].

In 1959 Kleene [13] and Kreisel [14] introduced what is now known as the *Kleene–Kreisel continuous functionals*. The motivation behind the two papers were different, the two approaches were different, and in fact, the two concepts of *countable functional* due to Kleene and of *continuous functional* due to Kreisel are not equivalent. Still, both authors claimed that the two approaches were essentially equivalent, without offering proofs. In Section 2 we will give a definition based on domain theory. We will let $Ct(\sigma)$ be the set of total continuous functionals of finite type σ.

Kleene's aim was to find a natural notion of computations relative to higher type objects. He observed that there is a natural sub-hierarchy, the *countable functionals*, of the full type structure that is closed under his notion of computability. Kreisel's motivation behind introducing the continuous functionals of finite type was to give an interpretation of the constructive content of a statement in second-order number theory. This should enable him to decide in an absolute way whether a statement in analysis is constructively true. We will not go into this analysis here.

In both approaches, the aim was to construct a hierarchy of total functionals of finite type, where the action of one functional Ψ on an input F is locally determined via finite "approximations" to Ψ and F. Kleene used *associates*, i.e. functions in $\mathbb{N}^{\mathbb{N}}$, to represent the functionals, whereas Kreisel used certain "total" ideals of formal neighborhoods. The Kleene–Kreisel density theorem was formulated in two ways:

Kleene: The set of finite sequences that may be extended to an associate for an object of type σ is computable.

Kreisel: Each formal neighborhood can be extended to a total ideal.

The density theorem was used to prove this now classical result:

Theorem 1 [Kreisel] *Let $A \subseteq \mathbb{N}^{\mathbb{N}}$ be Π_k^1 where $k \geq 1$. Then there is a primitive recursive relation R in $\mathbb{N}^{\mathbb{N}} \times Ct(k) \times \mathbb{N}$ (where k denotes the pure type at level k) such that*

$$f \in A \Leftrightarrow \forall \Phi \in Ct(k) \exists n \in \mathbb{N} R(f, \Phi, n).$$

The density theorem enables us to systematically replace a quantifier of the form $\exists F \in Ct(k-1)S(F)$ by $\exists n \in \mathbb{N}S(F_n)$ where $\{F_n\}_{n \in \mathbb{N}}$ is an effectively enumerated dense subset and S is a predicate that is computable in some parameters.

2 A modern view of the Kleene–Kreisel functionals

The two approaches to the continuous functionals discussed in Section 1 both belong to the CT-tradition. There is, however, an alternative approach more natural from the point of view of TCS, the approach via domains. In reality, this approach is not far from Kreisel's, but in spirit there is a certain gap. In CS it is important to model partiality, so the main structure is a hierarchy of partial continuous functionals. In order to define this hierarchy, we use domain theory, initiated by Scott [31] and by Ershov [9]. We will start with a brief introduction to domain theory. See Stoltenberg-Hansen, Lindström, and Griffor [33], Abramsky and Jung [1], or Amadiou and Curien [2] for detailed introductions.

2.1 Scott domains

A *complete partial ordering*, a *cpo* for short, is a partial ordering $(X, \sqsubseteq)$ such that each directed subset has a least upper bound $\sqcup A$. A *cpo* is *bounded complete* if each bounded set will have a least upper bound. Since the empty set is directed, a *cpo* X will have a least element, named $\perp_X$ or just $\perp$. $x \in X$ is *compact* or *finitary* if for each nonempty directed set A with $x \sqsubseteq \sqcup A$ there will be an $a \in A$ such that $x \sqsubseteq a$. A *cpo* $(X, \sqsubseteq)$ is an *algebraic domain* if for each $x \in X$, the set

$$C_x = \{x_0 \in X \mid x_0 \text{ is compact and } x_0 \sqsubseteq x\}$$

is directed and $x = \sqcup C_x$.

If X is bounded complete, C_x as defined above will always be directed.

In this paper we will restrict ourselves to *Scott domains*, i.e. bounded complete algebraic domains where the set of compacts is countable.

If $(X, \sqsubseteq_X)$ and $(Y, \sqsubseteq_Y)$ are two *cpo*s, we define a function $f : X \to Y$ to be *continuous* if f is monotone, and for each directed set $A \subseteq X$, we have that

$$f(\sqcup_X A) = \sqcup_Y \{f(a) \mid a \in A\}.$$

If we use the pointwise ordering of the continuous functions from X to Y, we obtain a new *cpo*, and if we in addition are dealing with Scott domains, the function space will be a Scott domain. We are not going to prove this here. The key to the argument for Scott domains is to characterize the compacts in $X \to Y$ as the least upper bounds of finite bounded sets of step functions, where whenever p is a compact in X and q is a compact in Y, we define the step function $f_{p,q}$ by

$f_{p,q}(x) = q$ if $p \sqsubseteq_X x$,

$f_{p,q}(x) = \perp_Y$ if $p \not\sqsubseteq_X x$.

If we use the continuous functions as morphisms, the Scott domains form a category that is cartesian closed.

There is a natural, non-Hausdorff topology on a Scott domain, known as the Scott topology:

Definition 2 The Scott topology on a Scott domain $(X, \sqsubseteq)$ is the topology generated from the base consisting of all

$$\{x \in X \mid x_0 \sqsubseteq x\},$$

where x_0 varies over all compacts in X.

A subset $H \subseteq X$ will be *dense* if it is dense with respect to the Scott topology, which means that every compact will have an extension in H. We justify our definition of continuous function from X to Y by observing that this means continuous with respect to the Scott topologies on X and Y.

In a bounded complete *cpo*, a set A is bounded if and only if each finite subset A_0 is bounded. If in addition, a set A is bounded whenever all subsets A_0 with at most two elements are bounded, the *cpo* is a *coherence space*. It is a basic and easy fact of domain theory that $X \to Y$ is a coherence space whenever X is an algebraic domain and Y is a coherence space.

2.2 Partial continuous functionals of finite type

Scott [31] introduced a formal logic LCF. In the language of LCF we have terms expressing functionals of higher types, and in the formal theory, we can reason about the relationship between various terms. His language is based on typed combinators, and he introduced a set of typed constants to be interpreted as the least fixed point operators at each type.

In this (for a long time) unpublished paper, Scott also gave a set-theoretical model for LCF. One important motivation at the time was that untyped λ-calculus lacked a set-theoretical model, and Scott suggested that LCF could replace λ-calculus for any practical purpose. We are not going to discuss the motivation of Scott further. To us, what is of interest is that Scott gave an interpretation $D(\sigma)$ for each finite type σ and by that laid the foundation of domain theory. For example, the natural numbers will be interpreted as the set $D(\iota) = \{\bot, 0, 1, \ldots\}$, where $\bot$ signifies *the undefined* and $a \sqsubseteq b \Leftrightarrow a = b \vee a = \bot$. This domain is known as the *flat domain* of natural numbers and is one of the base domains in the semantics for LCF. Another base domain will be the similar flat domain $D(o)$ of boolean values $\{\bot, tt, ff\}$. Then each finite type σ over the base type ι for the natural numbers and o for the booleans is interpreted as $D(\sigma)$ in the cartesian closed category of Scott domains. $D(\sigma)$ will also be a coherence space.

It is in order to make the interpretation of the constants for the fixed point operators possible that we need *cpos*; if $f : X \to X$ is continuous, then $\bigsqcup\{f^n(\bot) \mid n \in \mathbb{N}\}$ actually will be a least fixed point of f.

Developing the theory of enumerations, Ershov came up with concepts equivalent to Scott domains. Since it is irrelevant to our story, we will not give any details.

Ershov's characterization of the Kleene–Kreisel continuous functionals can be found in [9].

Although methodologically inspired from CT, LCF is a contribution to TCS, and [31] turned out to be an influential paper beyond its CT-content. Plotkin [28] reformulated LCF into the typed λ-calculus PCF. Special PCF is typed λ-calculus with constants for all natural numbers and boolean values, constants for the successor function, the predecessor function, the boolean test of zero-hood, and the conditionals over the two base types ι and o. In addition there are constants for the fixed point operator at every type. There are conversion rules from λ-calculus together with special rules for all the special constants.

Later we will discuss an application of the density theorem to a problem concerning PCF; see Section 3. For now, let us mention three important results:

1. To the extent it makes sense, Kleene's $S1$–$S9$-computability interpreted over the partial computable functionals is equivalent to PCF.

2. There is a compact object of mixed type 1 that is not the interpretation of any PCF-term.

3. If a closed PCF-term t of base type is interpreted as an element $\neq \perp$, then we may rewrite t to the constant for the interpretation using the conversion rules.

The proof of 1. goes back to Platek's thesis [27], where it is proved in the discontinuous case. A proof can also be found in Moldestad [18]. 2. is also known since Platek [27] and is observed in Scott [31] and Plotkin [28]. It is relevant to us to observe that this compact, an interpretation of *parallel or*, cannot be extended to a PCF-definable object. Thus the PCF-definable objects do not form a dense subset of the underlying domains seen as topological spaces. 3. is one of the main results of Plotkin [28] and is known as the *Plotkin adequacy theorem*.

2.3 Hereditarily total functionals

We may construct the Kleene–Kreisel continuous functionals from the partial continuous functionals via the so-called *hereditarily total functionals*.

Definition 3 a) For each finite type σ over the base types ι and o, we are going to define the set $H(\sigma) \subset D(\sigma)$ of hereditarily total elements:

1. $H(\iota) = \mathbb{N}$ and $H(o) = \mathbb{B} = \{tt, ff\}$.

2. Let $\sigma = \tau \to \delta$, and assume that $H(\tau)$ and $H(\delta)$ are both defined. Let $f \in D(\tau \to \delta)$.
   Then
   $$f \in H(\sigma) \Leftrightarrow \forall x \in H(\tau)\ (f(x) \in H(\delta)).$$

b) By recursion on the type σ, we define an equivalence relation $\approx_\sigma$ on $H(\sigma)$ as follows:

1. $\approx_\iota$ and $\approx_o$ are the identity relations on $\mathbb{N}$ and $\mathbb{B}$, respectively.

2. If $\sigma = \tau \to \delta$ and f and g are in $H(\sigma)$, we let

$$f \approx_\sigma g \Leftrightarrow \forall x \in H(\tau) \forall y \in H(\tau)(x \approx_\tau y \Rightarrow f(x) \approx g(y)).$$

Longo and Moggi [16] observed that we with ease may prove that

$$x \approx_\sigma y \Leftrightarrow x \sqcap y \in H(\sigma)$$

when x and y are in $H(\sigma)$. It then follows that $\approx_\sigma$ is an equivalence relation, something that had been seen as a consequence of the density theorem until then.

The best we can do within PCF with respect to the density theorem is to prove

Proposition 4 *Let σ be a finite type, and let $x_0 \in D(\sigma)$ be compact. Then there is a hereditarily total PCF-definable $x \in H(\sigma)$ such that x and x_0 are consistent.*

Although we have dropped every detail that could verify our claims, we are now in the position of giving a precise definition of the continuous functionals of finite type:

Definition 5 By recursion on the finite type σ we define the set $Ct(\sigma)$ of *continuous functionals of type σ* together with the surjective map $\rho_\sigma : H(\sigma) \to Ct(\sigma)$ as follows:

1. $Ct(\iota) = \mathbb{N}$ and $Ct(o) = \mathbb{B}$. ρ_ι and ρ_o are the respective identity maps.

2. Let $\sigma = \tau \to \delta$, and assume that $Ct(\tau)$, ρ_τ, $Ct(\delta)$, and ρ_δ are defined. If $F : Ct(\tau) \to Ct(\delta)$ and $f \in H(\sigma)$, we let $F = \rho_\sigma(f)$ if for all $x \in H(\tau)$, we have that
$$F(\rho_\tau(x)) = \rho_\delta(f(x)).$$

We then define $Ct(\sigma)$ as the image of ρ_σ.

We need to establish a few facts before we can claim that this definition makes sense for all types. The following lemma is easy to prove by induction on the type:

Lemma 6 *For every type σ we have that $\rho_\sigma(f)$ is defined for each $f \in H(\sigma)$ and that when f and g are in $H(\sigma)$, we have that*

$$f \approx_\sigma g \Leftrightarrow \rho_\sigma(f) = \rho_\sigma(g).$$

Each set $H(\sigma)$ will have a topology inherited from the Scott topology on $D(\sigma)$, and thus induce a quotient topology on $Ct(\sigma)$ via the identification map ρ_σ.

There are numerous characterizations of the Kleene–Kreisel functionals. The characterizations are mainly of two kinds:

1. We choose a way to model partial computable functionals and then we extract the hereditarily total objects.

2. We construct the hierarchy directly by imposing a superstructure, like a limspace structure or a topology, at each step.

The interesting fact is that, in particular for the first category, most conceptually well-based approaches lead to the same hierarchy of total functionals, even though the philosophy behind the superstructure (partial objects, limstructure, etc.) may differ.

2.4 The density theorem

Definition 7 By recursion on the type σ we will define the n'th approximation $(a)_n$ to any element a of $D(\sigma)$:

1. If $\sigma = \iota$ and $m \in \mathbb{N}$, we let $(m)_n = \min\{n, m\}$. $\perp_n = \perp$.

2. If $\sigma = o$ and $a \in \mathbb{B}_\perp$, we let $(a)_n = a$.

3. If $\sigma = \tau \to \delta$, $f \in D(\sigma)$, and $a \in D(\tau)$, we let $(f)_n(a) = (f((a)_n))_n$.

The following lemmas are trivial:

Lemma 8 For each type σ and each $n \in \mathbb{N}$, $\{(a)_n \mid a \in D(\sigma)\}$ is finite.

Lemma 9 For each type σ, each $a \in H(\sigma)$ and $n \in \mathbb{N}$, $(a)_n \in H(\sigma)$.

Lemma 10 Let σ be a type, and let $a \in D(\sigma)$ be compact. Then there is a number n such that for all $m \geq n$ we have

$$a \sqsubseteq (a)_m.$$

We will then have

Theorem 11 [Density Theorem] Let σ be a finite type.

a) Consistency (boundedness) is an equivalence relation on pairs from $H(\sigma)$.

b) Each compact $a \in D(\sigma)$ has an extension to an element of $H(\sigma)$.

The density theorem is proved over and over again in the literature, [4, 9, 13, 14] and in surveys in general. In the proof, a) and b) are proved by simultaneous induction on the type. In order to prove b) one proves that for each compact a of type σ there is a total Φ of type σ such that a is consistent with $(\Phi)_m$, where $m \geq n$ is as in Lemma 10.

2.5 Kleene schemes

The Kleene schemes S1–S9 were introduced in [12]. They can be seen as nine clauses in a grand inductive definition defining the relation

$$\{e\}(\phi_1, \dots, \phi_k\} \simeq a,$$

where each ϕ_i will be a functional of pure type and $a \in \mathbb{N}$. The interpretation of these schemes will depend on the typed structure at hand. To us, three such structures are of interest, the full type structure $\{Tp(n)\}_{n \in \mathbb{N}}$ of total functionals, the Kleene–Kreisel functionals $\{Ct(n)\}_{n \in \mathbb{N}}$, and the Scott hierarchy $\{D(n)\}_{n \in \mathbb{N}}$. We will not give a detailed introduction to S1–S9; see the original paper [12] or any later survey. The three first schemes introduce basic arithmetical functions like identity, successor, and constants. S4 is composition, S5 primitive recursion, S6 permutation of variables, and S7 the application operator on $(\mathbb{N} \to \mathbb{N}) \times \mathbb{N}$. S8 is a combination of application and composition at higher types:

$$\{e\}(\Phi^{k+2}, \phi_1, \dots, \phi_n) \simeq \Phi(\lambda \xi^k.\{d\}(\xi, \Phi, \phi_1, \dots, \phi_n)),$$

where e depends on d.

If we interpret this scheme over the full type structure or over the Kleene–Kreisel type structure, we require that $\lambda \xi^k.\{d\}(\xi, \Phi, \phi_1, \dots, \phi_n)$ is total, whereas this is relaxed in the Scott hierarchy since non-total functionals are present there. As we will see, this has a dramatical effect on the computational power of S1–S9.

Tait [34] observed that the *fan functional* is not S1–S9-computable over the Kleene–Kreisel functionals. Later, Martin Hyland [10, 11] showed that the functional Γ, defined by Gandy, is not S1–S9-computable relative to the fan functional. Finally, Normann [19] showed that for any type $k \geq 3$ and any continuous Φ of type k, there is a continuous Ψ of type k that is not S1–S9-computable relative to Φ and any continuous functional of lower type.

However, if we move to type $k + 1$ we can find a functional Φ with a computable associate such that every $\psi \in Ct(k)$ is uniformly μ-computable relative to Φ and any associate for ψ. This was also proved in [19].

3 The Cook–Berger Problem

Although S1–S9 interpreted over the Kleene–Kreisel continuous functionals is an interesting example of a computation theory, the relevance to TCS is rather meager. In modeling real computations, partiality is an important aspect. Moreover, the requirement in the traditional interpretation of S8:

$$\{e\}(\Phi, \vec{\phi}) \simeq \Phi(\lambda \xi \{d\}(\xi, \Phi, \vec{\phi}))$$

that $\lambda \xi \{d\}(\xi, \Phi, \vec{\phi})$ must be total, is rather unnatural. On the contrary, it is natural to imagine that we may have an algorithm for Φ that now and then consults $\lambda \xi \{d\}(\xi, \Phi, \vec{\phi})$ and after finitely many steps provides the value of $\Phi(\lambda \xi \{d\}(\xi, \Phi, \vec{\phi}))$.

PCF, with its operational semantics, is actually making this idea precise, and then, in a denotational semantics for PCF we have to accept a more liberal interpretation of what corresponds to S8. Again, this is obtained by interpreting S8 over the Scott hierarchy of partial continuous functionals.

What is interesting from a foundational point of view is that when we customize S8 to TCS, we increase the computational power of S1–S9 considerably. One simple example is the functional Γ defined by the equation

$$\Gamma(F) = F_0(\lambda n.\Gamma(F_{n+1})).$$

Using the fixed point operator in PCF, we see that Γ is a well-defined functional, and that actually $\Gamma \in H(3)$. If we interpret the index obtained by the recursion theorem in the Kleene–Kreisel hierarchy, we just get the nowhere defined functional. As remarked above, Γ is not S1–S9-computable in this latter sense. This shows that relaxing the requirements on S8 increases the computational power.

Apparently, Robin Gandy (unpublished) had made a similar observation for the fan functional. The fan functional essentially evaluates the modulus of uniform continuity of a continuous $F \in Ct(2)$ over a compact $C_f = \{g \in \mathbb{N}^{\mathbb{N}} \mid g \leq f\}$ in $\mathbb{N}^{\mathbb{N}}$. Independently, Berger [4] showed that there is a representative $\Phi \in H(3)$ for the fan functional that is S1–S9-computable in the Scott-hierarchy sense. Simpson [32] used this to show that we may write a PCF-program for Riemann integration if the reals are represented by the elements of a given σ-compact subset of $\mathbb{N}^{\mathbb{N}}$. We will not go into detail, but the point is that we may write programs for functions of interest in mathematics using PCF or S1–S9 and suitable representations of the data as domain elements. Thus it is of foundational interest to see to which extent we are able to write programs for functionals. Cook asked if, and Berger suggested that all Kleene–Kreisel functionals that have computable associates also have hereditarily total representatives that are PCF-computable. It turned out that this is true in a rather strong sense:

Theorem 12 (Normann [21]) *Let σ be a finite type. There is a PCF-definable function $EVAL$ of type $(\iota \to \iota) \to \sigma$ such that whenever $a \in H(\sigma)$ and f enumerates (the Gödel numbers of) the set of compacts in $D(\sigma)$ bounded by a, then $EVAL(f) \in H(\sigma)$ and $EVAL(f) \sqsubseteq a$.*

It will lead too far to give the details of the construction; see [21] or [22]. In [22] we only prove the theorem for type 3. There are some obstacles that are not visible at this level. Here we will explain how the density theorem is used.

For types ≤ 2 the theorem is quite easy, and the standard way of computing a total functional of type 2 from an enumeration of its approximations actually will give a $G \sqsubseteq F$ but not F itself in all cases. If we know that $\{f_{p_n,b_n}\}_{n\in\mathbb{N}}$ are step functions approximating F and f is given, we may for each n ask whether f is consistent with p_n. If it is, we let $G(f) = b_n$, whereas if f is not consistent with p_n, we go to the next step function approximating F for help.

If ϕ is of type 3, we may try a similar strategy. The problem is that consistency of an $F \in H(2)$ and a compact $p \in D(2)$ is not decidable; we just have that inconsistency is semi-decidable. In order to semi-decide inconsistency between $F \in H(2)$ and a compact $p \in D(2)$, we search through an effectively enumerated dense subset of the domain of p for an argument ξ such that $p(\xi)$ and $F(\xi)$ differ.

Now we let $\{f_{p_n,b_n}\}_{n \in \mathbb{N}}$ be an enumeration of all step functions bounded by $\phi \in H(3)$, and we let $F \in H(2)$. One ingredient in the algorithm for $\psi(F)$ is that we for each n search for a witness to the fact that p_n is inconsistent with F. If we find a witness, we just go on to $n + 1$. If we do not find a witness, and $p_n \sqsubseteq F$, the result of the search will be a partial object q for which $p_n(q) = F(q) \in \mathbb{N}$ and we conclude that $\psi(F) = b_n$. The problematic case is when neither of these occur, and then we have in parallel to search for a later stage actually verifying that $\psi(F) = b_n$ in any case, and then it does not matter that we found no witness either way. The functional ψ constructed this way will then be total, and $\psi \sqsubseteq \phi$. This argument explains where density is used, but it is of course not complete.

At type 3 we only use that there is an effectively enumerated dense set of objects of type 1. At higher types we have to use the full density theorem to the same effect. The obstacle is that not every functional constructed in order to verify the density theorem will be PCF-definable. Thus we have to use induction on the type, applying our construction to the effective enumerations of the approximations to the witnesses to the density theorem. We will not discuss the details here.

The Scott model is not fully abstract for PCF. This means that there are compacts in the Scott model that are not PCF-definable. Milner [17] showed that there exist one, and up to isomorphism, only one fully abstract model for PCF consisting of algebraic domains. Of course we may define the class of hereditarily total objects in Milner's model as well; consistency will be an equivalence relation on these objects, so we may even form the extensional collapse of the hereditarily total elements of Milner's model. Plotkin [29] showed as a corollary of Theorem 12 that this construction leads us to the Kleene–Kreisel continuous functionals. Thus Milner's construction is an alternative characterization. Recently Longley [15] showed that under reasonable general assumptions constructions of an extensional hierarchy of total continuous functionals containing $\mathbb{N}^{\mathbb{N}}$ at type 1 tends to end up with the Kleene–Kreisel functionals. Longley's proof is an elaboration on the argument from [21], where showing that we can simulate the effect of the density theorem is a nontrivial ingredient.

4 Replacing the natural numbers with the real numbers

Up to now, we have considered various models for computing relative to functionals of finite types over the natural numbers. Although we accept inputs to our algorithms that are infinite, the sets of natural numbers and boolean values are at the bottom, and these are discrete sets of data easily represented as digital data in a computer. As long as we are not involved in complexity issues, the nature of the representation is not important.

If we want to use other data-types, such as the real numbers, the set of differentiable functions on the complex numbers, certain Polish spaces, or even structured Polish spaces like Banach spaces, the situation is quite different. Then in addition to discussing which principles for forming valid algorithms we will accept and which superstructures we will find useful for the denotational semantics, we also have to discuss how to represent the basic data in digital form. Given a representation of basic data, there will be a conflict of interest:

1. If we write programs using the representation, we may get more efficient programs and we may be able to write programs solving more problems than we otherwise might.

2. If we hide the representation and write programs in a language based on the algebra of the data-type itself, it may be easier to analyze each program and to verify correctness.

It is well established that the decimal representation of real numbers is not suitable for modeling computability. Moreover, traditional constructions of $\mathbb{R}$ within set theory, like the set of Dedekind cuts or the set of equivalence classes of Cauchy sequences, do not lead to useful computational models either. As is customary in constructive analysis, Cauchy sequences with a prefixed rate of convergency works much better.

In this paper we will consider two ways of representing reals, one improvement of the Dedekind cut representation and one improvement of the Cauchy sequence representation. Although both approaches are good for analyzing what we mean by computable reals and computable functions on the reals, we will see that they may make a difference for the objects of higher types. In both cases there will be domains of partial computable functionals, and we will extract the hierarchies of total ones as substructures of particular interest.

4.1 The extensional hierarchy

Given the Dedekind cut of a real x, we may approximate x by half open rational intervals $(p, q]$ by first looking for a pair p_0, q_0 with distance ≤ 1, where p_0 is in the cut and q_0 is not, and then we recursively test whether $\frac{p_n + q_n}{2}$ is in the cut in order to decide whether this is p_{n+1} or q_{n+1}. The problem with cuts is that they are asymmetric, and it has turned out to be better to work with closed rational intervals as approximations. We let $R_0(0)$ be the set of closed rational intervals ordered by reverse inclusion (0 will denote the one base type now), and we let $R(0)$ be the set of ideals in this ordered set. An ideal $\mathcal{I}$ will determine a closed interval I of reals, the intersection of the ideal. We let an ideal be *total* if it determines an interval $[x, x]$ of length zero, and then we say that it *represents* x. Rational numbers can be represented in three ways, whereas irrational numbers are only represented by one ideal each.

Remark 13 We have chosen to use the algebraic domain version of the closed interval way of approximating reals. Even more common is the *continuous domain* of all closed intervals ordered by reverse inclusion. The point with both these domains is that the ordering of the domain elements only reflects the set of reals they approximate. Thus we call this an *extensional* way of representing reals via finite approximations.

We are now ready to define the full hierarchy of domains interpreting functionals of finite types over the reals. Replacing $\mathbb{N}_\perp$ with $R(0)$ and $\mathbb{N}$ with $\mathbb{R}$ we may copy the definition of the $Ct(\sigma)$-hierarchy defining $R(\sigma)$ in the category of algebraic domains for each finite type σ, the set $H_R(\sigma)$ of hereditarily total elements of type σ, the equivalence $\approx_\sigma^R$ on $H_R(\sigma)$, and the extensional collapse hierarchy $\{Ct_\mathbb{R}(\sigma)\}_{\sigma \text{ type}}$ with the corresponding identification functions ρ_σ^R. Details may be found, for example, in Normann [23].

It is easy to see that each $Ct_\mathbb{R}(\sigma)$ is organized into a topological vector-space in a natural way. By induction on the type, we also see that each compact p in $R(\sigma)$ will determine a closed, convex subset V_σ^p of $Ct_R(\sigma)$, the set of objects with a representative in $H_R(\sigma)$ extending p. We will use the convexity of V_σ^p in the following way:

Lemma 14 *Let $a_1, \ldots, a_n$ be elements in V_σ^p, and let μ be a probability distribution on $\{1, \ldots, n\}$. Then*

$$\sum_{i=1}^{n} \mu(i)a_i \in V_\sigma^p.$$

We are now going to discuss the density theorem, originally due to Normann [23] but with an alternative proof in DeJaeger [5]:

Theorem 15 *Let p be a compact in $R(\sigma)$. Then there is an element $a \in H_\mathbb{R}(\sigma)$ extending p.*

Since we have not given any detailed definitions, we will not give a detailed proof. We will, however, indicate how a proof goes, and our approach here is closer to DeJaeger's proof than to our original argument.

Definition 16 By recursion on the type σ for each $n \in \mathbb{N}$ we define a finitary type-structure $\{X_n(\sigma)\}_{\sigma \text{ type}}$ as follows:

$$X_n(0) = \{\tfrac{k}{n!} \mid -(n+1)! \leq k \leq (n+1)!\},$$
$$X_n(\tau \to \delta) = \{h \mid h : X_n(\tau) \to X_n(\delta)\}.$$

We are now, by simultaneous recursion for each n, going to define an embedding $\nu_{n,\sigma} : X_n(\sigma) \to H_R(\sigma)$ and for each $a \in H_R(\sigma)$ a probability distribution $\mu_{n,\sigma}(a)$ on $X_n(\sigma)$. The prime objects are the embeddings $\nu_{n,\sigma}$, and the probability distributions will replace projections that are used for similar purposes in the discrete case ($\mathbb{N}$ instead of $\mathbb{R}$).

Definition 17

1. Let $\sigma = 0$.

   If $x \notin (-(n+1), n+1)$, we let $\mu(\frac{k}{n!}) = 1$ if $\frac{k}{n!}$ is the object in $X_n(0)$ closest to x; otherwise we let $\mu(\frac{k}{n!}) = 0$.

   If $x \in (-(n+1), n+1)$, there are unique $k \in \mathbb{Z}$ and $y \in [0, 1)$ such that $x = \frac{k}{n!} + \frac{1-y}{n!}$.
   Let $\mu_{n,0}(\frac{k}{n!}) = y$, $\mu_{n,0}(\frac{k+1}{n!}) = 1 - y$ and $\mu_{n,0}(\frac{l}{n!}) = 0$ for all $l \neq k, k+1$.

2. Let $\sigma = \tau \rightarrow \delta$

   Let $h \in X_n(\tau \rightarrow \delta)$, $a \in H(\tau)$.
   Let

   $$\nu_{n,\sigma}(h)(a) = \sum_{c \in X_n(\tau)} \mu_{n,\tau}(a)(c) \cdot \nu_{n,\delta}(h(c)).$$

   Now let $f \in H(\tau \rightarrow \delta)$ and $h \in X_n(\tau \rightarrow \delta)$.
   Let

   $$\mu_{n,\sigma}(f)(h) = \prod_{c \in X_n(\tau)} \mu_{n,\delta}(f(\nu_{n,\tau}(c)))(h(c)).$$

We observe that at base type we are constructing explicit probability distributions, whereas at higher types we are using finite products of probability distributions, which again will be probability distributions. Thus $\mu_{n,\sigma}(a)$ is a probability distribution for all n, σ, and $a \in H(\sigma)$. We also observe that since all types are topological vector spaces, we may view products with scalars and sums as partial continuous operations on the underlying domains, and thus our constructions are sound and continuous over the underlying domains.

In an embedding-projection pair it is important that

$$\text{projection} \circ \text{embedding} = \text{identity}.$$

In this setting, we formulate the similar phenomenon as

Lemma 18 *For all types σ and each $h \in X_n(\sigma)$, we have that*

$$\mu_{n,\sigma}(\nu_{n,\sigma}(h))(h) = 1.$$

The proof is by induction on σ using a tedious but in principle simple calculation at the induction step.

The density theorem will follow from

Lemma 19 *Let $p \in R(\sigma)$ be compact. Then there is a number n_p such that*

a) For all $n \geq n_p$ there is a $b \in X_n(\sigma)$ such that p is consistent with $\nu_{n,\sigma}(b)$.

b) *If $a \in H_R(\sigma)$ and p is consistent with a, then*

$$\mu_{n,\sigma}(a)(\{b \in X_n(\sigma) \mid p \text{ is consistent with } \nu_{n,\sigma}(b)\}) = 1.$$

c) *If $a \in H_R(\sigma)$, then a is* pre-maximal *in the sense that a has a unique maximal extension.*

Proof. We use induction on σ.

Let $\sigma = 0$.

Let $p = [q, r]$, where q and r are rational numbers. Choose n_p such that $|q|$, $|r|$ and the denominators of q and r are bounded by n_p. This will do the trick for a) and b). c) is trivial.

Let $\sigma = \delta \rightarrow \tau$.

c) follows by the density theorem for τ and c for δ.

Let p be the least upper bound of the step functions $\{f_{q_i, r_i}\}_{i \in I}$ where I is finite. Let

$$n_p = \max\{n_{\sqcup_{i \in K} q_i}, n_{\sqcup_{i \in K} r_i} \mid K \subseteq I \wedge \{q_i \mid i \in K\} \text{ is consistent}\}.$$

We now prove a) and b), assuming both a) and b) for τ and δ.

Proof of a):

By c) for τ we have that q_i is consistent with $a \in H_R(\tau)$ if and only if q_i is in the maximal extension of a, so the set of q_i's consistent with a is itself consistent. Let $n \geq n_p$, and by the induction hypothesis, let $h \in X_n(\sigma)$ be such that for all $b \in X_n(\tau)$, r_i is consistent with $\nu_{n,\delta}(h(b))$ whenever q_i is consistent with $\nu_{n,\tau}(b)$. So let $a \in H_R(\tau)$ and let $K = \{i \in I \mid p_i \text{ is consistent with } a\}$. By definition,

$$\nu_{n,\sigma}(h)(a) = \sum_{c \in X_n(\tau)} \mu_{n,\tau}(a)(c) \cdot \nu_{n,\delta}(h(c)).$$

By the induction hypothesis, we only have to consider those $c \in X_n(\tau)$ where $\nu_{n,\tau}(c)$ is consistent with $\bigcup\{q_i \mid i \in K\}$, and for those c, $\nu_{n,\delta}(h(c))$ will be in $V_\delta^{\sqcup_{i \in K} r_i}$. By the convexity of this set, we see that $\nu_{n,\sigma}(h)(a) \in V_\delta^{\sqcup_{i \in K} r_i}$. This verifies a.

Proof of b):

Let $n \geq n_p$, $h \in X_n(\sigma)$, and $f \in H_R(\sigma)$ such that f is consistent with p, but $\nu_{n,\sigma}(h)$ is not consistent with p. We verify b) by showing that $\mu_{n,\sigma}(f)(h) = 0$. $\mu_{n,\sigma}(f)(h)$ is defined to be a product, and it is sufficient to prove that one of the factors is zero.

Since $\nu_{n,\sigma}(h)$ is not consistent with p, there must be an $i \in I$ and $a \in V_\tau^{q_i}$ such that $\nu_{n,\sigma}(h)(a)$ is inconsistent with r_i.

By the argument of a), we can find $b \in X_n(\tau)$ such that $\nu_{n,\tau}(b)$ is consistent with q_i while $\nu_{n,\delta}(h(b))$ is inconsistent with r_i.

Since f is consistent with p, we have that $f(\nu_{n,\tau}(b))$ is consistent with r_i and $f(\nu_{n,\tau}(b)) \in H_R(\delta)$. Then it follows by b) for δ that $\mu_{n,\delta}(f(\nu_{n,\tau}(b)))(h(b)) = 0$. Since this is one of the factors in $\mu_{n,\sigma}(f)(h)$, we are through.

If we consider a typed hierarchy with both connected and discrete base types, e.g., domain representations of $\mathbb{R}$ and $\mathbb{N}$, the density-theorem fails for trivial reasons. We may find a compact approximation to a partial continuous function $f : \mathbb{R} \rightarrow \mathbb{N}$ that is not constant, but all total such functions will be constant. In Normann [23] we define an alternative hierarchy of domains with totality satisfying the density theorem where the quotient spaces of hereditarily total functionals will be the same as with the traditional Scott domain approach. The drawback is that we will leave some interesting partial functionals out of this model.   □

4.2 The intensional hierarchy

In our first construction, we started with the closed interval domain where the elements represent approximations to the reals. This is called the *extensional approach* and may be viewed as improvements of Dedekind cuts. Alternative ways of representing the reals are via data-streams, or more mathematically, via infinite words in some alphabet. We will consider one example, the so-called *negative binary digit representation*.

Definition 20 Let $\alpha \in \mathbb{Z} \times \{-1, 0, 1\}^{\mathbb{N}^+}$, where $\mathbb{N}^+ = \{1, 2, 3, \ldots\}$. We consider α as a function defined on $\mathbb{N}$.
Let the real $r(\alpha)$ *represented by* α be

$$r(\alpha) = \alpha(0) + \sum_{n \in \mathbb{N}^+} \alpha(n) \cdot 2^{-n}.$$

Let REP_0 be the set of such α's.

We may consider REP_0 as the total objects in the domain $R^I(0)$ (I for intensional) of finite or infinite sequences $a b_1 b_2 \ldots$, where $a \in \mathbb{Z}$ and each $b_i \in \{-1, 0, 1\}$. This domain is again homeomorphic to a sub-domain of $D(\iota \rightarrow \iota)$, the domain of partial continuous functions from $\mathbb{N}_\bot$ to $\mathbb{N}_\bot$ such that both the embedding and the projection are total. This also suggests that based on this model, we may use this representation, its higher type semantics, and a PCF-like language to give a semantics to typed computations involving reals. This idea is discussed in more detail in DiGianantonio [6, 7, 8] and Simpson [32].

Let us define the hereditarily total functionals REP_σ with the equivalence relation $\sim_\sigma$ for each type σ:

Definition 21 We let $R^I(0)$ be as above, and we define $R^I(\sigma)$ for all finite types σ over the base type 0 in the category of algebraic domains. By recursion on σ, we define the set $REP_\sigma \subseteq R(\sigma)$ and the binary relation $\sim_\sigma$ on REP_σ

$\sigma = 0$:
If f and g are in REP_0, we let $f \sim_0 g$ if they represent the same real.

$\sigma = \tau \rightarrow \delta$:
Let $\phi \in R^I(\sigma)$. We let $\phi \in REP_\sigma$ if

i) $a \in REP_\tau \Rightarrow \phi(a) \in REP_\delta$ for all $a \in R(\tau)$.

ii) $a \sim_\tau b \Rightarrow \phi(a) \sim_\delta \phi(b)$ whenever $a, b \in REP_\delta$.

If ϕ and ψ are in REP_σ we let $\phi \sim_\sigma \psi$ if $\phi(a) \sim_\delta \psi(b)$ whenever $a \sim_\tau b$.

Except for the base type, REP_σ will not be dense in $R^I(\sigma)$. There will be inconsistent compacts in $R^I(0)$ that can be extended to inconsistent but equivalent total objects. This can be used to construct two consistent step-functions in $R^I(0 \to 0)$ that cannot be extended to any element in $REP_{0\to0}$.

It has actually been left open if the set of compacts that can be extended to a total object is decidable, but this is probably because no one tried hard to prove it. In Normann [24], an alternative hierarchy of effective domains with totality, satisfying density and leading to the same coefficient spaces of total objects, was constructed. The idea is that we introduce an extra relation representing "can be extended to equivalent total objects" on the compacts, and we only accept the compacts in a function space respecting this relation.

4.3 The coincidence problem

We have considered two hierarchies of total continuous functionals of finite types over the reals, both constructed as quotients of hereditarily total objects in a hierarchy of algebraic domains, one based on the closed interval domain and one on the negative binary representation of reals. There is a third approach, originating from Weihrauch's TTE [35]. In this approach one uses admissible representations at each type. The *coincidence problem* is whether these approaches coincide, i.e., if the typed structures defined via these approaches are the same.

Due to characterizations of the extensional hierarchy (Normann [23]) and the TTE-based hierarchy (Schröder [30]) as the one obtained from the reals in the category of limit spaces, these two approaches are known to be equivalent. Thus the real coincidence problem is whether the domain theoretical approaches based on extensional and intensional representations of the reals coincide. This problem is of course more of a foundational than of a practical nature. In its nontechnical form, the question is whether the choice of representation of reals as data will have any influence on what is considered to be a continuous, total functional of higher type.

The problem was first addressed by Bauer, Escardó and Simpson [3], and they obtained some partial results. Normann [26] showed that the coincidence problem is equivalent to a topological problem about the Kleene–Kreisel functionals:

> *Is the topology generated by all continuous functions $\phi : Ct(n) \to \mathbb{R}$ zero dimensional for all n?*

Theorem 22 (see below) is used to establish this equivalence.

5 Density and probability

In $Ct(\sigma)$, two distinct objects of the same type can be separated by a clopen set, i.e., a set that i both closed and open. This is of course useful in algorithm design; definitions by cases are handy. Each space $Ct_{\mathbb{R}}(\sigma)$ is path connected [23], so definitions by cases have to fail for some inputs. It has turned out that in some cases, elementary methods from probability theory can be combined with the density theorem to overcome this lack of separation. The proof of the density theorem itself is one such example. We will mention two more results here.

5.1 The embedding theorem

Theorem 22 *Let $Ct_{\mathbb{R}}(\sigma)$ be the quotient space of the hereditarily total functionals over $\mathbb{R}$ in the extensional hierarchy.*
For each type σ there is a continuous map $\pi_\sigma : Ct(\sigma) \to Ct_{\mathbb{R}}(\sigma)$ such that π_0 is the standard embedding of $\mathbb{N}$ into $\mathbb{R}$, and for each type $\sigma = \tau \to \delta$, each $\phi \in Ct(\sigma)$ and $a \in Ct(\tau)$, we have that

$$\pi_\sigma(\phi)(\pi_\tau(a)) = \pi_\delta(\phi).$$

This theorem, and an analog result for the hierarchy based on the intensional representation, is proved in [24].

We will not give the proof here, but we will discuss where the density theorem is used. Ideally we would like to replace the (possible) use of a (nonexistent) continuous projection from $Ct_{\mathbb{R}}(\sigma)$ to $Ct(\sigma)$ with a continuous function μ_σ defined on $Ct_{\mathbb{R}}(\sigma)$ where $\mu(\phi)$ then would be a probability measure on $Ct(\sigma)$. Then, if $\phi \in Ct(\tau \to \delta)$ and $b \in Ct_{\mathbb{R}}(\tau)$, we could define

$$\pi_{\tau \to \delta}(\phi)(b) = \int_{a \in Ct(\tau)} \phi(a) d\mu_\tau(b).$$

Unfortunately, the complexity of these spaces makes it impossible to define decent probability measures on them. The alternative is to use approximations via probability distributions on the sets $Z_n(\sigma) = \{(a)_n \mid a \in H(\sigma)\}$ and to use that every total object in $Ct(\tau)$ can be approximated in a uniform way by these finitary elements.

5.2 Representation theorems

It is well known that topological spaces can be represented as quotient spaces over domains; old and contemporary literature is too vast for us to be specific about it. It is also well known that Polish spaces will be homeomorphic to G_δ-subsets of $[0, 1]^\omega$. In Normann [25] we consider finite types where arbitrary Polish spaces may be used as base types, interpreted in the category of limit spaces. We then show that the hierarchy

$$\{Ct_{\mathbb{R}}(\sigma)\}_{\sigma \text{ type}}$$

is adequate for interpreting all these spaces of objects of higher types, in the sense that each such space will be homeomorphic to a subspace of some $Ct_{\mathbb{R}}(\sigma)$.

In finding these subspaces we use a concept of *representation* developed from an intermediate step in the proof of Theorem 1:

Definition 23 Let $A \subseteq Ct_R(\sigma)$. A *representation of A in $Ct_R(\pi)$* is a continuous map

$$\phi : R(\sigma) \to R(\pi)$$

such that for all $a \in REP_\sigma$, we have that

$$\phi(a) \in REP_\pi \Leftrightarrow \rho_\sigma^R(a) \in A.$$

We show that each space of finite type over Polish spaces is homeomorphic to a subspace of some $Ct_R(\sigma)$ with a representation. This is proved by induction on the type. We give a highly incomplete sketch of the induction step, omitting that the set we construct will be representable:

- By the density theorem for $Ct_R(\pi)$ there is a total map $h : R(\pi) \to (\mathbb{N}_\perp \to R(0))$ such that h identifies exactly the consistent elements.

- Let $A \subseteq Ct_R(\tau)$ and $B \subseteq_R (\delta)$, and let $h : R(\tau) \to R(\pi)$ be a representation of A. Let $f : A \to B$ be continuous. Uniformly (but nontrivial) in f there is a sequence $\{f_n\}_{n \in \mathbb{N}}$ of continuous functions from $Ct_R(\tau)$ to $Ct(\delta)$ such that f will be the limit of the restrictions of f_n to A. By linear interpolation, we define f_x for each non-negative real x.

- If $f : A \to B$, $a \in Ct(\tau)$ and $b \in Ct(\pi)$, we let $F(f)(a,b) = f(a)$ if $h(\phi(a)) = h(b)$, whereas it is $f_x(a)$ for some x continuously witnessing that $h(\phi(a)) \neq h(b)$ otherwise.

- We may recover f from $F(f)$ by $f(a) = F(a, \phi(a))$. Thus the set $\{F(f) \mid f \in A \to B\}$ will be homeomorphic to $A \to B$.

References

1. Abramsky, S. and Jung, A., *Domain Theory*, in S. Abramsky, D.M. Gabbay and T.S.E. Maibaum (eds.), Handbook of Logic in Computer Science, vol. 3, Clarendon Press (1994)
2. Amadio, R. M. and Curien, P. -L. , *Domains and Lambda-Calculi*, Cambridge University Press (1998)
3. Bauer, A., Escardó, M.H. and Simpson, A., *Comparing Functional Paradigms for Exact Real-number Computation*, in Proceedings ICALP 2002, Springer LNCS 2380, 488–500 (2002)

4. Berger, U., *Totale Objekte und Mengen in der Bereichtheorie* (in German), Thesis, München (1990)
5. DeJaeger, F., *Calculabilité sur les Réels*, Thesis, Paris VII (2003)
6. DiGianantonio, P., *A Functional Approach to Computability on Real Numbers*, Thesis, Università di Pisa - Genova - Udine, (1993)
7. DiGianantonio, P., *Real number computability and domain theory*, Inform. and Comput. **127**, 11–25 (1996)
8. DiGianantonio, P., *An abstract data type for real numbers*, Theoret. Comput. Sci. **221**, 295–326 (1999)
9. Ershov, Yu. L., *Maximal and everywhere defined functionals*, Algebra and Logic **13**, 210–225 (1974)
10. Gandy, R.O. and Hyland, J.M.E., *Computable and recursively countable functions of higher type*, Logic Colloquium '76, 407-438, North-Holland (1977)
11. Hyland, J.M.E., *Recursion on the Countable Functionals*, Dissertation, Oxford (1975)
12. Kleene, S. C., *Recursive functionals and quantifiers of finite types I*, Trans. Amer. Math. Soc. **91**, 1–52 (1959)
13. Kleene, S. C., *Countable Functionals*, in A. Heyting (ed.) Constructivity in Mathematics, 81–100, North-Holland (1959)
14. Kreisel, G. , *Interpretation of Analysis by Means of Functionals of Finite Type*, in A. Heyting (ed.) Constructivity in Mathematics, North-Holland, 101–128 (1959)
15. Longley, J.R., *On the ubiquity of certain total type structures*, Electr. Notes Theor. Comput. Sci. **73**, 87–109 (2004)
16. Longo, G. and Moggi, E., *The hereditary partial effective functionals and recursion theory in higher types*, J. Symbolic Logic **49**, 1319–1332 (1984)
17. Milner, R., *Fully abstract models for typed λ-calculi*, Theoret. Comput. Sci. **4**, 1–22 (1977)
18. Moldestad, J., *Computations in Higher Types*, Springer Lecture Notes in Mathematics No. 574, Springer Verlag (1977)
19. Normann, D., *The continuous functionals; computations, recursions and degrees*, Annals of Mathematical Logic **21**, 1–26 (1981)
20. Normann, D., *Computing with functionals - computability theory or computer science?*, Bull. Symbolic Logic **12**(1), 43–59, (2006)
21. Normann, D., *Computability over the partial continuous functionals*, J. Symbolic Logic **65**, 1133–1142 (2000)
22. Normann, D., *The Cook-Berger problem - A guide to the solution*, Electr. Notes Theor. Comput. Sci. **35**, (2000)
23. Normann, D., *The Continuous Functionals of Finite Types over the Reals*, in K. Keimel, G.Q. Zhang, Y. Liu and Y. Chen (eds.) Domains and Processes, Kluwer Academic Publishers, 103–124, (2001)
24. Normann, D., *Hierarchies of total functionals over the reals*, Theoret. Comput. Sci. **316**, pp. 137–151 (2004)
25. Normann, D., *Definability and reducibility in higher types over the reals*, to appear in the proceedings og Logic Colloquium '03
26. Normann, D., *Comparing hierarchies of total functionals*, Logical Methods in Computer Science **1**(2), (2005)
27. Platek, R. A., *Foundations of Recursion Theory*, Thesis, Stanford University (1966)
28. Plotkin, G., *LCF considered as a programming language*, Theoret. Comput. Sci. **5**, 223–255 (1977)
29. Plotkin, G., *Full abstraction, totality and PCF*, Math. Struct. in Comp. Science, **11**, 1–20 (1999)

30. Schröder, M., *Admissible representations of limit spaces*, in J. Blanck, V. Brattka, P. Hertling and K. Weihrauch (eds.), Computability and Complexity in Analysis, vol. 237, Informatik Berichte, 369–388, (2000)

31. Scott, D., *A type-theoretical alternative to ISWIM, CUCH, OWHY*, Unpublished notes, Oxford (1969). Printed with suplementary comments in Theoret. Comput. Sci. **121**, 411–440 (1993)

32. Simpson, A., *Lazy functional Algorithms for Exact Real Functionals*, Mathematical Foundations of Computer Science 1998, Springer LNCS **1450**, 456–464, (1998)

33. Stoltenberg-Hansen, V., Lindström, I. and Griffor, E.R., *Mathematical Theory of Domains*, Cambridge Tracts in Theor. Comp. Sci. 22, Cambridge University Press (1994)

34. Tait, W.W., *Continuity properties of partial recursive functionals of finite type*, unpublished notes (1958)

35. Weihrauch, K., *Computable Analysis*, in Texts in Theoretical Computer Science, Springer Verlag, Berlin, (2000)

Church Without Dogma: Axioms for Computability

Wilfried Sieg

Department of Philosophy, Carnegie Mellon University, Pittsburgh, PA 15213, USA
sieg@cmu.edu

Summary. Church's and Turing's theses assert dogmatically that an informal notion of effective calculability is captured adequately by a particular mathematical concept of computabilty. I present analyses of calculability that are embedded in a rich historical and philosophical context, lead to precise concepts, and dispense with theses.

To investigate effective calculability is to analyze processes that can in principle be carried out by calculators. This is a philosophical lesson we owe to Turing. Drawing on that lesson and recasting work of Gandy, I formulate boundedness and locality conditions for two types of calculators, namely, human computing agents and mechanical computing devices (or discrete machines). The distinctive feature of the latter is that they can carry out parallel computations.

Representing human and machine computations by discrete dynamical systems, the boundedness and locality conditions can be captured through axioms for Turing computors and Gandy machines; models of these axioms are all reducible to Turing machines. Cellular automata and a variety of artificial neural nets can be shown to satisfy the axioms for machine computations.

Background

The title of this essay promises axioms for computability. Such axioms will emerge from a *conceptual analysis* that begins with a straightforward observation: whatever we consider to be computable must be associated with computations that are carried out by some device or other. Consequently, we have to pay close attention to the nature of the device at hand, when thinking through the characteristic features that determine (the extension of) its notion of computability. My analysis builds on work by Turing and Gandy concerning computations that are carried out by human calculators and discrete machines, respectively.

I sharpen the informal concepts of computation for these two devices, specify rigorously their characteristic features, and formulate a representation theorem for the resulting systems of axioms. A broad methodological point can be immediately inferred: theses in the standard Church–Turing form are not needed to connect rigorously defined notions of computability with informally grasped concepts. It is however crucial to gain a proper understanding of these canonized connections, because the significance of logical results like Gödel's incompleteness theorems depends on it, as does the centrality of related issues in the philosophy of mind.

Part 1 articulates three principal *Church canons*[1] supporting the thesis. For the canonical argument from confluence, I distinguish between support that derives from examining the effective calculability of number theoretic functions and support that is obtained through analyzing mechanical operations on symbolic configurations. The analysis of such operations when carried out by a human calculator leads to Turing's claims in 1936. The arguments for these claims exploit *boundedness* and *locality conditions* that are presented in Part 2. Against this background I introduce in Part 3 axioms for *Turing computors* and *Gandy machines*, list models, and formulate a representation theorem. That completes the conceptual analysis. I will conclude with remarks on Gödel, Turing, and philosophical errors.

Note: This essay is based on two papers I published in 2002, but whose methodological considerations I would like to bring out more directly. I presented versions of this essay under the title *Beyond Church Canons* in the Distinguished Lecture Series (Haverford College, October 2002), in the Annual Lecture Series at the Center for Philosophy of Science (University of Pittsburgh, January 2004), at the Colloquium of the IHPST (Sorbonne, May 2004), as well as at the Colloquium of the Department of Philosophy (University of Florence, November 2004) and at the conference *Computability in Europe* (Amsterdam, July 2005). For detailed discussions of the origins and developments of computability, see also (Sieg 1994, 1997) and the rich literature that is referred to in those papers.

1 Church Canons

In a sense we have to untangle the relation between the concept of computability and the concept of computability, understanding the first concept as informally grasped and the second as rigorously defined. If one takes Gödel's notion of general recursiveness as the rigorously defined concept and effective calculability as the informally grasped one, then Church's Thesis expresses the relation between this and that concept of computability for number-theoretic functions: they are co-extensional. To provide a proper perspective for the broader investigation, I will examine the early history of computability hinted at in these remarks.

---

[1] According to the fifth edition of the Shorter OED, *canon* does not cover just ecclesiastical laws and decrees, but it has also the meaning of "a general law, rule, or edict; a fundamental principle" since the late middle ages, and that of "a standard of judgement; a criterion" since the early 17th century.

1.1 The Thesis

Gödel introduced general recursiveness for number-theoretic functions in his 1934 Princeton Lectures via his equational calculus; he viewed it as a heuristic principle that the informal concept of *finite computation* can be captured by suitably general *recursions*. Refining and generalizing a notion of finitistically calculable functions due to Herbrand, Gödel defined a number-theoretic function to be general recursive just in case it satisfies certain recursion equations and its values can be determined from the equations by simple steps, namely, replacement of variables by numerals and substitution of complex closed terms by their numerical values. When he gave this definition in 1934 Gödel was not convinced, however, that the underlying precise concept of recursion was the most general one, and he expressed his doubts in conversation with Church. Nevertheless, Church formulated the thesis a year later for the first time in print. Here is the classical statement found in the abstract for Church's talk to the American Mathematical Society in December 1935:

> ...Gödel has proposed ... a definition of the term recursive function, in a very general sense. In this paper a definition of recursive function of positive integers which is essentially Gödel's is adopted. And it is maintained that the notion of an effectively calculable function of positive integers should be identified with that of a recursive function, since other plausible definitions of effective calculability turn out to yield notions that are either equivalent to or weaker than recursiveness.

Between Church's conversations with Gödel in 1934, and the formulation of the above abstract in 1935, some crucial developments had taken place in Princeton. Kleene and Rosser had done significant quasi-empirical work, convincing themselves and Church that all known effective procedures are λ-definable. Kleene had discovered his normal-form theorem and had established the equivalence of Gödel's general recursiveness with μ-recursiveness. Finally, Church and Kleene had proved the equivalence of λ-definability and general recursiveness. All these developments are alluded to in Church's abstract, and they are interpreted as supporting the thesis, which was then, and is still now, principally defended on two grounds. First, there is the quasi-empirical reason: all known calculable functions are general recursive. This point, although important, is clearly not decisive and will be taken up in the broader context of section 2.3. Second, there is the argument from confluence: a variety of mathematical computability notions all turn out to be equivalent. This second important point is however only really convincing, if the "confluent" notions are of a quite different character and if there are independent reasons for believing that they capture the informal concept. Both Church and Gödel tried to give such independent reasons in 1936. Let me sketch their considerations.

1.2 Semi-circles

Church and Gödel took the evaluation of a function in some form of the equational calculus as the starting point for explicating the effective calculability of number-theoretic functions. Church generalized broadly: an evaluation is done in some logical calculus through a step-by-step process, and the steps must be elementary. Church argued that functions whose values can be computed in this way must be general recursive. Gödel, in contrast, just made a penetrating observation without giving an argument: the rules of the equational calculus are part of any adequate formal system of arithmetic, and the class of calculable functions is not enlarged beyond the general recursive ones, if the formal system is strengthened. This *absoluteness* of the notion was pointed out in a Postscriptum to (Gödel 1936) for transfinite extensions of type theory and in the Princeton Bicentennial lecture ten years later for extensions of formal set theory. Gödel formulated the significance of his observation in the lecture (Gödel 1946, p. 150) as follows:

> Tarski has stressed ... the great importance of the concept of general recursiveness (or Turing computability). It seems to me that this importance is largely due to the fact that with this concept one has for the first time succeeded in giving an absolute definition of an interesting epistemological notion, i.e., one not depending on the formalism chosen.

But what is the argument for Church's claim, and what could it be for Gödel's? If one uses the strategic considerations underlying the proof of Kleene's normal-form theorem, it is in both cases easily established that the functions calculable in the broader frameworks are general recursive, as long as the steps in the logical systems are elementary, formal, ... well, general recursive. Church turned the elementary steps explicitly into general recursive ones, whereas Gödel could not but exploit the formal character of the theories at hand through their recursive presentation.

Taken as principled arguments for the thesis, Gödel's and Church's considerations rely on a hidden and semi-circular condition for steps. Hilbert and Bernays moved this step-condition into the foreground when investigating calculations in deductive formalisms and reckonable functions ("regelrecht auswertbare Funktionen"). They imposed explicitly *recursiveness conditions* on deductive formalisms and showed that formalisms satisfying these conditions have as their calculable functions exactly the general recursive ones. In this way they provided mathematical underpinnings for Gödel's absoluteness claim and for Church's argument, but only *relative* to the recursiveness conditions: the crucial one requires the proof predicate of deductive formalisms, and thus the steps in formal calculations, to be primitive recursive.[2]

The work of Gödel, Church, Kleene, and Hilbert & Bernays had intimate historical connections and is still of deep interest. It explicated calculability of functions by exactly *one core notion*, namely, calculability of their values in logical calculi via (a finite number of) elementary steps. But no one gave convincing and non-circular reasons for the proposed rigorous restrictions on steps permitted in calculations. The question is, whether this stumbling block for a deeper analysis can be overcome. The answer lies in a motivated and general formulation of constraints on steps.

---

[2] These investigations are carried out in the second supplement of their *Grundlagen der Mathematik II*.

1.3 Symbolic Processes

Church reviewed in 1937 the two classical papers by Turing and Post, which had been published in 1936. When comparing Turing computability, general recursiveness, and λ-definability he claimed "the first [of these notions] has the advantage of making the identification with effectiveness in the ordinary (not explicitly defined) sense evident immediately. . . ." After all, Church reasoned, "To define effectiveness as computability by an arbitrary machine, subject to restrictions of finiteness, would seem to be an adequate representation of the ordinary notion, ..." The finiteness restrictions require that machines occupy only a finite space and that their working parts have finite size. Turing machines are obtained from such finite machines by further "convenient restrictions," but "these are of such a nature as obviously to cause no loss of generality." Church then observed, completely reversing Turing's sequence of analytic steps, "a human calculator, provided with pencil and paper and explicit instructions, can be regarded as a kind of Turing machine." He was obviously captured by the machine image and saw in it the reason for the deep interest of Turing's computability notion. In sum, we have arrived at three *Church canons* in support of the thesis, namely, (i) the confluence of notions, (ii) the step-by-recursive-step argument, and (iii) the immediate evidence of the adequacy of Turing's notion.

In his reviews Church failed to recognize two crucial aspects of a dramatic shift in perspective. One aspect underlies the work of both Turing and Post, whereas the other is distinctively Turing's. The first aspect becomes visible when Turing and Post, instead of considering schemes for computing the values of number-theoretic functions, look at identical symbolic processes that serve as building blocks for calculations. In order to specify such processes Post uses a human worker who operates in a symbol space and carries out, over a two-letter alphabet, exactly the kind of operations a Turing machine can perform. Post expects that his formulation will turn out to be equivalent to the Gödel–Church development. Given Turing's proof of the equivalence of his computability notion with λ-definability, Post's formulation is indeed equivalent.

Post asserts that "Church's identification of effective calculability with recursiveness" should be viewed as a "working hypothesis" in need of "continual verification." In sharp contrast, Turing attempts to give an analytic argument for the claim that these simple processes are sufficient to capture all human mechanical calculations. Turing exploits for his reductive argument broad constraints that are grounded in limitations of relevant capacities of the human computing agent. This is the second aspect of the novel perspective that made for genuine progress, and it is unique to Turing's work.

2 Computors

It is ironic that Post, when proposing his worker model, at no place used the fact that a human worker does the computing, whereas Turing who seems to emphasize machine computations explicitly examined *human* computations. Call a human computing agent who proceeds mechanically a *computor*; such a computor operates on finite configurations of symbols and, for Turing, deterministically so. The computer hovering about in Turing's paper is such a computor; computers in our contemporary sense are always called machines. Wittgenstein appropriately observed about Turing's machines that *these machines are humans who calculate*.[3] But how do we step from the calculations of computors to computations of Turing machines?

2.1 Preliminary Step

When Turing explores the extent of the computable numbers (or, equivalently, of the effectively calculable functions), he starts out by considering two-dimensional calculations "in a child's arithmetic book." Such calculations are first reduced to computations of *string machines*, and the latter are then shown to be equivalent to computations of a *letter machine*. Letter machines are ordinary Turing machines operating on one letter at a time, whereas string machines operate on finite sequences of letters. In the course of his reductive argument, Turing formulates and uses broadly motivated constraints. The argument concludes as follows: "We may now construct a machine to do the work of the computer [computor in our terminology]. . . . The machines just described [string machines] do not differ very essentially from computing machines as defined in §2 [letter machines], and corresponding to any machine of this type a computing machine can be constructed to compute the same sequence, that is to say the sequence computed by the computer." (Turing 1936, pp. 137–8)

For the presentation of Turing's argument, it is best to consider the description of Turing machines as Post production systems. This is most appropriate for a number of reasons. Post introduced this description in 1947 to establish that the word-problem of certain Thue-systems is unsolvable. Turing adopted it in 1950 when extending Post's results, but also in 1954 when writing a wonderfully informative and informal essay on solvable and unsolvable problems. In addition, this description reflects directly the move in Turing's (1936) to eliminate states of mind for computors[4] in favor of "more physical counterparts." Finally and most importantly, it makes perfectly clear that Turing is dealing with general symbolic processes, whereas the restricted machine model that results from his analysis almost obscures that fact.

---

[3] It is exactly right for Turing to look at human computations given the intellectual context that reaches back to at least Leibniz: the *Entscheidungsproblem* in the title of his (1936) paper asked for a procedure that can be carried out by humans; the restrictive formal conditions on axiomatic theories were imposed in mathematical logic to ensure intersubjectivity for humans on a minimal cognitive basis.

[4] Turing attributes states of mind only to *human computers*; machines have corresponding "m-configurations."

2.2 Boundedness and Locality

The constraints Turing imposes on symbolic processes derive from his central goal of isolating the most basic steps of computations, that is, steps that need not be further subdivided. This objective leads to the normative demand that the configurations, which are directly operated on, must be *immediately recognizable* by the computor. This demand and the evident limitation of the computor's sensory apparatus motivate most convincingly two central restrictive conditions:

(B) (*Boundedness*) A computor can immediately recognize only a bounded number of configurations.

(L) (*Locality*) A computor can change only immediately recognizable configurations.[5]

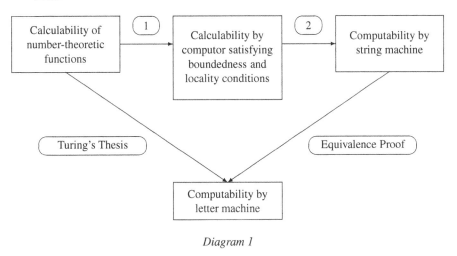

Diagram 1

Turing's considerations leading from operations of a computor on a two-dimensional piece of paper to operations of a letter machine on a linear tape are represented schematically in *Diagram 1*: Step 1 indicates Turing's analysis, whereas 2 refers to Turing's *central thesis* asserting that the calculations of a computor can be carried out by a string machine.

This remarkable progress has been achieved by bringing in, crucially and correctly, the computing agent who carries out the mechanical processes. Yet Turing finds the argument mathematically unsatisfactory as it involves an appeal to *intuition* in support of the central thesis, i.e., the ability of "making spontaneous judgments, which are not the result of conscious trains of reasoning." (Turing 1939, pp. 208–9) What more can be done?

---

[5] The boundedness and locality conditions are violated in Gödel's equational calculus: the replacement operations naturally involve terms of arbitrary complexity. That is, the shift from arithmetic calculations to symbolic processes is absolutely crucial in Turing's analysis.

2.3 Generalizations

At least two kinds of inductive support can be given for the quasi-empirical claim that all known effective procedures are general recursive or Turing computable. Turing provided in his paper one kind, by showing that large classes of numbers are indeed machine computable; Post suggested providing in his (1936) a second kind, by reducing ever-wider formulations of combinatory processes (as production systems) to his worker model.[6] This inductive support can be strengthened further through considering more general symbolic configurations with associated complex substitution operations.[7] In the spirit of this approach we can ask with Post, when have we gathered sufficient support to view the thesis as a *natural law*?

Gödel and Church faced in their analysis of effective calculability the stumbling block of having to define the elementary character of steps, rigorously and without semi-circles. Turing and Post faced at this point, it seems, a problem akin to that of induction. However, their fundamental difficulties are really the same and can be pinpointed more relevantly and quite clearly, as they are related to the looseness of the above restrictive conditions and the corresponding vagueness of the central thesis. These difficulties would be addressed by answering the questions, What are symbolic configurations? What changes can mechanical operations effect? Even without giving rigorous answers, some well-motivated ideas can be formulated for computors: (i) they operate deterministically on finite configurations; (ii) they recognize in each configuration exactly one pattern (from a bounded number of different kinds of such); (iii) they operate locally on the recognized pattern; and (iv) they assemble the next configuration from the original one and the result of the local operation. Exploiting these ideas I will attack the problem with a familiar tool, the axiomatic method.

However, before formulating the axioms for Turing computors, I discuss yet another sense of generalization that is relevant here. Gandy proposed in his (1980) a characterization of *machines* or, more precisely, *discrete mechanical devices*. The latter clause was to exclude analog machines from consideration. The novel aspect of Gandy's proposal was the fact that it incorporated parallelism in perfect generality. Gandy used, as Turing did, a *central thesis*: any discrete mechanical device satisfying some informal restrictive conditions can be represented as a particular kind of dynamical system. Instead, I characterize a *Gandy machine* axiomatically based on the following idea: the machine has to recognize all the patterns contained in a given finite configuration, act on them locally in parallel, and assemble the results of these local computations into the next configuration. As in the case of Turing computors, the configurations are finite, but unbounded; the generalization is simply this: there is no fixed bound on the number of patterns that such configurations may contain.

---

[6] Post of course did provide such reductions in his (1943) whose origins go back to investigations in the very early 1920s; see note 18 of Post's paper.

[7] In (Sieg and Byrnes 1996) that is done for K-graphs and K-graph machines; this is a generalization of the work on algorithms by Kolmogorov and Uspensky.

To help the imagination a bit, the reader should think of the Post-presentation of a Turing machine and the Game of Life as typical examples of a Turing computor and Gandy machine, respectively.

3 Axiomatics

The axioms are formulated for discrete dynamical systems and capture the above general ideas precisely. In the first subsection the broad mathematical setup for the axioms is discussed, whereas the specific principles for *Turing computors* and *Gandy machines* are formulated in the second subsection. The axioms for Turing computors are motivated by the restrictive conditions for human computing, i.e., the limitations of the human sensory apparatus. The axioms for Gandy machines are to capture the characteristic features of finite machines performing parallel computations. The restrictive conditions are in this case motivated by purely physical considerations: the uncertainty principle of quantum mechanics justifies a lower bound on the size of distinguishable "atomic" components, and the theory of special relativity yields an upper bound on signal propagation. Together, these conditions justify boundedness and locality conditions for machines in the very same way sensory limitations do for computors.[8]

3.1 Patterns and Local Operations

We consider pairs $\langle \mathbf{D}, \mathbf{F} \rangle$ where $\mathbf{D}$ is a *class of states* and $\mathbf{F}$ is an *operation* from $\mathbf{D}$ to $\mathbf{D}$ transforming a given state into the next one. States are finite objects and are represented by non-empty hereditarily finite sets over an infinite set of atoms. Such sets reflect states of computing devices just as other mathematical structures represent states of nature. Obviously, any ϵ-isomorphic set can replace a given one in this reflective role, and so we consider *structural classes* $\mathbf{D}$, i.e., classes of states that are closed under ϵ-isomorphisms. What invariance properties should the state transforming operations $\mathbf{F}$ have; i.e., how should the $\mathbf{F}$-images of ϵ-isomorphic states be related? These and other structural issues will be addressed now.

For the general setup we notice that any ϵ-isomorphism between states is an extension of some permutation π on atoms. Letting $\pi(\mathbf{x})$ stand for the result of applying the ϵ-isomorphism determined by a permutation π to the state $\mathbf{x}$, the requirement on $\mathbf{F}$ fixes the dependence of values on just structural features of a set, not the nature of its atoms: $\mathbf{F}(\pi(\mathbf{x}))$ is ϵ-isomorphic to $\pi(\mathbf{F}(\mathbf{x}))$, and this isomorphism must be the identity on the atoms occurring in $\pi(\mathbf{x})$; we say that $\mathbf{F}(\pi(\mathbf{x}))$ and $\pi(\mathbf{F}(\mathbf{x}))$ are *ϵ-isomorphic over* $\pi(\mathbf{x})$ and write $\mathbf{F}(\pi(\mathbf{x})) \cong_{\pi(\mathbf{x})} \pi(\mathbf{F}(\mathbf{x}))$. Note that we do not require $\mathbf{F}(\pi(\mathbf{x})) = \pi(\mathbf{F}(\mathbf{x}))$; that would be far too restrictive as new atoms may

---

[8] I hope the overall structure of the considerations will be clear from this informal presentation; for mathematical details (Gandy 1980) and (Sieg 2002*b*) should be consulted.

expand the state $\mathbf{x}$, and it should not matter which new atoms are chosen. The requirement $\mathbf{F}(\pi(\mathbf{x})) \cong \pi(\mathbf{F}(\mathbf{x}))$, on the other hand, would be too loose, as we want to guarantee the physical persistence of atomic components.

Now we turn to *patterns* and *local* operations. If $\mathbf{x}$ is a given state, regions of the next state are determined *locally* from particular *parts for* $\mathbf{x}$ on which the computor can operate.[9] *Boundedness* requires that there are only finitely many different kinds of such parts; i.e., each part lies in one of a finite number of isomorphism types or, using Gandy's terminology, *stereotypes*. A maximal part $\mathbf{y}$ for $\mathbf{x}$ of a certain stereotype is a *causal neighborhood* for $\mathbf{x}$, briefly $\mathbf{y} \in \mathrm{Cn}(\mathbf{x})$; we call the elements of $\mathrm{Cn}(\mathbf{x})$ also *patterns*. Finally, the local change is effected by a structural operation $\mathbf{G}$ that works on unique causal neighborhoods. The values of $\mathbf{G}$ are in general not exactly what we need in order to assemble the next state, because the configurations may have to be expanded and that expansion involves the addition and coordination of new atoms. To address that issue we introduce *determined regions* $\mathrm{Dr}(\mathbf{z}, \mathbf{x})$ of a state $\mathbf{z}$; they are ϵ-isomorphic to $\mathbf{G}(\mathbf{y})$ for some causal neighborhood $\mathbf{y}$ for $\mathbf{x}$ (and must satisfy a technical condition on the "newness" of atoms).

3.2 Axioms and a Theorem

Recalling the boundedness and locality conditions for computors, we define $\mathbf{M} = \langle \mathbf{S}; \mathbf{T}, \mathbf{G} \rangle$ to be a *Turing Computor on* $\mathbf{S}$, where $\mathbf{S}$ is a structural class, $\mathbf{T}$ a finite set of stereotypes, and $\mathbf{G}$ a structural operation on $\bigcup \mathbf{T}$, if and only if, for every $\mathbf{x} \in \mathbf{S}$, there is a $\mathbf{z} \in \mathbf{S}$, such that

$$(\mathbf{L.0}) : (\exists ! \mathbf{y}) \mathbf{y} \in \mathrm{Cn}(\mathbf{x});$$
$$(\mathbf{L.1}) : (\exists ! \mathbf{v} \in \mathrm{Dr}(\mathbf{z}, \mathbf{x})) \mathbf{v} \cong_{\mathbf{x}} \mathbf{G}(\mathbf{cn}(\mathbf{x}));$$
$$(\mathbf{A.1}) : \mathbf{z} = (\mathbf{x} \setminus \mathrm{Cn}(\mathbf{x})) \cup \mathrm{Dr}(\mathbf{z}, \mathbf{x}).$$

$(\exists ! \mathbf{y})$ is the existential quantifier expressing uniqueness; in $(\mathbf{L.1})$, $\mathbf{cn}(\mathbf{x})$ denotes the unique causal neighborhood guaranteed by $(\mathbf{L.0})$. As in the case of Gandy Machines below, $\mathbf{L}$ abbreviates locality and $\mathbf{A}$ stands for assembly. The state $\mathbf{z}$ is determined uniquely up to ϵ-isomorphism over $\mathbf{x}$. An $\mathbf{M}$-computation is a finite sequence of transition steps involving $\mathbf{G}$ that is halted when the operation on state $\mathbf{z}$ yields $\mathbf{z}$ as the next state. A function $\mathbf{F}$ is (Turing) *computable* if and only if there is a Turing computor $\mathbf{M}$ whose computation results determine, under a suitable encoding and decoding, the values of $\mathbf{F}$ for any of its arguments. A Turing machine is easily seen to be a Turing computor.

---

[9] A part $\mathbf{y}$ for $\mathbf{x}$ used to be in my earlier presentations a connected subtree $\mathbf{y}$ of the $\in$-tree for $\mathbf{x}$, briefly $\mathbf{y} <^* \mathbf{x}$, if $\mathbf{y} \neq \mathbf{x}$ and $\mathbf{y}$ has the same root as $\mathbf{x}$ and its leaves are also leaves of $\mathbf{x}$. More precisely, $\mathbf{y} \neq \mathbf{x}$ and $\mathbf{y}$ is a non-empty subset of $\{\mathbf{v} \mid (\exists \mathbf{z})(\mathbf{v} <^* \mathbf{z} \ \& \ \mathbf{z} \in \mathbf{x})\} \cup \{\mathbf{r} \mid \mathbf{r} \in \mathbf{x}\}$. Now it is just a subset, but I will continue to use the term "part" to emphasize that we are taking the whole $\in$-structure into account.

Generalizing these considerations to graph machines, for example, one notices quickly complications. When several new atoms are being introduced in the image of some causal neighborhood as well as in the next state, the new atoms have to be structurally coordinated; cf. (Sieg and Byrnes 1996). This issue is clearly even more pressing, when parallel computations are carried out. There the coordination can be achieved by a second local operation and a second set of stereotypes. Causal neighborhoods of type 1 are parts of larger neighborhoods of type 2, and the overlapping determined regions of type 1 must be parts of determined regions of type 2, so that they fit together appropriately. (Determined regions "overlap" if the intersection of their sets of new atoms is non-empty.)

For machines that carry out parallel computations we consequently need in addition to the finitely many stereotypes and the structural operation working on them a second set of stereotypes together with a second structural operation, which allow the machine to assemble the determined regions. This is reflected by separating the Assembly principle for Gandy machines into two kinds, where the principle of the first kind captures the idea expressed at the end of the last paragraph; the principle of the second kind is a more general form of the A-principle for Turing computors. Finally, we can define the central concept here: $\mathbf{M} = \langle \mathbf{S}; \mathbf{T}_1, \mathbf{G}_1, \mathbf{T}_2, \mathbf{G}_2 \rangle$ is a *Gandy machine on* $\mathbf{S}$, where $\mathbf{S}$ is a structural class, $\mathbf{T}_i$ a finite set of stereotypes, $\mathbf{G}_i$ structural operation on $\bigcup \mathbf{T}_i$, if and only if, for every $\mathbf{x} \in \mathbf{S}$ there is a $\mathbf{z} \in \mathbf{S}$, such that

$(\mathbf{L.1}) : (\forall \mathbf{y} \in \mathrm{Cn}_1(\mathbf{x}))(\exists! \mathbf{v} \in \mathrm{Dr}_1(\mathbf{z}, \mathbf{x}))\mathbf{v} \cong_{\mathbf{x}} \mathbf{G}_1(\mathbf{y});$

$(\mathbf{L.2}) : (\forall \mathbf{y} \in \mathrm{Cn}_2(\mathbf{x}))(\exists \mathbf{v} \in \mathrm{Dr}_2(\mathbf{z}, \mathbf{x}))\mathbf{v} \cong_{\mathbf{x}} \mathbf{G}_2(\mathbf{y});$

$(\mathbf{A.1}) : (\forall \mathbf{C})[\mathbf{C} \subseteq \mathrm{Dr}_1(\mathbf{z}, \mathbf{x}) \& \bigcap\{\mathrm{Sup}(\mathbf{v}) \cap A(\mathbf{z}, \mathbf{x}) \mid \mathbf{v} \in \mathbf{C}\} \neq \emptyset \rightarrow$
$(\exists \mathbf{w} \in \mathrm{Dr}_2(\mathbf{z}, \mathbf{x}))(\forall \mathbf{v} \in \mathbf{C})\mathbf{v} <^* \mathbf{w}];$

$(\mathbf{A.2}) : \mathbf{z} = \bigcup \mathrm{Dr}_1(\mathbf{z}, \mathbf{x}).$

$A(\mathbf{z}, \mathbf{x}) = \mathrm{Sup}(\mathbf{z}) \setminus \mathrm{Sup}(\mathbf{x});$ i.e., it consists of the new atoms that have been introduced into $\mathbf{z}$. Thus, the condition $\bigcap\{\mathrm{Sup}(\mathbf{v}) \cap A(\mathbf{z}, \mathbf{x}) \mid \mathbf{v} \in \mathbf{C}\} \neq \emptyset$ in $(\mathbf{A.1})$ expresses that the determined regions $\mathbf{v}$ in $\mathbf{C}$ have common new atoms; i.e., they overlap. The restrictions for Gandy machines, as those for Turing computors, amount to boundedness and locality conditions. They are justified *directly* by two physical bounds, namely, a lower bound on the size of atoms and an upper bound on the speed of signal propagation. On account of these bounds only boundedly many different configurations can be physically realized (within a unit time interval); cf. (Mundici and Sieg 1995). With these remarks I actually completed the foundational analysis, and I can describe now some important mathematical facts for Gandy machines.

The central facts are as follows: (i) the state $\mathbf{z}$ following $\mathbf{x}$ is determined uniquely up to ϵ-isomorphism over $\mathbf{x}$, and (ii) Turing machines can effect such transitions. The proof of the first fact contains the combinatorial heart of matters and uses crucially the first assembly condition. The proof of the second fact is rather direct. Only finitely many finite objects are involved in the transition, and all the axiomatic conditions are decidable. Thus, a search will allow us to find **z**. This can be understood as a Representation Theorem: any particular Gandy machine is computationally reducible to a two-letter Turing machine. Conversely, any Turing machine is a Gandy machine. Indeed, there is a rich variety of additional models, as the game of life, other cellular automata, and many artificial neural nets are Gandy machines. (cf. DePisapia 2000)

4 Adequacy and Philosophical Errors

So what? What have we gained? In very broad terms, taken from Hilbert, we have gained *eine Tieferlegung der Fundamente* (a deepening of the foundations) via the axiomatic method. In a conversation with Church in early 1934, Gödel found Church's proposal to identify effective calculability with λ-definability "thoroughly unsatisfactory." As a counter-proposal he suggested "to state a set of axioms which would embody the generally accepted properties of this notion [i.e., effective calculability], and to do something on that basis." Perhaps, the remarks in the 1964 Postscriptum to the Princeton Lectures of 1934 echo those earlier considerations. "Turing's work gives," according to Gödel, "an analysis of the concept of 'mechanical procedure' .... This concept is shown to be equivalent with that of a 'Turing machine'." Gödel did neither elucidate these remarks nor did he articulate what the generally accepted properties of effective calculability might be or what might be done on the basis of an appropriate set of axioms.

The work on which I reported substantiates Gödel's remarks in the following sense: it formulates axioms for the concept "mechanical procedure," and it shows that this axiomatically characterized concept is indeed equivalent to that of a Turing machine. As a matter of fact it does so for two such concepts, namely, when the computing agents are computors, respectively discrete machines. These considerations use only "generally accepted properties" of the informal concepts and avoid any appeal to theses, whether central or not. As to the correctness of the underlying analyses, an appeal to some understanding can no more be avoided in this case than in any other case of an axiomatically characterized (class of) mathematical structure(s) intended to mirror broad aspects of physical or intellectual reality. The general point is as follows: we do not have to face anything mysterious surrounding the concept of calculability; rather, we have to face the ordinary issues for the adequacy of mathematical concepts, and these are of course non-trivial.[10] From a slightly different and complementary perspective, the function of the axiom systems for computing devices can be seen as being similar to that of the axiom systems for the classical algebraic structures like groups, rings, or fields, namely, to abstract the essential aspects from a wide variety of instances and point to deep structural analogies. They explain here,

---

[10] Other examples of such analyses are provided by Dedekind's work on continuous domains (the reals) and simply infinite systems (natural numbers).

by way of the representation theorem, the computational reducibility of their models to Turing machines.

In the central case under discussion, Turing computability, its adequacy is still fraught with controversy and often misunderstanding. The controversy begins with the very question of what is the intended informal concept. For example, Gödel spotted in 1972 a "philosophical error" in Turing's work, *assuming* that Turing's argument in the 1936 paper was to show that "mental procedures cannot go beyond mechanical procedures." He considered the argument as inconclusive. Indeed, Turing does not give a conclusive argument for Gödel's claim, but it has to be added that he did not intend to argue for it. Even in his work of the late 1940s and early 1950s that deals explicitly with mental processes, Turing does not argue, "mental procedures cannot go beyond mechanical procedures."

Mechanical processes are, in this later work, still made precise as Turing machine computations; machines that might exhibit intelligence have in contrast a more complex structure than Turing machines. Conceptual idealization and empirical adequacy are being sought for quite different purposes, and Turing is trying to capture clearly what Gödel found missing in the would-be analysis of a broad concept of humanly effective calculability, namely, ". . . that mind, in its use, is not static, but constantly developing." The real difference between Turing's and Gödel's views, it seems, is Gödel's belief that it is "a prejudice of our time" that "[t]here is no mind separate from matter." This is reported by Wang. Gödel expected, also according to Wang, that this prejudice "will be disproved scientifically (perhaps by the fact that there aren't enough nerve cells to perform the observable operations of the mind)." Clearly, Turing did not share these expectations.

Many fascinating issues concerning physical and mental processes may or may not have adequate computational models. They are empirical, conceptual, mathematical ... well, indeed, richly interdisciplinary. Steps toward their clarification or resolution will be most illuminating. Why, let me ask, are we interested so deeply in computations?—One answer is, we want to determine states from other states, be they mathematical, physical, or mental; and we want to do that effectively and in a sharply intersubjective way that makes use of adequate symbolic representations.

References

[1936] Church, A. (1936) An unsolvable problem of elementary number theory; American Journal of Mathematics 58, 345–363; reprinted in (Davis 1965).
[1937] — (1937) Review of (Turing 1936); The Journal of Symbolic Logic 2, 40–41.
[1965] Davis, M., (ed.) (1965) *The Undecidable*, Basic papers on undecidable propositions, unsolvable problems and computable functions; Raven Press, Hewlett, New York.
[2000] De Pisapia, N. (2000) *Gandy Machines: an abstract model of parallel computation for Turing Machines, the Game of Life, and Artificial Neural Networks*; M.S. Thesis, Carnegie Mellon University, Pittsburgh.

[1980] Gandy, R. (1980) Church's Thesis and principles for mechanisms; in: *The Kleene Symposium* (edited by J. Barwise, H. J. Keisler and K. Kunen, North-Holland, 123–148.

[1934] Gödel, K. (1934) On undecidable propositions of formal mathematical systems; in: *Collected Works I*, 346–369.

[1936] — (1936) Über die Länge von Beweisen; in: *Collected Works I*, 396–399.

[1946] — (1946) Remarks before the Princeton bicentennial conference on problems in mathematics; in: *Collected Works II*, 150–153.

[1986–2003] — (1986–2003) *Collected Works*, volumes I–V; Oxford University Press.

[1939] Hilbert, D. and P. Bernays (1939) *Die Grundlagen der Mathematik II*; Springer Verlag, Berlin.

[1958] Kolmogorov, A. N and V. A. Uspensky (1958) On the definition of an algorithm; Uspekhi Mat. Nauk 13 (Russian), 1958; English translation in: *AMS Translations*, 2, 21 (1963), 217–245.

[1995] Mundici, D. and W. Sieg (1995) Paper Machines; Philosophia Mathematica 3, 5–30.

[1936] Post, E. (1936) Finite combinatory processes. Formulation I; Journal of Symbolic Logic 1, 103–105.

[1943] — (1943) Formal reductions of the general combinatorial decision problem; American Journal of Mathematics, 65 (2), 197–215.

[1947] — (1947) Recursive unsolvability of a problem of Thue; The Journal of Symbolic Logic 12, 1–11.

[1994] Sieg, W. (1994) Mechanical procedures and mathematical experience; in: *Mathematics and Mind* (A. George, ed.), Oxford University Press, 71–117.

[1997] — (1997) Step by recursive step: Church's analysis of effective calculability; The Bulletin of Symbolic Logic 3 (2), 154–180.

[2002a] — (2002a) Calculations by man and machine: conceptual analysis; Lecture Notes in Logic 15, 390–409.

[2002b] — (2002b) Calculations by man and machine: mathematical presentation; in: *In the Scope of Logic, Methodology and Philosophy of Science*, volume one of the 11th International Congress of Logic, Methodology and Philosophy of Science, Cracow, August 1999 (P. Gärdenfors, J. Wolenski and K. Kijania-Placek, eds.), Synthese Library volume 315, Kluwer, 247–262.

[1996] Sieg, W. and J. Byrnes (1996) K-Graph machines: generalizing Turing's machines and arguments; in: *Gödel '96* (P. Hajek, ed.), Lecture Notes in Logic 6, Springer Verlag, 98–119.

[1936] Turing, A. (1936) On computable numbers, with an application to the *Entscheidungsproblem*; Proceedings of the London Mathematical Society (Series 2) 42, 230–265.

[1939] — (1939) Systems of logic based on ordinals; Proceedings of the London Mathematical Society (Series 2) 45, 161–228; reprinted in (Davis 1965).

[1950] — (1950) The word problem in semi-groups with cancellation; Ann. of Math. 52, 491–505.

[1954] — (1954) Solvable and unsolvable problems; Science News 31, 7–23; reprinted in *Collected Works of A. M. Turing: Mechanical intelligence*, (D. C. Ince, ed.), North-Holland, 1992.

Computability on Topological Spaces via Domain Representations

Viggo Stoltenberg-Hansen[1] and John V. Tucker[2]

[1] Department of Mathematics, Uppsala University, S-75106 Uppsala, Sweden
`viggo@math.uu.se`
[2] Department of Computer Science, University of Wales Swansea, SA2 8PP, Wales
`J.V.Tucker@swansea.ac.uk`

Summary. Domains are ordered structures designed to model computation with approximations. We give an introduction to the theory of computability for topological spaces based on representing topological spaces and algebras using domains. Among the topics covered are different approaches to computability on topological spaces; orderings, approximations, and domains; making domain representations; effective domains; classifying representations; type two effectivity and domains; and special representations for inverse limits, regular spaces, and metric spaces. Lastly, we sketch a variety of applications of the theory in algebra, calculus, graphics, and hardware.

1 Introduction

The theory of topological spaces and continuous functions is about the approximation of data and the functions that preserve those approximations. Approximation is expressed by open subsets of the set of data. The primary intuitions are geometric, and its original applications were in geometry, differential equations, and functional analysis, where data are made from real and complex numbers, functions, and operators. A century of research has made topology essential to mathematics and physical science (see Aull and Lowen [2, 3] and James [39]). The question arises:

> How does one compute with data such as real and complex numbers, functions, and operators? More generally, how does one compute with data from a topological space?

Understanding computation on topological spaces is important. For example, it is needed to improve practical computation with continuous data; to compare and unify digital and analog computation; to explore computation in analysis and geometry; and to establish the computational and logical nature of physical systems—to name just four research problems of contemporary interest.

There are several answers to the computability question above: some methods are specific to the real numbers, and some are general for a class of topological spaces. Here we will explain one answer:

> *Represent topological spaces of data by domains, and reduce computation on those spaces to computation on domains.*

We will summarise other approaches to computability shortly (Section 2).

The theory of domains and order-preserving functions is also about the approximation of data and the functions that preserve approximations. Approximation is expressed by an ordering on the set of data. The primary intuitions are computational, and its primary applications are in computability theory and the semantics of programming languages and logics, where computations are defined using recursion equations on functions, memory states and environments, data, processes, formulae, and types. The foundations of the subject were laid by D. S. Scott [60, 61, 62] and Yu. L. Ershov [32, 33].

Domains are ordered algebraic structures containing both approximations and the data they approximate. The ordering $\sqsubseteq$ on a domain D formulates the idea that for $a, b \in D$,

$$a \sqsubseteq b \iff \text{'datum } b \text{ is a better approximation than datum } a\text{'}.$$

The limits of sequences of such approximations are the data to be approximated. Computations are modelled as a process of finding better and better approximations.

Domains are designed to solve equations. Their orderings are used to capture some of the features of using iterative algorithms to approximate solutions. The equations are formulated as fixed point equations; i.e., for a given function $f \colon D \to D$, find a in D such that $f(a) = a$. The fixed point methods build the solutions from their approximations. The inspiration of these essential features of domains and equation solving are the *complete partially ordered set* (cpo) and the equation solving methods of the *Tarski-Knaster Fixed Point Theorem*, proved in 1927; (see, e.g., Tarski [77]). These methods found their way into computability theory via theorems such as Kleene's recursion theorems. The methods were applied on particular cpos of functions on natural numbers to explain recursion. Through the theory of domains and domain representations, the wide applicability of fixed point methods to computational problems became evident.

The theory of domain representations of topological spaces is a general theory about how to:

(i) represent topological spaces using domains;

(ii) analyse computation on spaces via their representations;

(iii) compare and classify different domain representations;

(iv) compute the solutions of equations on spaces; and

(v) make applications.

In this chapter we will introduce these topics, sketch their development, and point out connections with other theories that answer the question posed above. Many kinds of domains have been discovered; we will focus on so-called algebraic domains that we consider to be the most simple and useful for computability.

The structure of the chapter is as follows. In Section 2 we will summarise the approaches to computability and sketch their origins. In Section 3 we introduce the idea of using orderings to formulate basic ideas about approximations. This leads directly to the concept of an algebraic domain. In Section 4 we introduce the continuous functions on domains. In Section 5 we define domain representations for spaces that are the structures within which computations take place. In Section 6 we add algorithms and define what is actually computable on the approximations that make up the domain. In Section 7 we introduce some simple types of domain representations (retract, dense, etc.), and we use reductions between domains that allow us to compare representations of topological spaces and discuss the stablility or invariance of computational properties of the representations. In Section 8 we examine a special form of algebraic domain representation that we derive from K. Weihrauch's approach to computability on spaces called Type 2 Theory of Effectivity (TTE): see Weihrauch [84]. In Section 9 we look at some standard constructions of domain representations, including metric spaces. In Section 10 we sketch some applications of the theory to studies of computation on different spaces, including real numbers, local rings, Banach spaces, process algebras, distributions, etc.

We thank Jens Blanck and Fredrik Dahlgren for useful comments on this article.

2 Computability on topological spaces: some principles, approaches, and history

To compute in a topological space we will choose some representation of the space, made from a domain, and compute on the domain representation. There are other ways to compute on spaces, not all of which are equivalent, and so before examining domain representations we will view the wider technical landscape.

2.1 Principles: Concrete versus abstract computability

By a *computability theory* we mean a theory of functions and sets that are definable using a model of computation. By a *model of computation* we mean some general method of calculating the value of a function or of deciding, or enumerating, the elements of a set. The functions and sets can be made from any kind of data.

With this terminology, Classical Computability Theory on the set $\mathbb{N}$ of natural numbers is made up of many computability theories, derived from different ideas about algorithms. The fact that different computability theories lead to equivalent theories of functions and sets on $\mathbb{N}$ gives the classical theory on $\mathbb{N}$ its unity, which was epitomised by the Church–Turing Thesis and was an early discovery.

Since the 1940s, computability theories have been created for other sets of data, including higher types over the natural numbers, real numbers, and spaces of real-valued functions. More generally, computability theories have been created for classes of structures, such as groups, rings, fields, and topological and metric spaces. However, the classification and the proofs of equivalences of models of computation—and, hence, the search for generalised Church–Turing Theses and the theoretical unity they represent—have proved much more difficult to achieve for these data types. An early example of different models of computation that are of equal conceptual value but are known not to be equivalent is provided by Tait's Theorem in higher types: the fan functional on total functions is recursively continuous but not computable in Kleene's schemes S1–S9; see Normann [51].

Some general insight into the phenomenon of inequivalent computability theories is to be found by examining treatment of data in models of computation. Computability theories can be classified into two types by introducing the following concepts.

Definition 2.1 In an *abstract computability theory* the computations are independent of all the representations of the data. Computations are uniform over all representations and are isomorphism invariants.

In a *concrete computability theory*, the computations are dependent on some representation of the data. Computations are not uniform, and different representations can yield different results. Computations are not isomorphism invariants.

Models of computation that are based on abstract ideas of program, equation, scheme, or logical formula are typical of abstract models. Models of computation that are based on concrete ideas of coding, numbering, or representing data using any other kind of data are typical of concrete models. Now, the distinction of abstract versus concrete is helpful in comparing models of computation. There is a need for both abstract and concrete models and an understanding of their relationship.

Clearly, within the concrete, there is great scope for variations in models of computation and we may expect different representations to lead to different computability theories. Abstract models, too, can vary, since the choice of operations, program constructs, and kinds of formulae can vary. Can there be concrete models that are sufficiently canonical to be equivalent to an abstract model?

A full general discussion of the distinction is given in Tucker and Zucker [81], motivated by their theory of abstract computation on topological spaces (see Tucker and Zucker [78, 79, 80]). The distinction is also directly relevant to the seemingly stable and unified classical world of computability on countable algebras, as pointed out in Tucker and Zucker [81]).

The theory of computable sets and functions is based on data that may be represented by finite discrete symbols. For Turing's analysis of human computation, the symbols came from the set $\mathbb{B} = \{0, 1\}$, or for Kleene's theory of recursive functions, the symbols were from $\mathbb{N}$. For Gödel's computations on syntax the symbols came from the set $\mathbb{N}$ of natural numbers via Gödel numberings. The early development of computability theory did not interest itself in ideas about data and how it was represented. What, after all, was worth saying about $\mathbb{B}$ and $\mathbb{N}$ other than they are *so* fundamental and an obvious place to start? In the 1950s computability theory was extended by advanced applications in logic, algebra, and analysis. Studying computability now required an interest in nature of data and how it was represented because what was computable depended upon the data. In algebra, rings and fields had to be considered as structures unique up to isomorphism, not just as specific representations. In Fröhlich and Shepherdson [36] we see great attention to representations and their equivalence. In the Mal'cev-Ershov theory of numberings of countable sets and structures [34, 35, 47], representations are studied in depth: a numbering $\alpha \colon \mathbb{N} \to A$ makes explicit the idea that one chooses a numerical representation of the data in A and computes on $\mathbb{N}$. The theory of numberings plays a role in the development of domain representations.

2.2 Computability theories for topological spaces

Most computability theories for topological spaces are developed using concrete models of computation. The study of computability on the reals began with Turing in 1936, but only later was it taken up in a systematic way, e.g., in Rice [57], Lacombe [45], and Grzegorczyk [37]. For example, to compute on the set $\mathbb{R}$ of real numbers with a concrete model of computation, we choose an appropriate concrete representation of the set $\mathbb{R}$, such as computable Cauchy sequences.

In the case of concrete computability, there have been a several general approaches to the analysis and classification of metric and topological structures since the 1950s:

(i) Effective metric spaces (Ceitin [17], Moschovakis [50]);

(ii) Computable sequence structures for Banach spaces (Pour El and Richards [56]);

(iii) Type 2 Theory of Effectivity or TTE (Weihrauch [83, 84]);

(iv) Algebraic domain representations (Stoltenberg-Hansen and Tucker [69, 70, 73]);

(v) Continuous domain representations (Edalat [23, 24, 25]);

(vi) Numbered spaces (Spreen [64, 65, 66, 67]).

Computable analysis has been greatly extended over the past decade using these models, which have been seen as competing. This has made the exciting rapid growth of the subject seem messy. In fact, for certain basic topological algebras, most of these concrete computability theories have been shown to be essentially equivalent in Stoltenberg-Hansen and Tucker [75].

We should say a word about the abstract approach. Analysis makes heavy use of algebraic structures, such as topological groups and vector spaces, Banach spaces, Hilbert spaces, C* algebras, and many more. These many sorted topological algebras specify (i) some basic continuous operations; (ii) normal forms for the algebraic representation of elements (e.g., using bases); (iii) structure-preserving operators (i.e., homomorphisms such as linear operators); and (iv) approximations, through inner products, norms, metrics, and topologies.

Abstract computability theories are created by simply applying the abstract models to these algebras. These models can be defined by programming languages whose programs are based on the operations of the algebras. However, thanks to approximation (iv), we obtain two classes of functions: the *computable functions* and the *computably approximable functions*.

A full account of the theory on general metric algebras, together with a detailed discussion of the bridge between abstract and concrete models, can be found in Tucker and Zucker [80, 81]. The most publicised abstract computability theory for $\mathbb{R}$ is that developed in Blum, Cucker, Shub, and Smale [16], but it is a theory that does not fit the concrete models because of its use of noneffective operations such as $=$.

2.3 Domain representation theory

The idea of representing topological spaces and algebras using effective domains was, as far as we know, first made explicit in a widely circulated report from Stoltenberg-Hansen and Tucker [69], which was later published as [70]. In this report a general methodology was described for topological algebras and applied to study the effective content of the completion of a computable Noetherian local ring. It was extended further to ultrametric spaces and locally compact regular spaces in [71, 72, 73] and to metric spaces in the thesis [8]. We will meet these constructions in Section 9.

A precursor to some of the central ideas of domain representability is Weihrauch and Schreiber [85], where embeddings of metric spaces into complete partial orders equipped with weight and distance are considered.

It was clear from the beginning of the development of domain theory that, in addition to the ease of building type structures, it is a theory of approximation and computation, and that computability often implies continuity. This was exploited in [33] where Ershov gave a domain representation of the Kleene–Kreisel continuous functionals. An effective and adequate domain model of Martin-Löf partial type theory is given in Palmgren and Stoltenberg-Hansen [55], which has been extended in Waagbø [82] to provide a domain representation of Martin-Löf total type theory (see also Berger [6] and Normann [52]).

3 Approximations, orderings, and domains

Suppose we want to compute on a possibly uncountable structure such as the field of real numbers $\mathbb{R}$. The elements of $\mathbb{R}$ are in general truly infinite objects (Cauchy sequences or Dedekind cuts) with no finite description. However, computations that can be performed by a digital computer or Turing machine must operate on 'finite' objects. By a finite object we mean that it is finitely describable or, equivalently, coded by a natural number. In particular, the structure on which computations are to be performed must be countable. Therefore it is not possible to compute directly on $\mathbb{R}$; we can at best compute on finite approximations of elements in $\mathbb{R}$. If the approximations are such that each real number is the limit of its approximations, then we can extend a computation on approximations to $\mathbb{R}$ by interpreting a computation on a real number as the 'limit' of the computations on its approximations, where such a limit exists. It follows, intuitively, that computations are continuous processes.

In this section we show that a simple analysis of the notion of approximation leads naturally to the class of *algebraic cpos*.

3.1 Approximations and orderings

Let us consider the problem of approximation abstractly. Suppose that X is a set, or, more generally, a structure. To say that a set P is an approximation for X should mean that elements of P are approximations for elements of X. That is, there is a relation $\prec$, the *approximation relation*, from P to X with the intended meaning for $p \in P$ and $x \in X$,

$$p \prec x \iff \text{``}p \text{ approximates } x\text{''}.$$

We illustrate this with a few relevant examples.

Example 3.1 Let $P = \{[a,b] : a \le b, a, b \in \mathbb{Q}\}$ and $X = \mathbb{R}$. Define

$$[a,b] \prec x \iff x \in [a,b].$$

Note that P is countable and consists of finite elements in the sense that an interval $[a,b]$ is finitely describable from finite descriptions of the rational numbers a and b and the symbols "[", "]" and ",". Furthermore, each $x \in \mathbb{R}$ is the 'limit' (intersection) of its approximations.

Example 3.2 Let $P = \mathbb{Q}$ and $X = \mathbb{R}$. For $a \in \mathbb{Q}$ and $x \in \mathbb{R}$, define

$$a \prec x \iff a < x,$$

where $<$ is the usual order on $\mathbb{R}$.

Note that Example 3.1 provides a better approximation of $\mathbb{R}$ than Example 3.2 in that $[a, b] \prec x$ gives more information than $a \prec x$.

Example 3.3 Let X be a topological space with a topological base $\mathcal{B}$. For $B \in \mathcal{B}$ and $x \in X$, define

$$B \prec x \Longleftrightarrow x \in B.$$

Let P and X be sets, and let $\prec$ be a relation from P to X. Then $\prec$ induces in a natural way a relation $\sqsubseteq$ on P, the *refinement (pre-)order* obtained from or induced by $\prec$: for $p, q \in P$, let

$$p \sqsubseteq q \Longleftrightarrow (\forall x \in X)(q \prec x \Longrightarrow p \prec x).$$

Thus $p \sqsubseteq q$ expresses that q *is a better approximation than* p, or q *refines* p, in the sense that q approximates fewer elements in X than does p. Note that the induced refinement order indeed is a *preorder*; i.e., it is reflexive and transitive.

We now put some reasonable requirements on P and $\prec$ in order to obtain an *approximation structure* for X. We require that

- each element $x \in X$ is determined uniquely by its approximations, and

- each element $x \in X$ is the 'limit' of its approximations.

In addition, for domain theoretic reasons guaranteeing the existence of fixed points, it is useful to require P to have a trivial approximation, i.e., an approximation that approximates all elements of X (and hence contains no information about elements of X). This leads us to

Definition 3.4 Let P and X be sets, $\prec$ a relation from P to X, and $\sqsubseteq$ the refinement preorder obtained from $\prec$. Then $(P, \sqsubseteq)$ is an *approximation structure* for X with respect to $\prec$ if

(i) $\{p \in P : p \prec x\} = \{p \in P : p \prec y\} \implies x = y$ *(uniqueness)*;

(ii) $p \prec x$ and $q \prec x \implies (\exists r \prec x)(p \sqsubseteq r$ and $q \sqsubseteq r)$ *(refinement)*;

(iii) $(\exists p \in P)(\forall x \in X)(p \prec x)$ *(trivial approximation)*.

Examples 3.1 and 3.2 are approximation structures when we add a trivial approximation. Example 3.3 gives an approximation structure precisely when the space X is T_0. In this sense, (i) in Definition 3.4 is a T_0 property.

3.2 Ideals and domains

Let $(P, \sqsubseteq)$ be an approximation structure for X with respect to $\prec$. Then each $x \in X$ is identified uniquely with the set $\{p \in P : p \prec x\}$. Note that if $p \sqsubseteq q \prec x$, then $p \prec x$. Together with (ii) and (iii) in Definition 3.4 we see that $\{p \in P : p \prec x\}$ is an *ideal* over $(P, \sqsubseteq)$. In particular, it is a (canonical) net 'converging' to x.

Let us recall the definitions. For a preorder $P = (P, \sqsubseteq)$, a set $A \subseteq P$ is said to be *directed* if A is non-empty, and if $p, q \in A$, then there is an $r \in A$ such that $p, q \sqsubseteq r$; i.e., every finite subset of A has an upper bound in A. A subset $I \subseteq P$ is an *ideal* over P if I is directed and whenever $x \sqsubseteq y$ and $y \in I$ then also $x \in I$; that is, I is *downwards closed*.

We often use the notation $\downarrow p = \{q \in P : q \sqsubseteq p\}$ and $\uparrow p = \{q \in P : p \sqsubseteq q\}$. Note that $\downarrow p$ is an ideal, the *principal ideal* generated by p. We denote by $\mathrm{Idl}(P, \sqsubseteq)$, or just $\mathrm{Idl}(P)$, the set of all ideals over $(P, \sqsubseteq)$.

Given an approximation structure $(P, \sqsubseteq)$ of X with respect to $\prec$, we obtain an injection of X into $\mathrm{Idl}(P)$; i.e., X 'lives' in $\mathrm{Idl}(P)$. In addition, $\mathrm{Idl}(P)$ contains the approximations P that we started with by means of the principal ideals $\downarrow p$. Thus

> $\mathrm{Idl}(P)$ *is a structure that contains both the original space* and *its approximations.*

$\mathrm{Idl}(P)$ is naturally ordered by inclusion $\subseteq$. For if ideals $I \subseteq J$, then J contains more approximations and hence more information about the elements approximated than does I. We consider $\mathrm{Idl}(P)$ as a structure ordered by inclusion.

Definition 3.5 Let $P = (P, \sqsubseteq)$ be a preorder. The *ideal completion* of P is the structure $\bar{P} = (\mathrm{Idl}(P), \subseteq)$.

It is verified easily that $\bar{P}$ is an *algebraic cpo* where the *compact elements* are precisely the principal ideals. We recall the definitions.

Let $D = (D, \sqsubseteq, \bot)$ be a partially ordered set with least element $\bot$. Then D is a *complete partial order* (abbreviated *cpo*) if whenever $A \subseteq D$ is directed, then $\bigsqcup A$ (the least upper bound or supremum of A) exists in D. An element $a \in D$ is said to be *compact* or *finite* if whenever $A \subseteq D$ is a directed set and $a \sqsubseteq \bigsqcup A$, then there is $x \in A$ such that $a \sqsubseteq x$. The set of compact elements in D is denoted by D_c. A cpo D is *algebraic* if for each $x \in D$, the set

$$\mathrm{approx}(x) = \{a \in D_c : a \sqsubseteq x\}$$

is directed and $x = \bigsqcup \mathrm{approx(x)}$.

Algebraic cpos have the following representation theorem. For its simple proof, see Stoltenberg-Hansen, Lindström, and Griffor [68].

Theorem 3.6 *Let* $D = (D, \sqsubseteq, \bot)$ *be an algebraic cpo, and let* $\overline{D_c}$ *be the ideal completion of* $D_c = (D_c, \sqsubseteq)$. *Then* $D \simeq \overline{D_c}$.

Note that if D is an algebraic cpo, then $(D_c, \sqsubseteq)$ is an approximation structure for D with respect to $\prec$, where for $a \in D_c$ and $x \in D$,

$$a \prec x \iff a \sqsubseteq x.$$

Thus we have

Corollary 3.7 *Algebraic cpos are precisely the ideal completions of approximation structures.*

Algebraic cpos are determined completely by their sets of compact elements. Also continuous functions between algebraic cpos are determined completely by their action on compact elements. Therefore, as we shall see in Section 6, algebraic cpos carry a natural theory of effectivity by computing on the set of compact elements, and a large subclass of them is effectively closed under various constructions, including the function space construction.

We say that an algebraic cpo D is κ-*based* if the cardinality of D_c equals κ, where κ is a cardinal. D is *countably based* if D_c is countable. When considering effective algebraic cpos, we are thus restricted to countably based algebraic cpos.

3.3 Methodology of domain representability

Assume the task is to study computability on a set or structure X. We find a suitable set P of approximations and then form the ideal completion $\bar{P}$ of the induced approximation structure. Then $\bar{P}$ contains both the structure X and the set of approximations for X. Furthermore, the effectivity of $\bar{P}$ and hence of X is determined completely by the computability of the set P of approximations. Now we use the general theory of domains to study the structure X, including

- fixed point theorems to compute solutions to equations;

- ease in building higher type objects (e.g., streams and stream transformers, see [14], and higher type operations such as integrals and distributions);

- computability, inherited from the computability of P.

Our claim is that the use of domains (of various kinds) provides a general, uniform, and useful way to study computability via approximations on a large class of structures.

4 Continuous functions and algebraic domains

Since computations are based on approximations, an approximation of the value of a computable function should depend only on an approximation of its argument. This property gives rise to a notion of continuity.

Let D and E be cpos. Then $f \colon D \to E$ is *continuous* if f is monotone, and for each directed set $A \subseteq D$, $f(\bigsqcup A) = \bigsqcup f[A]$. Thus f is continuous if it preserves information and, regarding $\bigsqcup A$ as the limit of the 'net' A, preserves limits. In case D and E are algebraic, then f is continuous if, and only if, f is monotone and for each $x \in D$,

$$(\forall b \in \text{approx}(f(x)))(\exists a \in \text{approx}(x))(b \sqsubseteq f(a)).$$

This says that for each concrete approximation b of $f(x)$ there is a concrete approximation a of x such that f applied to a 'computes' at least as much information as b.

The topology corresponding to this notion of continuity is the *Scott topology*. For an algebraic cpo, it is generated by the topological base $\{\uparrow a : a \in D_c\}$.

For cpos D and E, we define the *function space* $[D \to E]$ of D and E by

$$[D \to E] = \{f \colon D \to E \mid f \text{ is continuous}\},$$

and we give $[D \to E]$ the pointwise ordering:

$$f \sqsubseteq g \iff (\forall x \in D)(f(x) \sqsubseteq g(x)).$$

It is easy to see that $[D \to E]$ is a cpo where for a directed set $\mathcal{F} \subseteq [D \to E]$ and $x \in D$,

$$\left(\bigsqcup \mathcal{F}\right)(x) = \bigsqcup \{f(x) \colon f \in \mathcal{F}\}.$$

It is well known that the class of algebraic cpos is *not* closed under the function space construction. The usual additional requirement (though not the finest) is to assume consistent completeness. An algebraic cpo D is said to be *consistently complete* if each consistent (i.e., bounded) pair a and b of compact elements has a supremum (denoted $a \sqcup b$). It follows that $\bigsqcup A$ exists for each consistent set $A \subseteq D$.

Definition 4.1 An *algebraic domain* is a consistently complete algebraic cpo.

Proposition 4.2 *The class of algebraic domains is closed under the function space construction.*

Let D and E be algebraic domains. Then the compact elements of $[D \to E]$ are suprema of finite consistent sets of step functions $\langle a; b \rangle$, where the latter are defined as follows for $a \in D_c$ and $b \in E_c$:

$$\langle a; b \rangle(x) = \begin{cases} b & \text{if } a \sqsubseteq x, \\ \bot & \text{otherwise.} \end{cases}$$

The class of algebraic domains is quite robust in that it is closed under all the usual constructions with the exception of the Plotkin power domain construction. The category of algebraic domains along with continuous functions is cartesian closed. In addition, the fixed point operator $\text{fix} \colon [D \to D] \to D$, defined by $\text{fix}(f) = \text{least } x$ such that $f(x) = x$, is continuous.

5 Domain representations

Here we define the concept of a domain representation. We begin by considering the canonical example of the reals and conclude with some comments on using different kinds of domains.

5.1 Representing the reals

Recall Example 3.1 of the interval approximation structure for the reals $\mathbb{R}$. The set of approximations P consists of all finite closed intervals with rational end points. We also add $\mathbb{R}$ to P and say $\mathbb{R} \prec r$ for each $r \in \mathbb{R}$. The induced refinement order is $[a, b] \sqsubseteq [c, d] \iff [a, b] \supseteq [c, d]$. Let $r \in \mathbb{R}$. Then, as discussed in Section 3, the ideal

$$I^r = \{[a, b] \in P : a \leq r \leq b\} \cup \{\mathbb{R}\}$$

represents r. Note that I^r has the property that $\bigcap I^r = \{r\}$. Now consider the ideal

$$I_r = \{[a, b] \in P : a < r < b\} \cup \{\mathbb{R}\}.$$

Also in this case $\bigcap I_r = \{r\}$, but $I^r \neq I_r$ in case r is a rational number. Both ideals give complete information about r and can be considered to represent r. We say that an ideal $I \in \bar{P}$ *represents* a real number r just in case $\bigcap I = \{r\}$. Let $\bar{P}^R$ be the set of ideals whose intersection is a singleton and define a function $\nu : \bar{P}^R \to \mathbb{R}$ by

$$\nu(I) = r \iff \bigcap I = \{r\}.$$

Proposition 5.1 *The function* $\nu \colon \bar{P}^R \to \mathbb{R}$ *is a continuous surjection with respect to the Scott and Euclidean topologies.*

We have the following picture:

$$P \hookrightarrow \bar{P} \hookleftarrow \bar{P}^R \xrightarrow{\nu} \mathbb{R}.$$

Thus, computability on $\mathbb{R}$ can be induced via the continuous function ν from computability considerations on $\bar{P}$, which in turn depends on computations on P. The tuple $(\bar{P}, \bar{P}^R, \nu)$ is a canonical example of a domain representation of $\mathbb{R}$. It is *upwards closed* in the sense that if $I \in \bar{P}^R$ and $I \subseteq J \in \bar{P}$, then $J \in \bar{P}^R$. Furthermore $\nu(I) = \nu(J)$ for such I and J. Note that if we instead choose P to consist of open intervals, then $\bar{P}^R$ consisting of the ideals whose intersection is a singleton will not be upwards closed.

5.2 General definitions

We now generalise to an arbitrary topological space.

Definition 5.2 Let X be a topological space, and let D be a domain and D^R a subset of D. Then (D, D^R, ν) is a *domain representation* of X in case $\nu \colon D^R \to X$ is a surjective continuous map when D^R is given the (relativised) Scott topology.

We have on purpose used the generic term 'domain' since the definition makes sense for any type of ordered structure or, for that matter, topological space. Commonly used ordered structures are algebraic cpos, algebraic domains, continuous domains, and bifinite domains. We return to this point below and in Section 10.

Suppose we have a domain representation of a *set* X, where we thus only require ν to be a surjection. Then the domain representation induces a topology on X by giving X the quotient topology. That is, $U \subseteq X$ is open $\Longleftrightarrow \nu^{-1}[U]$ is open in D^R. This may at times be a useful way to topologise function spaces and thus build type structures of topological spaces.

It is quite common when constructing domain representations that the obtained mapping ν is a quotient mapping. For example, this is the case for the representation of $\mathbb{R}$ given above. Thus $\mathbb{R}$ is a quotient of $\bar{P}^R$; that is,

$$\mathbb{R} \simeq \bar{P}^R / \sim,$$

where $I \sim J$ if $\bigcap I = \bigcap J$.

The next step is to represent functions between and operations on topological spaces.

Definition 5.3 Let (D, D^R, ν) and (E, E^R, μ) be domain representations of X and Y, respectively. A function $f \colon X \to Y$ is *represented* by (or *lifts* to) a continuous function $\bar{f} \colon D \to E$ if $\bar{f}[D^R] \subseteq E^R$ and $\mu(\bar{f}(x)) = f(\nu(x))$, for all $x \in D^R$.

Note that $\bar{f}$ is required to be defined on all of D. In certain situations, when considering computability aspects, it may be useful to allow partial functions on D. This is developed in Dahlgren [19].

Suppose $\bar{f} \colon D \to E$ is such that $\bar{f}[D^R] \subseteq E^R$ and such that $\nu(x) = \nu(y) \Longrightarrow \mu(\bar{f}(x)) = \mu(\bar{f}(y))$. Then $\bar{f}$ induces a unique function $f \colon X \to Y$ defined by $f(\nu(x)) = \mu(\bar{f}(x))$.

Proposition 5.4 *Let* (D, D^R, ν) *and* (E, E^R, μ) *be domain representations of* X *and* Y, *respectively, and assume* ν *is a quotient map. If* $f \colon X \to Y$ *is represented by a continuous function* $\bar{f} \colon D \to E$, *then* f *is continuous.*

The proposition is a trivial topological fact. The converse is more interesting. When does a continuous function $f \colon X \to Y$ have a continuous lifting $\bar{f} \colon D \to E$? We will return to this kind of question in Section 7.

Domain representability is naturally extended to topological algebras. Recall that a topological Σ-algebra is a topological space with continuous operations specified by the signature Σ. The field $\mathbb{R} = (\mathbb{R}, +, \times, 0, 1)$ of real numbers is a relevant example here.

Definition 5.5 Let $A = (A, \sigma_1, \ldots, \sigma_n)$ be a topological Σ-algebra. Then A is *domain representable* by $D = (D, D^R, \nu; \bar{\sigma}_1, \ldots, \bar{\sigma}_n)$ when (D, D^R, ν) is a domain representation of the topological space A, and each $\bar{\sigma}_i \colon D^{n_i} \to D$ is a continuous operation on D representing the operation σ_i. The domain with operations $(D, \bar{\sigma}_1, \ldots, \bar{\sigma}_n)$ is called a Σ-*domain.*

Note that the mapping ν in the definition is a Σ-homomorphism.

5.3 Other domains

One can represent topological spaces using other kinds of domains and ordered structures. Examples are *Baire–Cantor domains* and *continuous domains*. Each yields a theory of computability on spaces with an extensive set of applications.

The Cantor–Weihrauch domains are simply the Baire and Cantor spaces of functions on $\mathbb{N}$ seen as domains; we will meet them in Section 8. K. Weihrauch created the theory of TTE computability independently of notions of domain theory. Indeed after having considered cpos as a general approximating structure, he chose to use Baire and Cantor spaces and their computability theories based on relativised Turing computability, to represent spaces. TTE has an extensive theory and a huge range of applications; see Weihrauch [84]. It is possible to view TTE as a theory of *Baire–Cantor–Weihrauch domain representations*; see Blanck [13].

The continuous domains have a different axiomatisation of the intuitions behind domains and form a larger class of structures containing the algebraic domains. They were first used for the representation of the real numbers and other topological spaces by A. Edalat, who has also created an extensive set of applications; see Section 10.9.

The relationship between the use of these various kinds of domains in representation theory has been discussed in Stoltenberg-Hansen and Tucker [75] and in Blanck [13].

6 Effectivity

In this section we impose and study notions of computability or effectivity on domains in order to study computability on the represented structure. The type of effectivity we consider is, in the terminology of Definition 2.1, concrete computability. Our computability theory is driven by the partial recursive functions. We use the Mal'cev–Ershov theory of numberings in order to extend computability from the natural numbers to other structures, such as domains.

We assume some very basic knowledge of recursion theory that can be found in any basic text. Our notation is standard. In particular we let $\{W_e\}_{e\in\mathbb{N}}$ be a standard numbering of the recursively enumerable (r.e.) sets.

Let A be a set. A *numbering* of A is a surjective function $\alpha: \Omega_A \rightarrow A$, where $\Omega_A \subseteq \mathbb{N}$. It should be thought of as a coding of A by natural numbers. In case Ω_A is recursive, we say that a subset $S \subseteq A$ is α-*semidecidable* if $\alpha^{-1}(S)$ is r.e. and S is α-*decidable* if $\alpha^{-1}(S)$ is recursive.

Let B be a set with a numbering β. Then a function $f: A \rightarrow B$ is said to be (α, β)-*computable* if there is a partial recursive function $\bar{f}$ such that for each $n \in \Omega_A$, $\bar{f}(n)$ is defined and

$$f(\alpha(n)) = \beta(\bar{f}(n)).$$

We say that $\bar{f}$ *tracks* f.

6.1 Effective domains

At the heart of an algebraic cpo are the compact elements that play the role of the finite approximations. All computations will take place on the compact elements. Moreover continuous functions between algebraic cpos are determined completely by their behaviour on the compact elements. Thus it suffices to have a numbering of the compact elements of an algebraic cpo.

The following weak notion of effectivity suffices for many basic results with the important exception of the function space construction.

Definition 6.1 An algebraic cpo $D = (D, \sqsubseteq, \perp)$ is *weakly effective* if there is a numbering

$$\alpha: \mathbb{N} \to D_c$$

of D_c such that the relation $\alpha(n) \sqsubseteq \alpha(m)$ is a recursively enumerable relation on $\mathbb{N}$ (i.e., the relation $\sqsubseteq$ is α-semidecidable).

We denote an algebraic cpo D, which is weakly effective under a numbering α, by (D, α).

Computable elements of a weakly effective cpo are those that can be effectively approximated, and effective functions are those whose values can be effectively approximated from effective approximations of the arguments.

Making this precise, given weakly effective (D, α) and (E, β), we say that an element $x \in D$ is α-*computable* if the set approx(x) is α-semidecidable. The set of computable elements in (D, α) is denoted by $D_{k,\alpha}$.

A continuous function $f: D \to E$ is (α, β)-*effective* if the relation $b \sqsubseteq f(a)$ is α-semidecidable on $D_c \times E_c$. The intuition for the latter is that the approximations of $f(x)$ are generated effectively and simultaneously with the approximations of x. (Recall the characterisation of a continuous function between algebraic cpos from Section 4.)

It is straightforward to show that an effective function takes a computable element to a computable element and that the composition of effective functions is effective.

The set $D_{k,\alpha}$ has a natural numbering.

Theorem 6.2 *Let (D, α) be a weakly effective algebraic cpo. Then there is a numbering $\bar{\alpha}: \mathbb{N} \to D_{k,\alpha}$ such that*

(i) *the inclusion mapping $\iota: D_c \to D_{k,\alpha}$ is $(\alpha, \bar{\alpha})$-computable;*

(ii) *the relation $\alpha(n) \sqsubseteq \bar{\alpha}(m)$ is r.e.; i.e., approx$(\bar{\alpha}(m))$ is α-semidecidable uniformly in m; and*

(iii) *there is a total recursive function h such that for each e,*

$$\bar{\alpha}[W_e] \text{ directed} \implies \bar{\alpha}h(e) = \bigsqcup \bar{\alpha}[W_e].$$

A numbering satisfying (i) and (ii) of the theorem is said to be a *constructive numbering* of $D_{k,\alpha}$. It is *recursively complete* if it also satisfies (iii). It is a fact that all recursively complete constructive numberings of $D_{k,\alpha}$ are *recursively equivalent* as numberings. In general, two numberings μ and ν of a set A are recursively equivalent if id: $A \to A$ is (μ, ν)-computable and (ν, μ)-computable.

To relate our domain theoretic notions to classical recursion theory, let $\mathcal{P}$ be the algebraic domain of all partial functions from $\mathbb{N}$ into $\mathbb{N}$ ordered by graph inclusion. Let α be a standard numbering of the set $\mathcal{P}_c$ of finite functions. Then $\mathcal{P}_{k,\alpha}$ is the set of partial recursive functions. The numbering $\bar{\alpha}$ is a standard numbering of the partial recursive functions in the sense of Hartley Rogers in that it satisfies the universal property and the s-m-n theorem.

Given weakly effective algebraic cpos (D, α) and (E, β), we have the notion of an (α, β)-effective function from D to E and of an $(\bar{\alpha}, \bar{\beta})$-computable function from $D_{k,\alpha}$ to $E_{k,\beta}$. They are related by the following deep theorem due to Ershov [33], a generalisation of the Myhill–Shepherdson theorem.

Theorem 6.3 *Let (D, α) and (E, β) be weakly effective domains, and let $f: D_{k,\alpha} \to E_{k,\beta}$. Then f is $(\bar{\alpha}, \bar{\beta})$-computable if, and only if, there is an (α, β)-effective function $\bar{f}: D \to E$ such that $\bar{f} \restriction D_{k,\alpha} = f$.*

For the function space construction, a stronger form of effectivity is needed.

Definition 6.4 An algebraic domain $D = (D, \sqsubseteq, \bot)$ is *effective* if there is a numbering $\alpha: \mathbb{N} \to D_c$ such that the following relations are α-decidable for $a, b, c \in D_c$:

(i) $a \sqsubseteq b$;

(ii) $\exists d \in D_c(a, b \sqsubseteq d)$; and

(iii) $a \sqcup b = d$.

Proposition 6.5 *The category of effective domains with effective functions as morphisms is cartesian closed.*

The proof uses the intuitively effective criterion for determining whether a finite set of step functions is consistent, namely

$$\{\langle a_1; b_1 \rangle, \ldots, \langle a_n; b_n \rangle\} \text{ is consistent in } [D \to E]$$

if, and only if,

$$\forall I \subseteq \{1, \ldots, n\}(\{a_i : i \in I\} \text{ consistent} \implies \{b_i : i \in I\} \text{ consistent}).$$

6.2 Effective domain representations

The method we pursue to study effective properties of a topological algebra A is to find an *effective* domain D representing A in the sense of Definition 5.2 and then measure the effectivity of A by means of the effectivity of the representing domain D. Thus, the effectivity of A is dependent on the domain representation D and its effectivity. In practice, as described in Section 3.3, given an algebra A one finds a *computable* or *effective* structure P of approximations for A that is such that the ideal completion $\bar{P}$ of the approximation structure P is a domain representation of A.

Definition 6.6 Let X be a topological space. Then X is *(weakly) effectively domain representable* by (D, D^R, ν, α) when (D, D^R, ν) is a domain representation of X and (D, α) is a (weakly) effective domain.

The computable elements of X are induced by the computable elements of D. More precisely, the set $X_{k,\alpha}$ of *computable elements* of X is the set

$$X_{k,\alpha} = \{x \in X : \nu^{-1}(x) \cap D_{k,\alpha} \neq \emptyset\}.$$

The above notions are easily extended to topological Σ-algebras. Let $(A, \sigma_1, \ldots, \sigma_n)$ be a topological Σ-algebra. Then $(D, D^R, \nu, \alpha; \bar{\sigma}_1, \ldots, \bar{\sigma}_n)$ is a (weakly) effective domain representation of $(A, \sigma_1, \ldots, \sigma_n)$ if the operations $\bar{\sigma}_i$ are α-effective, and $(D, D^R, \nu; \bar{\sigma}_1, \ldots, \bar{\sigma}_n)$ is a domain representation of $(A, \sigma_1, \ldots, \sigma_n)$.

A Σ-algebra A is said to have a *numbering with recursive operations* if there is a numbering $\beta : \Omega_A \to A$, such that each operation in A is β-computable. And we say that (A, β) is a *numbered algebra with recursive operations* if β is a numbering of A with recursive operations. Note that we put *no* requirement on the complexity of the code set Ω_A nor on the (relative) complexity of the equality relation.

Proposition 6.7 *Let $(A, \sigma_1, \ldots, \sigma_q)$ be a topological Σ-algebra weakly effective domain representable by $(D, D^R, \nu, \alpha; \bar{\sigma}_1, \ldots, \bar{\sigma}_q)$.*

(i) *$A_{k,\alpha}$ is a subalgebra of A.*

(ii) *$A_{k,\alpha}$ is a numbered algebra with recursive operations with a numbering $\tilde{\alpha}$ induced by α.*

The first part of the proposition follows immediately since an effective domain function takes computable elements to computable elements. For the second part, let $\Omega_A = \bar{\alpha}^{-1}(D_{k,\alpha} \cap D^R)$ and define $\tilde{\alpha} : \Omega_A \to A_{k,\alpha}$ by

$$\tilde{\alpha}(n) = \nu(\bar{\alpha}(n))$$

for $n \in \Omega_A$, where $\bar{\alpha}$ is the canonical numbering of $D_{k,\alpha}$ obtained from α as in Theorem 6.2.

Finally we introduce two notions of effectivity for functions between weakly effective, domain-representable topological spaces.

Definition 6.8 Let A and B be topological spaces, weakly effective domain representable by (D, D^R, ν, α) and (E, E^R, μ, β), respectively.

(i) A continuous function $f \colon A \rightarrow B$ is said to be (α, β)-*effective* if there is an (α, β)-effective continuous function $\bar{f} \colon D \rightarrow E$ representing f; that is, $\bar{f}[D^R] \subseteq E^R$ and for each $x \in D^R$, $f(\nu(x)) = \mu(\bar{f}(x))$.

(ii) A function $f \colon A_{k,\alpha} \rightarrow B_{k,\beta}$ is $(\tilde{\alpha}, \tilde{\beta})$-*computable*, where $\tilde{\alpha}$ and $\tilde{\beta}$ are the numberings obtained in Proposition 6.7, if there is a partial recursive function $\tilde{f}$ such that $\Omega_A \subseteq \mathrm{dom}(\tilde{f})$ and for all $n \in \Omega_A$,

$$f(\tilde{\alpha}(n)) = \tilde{\beta}(\tilde{f}(n));$$

that is $\tilde{f}$ *tracks* f with respect to $\tilde{\alpha}$ and $\tilde{\beta}$.

It is not difficult to see from Theorem 6.3 that if $f \colon A \rightarrow B$ is (α, β)-effective, then $f{\restriction}A_{k,\alpha} \colon A_{k,\alpha} \rightarrow B_{k,\beta}$ is $(\tilde{\alpha}, \tilde{\beta})$-computable (and continuous). The converse direction is more difficult. It is related to the Kreisel–Lacombe–Shoenfield theorem [41] and Ceitin's theorem [17]. Note that continuity is not assumed in (ii).

7 Classes of domain representations

Recall that we only required of a domain representation (D, D^R, ν) of a space X that the function $\nu \colon D^R \rightarrow X$ be continuous. In most cases, but not all, we have stronger representation. Here are some common and useful additional properties.

Definition 7.1 Let (D, D^R, ν) be a domain representation of the topological space X.

(i) The representation is a *quotient representation* if ν is a quotient map.

(ii) The representation is a *retract representation with respect to* $\rho \colon X \rightarrow D^R$ if ρ is continuous and $\nu\rho = \mathrm{id}_X$.

(iii) The representation is a *homeomorphic representation* if ν is a homeomorphism.

It is straightforward to see that (iii) $\implies$ (ii) $\implies$ (i). Recall that if we restrict ourselves to quotient representations, then representable functions are continuous. For a retract representation (D, D^R, ν) with respect to ρ of X, we have that $\rho\nu$ is a retract; that is, $(\rho\nu)^2 = \rho\nu$, and hence that $(D, \rho\nu[D^R], \nu \restriction \rho\nu[D^R])$ is a homeomorphic representation of X.

Consider the standard representation of $\mathbb{R}$ obtained from the approximations in Example 3.1. It is easy to see that this is a retract representation with respect to the function sending each $r \in \mathbb{R}$ to the ideal $I_r = \{[a, b] \in P : a < r < b\} \cup \{\mathbb{R}\}$. Thus we obtain a homeomorphic representation of $\mathbb{R}$. Note, however, that this representation is not upwards closed. In fact, there is no homeomorphic domain representation (D, D^R, ν) of $\mathbb{R}$ where D^R is the set of maximal elements of an algebraic domain D. (There is, however, a homeomorphic continuous domain representation consisting of the maximal elements; see Section 10.9.)

Theorem 7.2 *Every T_0 topological space X has a homeomorphic algebraic domain representation.*

The construction is as follows. Let $\mathcal{B}$ be a topological base of non-empty sets closed under finite intersections as in Example 3.3. Taking the ideal completion of the approximation structure $(\mathcal{B}, \supseteq)$ and letting the representing ideals be $I_x = \{B \in \mathcal{B} : x \in B\}$, we obtain a homeomorphic representation. Thus, every T_0 space X has a homeomorphic κ-based domain representation, where κ is the *weight* of X, that is, the smallest infinite cardinality of a topological base for X. In particular, each second countable T_0-space has a countably based homeomorphic domain representation. However, it is not the case that countably based domain representations are restricted to second countable spaces. As we shall see in Section 10.6, there are good effective and hence countably based domain representations of important spaces that are not second countable.

The set D^R in a domain representation (D, D^R, ν) is often referred to as the set of *total* elements in the sense that its elements give total information about the elements of the represented space. There is an abstract theory of *domains with totality*, i.e., pairs (D, D^t) where $D^t \subseteq D$ and (often) satisfies some trivial properties. We will not pursue this theory here, but we will use the concept.

We will mainly restrict ourselves to dense representations. A domain with totality (D, D^R) is *dense* if D^R is dense in D with respect to the Scott topology. And a domain representation (D, D^R, ν) of a space X is *dense* if (D, D^R) is dense.

The advantage of a dense representation (D, D^R, ν) is the relative ease with which a continuous function from D^R can be lifted or extended to the whole of D. It is always possible to obtain an equivalent dense representation from any given representation (D, D^R, ν) by considering the domain generated by all compact approximations lying below some element of D^R. This construction, however, is in general far from being effective. One way to deal with this problem is to use *partial* continuous functions [19]. There are important situations where liftings can be achieved also for non-dense representations [13, 43, 54].

Definition 7.3 Let $D = (D, D^R, \nu)$ and $E = (E, E^R, \mu)$ be domain representations of a topological space X. The representation D *reduces* (continuously) to E, denoted by $D \le E$, if there is a continuous function $\phi \colon D \to E$ such that $\phi[D^R] \subseteq E^R$ and $(\forall x \in D^R)(\nu(x) = \mu\phi(x))$, i.e., ν *factors* through μ via ϕ on the representing elements D^R. We say that $D \equiv E$ when $D \le E$ and $E \le D$.

Like the definition of domain representability, this notion of reducibility works with many types of ordered structures.

Let C be a class of domains with totality. We will in this connection say, e.g., that C is the class of dense algebraic cpos, thus suppressing the 'with totality'. Then we let $\mathbf{Spec}_C(X)$ denote the equivalence classes of $\equiv$ over the class of domain representations (D, D^R, ν) of X, where $(D, D^R) \in C$. Note that if C is the class of dense algebraic domains, then $\mathbf{Spec}_C(X)$ contains a largest element, assuming X is a T_0-space, by considering the homeomorphic representation obtained from a topological base.

Theorem 7.4 (Blanck [13]) *Let C be the class of dense algebraic domains and assume X is a T_0-space. Then the largest degree of $\mathbf{Spec}_C(X)$ contains precisely the retract representations of X over C.*

In particular we know that the standard representation of $\mathbb{R}$ is the largest representation over dense algebraic domains and is equivalent to the standard homeomorphic representation of $\mathbb{R}$. In fact, Blanck shows that if (D, D^R, ν) is a retract algebraic domain representation of X, then (D, D^R, ν) is the largest representation over the class C of dense algebraic cpos. This then applies to the standard representation of $\mathbb{R}$.

Another related but important concept is that of an admissible domain representation. The analogous notion for TTE was first formulated by Schröder [59], whereas Weihrauch considered a similar notion for second countable spaces.

Definition 7.5 Let $D = (D, D^R, \nu)$ be a domain representation of a topological space X. Then D is an *admissible* representation of X over a class C of domains with totality if whenever $(E, E^R) \in C$ and $\phi: E^R \to X$ is continuous, then there is a continuous function $\bar{\phi}: E \to D$ such that $\bar{\phi}[E^R] \subseteq D^R$ and for each $w \in E^R$, $\phi(w) = \nu\bar{\phi}(w)$.

Again the term 'domain' is generic. We will, as usual, restrict ourselves to algebraic cpos or algebraic domains.

We have the following relation between admissibility and the reduction ordering of domain representations.

Theorem 7.6 *Let C be the class of dense algebraic cpos. Then $D = (D, D^R, \nu)$ is the largest representation of X with respect to $\leq$ over C if, and only if, D is an admissible representation of X over C.*

The theorem is true for any reasonable class C. It is proved by considering a direct sum of a largest representation of X and $(E, E^R) \in C$.

Admissibility has implications on the nature of the coding function of the representation. The following is observed in Hamrin [38].

Theorem 7.7 *Let D be an algebraic cpo, and assume (D, D^R, ν) is an admissible domain representation of X over the class of dense algebraic cpos. Then ν is a quotient mapping.*

The key point here is that open sets can be characterised using nets of arbitrary large cardinalities. On the other hand we are interested primarily in effective representations D, and hence, D_c must be countable. It is therefore interesting to introduce cardinality restrictions to the notion of admissibility.

Definition 7.8 Let κ be an infinite cardinal, and let $\mathcal{C}$ be a class of algebraic cpos with totality. Let $D = (D, D^R, \nu)$ be a domain representation of a topological space X. Then D is a κ-*admissible* representation of X over $\mathcal{C}$ if whenever $(E, E^R) \in \mathcal{C}$, the cardinality of E_c is less than or equal to κ, and $\phi \colon E^R \to X$ is continuous, then there is a continuous function $\bar{\phi} \colon E \to D$ such that $\bar{\phi}[E^R] \subseteq D^R$, and for each $w \in E^R$, $\phi(w) = \nu\bar{\phi}(w)$.

Recall that if the coding function ν was a quotient, then every representable function is a continuous. For κ-admissible, κ-based representations, we have a precise characterisation of the representable functions. We formulate it here for $\kappa = \omega$ so as not to introduce the notion of a κ-continuous function.

Theorem 7.9 *Let $\mathcal{C}$ be the class of dense algebraic cpos with totality or the class of dense domains with totality. Suppose that $(D, D^R) \in \mathcal{C}$ and $D = (D, D^R, \nu)$ is a countably based representation of X such that D is ω-admissible over $\mathcal{C}$. Let $E = (E, E^R, \mu)$ be a representation of Y that is ω-admissible over $\mathcal{C}$. Then a function $f \colon X \to Y$ is representable over D and E if, and only if, f is sequentially continuous.*

Recall that continuous functions are sequentially continuous.

Finally, we mention a theorem from Hamrin [38] characterising the spaces representable by κ-admissible and κ-based domains.

We say that a topological space X has a κ-*pseudobase* if there is a family $\mathcal{B} \subseteq \wp(X)$ such that for each open set $U \subseteq X$ and each κ-net $S \to x \in U$ there is $B \in \mathcal{B}$ such that $x \in B \subseteq U$ and S is eventually in B. A κ-net is a net of cardinality at most κ. Thus a space X has an ω-pseudobase $\mathcal{B}$ if the condition holds for each open set U and each sequence $(x_n)_n$ approaching $x \in U$.

The space of test functions used in distribution theory is an example of a topological space that is not second countable but has a countable pseudobase. (See Section 10.6.)

Theorem 7.10 *A topological space X has a κ-based and κ-admissible domain representation if, and only if, X is a T_0-space and has a pseudobase of size at most κ.*

It has been shown by Schröder [59] using TTE that the category of spaces representable by ω-based and ω-admissible domains is cartesian closed. In fact, this category coincides with the category QCB consisting of topological spaces that are quotients of second countable spaces; see Menni and Simpson [48]. For $\kappa > \omega$, the question of finding a large cartesian closed category of topological spaces is unclear.

8 TTE and domain representability

An important and successful approach to computability on topological algebras and to Computable Analysis is Type 2 Theory of Effectivity, abbreviated as TTE. A large amount of work has been done using this approach by K. Weihrauch, his students and collaborators, and others. The idea is to generalise the basic definition of a numbering, replacing the natural numbers $\mathbb{N}$ with the Baire space $\mathbb{F} = \mathbb{N} \to \mathbb{N}$ and giving $\mathbb{F}$ the Baire topology, or, more generally, with Σ^ω, where Σ is a finite or countable set. Then the established computability theory on Σ^ω induces computability on the represented space via the numbering.

We will relate TTE to (effective) domain representability. For simplicity we restrict ourselves to the Baire space, leaving the simple coding necessary when going to finite Σ.

Let X be a topological space. We say that a partial surjective function $\rho\colon \mathrm{dom}(\rho) \subseteq \mathbb{F} \to X$ is a *TTE-representation* of X if ρ is continuous. An element $x \in X$ is ρ-*computable* if there is a recursive function in $\mathrm{dom}(\rho)$ such that $\rho(f) = x$.

Suppose $\eta\colon \mathrm{dom}(\eta) \subseteq \mathbb{F} \to Y$ is a TTE-representation of Y. Then a function $f\colon X \to Y$ is TTE-*representable* with respect to ρ and η if there is a continuous partial function $\bar{f}\colon \mathrm{dom}(\bar{f}) \subseteq \mathbb{F} \to \mathbb{F}$ tracking f; i.e., $f(\rho(x)) = \eta(\bar{f}(x))$ for each $x \in \mathrm{dom}(\rho)$. The function f is (ρ, η)-*effective* if there is a computable tracking function $\bar{f}$ for f.

A first observation is that the Baire space $\mathbb{F}$ naturally extends to an algebraic domain $\mathbb{B} = \mathbb{N}^{<\omega} \cup \mathbb{F}$ with the ordering $w \preceq v \iff w$ is an initial segment of v. The usual Baire topology on $\mathbb{F}$ is the subspace topology obtained from the Scott topology on $\mathbb{B}$. We call $\mathbb{B}$ the *Baire domain*.

It is well known that each partial continuous function $f\colon \mathrm{dom}(f) \subseteq \mathbb{F} \to \mathbb{F}$ extends to a total continuous function $\bar{f}\colon \mathbb{B} \to \mathbb{B}$. Furthermore, if f is computable, then $\bar{f}$ can be chosen to be effective in a uniform way from f. From these observations we have the following equivalence theorem:

Theorem 8.1 *Let $\rho\colon \mathrm{dom}(\rho) \subseteq \mathbb{F} \to X$ be a TTE-representation of X. Then $(\mathbb{B}, \mathrm{dom}(\rho), \rho)$ is an effective domain representation of X. An element $x \in X$ is ρ-computable in the TTE sense if, and only if, it is computable in the Baire domain representation sense.*

Furthermore, if $\eta\colon \mathrm{dom}(\eta) \subseteq \mathbb{F} \to Y$ is a TTE representation of Y, then $f\colon X \to Y$ is TTE-representable (and effective) with respect to ρ and η in the TTE sense if, and only if, f is representable (and effective) with respect to ρ and η in the Baire domain representation sense.

For the converse reduction, we have the following observation.

Lemma 8.2 *If D is a countably based algebraic domain, then there is a surjective quotient map $\varphi\colon \mathbb{B} \to D$. Furthermore, $\varphi[\mathbb{F}] = D$.*

Proof. Let (a_i) be an enumeration of D_c. For $w \in \mathbb{N}^{<\omega}$ we define $\varphi(w)$ as follows. Let $v \preceq w$ be the largest initial segment such that $\{a_{v(i)} : i < \text{length of } w\}$ is consistent, and let $\varphi(w) = \bigsqcup\{a_{v(i)} : i < \text{length of } v\}$. Then φ is monotone on $\mathbb{N}^{<\omega}$ and hence extends uniquely to a continuous function on $\mathbb{B}$, which is easily seen to be a quotient. For $x \in D$, let $w \in \mathbb{F}$ be such that $\text{approx}(x) = \{a_{w(i)} : i \in \mathbb{N}\}$. Then clearly $\varphi(w) = x$. $\qquad\square$

It follows from the proof that if (D, α) is an effective domain, then φ is effective, using the numbering of D_c given by α. Furthermore, $D_k = \varphi[\mathbb{F}_k]$. Thus we obtain

Theorem 8.3 *Let (D, D^R, ν, α) be an effective domain representation of X. Then there is a TTE-representation $\rho\colon \text{dom}(\rho) \subseteq \mathbb{F} \to X$ such that the sets of computable elements of X with respect to the two representations coincide.*

Proof. Let $\varphi\colon \mathbb{B} \to D$ be as in Lemma 8.2, and define $\rho\colon \varphi^{-1}[D^R] \cap \mathbb{F} \to X$ by $\rho(x) = \nu\varphi(x)$. $\qquad\square$

We now consider representable functions. Let (D, α) and (E, β) be countably based domains, and let $\varphi\colon \mathbb{B} \to D$ and $\psi\colon \mathbb{B} \to E$ be the effective surjections obtained from Lemma 8.2. The following can be proved along similar lines.

Lemma 8.4 *Suppose $f\colon D \to E$ is (α, β)-effective. Then there is an effective function $\tilde{f}\colon \mathbb{B} \to \mathbb{B}$, obtained uniformly from f, such that $\tilde{f}[\mathbb{F}] \subseteq \mathbb{F}$, and for each $x \in \mathbb{F}$, $\psi\tilde{f}(x) = f\varphi(x)$.*

Theorem 8.5 *Let $D = (D, D^R, \nu, \alpha)$ and $E = (E, E^R, \mu, \beta)$ be effective domain representations of topological spaces X and Y, respectively. There are TTE-representations ρ and η of X and Y, respectively, such that if $f\colon X \to Y$ is effectively representable over (D, α) and (E, β), then f is effectively representable with respect to ρ and η.*

Proof. Let $\varphi\colon \mathbb{B} \to D$ and $\psi\colon \mathbb{B} \to E$ be the effective surjections obtained from Lemma 8.2, and let $\bar{f}\colon D \to E$ be an (α, β)-effective representation of $f\colon X \to Y$. Let $\tilde{f}\colon \mathbb{B} \to \mathbb{B}$ be the effective function obtained from $\bar{f}$ as in Lemma 8.4. Then we define $\rho\colon \varphi^{-1}[D^R] \cap \mathbb{F} \to X$ by $\rho = \nu\varphi$, and, similarly, $\eta\colon \psi^{-1}[E^R] \cap \mathbb{F} \to Y$ by $\eta = \mu\psi$. These are clearly continuous surjections and hence TTE-representations. Furthermore, for each $x \in \varphi^{-1}[D^R] \cap \mathbb{F}$,

$$f\rho(x) = f\nu\varphi(x) = \mu\bar{f}\varphi(x) = \mu\psi\tilde{f}(x) = \eta\tilde{f}(x),$$

which shows that f is effectively representable with respect to ρ and η. $\qquad\square$

A detailed analysis of the relationship between domain representability, using the category EQU of equilogical spaces [5], and TTE is given in Bauer [4]. Dahlgren [20] shows that there is an adjoint pair of effective functors taking a TTE-representation of a topological space X to an effective domain representation of X and, conversely, taking an effective domain representation of X to a TTE-representation of X.

9 Standard constructions

In this section we consider various standard ways to obtain algebraic domain representations.

9.1 Representation of inverse limits and ultrametric algebras

We introduced domain representations to analyse the computability of topological algebras. We wanted to study the completions of local rings and algebras of infinite processes. Both algebras were constructed as countable inverse limits of algebras; such limits posessed ultrametrics and were therefore topological algebras. Many algebras of interest in computing have this form. The following special construction for countable inverse limits was introduced in Stoltenberg-Hansen and Tucker [69, 70, 71].

Let $A = (A, \sigma_1, \ldots, \sigma_k)$ be a Σ-algebra, and let $\{\equiv_n\}_{n \in \mathbb{N}}$ be a family of congruences on A. We say that $\{\equiv_n\}_{n \in \mathbb{N}}$ is *separating* if $n \geq m$ and $x \equiv_n y \implies x \equiv_m y$, and if $\bigcap_{n \in \mathbb{N}} \equiv_n = \{(x, x) : x \in A\}$.

There is an abundance of natural examples of algebras with a family of separating congruences. For a simple example, let $T(\Sigma, X)$ be the term algebra over a signature Σ and a set of variables X. Then, for $t, t' \in T(\Sigma, X)$, let $t \equiv_n t'$ if t and t' are identical up to height $n - 1$, for $n \in \mathbb{N}$. Further examples will be given in Section 10.

Given a Σ-algebra A together with a family $\{\equiv_n\}_{n \in \mathbb{N}}$ of separating congruences, we define a metric d on A by

$$d(x, y) = \begin{cases} 0 & \text{if } x = y, \\ 2^{-n} & \text{if } x \neq y, \text{ where } n \text{ is least s.t. } x \not\equiv_n y. \end{cases}$$

The metric d is an *ultrametric*; i.e., d satisfies the stronger triangle inequality

$$d(x, y) \leq \max\{d(x, z), d(z, y)\}.$$

Furthermore, each operation σ on A is *non-expansive*, i.e., satisfies

$$d(\sigma(x_1, \ldots, x_n), \sigma(y_1, \ldots, y_n)) \leq \max\{d(x_i, y_i) : 1 \leq i \leq n\}.$$

Conversely, suppose (A, d) is an ultrametric algebra with non-expansive operations. Then we define a family $\{\equiv_n\}_{n \in \mathbb{N}}$ by $x \equiv_n y \iff d(x, y) \leq 2^{-n}$.

Given a Σ-algebra A with a family of separating congruences $\{\equiv_n\}_{n \in \mathbb{N}}$, we form the Σ-algebra

$$\hat{A} = \varprojlim A/\equiv_n,$$

the inverse limit of the $A/\equiv_n$ with respect to the homomorphisms $\phi_m^n : A/\equiv_n \to A/\equiv_m$ defined by $\phi_m^n([a]_n) = [a]_m$, for $n \geq m$. Here $[a]_n$ denotes the equivalence class of a with respect to $\equiv_n$.

The inverse limit $\hat{A} = \lim_{\leftarrow} A/\equiv_n$ is a *completion* of A. The completion of $T(\Sigma, X)$ is the set $T^\infty(\Sigma, X)$ of all finite and infinite terms. The (metric) completion of an ultrametric algebra A with non-expansive operations is isomorphic *as topological algebras* to the inverse limit $\hat{A} = \lim_{\leftarrow} A/\equiv_n$, where $\equiv_n$ is obtained from the metric as above.

To construct a domain representation of $\lim_{\leftarrow} A/\equiv_n$, let

$$\mathcal{C} = \overset{\cdot}{\bigcup}\{A/\equiv_n : n \in \mathbb{N}\},$$

the disjoint union of the $A/\equiv_n$. Order $\mathcal{C}$ by

$$[a]_m \sqsubseteq [b]_n \iff m \leq n \text{ and } a \equiv_m b.$$

Let $D(A) = \bar{\mathcal{C}}$, which is the ideal completion of $\mathcal{C}$. Then $D(A)$ is an algebraic domain of a rather simple kind. It is a tree of height ω, where the maximal elements of the domain correspond to the infinite branches of the tree.

There is an embedding of $\hat{A} = \lim_{\leftarrow} A/\equiv_n$ into $D(A)_m$, the maximal elements of $D(A)$, given by $\psi(x) = \{[\phi_n(x)]_n : n \in \mathbb{N}\}$, where $\phi_n : \hat{A} \to A/\equiv_n$ is the mapping obtained from the inverse limit construction.

Let σ be a k-ary operation on A. We define $\phi_\sigma : D(A)_c^k \to D(A)$ by

$$\phi_\sigma([a_1]_{n_1}, \ldots, [a_k]_{n_k}) = [\sigma(a_1, \ldots, a_k)]_{\min\{n_1, \ldots, n_k\}}.$$

Then ϕ_σ is well defined and monotone and hence extends to a continuous function $\phi_\sigma : D(A) \to D(A)$ representing σ on $\hat{A}$.

As a final remark we mention that the Banach fixed point theorem for an ultrametric space A is a direct consequence of the fixed point theorem for $D(A)$.

9.2 Standard representation of regular spaces

In the previous section we described how ultrametric spaces and certain inverse limit spaces have homeomorphic domain representations using the maximal elements of the domain. However, it is an easy fact that the set of maximal elements of an algebraic domain is totally disconnected, whereas essentially all spaces used in analysis are not. If one wants to keep dealing with homeomorphic representations using maximal elements, one is forced to consider a larger class of domains such as continuous cpos. Here we continue to consider the simpler structures of algebraic domains and drop the wish for a homeomorphic representation. From a computational viewpoint, this is not as problematic as it may seem since the computations take place on the representing structure.

Many spaces, such as the real numbers, cannot be constructed as inverse limits. Thus we must find other constructions when representing a wider class of spaces. In Stoltenberg-Hansen and Tucker [73], we introduced the following general method to represent regular spaces.

Definition 9.1 Let X be a topological space. Then a family P of non-empty subsets of X is a *neighbourhood system* if $X \in P$ and

(i) if $F, G \in P$ and $F \cap G \neq \emptyset$, then $F \cap G \in P$; and

(ii) if $x \in U$, where U is open, then $(\exists F \in P)(x \in F^\circ \subseteq \bar{F} \subseteq U)$.

For $F \subseteq X$, F° denotes the interior of F and $\bar{F}$ denotes the closure of F. Note that (ii) forces the space X to be regular.

Examples are topological bases of non-empty open (or closed) sets of a regular space X. Another example is a sufficiently rich family of non-empty compact sets in a locally compact space. The set of approximations for $\mathbb{R}$ in Example 3.1 is a countable and effective neighbourhood system.

Let P be a neighbourhood system for X. Then $P = (P, \supseteq, X)$ is an approximation structure for X via the approximation

$$F \prec x \Longleftrightarrow x \in F.$$

Let $\bar{P}$ be the ideal completion of P. It is an algebraic domain. (Condition (i) is only used to show consistent completeness.)

An ideal $I \in \bar{P}$ *converges* to a point $x \in X$ if for every open set U containing x there is $F \in I$ such that $x \in F \subseteq U$. I converges to x is denoted by $I \to x$. Note that a converging ideal converges to a unique point for a T_1 space X (which we include in our definition of regularity).

We let $\bar{P}^R = \{I \in \bar{P} : I \text{ convergent}\}$ and define $\nu \colon \bar{P}^R \to X$ by

$$\nu(I) = x \iff I \to x.$$

For $x \in X$ we define the ideal I_x by

$$I_x = \{F \in P : x \in F^\circ\}.$$

Note that $I_x \to x$ and that $J \to x \iff I_x \subseteq J$.

Theorem 9.2 *Let X be a regular space and P be a neighbourhood system for X. Then $\bar{P}$ is an algebraic domain and $(\bar{P}, \bar{P}^R, \nu)$ is a retract representation of X.*

Proof. Suppose $U \subseteq X$ is open and $\nu(I) = x \in U$. By Definition 9.1 (ii) there is $F \in P$ such that $x \in F^\circ \subseteq \bar{F} \subseteq U$. Thus $F \in I_x$ and hence $F \in I$. Suppose

$J \in \bar{P}^R$ and $F \in J$. Then, clearly $\nu(J) \in \bar{F}$; i.e., $\nu(\uparrow F \cap \bar{P}^R) \subseteq U$. (As usual F is identified with its principal ideal $\downarrow F$.) Thus ν is continuous.

Define $\eta \colon X \to \bar{P}^R$ by $\eta(x) = I_x$. Then $\nu \circ \eta = \mathrm{id}_X$. Furthermore η is continuous since for $F \in P$,

$$\eta^{-1}(\uparrow F \cap \bar{P}^R) = \{x \in X \colon F \in I_x\} = \{x \in X \colon x \in F^{\circ}\} = F^{\circ}.$$

$\square$

Next we consider the problem of lifting continuous functions to the representing domains.

Theorem 9.3 *Let X and Y be regular spaces with neighbourhood systems P and Q, respectively. Let $(\bar{P}, \bar{P}^R, \nu)$ and $(\bar{Q}, \bar{Q}^R, \mu)$ be the domain representations of X and Y obtained from P and Q. Suppose $f \colon X \to Y$ is a continuous function. Then there is a continuous function $\bar{f} \colon \bar{P} \to \bar{Q}$ such that for all $I \in \bar{P}^R$,*

$$\mu(\bar{f}(I)) = f(\nu(I));$$

i.e., $\bar{f}$ is a lifting or representation of f.

Proof. Given continuous $f \colon X \to Y$, define $\bar{f} \colon P \to \bar{Q}$ by

$$\bar{f}(F) = \{G \in Q \colon f[F] \subseteq G^{\circ}\}.$$

It is easily verified that $\bar{f}(F)$ is an ideal and that $\bar{f}$ is monotone. We also denote by $\bar{f}$ its unique continuous extension to all of $\bar{P}$. In fact for $I \in \bar{P}$,

$$\bar{f}(I) = \{G \in Q \colon (\exists F \in I)(f[F] \subseteq G^{\circ})\}.$$

Suppose $I \in \bar{P}^R$ and $\nu(I) = x$. Then $I_x \subseteq I$ and hence $\bar{f}(I_x) \subseteq \bar{f}(I)$. Thus it suffices to show that $I_{f(x)} \subseteq \bar{f}(I_x)$.

Let $G \in I_{f(x)}$. Then $f(x) \in G^{\circ}$ and $x \in f^{-1}[G^{\circ}]$. But then there is $F \in I_x$ such that $F \subseteq f^{-1}[G^{\circ}]$. This shows that $G \in \bar{f}(I_x)$. $\square$

In case $\bar{P}$ and $\bar{Q}$ are effective representations for X and Y, respectively, we see from the proof that the crucial point for knowing that a continuous function $f \colon X \to Y$ is effective with respect to the representations is that the relation $f[F] \subseteq G^{\circ}$ for $F \in P$ and $G \in Q$ is semidecidable.

The standard notion of a computable function on $\mathbb{R}$ is the one by Grzegorczyk [37]. Applying the above to the neighbourhood system P for $\mathbb{R}$ from Example 3.1 with a standard numbering, we have the following theorem proved in [73].

Theorem 9.4 *A function $f \colon \mathbb{R} \to \mathbb{R}$ is computable in the sense of Grzegorczyk if, and only if, it is effective with respect to the above representation of $\mathbb{R}$.*

9.3 Representation of metric spaces

Most topological spaces of interest possess useful metrics that can define their open sets. In analysis these metrics typically come from norms whose general theory involves Banach spaces and Hilbert spaces, for example. The effective content of metric spaces was analysed early on in a constructive framework in Ceitin [17]—see also the monograph Kushner [42]. Fundamental early contributions based on computability theory are Lacombe [45] and Moschovakis [50]. Banach spaces have received special attention in Pour El and Richards [56], where the computability of linear operators was classified. The computability of homomorphisms between metric algebras in general is studied in Stoltenberg-Hansen and Tucker [76]. We will now consider effectivity in metric spaces using domain representations following Blanck [8, 9].

We say that a metric space (X, d) is *recursive* in the sense of Moschovakis if there is a numbering $\alpha \colon \Omega_\alpha \to X$ such that the metric $d \colon X \times X \to \mathbb{R}_k$ is (α, ρ)-computable, where ρ is a standard numbering of the recursive reals $\mathbb{R}$.

This is a very general definition. The difficulty from a computational point of view is that calculations with distances are limited to those possible with recursive reals. Nonetheless it is possible to give a weakly effective domain representation to (the completion of) a recursive metric space along the lines given below. We shall not pursue this here. Instead we give an alternative definition that strengthens the computability of the space while still covering important examples.

By an ordered field K we mean a field $K = (K, +, -, \times, 0, 1, \leq)$. The field K is *computable* if there is a numbering $\gamma \colon \mathbb{N} \to K$ such that all the operations and the relation $\leq$ (and hence $=$) are γ-computable. It is known that if K is a computable ordered field, then its real closure is computable as an ordered field (Madison [46]). Furthermore, if K is archimedian, then K can be computably embedded into $\mathbb{R}_k$, (Lachlan and Madison [44]).

Now we say that the metric space (A, d) is *computable* if there is a numbering $\alpha \colon \mathbb{N} \to A$ and a computable archimedian ordered field (K, γ) such that d takes values in K and d is $(\alpha \times \alpha, \gamma)$-computable. We extend this to a possibly uncountable metric space (X, d) by saying that (X, d) is *effective* if there is a dense subset $A \subseteq X$ such that (A, d) is computable. Examples of effective metric spaces are the Euclidean spaces $\mathbb{R}^n$, the space $C[0, 1]$ of continuous functions $[0, 1] \to \mathbb{R}$ with the sup norm, and L^p spaces for rational $p \geq 1$.

Let (X, d) be a metric space with a dense subset A. A *formal closed ball* is a 'notation' $F_{a,r}$, where $a \in A$ and $r \in \mathbb{Q}_+$, the set of non-negative rational numbers. The formal ball is a name or syntax for a closed ball, and we may write it semantically by

$$F_{a,r} = \{x \in X : d(a, x) \leq r\}.$$

Two formal balls are *consistent*, $F_{a,r} \uparrow F_{b,s}$, if $d(a, b) \leq r + s$. And $F_{a,r}$ *formally contains* $F_{b,s}$, $F_{a,r} \sqsubseteq F_{b,s}$, if $d(a, b) + s \leq r$.

A set $\{F_{a_1,r_1}, \ldots, F_{a_n,r_n}\}$ of formal balls is *permissible* if the balls are pairwise consistent and no ball is contained within another; i.e., for $1 \le i < j \le n$, $F_{a_i,r_i} \uparrow F_{a_j,r_j}$ and it is not the case that $F_{a_i,r_i} \sqsubseteq F_{a_j,r_j}$ or $F_{a_j,r_j} \sqsubseteq F_{a_i,r_i}$. We use the notation σ, τ for permissible sets.

Let P be the set of all permissible sets of formal balls. We need to extend the relation $\sqsubseteq$ to permissible sets:

$$\sigma \sqsubseteq \tau \iff (\forall F_{a,r} \in \sigma)(\exists F_{b,s} \in \tau)(F_{a,r} \sqsubseteq F_{b,s}).$$

Note that consistency is characterised by

$$\sigma \uparrow \tau \iff (\forall F_{a,r} \in \sigma)(\forall F_{b,s} \in \tau)(F_{a,r} \uparrow F_{b,s}).$$

Given consistent permissible sets σ and τ, the supremum $\sigma \sqcup \tau = g(\sigma, \tau)$, where g removes those formal balls in $\sigma \cup \tau$ formally containing others.

The following is immediate from the construction above. But note that we need to consider sets of formal balls in order to be able to compute the supremum operation.

Lemma 9.5 *If (A, d) is a computable metric space, then the obtained structure $P = (P, \sqsubseteq, \uparrow, \sqcup, \bot)$ is computable with a numbering α obtained from the numbering of A.*

We now let $D = \bar{P}$, which is the ideal completion of P. Thus (D, α) is an effective domain.

An ideal $I \in D$ is *converging* if for any $\varepsilon > 0$, there exists $\{F_{a,r}\} \in I$ such that $r < \varepsilon$. An element $x \in \bar{A}$, the metric completion of A, is *approximated* by the ideal I if $(\forall \sigma \in I)(\forall F_{a,r} \in \sigma)(x \in F_{a,r})$. A convergent ideal I approximates exactly one element x in $\bar{A}$; we write $I \to x$. Let $D^R = \{I \in D : I \to x \in X\}$. The function $\nu \colon D^R \to X$ defined by

$$\nu(I) = x \iff I \to x$$

is a quotient mapping.

In this way we have obtained an effective domain representation of $\bar{A}$ and hence of X.

Theorem 9.6 *Each effective metric space (X, d) has an effective domain representation (D, D^R, ν, α) such that the set $X_{k,\alpha}$ of computable elements in X induced by (D, D^R, ν, α) is a recursive metric space in the sense of Moschovakis.*

The situation with computable functions between effective metric spaces is more difficult. We state the following theorem, which is, essentially, Theorem 3.4.33 in Blank [8]. It uses Berger's generalisation in [6] of the Kreisel–Lacombe–Shoenfield theorem.

By a semieffective domain we mean one where the consistency relation on the compact elements need not be decidable. A semieffective domain representation of X in the theorem below is obtained by taking the dense part of a standard effective formal ball representation of X.

Theorem 9.7 *Let X and Y be effective metric spaces. Then there exists a semieffective domain representation (D, D^R, ν, α) of X consisting of permissible sets of formal balls such that together with a standard effective formal ball representation (E, E^R, μ, β) of Y, the following are equivalent.*

(i) *The function $f: X_{k,\alpha} \to Y_{k,\beta}$ is computable in the sense of Definition 6.8;*

(ii) *There is a continuous extension of f to $f: X \to Y$ that is effective with respect to the domain representations (D, D^R, ν, α) and (E, E^R, μ, β).*

Note that the function f in (i) is not assumed to be continuous. The implication (i) implies (ii) has the form of Ceitin's Theorem, that computability implies effective continuity, as a corollary.

9.4 Representation of partial and discontinuous functions

There are important phenomena in computing that are not continuous. For example, suppose we model a *stream* of data as a function from time into a set of data, where time is thought of as continuous and data is a discrete set. It is reasonable to model time by the real number line $\mathbb{R}$ or a final segment of $\mathbb{R}$ and give the data set the discrete topology. However, the only continuous functions from $\mathbb{R}$ into a discrete set are the constant functions (since $\mathbb{R}$ is a connected space). Thus transmission of discrete data in continuous time cannot be modelled by continuous functions.

Given domain representations (D, D^R, ν) of $\mathbb{R}$ and (E, E^R, μ) of the data set A, the domain $[D \to E]$ will contain approximations to arbitrary functions from $\mathbb{R}$ to A. There is no hope of having exact continuous representations of discontinuous functions. But there are best possible *approximate* representations.

Let (D, D^R, ν) and (E, E^R, μ) be domain representations of the topological spaces X and Y, respectively. Then we say that a function $f: X \to Y$ (not necessarily continuous) is *represented approximately* by (or *lifts approximately* to) $\bar{f} \in [D \to E]$ if for each $x \in D^R$,

(i) f continuous at $\nu(x) \implies \bar{f}(x) \in E^R$ and $f\nu(x) = \mu\bar{f}(x)$; and

(ii) f not continuous at $\nu(x) \implies (\exists y \in \mu^{-1}[f\nu(x)])(\bar{f}(x) \sqsubseteq y)$.

To illustrate we consider the simple example of the floor function $\lfloor \cdot \rfloor : \mathbb{R} \to \mathbb{Z}$, which is discontinuous at precisely the integer points. Let (D, D^R, ν) be the standard closed interval representation of $\mathbb{R}$ from Example 3.1. For $\mathbb{Z}$ we could have chosen the flat domain $\mathbb{Z}_\perp$. This, however, would give no information at points of discontinuities. Instead we let $E = \wp_f(\mathbb{Z}) \cup \{\mathbb{Z}\}$ ordered by reverse inclusion $\supseteq$. In fact, E is the upper (or Smyth) power domain of $\mathbb{Z}_\perp$. Letting E^R be the set of maximal elements in E, i.e., the set of singletons $\{n\}$, we obtain a domain representation (E, E^R, μ) by mapping $\{n\}$ to n.

Define $f \colon D_c \to E$ by $f([a, b]) = \{n \in \mathbb{Z} : \lfloor a \rfloor \leq n \leq \lfloor b \rfloor\}$ and extend f continuously to D. Then clearly f represents the floor function approximately. But note that at the discontinuity $n \in \mathbb{R}$ we have for $\nu(I) = n$ that $f(I) \subseteq \{n - 1, n\}$. Thus, by choosing our representations with some care, we are able to recover much information also at points of discontinuities.

If a function f has an approximate representation, then it has a best approximate representation.

Theorem 9.8 ([14]) *Let (D, D^R, ν) and (E, E^R, μ) be algebraic domain representations of X and Y, respectively. Assume that D^R is dense in D, and that (E, E^R, μ) satisfies the following local property: if $x \sqsubseteq y$ and $x \in E^R$, then $y \in E^R$ and $\mu(x) = \mu(y)$. Let $f \colon X \to Y$ be a function, and assume that f has one approximate representation in $[D \to E]$. Then there is a best approximate representation $\bar{f} \in [D \to E]$ in the sense of the domain ordering.*

10 Applications

A theory of computability on topological spaces can be used to analyse computation in many application areas, including analysis, algebra, semantics of data types and programming, graphics, and hardware.

10.1 More on real numbers

Throughout the paper we have chosen the field $\mathbb{R}$ of real numbers together with the closed interval domain representation of Example 3.1, which we here denote by $\mathcal{R}$, as a canonical example. We have observed that this representation is an effective dense retract representation, that the elements in $\mathbb{R}$ computable from the representation are precisely the recursive reals, and that the effective functions from $\mathbb{R}$ to $\mathbb{R}$ are precisely the functions computable in the sense of Grzegorczyk. In addition, the representation is ω-admissible and it is a largest representation of $\mathbb{R}$ with respect to the reduction $\leq$ from Definition 7.3.

Now consider the set $C(\mathbb{R}, \mathbb{R})$ of continuous functions from $\mathbb{R}$ to $\mathbb{R}$. This space has a natural topology, namely the compact-open topology. The set $C(\mathbb{R}, \mathbb{R})$ has a natural domain representation $[\mathcal{R} \to \mathcal{R}]$ where the representing elements $[\mathcal{R} \to \mathcal{R}]^R$ are those continuous domain functions representing functions in $C(\mathbb{R}, \mathbb{R})$. The obtained function

$$\nu \colon [\mathcal{R} \to \mathcal{R}]^R \to C(\mathbb{R}, \mathbb{R})$$

induces a topology on $C(\mathbb{R}, \mathbb{R})$ from the relativised Scott topology on $[\mathcal{R} \to \mathcal{R}]^R$ which coincides with the compact-open topology (see Blanck [10]); i.e.,

$$([\mathcal{R} \to \mathcal{R}], [\mathcal{R} \to \mathcal{R}]^R, \nu)$$

is a domain representation of $C(\mathbb{R}, \mathbb{R})$. Di Gianantonio [22] used signed digit representations of real numbers to construct another domain representation capturing the compact-open topology of $C(\mathbb{R}, \mathbb{R})$.

It is well known that $C(\mathbb{R}, \mathbb{R})$ is not locally compact, and hence, there is no 'natural' topology on $C(C(\mathbb{R}, \mathbb{R}), \mathbb{R})$. On the other hand the type structure over $\mathcal{R}$ is well behaved, and therefore, we can construct a type structure also over $\mathbb{R}$, including $C(C(\mathbb{R}, \mathbb{R}), \mathbb{R})$, and give each such type a topology.

To make this precise we define the set of finite type symbols as follows: ι is a type symbol, and if σ and τ are type symbols, then $(\sigma \to \tau)$ is a type symbol. The pure type symbols are $t_0 = \iota$ and $t_{n+1} = (t_n \to \iota)$. For each type symbol σ we define a domain with totality $(\sigma(\mathcal{R}), \sigma(\mathcal{R})^R)$. Simultaneously we define the type $\sigma(\mathbb{R})$ over $\mathbb{R}$ and a surjective map $\nu_\sigma : \sigma(\mathcal{R})^R \to \sigma(\mathbb{R})$ such that $(\sigma(\mathcal{R}), \sigma(\mathcal{R})^R, \nu_\sigma)$ is a domain representation of $\sigma(\mathbb{R})$. For the base case we use the standard closed interval domain representation $(\mathcal{R}, \mathcal{R}^R, \nu)$. Thus we let $\iota(\mathcal{R}) = \mathcal{R}$, $\iota(\mathcal{R})^R = \mathcal{R}^R$, $\iota(\mathbb{R}) = \mathbb{R}$, and $\nu_\iota = \nu$.

Inductively let $(\sigma \to \tau)(\mathcal{R}) = [\sigma(\mathcal{R}) \to \tau(\mathcal{R})]$ and let $(\sigma \to \tau)(\mathcal{R})^R$ be the set of functions in $(\sigma \to \tau)(\mathcal{R})$ representing a function from $\sigma(\mathbb{R})$ into $\tau(\mathbb{R})$ via ν_σ and ν_τ. Then let $(\sigma \to \tau)(\mathbb{R})$ be the set of functions from $\sigma(\mathbb{R})$ into $\tau(\mathbb{R})$ having a representing function in $(\sigma \to \tau)(\mathcal{R})$. Finally let $\nu_{(\sigma \to \tau)} : (\sigma \to \tau)(\mathcal{R})^R \to (\sigma \to \tau)(\mathbb{R})$ be the map taking a representing function in $(\sigma \to \tau)(\mathcal{R})^R$ to the function in $(\sigma \to \tau)(\mathbb{R})$ that it represents.

By the fact that the category of effective algebraic domains is cartesian closed, the domain representations $(\sigma(\mathcal{R}), \sigma(\mathcal{R})^R, \nu_\sigma)$ induce a topology (the quotient topology) and effectivity on each type $\sigma(\mathbb{R})$.

D. Normann shows in [53] that each representation $(\sigma(\mathcal{R}), \sigma(\mathcal{R})^R, \nu_\sigma)$ is dense. This is analogous to the density theorem for the finite type structure over the discrete space $\mathbb{N}$ of natural numbers proved by Berger [6], but it uses by necessity a different proof. Normann also observes some 'anomalies' of the type structures $\sigma(\mathbb{R})$, e.g., that the space $t_2(\mathbb{R})$ is not metrizable.

The natural continuous domain representation for real numbers is the interval domain consisting of real intervals; this suggests strong connections to Interval Analysis [1, 49]. For example, an often-used notion in Interval Analysis is the *monotone interval function*, which is nothing more than a monotone map on the interval domain. Interval Analysis has traditionally used the topology induced by the Moore metric, whereas the Scott topology has been used for the interval domain. It is easy to construct interval functions that are continuous with respect to either topology but not both. In [58] it is shown that for a continuous function f the optimal interval representation of f is continuous with respect to both topologies.

Interval Analysis is an established approach to practical exact computation. The interval domain and certain substructures thereof have also been used to investigate and reason about the practical implementation of exact real arithmetic [11, 12]. Thus, domain representations can be used to reason abstractly about the computability of functions, and to model concretely the exact steps taken in making exact real computations. Thus, there is evidence that domain representations may be a powerful tool towards practical exact computation on many forms of continuous data.

10.2 Local rings

In 1983 we knew a great deal about computable algebra (see, e.g., our later survey [74]), and our interest in domain representability began with the problem of investigating the computability of local rings. Thinking about the completions of local rings, we wanted a *general* method of introducing computability into uncountable algebras. There were four algebras in view: complete local rings, algebras of infinite processes (satisfying Bergstra and Klop's laws for ACP), algebras of infinite terms, and the field of real numbers. The first three had a common structure: they were inverse limits of countably many factor algebras and looked like domains!

Let R be a local commutative Noetherian ring whose unique maximal ideal is $\mathbf{m}$. We showed that $\mathbf{m}$ is decidable when R is computable as a ring. Define for $x, y \in R$ and $n \in \mathbb{N}$,

$$x \equiv_n y \Leftrightarrow x - y \in \mathbf{m}^n,$$

which is decidable. By Krull's Theorem, $\{\equiv_n\}_{n \in \mathbb{N}}$ is a family of separating congruences with respect to the ring operations, and the general constructions of Section 9.1 can be applied to obtain an effective domain representation of the completion of R.

The local ring and the general method was circulated in Stoltenberg-Hansen and Tucker [69] and later published in Stoltenberg-Hansen and Tucker [70]).

10.3 Process algebra

Think of a process made of atomic actions that can be performed sequentially or in parallel, can be independent or communicate, and can branch deterministically or non-deterministically. Such processes abound in both computers and machines and in nature too. In process algebra such intuitive ideas are analysed very abstractly: processes are modelled and classified by postulating operations on processes, such as

$$p \cdot q, p \| q \text{ and } p + q,$$

and axioms that they should satisfy. There are many kinds of semantic ideas to be found in systems so there are many operations and axioms—see Bergstra, Ponse and Smolka [7]. In modelling a particular system, the idea is to devise a specification that is a set of equations, based on some choice of operations. The semantics of the specification is given by solving the equations in process algebras satisfying axioms appropriate to the problem.

It is common to need complicated infinite processes in the semantic modelling of systems, and so the process algebras used are complicated uncountable structures. In particular, with some process algebra methods, the algebras of infinite processes have the beautiful structure of inverse limits of finite models of equational theories. This means that algebras of infinite processes are algebras with ultrametric topologies, and the methods of Section 9.1 can be used to study processes. Applications of solving finite systems of equations in process algebras are given in [71] and infinite systems of equations in [72].

10.4 Banach spaces

Functions and functionals on $\mathbb{R}$ and $\mathbb{C}$ can be approximated in many different ways. However, the methods used have been found to have two fundamental properties in common: they use linear combinations of basic functions, and they measure the accuracy of approximations by metrics derived form *norms*. Theories of these methods have been created using vector spaces equipped with norms and other operations, such as Banach spaces, Banach algebras, Hilbert spaces, and C*-algebras. Since all of these topological algebras are special kinds of metric spaces, the method for metric spaces, given in Section 9.3, can be used to make domain representations for them. Algebraic domain representations for Banach spaces were made in Stoltenberg-Hansen and Tucker [75], in order to prove the equivalence of various models of computation, including that of Pour El and Richards designed for Banach spaces.

10.5 C^∞ functions

A common way to approximate a continuous function on the real numbers is by finite collections of compact boxes enclosing the graph of the function. Tighter boxes covering a larger segment of the graph naturally yield more information about the function we wish to approximate. This idea can be generalised to approximations of C^k and C^∞ functions on the reals in a natural way: an approximation of a C^k function f is a finite set of approximations of the the function f and the first k derivatives of f (as continuous functions from $\mathbb{R}$ to $\mathbb{R}$). Similarly, an approximation of a C^∞ function f on $\mathbb{R}$ is a finite set of approximations of the function f and the first k derivatives of f *for some $k \geq 0$*.

A C^∞ function from $\mathbb{R}$ to $\mathbb{C}$ can be thought of as a pair of smooth functions from $\mathbb{R}$ to $\mathbb{R}$ (corresponding to the real and imaginary parts of the function). Thus, we can approximate smooth functions from $\mathbb{R}$ to $\mathbb{C}$ by approximating the real and imaginary parts separately. In this way we get an effective domain representation of the space of smooth functions from $\mathbb{R}$ to $\mathbb{C}$.

10.6 Test functions and distributions

An interesting class of functions in this context is the space $\mathcal{D}$ of test functions considered in distribution theory. If we restrict ourselves to one variable, a test function is simply a smooth function from $\mathbb{R}$ to $\mathbb{C}$ with compact support. Formally, the space of test functions is constructed as an inductive limit of metrisable spaces but is itself not metrisable. In fact, it is not even first countable. Nevertheless, we may construct an effective domain representation of $\mathcal{D}$ and study computable processes on the space of test functions. This is interesting from a purely computability theoretic point of view since it has sometimes been argued that the stronger property of second countability is needed to develop a viable computability theory on a topological space (cf. Smyth [63]). To approximate a test function f we simply add information about (i.e., bounds on) the support of f to a $C^\infty(\mathbb{R})$-approximation of f. This idea yields an ω-admissible effective domain representation of the space of test functions and thus allows us to introduce a notion of computability on $\mathcal{D}$. We note that standard operations on $\mathcal{D}$ such as integration, differentiation, regularisation, addition, and scalar multiplication are all effective with respect to this representation.

A distribution is a continuous linear functional on the space of test functions. Since we have effective representations of the spaces $\mathcal{D}$ and $\mathbb{C}$, general domain theory yields an effective domain representation of the space of distributions. Moreover, similar methods may be applied to construct effective representations of the spaces of tempered distributions and distributions with compact support. This allows us to introduce a notion of computability on the space of distributions in the spirit of Weihrauch and Zhong [86], and to study computable processes on spaces of distributions. In particular, the space of distributions, the space of tempered distributions, and the space of distributions with compact support are all effective vector spaces, the standard embedding theorems effectivise, and the Fourier transform and its inverse lift to effective functions on the space of tempered distributions. For details, see Dahlgren [21].

10.7 Volume graphics

In volume graphics, objects are defined in three dimensions. Objects can be regular, like buildings and crockery, or irregular but structured, like 3D body scans, or amorphous like clouds and fire. The objects may be combined to create 3D scenes. In volume graphics, objects and scenes must be created, transformed, and rendered in 2D.

In practice, different objects can have quite different representations, ranging from a collection of simple mathematical functions to large 3D arrays of physical data. *Constructive volume geometry* (CVG) is a high-level approach to volume graphics that abstracts from specific representations by focussing on high level operations on volume objects. First, to unify representations, each spatial object is required to assign data, called attributes, to every point in 3D. Thus, spatial objects are modelled by vectors

$$\phi_1, \ldots, \phi_k$$

of scalar fields of the form:

$$\phi \colon \mathbb{R}^3 \to [0, 1] \text{ or } \phi \colon \mathbb{R}^3 \to \mathbb{R}.$$

The attributes chosen depend on the application. For example, a simple graphics application is the RGB model, which has $k = 4$ and attributes of opacity, measured by the interval $[0, 1]$, and colours red (R), green (G), and blue (B), measured by $\mathbb{R}$.

Then operations on these objects are defined to make algebras of spatial objects. There are lots of simple operations to create RGB algebras, with attributes opacity and three colours. CVG algebras are as varied as the applications of computer graphics.

CVG was first proposed in Chen and Tucker [18], where various operations and their laws were given, the high-level representation of graphics objects using CVG terms explained, and recursive rendering via structural induction on terms introduced. In [18], the scalar fields are total functions, which simplifies the algebra. A fuller mathematical treatment of CVG, including approximation, is in Johnson [40].

Computation in CVG involves computation on real numbers, real-valued functions, and operators. To understand the semantics of the CVG programming, the framework needs to be analysed by a computabilty theory for topological spaces. In Blanck, Stoltenberg-Hansen, and Tucker [15] we consider computability with partial functions and apply the theory to the computability of CVG algebras, such as the RGB algebras.

10.8 Analog and digital systems

In computer science the interfaces between continuous and discrete data types are not well understood. Domains and topological spaces are designed to model continuous data, but they can also model discrete data. Can domain representations model computation with continuous and discrete data in a uniform way? Yes.

Consider analog and digital data and the interface between them. A *data stream* is a sequence of data indexed by time. Mathematically, we model data streams by functions

$$s \colon T \to A,$$

where $s(t) = $ datum or measurement from A at time t. The functions may be total or partial.

There are several cases of practical importance to consider, especially the purely digital case:

discrete time $T = \mathbb{Z}$ and discrete data $A = \{0, 1\}$;

and the purely analog case:

continuous time $T = \mathbb{R}$ and continuous data $A = \mathbb{R}$.

We model computation with these streams by mappings of the form

$$F: [T \to A] \to [T' \to B],$$

where T, T' are time scales and A, B are data types. The stream transformations include analog-to-digital and digital-to-analog transformations.

We have seen a number of mathematical tools to tackle the problem of analysing the semantics of analog *versus* digital computing and signal processing, starting with domain representations of the reals. In applying domain representations and computability theory we focus on streams and stream transformers that are continuous functions. (The functions may be partial to help model discontinuities in streams.) The interface between analog and digital computation is studied in [14], using domain representations of spaces with the compact-open topology.

10.9 Applications using continuous domains

Let us remind the reader that in this introduction to domain representation theory we have used algebraic domains exclusively and have concentrated on our own interests. As emphasised earlier, one can use many types of ordered structure for representation. In particular, A. Edalat has used continuous domains to represent topological spaces in many applications, including several areas we have not discussed here.

The early applications of continuous domain representations focussed on semantic modelling of case studies of mathematical approximation, including iterative maps and integration; see Edalat [23, 24, 25]. This was done without emphasis on computability. A great deal of effort was devoted to using domains to develop software for exact arithmetic on computers.

With the rise of Computable Analysis, later studies of metric spaces in Edalat and Heckmann [26], real numbers in Edalat and Sünderhauf [30], and Banach spaces in Edalat and Sünderhauf [31] looked at computability and may be compared with approaches based on algebraic domains mentioned above.

Recently, new subjects have been started. There is extensive work on computational geometry and Constructive Solid Geometry (CSG), which is a modelling technique well established in CAD; see Edalat and Lieutier [27]. CSG is a precursor to CVG mentioned in Subsection 10.7. Some first steps into the rich and vast subject of calculus and solving differential equations have been also taken in [28, 29].

The use of continuous domains has the advantage that often (but not always) D^R may be chosen as the set of maximal elements of D and that the definition of representability is then reformulated in these terms. A disadvantage is that the theory of continuous domains is more involved. We have stuck to algebraic cpos and domains because of their simplicity and the fact that they arise from our consideration of approximation structures. Moreover, it is well known that every continuous cpo is a retract of an algebraic cpo. It follows that the two approaches of using continuous representations or merely algebraic representations are essentially equivalent.

References

1. G. ALEFELD AND J. HERZBERGER, *Introduction to Interval Computations*, Academic Press, New York, 1983.
2. C. E. AULL AND R. L. OWEN (editors), *Handbook of the History of General Topology*, Volume 1, Kluwer Academic Publishers, Dordrecht, 1997.
3. C. E. AULL AND R. L. OWEN (editors), *Handbook of the History of General Topology*, Volume 2, Kluwer Academic Publishers, Dordrecht, 1998.
4. A. BAUER, A relationship between equilogical spaces and type two effectivity, *Mathematical Logic Quarterly* 48 (2002), 1 – 15.
5. A. BAUER, L. BIRKEDAL AND D. S. SCOTT, Equilogical spaces, *Theoretical Computer Science* 315 (2004), 35 – 59.
6. U. BERGER, Total sets and objects in domain theory, *Annals of Pure and Applied Logic* 60 (1993), 91 – 117.
7. J. A. BERGSTRA, A. PONSE AND S. A. SMOLKA, *Handbook of Process Algebra*, Elsevier, Amsterdam, 2001.
8. J. BLANCK, *Computability on Topological Spaces by Effective Domain Representations*, Uppsala Dissertations in Mathematics 7, 1997.
9. J. BLANCK, Domain representability of metric spaces, *Annals of Pure and Applied Logic* 83 (1997), 225 – 247.
10. J. BLANCK, Domain representations of topological spaces, *Theoretical Computer Science* 247 (2000), 229 – 255.
11. J. BLANCK, Efficient exact computation of iterated maps, *Journal of Logic and Algebraic Programming* 64 (2005), 41 – 59.
12. J. BLANCK, Exact real arithmetic using centered intervals and bounded error, *Journal of Logic and Algebraic Programming* 66 (2006), 50 – 67.
13. J. BLANCK, Reducibility of Domain Representations and Cantor-Weihrauch Domain Representations, Report CSR 15-2006, Department of Computer Science, Swansea University.
14. J. BLANCK, V. STOLTENBERG-HANSEN AND J. V. TUCKER, Streams, stream transformers and domain representations, in B Moller and J. V. Tucker (eds.), *Prospects for Hardware Foundations*, Lecture Notes in Computer Science, volume 1546, Springer Verlag, New York, 1998, 27 – 68.
15. J. BLANCK, V. STOLTENBERG-HANSEN AND J. V. TUCKER, Domain representations of partial functions, with applications to spatial objects and constructive volume geometry, *Theoretical Computer Science* 284 (2002), 207 – 224.
16. L. BLUM, F. CUCKER, M. SHUB, AND S. SMALE, *Complexity and Real Computation*, Springer-Verlag, New York, 1998.
17. G. S. CEITIN, Algorithmic operators in constructive complete separable metric spaces, *Doklady Akademii Nauk SSSR* 128 (1959), 49 – 52.

18. M. CHEN AND J. V. TUCKER, Constructive volume geometry, *Computer Graphics Forum* 19 (2000), 281 – 293.
19. F. DAHLGREN, Partial continuous functions and admissible domain representations (extended abstract), in A. Beckman et al. (eds.), *Logical Approaches to Computational Barriers*, Lecture Notes in Computer Science, volume 3988, Springer-Verlag, New York, 2006, 94 – 104.
20. F. DAHLGREN, Effective domain representability vs. TTE representability, manuscript, 2006.
21. F. DAHLGREN, Effective distribution theory, manuscript, 2006.
22. P. DI GIANANTONIO, Real number computability and domain theory, *Information and Computation* 127 (1996), 11 – 25.
23. A. EDALAT, Dynamical systems, measures, and fractals via domain theory, *Information and Computation* 120 (1995), 32 – 48.
24. A. EDALAT, Power domains and iterated function systems, *Information and Computation* 124 (1996), 182 – 197.
25. A. EDALAT, Domains for computation in mathematics, physics and exact real arithmetic, *Bulletin of Symbolic Logic* 3 (1997), 401 – 452.
26. A. EDALAT AND R. HECKMANN, A computational model for metric spaces, *Theoretical Computer Science* 193 (1998), 53 –73.
27. A. EDALAT AND A. LIEUTIER, Foundation of a computable solid modeling, *Theoretical Computer Science* 284 (2002), 319–345.
28. A. EDALAT AND A. LIEUTIER, Domain theory and differential calculus (functions of one variable), *Mathematical Structures in Computer Science* 14 (2004), 771 – 802.
29. A. EDALAT AND A. LIEUTIER, A Domain Theoretic Account of Picard's Theorem, in J. Diaz et al. (eds.), *Automata, Languages and Programming*, Lecture Notes in Computer Science, volume 3142, Springer, Berlin, 2004, 494 – 505.
30. A. EDALAT AND P. SÜNDERHAUF, A domain-theoretic approach to computability on the real line, *Theoretical Computer Science* 210 (1999), 73 – 98.
31. A. EDALAT AND P. SÜNDERHAUF, Computable banach spaces via domain theory, *Theoretical Computer Science* 219 (1999), 169 – 184.
32. YU. L. ERSHOV, The theory of A-spaces, *Algebra and Logic* 12 (1973), 209 – 232.
33. YU. L. ERSHOV, The model C of the partial continuous functionals, in R. O. Gandy and J. M. E. Hyland (eds.), *Logic Colloquium 76*, North-Holland, Amsterdam, 1977, 455 – 467.
34. YU. L. ERSHOV, *Theory of Numerations*, Monographs in Mathematical Logic and the Foundation of Mathematics, 'Nauka', Moscow, 1977.
35. YU. L. ERSHOV, Theorie der Numerierungen III, *Zeitschrift für Mathematische Logik und Grundlagen der Mathematik* 23 (1977), 289 – 371.
36. A. FRÖLICH AND J. C. SHEPHERDSON, Effective procedures in field theory, *Philosophical Transactions of the Royal Society London. Ser. A.* 248 (1956), 407 – 432.
37. A. GRZEGORCZYK, On the definitions of computable real continuous functions, *Fundamenta Mathematicae* 44 (1957), 61 – 71.
38. G. HAMRIN, Effective Domains and Admissible Domain Representations, Uppsala Dissertations in Mathematics 42, 2005.
39. I. M. JAMES (editor), *History of Topology*, North-Holland, Amsterdam, 1999.
40. K. JOHNSON, The algebraic specification of spatial data types with applications to constructive volume geometry, PhD Thesis, Department of Computer Science, Swansea University, 2006.

41. G. KREISEL, D. LACOMBE, AND J. R. SHOENFIELD, Partial recursive functionals and effective operations, in A. Heyting (ed.), *Constructivity in Mathematics*, North-Holland, Amsterdam, 1959, 195 – 207.

42. B. A. KUSHNER, *Lectures on Constructive Mathematical Analysis*, Translations of Mathematical Monographs, v. 60, AMS, Providence, 1984.

43. P. KØBER, Uniform domain representations of l^p-spaces, *Mathematical Logic Quarterly* 53 (2007), 180 – 205.

44. A. H. LACHLAN AND E. W. MADISON, Computable fields and arithmetically definable ordered fields, *Proceedings of the American Mathematical Society* 24 (1970), 803 – 807.

45. D. LACOMBE, Extension de la notion de fonction récursive aux fonctions d'une ou plusieurs variables réelles, I, II, III. *Comptes Rendus* 240, 241 (1955), 2478 – 2480, 13 – 14, 151 – 155.

46. E. W. MADISON, A note on computable real fields, *Journal of Symbolic Logic* 35 (1970), 239 – 241.

47. A. I. MAL'CEV, Cconstructive algebras, I, *The Metamathematics of Algebraic Systems. Collected papers: 1936 – 1967*, North-Holland, Amsterdam, 1971, 148 – 212.

48. M. MENNI AND A. SIMPSON, Topological and limit-space subcategories of countably-based equilogical spaces, *Mathematical Structures in Computer Science* 12 (2002), 739 – 770.

49. R. E. MOORE, *Interval Analysis*, Prentice-Hall, Englewood Cliffs, 1966.

50. Y. N. MOSCHOVAKIS, Recursive metric spaces, *Fundamenta Mathematicae* 55 (1964), 215 – 238.

51. D. NORMANN, *Recursion on the Countable Functionals*, Springer Lecture Notes in Mathematics 811, 1980.

52. D. NORMANN, A hierarchy of domains with totality but without density, in B. Cooper, T. Slaman and S. S. Wainer (eds.), *Computability, Enumerability, Unsolvability*, Cambridge University Press, 1996, 233 – 257.

53. D. NORMANN, The continuous functionals of finite types over the reals, *Elkectronic Notes in Theoretical Computer Science* 35 (2000).

54. D. NORMANN, The continuous functionals of finite types over the reals, in K. Keimel, G. Q. Zhang, Y. Liu and Y. Chen (eds.), *Domains and Processes*, Proc. 1st Intern. Symp. on Domain Theory, Shanghai, China, 1999, Kluwer, Boston, 2001, 103 – 124.

55. E. PALMGREN AND V. STOLTENBERG-HANSEN, Domain interpretations of Martin-Löf's partial type theory, *Annals of Pure and Applied Logic* 48 (1990), 135 – 196.

56. M. B. POUR-EL AND J. I. RICHARDS, *Computability in Analysis and Physics*, Perspectives in Mathematical Logic, Springer-Verlag, Berlin, 1989.

57. H. RICE, Recursive real numbers, *Proceedings of the American Mathematical Society* 5 (1954), 784 – 791.

58. R. SANTIAGO, B. BEDREGAL AND B. ACIÓLY, Formal aspects of correctness and optimality of interval computations, *Formal Aspects of Computing* 18 (2006), 231 – 243.

59. M. SCHRÖDER, Extended admissibility, *Theoretical Computer Science* 284 (2002), 519 – 538.

60. D. S. SCOTT, A theory of computable functionals of higher type, Unpublished notes, Oxford University, 1969.

61. D. S. SCOTT, Continuous lattices, in F. W. Lawvere (ed.), *Toposes, Algebraic Geometry and Logic*, Springer Lecture Notes in Mathematics, volume 274, Springer–Verlag, Berlin, 1972, 97 – 136.

62. D. S. SCOTT, A type-theoretical alternative to ISWIM, CUCH, OWHY, *Theoretical Computer Science* 121 (1993), 411 – 440.

63. M. B. SMYTH, Topology, in S. Abramsky, D. Gabbay, and T. S. E. Maibaum (eds.), *Handbook of Logic in Computer Science*, Volume 1, Oxford University Press, 1992, 641 – 751.

64. D. SPREEN, Effective inseparability in a topological setting, *Annals of Pure and Applied Logic* 80 (1996), 257 – 275.

65. D. SPREEN, On effective topological spaces, *Journal of Symbolic Logic* 63 (1998), 185 – 221.

66. D. SPREEN, Representations versus numberings: on the relationship of two computability notions, *Theoretical Computer Science* 262 (2001), 473 – 499.

67. D. SPREEN AND H. SCHULZ, On the Equivalence of some approaches to computability on the real line, in K. Keimel, G. Q. Zhang, Y. Liu and Y. Chen (eds.), *Domains and Processes*, Proc. 1st Intern. Symp. on Domain Theory, Shanghai, China, 1999, Kluwer, Boston, 2001, 67 – 101.

68. V. STOLTENBERG-HANSEN, I. LINDSTRÖM AND E. R. GRIFFOR, *Mathematical Theory of Domains*, Cambridge University Press, 1994.

69. V. STOLTENBERG-HANSEN AND J. V. TUCKER, Complete local rings as domains, Report 1.85, Centre for Theoretical Computer Science, University of Leeds, Leeds, 1985.

70. V. STOLTENBERG-HANSEN AND J. V. TUCKER, Complete local rings as domains, *Journal of Symbolic Logic* 53 (1988), 603 – 624.

71. V. STOLTENBERG-HANSEN AND J. V. TUCKER, Algebraic equations and fixed-point equations in inverse limits, *Theoretical Computer Science* 87 (1991), 1 – 24.

72. V. STOLTENBERG-HANSEN AND J. V. TUCKER, Infinite systems of equations over inverse limits and infinite synchronous concurrent algorithms, in J. W. de Bakker, W. P. de Roever and G. Rozenberg (eds.), *Semantics – Foundations and Applications*, Lecture Notes in Computer Science, volume 666, Springer Verlag, Berlin, 1993, 531 – 562.

73. V. STOLTENBERG-HANSEN AND J. V. TUCKER, Effective algebra, in S. Abramsky, D. Gabbay, and T. S. E. Maibaum (eds.), *Handbook of Logic in Computer Science*, Volume 4, Oxford University Press, 1995, 357 – 526.

74. V. STOLTENBERG-HANSEN AND J. V. TUCKER, Computable rings and fields, in E. Griffor (ed.), *Handbook of Computability Theory*, Elsevier, 1999, 363 – 447.

75. V. STOLTENBERG-HANSEN AND J. V. TUCKER, Concrete models of computation for topological algebras, *Theoretical Computer Science* 219 (1999), 347 – 378.

76. V. STOLTENBERG-HANSEN AND J. V. TUCKER, Computable and continuous partial homomorphisms on metric partial algebras, *Bulletin of Symbolic Logic* 9 (2003), 299 – 334.

77. A. TARSKI, A lattice-theoretical fixed point theorem and its applications, *Pacific Journal of Mathematics* 5 (1955), 285 – 309.

78. J. V. TUCKER AND J. I. ZUCKER, Computation by while programs on topological partial algebras, *Theoretical Computer Science* 219 (1999), 379 – 421.

79. J. V. TUCKER AND J. I. ZUCKER, Computable functions and semicomputable sets on many sorted algebras, in S. Abramsky, D. Gabbay and T. Maibaum (eds.), *Handbook of Logic for Computer Science*, Volume 5, Oxford University Press, 2000, 317 – 523.

80. J. V. TUCKER AND J. I. ZUCKER, Abstract versus concrete computation on metric partial algebras, *ACM Transactions on Computational Logic* 5 (4) (2004), 611–668.

81. J. V. TUCKER AND J. I. ZUCKER, Abstract versus concrete computability: The case of countable algebras, in V. Stoltenberg-Hansen and J. Väänänen (eds.), *Logic Colloquium 03, Proceedings of Annual European Summer Meeting of Association for Symbolic Logic, Helsinki, 2003*, Lecture Notes in Logic 24, Association for Symbolic Logic, 2006, 377 – 408.

82. G. WAAGBØ, Denotational semantics for intuitionistic type theory using a hierarchy of domains with totality, *Archive for Mathematical Logic* 38 (1999), 19 – 60.

83. K. WEIHRAUCH, *Computability*, Springer Verlag, New York, 1987.

84. K. WEIHRAUCH, *Computable Analysis*, Springer Verlag, New York, 2000.

85. K. WEIHRAUCH AND U. SCHREIBER Embedding metric spaces into cpo's, *Theoretical Computer Science* 16 (1981), 5 – 24.

86. K. WEIHRAUCH AND N. ZHONG, Computability theory of generalized functions, *Journal of the ACM* 50 (2003), 469 – 505.

On the Power of Broadcasting in Mobile Computing

Jiří Wiedermann[1]* and Dana Pardubská[2]†

[1] Institute of Computer Science, Academy of Sciences of the Czech Republic,
182 07 Prague 8, Czech Republic
jiri.wiedermann@cs.cas.cz
[2] Department of Computer Science, Comenius University,
842 48 Bratislava, Slovakia
pardubska@fmph.uniba.sk

Summary. A computational model reflecting fundamental computational aspects of wirelessly communicating mobile processors is presented. In essence, our model is a deterministic Turing machine that can launch new processes among which a wireless communication via explicitly assigned channels must be programmed. We show that computations of such machines are polynomially time- and space-equivalent to the synchronized alternating Turing machines studied previously in the literature. This shows that nondeterminism can be completely eliminated from synchronized alternation at the price of introducing a program-driven communication among the respective processors.

1 Introduction

Recent progress in wireless and mobile information technologies has caused an increased interest in algorithmic aspects of the underlying computing and communicating mechanisms. It seems that so far the respective research has mainly concentrated on the concrete algorithmic issues, neglecting almost completely the computational complexity aspects in that kind of computing. This might be due to the lack of formal computational models underlying the wireless mobile computing. If one tries to identify the basic computational and communication properties of the respective mode of computing, after a considerable simplification, one arrives at a notion of dynamically reconfigurable nets of mobile processors that can communicate one with each other via a "radio." On a sufficiently high level of abstraction, this can be modelled as though the processes communicated over a set of channels, with messages

* The research was carried out within the institutional research plan AV0Z10300504 and partially supported by Grant 1ET100300517.
† The research was partially supported by Grant VEGA 1/3106/06.

broadcasted over different channels being "heard" by any other processes tuned to the respective channels.

What can be said about the computational power and efficiency of the "computational model" we have just sketched? Does wireless mobile computing as captured by the simplified model bring new quality into computing when compared with the classical ways of computing? Answers to these questions will present the main motivation for this paper. In order to get such answers we will devise a recursion-theoretic model capturing the elementary features of mobile wireless computing and we will investigate its computational efficiency. Our model is a parallel deterministic Turing machine that can spawn parallel processes, which can create communicating groups. The main design idea of our model has been its transparency as far as the communication mechanism among the processes is concerned. That is, in our model establishing a connection within different subsets of processors requires explicit allocation of different communication channels to the respective processes. Moreover, the same mechanism of inter-processor communication is also used for detecting termination condition. The above-mentioned transparency makes visible (and thus, chargeable) the activities that in the case of alternating Turing machines occur "behind the scene" and bring free benefits.

In our investigations we have been primarily inspired by the concept of alternation (cf. [1]). In complexity theory this concept is seen as the theoretically neatest framework for studying parallel computations. It might look as quite far-fetched to try to see the alternating processes as wirelessly communicating mobile processes, but it is not entirely so. In a classical alternating Turing machine, the "wireless" communication among its running processes is used in order to determine the termination of the whole computation. The mobility of processes is captured by their dynamic emergence and extinction and by the fact that they are not associated with any concrete location.

The second source of inspiration to our work has been offered by so-called synchronized alternation, which is a more general concept that the classical alternation. Synchronized alternating Turing machines have been investigated since the end of the 1990s ([3], [5]). The motivation behind these studies was similar to ours: what is the benefit of an additional communication among processes of a running alternating machine? In our terms, in the original works on synchronized alternation (cf. [3] and [5]), only a single channel communication was considered. It appeared that the respective machines had the same time efficiency as the classic alternating Turing machines, but they were more space efficient than the latter machines: their logarithmic space had the same power as their polynomial time. This property is not known to hold for the classical alternating Turing machines.

In our modeling, we have disposed both of nondeterminism and of the acceptance mechanism, which both are crucial ingredients of alternation (either a classical or a synchronized one). Instead, we introduced a versatile deterministic inter-processor message exchange mechanism. Surprisingly, our main result states that these changes compensate for the loss of the respective abilities. In fact, our entirely deterministic parallel machines are equivalent to the synchronized alternating Turing machines.

We are far from claiming that our model represents an ideal model of mobile wireless computing. Nevertheless, we believe that the benefit of having such a model is threefold. First, our model characterizes the computational power of a certain type of wireless mobile computing. Second and perhaps more importantly, the model shows that nondeterminism and the acceptance mechanism of classical alternating machines are not necessary in order to get the full computational power of synchronized Turing machines. Last but not least, the new model leads to an alternative machine characterization of PSPACE and EXPSPACE, or APTIME and AEXPTIME, respectively. Namely, it shows that synchronized alternation is equivalent to deterministic parallelism enhanced by an explicit communication mechanism.

The paper is structured as follows. In Section 2 we present our model of so-called wireless parallel Turing machine and introduce the respective complexity measures. Next, in Section 3 we state the main complexity characterizations of computations by such a machine. Finally, in Section 4 we review the achievements of the paper.

2 Wireless Parallel Turing Machine

In order to arrive at a computational model in which processes communicate via broadcasting we use the multiplicative ability of processes of an alternating Turing machine in universal states and enhance the latter machine by a special mechanism that enables "wireless" information transfer among the processes that have "tuned in" to the same channel. We will also dispose of the acceptance mechanism of alternating machines whose activity will be substituted by a versatile communication mechanism.

Definition 2.1. A k-tape wireless parallel Turing machine (WPTM) with a separate read-only input tape and a separate channel tape is an 11-tuple $M = (k, Q, R, \Sigma, \Gamma, \Delta, q_0, r_0, \varepsilon, q_{\text{accept}}, q_{\text{reject}})$, where

- k is the number of work tapes;

- $Q \times R$ is the finite set of states;

  - Q is the set of working states with the initial state $q_0 \in Q$;

  - R is the set of communication states also containing four distinguished states: initial communication state r_0, empty communication state ε, and states $q_{\text{accept}}, q_{\text{reject}}$, which are *accepting* and *rejecting* states, respectively;

- Σ is a finite input alphabet ($\$ \notin \Sigma$ is an endmarker);

- Γ is a finite work tape alphabet ($\natural \in \Gamma$ is the blank symbol, $\natural \notin \Sigma$);

- $\Delta \subseteq Q \times R \times (\Sigma \cup \{\$\}) \times \Gamma^{k+1} \times Q \times R \times (\Gamma - \{\sharp\})^{k+1} \times \{-1, 1\}^{k+2}$ is the next move relation.

The elements of Δ are called transitions. The machine has a read-only input tape with endmarkers, k work tapes, and one channel tape. The work tapes and the channel tape, jointly referred to as tapes, are initially blank. The tapes are unbounded to the right, with their cells numbered from 0.

Let $\delta = \langle q, r, x, a_1, \ldots, a_{k+1}, q', r', a_1', \ldots, a_{k+1}', d_1, \ldots, d_{k+2} \rangle \in \Delta$ be a transition of M. According to this transition in a single *step* M finding itself in working state q, in communication state r, reading symbol x from the input tape and a_i from the i-th tape, for $i = 1, 2, \ldots, k + 1$, enters a new working state q', new communication state r', writes symbol a_i' on the i-th tape and moves each of the $k + 2$ heads in direction d_j (left or right) one tape cell, for $j = 1, 2, \ldots, k + 2$.

A *configuration* of a WPTM M is an element of $Q \times R \times \Sigma^* \times ((\Gamma - \{\sharp\})^*)^{k+1} \times \mathbb{N}^{k+2}$, representing the working and communication state of the finite control, the input, the nonblank contents of $k + 1$ tapes, and $k + 2$ head positions.

A *head configuration* of M is an element of $Q \times R \times (\Sigma \cup \{\$\}) \times \Gamma^{k+1}$ representing the working and communication state of the finite control and the contents of cells scanned by each head. Note that for a given head configuration, there can exist several transitions in Δ applicable in that configuration. We will later see that this possibility provides the parallelism to our model.

We say that a transition with the new communication state $r' \neq \varepsilon$ broadcasts state $r' \in R$. Such a transition is called a *broadcasting transition*. There is one syntactic restriction holding for broadcasting transitions called the "unanimous broadcast rule": *the broadcasting transitions pertinent to the same head configuration must all broadcast the same communication state*. They can differ in the remaining parts; i.e., they can prescribe entering different working states and different rewritings and movements. We say a configuration is *tuned to channel c* if it has string c written to the left from the current channel tape head position. If c is a nonempty string, then it is also called a *channel number*. A configuration tuned to c to which a transition with the new communication state $r' \neq \varepsilon$ applies is said to broadcast r' on channel c. A configuration broadcasting ε is considered effectively as a no-broadcasting (or *silent*) configuration (cf. the definition of a transition modified by a broadcasting below). The silent transitions are useful in situations when a process "does not want" to broadcast any information, e.g., when retuning its channel.

A configuration β is a δ-successor of a configuration α with respect to transition $\delta \in \Delta$ (written as $\alpha \vdash^\delta \beta$) if β follows from α in one step, according to transition δ. The move $\alpha \vdash^\delta \beta$ is called a *simple step* of M. A configuration without successors is called a *terminal configuration*.

In order to define a computation of M, we need a couple of further preliminary definitions.

Function $Tuned : Q \times R \times \Sigma^* \times ((\Gamma - \{\sharp\})^*)^{k+1} \times \mathbb{N}^{k+2} \to (\Gamma - \{\sharp\})^*$ assigns to each configuration its channel number.

In a similar vein we define function $Broadcast : Q \times R \times \Sigma^* \times ((\Gamma - \{\sharp\})^*)^{k+1} \times \mathbb{N}^{k+2} \to R$ returning to each configuration the unique state broadcasted by the transition applicable to that configuration (remember the "unanimous broadcast rule"). With a slight abuse of notation this function will naturally be extended to a set $L \neq \emptyset$ of configurations as follows:

$$Broadcast(L) = \begin{cases} b & \text{iff for all } \alpha, \beta \in L, Tuned(\alpha) = Tuned(\beta) \\ & \text{and } Broadcast(\alpha) = b \\ \bot & \text{otherwise} \end{cases}$$

(symbol $\bot$ denotes an undefined value).

Finally, we define projection $Comm : Q \times R \times \Sigma^* \times ((\Gamma - \{\sharp\})^*)^{k+1} \times \mathbb{N}^{k+2} \to R$, assigning to each configuration its communication state.

For any communication states $u, v \in R$ and any configuration α in communication state u, the notation $\alpha|_{u:=v}$ denotes configuration α in which state u is changed to v.

Let L be a set of configurations, $L_c \subseteq L$ a subset of configurations tuned to c, and $\alpha \vdash^\delta \beta$ a simple step. Then configuration γ is a so-called δ_L-*successor* of α w.r.t. transition δ *modified by broadcasting* from L (denoted as $\alpha \vdash^{\delta_L} \gamma$) if γ is defined as follows:

$$\gamma := \begin{cases} \beta|_{Comm(\beta):=b} & \text{iff } L_{Tuned(\beta)} \neq \emptyset \text{ and } Broadcast(L_{Tuned(\beta)}) = b \\ \bot & \text{iff } L_{Tuned(\beta)} \neq \emptyset \text{ and } Broadcast(L_{Tuned(\beta)}) = \bot \\ \beta & \text{iff } L_{Tuned(\beta)} = \emptyset \text{ or } Broadcast(L_{Tuned(\beta)}) = \varepsilon. \end{cases}$$

The previous three items describe the effect of broadcasting from L on a configuration β. The items correspond to all possible situations that might occur: unanimous broadcast, conflict and no broadcast on the channel of interest, respectively.

Note that there can be several computational steps possible from a given α. This occurs when there are several transitions applicable to α. If this is the case, we say that α *spawns* all configurations γ for which there exists $\delta \in \Delta$ such that $\alpha \vdash^{\delta_L} \gamma$.

For any configuration ϱ, the *computational graph* $T(\varrho)$ w.r.t. the transition relation Δ of M is a rooted, directed, possibly infinite acyclic multigraph whose nodes are configurations of M and edges correspond to transition and communication links. This graph is defined inductively:

1. ϱ is the root of $T(\varrho)$ at depth $d = 0$;

2. Let C_d be the set of configurations at depth $d \geq 0$. Then for all nonterminal configurations $\alpha \in C_d$, set C_{d+1} contains all δ_{C_d}-successors of α; i.e., set

$C_{d+1} = \{\gamma \mid \exists \text{ nonterminal } \alpha \in C_d : \alpha \vdash^{\delta_{C_d}} \gamma\}$. We also say that α *spawns* its all δ_{C_d}-successors. If some of the δ_{C_d}-successors of α is undefined, then the whole graph $T(\varrho)$ is undefined.

3. In $T(\varrho)$ there are two kinds of edges:

   - so-called *transition edges* leading from each $\alpha \in C_d$ to each of its δ_{C_d}-successors $\gamma \in C_{d+1}$;

   - so-called *broadcasting edges* leading from each broadcasting configuration $\alpha \in C_d$ to each configuration $\beta \in C_{d+1}$, with $Tuned(\alpha) = Tuned(\beta)$.

Thus, the parallelism in M's computations is enforced by insisting, in condition 2, that in $T(\varrho)$ for a given configuration all its successor configurations (with their communication states possibly modified by broadcasting) spawned by that configuration are included. Note that the successor configurations in $T(\varrho)$ are spawned by a mechanism similar to universal branching used in the case of alternating Turing machines. The schema of a computational graph of a WPTM computation is depicted in Fig. 1.

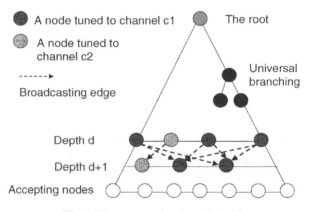

Fig. 1. The computational graph of M

From the previous description it is seen that the computational graph is built in a completely deterministic manner: in this graph, all transition and broadcasting edges are defined uniquely. Formally, it is a multigraph since there can be one transition and one broadcasting edge between some nodes of this graph.

A *computational path* starting in configuration ϱ is a (possibly infinite) sequence of configurations of M that are encountered during a traversal along the transition edges down any path in $T(\varrho)$. Any computational path represents a computation of M along that path or a *process* corresponding to that path.

The nodes of $T(\varrho)$ without successors are called the *leaves* of $T(\varrho)$.

A computational graph $T(\varrho)$ of M is a *computational graph accepting input w* if it satisfies the following conditions:

1. *Finiteness:* $T(\varrho)$ is a finite graph.

2. *Initial condition:* $\varrho = (q_0, r_0, w, \underbrace{\nu, \ldots, \nu}_{k+1}, \underbrace{0, \ldots, 0}_{k+2})$ is the *initial configuration* where ν is the null string.

3. *Acceptance agreement:* All leaves of $T(\varrho)$ are terminal configurations at the same depth, in communication state q_{accept} and tuned to the same channel.

In a similar way, the notion of a computational graph rejecting the input string is defined.

From a practical point of view the acceptance agreement means that all processes share the information that all of them have accepted the input. If the termination time is unknown, then the acceptance in the same depth can be achieved, e.g., via barrier synchronization. The idea is to periodically check, synchronously in all processes, their "readiness" to terminate the computation (cf. [4]).

We say that M *accepts w* if M's computational graph accepts w; we define $L(M)$ to be the set of strings accepted by M.

The working space (channel space) complexity of a configuration is the sum of lengths of the nonblank contents of corresponding work tapes (the length of the channel tape). The working space of a computational graph T is the maximum work space of any configuration in T; the channel space of T is defined similarly. The time of T is the maximum length of any path in T.

A WPTM M *operates in work space $S(n)$ (channel space $C(n)$)* if for every string $w \in L(M)$ of length n there is a computational graph of M of working space at most $S(n)$ (channel space $C(n)$) that accepts w. Similarly, M *operates in time $T(n)$* if for every string $w \in L(M)$ of length n there is a computational graph of M of time at most $T(n)$ that accepts w.

The introduction of a separate space measure based on the size of the channel tape is motivated by the wireless mobile computing. The channel space measure hints to the size of the respective communication mechanism.

The resulting model can also be characterized as a co-nondeterministic machine (i.e., an alternating Turing machine with universal states only) enhanced by a certain processor inter-communication mechanism and without the acceptance mechanism of alternating Turing machines: the acceptance criterion must be programmed in the machine's instructions. The deterministic nature of a WPTM contributes to the "realness" of this model as far as the wireless mobile computations are concerned.

When the transition relation becomes a function it is easily seen that a WPTM turns into the classical deterministic Turing machine. Similarly, it is an easy exercise to realize that a WPTM can simulate in linear time a nondeterministic or a co-nondeterministic Turing machine. Although for both machines the acceptance criteria are different, the WPTM can easily accommodate any of them. In fact, we will show that the WPTMs and alternating Turing machines are polynomially time related.

3 The Power of the Wireless Communication

We start by comparing the WPTMs with the classical ATMs.

Theorem 3.1. *Let $T(n)$ be a time constructible function with $T(n) \geq n$. Let A be an alternating Turing machine of time complexity $T(n)$. Then there is a WPTM M simulating A in time $O(T(n))$ and in working and channel space $O(T(n))$.*

Sketch of the proof: W.l.o.g. assume that in the computation tree T_A of A all branchings are of degre at most 2. Design M as follows: starting from the initial configuration of A, M applies to the emerging configurations all applicable transitions of A irrespective of whether the configuration was an existential or an universal one, and in this manner proceeds as though descending the levels in T_A. However, in their working memory, the processes in M also remember the so-called *routing information* representing the computational path in T_A from its root to a configuration c at depth d describing the respective process at time d. The routing information for a configuration c stored at v is of form $\langle (b_1, \square_1), \ldots, (b_d, \square_d) \rangle$, with $b_i \in \{0, 1\}$ and $\square_i \in \{\exists, \forall, |.\}$ The binary number $(b_1, \ldots, b_d)_2$ is also called the *path number*. The quantifier $\exists$ denotes the existential branching at v, $\forall$ denotes the universal branching (with two successors), and $|$ denotes "no branching"—a deterministic configuration having but one successor. Routing information description is at most of length $O(T(n))$.

After simulating $T(n)$ steps (remember that $T(n)$ has been a time constructible function) of A, the processes of M start to verify the acceptance condition for A. Essentially, this condition states how the "answers" from the individual terminated processes must be combined while "climbing" toward the root of T_A (cf. [1]).

Evaluation phase starts by assigning values Y, N, or $\bot$ to local variable $qual$ in each (leaf-)process of M, which has accepted, rejected, or has not finished its task, respectively. Then, the processes simulate "climbing" up the T_A so that at the end of the so-called d-round the value of $qual$ in the process corresponds to a value of a vertex at depth d identified by the path number $(b_1, \ldots, b_d)_2$ stored in that process.

Let $S_{\langle b_1, \ldots, b_d \rangle}$ denotes the set of processes of M whose path numbers share the prefix $(b_1, \ldots, b_d)_2$. Note that the values of $\square_i$ for $i = 1, \ldots, d$, resp. $qual$ for all processes from $S_{\langle b_1, \ldots, b_d \rangle}$, coincide.

At the beginning of the d-round, each process in $S_{\langle b_1, \ldots, b_d \rangle}$ inspects the value of $\square_d$ in its routing information.

If it was "|", then the process does not change the value of $qual$.

If it was "∀" or "∃", then there exist two "brother" sets of processes $S_{\langle b_1,...,b_d,0\rangle} = S_1$ and $S_{\langle b_1,...,b_d,1\rangle} = S_2$ within $S_{\langle b_1,...,b_d\rangle} = S$. Clearly, S_1 and S_2 are disjoint, $S_1 \cup S_2 = S$, and for all processes in S_1, $qual$ has the same value, and the same holds for S_2. In order to compute the new value $qual$ in S from knowing old $qual$ both in S_1 and S_2, respectively, both sets must broadcast the values of their respective $qual$ to all processes in S. The broadcasting is performed over channel no. $(b_1 \ldots b_d)_2$: first, the processes from S_1 broadcast their value to processes of S, and the same is then done by processes of S_2. Due to the fact that the values of $qual$ are the same within each S_1 and S_2, respectively, no broadcasting conflict can happen. As a result, in accordance with the rules for quality assignment in alternation trees (cf. [1]), each process in S can compute its quality at depth d. Obviously, all processes in S will get the same value for $qual$. Note that for all sets $S_{\langle b_1,...,b_d\rangle}$, all the necessary communication is carried out over the disjoint channels. The round is finished by setting $d := d - 1$.

In a similar way we proceed level by level, until we get $d = 1$, and in all processes, $qual$ gets the acceptance value of the root of T_A. Based on this, a broadcasting transition on channel 0 into the terminating communication state q_{accept} or q_{reject} can be realized in all processes and the simulation can terminate.

We conclude that in $O(T(n))$ steps, the acceptance condition for A can be verified by M, indeed.   □

Theorem 3.2. *Let $T(n)$ be a time constructible function with $T(n) \geq n$. Let M be a WPTM of time complexity $T(n)$. Then M can be simulated by an alternating Turing machine A of time complexity $O(T^2(n))$ and space complexity $O(T(n))$.*

Sketch of the proof: Let T_M be the computational graph of M on input w. T_M is a superposition of two graphs: w.l.o.g. we can assume that the first is the binary tree—called the *spawning tree* hereafter—underlying T_M and capturing the processes spawning history. The second graph is the *broadcasting graph* capturing the broadcasting of communication states from the broadcasting configurations to the respective target configurations. If T_M is an accepting computational graph, then, w.l.o.g., the leaf configurations are tuned to channel 0.

The idea of the simulation algorithm is to guess accepting configurations and to check that these configurations are leaves of an accepting computational graph T_M for input w. Note that the leaf configurations are determined uniquely by M and w thanks to the fact that any WPTM is a fully deterministic device.

To verify the before-mentioned property we will first guess the accepting configurations c_v in the leaves of T_M by creating pairs (path number p_v, accepting configuration c_v). Then, we will reconstruct T_M in a bottom-up manner by checking in parallel for each leaf v in T_M that the computational path with path number p_v ends up in configuration c_v. In fact, we will traverse the computational path in a bottom-up manner, from the leaf to the root and verify that the path ends up in the initial configuration. Within this process we will have to guess repeatedly and verify the subgraphs of T_M consisting of all paths leading from the root to a given node. Due to the deterministic nature of M's computation such paths are unique, and therefore their repeated bottom-up traversals will be possible.

To better understand how the bottom-up traversal is done, let us focus on one computational step of M first. Let v be a vertex of T_M at depth d, and let p_v, c_v be the corresponding path number and configuration, respectively. Note that $d = |p_v|$, with $|p_v|$ denoting the length of the (binary) representation of p_v. With the exception of the communication state, configuration c_v is determined uniquely by the configuration in the predecessor v_1 of v in the spawning tree. The communication state of v is determined by the set $S_{(|p_v|-1,Tuned(c_v))}$ of v's predecessor vertices in the broadcasting graph (see Fig. 2). For a correct realization of a computational step, the condition of unanimous broadcasting must be satisfied; i.e., all vertices from $S_{(|p_v|-1,Tuned(c_v))}$ must broadcast the same communication state $Comm(c_v)$.

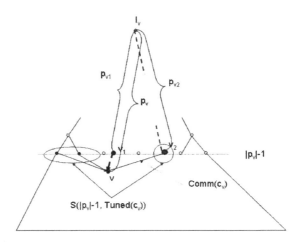

Fig. 2. The schema of broadcasting in T_M

In order to proceed from v to v_1 we will (i) *guess* a configuration c_{v_1}, (ii) *guess* a vertex $v_2 \in S_{(|p_v|-1,Tuned(c_v))}$, and (iii) verify that all vertices from $S_{(|p_v|-1,Tuned(c_v))}$ broadcast the same communicating state $Comm(c_v)$.

The simulation algorithms starts in a so-called *preparatory phase*. In this phase, in universal mode, A creates $2^{T(n)}$ processes at depth $T(n)$ making use of the constructibility of $T(n)$. For each such a process A computes its path number and guesses the respective terminal leaf configuration in T_M with channel number $cons$ set to 0 (for the acceptance agreement) and communication state q_{accept}. Thanks to the fact that all configurations are accepting ones, no broadcasting conflict can occur at depth $T(n)$. Then, in every process just created, procedure $Reachability(p_v, c_v)$ is issued to verify whether a configuration c_v is reachable from the root via a specified computational path p_v assuming no broadcasting conflicts occur in T_M whenever necessary.

$Reachability(p_v, c_v)$ works as follows.

If $|p_v| = 0$, then there cannot be any broadcasting from configurations at a higher level and hence $Reachability(\lambda, c_v)$, with λ denoting the empty string of length 0, straightforwardly verifies whether c_v equals the initial configuration and returns the corresponding answer.

If $|p_v| > 0$, from p_v the process computes $p_{v_1} = p_v \div 2$, guesses c_{v_1}, and then existentially splits into two branches.

The first branch corresponds to the case when $c_{v_1} \vdash^\delta c_v$, i.e., when the transition δ from v_1 to v was a simple move (not using any broadcasting). Thus we have to verify the existence of $\delta \in \Delta$ such that $c_{v_1} \vdash^\delta c_v$, that c_{v_1} is reachable (via path p_{v_1}), and that there was no broadcasting on channel $Tuned(c_v)$. The first condition is checked by existentially trying all δ's for $c_{v_1} \vdash^\delta c_v$. Once a matching δ is found we verify the remaining two conditions by splitting universally and calling $Reachability(p_{v_1}, c_{v_1})$ on one branch and procedure $UndisturbedBroadcast(|p_{v_1}|, Tuned(c_v), \varepsilon)$ on the other branch (see the description of procedure $UndisturbedBroadcast$ in the sequel). If the move $c_{v_1} \vdash^\delta c_v$ could not have been verified, then the procedure rejects.

The second branch corresponds to the case when state $Broadcast(c_{v_2}) = b$ (say) was broadcasted to c_v on channel $h = Tuned(c_v)$. In this case $Reachability$ guesses p_{v_2} and c_{v_2} such that $Tuned(c_{v_2}) = h$. Then it universally creates a process for every δ such that $v_1 \vdash^{\delta\{v_2\}} v$. If no such δ exists, procedure $Reachability$ rejects. Otherwise, the procedure splits universally into three processes performing in parallel procedure $Reachability(p_{v_1}, c_{v_1})$, procedure $Reachability(p_{v_2}, c_{v_2})$, and procedure $UndisturbedBroadcast(|p_{v_2}|, h, b)$, respectively.

$UndisturbedBbroadcast(d, h, b)$ verifies that no other state than b is broadcasted on channel h at depth d in T_M. The other way around, no configuration tuned to h and broadcasting $b' \neq b$ is reachable in T_M at depth d.

To check the latter property the procedure splits universally, systematically creating a process for each path number p between $\underbrace{(0, \ldots, 0)_2}_{d\ times}$ and $\underbrace{(1, \ldots, 1)_2}_{d\ times}$, each communicating state $b' \in R$, $b' \neq b$ (or $b' \in R$ iff $b = \varepsilon$, respectively), and configuration c of size at most d such that c broadcasts b' on channel h. On each, such a branch $Nonreachability(p, c)$ (described in the sequel) is called in order to verify that in T_M configuration c is not reachable via path p.

$Nonreachability(p, c)$ checks that the statement "in T_M, path p starting in initial configuration leads to c" is false. In fact, we want $Nonreachability(p, c) = \neg(Reachability(p, c))$ to hold. Hence, $Nonreachability(p, c)$ works much like

$Reachability(p, c)$ with the substantial difference that existential states are replaced by universal ones and vice versa, and instead of $UndisturbedBroadcast$ procedure, $Broadcast$ (described in the sequel) is called. $Nonreachability(p, c)$ accepts if and only if $Reachability(p, c)$ called with the same parameters rejects or is not defined.

$Broadcast(d, h, b)$ verifies that state b is broadcasted on channel h by some configuration at depth d in T_M. For that purpose a process is existentially created for every path number $p \in \{0, 1\}^d$ and configuration c, which could in principle occur in T_M at depth d tuned to h broadcasting a communication symbol b. Then, $Reachability(p, c)$ is called in each such process.

The simulation of M by A ends successfully by accepting the input if and only if all calls of checking procedures issued in the verification phase end successfully.

Now a few words regarding the correctness of the simulation algorithm are in order. The preparatory phase obviously serves its purpose. Subsequently, procedure $Reachability$ was used to ensure the reachability of nodes of T_M from the root. However, this property has been checked under the condition that there were no broadcasting conflicts in T_M. Such conflicts, if any, were detected by $UndisturbedBraodcast$ by inspecting all configurations in T_M possibly "jamming" a state broadcasted on a given channel at the given level. In this way the correctness of guesses of $Reachbility$ concerning configurations in T_M modified by broadcasting on the given channel has been either verified or a "mismatch" in broadcasted communication states has been detected, if any.

Finally we determine the complexity of A. Obviously, the preparatory phase is of time complexity $O(T(n))$. The complexity of the next phase is quadratic w.r.t. $T(n)$ since by each call of the checking procedures, the depth parameter in the calls decreases by 1 until the calls terminate at depth 0. However, preparing the necessary parameters for the calls also takes time $O(T(n))$, which leads to the quadratic time complexity of the verification phase.

A more detailed proof including the formal description of all before-mentioned procedures is given in the technical report [6].   □

In what follows, in addition to the standard deterministic complexity classes such as **LOGSPACE**, **PTIME**, **PSPACE**, and **EXPTIME**, we will also use their analogs defined for the alternating and wireless parallel Turing machines. These classes will be denoted by prefixing the standard complexity classes listed before by **A** or **W**, respectively.

The following corollary characterizing the power of WPTM's polynomial time is the consequence of both previous theorems and of known properties of alternating complexity classes (cf. [1]).

Corollary 3.3. *For any time constructible function* $T(n)$ *with* $T(n) \geq n$,

$$\bigcup_{k>0} \text{WTIME}(\text{T}^k(n)) = \bigcup_{k>0} \text{ATIME}(\text{T}^k(n));$$

i.e.,

$$\text{WPTIME} = \text{APTIME} = \text{PSPACE}.$$

Next, we turn our attention to the relation between deterministic and wireless mobile space.

Theorem 3.4. *Let* D *be a deterministic Turing machine of space complexity* $S(n)$. *Then* D *can be simulated by a WPTM* M *of channel space complexity* $O(\log S(n))$ *and of working tape complexity* $O(1)$.

Sketch of the proof: The proof mirrors the proofs of similar theorems known in the theory of synchronized alternation (cf. [2], [5]). That is, the contents of cells of a single tape of D are kept in processes of M. A new process is spawned each time when the head of D on its work tape enters a blank cell. The cells are numbered by their position number and so are the processes. Thus, on their working tape, process no. i remembers the symbol written in the i-th cell, the presence of the working head of D at this cell (yes or no), and the state of D if head scans the i-th cell of D. Number i in binary is stored at the channel tape of the i-th process. M simulates D by following its moves and updating the information in the processes corresponding to the respective updated cells and state changes of D. The "head movements" from the i-th process (cell) are realized by broadcasting the respective "message" to the cell's left or right neighbor on channel $(i-1)$ or $(i+1)$ whose number equals the index of that cell. Thus, retuning a channel amounts to adding or subtracting 1 from the current channel number. Clearly, channel space complexity of the simulating machine is $O(\log S(n))$, whereas working space complexity is $O(1)$.   □

Theorem 3.5. *Let* $S(n) \geq \log n$. *Let* M *be a WPTM of both working and channel space complexity* $O(S(n))$. *Then there exists* $c > 0$ *such that* M *can be simulated by a deterministic Turing machine* D *in space* $O(c^{S(n)})$.

Sketch of the proof: Consider the computational graph T_M of M on input w. Each configuration in T_M is of size $O(S(n))$; in this estimate the input head position is included thanks to our assumption on the size of $S(n)$. If w is accepted by M, then at most $O(c_1^{S(n)})$ different processes can occur in M. It follows that for a suitable $c > c_1$ D has enough space to write down all the respective configurations, and within this space, it can easily keep track of all M's actions. D will accept if and only if its computation will correspond to an accepting tree of M on w.   □

From theorems 3 and 4 and from Corollary 1, we get further characterizations of WPTM complexity classes.

Corollary 3.6. *For any* $S(n) \geq \log n$

$$\mathsf{WSPACE}(S(n)) = \bigcup_{c>0} \mathsf{DSPACE}(c^{S(n)}),$$

WLOGSPACE = PSPACE = WPTIME, WPSPACE
= EXPSPACE = WEXPTIME.

From corollaries 1 and 2, we see that the fundamental complexity classes, viz. logarithmic space and polynomial time, coincide both for synchronized alternating (cf. [5] and [3]) and wireless parallel Turing machines. Thus, w.r.t. these classes both models are equivalent. Especially note that similarly as synchronized alternating Turing machines, WPTMs use their space in an optimal manner—e.g., "wireless" polynomial time equals "wireless" logarithmic space.

4 Conclusion

The effort of designing a computational model of wireless mobile nets has brought an unexpected result. In addition to the original goal of characterizing the computational power of a certain kind of such nets, our results have also thrown new light on the nature of computational resources used by the classical and synchronized alternation. Although the (synchronized) alternation machines with their unrestricted use of nondeterminism and their fancy acceptance mechanism seem to be an unrealistic computational model, we have shown that these machines are in fact equivalent to deterministic parallel devices possessing the ability of information exchange via broadcasting on unbounded number of different channels. On the one hand, these results rank the wireless mobile nets under our consideration among the very powerful computational devices. On the other hand, the same results confirm the fact that the very idea of alternation is not so far from the reality as it might appear at first glance.

In the future it would be of interest to study the computational power of nondeterministic wireless parallel Turing machines.

References

1. Chandra, A., Kozen, D., Stockmeyer, L.: Alternation. Journal of the ACM, **28**, 114–133, 1981.
2. Geffert, V.: A Communication Hierarchy of Parallel Computations. Theor. Comput. Sci., **198**, No. 1-2, 99–130, 1998.
3. Hromkovič, J., Karhumäki, J., Rovan, B., Slobodová, A.: On the power of synchronization in parallel computations. Discrete Appl. Math. **32**, 155–182, (1991).

4. Valiant, L. G.: A Bridging Model for Parallel Computation, Comm. of the ACM, **33**, No. 8, 103–111, 1990.
5. Wiedermann, J.: On the Power of Synchronization. Elektronische Informationsverarbeitung und Kybernetik (EIK), **25**, No. 10, 499–506, 1989.
6. Wiedermann, J., Pardubská, D.: On the power of broadcasting in mobile computing. Technical report V–944, Institute of Computer Science, Academy of Sciences of the Czech Republic, Prague 2005, 17 p.

Part II

Logic, Algorithms and Complexity

The Computational Power of Bounded Arithmetic from the Predicative Viewpoint

Samuel R. Buss*

Department of Mathematics
University of California, San Diego
La Jolla, CA 92093-0112, U.S.A.
sbuss@math.ucsd.edu

Summary. This paper considers theories of bounded arithmetic that are predicative in the sense of Nelson, that is, theories that are interpretable in Robinson's Q. We give a nearly exact characterization of functions that can be total in predicative bounded theories. As an upper bound, any such function has a polynomial growth rate and its bit-graph is in nondeterministic exponential time and in co-nondeterministic exponential time. In fact, any function uniquely defined in a bounded theory of arithmetic lies in this class. Conversely, any function that is in this class (provably in $I\Delta_0 + \exp$) can be uniquely defined and total in a (predicative) bounded theory of arithmetic.

1 Introduction

Theories of bounded arithmetic and their associated provable total functions have been studied extensively for over two decades. Bounded arithmetic arose originally from the definition of $I\Delta_0$ by Parikh. A subsequent development by Nelson of his "predicative" theories gave an alternate route to bounded arithmetic. The present author's thesis [2] introduced the fragments S_2^i and T_2^i of bounded arithmetic (with Nelson's smash function present), and these have been extended over the years to a proliferation of theories of bounded arithmetic that have good characterizations of their provably total functions in terms of computational complexity. To mention only a few such characterizations, the Δ_0-definable predicates of $I\Delta_0$ are precisely the functions in the linear time hierarchy [7, 8], the provably total functions of S_2^1 are precisely the polynomial time computable functions [2], the provably total functions of T_2^1 are precisely the projections of polynomial local search (PLS) functions [4], Clote and Takeuti [5] gave formal theories that capture log space functions and alternating log time, and Arai [1] defined a system AID that better captures alternating log time. A large number of further bounded theories of arithmetic have been

---

* Supported in part by NSF grant DMS-0400848.

214   Samuel R. Buss

formulated by others, including Zambella, Cook, and several students of Cook; see Cook and Nguyen [6] for a partial survey.

This paper returns to (one of) the historical motivations for bounded arithmetic by considering the computational complexity of functions that are definable in predicative theories of arithmetic. By "predicative" is meant in the sense of Nelson, namely, interpretable in Robinson's theory Q. Nelson introduced this notion of predicative because of his finitistic formalist philosophy of mathematics. Remarkably, another group of researchers were independently investigating mathematically equivalent notions of interpretability in Q, not for philosophical reasons, but for investigations into independence results for arithmetic and into computational complexity. This latter line of research included the foundational results of Solovay [11], Paris and Dimitracopoulos [9], Wilkie and Paris [13], Wilkie [12], and Pudlák [10].

The goal of this paper is to study functions f that are defined in a predicative theory of arithmetic, and to characterize such functions in terms of computational complexity. We shall consider only bounded predicative theories and only theories that are interpretable with cut-interpretations. These restrictions are quite natural, since only cut-interpretations have been used for predicative theories to date, and at the current state of the art, we have essentially no ideas for what kinds of non-cut-interpretations could be defined, much less be useful.

The general outline of the paper is as follows. Although we presume the reader is familiar with both bounded arithmetic and predicative arithmetic, the next section gives some technical preliminaries necessary for our exposition. Then section 3 proves an upper bound on the complexity of predicative functions. This upper bound is actually a upper bound on the complexity of any function that is uniquely determined by a definition over a bounded theory of arithmetic (not necessarily a predicative theory). Section 4 gives lower bounds. These lower bounds state that any function within a certain complexity class can be defined by a function symbol in some bounded predicative theory. For nonpredicative theories, the upper and lower bounds match; namely, it is the class of functions whose bit-graph is in both nondeterministic exponential time and co-nondeterministic exponential time (see below for the exact definitions). For predicative theories, the lower bound further requires provability in the theory $I\Delta_0 + \exp$ of membership in this class.

2 Definitions

2.1 Cut-interpretability

When considering interpretability of a theory T in Q, we shall restrict our attention to theories T that (a) are Δ_0-axiomatizable and (b) are interpreted in an inductive cut. The first condition (a) means that T is axiomatized with a set of (universal closures of) bounded formulas over some language that extends the language of Q.

The second condition (b) means that the interpretation is relative to some *inductive* formula $\theta(x)$, for which Q proves the closure properties:

$$\theta(0) \;\wedge\; (\forall x)(\forall y)(\theta(x) \wedge y < x \rightarrow \theta(y))$$

and

$$(\forall x)(\theta(x) \rightarrow \theta(Sx)).$$

By compactness considerations, we may always assume without loss of generality that T has a finite language and is finitely axiomatized. Thus we do not need to worry about the distinction between local interpretability versus global interpretability.

We also may assume without loss of generality that any theory T contains S_2^1 (for instance) as a subtheory, and that any subset of the polynomial time functions and bounded axioms defining these functions and their properties are included in the theory T.

We frequently need the theory T to include induction for all bounded formulas in the language of T. In particular, if the theory T has non-logical language L, we write $I\Delta_0(L)$ to denote the set of induction axioms for all Δ_0-formulas over the language L. It is known (see Pudlák [10]) that if T is interpretable in Q, then so is $T + I\Delta_0(L)$. Thus, we may assume w.l.o.g. that any bounded theory T interpretable in Q includes the $I\Delta_0(L)$ axioms. For similar reasons, the theory T may be assumed to include the smash function $\#$ and its defining axioms; in this case, the $I\Delta_0(L)$ axioms imply all of $S_2(L)$.

An important fact about interpretability in Q is that the growth rate of functions can be characterized exactly. We define $|x|$ as usual to equal the length of the binary representation of x. Then we define

$$\Omega_0(x) \;=\; 2^{2|x|}$$

so that $\Omega_0(x) \approx x^2$, and we further define

$$\Omega_{i+1}(x) \;=\; 2^{\Omega_i(|x|)}.$$

We have $\Omega_1(x) = 2^{2^{2||x||}} \approx 2^{|x|^2} = x\#_2 x$, $\Omega_2(x) = 2^{2^{2^{2|||x|||}}} \approx x\#_3 x$, etc.

Solovay [11] proved that the functions $\Omega_i(x)$ are interpretable with inductive cuts in Q. Conversely, Wilkie [12] proved that any function f that is interpretable in an inductive cut in Q is bounded by some Ω_i; i.e., that for some i, $f(x) < \Omega_i(x)$ for all x. More generally, for a multivariable function symbol $f(x_1, \ldots, x_n)$, we say that f *is dominated eventually* by Ω_i provided that

$$(\exists \ell)(\forall x_1)(\forall x_2) \cdots (\forall x_n)(\ell \leq x_1 + x_2 + \cdots + x_n$$

$$\rightarrow f(x_1, x_2, \ldots, x_n) < \Omega_i(x_1 + x_2 + \cdots + x_n)).$$

Then, if f is interpretable in an inductive cut in Q, f is dominated eventually by Ω_i for some i. Furthermore, the fact that f is dominated by Ω_i is provable in a suitable theory (which is interpreted in Q). This is expressed by the following theorem.

Theorem 1 *Suppose T is a finite Δ_0-axiomatized theory, interpretable in Q with an inductive cut. Then a finite extension T' of T is also Δ_0-axiomatized and interpretable in Q with an inductive cut such that T' proves that every function symbol is eventually dominated by some Ω_i.*

Theorem 1 is a strengthening of the theorem of Wilkie [12]: its proof is beyond the scope of this paper, but the crucial point of the proof is that Lemmas 8 and 9 and Corollary 10 of [12] can be formalized in $I\Delta_0 + \exp$.

In view of the restriction on the growth rate of functions in inductive cuts in Q, we define the computational complexity class Ω_i-TIME to be the class

$$\Omega_i\text{-TIME} = \Omega_{i-1}(n^{O(1)})\text{-TIME}.$$

Here n indicates the length of an input, and it is easy to check that this means that the runtime of an Ω_i-TIME function on an input x of length $n = |x|$ is bounded by $|t(x)|$ for some term $t = \Omega_i^{(k)}(x)$, where $k \in \mathbb{N}$ and $\Omega_i^{(k)}$ indicates the k-fold composition of Ω_i.

Likewise we define analogs of exponential time by

$$\text{EXP}^i\text{-TIME} = 2^{\Omega_{i-1}(n^{O(1)})}\text{-TIME}.$$

The nondeterministic and co-nondeterministic time classes NEXP^i-TIME and coNEXP^i-TIME are defined similarly.

The bit-graph of a function f is the binary relation $BG_f(x, b, j)$ that is true exactly when the b-th bit of the binary representation of f is equal to j. Letting C be any of the time classes defined above, we define f to be in the complexity class C provided its bit-graph is in C. Note that, assuming f is a single-valued function (rather than a multifunction), f is in NEXP^i-TIME iff f is in coNEXP^i-TIME. In this case, we can say that f is in NEXP^i-TIME $\cap$ coNEXP^i-TIME.

2.2 Definition of Δ_0-interpretable function

This section presents the crucial definition of what is meant by a function being interpretable in Q by a bounded theory. The intuition is that there should be a theory T with language $L \cup \{f\}$ that is Δ_0-axiomatized and is interpretable in Q and that uniquely specifies the function f. The only tricky part of the definition is what it means to specify uniquely f: for instance, it would be cheating to have a function symbol $g \in L$ and an axiom $(\forall x)(f(x) = g(x))$, since this would merely beg the question of whether g is specified uniquely.

In order to formalize this properly, we let L^* be the language that is obtained by making a "copy" of L: for each symbol $g \in L$, there is a symbol $g^* \in L^*$ (g may be a function symbol, a constant, or a predicate symbol). The function symbol f^* is defined similarly. The theory T^* is obtained from T be replacing all the symbols in the language $L \cup \{f\}$ with the corresponding symbol from $L^* \cup \{f^*\}$.

Definition A Δ_0-*interpretation in Q* of a function f consists of a theory T as above that is Δ_0-axiomatizable, is interpretable in Q with a cut interpretation, and for which

$$I\Delta_0(f, f^*, L, L^*) + T + T^* \vdash (\forall x)(f(x) = f^*(x)). \tag{1}$$

It is obvious that any Δ_0-interpretable function defines a function $f : \mathbb{N} \to \mathbb{N}$; that is to say, it defines an "actual" function on the integers. At the risk of confusing syntax and semantics, we define that any actual function defined by a symbol f of a theory T satisfying the conditions of the definition is Δ_0-interpretable in Q.

3 Upper bound

This section gives an upper bound on the computational complexity of functions that are Δ_0-interpretable in Q. The upper bound will not use the interpretability at all, but rather, it will depend only on the fact that the function is defined uniquely in a bounded theory with the right growth rate functions.

Theorem 2 *Let T be Δ_0 axiomatized, with language $L \cup \{f\}$, f a unary function symbol. Suppose $T \supset I\Delta_0(f, L)$ and that equation (1) holds so that T defines f uniquely. Then suppose that there is a $i > 0$ such that $\Omega_i \in L$ and the defining axioms of Ω_i are in T, and such that for each function symbol g in the language of T, T proves that g is dominated by Ω_i. Then, f is in $\mathrm{NEXP}^i\text{-TIME} \cap \mathrm{coNEXP}^i\text{-TIME}$.*

Proof. We may assume w.l.o.g. that T is axiomatized finitely and that T contains as many $I\Delta_0$ axioms as is helpful. In fact, we may suppose T is axiomatized by a single $\forall \Delta_0$-sentence $(\forall x)\Theta(x)$. Also without loss of generality, we may assume that the axiom contains only terms of depth 1; that is, that no function symbols are nested. (This is easily accomplished at the expense of making the formula Θ more complicated with additional bounded quantifiers.) In addition, we may assume that every bounded quantifier in Θ is of the form $(\forall y \leq x)$ or $(\exists y \leq x)$; i.e., the universal quantified variable x effectively bounds all variables in the axiom. We let Θ^* denote the formula obtained from Θ by replacing each nonlogical symbol g with g^*.

Let c be a new constant symbol. By (1), we have

$$I\Delta_0(f, L, f^*, L^*) + f(c) \neq f^*(c) \vdash (\exists x) [\neg\Theta(x) \vee \neg\Theta^*(x)] .$$

By Parikh's thoerem, there is a term $t(c)$ such that the quantifier $(\exists x)$ may be replaced by $(\exists x \leq t(c))$. It follows that there is a $k \in \mathbb{N}$ such that $t(x)$ is eventually dominated by $\Omega_i^{(k)}(x)$, provably in $I\Delta_0(f, L, f^*, L^*)$, where the superscript "(k)" indicates k-fold iterated function composition. It follows that

$$I\Delta_0(f, L, f^*, L^*) \vdash \text{"c is sufficiently large"} \wedge f(c) \neq f^*(c)$$

$$\to (\exists x \leq \Omega_i^{(k)}(c)) \left[\neg\Theta(x) \vee \neg\Theta^*(x)\right].$$

The algorithm to compute the bit-graph of f can now be described: On input an integer c and integers b and j, the algorithm nondeterministically guesses and saves the following values:

1. For each n-ary function symbol $g(x_1, \ldots, x_n)$ of L and all values of $x_1, \ldots, x_n$ $\leq \Omega_i^{(k)}(c)$, a value of $g(x_1, \ldots, x_n)$, which is $\leq \Omega_i(\Omega_i^{(k)}(c)) = \Omega_i^{(k+1)}(c)$,[2] and

2. For each n-ary predicate P and all values of $x_1, \ldots, x_n \leq \Omega_i^{(k+1)}(c)$, a truth value of $P(x_1, \ldots, x_n)$.

After nondeterministically guessing these values, the algorithm verifies that the axiom $\Theta(x)$ holds for all $x < \Omega_i(c)$. If they all hold, the algorithm accepts if the bth bit of the guessed value of $f(c)$ is equal to j. Otherwise, the algorithm rejects.

It is now straightforward to check that the nondeterministic algorithm correctly recognizes the bit-graph of f. Furthermore, the run time of f is bounded clearly by $\Omega_i^{(s)}(c)$ for some $s \in \mathbb{N}$. Thus, the the algorithm is in $\text{NEXP}^i\text{-TIME}$. Since the function f is single-valued, the bit-graph is also in $\text{coNEXP}^i\text{-TIME}$. $\square$

Corollary 3 *Suppose that* $f : \mathbb{N} \to \mathbb{N}$ *is* Δ_0-*interpretable in* Q. *Then* f *is in* $\text{NEXP}^i\text{-TIME} \cap \text{coNEXP}^i\text{-TIME}$ *for some* $i \geq 0$.

The corollary is an immediate consequence of Theorem 2 because of the results discussed in Section 2.

4 Lower bound

This section gives lower bounds for the definability of functions in bounded theories that match the upper bounds of the earlier section. Theorem 4 applies to arbitrary bounded theories, and Theorem 5 applies to predicative bounded theories.

---

[2] This case covers constant symbols, since they may be viewed as 0-ary function symbols. In addition, the symbol f is of course one of the functions g, so the value of $f(c)$ is guessed as part of this process.

Theorem 4 *Suppose $f(x)$ is dominated by $\Omega_i(x)$ for some $i \geq 0$ and that the bit-graph of $f(x)$ is in $\mathrm{NEXP}^i\text{-TIME} \cap \mathrm{coNEXP}^i\text{-TIME}$. Then there is a bounded theory T in a language $L \cup \{f\}$ such that $T \vdash I\Delta_0(f, L)$ and such that T proves f is total and defines $f(x)$ uniquely.*

Theorem 5 *Let $f : \mathbb{N} \to \mathbb{N}$ be in $\mathrm{NEXP}^i\text{-TIME} \cap \mathrm{coNEXP}^i\text{-TIME}$ and be dominated by Ω_i for some $i \geq 0$. Suppose that $I\Delta_0 + \exp$ can prove those facts; namely, there are predicates $A(x, b, j)$ and $B(x, b, j)$ such that $I\Delta_0 + \exp$ can prove that*

(a) *$A(x, b, j)$ and $B(x, b, j)$ are equivalent for all x, b, j,*

(b) *A is computable by a $\mathrm{NEXP}^i\text{-TIME}$ Turing machine, and*

(c) *B is computable by a $\mathrm{coNEXP}^i\text{-TIME}$ machine,*

and such that the predicates A and B each define the bit-graph of f.

Then f is Δ_0-interpretable in Q.

We already gave a sketch of a proof of a weakened form of these two theorems in the appendix to [3]. That proof was based on the equivalence of alternating polynomial space and exponential time. Our proofs below, however, are based on directly representing nondeterministic exponential time computation with function values.

Proof (of Theorem 4). Consider two $\mathrm{NEXP}^i\text{-TIME}$ Turing machines, M and N, such that the language accepted by M is the complement of the language of N. Without loss of generality, the machines accept a single integer as input (in binary notation, say), use a single half-infinite work tape, and halt after exactly $\Omega_i(n)$ steps on an input z of length $n = |z|$.

We describe an execution of the machine M with a trio of functions $T_M(z, i, j)$, $H_M(z, i)$, and $S_M(z, i)$. The intended meaning of $T_M(z, i, j) = c$ is that in the execution of M on input z, after i steps, the jth-tape square of M contains the symbol c. The intended meaning of $S_M(z, i) = q$ is that, M on input z after i steps, is in state q. The intended meaning of $H_M(z, i) = j$ is that M's tape head is positioned over tape square j after i steps. Now, since M is nondeterministic, there is more than one possible execution of M on input z, and this means the above "intended meanings" are under-specified. To clarify, the real intention is that if there is an execution of M on input z that leads to an accepting state, then we choose an arbitrary accepting computation and let $T_M(z, i, j)$, $H_M(z, i)$, and $S_M(z, i)$ be defined according to that accepting computation. On the other hand, if there is no accepting computation, we just choose any computation and set the values of $T_M(z, i, j)$, $H_M(z, i)$, and $S_M(z, i)$ accordingly.

These conditions can be represented by Δ_0-axioms that express the following conditions stating that T_M, H_M, and S_M correctly define an execution of M:

(1) For all j, $T_M(z, 0, j)$ has the correct value for the initial state of M with z written on its input tape.

(2) $S_M(z, 0)$ is equal to the initial state of M.

(3) $H_M(z, 0) = 0$, where w.l.o.g., M starts at tape square zero.

(4) For all $i \geq 0$, if $H_M(z, i) \neq j$, then $T_M(z, i, j) = T_M(z, i+1, j)$.

(5) For all $i \geq 0$, the transition rules for the machine M include a rule that allows M when reading symbol $T_M(z, i, H_M(z, i))$ in state $S_M(z, i)$ to write symbol $T_M(z, i+1, H_M(z, i))$, enter state $S_M(z, i+1)$, and either (i) move right one tape square or (ii) move left one tape square. In case (i), it is required that $H_M(z, i+1) = H_M(z, i) + 1$, and in case (ii), it is required that $H_M(z, i+1) = H_M(z, i) - 1$.

Clearly, the conditions (1)–(5) are satisfied only if T_M, H_M, and S_M describe a correct execution of M (which may be either accepting or rejecting). Furthermore, there are natural exponential bounds on i and j given that M is an exponential time machine. It is clear that the conditions (1)–(5) can be expressed by $\forall \Delta_0$ statements, Γ_M.

Similar conditions Γ_N can be defined for the machine N using symbols T_N, H_N, and S_N.

The theory T defining f can now be defined. Let M be the $\mathrm{NEXP}^i\text{-TIME}$ machine that accepts an input $z = \langle x, b \rangle$ precisely when the bth bit of $f(x)$ is equal to 1. Likewise, let N be the NEXP^i machine that accepts the complement of the set accepted by M. The language of T includes the function symbol f (a sufficiently large subset of), the language of PV, and the symbols T_M, H_M, S_M, T_N, H_N, and S_N. The axioms of T include induction for all Δ_0-formulas of T, plus the axioms Γ_M and Γ_N, and axioms expressing the following two conditions:

(a) For each z, either $S_M(z, \Omega_i(z))$ is an accepting state or $S_N(z, \Omega_i(z))$ is an accepting state, but not both, and

(b) The bth bit of $f(x)$ is equal to 1 if and only $S_M(\langle x, b \rangle, \Omega_i(\langle x, b \rangle))$ is an accepting state.

By the fact that M accepts the complements of the set accepted by N, we see that (a) is a true condition, and the condition (b) is a correct definition of a function f. It is clear from the construction that the theory T correctly defines f.   □

Proof (of Theorem 5, sketch). The idea of the proof is to formalize the proof of Theorem 4 in $I\Delta_0 + \exp$. For this reason, recall from [13] that if $I\Delta_0 + \exp$ can prove a Δ_0-formula $\theta(x)$, then there is some $k > 0$ such that $I\Delta_0$ can prove "if the k-fold exponential of x exists, then $\theta(x)$ holds." Thus if the hypotheses of Theorem 5 hold, there is some $k > i$ such that $I\Delta_0 + $ "k-fold exponential of x exists" can prove that the predicates $A(x, b, j)$ and $B(x, b, j)$ accept the same set and are in $\mathrm{NEXP}^i\text{-TIME}$ and $\mathrm{coNEXP}^i\text{-TIME}$ (respectively).

It is well known [11, 13] that Q can define inductive cuts $I(a)$ and $J(a)$ such that $J \subseteq I$ and $I \vDash I\Delta_0$, and such that for all $x \in J$, the k-fold iterated exponential of x

exists in I. Working in the cut J, Q can formalize the construction of the proof of Theorem 5 (with the aid of the k-fold exponentials of elements of J that exist in I). Then, introducing the function symbol f and the function symbols T_M, H_M, S_M, T_N, H_N, and S_N for the NEXP^i-TIME Turing machines M and N, which accept the set $\{z = \langle x, b \rangle : A(x, b, 1)\}$ and its complement (respectively), and restricting to the cut J, we have interpreted the definition of f into Q.   □

The above theorems give an essentially exact characterization of the computational complexity of the functions that are Δ_0-interpretable in Q. It is clear that Theorems 2 and 4 give matching upper and lower bounds on the computational complexity of functions that are uniquely definable in Δ_0-theories. For Δ_0-interpretability in Q, Corollary 3 and Theorem 5 differ in the bounds on the function since the latter mentions provability in $I\Delta_0 + \exp$, whereas Corollary 3 does not mention provability explicitly. However, already the definition of Δ_0-interpretability, especially the provability in Q of the uniqueness condition (1), essentially implies the provability in $I\Delta_0 + \exp$ of the fact that the bit-graph of f is in NEXP^i-time (via the construction of the proof of Theorem 2).

We conclude with a few open problems. The first problem is the question of what multifunctions are interpretable in Q. A multifunction is a multiple-valued function (i.e., a relation); that is to say, there may be several values y such that $f(x) = y$ for a fixed x. The question is as follows a multifunction is interpretable in Q, what is the minimum computational complexity of a total multifunction that satisfies the axioms of the theory?

A second line of research is to answer some questions left open from the work of Wilkie [12]. One such question is whether (now working over the base theory $I\Delta_0$ rather than Q) it is possible for a Σ_{n+1}-formula to define an inductive cut closed under Ω_n. Wilkie [12] shows that it is not possible for a Σ_n or Π_n formula to define such a cut in $I\Delta_0$ and that it is possible for a Π_{n+1} to define one, but leaves the Σ_{n+1} case open. Likewise, it appears no one has studied the corresponding questions over the base theory Q. Another question is whether Theorem 1 can be proved by a direct proof-theoretic argument. The game-theoretic argument in [12] is combined with a model-theoretic proof. The model-theoretic part can be avoided, but it would nice to give a more direct proof-theoretic proof.

References

1. T. ARAI, *Frege system, ALOGTIME, and bounded arithmetic*. Manuscript, 1992.
2. S. R. BUSS, *Bounded Arithmetic*, Bibliopolis, 1986. Revision of 1985 Princeton University Ph.D. thesis.
3. ——, *Nelson's Work on Logic and Foundations and Other Reflections on Foundations of Mathematics*, Princeton University Press, 2006, pp. 183–208. edited by W. Faris.

4. S. R. BUSS AND J. KRAJÍČEK, *An application of Boolean complexity to separation problems in bounded arithmetic*, Proc. London Math. Society, 69 (1994), pp. 1–21.

5. P. CLOTE AND G. TAKEUTI, *Bounded arithmetics for NC, ALOGTIME, L and NL*, Annals of Pure and Applied Logic, 56 (1992), pp. 73–117.

6. S. COOK AND P. NGUYEN, *Foundations of Proof Complexity: Bounded Arithmetic and Propositional Translations*. Book in preparation. Draft manuscipt available on web.

7. R. J. LIPTON, *Model theoretic aspects of computational complexity*, in Proceedings of the 19th Annual Symposium on Foundations of Computer Science, IEEE Computer Society, 1978, pp. 193–200.

8. R. J. PARIKH, *Existence and feasibility in arithmetic*, The Journal of Symbolic Logic, 36 (1971), pp. 494–508.

9. J. B. PARIS AND C. DIMITRACOPOULOS, *A note on the undefinability of cuts*, Journal of Symbolic Logic, 48 (1983), pp. 564–569.

10. P. PUDLÁK, *Cuts, consistency statements and interpretation*, The Journal of Symbolic Logic, 50 (1985), pp. 423–441.

11. R. M. SOLOVAY, Letter to P. Hájek, August 1976.

12. A. J. WILKIE, *On sentences interpretable in systems of arithmetic*, in Logic Colloquium '84, North-Holland, 1986, pp. 329–342, edited by J. B. Paris, A. J. Wilkie and G. M. Wilmers.

13. A. J. WILKIE AND J. B. PARIS, *On the scheme of induction for bounded arithmetic formulas*, Annals of Pure and Applied Logic, 35 (1987), pp. 261–302.

Effective Uniform Bounds from Proofs in Abstract Functional Analysis

Ulrich Kohlenbach

Department of Mathematics
Technische Universität Darmstad
D-64289 Darmstadt, Germany
kohlenbach@mathematik.tu-darmstadt.de
http://www.mathematik.tu-darmstadt.de/~kohlenbach

1 Introduction

In recent years (though influenced by papers of G. Kreisel going back to the 1950s, e.g., [74, 75, 76], as well as subsequent work by H. Luckhardt [82, 83] and others) an applied form of proof theory systematically evolved that is also called 'Proof Mining' ([72]). It is concerned with transformations of prima facie ineffective proofs into proofs from that certain quantitative computational information as well as new qualitative information can be read off that was not visible beforehand. Applications have been given in the areas of number theory [82], combinatorics [2, 35, 97, 98], algebra [19, 20, 21, 22, 23, 24, 25], and most systematically, in the area of functional analysis (see the references below). In particular, general logical metatheorems [34, 54, 64] have been proved that guarantee a-priorily for large classes of theorems and proofs in analysis the extractability of effective bounds that are independent from parameters in general classes of metric, hyperbolic, and normed spaces if certain local boundedness conditions are satisfied. Unless separability assumptions on the spaces involved are used in a given proof, the independence results from parameters only need metric bounds but no compactness [34, 64]. The theorems treat results involving concrete Polish metric spaces P (such as $\mathbb{R}^n$ or $C[0, 1]$) as well as **abstract** structures (metric, hyperbolic, normed spaces, etc.) that are axiomatically added to the formal systems as a kind of 'Urelements'. It is for the latter structures that we can replace the dependency of the bounds from inputs involving elements of these spaces by hereditary bounds ('majorants') of such elements that in our applications will be either natural numbers or number-theoretic functions. So we can apply the usual notions of computability and complexity for type-2 functionals and do not have to restrict ourselves to instances of these structures that are representable in some effective way or would carry a computability structure. The latter is only required for the **concrete** Polish metric spaces where we rely on the usual 'standard (Cauchy) representation'.

Obviously, certain restrictions on the logical form of the theorems to be proved as well as on the axioms to be used in the proofs are necessary (for a large class of semi-constructive proofs, the restrictions on the form of the theorems can largely be avoided; see [33]). These restrictions in turn depend on the language of the formal systems used as well as on the representation of the relevant mathematical objects such as general function spaces. The correctness of the results, moreover, depends in subtle ways on the amount of extensionality properties used in the proof, which has a direct analytic counterpart in terms of uniform continuity conditions.

The applications that we discuss in this survey include a number of new qualitative existence results in the area of nonlinear functional analysis that follow from the metatheorems but so far did not have a functional analytic proof. Applying the extraction algorithm provided by the proofs of the metatheorems to these results yields the explicit quantitative versions stated below and at the same time direct proofs that no longer rely on the logical metatheorems themselves [9, 32, 61, 63, 65, 68, 69, 79, 80].

The page limitations of this paper prevent us from formulating precisely the various logical metatheorems and the formal systems involved, and we refer to [34, 64, 66]. We will rather give a comprehensive presentation of the effective bounds obtained with the help of this logical approach in analysis (often all the qualitative features of the bounds concerning the (in)dependence from various parameters as well as some crude complexity estimates are guaranteed a-priorily by logical metatheorems) and refer for information on the logical background as well as for the proofs of these bounds to the literature.

Notations: $\mathbb{N}$ denotes the set of natural numbers, $\mathbb{N} = \{0, 1, 2, \ldots\}$. $\mathbb{Q}_+^*$ and $\mathbb{R}_+^*$ denote the sets of strictly positive rational and real numbers, respectively.

The bounds presented below are all obviously effective if stated for $\varepsilon \in \mathbb{Q}_+^*$. Sometimes it is more convenient to state them (and to formulate the various moduli involved) for $\varepsilon \in \mathbb{R}_+^*$. It will, nevertheless, always be straightforward to make the use of, e.g., $\lceil x \rceil$ effective by restricting things to rational ε (and corresponding moduli formulated for rationals).

2 Logical metatheorems

In this section we give an informal presentation of the main metatheorems on which the applications reported in this paper are based (details can be found in [34, 64]).

Definition 2.1.

1) The set **T** of all finite types over 0 is defined inductively by the clauses

$$(i)\ 0 \in \mathbf{T},\ (ii)\ \rho, \tau \in \mathbf{T} \Rightarrow (\rho \to \tau) \in \mathbf{T}.$$

2) The set $\mathbf{T}^X$ of all finite types over the two ground types 0 and X is defined by

$$(i)\ 0, X \in \mathbf{T}^X,\ (ii)\ \rho, \tau \in \mathbf{T}^X \Rightarrow (\rho \to \tau) \in \mathbf{T}^X.$$

3) A type is called small if it is of degree 1 (i.e., $0 \to \cdots \to 0 \to 0$) or the form $\rho_1 \to \cdots \to \rho_k \to X$ with the ρ_i being 0 or X.[1]

The theory $\mathcal{A}^\omega$ for classical analysis is the extension of the weakly extensional Peano arithmetic in all types of WE-PA$^\omega$ by the schemata of quantifier-free choice QF-AC and dependent choice DC for all types in $\mathbf{T}$ (formulated for tuples of variables).

The theories $\mathcal{A}^\omega[X, d]_{-b}$ and $\mathcal{A}^\omega[X, d, W]_{-b}$ result[2] by extending $\mathcal{A}^\omega$ to all types in $\mathbf{T}^X$ and by adding axioms for an abstract metric (in the case of $\mathcal{A}^\omega[X, d]_{-b}$), resp. hyperbolic (in the case of $\mathcal{A}^\omega[X, d, W]_{-b}$), space. $\mathcal{A}^\omega[X, d, W, \mathrm{CAT}(0)]_{-b}$ is the extension by an abstract CAT(0)-space. Analogously, one has theories $\mathcal{A}^\omega[X, \|\cdot\|]$ with an abstract, non-trivial, real normed space added (as well as further extensions $\mathcal{A}^\omega[X, \|\cdot\|, C]$ resp. $\mathcal{A}^\omega[X, \|\cdot\|, C]_{-b}$ with bounded, resp. general, convex subsets $C \subseteq X$, which we will, however, due to lack of space not formulate here). Our theories also contain a constant 0_X of type X that in the normed case represents the zero vector and in the other cases stands for an arbitrary element of the metric space. For details on all this, see [34, 64].

Real numbers are represented as Cauchy sequences of rationals with a fixed rate 2^{-n} of convergence that in turn are encoded as number-theoretic functions f^1, where an equivalence relation $f =_{\mathbb{R}} g$ expresses that f^1, g^1 denote the same real numbers, and $\leq_{\mathbb{R}}, <_{\mathbb{R}}, |\cdot|_{\mathbb{R}}$ express the obvious relations and operations on the level of these codes. Here $=_{\mathbb{R}}, \leq_{\mathbb{R}} \in \Pi^0_1$, whereas $<_{\mathbb{R}} \in \Sigma^0_1$. Again details can be found in [64].

'Weakly extensional' means that we only have Spector's quantifier-free extensionality rule. In particular, for the defined equality $x =_X y :\equiv (d_X(x, y) =_{\mathbb{R}} 0_{\mathbb{R}})$, we do not have

$$x =_X y \to f^{X \to X}(x) =_X f(y)$$

but only from a proof of $s =_X t$ can infer that $f(s) =_X f(t)$. This is of crucial importance for our metatheorems to hold. Fortunately, we can in most cases prove the extensionality of f for those functions we consider, e.g., for nonexpansive functions, so that this only causes some need for extra care in a few cases (for an extensive discussion of this point, see [64]).

Definition 2.2. For $\rho \in \mathbf{T}^X$, we define $\widehat{\rho} \in \mathbf{T}$ inductively as follows:

$$\widehat{0} := 0,\ \widehat{X} := 0,\ \widehat{(\rho \to \tau)} := (\widehat{\rho} \to \widehat{\tau});$$

i.e. $\widehat{\rho}$ is the result of replacing all occurrences of the type X in ρ by the type 0.

---

[1] In [34] a somewhat bigger class of types of so-called degree $(1, X)$ is allowed. However, for the applications presented in this paper, the small types suffice, which simplifies the statement of the metatheorem below.

[2] The index '$-b$' indicates that in contrast to the corresponding theories in [64], we (following [34]) do not require the metric space to be bounded.

Definition 2.3 ([34]). We define a ternary majorization relation $\gtrsim_\rho^a$ between objects x, y, and a of type $\widehat{\rho}, \rho$, and X, respectively, by induction on ρ as follows[3]:

- $x^0 \gtrsim_0^a y^0 :\equiv x \geq_{\mathbb{N}} y$,

- $x^0 \gtrsim_X^a y^X :\equiv (x)_{\mathbb{R}} \geq_{\mathbb{R}} d_X(y, a)$,

- $x \gtrsim_{\rho \to \tau}^a y :\equiv \forall z', z(z' \gtrsim_\rho^a z \to xz' \gtrsim_\tau^a yz) \wedge \forall z', z(z' \gtrsim_{\widehat{\rho}}^a z \to xz' \gtrsim_{\widehat{\tau}}^a xz)$.

For normed linear spaces, we choose $a = 0_X$.

Definition 2.4. A formula F in $\mathcal{L}(\mathcal{A}^\omega[X, \ldots]_{-b})$ is called $\forall$-formula (resp., $\exists$-formula) if it has the form $F \equiv \forall \underline{a}^{\underline{\sigma}} F_{qf}(\underline{a})$ (resp., $F \equiv \exists \underline{a}^{\underline{\sigma}} F_{qf}(\underline{a})$), where F_{qf} does not contain any quantifier and the types in $\underline{\sigma}$ are small.

In the following $\mathcal{S}^\omega = \langle S_\rho \rangle_{\rho \in \mathbf{T}}$ refers to the full set-theoretic-type structure of all set-theoretic functionals of finite type.

Theorem 2.5 ([34]).

1) Let ρ be a small type, and let $B_\forall(x, u)$, resp. $C_\exists(x, v)$, be $\forall$- and $\exists$-formulas that contain only x, u free, resp. x, v free. Assume that the constant 0_X does not occur in $B_\forall, C_\exists$ and that

$$\mathcal{A}^\omega[X, d]_{-b} \vdash \forall x^\rho (\forall u^0 B_\forall(x, u) \to \exists v^0 C_\exists(x, v)).$$

Then there exists a computable functional[4] $\Phi : S_{\widehat{\rho}} \to \mathbb{N}$ such that the following holds in all nonempty metric spaces (X, d): for all $x \in S_\rho$, $x^* \in S_{\widehat{\rho}}$, if there exists an $a \in X$ s.t. $x^* \gtrsim^a x$ then

$$\forall u \leq \Phi(x^*) B_\forall(x, u) \to \exists v \leq \Phi(x^*) C_\exists(x, v).$$

If 0_x does occur in $B_\forall$ and/or $C_\exists$, then the bound Φ depends (in addition to x^*) on an upper bound $\mathbb{N} \ni n \geq d(0_X, a)$.

2) The theorem also holds for nonempty hyperbolic spaces $\mathcal{A}^\omega[X, d, W]_{-b}$, (X, d, W), and for $\mathcal{A}^\omega[X, d, W, \mathrm{CAT}(0)]_{-b}$, where (X, d, W) is a $\mathrm{CAT}(0)$ space.

3) The theorem also holds for nontrivial, real normed spaces $\mathcal{A}^\omega[X, \|\cdot\|]$, $(X, \|\cdot\|)$, where then 'a' has to be interpreted by the zero vector 0_X in $(X, \|\cdot\|)$ and 0_X is allowed to occur in $B_\forall, C_\exists$.

Instead of single variables x, u, v and single premises $\forall u B_\forall(x, u)$ we may have tuples of variables and finite conjunctions of premises. In the case of a tuple $\underline{x}$, we then have to require that we have a tuple $\underline{x}^*$ of a-majorants for a common $a \in X$ for all the components of the tuple $\underline{x}$.

---

[3] Here $(x)_{\mathbb{R}}$ refers to the embedding of $\mathbb{N}$ into $\mathbb{R}$ in the sense of our representation of $\mathbb{R}$.

[4] Note that for small types ρ the type $\widehat{\rho}$ is of degree 1. So Φ essentially is a type-2 functional $: \mathbb{N}^{\mathbb{N}} \to \mathbb{N}$.

Remark 2.6. From the proof of Theorem 2.5, two further extensions follow:

1) The language may be extended by a-majorizable constants (in particular constants of types 0 and 1, which always are uniformly majorizable) where the extracted bounds then additionally depend on (a-majorants for) the new constants.

2) The theory may be extended by purely universal axioms or, alternatively, axioms that can be reformulated into purely universal axioms using new majorizable constants if the types of the quantifiers are small.

Using these extensions, the theorem above can be adapted to other structures such as uniformly convex normed spaces or inner product spaces [34] as well as to uniformly convex hyperbolic spaces, δ-hyperbolic spaces (in the sense of Gromov), and $\mathbb{R}$-trees in the sense of Tits (see [80]).

A crucial aspect of theorem 2.5 is that the bound Φ operates on objects of degree ≤ 1, i.e., natural numbers or n-ary number-theoretic functions, so that the usual type-2 computability theory as well as well-known subrecursive classes of such functionals apply here irrespective of whether the metric and normed spaces to which the bounds are applied come with any notion of computability. Since we included the axiom of dependent choice (and so also countable choice and hence full comprehension over numbers) in our systems, the functional Φ extracted will be in general a bar recursive functional in the sense of Spector [94]. However, if (as usually is the case) only small fragments of this are used, e.g., if in addition to basic arithmetic only the weak König's lemma WKL is used, then the bound will be primitive recursive in the sense of Gödel's T ([39]) if full induction is used, resp. primitive recursive in the ordinary sense of Kleene if only Σ_1^0-induction is used. If not even full Σ_1^0-induction is used, then in many cases even polynomial bounds (in the data) can be expected (see [54, 57, 58, 59]).

The proof of theorem 2.5 provides an algorithm (based on (monotone) functional ('Dialectica') interpretation [39, 42, 56, 94]) for the extraction of Φ.

In the concrete applications, theorem 2.5 is used via various applied corollaries of which we give an example now:

Definition 2.7. Let (X, d) be a metric space. A mapping $f : X \to X$ is called nonexpansive (short 'n.e.') if

$$\forall x, y \in X (d(f(x), f(y)) \leq d(x, y)).$$

Corollary 2.8 ([34]). *Let $C_\exists$ be an $\exists$-formula and P, K Polish, resp. compact, metric spaces in standard representation by $\mathcal{A}^\omega$-definable terms (see [54] for a precise definition). If $\mathcal{A}^\omega[X, d, W]_{-b}$ proves a sentence*

$$\forall x \in P \forall y \in K \forall z^X, \tilde{z}^X, c^{0 \to X}, f^{X \to X} \left(f \text{ nonexpansive} \to \exists v^{\mathbb{N}} C_\exists \right),$$

then there is a computable functional $\Phi(g_x, b, h)$ s.t. for all $x \in P, g_x \in \mathbb{N}^{\mathbb{N}}$ representative of x, $b \in \mathbb{N}, h \in \mathbb{N}^{\mathbb{N}}$

$$\forall y \in K \forall z, \tilde{z} \in X \forall c : \mathbb{N} \to X \forall f : X \to X \big(f \text{ n.e.} \wedge d(z, f(z)), d(z, \tilde{z}) \leq b$$

$$\wedge \forall n(d(z, c(n)) \leq h(n)) \to \exists v \leq \Phi(g_x, b, h) \, C_{\exists}\big)$$

holds in **any** *nonempty hyperbolic space* (X, d, W).

Proof (sketch). The fact that P, K have a standard representation by $\mathcal{A}^\omega$-terms essentially means that $\forall$-quantification over P, resp. K, can be expressed as quantification $\forall x^1$, resp. $\forall y \leq_1 N$, where N is a fixed simple (primitive recursive) function depending on K. Here the number-theoretic functions encode Cauchy sequences (with a fixed rate of convergence) of elements from the countable dense subset of P, resp. K, on which the standard representations are based. We now apply theorem 2.5 with $a := z$. For this reason, we have to construct $\gtrsim^z$-majorants for $x^1, y^1, z^X, \tilde{z}^X, c^{0 \to X}$, and $f^{X \to X}$:

$$x^* := x^M := \lambda n. \max\{x(i) : i \leq n\}, y^* := N^M, z^* := 0^0, \tilde{z}^* := b, c^* := h^M,$$

$$f^* := \lambda n^0 . n + b.$$

For f^* we use that

$$d(x, z) \leq n \to d(f(x), z) \leq d(f(x), f(z)) + d(f(z), z)$$

$$\leq d(x, z) + d(f(z), z)$$

$$\leq n + b.$$

Note that the majorants only depend on x, b, h. □

3 Applications of proof mining in approximation theory

Let $(X, \|\cdot\|)$ be a (real) normed linear space and $E \subseteq X$ a finite-dimensional subspace. By a standard (ineffective) compactness argument, each $x \in X$ possesses at least one element $y_b \in E$ of best approximation; i.e.,

$$\|x - y_b\| = \inf_{y \in E} \|x - y\| =: \text{dist}\,(x, E).$$

In some important cases (see further below), y_b is determined uniquely

$$\forall x \in X \forall y_1, y_2 \in E(\|x - y_1\|, \|x - y_2\| = \text{dist}\,(x, E) \to y_1 = y_2),$$

which can be written as follows:

$$\forall x \in X \forall y_1, y_1 \in E \forall k \in \mathbb{N} \exists n \in \mathbb{N}(\|x - y_1\|, \|x - y_2\| \leq \text{dist}\,(x, E) + 2^{-n}$$

$$\to \|y_1 - y_2\| < 2^{-k}),$$

where (using the representation of real numbers mentioned above)

$$\|x - y_1\|, \|x - y_2\| \le \operatorname{dist}(x, E) + 2^{-n} \to \|y_1 - y_2\| < 2^{-k}$$

is equivalent to a Σ_1^0-formula.

Every best approximation $y_b \in E$ clearly satisfies $\|y_b\| \le 2\|x\|$ (since otherwise $0 \in E$ would be a better appoximation). Hence we can replace above the space E by the compact subset $K_x := \{y \in E : \|y\| \le 2\|x\|\}$. Now suppose that one has a computable bound $\Phi(x, k)$ (depending on a suitable representation of x) for '$\exists n \in \mathbb{N}$' that is independent of $y_1, y_2 \in K_x$; i.e.,

$$\forall x \in X \forall y_1, y_1 \in K_x \forall k \in \mathbb{N}(\|x - y_1\|, \|x - y_2\| \le \operatorname{dist}(x, E) + 2^{-\Phi(x,k)}$$

$$\to \|y_1 - y_2\| < 2^{-k}).$$

We call such a Φ a modulus of uniqueness. Then any algorithm for computing 2^{-n}-best approximations $y_n \in K_x$; i.e., $\|x - y_n\| \le \operatorname{dist}(x, E) + 2^{-n}$, can be used to compute y_b with any prescribed precision since

$$\forall k \in \mathbb{N} \ (\|y_{\Phi(x,k)} - y_b\| < 2^{-k}).$$

If we use $\tilde{K}_x := \{y \in E : \|y\| \le \frac{5}{2}\|x\|\}$ instead of K_x, then by an easy argument a modulus of uniqueness on $\tilde{K}_x$ can be extended effectively to the whole space E. So we now always refer to moduli of uniqueness on all of E and—for convenience—use $q \in \mathbb{Q}_+^*$ instead of 2^{-k} with $\Phi(x, q) \in \mathbb{Q}_+^*$. The next proposition further indicates the relevance of this notion:

Proposition 3.1 ([54]). *Let $(X, \|\cdot\|)$ be a real normed linear space, $E \subseteq X$ a finite-dimensional subspace. Assume that every $x \in X$ possesses a uniquely determined best approximation in E and that the operation Φ is a modulus of uniqueness. Then the following holds:*

1) $\frac{1}{2} \cdot \Phi$ is a modulus of pointwise continuity for the projection $\mathcal{P} : X \to E$, which maps $x \in X$ to its best approximation $y_b \in E$; i.e.,

$$\forall x, x_0 \in X, q \in \mathbb{Q}_+^* \left(\|x - x_0\| \le \frac{1}{2}\Phi x_0 q \to \|\mathcal{P}x - \mathcal{P}x_0\| \le q\right).$$

2) If Φ is linear in q, i.e., $\Phi x q = q \cdot \gamma(x)$, then $\gamma(x)$ is a 'constant of strong unicity'; i.e.,

$$\forall x \in X, y \in E \big(\|x - y\| \ge \|x - y_b\| + \gamma(x) \cdot \|y - y_b\|\big),$$

where y_b is the best approximation of x in E.

3) For $\gamma(x)$ as in '2)', we get that $\lambda(x) := \frac{2}{\gamma(x)}$ is a pointwise Lipschitz constant for $\mathcal{P}$; i.e.,

$$\forall x, x_0 \in X \big(\|\mathcal{P}x - \mathcal{P}x_0\| \le \lambda(x_0) \cdot \|x - x_0\|\big).$$

In the following, we discuss two specific best approximation problems. Let $C[0,1]$ be the space of all continuous real-valued functions on $[0,1]$ and P_n the subspace of all polynomials of degree $\leq n$. We consider best approximations of $f \in C[0,1]$ by polynomials in P_n w.r.t. the maximum norm $\|f\|_\infty := \sup_{x \in [0,1]} |f(x)|$ (called best Chebycheff approximation) as well as w.r.t. the L_1-norm $\|f\|_1 := \int_0^1 |f|$ (also called 'approximation in the mean'). Even in the latter case we represent $C[0,1]$ as a Polish space w.r.t. the metric induced by $\| \cdot \|_\infty$ since it is not complete w.r.t. $\| \cdot \|_1$. The usual so-called standard representation of $(C[0,1], \| \cdot \|_\infty)$ is constructively equivalent to the representation of f via its restriction to the rational numbers in $[0,1]$ and a modulus $\omega : \mathbb{Q}_+^* \to \mathbb{Q}_+^*$ of uniform continuity of f; i.e.,

$$\forall x, y \in [0,1] \forall \varepsilon \in \mathbb{Q}_+^* (|x - y| < \omega(\varepsilon) \to |f(x) - f(y)| < \varepsilon)$$

so that the bounds will depend on ω.

Since in this section we do not use abstract classes of metric spaces but (in addition to $\mathbb{R}$), only the concrete Polish metric space $(C[0,1], \| \cdot \|_\infty)$, the applications in this section are instances already of the older metatheorems from [54].

We first consider the case of best Chebycheff approximation: A well-known theorem in so-called Chebycheff approximation theory states that every $f \in C[0,1]$ possesses a unique polynomial $p_b \in P_n$ of best approximation in the $\| \cdot \|_\infty$-norm, i.e., a polynomial in P_n such that $\|f - p_b\|_\infty = \mathrm{dist}_\infty(f, P_n) := \inf_{p \in P_n} \|f - p\|_\infty$. Both the existence as well as the uniqueness of p_b are established by classical arguments that make use of the theorem that continuous real-valued functions attain their minimum on compact spaces, i.e., use the ineffective weak König's lemma WKL (see [92]).

By (the algorithm implicit in) our general metatheorems from [54], it is guaranteed that the uniqueness proof, nevertheless, allows one to extract a (primitive recursively) computable modulus of uniqueness (even of relatively low complexity), a concept that—under the name of strong unicity—plays an important role in approximation theory (see [17]). By proposition 3.1, such a modulus of uniqueness provides a stability rate for the Chebycheff projection that assigns to $f \in C[0,1]$ the unique polynomial p_b of best approximation in P_n. Furthermore, it can be used to compute p_b and to (upper) estimate its computational complexity (see [54] for all this). In [55] the following explicit moduli (also for the case of general Haar spaces) were extracted from the classical uniqueness proof due to [99]:

Theorem 3.2 ([55]). *Let*

$$\Phi(\omega, n, \varepsilon) := \min \left\{ \varepsilon/4, \frac{\lfloor \frac{n}{2} \rfloor! \lceil \frac{n}{2} \rceil!}{2(n+1)} \cdot (\omega_n(\varepsilon/2))^n \cdot \varepsilon \right\},$$

with

$$\omega_n(\varepsilon) := \begin{cases} \min \left\{ \omega \left(\frac{\varepsilon}{2} \right), \frac{\varepsilon}{8n^2 \lceil \frac{1}{\omega(1)} \rceil} \right\}, & \text{if } n \geq 1, \\ 1, & \text{if } n = 0. \end{cases}$$

Then Φ is a common modulus of uniqueness for all $f \in C[0,1]$ that have the modulus of uniform continuity ω, i.e., for all $n \in \mathbb{N}$. More precisely, we have

$$\forall p_1, p_2 \in P_n; \varepsilon \in \mathbb{Q}_+^* \big(\bigwedge_{i=1}^{2} (\|f - p_i\|_\infty - \mathrm{dist}_\infty(f, P_n) < \Phi(\omega, n, \varepsilon))$$

$$\rightarrow \|p_1 - p_2\|_\infty \leq \varepsilon \big).$$

Moreover if $\mathrm{dist}_\infty(f, P_n) > 0$ *and* $l \in \mathbb{Q}_+^*$ *such that* $l \leq \mathrm{dist}_\infty(f, P_n)$ *and*

$$\tilde{\Phi}(\omega, n, l) := \frac{\lfloor \frac{n}{2} \rfloor! \lceil \frac{n}{2} \rceil!}{2(n+1)} \cdot (\omega_n(2l))^n,$$

then $\tilde{\Phi}(\omega, n, l) \cdot \varepsilon$ is a modulus of uniqueness for f that is linear in ε, and so $\tilde{\Phi}(\omega, n, l)$ (by proposition 3.1) is a 'constant of strong unicity'.

Remark 3.3.   1) The most important aspect of $\Phi, \tilde{\Phi}$ above is that these bounds do not depend on p_1, p_2. This is guaranteed by the metatheorems in [54] since one can—as discussed above—restrict things to the bounded (and hence compact) subset $\tilde{K}_{f,n} := \{p \in P_n : \|p\|_\infty \leq \frac{5}{2}\|f\|_\infty\}$ of the finite-dimensional space P_n.

2) Instead of the term $\lceil \frac{1}{\omega(1)} \rceil$ in the definition of ω_n, we may use an arbitrary upper bound $M \geq \|f\|_\infty$. Actually the result is proved in this form in [55]. Using the construction $f \mapsto \tilde{f}, \tilde{f}(x) := f(x) - f(0)$ (using that $\mathrm{dist}_\infty(f, P_n) = \mathrm{dist}_\infty(\tilde{f}, P_n)$), one sees that one may assume without loss of generality that $f(0) = 0$. With this assumption $\lceil \frac{1}{\omega(1)} \rceil$ is an upper bound of $\|f\|_\infty$ that reduces the dependence of the bound on f to just ω.

3) Our constant of strong unicity tends to 0 as $n \rightarrow \infty$. Except for the trivial case where $f \in P_n$, this is unavoidable by a deep result in [31].

The modulus of uniqueness in theorem 3.2 is significantly better than the one implicit in [52, 53] (see [55] for a comparison).

The existence of a unique element of best approximation to $f \in C[0,1]$ extends from $P_{n-1} := Lin_\mathbb{R}\{1, x, \ldots, x^{n-1}\}$ to general so-called Haar spaces $H := Lin_\mathbb{R}\{\phi_1, \ldots, \phi_n\}$, i.e., n-dimensional subspaces of $C[0,1]$ that have the unique interpolation property; i.e.,

$$\forall \phi \in H \forall \underline{x} \in [0,1] \big(\bigwedge_{i=1}^{n-1} (x_i < x_{i+1}) \wedge \bigwedge_{i=1}^{n} (\phi(x_i) = 0) \rightarrow \phi \equiv 0 \big).$$

The tuple $(\phi_1, \ldots, \phi_n)$ of functions in $C[0,1]$ is called a Chebycheff system over $[0,1]$.

Let $\phi := (\phi_1, \ldots, \phi_n)$ be a Chebycheff system over $[0,1]$,

$$\underline{\phi}(x) := \big(\phi_1(x), \ldots, \phi_n(x)\big) \in \mathbb{R}^n,$$

$$\|\underline{\phi}\| := \sup_{x \in [0,1]} \|\underline{\phi}(x)\|_2,$$

where $\| \cdot \|_2$ denotes the Euclidean norm on $\mathbb{R}^n$.

$\beta, \gamma, \kappa : (0, \frac{1}{n}] \to \mathbb{R}_+^*$ are defined by

$$\beta(\alpha) := \begin{cases} \inf_{x \in [0,1]} |\phi_1(x)|, & \text{if } n = 1 \\[2mm] \inf\left\{ |\det(\phi_j(x_i))| : 0 \le x_1, \ldots, x_n \le 1, \bigwedge_{i=1}^{n-1} (x_{i+1} - x_i \ge \alpha) \right\}, \\[2mm] \text{if } n > 1 \end{cases}$$

and

$$\gamma(\alpha) := \min\left\{ \|\underline{\phi}\|, \frac{\beta(\alpha)}{n^{\frac{1}{2}}(n-1)! \prod_{i=1}^{n}(1 + \|\phi_i\|_\infty)} \right\}, \quad \kappa(\alpha) := \gamma(\alpha)^{-1} \cdot \|\underline{\phi}\|$$

for $\alpha \in (0, \frac{1}{n}]$. Since ϕ is a Chebycheff system, it follows that $\beta(\alpha) > 0$.
$H := Lin_{\mathbb{R}}(\phi_1, \ldots, \phi_n)$; ω_ϕ denotes a modulus of uniform continuity of $\underline{\phi}$.
$E_{H,f} := \mathrm{dist}_\infty(f, H)$.

Lemma 3.4 ([5, 6]).

1) Suppose that $A \subset C[0,1]$ is totally bounded, ω_A is a common modulus of uniform continuity for all $f \in A$, and $M > 0$ is a common bound $M \ge \|f\|_\infty$ for all $f \in A$. Then

$$\omega_{A,H}(\varepsilon) := \min\left\{ \omega_A\Big(\frac{\varepsilon}{2}\Big), \omega_\phi\left(\frac{\varepsilon \cdot \beta(\frac{1}{n})}{4Mn^{\frac{3}{2}}(n-1)! \prod_{i=1}^{n}(1 + \|\phi_i\|_\infty)} \right) \right\}$$

is a common modulus of uniform continuity for all $\psi_b - f$, where $f \in A$ and ψ_b is the best approximation of f in H.

2) Assume $0 < \alpha \le \frac{1}{n}$ and $\bigwedge_{i=1}^{n-1} (x_{i+1} - x_i \ge \alpha)$ $(x_1, \ldots, x_n \in [0,1])$ for $n \ge 2$. Then

$$\forall \psi \in H, \varepsilon > 0 \Big(\bigwedge_{i=1}^{n} |\psi(x_i)| \le \frac{\gamma(\alpha)}{n \cdot \|\underline{\phi}\|} \cdot \varepsilon \to \|\psi\|_\infty \le \varepsilon \Big).$$

Theorem 3.5 ([55]). *Let* $A, \omega_{A,H}, \gamma, \kappa$ *be as in lemma 3.4 and* $E_{H,A} := \inf\limits_{f \in A} E_{H,f}$.
Then

$$\Phi_A \varepsilon := \min\left\{ \frac{\varepsilon}{4}, \frac{1}{2} \frac{\gamma\left(\min\{\frac{1}{n}, \omega_{A,H}(\frac{\varepsilon}{2})\}\right)}{n \cdot \|\underline{\phi}\|} \cdot \varepsilon \right\}$$

$$= \min\left\{ \frac{\varepsilon}{4}, \frac{\varepsilon}{2n\kappa\left(\min\{\frac{1}{n}, \omega_{A,H}(\frac{\varepsilon}{2})\}\right)} \right\}$$

is a common modulus of uniqueness (and a common modulus of continuity for the Chebycheff projection in f*) for all* $f \in A$.
For $l_{H,A} \in \mathbb{Q}_+^*$ *such that* $l_{H,A} < E_{H,A}$ *and* $0 < \alpha \leq \min\{\frac{1}{n}, \omega_{A,H}(2 \cdot l_{H,A})\}$, *we have* $\frac{\gamma(\alpha)}{n \cdot \|\underline{\phi}\|}$ *(resp.,* $2n\kappa(\alpha)$*) as a uniform constant of strong unicity (resp. Lipschitz constant) for all* $f \in A$.

The bounds in theorem 3.5 are significantly better than the ones obtained in [5, 6, 7] (see [55] for a detailed comparison). The (ineffective) existence of a constant of strong unicity was proved first in [84]. The existence of a uniform such constant (in the sense above) was established (again ineffectively) first in [41]. The local Lipschitz continuity of the projection is due to [30].

If the Haar space contains the constant-1 function, then using again the transformation $f \mapsto \tilde{f}$, with $\tilde{f}(x) := f(x) - f(0)$, one can even eliminate the dependence of the bounds on $M \geq \|f\|_\infty$ and conclude:

Theorem 3.6. *Let* $\{\phi_1, \ldots, \phi_n\}$ *be a Chebycheff system such that* $1 \in H := Lin_{\mathbb{R}}(\phi_1, \ldots, \phi_n)$, *and let* $\omega : \mathbb{R}_+^* \to \mathbb{R}_+^*$ *be any function. Then*

$$\Phi_H(\omega, \varepsilon) := \min\left\{ \frac{\varepsilon}{4}, \frac{\varepsilon}{2n\kappa\left(\min\{\frac{1}{n}, \omega^H(\frac{\varepsilon}{2})\}\right)} \right\}$$

with

$$\omega^H(\varepsilon) := \min\left\{ \omega(\frac{\varepsilon}{2}), \omega_{\underline{\phi}}\left(\frac{\varepsilon \cdot \beta(\frac{1}{n})}{4\lceil\frac{1}{\omega(1)}\rceil n^{\frac{3}{2}}(n-1)! \prod\limits_{i=1}^{n}(1 + \|\phi_i\|_\infty)} \right) \right\}$$

is a common modulus of uniqueness (and a common modulus of continuity for the Chebycheff projection) for all functions $f \in C[0, 1]$ *that have* ω *as a modulus of uniform continuity.*

As a corollary we obtain that for arbitrary Haar spaces having the constant function 1, the continuity behavior of the Chebycheff projection is uniform for any class of equicontinuous functions that generalizes a result of [78] for the case of (trigonometric) polynomials.

We now move to best approximations of f by polynomials in P_n w.r.t. the L_1-norm $\|f\|_1 := \int_0^1 |f(x)|dx$, so-called best 'approximation in the mean'.

Theorem 3.7 ([44]). Let $f \in C[0,1]$ and $n \in \mathbb{N}$. There exists a unique polynomial $p_b \in P_n$ of degree $\leq n$ that approximates f best in the L_1-norm; i.e.,

$$\|f - p_b\|_1 = \inf_{p \in P_n} \|f - p\|_1 =: \text{dist}_1(f, P_n).$$

Since $C[0,1]$ is not complete w.r.t. the norm $\|f\|_1$, we still use the representation w.r.t. $\|f\|_\infty$ in which the norm $\|f\|_1$ can be computed easily. As a result, we again have to expect our modulus of uniqueness to depend on a modulus ω of uniform continuity of f. Again, both the existence and the uniqueness part are proved using compactness arguments that are equivalent to WKL. Despite this ineffectivity, using the algorithm implicit in the logical metatheorems from [54] the following result was extracted from the ineffective uniqueness proof due to [16] (the extractability of a primitive recursive modulus of uniqueness again is a-priorily guaranteed by logical metatheorems; see [54]):

Theorem 3.8 ([71]). Let

$$\Phi(\omega, n, \varepsilon) := \min\{\frac{c_n \varepsilon}{8(n+1)^2}, \frac{c_n \varepsilon}{2}\omega_n(\frac{c_n \varepsilon}{2})\},$$

where

$$c_n := \frac{\lfloor n/2 \rfloor! \lceil n/2 \rceil!}{2^{4n+3}(n+1)^{3n+1}} \quad \text{and} \quad \omega_n(\varepsilon) := \min\{\omega(\frac{\varepsilon}{4}), \frac{\varepsilon}{40(n+1)^4 \lceil \frac{1}{\omega(1)} \rceil}\}.$$

Then $\Phi(\omega, n, \varepsilon)$ is a modulus of uniqueness for the best L_1-approximation of any function f in $C[0,1]$ having modulus of uniform continuity ω from P_n; i.e., for all n and $f \in C[0,1]$:

$$\forall p_1, p_2 \in P_n; \varepsilon \in \mathbb{Q}_+^* \left(\bigwedge_{i=1}^{2} (\|f - p_i\|_1 - \text{dist}_1(f, P_n) \leq \Phi(\omega, n, \varepsilon)) \right.$$

$$\|p_1 - p_2\|_1 \leq \varepsilon),$$

where ω is a modulus of uniform continuity of the function f. Note that again Φ only depends on f only via the modulus ω.

The uniqueness of the best L_1-approximation was proved already in 1921 [44]. In 1975, Björnestål [3] proved ineffectively the existence of a rate of strong unicity Φ having the form $c_{f,n} \varepsilon \omega_n(c_{f,n} \varepsilon)$, for some constant $c_{f,n}$ depending on f and n. In 1978, Kroó [77] improved Björnestål's results by showing—again ineffectively— that a constant $c_{\omega,n}$ depending only on the modulus of uniform continuity of f and n exists. Moreover, Kroó proved that the ε-dependency established by Björnestal is optimal. Note that the effective rate given above has this optimal dependency.

The effective rate of strong unicity given above allows one for the first time to compute effectively the best approximation. An upper bound on the complexity of that procedure is given in [85].

4 Effective computation of fixed points for functions of contractive type

There is a long history of extensions of Banach's well-known fixed point theorem for contractions to various more liberal notions of contractive type functions. The results usually are of the same shape as Banach's theorem; i.e., they state that the functions under consideration have a unique fixed point and that the Picard iteration $(f^n(x))_{n \in \mathbb{N}}$ of an arbitrary starting point converges to this fixed point. However, in contrast to Banach's theorem, in general no explicit rates of convergence can be read off from the (often ineffective) proofs.

The oldest of these results are due to Edelstein [27] and Rakotch [87].

Definition 4.1 ([27]). A self-mapping f of a metric space (X, d) is contractive if

$$\forall x, y \in X (x \neq y \rightarrow d(f(x), f(y)) < d(x, y)).$$

Theorem 4.2 ([27]). *Let* (X, d) *be a complete metric space, let* f *be a contractive self-mapping on* X, *and suppose that for some* $x_0 \in X$ *the sequence* $(f^n(x_0))$ *has a convergent subsequence* $(f^{n_i}(x_0))$. *Then* $\xi = \lim_{n \to \infty} f^n(x_0)$ *exists and is a unique fixed point of* f.

Rakotch observed that when contractivity is formulated in the following uniform way (which in the presence of compactness is equivalent to Edelstein's definition but in general is a strictly stronger condition), then it is possible to drop the assumption of the existence of convergent subsequences.

Definition 4.3 ([87]). [5] A self-mapping $f : X \to X$ of a metric space is called uniformly contractive with modulus $\alpha : \mathbb{Q}_+^* \to (0, 1) \cap \mathbb{Q}$ if

$$\forall \varepsilon \in \mathbb{Q}_+^* \forall x, y \in X (d(x, y) > \varepsilon \rightarrow d(f(x), f(y)) \leq \alpha(\varepsilon) \cdot d(x, y)).$$

Theorem 4.4 ([87]). *Let* (X, d) *be a complete metric space, and let* f *be a uniformly contractive self-mapping on* X *(i.e.,* f *has modulus of contractivity* α*); then, for all* $x \in X$, $\xi = \lim_{n \to \infty} f^n(x)$ *exists and is a unique fixed point of* f.

Example 4.5. The functions $f : [1, \infty) \to [1, \infty)$, $f(x) := x + \frac{1}{x}$, and $f : \mathbb{R} \to \mathbb{R}$, $f(x) := \ln(1 + e^x)$ are both contractive in the sense of Edelstein but not uniformly contractive in the sense of Rakotch. The function $f : [1, \infty) \to [1, \infty)$, $f(x) := 1 + \ln x$ is uniformly contractive in the sense of Rakotch but not a contraction.

---

[5] This definition is taken from [33] and is slightly more general than Rakotch's original definition.

From the essentially constructive proof in [87] one obtains (as predicted by a general logical metatheorem established in [33]) the following bound (see also [8] for a related result):

Theorem 4.6 ([33]). *With the conditions as in the previous theorem we have the following rate of convergence of the Picard iteration from an arbitrary point $x \in X$ towards the unique fixed point ξ of f :*

$$\forall x \in X \forall \varepsilon \in \mathbb{Q}_+^* \forall n \geq \delta(\alpha, b, \varepsilon)(d(f^n(x), \xi) \leq \varepsilon),$$

where

$$\delta(\alpha, b, \varepsilon) = \left\lceil \frac{\log \varepsilon - \log b'(\alpha, b)}{\log \alpha(\varepsilon)} \right\rceil \quad \text{for}$$

$$b'(\alpha, b) = \max(\rho, \frac{2 \cdot b}{1 - \alpha(\rho)}) \text{ with } \mathbb{N} \ni b \geq d(x, f(x)) \text{ and } \rho > 0 \text{ arbitrary.}$$

Remark 4.7. 1) Note that the rate of convergence depends on f, x only via α, and an upper bound for $d(x, f(x))$.

2) Instead of the multiplicative modulus of uniform contractivity α, one can also consider an additive modulus $\eta : \mathbb{Q}_+^* \to \mathbb{Q}_+^*$ s.t.

$$\forall \varepsilon \in \mathbb{Q}_+^* \forall x, y \in X \big(d(x, y) > \varepsilon \to d(f(x), f(y)) + \eta(\varepsilon) \leq d(x, y)\big)$$

and can construct a rate of convergence in terms of η (see [33]).

Instead of starting from a constructive proof one also could take an ineffective proof of $f^n(x) \to 0$ and first extract an effective bound Φ such that

$$\forall x \in X \forall \varepsilon \in \mathbb{Q}_+^* \exists n \leq \Phi(\alpha, b, \varepsilon)(d(f^n(x), f^{n+1}(x)) < \varepsilon)$$

using theorem 2.5 (which is possible since '$\exists n(d(f^n(x), f^{n+1}(x)) < \varepsilon)$' is purely existential).
Since the sequence $(d(f^n(x), f^{n+1}(x)))_n$ is nonincreasing, this yields

$$\forall x \in X \forall \varepsilon \in \mathbb{Q}_+^* \forall n \geq \Phi(\alpha, b, \varepsilon)(d(f^n(x), f^{n+1}(x)) < \varepsilon).$$

One then extracts (using again theorem 2.5) a modulus Ψ of uniqueness from the uniqueness proof. Similarly to our applications in approximation theory, these two moduli Φ, Ψ together then provide a rate of convergence towards the fixed point (for details, see [72]).

In the fixed point theorems due to Kincses/Totik ([45]) and Kirk ([48]) that we discuss next, only ineffective proofs were known so that an approach as outlined above had to be anticipated. However, due to the lack of monotonocity of $(d(f^n(x), f^{n+1}(x)))_n$ in these cases, this approach would not yield a full rate of convergence.

Nevertheless, this problem could be overcome, and in fact, recent work of E.M. Briseid ([14]) shows that under rather general conditions on the class of functions to be considered (satisfied in the two cases at hand for uniformly continuous functions), theorem 2.5 **can** be used to guarantee effective rates of convergence of $(f^n(x))_n$ towards a unique fixed point from a given ineffective proof of this fact.

In [89, 90], 25 different notions of contractivity are considered starting from Edelstein's definition. The most general one among those is called 'generalized contractivity' in [9, 10]. If only some iterate f^p for $p \in \mathbb{N}$ is required to satisfy this condition, the function is called 'generalized p-contractive':

Definition 4.8 ([89]). Let (X, d) be a metric space and $p \in \mathbb{N}$. A function $f : X \to X$ is called generalized p-contractive if

$$\forall x, y \in X \big(x \neq y \to d(f^p(x), f^p(y)) < \ \text{diam} \{x, y, f^p(x), f^p(y)\}\big).$$

Theorem 4.9 (Kincses/Totik,[45]). *Let (K, d) be a compact metric space and $f : K \to K$ a continuous function that is generalized p-contractive for some $p \in \mathbb{N}$. Then f has a unique fixed point ξ, and for every $x \in K$, we have*

$$\lim_{n \to \infty} f^n(x) = \xi.$$

Guided by the logical metatheorems from [54, 62, 64], Briseid ([9]) (i) generalized theorem 4.9 to the noncompact case (similar to Rakotch's form of Edelstein's theorem) and (ii) provided a fully effective quantitative form of this generalized theorem:

Definition 4.10 ([9, 10]). Let (X, d) be a metric space, $p \in \mathbb{N}$. $f : X \to X$ is called uniformly generalized p-contractive with modulus $\eta : \mathbb{Q}_+^* \to \mathbb{Q}_+^*$ if

$$\forall x, y \in X \forall \varepsilon \in \mathbb{Q}_+^* (d(x, y) > \varepsilon \to d(f^p(x), f^p(y)) + \eta(\varepsilon)$$

$$< \ \text{diam} \{x, y, f^p(x), f^p(y)\}).$$

It is clear that for compact spaces and continuous f the notions 'generalized p-contractive' and 'uniformly generalized p-contractive (with some modulus η)' coincide.

Theorem 4.11 ([9, 10]). *Let (X, d) be a complete metric space and $p \in \mathbb{N}$. Let $f : X \to X$ be a uniformly continuous and uniformly generalized p-contractive function with moduli of uniform continuity ω and uniform generalized p-contractivity η. Let $x_0 \in X$ be the starting point of the Picard iteration $(f^n(x_0))$ of f, and assume that $(f^n(x_0))$ is bounded by $b \in \mathbb{Q}_+^*$. Then f has a unique fixed point ξ and $(f^n(x_0))$ converges to ξ with rate of convergence $\Phi : \mathbb{Q}_+^* \to \mathbb{N}$; i.e.,*

$$\forall \varepsilon \in \mathbb{Q}_+^* \forall n \geq \Phi(\varepsilon)(d(f^n(x_0), \xi) \leq \varepsilon),$$

where

$$\Phi(\varepsilon) := \begin{cases} p\lceil (b - \varepsilon)/\rho(\varepsilon)\rceil & \text{if } b > \varepsilon, \\ 0, \text{otherwise}, \end{cases}$$

with

$$\rho(\varepsilon) := \min\left\{\eta(\varepsilon), \frac{\varepsilon}{2}, \eta(\frac{1}{2}\omega^p(\frac{\varepsilon}{2}))\right\}.$$

For a discussion of the logical background of this result, see [9].

Another notion of contractivity was recently introduced by Kirk and has received quite some interest in the last few years:

Definition 4.12 ([48]). Let (X, d) be a metric space. A self-mapping $f : X \to X$ is called an asymptotic contraction with moduli $\Phi, \Phi_n : [0, \infty) \to [0, \infty)$ if Φ, Φ_n are continuous, $\Phi(s) < s$ for all $s > 0$,

$$\forall n \in \mathbb{N} \forall x, y \in X (d(f^n(x), f^n(y)) \le \Phi_n(d(x, y))),$$

and $\Phi_n \to \Phi$ uniformly on the range of d.

Theorem 4.13 (Kirk, [48]). *Let (X, d) be a complete metric space and $f : X \to X$ a continuous asymptotic contraction. Assume that some orbit of f is bounded. Then f has a unique fixed point $\xi \in X$ and the Picard sequence $(f^n(x))$ converges to ξ for each $x \in X$.*

The following definition is essentially due to [32] (with a small generalization given by [11]) and was prompted by applying the method of monotone functional interpretation on which the logical metatheorems mentioned before are based on Kirk's definition.

Definition 4.14 ([11, 32]). A self-mapping $f : X \to X$ of a metric space (X, d) is called an asymptotic contraction in the sense of Gerhardy and Briseid if for each $b > 0$, there exist moduli $\eta^b : (0, b] \to (0, 1)$ and $\beta^b : (0, b] \times (0, \infty) \to \mathbb{N}$ such that the following hold:

1) There exists a sequence of functions $\phi_n^b : (0, \infty) \to (0, \infty)$ such that for each $0 < l \le b$ the function $\beta_l^b := \beta^b(l, \cdot)$ is a modulus of uniform convergence for $(\phi_n^b)_n$ on $[l, b]$; i.e.,

$$\forall \varepsilon > 0 \forall s \in [l, b] \forall m, n \ge \beta_l^b(\varepsilon)(|\phi_m^b(s) - \phi_n^b(s)| \le \varepsilon).$$

   Furthermore, if $\varepsilon < \varepsilon'$, then $\beta_l^b(\varepsilon) \ge \beta_l^b(\varepsilon')$.

2) For all $x, y \in X$, for all $\varepsilon > 0$, and for all $n \in \mathbb{N}$ with $\beta_\varepsilon^b(1) \le n$, we have that

$$b \ge d(x, y) \ge \varepsilon \to d(f^n(x), f^n(y)) \le \phi_n^b(\varepsilon)d(x, y).$$

3) For $\phi^b := \lim\limits_{n\to\infty} \phi^b_n$, we have

$$\forall \varepsilon \in (0,b] \forall s \in [\varepsilon, b](\phi^b(s) + \eta^b(\varepsilon) \le 1).$$

As shown in [32] (see also [11]) every asymptotic contraction in the sense of Kirk is also an asymptotic contraction in the sense of Gerhardy and Briseid (for suitable moduli). Moreover, as shown in [11], in the case of bounded and complete metric spaces, both notions coincide and are equivalent to the existence of a rate of convergence of the Picard iterations, which is uniform in the starting point (as the one presented below).

Guided by logical metatheorems, Gerhardy [32] not only developed the above explicit form of asymptotic contractivity but also extracted from Kirk's proof an effective so-called rate of proximity $\Psi(\eta, \beta, b, \varepsilon)$ such that

$$(f^n(x))_n \text{ bounded by } b \ \to \ \forall \varepsilon > 0 \exists n \le \Psi(\eta, \beta, b, \varepsilon)(d(f^n(x), \xi))$$

for the unique fixed point ξ of f. For functions f that in addition to being continuous asymptotic contractions (with moduli η, β) are quasi-nonexpansive (see the final section of this paper), this already yields a rate of convergence towards the fixed point since $(d(f^n(x), \xi))_n$ is nonincreasing in this case. Building upon Gerhardy's result, Briseid [11] gave an effective rate of convergence in the general case:

Theorem 4.15 ([11]). *Let (X, d) be a complete metric space and f a continuous asymptotic contraction (in the sense of Gerhardy and Briseid) with moduli η, β. Let, furthermore, $b > 0$. If for some $x_0 \in X$ the Picard iteration sequence $f^n(x_0)$ is bounded by b, then f has a unique fixed point ξ and*

$$\forall \varepsilon > 0 \forall n \ge \Phi(\eta, \beta, b, \varepsilon)\big(d(f^n(x_0), \xi) \le \varepsilon\big),$$

where

$$\Phi(\eta, \beta, b, \varepsilon) := \max\{k(2M_\gamma + \beta_{(\frac{\varepsilon}{2})}(\delta) + K_\gamma - 1),$$

$$(k-1) \cdot (2M_\gamma + \beta_{(\frac{\varepsilon}{2})}(\delta) + K_\gamma - 1) + M_\gamma + 1\},$$

with

$$k := \left\lceil \frac{\ln \varepsilon - \ln b}{\ln(1 - \frac{\eta(\gamma)}{2})} \right\rceil, \quad M_\gamma := K_\gamma \cdot \left\lceil \frac{\ln \gamma - \ln b}{\ln(1 - \frac{\eta(\gamma)}{2})} \right\rceil, \quad K_\gamma := \beta_\gamma\left(\frac{\eta(\gamma)}{2}\right),$$

$$\delta := \min\{\tfrac{\varepsilon}{2}, \tfrac{\eta(\frac{\varepsilon}{2})}{2}\} \quad \gamma := \min\{\delta, \tfrac{\delta\varepsilon}{4}\}.$$

Using results from [32] it is shown in [12] that Picard iteration sequences of asymptotic contractions always are bounded so that the corresponding assumption in Kirk's theorem 4.13 is superfluous (see also [13, 95]). Moreover, [12] gives an effective rate of convergence that does not depend on a bound b on (x_n) but instead on (strictly positive) lower and upper bounds on $d(x_0, f(x_0))$.

5 Fixed points and approximate fixed points of nonexpansive functions in hyperbolic spaces

Already for bounded metric spaces we cannot even hope that nonexpansive functions have approximate fixed points. This is due to the fact that (in contrast to functions of contractive type treated above) we can always change a given metric d to a bounded one by defining the truncated metric $D(x, y) := \max\{d(x, y), 1\}$ without destroying the property of nonexpansiveness: for example, consider the bounded metric space $(\mathbb{R}, D)$, where $D(x, y) := \max\{|x - y|, 1\}$ and the nonexpansive function $f(x) := x + 1$. Then $\inf\{D(x, f(x)) : x \in \mathbb{R}\} = 1$. In the case of bounded, closed, and convex subsets C of Banach spaces, nonexpansive mappings always have approximate fixed points (see [60] for an easy proof of this fact), but in general they have no fixed points (see [93]). Moreover, as the example $f = id_X$ shows, if a fixed point exists, it will in general no longer be unique, and even in cases where a unique fixed point exists, the Picard iteration will not necessarily converge to the fixed point and may even fail to produce approximate fixed points: consider e.g., $f : [0, 1] \to [0, 1], f(x) := 1 - x$. Then for each $x \in [0, 1] \setminus \{\frac{1}{2}\}$, the iteration sequence $f^n(x)$ oscillates between x and $1 - x$ and so stays bounded away from the unique fixed point $\frac{1}{2}$. This is the reason why one considers so-called Krasnoselski–Mann iterations $(x_n)_{n \in \mathbb{N}}$ (see below), which make use of a concept of a convex combination that exists in normed spaces and in so-called hyperbolic spaces. Even in cases where (x_n) converges to a fixed point, one can no longer hope for an effective rate of convergence. In fact it has been shown that already in almost trivial contexts, such effective rates do not exist (see [65]). This failure of effectivity is largely due to the nonuniqueness of the fixed point (and hence the absence of a modulus of uniqueness in the sense of section 3). However, in many cases, one can extract from the proofs effective rates on the so-called asymptotic regularity

$$d(x_n, f(x_n)) \to 0,$$

which holds under much more general conditions than the ones needed to guarantee the existence of fixed points. As mentioned above, we need somewhat more structure than just a metric space to define the Krasnoselski–Mann iteration:

Definition 5.1 ([37, 46, 64, 88]). (X, d, W) is called a hyperbolic space if (X, d) is a metric space and $W : X \times X \times [0, 1] \to X$ is a function satisfying

(i) $\forall x, y, z \in X \forall \lambda \in [0, 1]\big(d(z, W(x, y, \lambda)) \leq (1 - \lambda)d(z, x) + \lambda d(z, y)\big)$,

(ii) $\forall x, y \in X \forall \lambda_1, \lambda_2 \in [0, 1]\big(d(W(x, y, \lambda_1), W(x, y, \lambda_2)) = |\lambda_1 - \lambda_2| \cdot d(x, y)\big)$,

(iii) $\forall x, y \in X \forall \lambda \in [0, 1]\big(W(x, y, \lambda) = W(y, x, 1 - \lambda)\big)$,

(iv) $\begin{cases} \forall x, y, z, w \in X, \lambda \in [0, 1], \\ \big(d(W(x, z, \lambda), W(y, w, \lambda)) \leq (1 - \lambda)d(x, y) + \lambda d(z, w)\big). \end{cases}$

Remark 5.2. The definition (introduced in [64]) is slightly more restrictive than the notion of 'space of hyperbolic type' as defined in [37] (which results if (iv) is dropped) but somewhat more general than the concept of 'space of hyperbolic type' as defined in [46] and—under the name of 'hyperbolic space'—in [88]. Our definition was prompted by the general logical metatheorems developed in [64] and appears to be most useful in the context of proof mining (see [34, 64] for detailed discussions). Moreover, our notion comprises the important class of CAT(0)-spaces (in the sense of Gromov), whereas the concept from [46, 88] only covers CAT(0)-spaces having the so-called geodesic line extension property. With axiom (i) alone, the above notion coincides with the concept of 'convex metric space' as introduced in [96].

In the following we denote $W(x, y, \lambda)$ by $(1 - \lambda)x \oplus \lambda y$.

In this section (X, d, W) always denotes a hyperbolic space and (λ_n) a sequence in $[0, 1)$ that is bounded away from 1 (i.e., $\limsup \lambda_n < 1$) and divergent in sum (i.e., $\sum_{i=0}^{\infty} \lambda_i = \infty$). $f : X \to X$ is a self-mapping of X. Furthermore, given an $x \in X$, the sequence (x_n) refers (unless stated otherwise) to the so-called Krasnoselski–Mann iteration of f; i.e.,

$$x_0 := x, \quad x_{n+1} := (1 - \lambda_n)x_n \oplus \lambda_n f(x_n).$$

Theorem 5.3 ([37, 43]). *Let (X, d, W) be a hyperbolic space and $f : X \to X$ be nonexpansive. Then for all $x \in X$, the following holds:*

$$\text{If } (x_n) \text{ is bounded, then } d(x_n, f(x_n)) \to 0.$$

Theorem 5.4 ([4]). *Let (X, d, W) be a hyperbolic space and $f : X \to X$ be a nonexpansive function. Then for all $x \in X$, the following holds:*

$$d(x_n, f(x_n)) \to r_X(f) := \inf_{y \in X} d(y, f(y)).$$

The quantity $r_X(f)$ is often called 'minimal displacement of f on X'.

As shown in [34], corollary 2.8 a-priorily guarantees that the proofs of the previous two results allow one to extract effective bounds on both theorems depending only on those parameters the concrete bounds in theorems 5.5 and 5.10 below depend that are extracted in this way. We start with theorem 5.4: Since $(d(x_n, f(x_n)))$ is nonincreasing, theorem 5.4 formalizes as either

$$(a) \ \forall \varepsilon > 0 \exists n \in \mathbb{N} \forall x^* \in X (d(x_n, f(x_n)) < d(x^*, f(x^*)) + \varepsilon)$$

or

$$(b) \ \forall \varepsilon > 0 \forall x^* \in X \ \exists n \in \mathbb{N}(d(x_n, f(x_n)) < d(x^*, f(x^*)) + \varepsilon).$$

Trivially, (a) implies (b), but ineffectively (using the existence of $r_X(f)$) also the implication in the other direction holds. Only (b) meets the specification in the metatheorem.

In the following, let $\alpha : \mathbb{N} \times \mathbb{N} \to \mathbb{N}$ be such that[6]

$$\forall i, n \in \mathbb{N}\big(\alpha(i,n) \leq \alpha(i+1,n)\big) \text{ and}$$

$$\forall i, n \in \mathbb{N}\Big(n \leq \sum_{s=i}^{i+\alpha(i,n)-1} \lambda_s\Big).$$

Let $k \in \mathbb{N}$ be such that $\lambda_n \leq 1 - \frac{1}{k}$ for all $n \in \mathbb{N}$.
Corollary 2.8 predicts a uniform bound depending on x, x^*, f only via $b \geq d(x, x^*)$, on $d(x, f(x))$, and on (λ_k) only via k, α (see [34]):

Theorem 5.5 ([69]). *Let (X, d, W) be a hyperbolic space and $(\lambda_n)_{n \in \mathbb{N}}, k, \alpha$ as above. Let $f : X \to X$ be nonexpansive and $b > 0, x, x^* \in X$ with*

$$d(x, x^*), d(x, f(x)) \leq b.$$

Then for the Krasnoselski–Mann iteration (x_n) of f starting from x, the following holds:

$$\forall \varepsilon \in \mathbb{Q}_+^* \forall n \geq \Psi(k, \alpha, b, \varepsilon)\, (d(x_n, f(x_n)) < d(x^*, f(x^*)) + \varepsilon),$$

where

$$\Psi(k, \alpha, b, \varepsilon) := \widehat{\alpha}(\lceil 2b \cdot \exp(k(M+1)) \rceil \div 1, M),$$

$$\text{with } M := \left\lceil \frac{1+2b}{\varepsilon} \right\rceil \text{ and}$$

$$\widehat{\alpha}(0, M) := \tilde{\alpha}(0, M), \ \widehat{\alpha}(m+1, M) := \tilde{\alpha}(\widehat{\alpha}(m, M), M) \text{ with}$$

$$\tilde{\alpha}(m, M) := m + \alpha(m, M) \ (m \in \mathbb{N}).$$

Definition 5.6 ([47, 69]). If (X, d, W) is a hyperbolic space, then $f : X \to X$ is called directionally nonexpansive (short 'f d.n.e') if

$$\forall x \in X \forall y \in \text{seg}\,(x, f(x))\big(d(f(x), f(y)) \leq d(x, y)\big),$$

where

$$\text{seg}\,(x, y) := \{\, W(x, y, \lambda) : \lambda \in [0, 1] \,\}.$$

---

[6] One easily verifies that one could start with any function $\beta : \mathbb{N} \to \mathbb{N}$ satisfying $n \leq \sum_{s=0}^{\beta(n)} \lambda_s$ and then define $\alpha(i, n) := \max_{j \leq i}(\beta(n+j) - j + 1)$ to get an α satisfying these conditions. However, this would in general give less good bounds than when working with α directly. See [61, 69] for more information in this point.

Example 5.7. Consider the convex subset $[0,1]^2$ of the normed space $(\mathbb{R}^2, \| \cdot \|_{\max})$ and the function

$$f : [0,1]^2 \to [0,1]^2, \; f(x,y) := \begin{cases} (1,y), & \text{if } y > 0, \\ (0,y), & \text{if } y = 0. \end{cases}$$

f is directionally nonexpansive but discontinuous at $(0,0)$ and so, in particular, not nonexpansive.

Theorem 5.5 generalizes to directionally nonexpansive mappings. The additional assumption needed is redundant in the case of nonexpansive mappings:

Theorem 5.8 ([69]). *The previous theorem (and bound) also holds for directionally nonexpansive mappings if $d(x, x^*) \le b$ is strengthened to $d(x_n, x_n^*) \le b$ for all n.*

The next result is proved in [63] for the case of convex subsets of normed spaces, but the proof immediately extends to hyperbolic spaces. We include the proof for completeness. It applies corollary 2.8 to a formalization of theorem 5.4 that corresponds to the Herbrand normal form of (a) and constructively has a strength in between (a) and (b). Here x^* is replaced by a sequence (y_n) and we search for an n such that

$$d(x_n, f(x_n)) < d(y_n, f(y_n)) + \varepsilon;$$

i.e., (b) is just the special case with the constant sequence $y_n := x^*$. As predicted by corollary 2.8, we get a quantitative version of the following form:

Theorem 5.9. *Under the same assumptions as in theorem 5.5, the following holds: Let (b_n) be a sequence of strictly positive real numbers. Then for all $x \in X$, $(y_n)_{n \in \mathbb{N}} \subset X$ with*

$$\forall n \in \mathbb{N}(d(x, f(x)), d(x, y_n) \le b_n)$$

and all $\varepsilon > 0$, there exists an $i \le j(k, \alpha, (b_n)_{n \in \mathbb{N}}, \varepsilon)$ s.t.[7]

$$d(x_i, f(x_i)) < d(y_i, f(y_i)) + \varepsilon,$$

where (omitting the arguments k, α for better readability)

$$j((b_n)_{n \in \mathbb{N}}, \varepsilon) := \max_{i \le h((b_n)_{n \in \mathbb{N}}, \varepsilon)} \Psi(k, \alpha, b_i, \varepsilon/2)$$

with

$$h((b_n)_{n \in \mathbb{N}}, \varepsilon) := \max_{i < N} g^i(0), \; g(n) := \Psi(k, \alpha, b_n, \varepsilon/2), N := \left\lceil \frac{6b_0}{\varepsilon} \right\rceil.$$

Here Ψ is the bound from theorem 5.5 and $g^n(0)$ is defined primitive recursively: $g^0(0) := 0, \; g^{n+1}(0) := g(g^n(0))$.
Instead of N, we can take any integer upper bound for $6b_0/\varepsilon$.

---

[7] Recall that whereas (y_n) is an arbitrary sequence of points in X, (x_n) denotes the Krasnoselski–Mann iteration of f starting from x.

Proof. By theorem 5.5 we have that

$$(1) \; \forall n \in \mathbb{N}\big(d(x_{g(n)}, f(x_{g(n)})) < d(y_n, f(y_n)) + \frac{\varepsilon}{2}\big),$$

where $g(n) := \Psi(k, \alpha, b_n, \varepsilon/2)$. Let $N := \lceil \frac{6b_0}{\varepsilon} \rceil$ and $l := \max_{i < N} g^i(0)$. Using that

$$(2) \; d(y_0, f(y_0)) \leq d(y_0, x) + d(x, f(x)) + d(f(x), f(y_0))$$
$$\leq 2d(y_0, x) + d(x, f(x)) \leq 3b_0,$$

we now show that

$$(3) \; \exists i < N \big(d(y_{(g^i(0))}, f(y_{(g^i(0))})) \leq d(y_{(g^{i+1}(0))}, f(y_{(g^{i+1}(0))})) + \frac{\varepsilon}{2}\big) :$$

Suppose not, then for all $i < N$,

$$d(y_{(g^{i+1}(0))}, f(y_{(g^{i+1}(0))})) < d(y_{(g^i(0))}, f(y_{(g^i(0))})) - \frac{\varepsilon}{2},$$

and therefore,

$$d(y_{(g^N(0))}, f(y_{(g^N(0))})) < d(y_0, f(y_0)) - N\frac{\varepsilon}{2} \overset{(2)}{\leq} 3b_0 - N\frac{\varepsilon}{2} \leq 0,$$

which is a contradiction and finishes the proof of (3).
Let i be as in (3). Then by (1) we get for $p := g^i(0)$

$$(4) \; \forall n \in \mathbb{N}\big(d(x_{g(p)}, f(x_{g(p)})) < d(y_{g(p)}, f(y_{g(p)})) + \varepsilon\big),$$

where $p \leq l$. Hence the theorem is satisfied with $j((b_n)_n, \varepsilon) := \max_{i \leq l} g(i)$. □

The next theorem gives a uniform quantitative version of the theorem of Ishikawa [43] as generalized by Goebel and Kirk [37] to hyperbolic spaces.

Theorem 5.10 ([67, 69]). *Let (X, d, W) be a nonempty hyperbolic space and $f : X \to X$ a nonexpansive mapping, and $(\lambda_n)_{n \in \mathbb{N}}, \alpha$, and k be as before. Let $b > 0, x, x^* \in X$ be such that*

$$d(x, x^*) \leq b \wedge \forall n, m \in \mathbb{N}(d(x_n^*, x_m^*) \leq b),$$

where (x_n^*) is the Krasnoselski–Mann iteration starting from x^*. Then the following holds:

$$\forall \varepsilon > 0 \forall n \geq h(k, \alpha, b, \varepsilon)\big(d(x_n, f(x_n)) \leq \varepsilon\big),$$

where

$$h(k, \alpha, b, \varepsilon) := \widehat{\alpha}(\lceil 10b \cdot \exp(k(M+1)) \rceil - 1, M)), \; with$$

$$M \qquad := \lceil \tfrac{1+4b}{\varepsilon} \rceil \; and \; \widehat{\alpha} \; as \; before.$$

Next we generalize the previous theorem (for $x^* := x$) to directionally nonexpansive functions.

Theorem 5.11 ([69]). *Let (X, d, W) be a nonempty hyperbolic space and $f : X \to X$ a directionally nonexpansive mapping. Let $(\lambda_n)_{n \in \mathbb{N}}, \alpha, k$ be as before. Let $b > 0$ and $x \in X$ such that*

$$\forall n, k, m \in \mathbb{N}\big(d(x_n, (x_k)_m) \leq b\big),$$

where

$$(x_k)_0 = x_k, \qquad (x_k)_{m+1} = (1 - \lambda_m)(x_k)_m \oplus \lambda_k f((x_k)_m).$$

Then the following holds:

$$\forall \varepsilon > 0 \forall n \geq h(k, \alpha, b, \varepsilon)\big(d(x_n, f(x_n)) \leq \varepsilon\big),$$

where

$h(k, \alpha, b, \varepsilon) := \alpha(0, 1) + \widehat{\alpha^*}(\lceil 2b \cdot \alpha(0, 1) \cdot \exp(k(M + 1))\rceil - 1, M)$, *with*

$M \qquad := \lceil \frac{1+2b}{\varepsilon} \rceil$ *and* $\widehat{\alpha^*}(0, n) := \tilde{\alpha}^*(0, n)$, $\widehat{\alpha^*}(i+1, n) := \tilde{\alpha}^*(\widehat{\alpha^*}(i, n), n)$ *with*

$\tilde{\alpha}^*(i, n) \quad := i + \alpha^*(i, n),$

$\alpha^*(i, n) \quad := \alpha(i + \alpha(0, 1), n) \; (i, n \in \mathbb{N}).$

Remark 5.12. Note that for constant $\lambda_k := \lambda$ we have $(x_k)_m = x_{k+m}$ so that the assumption $d(x_n, x_m) \leq b$ for all m, n suffices.

Previously known existence and uniformity results in the bounded case[8]:

- Krasnoselski ([73]): Uniformly convex normed spaces X and special constant $\lambda_k = \frac{1}{2}$, no uniformity.

- Browder/Petryshyn ([15]): Uniformly convex normed spaces X and constant $\lambda_k = \lambda \in (0, 1)$, no uniformity.

- Groetsch ([40]): X uniformly convex, general (λ_k), no uniformity (see also below).

- Ishikawa ([43]): General normed space X and general (λ_k), no uniformity.

- Edelstein/O'Brien ([28]): General normed space X and constant $\lambda_k := \lambda \in (0, 1)$. Uniformity w.r.t. $x_0 \in C$ (and implicitly, though not stated, w.r.t. f).

- Goebel/Kirk ([37]): General hyperbolic X and general (λ_k). Uniformity w.r.t. x_0 and f.

- Kirk/Martinez-Yanez ([50]): Uniformity w.r.t. x_0, f for uniformly convex normed spaces X and special constant $\lambda_k := 1/2$.

---

[8] That is, the case of bounded convex subsets in the normed case, resp. bounded, hyperbolic spaces.

- Goebel/Kirk ([38]): Conjecture: no uniformity w.r.t. C.[9]

- Baillon/Bruck ([1]): Uniformity w.r.t. x_0, f, C for general normed spaces X and constant $\lambda_k := \lambda \in (0, 1)$.

- Kirk ([47]): Uniformity w.r.t. x_0, f for constant $\lambda_k := \lambda \in (0, 1)$ for directionally nonexpansive functions in normed spaces.

- Kohlenbach ([61]): Uniformity w.r.t. x_0, f, C for general (λ_k) for nonexpansive functions in the normed case.

- K./Leuştean ([69]): Uniformity w.r.t. x_0, f, C for general (λ_k) for directionally nonexpansive functions in the hyperbolic case.

Theorem 5.3 by Ishikawa [43] and Goebel and Kirk [37] has the following consequence in the compact case:

Theorem 5.13 ([37, 43]). *Let* (X, d, W) *be a compact hyperbolic space, and let* $(\lambda_n), f, (x_n)$ *be as in theorem 5.3. Then* $(x_n)_n$ *converges towards a fixed point of* f *for any starting point* $x_0 := x \in X$ *of the Krasnoselski–Mann iteration* (x_n).

By theorem 5.3, the completeness of the space, and the continuity of f, the conclusion of theorem 5.13 is equivalent to the property of (x_n) being a Cauchy sequence. That property is Π_3^0 and so of too complicated a logical form to allow for an effective bound in general. In fact, as shown in [65] there is no effective bound (uniformly in the parameters) even in the most simple cases. However, we can extract an effective bound on the Herbrand normal form

$$(H) \ \forall k \in \mathbb{N}, g \in \mathbb{N}^{\mathbb{N}} \exists n \in \mathbb{N} \forall i, j \in [n; n + g(n)](d(x_i, x_j) < 2^{-k})$$

of the Cauchy property that classically is equivalent to the latter. Here $[n; n + g(n)]$ denotes the set of all natural numbers j with $n \leq j \leq n + g(n)$. Note that '$\forall i, j \in [n; n + g(n)](d(x_i, x_j) < 2^{-k})$' is equivalent to a purely existential formula. Since

$$\lambda_n d(x_n, f(x_n)) = d(x_n, x_{n+1}),$$

the asymptotic regularity $d(x_n, f(x_n)) \to 0$ property is equivalent to the special case of (H) with $g \equiv 1$ (for seqences (λ_n) that are bounded away from 0). So (H) is a generalization of asymptotic regularity that for general g fails in the absence of compactness, whereas asymptotic regularity only needs the boundedness of X (or rather of the sequence (x_n)). Our effective bound on (H), therefore, will depend on a modulus of total boundedness of the space (see [67] for a detailed discussion).

Definition 5.14. Let (M, d) be a totally bounded metric space. We call $\gamma : \mathbb{N} \to \mathbb{N}$ a modulus of total boundedness for M if for any $k \in \mathbb{N}$, there exist elements $a_0, \ldots, a_{\gamma(k)} \in M$ such that

---

[9] By uniformity w.r.t. C, it is meant that the bound depends on C only via an upper bound on the diameter of C.

$$\forall x \in M \exists i \leq \gamma(k)\big(d(x, a_i) \leq 2^{-k}\big).$$

Definition 5.15. Let (M, d) be a metric space, $f : M \to M$ a self-mapping of M, and (x_n) an arbitrary sequence in M. A function $\delta : \mathbb{N} \to \mathbb{N}$ is called an approximate fixed point bound for (x_n) if

$$\forall k \in \mathbb{N} \exists m \leq \delta(k)\big(d(x_m, f(x_m)) \leq 2^{-k}\big).$$

Of course, an approximate fixed point bound only exists if (x_n) contains arbitrarily good approximate fixed points.

Theorem 5.16 ([65]). *Let* $(X, d, W), (\lambda_n), f, (x_n)$ *be as in theorem 5.3 and* $k \in \mathbb{N}, g : \mathbb{N} \to \mathbb{N}, \delta : \mathbb{N} \to \mathbb{N}$ *and* $\gamma : \mathbb{N} \to \mathbb{N}$. *We define a function* $\Omega(k, g, \delta, \gamma)$ *(primitive) recursively as follows:*

$$\Omega(k, g, \delta, \gamma) := \max_{i \leq \gamma(k+3)} \Psi_0(i, k, g, \delta),$$

where

$$\begin{cases} \Psi_0(0, k, g, \delta) := 0, \\ \Psi_0(n + 1, k, g, \delta) := \delta\left(k + 2 + \lceil \log_2(\max_{l \leq n} g(\Psi_0(l, k, g, \delta)) + 1) \rceil\right). \end{cases}$$

If δ *is an approximate fixed point bound for the Krasnoselski–Mann iteration* (x_n) *starting from* $x \in X$ *and* γ *a modulus of total boundedness for* X, *then*

$$\forall k \in \mathbb{N} \forall g : \mathbb{N} \to \mathbb{N} \exists n \leq \Omega(k, g, \delta, \gamma) \forall i, j \in [n; n + g(n)]\big(d(x_i, x_j) \leq 2^{-k}\big).$$

We now extend the previous theorem to asymptotically nonexpansive functions (though only in the context of convex subsets C of normed linear spaces $(X, \| \cdot \|)$:

Definition 5.17 ([36]). Let $(X, \| \cdot \|)$ be normed space and $C \subset X$ a nonempty convex subset. $f : C \to C$ is said to be asymptotically nonexpansive with sequence $(k_n) \in [0, \infty)^{\mathbb{N}}$ if $\lim_{n \to \infty} k_n = 0$ and

$$\forall n \in \mathbb{N} \forall x, y \in X\big(\|f^n(x) - f^n(y)\| \leq (1 + k_n)\|x - y\|\big).$$

In the context of asymptotically nonexpansive mappings $f : C \to C$, the Krasnoselski–Mann iteration starting from $x \in C$ is defined in a slightly different form as

$$(+)\ x_0 := x, \quad x_{n+1} := (1 - \lambda_n)x_n + \lambda_n f^n(x_n).$$

Definition 5.18. An approximate fixed point bound $\Phi : \mathbb{Q}_+^* \to \mathbb{N}$ is called monotone if

$$q_1 \leq q_2 \to \Phi(q_1) \geq \Phi(q_2), \quad q_1, q_2 \in \mathbb{Q}_+^*.$$

Remark 5.19. Any approximate fixed point bound Φ for a sequence (x_n) can effectively be converted into a monotone approximate fixed point bound for (x_n) by

$$\Phi_M(q) := \Phi_m(\min k[2^{-k} \leq q]), \text{ where } \Phi_m(k) := \max_{i \leq k} \Phi(2^{-i}).$$

We now assume that C is totally bounded.

Theorem 5.20 ([65]). *Let $k \in \mathbb{N}, g : \mathbb{N} \to \mathbb{N}, \Phi : \mathbb{Q}_+^* \to \mathbb{N}$, and $\gamma : \mathbb{N} \to \mathbb{N}$. Let $f : C \to C$ be asymptotically nonexpansive with a sequence (k_n) such that $\mathbb{N} \ni K \geq \sum\limits_{n=0}^{\infty} k_n$ and $N \in \mathbb{N}$ be such that $N \geq e^K$. We define a function $\Psi(k, g, \Phi, \gamma)$ (primitive) recursively as follows:*

$$\Psi(k, g, \Phi, \gamma) := \max_{i \leq \gamma(k + \log_2(N) + 3)} \Psi_0(i, k, g, \Phi),$$

where (writing $\Psi_0(l)$ for $\Psi_0(l, k, g, \Phi)$)

$$\begin{cases} \Psi_0(0) := 0, \\ \Psi_0(n+1) := \\ \Phi\left(2^{-k - \log_2(N) - 2} / (\max_{l \leq n}[g^M(\Psi_0(l))(\Psi_0(l) + g^M(\Psi_0(l)) + \log_2(N)) + 1])\right) \end{cases}$$

with $g^M(n) := \max\limits_{i \leq n} g(i)$.

If Φ is a monotone approximate fixed point bound for the Krasnoselski–Mann iteration (x_n) (defined by $(+)$) and γ a modulus of total boundedness for C, then

$$\forall k \in \mathbb{N} \forall g : \mathbb{N} \to \mathbb{N} \exists n \leq \Psi(k, g, \Phi, \gamma) \forall i, j \in [n; n + g(n)] \left(\|x_i - x_j\| \leq 2^{-k}\right).$$

Remark 5.21. The previous two theorems even hold for arbitrary sequences (λ_n) in $[0, 1]$. However, in order to construct approximate fixed point bounds, one will need extra conditions.

For uniformly convex spaces and (λ_n) bounded away from both 0 and 1, an approximate fixed point bound Φ for asymptotically nonexpansive mappings will be presented in the last section.

We will now show that the qualitative features of the bounds in theorem 5.5 and 5.10 can be used to obtain new information on the approximate fixed point property (AFPP) for product spaces. A metric space (M, ρ) is said to have the AFPP for nonexpansive mappings if every nonexpansive mapping $f : M \to M$ has arbitrarily good approximate fixed points, i.e., if $\inf\limits_{u \in M} \rho(u, f(u)) = 0$.

Let (X, d, W) be a hyperbolic space and (M, ρ) a metric space with AFPP for nonexpansive mappings. Let $\{C_u\}_{u \in M} \subseteq X$ be a family of convex sets such that there exists a nonexpansive *selection* function $\delta : M \to \bigcup_{u \in M} C_u$ with

$$\forall u \in M\big(\delta(u) \in C_u\big).$$

Consider subsets of $(X \times M)_\infty$ (with the metric $d_\infty((x,u),(y,v)) := \max\{d(x,y), \rho(u,v)\}$)

$$H := \{(x,u) : u \in M, x \in C_u\}.$$

If $P_1 : H \to \bigcup_{u \in M} C_u, P_2 : H \to M$ are the projections, then for any nonexpansive function $T : H \to H$ w.r.t. d_∞ satisfying

$$(*)\ \forall(x,u) \in H\ \big((P_1 \circ T)(x,u) \in C_u\big),$$

we can define for each $u \in M$ the nonexpansive function

$$T_u : C_u \to C_u, \quad T_u(x) := (P_1 \circ T)(x,u).$$

We denote the Krasnoselski–Mann iteration starting from $x \in C_u$ and associated with T_u by (x_n^u) $((\lambda_n)$ as in theorem 5.3).

$r_S(F)$ always denotes the minimal displacement of F on S.

Theorem 5.22 ([70]). *Assume that $T : H \to H$ is nonexpansive with $(*)$ and $\sup_{u \in M} r_{C_u}(T_u) < \infty$.*

Suppose there exists $\varphi : \mathbb{R}_+^* \to \mathbb{R}_+^*$ s.t.

$$\forall \varepsilon > 0\, \forall v \in M\, \exists x^* \in C_v\ \big(d(\delta(v), x^*) \leq \varphi(\varepsilon)\ \wedge$$

$$\wedge\, d(x^*, T_v(x^*)) \leq \sup_{u \in M} r_{C_u}(T_u) + \varepsilon\big).$$

Then

$$r_H(T) \leq \sup_{u \in M} r_{C_u}(T_u).$$

Theorem 5.23 ([70]). *Assume that there is a $b > 0$ s.t.*

$$\forall u \in M\, \exists x \in C_u\big(d(\delta(u), x) \leq b \wedge \forall n, m \in \mathbb{N}(d(x_n^u, x_m^u) \leq b).$$

Then $r_H(T) = 0$.

Corollary 5.24 ([70]). *Assume that there is a $b > 0$ with the property that*

$$\forall u \in M\big(diam(C_u) \leq b\big).$$

Then H has AFPP for nonexpansive mappings $T : H \to H$ satisfying $(*)$.

As a special case of the previous corollary we obtain a recent result of Kirk (note that for $C_u := C$ being constant, we can take as δ any constant function $: M \to C$):

Corollary 5.25 ([49]). *If $C_u := C$ constant and C bounded, then H has the approximate fixed point property.*

6 Bounds on asymptotic regularity in the uniformly convex case

Prior to Ishikawa's paper [43], the fixed point theory of nonexpansive mappings was essentially restricted to the case of uniformly convex normed spaces ([15, 73]). Although Ishikawa showed that the fundamental theorem 5.3 holds without uniform convexity, the case of uniformly convex spaces is still of interest for the following reasons (among others):

- As shown by Groetsch [40] (see below) in the uniformly convex case, the conditions on (λ_n) in theorem 5.3 can be weakened to

$$\sum_{i=0}^{\infty} \lambda_i (1 - \lambda_i) = \infty,$$

  which is known to be optimal even for the case of Hilbert spaces (for general normed spaces, it is still open whether this condition is sufficient).

- The bounds extracted from proofs using uniform convexity are often better than the ones known for the general case (see below; a notable exception is the optimal quadratic bound from [1] for the case of general normed spaces and constant $\lambda_k = \lambda \in (0, 1)$).

- In the uniformly convex case, only corresponding results for more general classes of functions such as asymptotically nonexpansive functions and (weakly) quasi-nonexpansive functions are known (see below).

Definition 6.1 ([18]). A normed linear space $(X, \| \cdot \|)$ is uniformly convex if for any $\varepsilon > 0$, there exists a $\delta > 0$ such that

$$\forall x, y \in X \big(\|x\|, \|y\| \leq 1 \wedge \|x - y\| \geq \varepsilon \to \|\tfrac{1}{2}(x + y)\| \leq 1 - \delta \big).$$

A mapping $\eta : (0, 2] \to (0, 1]$ providing such a $\delta := \eta(\varepsilon) > 0$ for given $\varepsilon \in (0, 2]$ is called a modulus of uniform convexity.

Theorem 6.2 ([40]). *Let C be a convex subset of uniformly convex Banach space $(X, \| \cdot \|)$, and let (λ_n) be a sequence in $[0, 1]$ with $\sum_{i=0}^{\infty} \lambda_i(1 - \lambda_i) = \infty$. If $f : C \to C$ is nonexpansive and has at least one fixed point, then for the Krasnoselski–Mann iteration (x_n) of f starting at any point $x_0 \in C$, the following holds:*

$$\|x_k - f(x_k)\| \stackrel{k \to \infty}{\to} 0.$$

We now give a quantitative version of a strengthening of Groetsch's theorem, which only assumes the existence of approximate fixed points in some neighborhood of x (see [64, 80] for a discussion on how this fits under the logical metatheorems):

Theorem 6.3 ([63]). *Let* $(X, \| \cdot \|)$ *be a uniformly convex normed linear space with modulus of uniform convexity* η, $d > 0$, $C \subseteq X$ *a (nonempty) convex subset,* $f : C \to C$ *nonexpansive, and* $(\lambda_k) \subset [0, 1]$ *and* $\gamma : \mathbb{N} \to \mathbb{N}$ *such that*

$$\forall n \in \mathbb{N}\left(\sum_{k=0}^{\gamma(n)} \lambda_k(1 - \lambda_k) \geq n \right).$$

Then for all $x \in C$, *which satisfy that for all* $\varepsilon > 0$, *there is a* $y \in C$ *with*

$$\|x - y\| \leq d \text{ and } \|y - f(y)\| < \varepsilon,$$

one has

$$\forall \varepsilon > 0 \forall n \geq h(\varepsilon, d, \gamma, \eta)\big(\|x_n - f(x_n)\| \leq \varepsilon\big),$$

where $h(\varepsilon, d, \gamma, \eta) := \gamma\left(\left\lceil \frac{3(d+1)}{2\varepsilon \cdot \eta(\frac{\varepsilon}{d+1})} \right\rceil\right)$ *for* $\varepsilon < 2d$ *and* $h(\varepsilon, d, \gamma, \eta) := 0$ *otherwise.*
Moreover, if $\eta(\varepsilon)$ *can be written as* $\eta(\varepsilon) = \varepsilon \cdot \tilde{\eta}(\varepsilon)$ *with*

$$\varepsilon_1 \geq \varepsilon_2 \to \tilde{\eta}(\varepsilon_1) \geq \tilde{\eta}(\varepsilon_2), \text{ for all } \varepsilon_1, \varepsilon_2 \in (0, 2], \tag{1}$$

then the bound $h(\varepsilon, d, \gamma, \eta)$ *can be replaced (for* $\varepsilon < 2d$*) by*

$$\tilde{h}(\varepsilon, d, \gamma, \tilde{\eta}) := \gamma\left(\left\lceil \frac{d+1}{2\varepsilon \cdot \tilde{\eta}(\frac{\varepsilon}{d+1})} \right\rceil\right).$$

For a Hilbert space one can take as modulus of uniform convexity $\eta(\varepsilon) := \varepsilon^2/8$, and hence, the bound in theorem 6.3 applies with $\tilde{\eta}(\varepsilon) := \varepsilon/8$. If, moreover, $\lambda_n := \lambda \in (0, 1)$ for all n, then we can take $\gamma(n) := \lceil n/(\lambda(1 - \lambda)) \rceil$. So for the case of Hilbert spaces and constant λ, we obtain a quadratic bound in ε.

In [81], Groetsch's theorem and its quantitative analysis from [63] is extended to uniformly convex hyperbolic spaces. The bounds obtained are roughly the same as in theorem 6.3 but now also apply, e.g., to the important class of CAT(0)-spaces that are uniformly convex with the same modulus as in the Hilbertian case. Hence, as a corollary, the following quadratic bound follows:

Theorem 6.4 ([81]). *Let* (X, d) *be a CAT(0)-space;* $C \subseteq X$ *is a nonempty convex subset whose diameter is bounded by* $d \in \mathbb{Q}_+^*$. *Let* $f : C \to C$ *be nonexpansive and* $\lambda \in (0, 1)$. *Then*

$$\forall \varepsilon \in \mathbb{Q}_+^* \forall n \geq g(\varepsilon, d, \lambda)(d(x_n, f(x_n)) < \varepsilon),$$

where (x_n) *is the Krasnoselski–Mann iteration starting from* $x_0 := x \in C$ *and*

$$g(\varepsilon, d, \lambda) := \begin{cases} \frac{1}{\lambda(1-\lambda)} \left\lceil \frac{4(d+1)^2}{\varepsilon^2} \right\rceil, & \text{for } \varepsilon < 2d, \\ 0, & \text{otherwise.} \end{cases}$$

In the following, $C \subseteq X$ is a convex subset of a normed linear space $(X, \|\cdot\|)$.

Definition 6.5 ([91]). $f : C \to C$ is said to be uniformly λ-Lipschitzian ($\lambda > 0$) if

$$\forall n \in \mathbb{N} \forall x, y \in C\big(\|f^n(x) - f^n(y)\| \le \lambda \|x - y\|\big).$$

Definition 6.6 ([26]). $f : C \to C$ is quasi-nonexpansive if

$$\forall x \in C \forall p \in Fix(f)\big(\|f(x) - p\| \le \|x - p\|\big),$$

where $Fix(f)$ is the set of fixed points of f.

Example 6.7. $f : [0, 1) \to [0, 1), f(x) := x^2$ is quasi-nonexpansive but not nonexpansive.

Definition 6.8 ([86]). $f : C \to C$ is asymptotically quasi-nonexpansive with $k_n \in [0, \infty)^{\mathbb{N}}$ if $\lim_{n \to \infty} k_n = 0$ and

$$\forall n \in \mathbb{N} \forall x \in X \forall p \in Fix(f)\big(\|f^n(x) - p\| \le (1 + k_n)\|x - p\|\big).$$

Definition 6.9 ([34, 68]).

1) $f : C \to C$ is weakly quasi-nonexpansive[10] if

$$\exists p \in Fix(f) \forall x \in C\big(\|f(x) - f(p)\| \le \|x - p\|\big)$$

or—equivalently—

$$\exists p \in C \forall x \in X\big(\|f(x) - p\| \le \|x - p\|\big).$$

2) $f : C \to C$ is asymptotically weakly quasi-nonexpansive if

$$\exists p \in Fix(f) \forall x \in C \forall n \in \mathbb{N}\big(\|f^n(x) - f^n(p)\| \le (1 + k_n)\|x - p\|\big).$$

Example 6.10. $f : [0, 1] \to [0, 1], f(x) := x^2$ is weakly quasi-nonexpansive but not quasi-nonexpansive.

For asymptotically (weakly) quasi-nonexpansive mappings $f : C \to C$, the Krasnoselski–Mann iteration with errors is

$$(++) \ x_0 := x \in C, \ x_{n+1} := \alpha_n x_n + \beta_n f^n(x_n) + \gamma_n u_n,$$

where $\alpha_n, \beta_n, \gamma_n \in [0, 1]$ with $\alpha_n + \beta_n + \gamma_n = 1$ and $u_n \in C$.

Relying on the previous results of Opial, Dotson, Schu, Rhoades, Tan, Xu, and—most recently—Qihou, we have

---

[10] The same class of mappings has recently been introduced also in [29] under the name of J-type mappings.

Theorem 6.11 ([68]). *Let $(X, \|\cdot\|)$ be a uniformly convex normed space and $C \subseteq X$ convex. $(k_n) \subset \mathbb{R}_+$ with $\sum k_n < \infty$. Let $k \in \mathbb{N}$ and $\alpha_n, \beta_n, \gamma_n \in [0,1]$ such that $1/k \leq \beta_n \leq 1 - 1/k$, $\alpha_n + \beta_n + \gamma_n = 1$, and $\sum \gamma_n < \infty$. $f : C \to C$ uniformly Lipschitzian and asymptotically weakly quasi-nonexpansive and (u_n) be a bounded sequence in C. Then the following holds for (x_n) as defined in $(++)$ for an arbitrary starting point $x \in X$:*

$$\|x_n - f(x_n)\| \to 0.$$

Unless f is nonexpansive we in general cannot conclude (in contrast to the situation in theorems 5.3 and 5.4) that $(\|x_n - f(x_n)\|)_{n \in \mathbb{N}}$ is nonincreasing, which is needed to reduce the logical complexity of the convergence statement from Π_3^0 to Π_2^0. That is why we can apply our metatheorems only to the Herbrand normal form to get the following result (see [68]) for an extended discussion on how the metatheorems apply here and to a large extent predict the general form of the result):

Theorem 6.12 ([68]). *Let $(X, \|\cdot\|)$ be uniformly convex with modulus of convexity η, $C \subseteq X$ convex, $x \in C, f : C \to C, k, \alpha_n, \beta_n, \gamma_n, k_n, u_n$ as before with $\sum \gamma_n \leq E$, $\sum k_n \leq K, \forall n \in \mathbb{N}(\|u_n - x\| \leq u)$, and $E, K, u \in \mathbb{Q}_+$. Let $d \in \mathbb{Q}_+^*$ and (x_n) as in theorem 6.11.*

If f is λ-uniformly Lipschitzian and

$$\forall \varepsilon > 0 \exists p_\varepsilon \in C \left(\begin{array}{l} \|f(p_\varepsilon) - p_\varepsilon\| \leq \varepsilon \wedge \|p_\varepsilon - x\| \leq d \wedge \\ \forall y \in C \forall n \in \mathbb{N}(\|f^n(y) - f^n(p_\varepsilon)\| \leq (1 + k_n)\|y - p_\varepsilon\|) \end{array} \right),$$

then

$$\forall \varepsilon \in (0,1] \forall g : \mathbb{N} \to \mathbb{N} \exists n \leq \Phi \forall m \in [n, n + g(n)] \left(\|x_m - f(x_m)\| \leq \varepsilon \right),$$

where

$$\Phi \quad := \Phi(K, E, u, k, d, \lambda, \eta, \varepsilon, g) := h^i(0), \text{ where}$$

$$h(n) := g(n+1) + n + 2,$$

$$i \quad = \left\lceil \frac{3(5KD + 6E(U+D) + D)k^2}{\tilde{\varepsilon}\eta(\tilde{\varepsilon}/(D(1+K)))} \right\rceil,$$

$$D \quad := e^K(d + EU), U := u + d,$$

$$\tilde{\varepsilon} \quad := \varepsilon/(2(1 + \lambda(\lambda+1)(\lambda+2))).$$

Remark 6.13.   1) Specializing theorem 6.12 to $g \equiv 0$ yields

$$\forall \varepsilon \in (0,1] \exists n \leq \Psi \left(\|x_n - f(x_n)\| \leq \varepsilon \right),$$

where

$$\Psi := \Psi(K, E, u, k, d, \lambda, \eta, \varepsilon) := 2 \left\lceil \frac{3(5KD + 6E(U+D) + D)k^2}{\tilde{\varepsilon}\eta(\tilde{\varepsilon}/(D(1+K)))} \right\rceil,$$

$$D := e^K(d + EU), U := u + d,$$

$$\tilde{\varepsilon} := \varepsilon/(2(1 + \lambda(\lambda+1)(\lambda+2))).$$

2) As in the quantitative analysis of Groetsch's theorem above, one can replace in the bound in theorem 6.12 η by $\tilde{\eta}$ if η can be written in the form $\eta(\varepsilon) = \varepsilon\tilde{\eta}(\varepsilon)$ with $\tilde{\eta}$ satisfying

$$0 < \varepsilon_1 \leq \varepsilon_2 \leq 2 \rightarrow \tilde{\eta}(\varepsilon_1) \leq \tilde{\eta}(\varepsilon_2).$$

3) For asymptotically nonexpansive mappings with sequence (k_n) in $\mathbb{R}_+$ such that $\sum k_n \leq K$, the assumption 'uniformly Lipschitzian' is automatically satisfied by $\lambda := 1 + K$ since $K \geq k_n$ for all n.

References

1. Baillon, J., Bruck, R.E., The rate of asymptotic regularity is $0(\frac{1}{\sqrt{n}})$. Theory and applications of nonlinear operators of accretive and monotone type, Lecture Notes in Pure and Appl. Math. 178, Dekker, New York, pp. 51–81 (1996).
2. Bellin, G., Ramsey interpreted: a parametric version of Ramsey's theorem. In: Logic and computation (Pittsburgh, PA, 1987), Contemp. Math., 106, Amer. Math. Soc., Providence, RI, pp. 17–37 (1990).
3. Björnestal, B.O., Continuity of the metric projection operator I-III. The preprint series of Department of Mathematics. Royal Institute of Technology. Stockholm, TRITA-MAT 17 (1974), 20 (1974), 12 (1975).
4. Borwein, J., Reich, S., Shafrir, I., Krasnoselski-Mann iterations in normed spaces. Canad. Math. Bull. 35, pp. 21–28 (1992).
5. Bridges, D.S., A constructive development of Chebychev approximation theory. J. Approx. Theory 30, pp. 99–120 (1980).
6. Bridges, D.S., Lipschitz constants and moduli of continuity for the Chebychev projection. Proc. Amer. Math. Soc. 85, pp. 557–561 (1982).
7. Bridges, D.S., Recent progress in constructive approximation theory. In: Troelstra, A.S./van Dalen, D. (eds.) The L.E.J. Brouwer Centenary Symposium. North-Holland, Amsterdam, pp. 41–50 (1982).
8. Bridges, D.S., Richman, F., Julian, W.H., Mines, R., Extensions and fixed points of contractive maps in R^n. J. Math. Anal. Appl. 165, pp. 438–456 (1992).
9. Briseid, E.M., Proof mining applied to fixed point theorems for mappings of contractive type. Master Thesis, Oslo, 70 pp. (2005).
10. Briseid, E.M., Fixed points of generalized contractive mappings. To appear in: J. Nonlinear and Convex Analysis.
11. Briseid, E.M., A rate of convergence for asymptotic contractions. J. Math. Anal. Appl. 330, pp. 364–376 (2007).
12. Briseid, E.M., Some results on Kirk's asymptotic contractions. Fixed Point Theory 8, No.1, pp. 17–27 (2007).
13. Briseid, E.M., Addendum to [12]. To appear in: Fixed Point Theory.
14. Briseid, E.M., Logical aspects of rates of convergence in metric spaces. In preparation.
15. Browder, F.E., Petryshyn, W.V., The solution by iteration of nonlinear functional equations in Banach spaces. Bull. Amer. Math. Soc. 72, pp. 571–575 (1966).
16. Cheney, E.W., An elementary proof of Jackson's theorem on mean-approximation. Mathematics Magazine 38, 189-191 (1965).
17. Cheney, E.W., Approximation Theory. AMS Chelsea Publishing, Providence RI, (1966).

18. Clarkson, J.A., Uniformly convex spaces. Trans. Amer. Math. Soc. **40**, pp. 396–414 (1936).
19. Coquand, Th., Sur un théorème de Kronecker concernant les variétés algébriques. C.R. Acad. Sci. Paris, Ser. I **338**, pp. 291–294 (2004).
20. Coquand, Th., Lombardi, H., Quitte, C., Generating non-Noetherian modules constructively. Manuscripta mathematica **115**, pp. 513–520 (2004).
21. Coste, M., Lombardi, H., Roy, M.F., Dynamical methods in algebra: effective Nullstellensätze. Ann. Pure Appl. Logic **111**, pp. 203–256 (2001).
22. Delzell, C., Continuous sums of squares of forms. In: Troelstra, A.S./van Dalen, D. (eds.) The L.E.J. Brouwer Centenary Symposium. North-Holland, Amsterdam, pp. 65–75 (1982).
23. Delzell, C., Case distinctions are necessary for representing polynomials as sums of squares. In: Stern, J. (ed.), Proc. Herbrand Symposium, pp. 87–103 (1981).
24. Delzell, C., A finiteness theorem for open semi-algebraic sets, with applications to Hilbert's 17th problem. Contemporary Math. **8**, pp. 79–97 (1982).
25. Delzell, C., Kreisel's unwinding of Artin's proof-Part I. In: Odifreddi, P., Kreiseliana, A K Peters, Wellesley, MA, pp. 113–246 (1996).
26. Dotson, W.G., Jr., On the Mann iterative process. Trans. Amer. Math. Soc. **149**, pp. 65–73 (1970).
27. Edelstein, M., On fixed and periodic points under contractive mappings. J. London Math. Soc. **37**, pp. 74–79 (1962).
28. Edelstein, M., O'Brien, R.C., Nonexpansive mappings, asymptotic regularity and successive approximations. J. London Math. Soc. **17**, pp. 547–554 (1978).
29. Garcia-Falset, J., Llorens-Fuster, E., Prus, S., The fixed point property for mappings admitting a center. To appear in: Nonlinear Analysis.
30. Freud, G., Eine Ungleichung für Tschebyscheffsche Approximationspolynome. Acta Scientiarum Math. (Szeged) **19**, pp. 162–164 (1958).
31. Gehlen, W., On a conjecture concerning strong unicity constants. J. Approximation Theory **101**, pp. 221–239 (1999).
32. Gerhardy, P., A quantitative version of Kirk's fixed point theorem for asymptotic contractions. J. Math. Anal. Appl. **316**, pp. 339–345 (2006).
33. Gerhardy, P., Kohlenbach, U., Strongly uniform bounds from semi-constructive proofs. Ann. Pure Appl. Logic **141**, pp. 89–107 (2006).
34. Gerhardy, P., Kohlenbach, U., General logical metatheorems for functional analysis. To appear in: Trans. Amer. Math. Soc.
35. Girard, J.-Y., Proof Theory and Logical Complexity Vol.I. Studies in Proof Theory. Bibliopolis (Napoli) and Elsevier Science Publishers (Amsterdam), (1987).
36. Goebel, K., Kirk, W.A., A fixed point theorem for asymptotically nonexpansive mappings. Proc. Amer. Math. Soc. **35**, pp. 171–174 (1972).
37. Goebel, K., Kirk, W.A., Iteration processes for nonexpansive mappings. In: Singh, S.P., Thomeier, S., Watson, B., eds., Topological Methods in Nonlinear Functional Analysis. Contemporary Mathematics **21**, AMS, pp. 115–123 (1983).
38. Goebel, K., Kirk, W.A., Topics in metric fixed point theory. Cambridge Studies in Advanced Mathematics **28**, Cambridge University Press (1990).
39. Gödel, K., Über eine bisher noch nicht benützte Erweiterung des finiten Standpunktes. Dialectica **12**, pp. 280–287 (1958).
40. Groetsch, C.W., A note on segmenting Mann iterates. J. of Math. Anal. and Appl. **40**, pp. 369–372 (1972).

41. Henry, M.S., Schmidt, D., Continuity theorems for the product approximation operator. In: Law, A.G., Sahney, B.N. (eds.), Theory of Approximation with Applications, Alberta 1975, Academic Press, New York, pp. 24–42 (1976).
42. Hernest, M.-D., Synthesis of moduli of uniform continuity by the monotone Dialectica interpretation in the proof-system MinLog. Electronic Notes in Theoretical Computer Science **174**, pp. 141–149 (2007).
43. Ishikawa, S., Fixed points and iterations of a nonexpansive mapping in a Banach space. Proc. Amer. Math. Soc. **59**, pp. 65–71 (1976).
44. Jackson, D., Note on a class of polynomials of approximation. Trans. Amer. Math. Soc. **22**, pp. 320–326 (1921).
45. Kincses, J., Totik, V., Theorems and counterexamples on contractive mappings. Mathematica Balkanica, New Series **4**, pp. 69–90 (1990).
46. Kirk, W.A., Krasnosel'skii iteration process in hyperbolic spaces, Numer. Funct. Anal. and Optimiz. **4**, pp. 371–381 (1982).
47. Kirk, W.A., Nonexpansive mappings and asymptotic regularity. Nonlinear Analysis **40**, pp. 323–332 (2000).
48. Kirk, W.A., Fixed points of asymptotic contractions. J. Math. Anal. Appl. **277**, pp. 645–650 (2003).
49. Kirk, W.A., Geodesic geometry and fixed point theory II. In: García Falset, J. Llorens Fuster, E. Sims B. (eds.), Proceedings of the International Conference on Fixed Point Theory and Applications, Valencia (Spain), July 2003, Yokohama Publishers, pp. 113–142, (2004).
50. Kirk, W.A., Martinez-Yanez, C., Approximate fixed points for nonexpansive mappings in uniformly convex spaces. Annales Polonici Mathematici **51**, pp. 189–193 (1990).
51. Kirk, W.A., Sims, B. (eds.), Handbook of Metric Fixed Point Theory. Kluwer Academic Publishers, Dordrecht, xi+703 pp. (2001).
52. Ko, K.-I., On the computational complexity of best Chebycheff approximation. J. of Complexity **2**, pp. 95–120 (1986).
53. Ko, K.-I., Complexity theory of real functions. Birkhäuser, Boston, x+309 pp., (1991).
54. Kohlenbach, U., Effective moduli from ineffective uniqueness proofs. An unwinding of de La Vallée Poussin's proof for Chebycheff approximation. Ann. Pure Appl. Logic **64**, pp. 27–94 (1993).
55. Kohlenbach, U., New effective moduli of uniqueness and uniform a–priori estimates for constants of strong unicity by logical analysis of known proofs in best approximation theory. Numer. Funct. Anal. and Optimiz. **14**, pp. 581–606 (1993).
56. Kohlenbach, U., Analysing proofs in analysis. In: W. Hodges, M. Hyland, C. Steinhorn, J. Truss, editors, *Logic: from Foundations to Applications. European Logic Colloquium* (Keele, 1993), Oxford University Press, pp. 225–260 (1996).
57. Kohlenbach, U., Mathematically strong subsystems of analysis with low rate of growth of provably recursive functionals. Arch. Math. Logic **36**, pp. 31–71 (1996).
58. Kohlenbach, U., Arithmetizing proofs in analysis. In: Larrazabal, J.M., Lascar, D., Mints, G. (eds.), Logic Colloquium '96, Springer Lecture Notes in Logic **12**, pp. 115–158 (1998).
59. Kohlenbach, U., Proof theory and computational analysis. Electronic Notes in Theoretical Computer Science **13**, Elsevier, 34 pp., (1998).
60. Kohlenbach, U., On the computational content of the Krasnoselski and Ishikawa fixed point theorems. In: Proceedings of the Fourth Workshop on Computability and Complexity in Analysis, (eds.), Blanck, J. Brattka, V. Hertling P., Springer LNCS **2064**, pp. 119–145 (2001).
61. Kohlenbach, U., A quantitative version of a theorem due to Borwein-Reich-Shafrir. Numer. Funct. Anal. and Optimiz. **22**, pp. 641–656 (2001).

62. Kohlenbach, U., Foundational and mathematical uses of higher types. In: Sieg, W., Sommer, R., Talcott, C. (eds.), Reflections on the foundations of mathematics. Essays in honor of Solomon Feferman, Lecture Notes in Logic **15**, A.K. Peters, pp. 92–120 (2002).

63. Kohlenbach, U., Uniform asymptotic regularity for Mann iterates. J. Math. Anal. Appl. **279**, pp. 531–544 (2003).

64. Kohlenbach, U., Some logical metatheorems with applications in functional analysis. Trans. Amer. Math. Soc. **357**, no. 1, pp. 89–128 (2005).

65. Kohlenbach, U., Some computational aspects of metric fixed point theory. Nonlinear Analysis **61**, pp. 823–837 (2005).

66. Kohlenbach, U., A logical uniform boundedness principle for abstract metric and hyperbolic spaces. Electronic Notes in Theoretical Computer Science **165** (Proc. WoLLIC 2006), pp. 81–93 (2006).

67. Kohlenbach, U., Applied Proof Theory: Proof Interpretations and their Use in Mathematics. Book in prepration for 'Springer Monographs in Mathematics'. Expected to appear: 2008.

68. Kohlenbach, U., Lambov, B., Bounds on iterations of asymptotically quasi-nonexpansive mappings. In: Falset, J.G., Fuster, E.L., Sims, B. (eds.), Proc. International Conference on Fixed Point Theory and Applications, Valencia 2003, Yokohama Publishers, pp. 143–172 (2004)

69. Kohlenbach, U., Leuştean, L., Mann iterates of directionally nonexpansive mappings in hyperbolic spaces. Abstr. Appl. Anal. vol. 2003, no. 8, pp. 449–477 (2003).

70. Kohlenbach, U., Leuştean, L., The approximate fixed point property in product spaces. Nonlinear Analysis **66** , pp. 806–818 (2007).

71. Kohlenbach, U., Oliva, P., Proof mining in L_1-approximation. Ann. Pure Appl. Logic **121**, pp. 1–38 (2003).

72. Kohlenbach, U., Oliva, P., Proof mining: a systematic way of analysing proofs in mathematics. Proc. Steklov Inst. Math. **242**, pp. 1–29 (2003).

73. Krasnoselski, M. A., Two remarks on the method of successive approximation. Usp. Math. Nauk (N.S.) **10**, pp. 123–127 (1955) (Russian).

74. Kreisel, G., On the interpretation of non-finitist proofs, part I. J. Symbolic Logic **16**, pp. 241–267 (1951).

75. Kreisel, G., On the interpretation of non-finitist proofs, part II: Interpretation of number theory, applications. J. Symbolic Logic **17**, pp. 43–58 (1952).

76. Kreisel, G., Macintyre, A., Constructive logic versus algebraization I. In: Troelstra, A.S./van Dalen, D. (eds.) The L.E.J. Brouwer Centenary Symposium. North-Holland, Amsterdam, pp. 217–260 (1982).

77. Kroó, A., On the continuity of best approximations in the space of integrable functions. Acta Math. Acad. Sci. Hungar. **32**, pp. 331–348 (1978).

78. Kroó, A., On the uniform modulus of continuity of the operator of best approximation in the space of periodic functions. Acta Math. Acad. Sci. Hungar. **34**, no. 1-2, pp. 185–203 (1979).

79. Lambov, B., Rates of convergence of recursively defined sequences. Electronic Notes in Theoretical Computer Science, **120**, pp. 125–133 (2005).

80. Leuştean, L., Proof mining in IR-trees and hyperbolic spaces. Electronic Notes in Theoretical Computer Science **165** (Proc. WoLLIC 2006), pp. 95–106 (2006).

81. Leuştean, L., A quadratic rate of asymptotic regularity for CAT(0)-spaces. J. Math. Anal. Appl. **325**, pp. 386–399 (2007).

82. Luckhardt, H., Herbrand-Analysen zweier Beweise des Satzes von Roth: Polynomiale Anzahlschranken. J. Symbolic Logic **54**, pp. 234–263 (1989).

83. Luckhardt, H., Bounds extracted by Kreisel from ineffective proofs. In: Odifreddi, P., Kreiseliana, A K Peters, Wellesley, MA, pp. 289–300, 1996.
84. Newman, D.J., Shapiro, H.S., Some theorems on Cebysev approximation. Duke Math. J. **30**, pp. 673–682 (1963).
85. Oliva, P., On the computational complexity of best L_1-Approximation. Math. Logic. Quart. **48**, suppl. I, pp. 66–77 (2002).
86. Qihou, L., Iteration sequences for asymptotically quasi-nonexpansive mappings. J. Math. Anal. Appl. **259**, pp. 1–7, (2001).
87. Rakotch, E., A note on contractive mappings. Proc. Amer. Math. Soc. **13**, pp. 459–465 (1962).
88. Reich, S., Shafrir, I., Nonexpansive iterations in hyperbolic spaces. Nonlinear Analysis, Theory, Methods and Applications **15**, pp. 537–558 (1990).
89. Rhoades, B.E., A comparison of various definitions of contractive mappings. Trans. Amer. Math. Soc. **226**, pp. 257–290 (1977)
90. Rhoades, B.E., Contractive definitions. In: Rassias, Th., M. editor, Nonlinear Analysis, World Sci. Publishing, Singapore, pp. 513–526, 1987.
91. Schu, J., Iterative construction of fixed points of asymptotically nonexpansive mappings. J. Math. Anal. Appl. **158**, pp. 407–413 (1991).
92. Simpson, S.G., Subsystems of Second Order Arithmetic. Perspectives in Mathematical Logic. Springer-Verlag, xiv+445 pp. 1999.
93. Sims, B., Examples of fixed point free mappings. In: [51], pp. 35–48 (2001).
94. Spector, C., Provably recursive functionals of analysis: a consistency proof of analysis by an extension of principles formulated in current intuitionistic mathematics. In: Recursive function theory, Proceedings of Symposia in Pure Mathematics, vol. 5 (J.C.E. Dekker (ed.)), AMS, Providence, RI, pp. 1–27 (1962).
95. Suzuki, T., Fixed-point theorem for asymptotic contractions of Meir-Keeler type in complete metric spaces. Nonlinear Analysis **64**, pp. 971–978 (2006).
96. Takahashi, W., A convexity in metric space and nonexpansive mappings, I. Kodai Math. Sem. Rep. **22**, pp. 142–149 (1970).
97. Weiermann, A., A classification of rapidly growing Ramsey functions. Proc. Amer. Math. Soc. **132**, no. 2, pp. 553–561 (2004).
98. Weiermann, A., Phasenübergänge in Logik und Kombinatorik. DMV-Mitteilungen **13**, no. 3, pp. 152–156 (2005).
99. Young, J.W., General theory of approximation by functions involving a given number of arbitrary parameters. Trans. Amer. Math. Soc. **8**, pp. 331–344 (1907).

Effective Fractal Dimension in Algorithmic Information Theory

Elvira Mayordomo*

Departamento de Informática e Ingeniería de Sistemas, Universidad de Zaragoza,
50018 Zaragoza, Spain
elvira@unizar.es

Summary. Effective fractal dimension was defined by Lutz (2003) in order to quantitatively analyze the structure of complexity classes, but then interesting connections of effective dimension with information theory were also found, justifying the long existent intuition that dimension is an information content measure. Considering different bounds on computing power that range from finite memory to constructibility, including time-bounded and space-bounded computations, we review all known characterizations of effective dimension that support the thesis that effective dimensions capture what can be considered the inherent information content of a sequence in each setting.

1 Introduction

Hausdorff dimension assigns a dimension value to each subset of an arbitrary metric space. In Euclidean space, this concept coincides with our intuition that smooth curves have dimension 1 and smooth surfaces have dimension 2, but from its introduction in 1918 [23] Hausdorff noted that many sets have noninteger dimension, what he called "fractional dimension." The development and applications of fractal geometry quickly outgrew the field of geometry and spread through many other areas [19, 56, 15, 16, 17, 13, 12, 49]. In the 1980s Tricot [73] and Sullivan [71] independently developed a dual of Hausdorff dimension called the *packing dimension* that is now widely used.

In this paper we will focus on the use of fractal dimensions in the Cantor space of infinite sequences over a finite alphabet. The results obtained since the 1990s, and in particular the effectivizations of dimension that we will review in this paper, have introduced the powerful tools of fractal geometry into computational complexity and information theory.

* Research was supported in part by Spanish Government MEC Project TIN 2005-08832-C03-02.

In 2000 Lutz [45] proved a new characterization of the Hausdorff dimension for the case of Cantor space that was based on gales. This characterization was the beginning of a whole range of effective versions of dimensions naturally based on bounding the computing power of the gale. Gales are a generalization of martingales, which are strategies for betting on the successive bits of infinite binary sequences with fair payoffs. Martingales were introduced by Ville [74] in 1939 (also implicit in [38, 39]) and used by Schnorr [61, 62, 63, 64] in his work on randomness. In the 1990s, Ryabko [59, 60] and Staiger [69] proved several connections of Hausdorff dimension and martingales, that included relating the Hausdorff dimension of a set X of binary sequences to the growth rates achievable by computable martingales betting on the sequences in X (see section 4 for more details).

The introduction of a resource-bounded dimension by Lutz [45] had the immediate motivation of overcoming the limitations of a resource-bounded measure, a generalization of the classical Lebesgue measure, in the quantitative analysis of complexity classes [43]. The resulting concepts of effective dimension have turned out to be robust, since they have been shown to admit several equivalent definitions that relate them to well-studied concepts in computation, and they have proven very fruitful in investigating not only the structure of complexity classes but also in the modeling and analysis of sequence information and, more recently, back in fractal geometry. See [30] for an updated bibliography on effective dimension.

There is a recent survey on the applications of effective dimension to the study of complexity classes by Hitchcock et al. [28]. The purpose of this paper will be centered on the information theory connections. In fact, as could be suspected from earlier results by Ryabko [57, 58], Staiger [68, 69], and Cai and Hartmanis [3], effective dimensions have very clear interpretations in terms of information content or compressibility of a sequence. Considering different bounds on computing power that range from finite memory to constructibility, including time-bounded and space-bounded computations, effective dimensions capture what can be considered the inherent information content of a sequence in the corresponding setting. We will present in this paper all known characterizations of effective dimension that support this thesis.

We start by developing very general definitions of dimension, including an extension of scaled dimension to a general metric space. Scaled dimension allows a rescaling of dimension that can give more meaningful results for dimension 0 sets, for instance. It was introduced in [27] for the particular case of Cantor space with the usual metric, based on the uniform probability distribution. We think this more general definition will allow further insight into the interest of scaled dimension with different metrics.

Next we review the different notions of effective dimension, starting with the finite-state dimension in which computation is restricted to finite-state devices. In this setting compression has been widely studied as a precursor of the Lempel–Ziv algorithm [37]. Dai et al. [6] proved that the finite-state dimension can be characterized in terms of information-lossless finite-state compressors, and Doty and Moser [10] remarked that a Kolmogorov-complexity-like characterization is also possible from earlier results by Sheinwald et al. [67].

In section 4 we will develop a constructive dimension that corresponds to the use of lower semicomputable strategies, and that has good properties inherited from the existence of a universal constructible semimeasure. Lutz introduced this notion in [46]. Athreya et al. [1] introduced the dual constructive strong dimension. We present a characterization of both notions of constructive dimension in terms of Kolmogorov complexity, present a correspondence principle stating that constructive dimension coincides with Hausdorff dimension for sufficiently simple sets, and summarize the main results. The open question of whether positive dimension sequences can substitute Martin–Löf random sequences as the randomness source of a computation has received recent attention from different areas. We present the main known results here and refer the reader to [11] and [51] for more information (these two references use the term "effective dimension" for Lutz's constructive dimension).

Our last section concerns resource-bounds on time and space. Polynomial-space bounded dimension has been well studied in terms of information content [25], but polynomial-time dimension seems harder to grasp. We know very little about time-bounded Kolmogorov complexity, but a compressibility characterization of polynomial-time dimension has been obtained in [41] via polynomial-time compression algorithms. We should consider polynomial-time dimension as an interesting alternative to time-bounded Kolmogorov complexity, expecting that we can import robustness properties from fractal dimension.

There are many related topics we chose not to cover in this paper, mainly due to lack of space for proper development. We mention very interesting recent results on the effective dimension on Euclidean space ([21], [47]) that would require a paper of its own.

2 Fractal dimensions and gale characterizations

In this section we first review the classical Hausdorff and packing dimensions and then we introduce a scaled dimension for a general metric space. We present the characterizations of these notions in terms of gales for the case of Cantor space. This characterization is crucial in the definition of effective dimensions that we will introduce in the rest of the paper.

2.1 Hausdorff and packing dimensions

Let ρ be a metric on a set $\mathcal{X}$. We use the following standard terminology. The *diameter* of a set $X \subseteq \mathcal{X}$ is $\text{diam}(X) = \sup \{\rho(x, y) \mid x, y \in X\}$ (which may be ∞). For each $x \in \mathcal{X}$ and $r \in \mathbb{R}$, the *closed ball* of radius r about x is the set

$B(x, r) = \{y \in \mathcal{X} \mid \rho(y, x) \leq r\}$, and the *open ball* of radius r about x is the set $B^o(x, r) = \{y \in \mathcal{X} \mid \rho(y, x) < r\}$. A *ball* is any set of the form $B(x, r)$ or $B^o(x, r)$. A ball B is *centered* in a set $X \subseteq \mathcal{X}$ if $B = B(x, r)$ or $B = B^o(x, r)$ for some $x \in X$ and $r \geq 0$.

For each $\delta > 0$, we let $\mathcal{C}_\delta$ be the set of all countable collections $\mathcal{B}$ of balls such that $\mathrm{diam}(B) \leq \delta$ for all $B \in \mathcal{B}$, and we let $\mathcal{D}_\delta$ be the set of all $\mathcal{B} \in \mathcal{C}_\delta$ such that the balls in $\mathcal{B}$ are pairwise disjoint. For each $X \subseteq \mathcal{X}$ and $\delta > 0$, we define the sets

$$\mathcal{H}_\delta(X) = \left\{ \mathcal{B} \in \mathcal{C}_\delta \,\middle|\, X \subseteq \bigcup_{B \in \mathcal{B}} B \right\},$$

$$\mathcal{P}_\delta(X) = \{\mathcal{B} \in \mathcal{D}_\delta \mid (\forall B \in \mathcal{B})B \text{ is centered in } X\}.$$

If $\mathcal{B} \in \mathcal{H}_\delta(X)$, then we call $\mathcal{B}$ a δ-*cover* of X. If $\mathcal{B} \in \mathcal{P}_\delta(X)$, then we call $\mathcal{B}$ a δ-*packing* of X. For $X \subseteq \mathcal{X}$, $\delta > 0$, and $s \geq 0$, we define the quantities

$$H_\delta^s(X) = \inf_{\mathcal{B} \in \mathcal{H}_\delta(X)} \sum_{B \in \mathcal{B}} \mathrm{diam}(B)^s,$$

$$P_\delta^s(X) = \sup_{\mathcal{B} \in \mathcal{P}_\delta(X)} \sum_{B \in \mathcal{B}} \mathrm{diam}(B)^s.$$

Since $H_\delta^s(X)$ and $P_\delta^s(X)$ are monotone as $\delta \to 0$, the limits

$$H^s(X) = \lim_{\delta \to 0} H_\delta^s(X),$$

$$P_0^s(X) = \lim_{\delta \to 0} P_\delta^s(X)$$

exist, though they may be infinite. Let

$$P^s(X) = \inf \left\{ \sum_{i=0}^\infty P_0^s(X_i) \,\middle|\, X \subseteq \bigcup_{i=0}^\infty X_i \right\}. \tag{2.1}$$

It is routine to verify that the set functions H^s and P^s are outer measures [16]. The quantities $H^s(X)$ and $P^s(X)$—which may be infinite—are called the s-*dimensional Hausdorff (outer) ball measure* and the s-*dimensional packing (outer) ball measure* of X, respectively. The optimization (2.1) over all countable partitions of X is needed because the set function P_0^s is *not* an outer measure.

Definition. Let ρ be a metric on a set $\mathcal{X}$, and let $X \subseteq \mathcal{X}$.

1. (Hausdorff [23]). The *Hausdorff dimension* of X with respect to ρ is

$$\dim^{(\rho)}(X) = \inf\{s \in [0, \infty) \mid H^s(X) = 0\}.$$

2. (Tricot [73], Sullivan [71]). The *packing dimension* of X with respect to ρ is

$$\mathrm{Dim}^{(\rho)}(X) = \inf\{s \in [0, \infty) \mid P^s(X) = 0\}.$$

When $\mathcal{X}$ is a Euclidean space $\mathbb{R}^n$ and ρ is the usual Euclidean metric on $\mathbb{R}^n$, $\dim^{(\rho)}$ and $\mathrm{Dim}^{(\rho)}$ are the ordinary Hausdorff and packing dimensions, also denoted by $\dim_H$ and $\dim_P$, respectively.

2.2 Scaled dimensions

This subsection introduces the notion of a scaled dimension for a general metric space. Our treatment is based on [27], which presents only the case of Cantor space and uses directly gales in the definition.

The notions of Hausdorff and packing dimensions introduced above depend on the expression "$\mathrm{diam}(B)^s$" that is used in both s-Haussdorff and s-packing measures (see definitions of $H_\delta^s(X)$ and $P_\delta^s(X)$ above). Here we consider alternative functions on s and the diameter.

Definition. A *scale* is a function $h(x, s)$, $h : [0, \infty) \times [0, \infty) \to [0, \infty)$, with the following two properties:

1. For every $s \in [0, \infty)$, $h(\_, s)$ is nondecreasing.

2. For every $s, \epsilon \in [0, \infty)$, $\lim_{x \to 0} \frac{h(x, s+\epsilon)}{h(x, s)} = 0$.

From each scale h we define scaled Hausdorff and packing measures generalizing the definitions in subsection 2.1.

For $X \subseteq \mathcal{X}$, $\delta > 0$ and $s \geq 0$, we define the quantities

$$SH_\delta^{h,s}(X) = \inf_{\mathcal{B} \in \mathcal{H}_\delta(X)} \sum_{B \in \mathcal{B}} h(\mathrm{diam}(B), s),$$

$$SP_\delta^{h,s}(X) = \sup_{\mathcal{B} \in \mathcal{P}_\delta(X)} \sum_{B \in \mathcal{B}} h(\mathrm{diam}(B), s).$$

Since $SH_\delta^s(X)$ and $SP_\delta^s(X)$ are monotone as $\delta \to 0$, the limits

$$SH^{h,s}(X) = \lim_{\delta \to 0} SH_\delta^{h,s}(X),$$

$$SP_0^{h,s}(X) = \lim_{\delta \to 0} SP_\delta^{h,s}(X)$$

exist, though they may be infinite. Let

$$SP^{h,s}(X) = \inf \left\{ \sum_{i=0}^{\infty} SP_0^{h,s}(X_i) \, \middle| \, X \subseteq \bigcup_{i=0}^{\infty} X_i \right\}. \tag{2.2}$$

In this case it is also routine to verify that the set functions $SH^{h,s}$ and $SP^{h,s}$ are outer measures. The optimization (2.2) over all countable partitions of X is needed because the set function $SP_0^{h,s}$ is *not* an outer measure.

Rogers introduced in [56] the generalized notion of the Hausdorff measure using a function $f(\text{diam}(B))$ in the place of $\text{diam}(B)^s$ in the definition of the Hausdorff measure. More recently Roger's approach was revisited by Reimann and Stephan [55] in the context of algorithmic randomness. In those references the authors did not consider dependence on a second parameter s or a dimension concept in this context.

Our first property is that for each $X \subseteq \mathcal{X}$ there is at most one s for which $0 < SH^{h,s}(X) < \infty$.

Proposition 2.1 *Let $X \subseteq \mathcal{X}$, let h be a scale, and let $s \in [0, \infty)$.*

1. *If $0 < SH^{h,s}(X) < \infty$, then for every $\epsilon > 0$, $SH^{h,s+\epsilon}(X) = 0$.*

2. *If $0 < SP^{h,s}(X) < \infty$, then for every $\epsilon > 0$, $SP^{h,s+\epsilon}(X) = 0$.*

Proof. The property follows from the fact that $\lim_{x \to 0} \frac{h(x,s+\epsilon)}{h(x,s)} = 0$, in the definition of scale. $\square$

Definition. Let ρ be a metric on a set $\mathcal{X}$, let $X \subseteq \mathcal{X}$, and let h be a scale.

1. The *h-scaled dimension* of X with respect to ρ is
$$\dim^{(h),(\rho)}(X) = \inf \left\{ s \in [0, \infty) \,\middle|\, SH^{h,s}(X) = 0 \right\}.$$

2. The *h-scaled packing dimension* of X with respect to ρ is
$$\text{Dim}^{(h),(\rho)}(X) = \inf \left\{ s \in [0, \infty) \,\middle|\, SP^{h,s}(X) = 0 \right\}.$$

The basic properties of scaled dimensions are monotonicity and countable stability, which also hold for Hausdorff and packing dimensions [16].

Proposition 2.2 *Let h be a scale.*

1. *For every $x \in \mathcal{X}$, $\text{Dim}^{(h),(\rho)}(\{x\}) = \dim^{(h),(\rho)}(\{x\}) = 0$.*

2. *For every $X \subseteq \mathcal{X}$, $0 \leq \dim^{(h),(\rho)}(X) \leq \text{Dim}^{(h),(\rho)}(X)$.*

3. *Let $X_i \subseteq \mathcal{X}$ for each $i \in \mathbb{N}$,*
$$\dim^{(h),(\rho)}(\cup_i X_i) = \sup_i \dim^{(h),(\rho)}(X_i), \text{ and}$$
$$\text{Dim}^{(h),(\rho)}(\cup_i X_i) = \sup_i \text{Dim}^{(h),(\rho)}(X_i).$$

4. *For every $X, Y \subseteq \mathcal{X}$ with $X \subseteq Y$,*
$$\dim^{(h),(\rho)}(X) \leq \dim^{(h),(\rho)}(Y)$$
and
$$\text{Dim}^{(h),(\rho)}(X) \leq \text{Dim}^{(h),(\rho)}(Y).$$

In particular, every countable set has a zero scaled dimension for any scale.

Notice that for $h_0(x, s) = x^s$, $\dim^{(h_0),(\rho)}(X) = \dim^{(\rho)}(X)$ and $\mathrm{Dim}^{(h_0),(\rho)}(X) = \mathrm{Dim}^{(\rho)}(X)$.

We can compare the scaled dimensions for different scales.

Proposition 2.3 *Let h, h' be scales such that $h(x, s) \leq h'(x, s)$ for every s and for every $x \in [0, \epsilon)$, where $\epsilon > 0$ may depend on s. Then for every $X \subseteq \mathcal{X}$,*

$$\dim^{(h),(\rho)}(X) \leq \dim^{(h'),(\rho)}(X)$$

and

$$\mathrm{Dim}^{(h),(\rho)}(X) \leq \mathrm{Dim}^{(h'),(\rho)}(X).$$

Proof. The property follows from the definition of scaled Hausdorff and packing measures. □

The next property concerns the scaled dimension of the whole space.

Proposition 2.4 *Let $\mathcal{X}$ be a metric space such that $0 < H^{\dim(\mathcal{X})}(\mathcal{X}) < \infty$. Let h be a scale, and let $s \in [0, \infty)$.*
If $h(x, s) = \Omega(x^{\dim(\mathcal{X})})$, then $\dim^{(h)}(\mathcal{X}) \geq s$.
If $h(x, s) = O(x^{\dim(\mathcal{X})})$, then $\dim^{(h)}(\mathcal{X}) \leq s$.

Hitchcock et al. consider in [27] the following scales, which are useful for dimension values up to 1.

Definition. For every $x \in [0, \infty)$, $s \in [0, \infty)$ we define

1. $h_0(x, s) = x^s$.

2. For each $i \geq 0$,

$$
\begin{aligned}
h_{i+1}(x, s) &= 2^{-\frac{1}{h_i(-1/\log(x), s)}} && \text{if } x \leq 1/2, s < 1, \\
h_{i+1}(x, s) &= 2^{-1/h_i(1, s)} && \text{if } x > 1/2, s < 1, \\
h_{i+1}(x, s) &= x^s && \text{if } s \geq 1.
\end{aligned}
$$

3. For each $i > 0$,

$$
h_{-i}(x, s) = \begin{cases} x/h_i(x, 1 - s) & \text{if } s < 1, \\ x^s & \text{if } s \geq 1. \end{cases}
$$

For $s < 1$, the above-defined scales are below the inverse of the logarithm, and for every k, h_k is asymptotically below h_{k+1}. This provides a fine family of scales that, for instance, can be used to distinguish different circuit size rates [27].

Proposition 2.5 *For every $k \in \mathbb{Z}$, the above-defined h_k is a scale.*

Notation. For each $k \in \mathbb{Z}$, we denote $\dim^{(h_k),(\rho)}(X)$ as $\dim^{(k),(\rho)}(X)$ and $\mathrm{Dim}^{(h_k),(\rho)}(X)$ as $\mathrm{Dim}^{(k),(\rho)}(X)$.

The relationship between scales h_k is as follows:

Proposition 2.6 *Let $k \in \mathbb{Z}$. Then for every $X \subseteq \mathcal{X}$,*

$$\dim^{(k),(\rho)}(X) \leq \dim^{(k+1),(\rho)}(X)$$

and

$$\mathrm{Dim}^{(k),(\rho)}(X) \leq \mathrm{Dim}^{(k+1),(\rho)}(X).$$

2.3 Gale characterizations

We now focus our attention on sequence spaces. Let Σ be a finite alphabet with $|\Sigma| \geq 2$. We will consider the following metric on Σ^{∞}:

$$\rho(S,T) = \inf \left\{ |\Sigma|^{-|w|} \mid w \sqsubseteq S \text{ and } w \sqsubseteq T \right\}$$

for all $S, T \in \Sigma^{\infty}$.

We fix the above ρ and denote $\dim^{(\rho)}(X)$ and $\mathrm{Dim}^{(\rho)}(X)$ as $\dim(X)$ and $\mathrm{Dim}(X)$, for $X \subseteq \Sigma^{\infty}$. Similarly for scaled dimension we use $\dim^{(h)}(X)$ and $\mathrm{Dim}^{(h)}(X)$ for $\dim^{(h),(\rho)}(X)$, $\mathrm{Dim}^{(h),(\rho)}(X)$. Recently Lutz and Mayordomo have considered alternative metrics on Σ^{∞} with interesting applications to dimension in Euclidean space [47].

Lutz [45] characterized the Hausdorff dimension in terms of gales, as presented next.

Definition. ([45]) Let Σ be a finite alphabet with $|\Sigma| \geq 2$, and let $s \in [0, \infty)$.

1. An *s-gale* is a function $d : \Sigma^* \to [0, \infty)$ that satisfies the condition

$$d(w) = |\Sigma|^{-s} \sum_{a \in \Sigma} d(wa) \tag{2.3}$$

for all $w \in \Sigma^*$.

2. A *martingale* is a 1-gale.

In fact Lutz [45] considered also supergales, which are functions for which equality (2.3) is substituted by the inequality

$$d(w) \geq |\Sigma|^{-s} \sum_{a \in \Sigma} d(wa).$$

Supergales give additional flexibility and in most interesting cases can be substituted by gales in the definitions and characterizations of different dimensions and effective dimensions. For the sake of readability, we will restrict to gales in this paper.

The following observation shows how gales are affected by variation of the parameter s.

Observation 2.7 ([46]) *Let $s, s' \in [0, \infty)$, and let $d, d' : \Sigma^* \to [0, \infty)$. Assume that*

$$d(w)|\Sigma|^{-s|w|} = d'(w)|\Sigma|^{-s'|w|}$$

holds for all $w \in \Sigma^*$. Then d is an s-gale if and only if d' is an s'-gale.

For example, a function $d : \Sigma^* \to [0, \infty)$ is an s-gale if and only if the function $d' : \Sigma^* \to [0, \infty)$ defined by $d'(w) = |\Sigma|^{1-s|w|}d(w)$ is a martingale.

Martingales were introduced by Lévy [39] and Ville [74]. They have been used extensively by Schnorr [62, 63, 65] and others in investigations of randomness, by Lutz [42, 44] and others in the development of resource-bounded measure, and by Ryabko [60] and Staiger [69] regarding exponents of increase. Gales are a convenient generalization of martingales introduced by Lutz [45, 46] in the development of effective fractal dimensions.

Intuitively, an s-gale d is a strategy for betting on the successive symbols in a sequence $S \in \Sigma^\infty$. We regard the value $d(w)$ as the amount of money that a gambler using the strategy d will have after betting on the symbols in w; w is a prefix of S. If $s = 1$, then the s-gale identity (2.3) ensures that the payoffs are fair in the sense that the conditional expected value of the gambler's capital after the symbol following w, given that w has occurred, is precisely $d(w)$, the gambler's capital after w. If $s < 1$, then (2.3) says that the payoffs are less than fair. If $s > 1$, then (2.3) says that the payoffs are more than fair. Clearly, the smaller s is, the more hostile the betting environment.

There are two important notions of success for a gale.

Definition. Let d be an s-gale, where $s \in [0, \infty)$, and let $S \in \Sigma^\infty$.

1. We say that d *succeeds* on S, and we write $S \in S^\infty[d]$, if
   $\limsup_{t \to \infty} d(S[0..t-1]) = \infty$.

2. We say that d *succeeds strongly* on S, and we write $S \in S^\infty_{\text{str}}[d]$, if
   $\liminf_{t \to \infty} d(S[0..t-1]) = \infty$.

The following theorem gives useful characterizations of the classical Hausdorff and packing dimensions on sequence spaces. The Hausdorff dimension part was proven by Lutz [45], and the packing dimension part was proven by Athreya et al. in [1].

Theorem 2.8 ([45] and [1]) *For all $X \subseteq \Sigma^\infty$,*

$$\dim(X) = \inf \{ s \in [0, \infty) \mid \text{there is an } s\text{-gale } d \text{ with } X \subseteq S^\infty[d] \}$$

and

$$\text{Dim}(X) = \inf \{ s \in [0, \infty) \mid \text{there is an } s\text{-gale } d \text{ with } X \subseteq S^\infty_{\text{str}}[d] \}.$$

The effectivization of both Hausdorff and packing (or strong) dimensions will be based on Theorem 2.8. By restricting the set of gales that are allowed to different classes of computable gales, we will obtain effective versions of dimension that will be meaningful in different subclasses of Σ^∞. This will be developed in the following sections.

Eggleston [14] proved the following classical result on the Hausdorff dimension of a set of sequences with a fixed asymptotic frequency.

The *frequency* of a nonempty binary string $w \in \{0,1\}^*$ is the ratio $\mathrm{freq}(w) = \frac{\#(1,w)}{|w|}$, where $\#(b,w)$ denotes the number of occurrences of the bit b in w. For each $\alpha \in [0,1]$, we define the set

$$\mathrm{FREQ}(\alpha) = \left\{ S \in \{0,1\}^\infty \,\middle|\, \lim_{n\to\infty} \mathrm{freq}(S\,[0..n-1]) = \alpha \right\}.$$

The binary Shannon *entropy* function $\mathcal{H} : [0,1] \to [0,1]$ is defined as $\mathcal{H}(x) = x \log \frac{1}{x} + (1-x) \log \frac{1}{1-x}$, with $\mathcal{H}(0) = \mathcal{H}(1) = 0$.

Theorem 2.9 ([14]) *For each real number $\alpha \in [0,1]$,*

$$\dim_\mathrm{H}(\mathrm{FREQ}(\alpha)) = \mathcal{H}(\alpha).$$

We will reformulate this last result in the contexts of the dimensions defined in sections 3, 4, and 5.

We finish this section with the fact that scaled dimension in Σ^∞ admits a similar characterization.

The notion of scaled gales is introduced in [27].

Definition. Let Σ be a finite alphabet with $|\Sigma| \geq 2$, let h be a scale, and let $s \in [0,\infty)$. An *h-scaled s-gale* (briefly, an $s^{(h)}$-gale) is a function $d : \Sigma^* \to [0,\infty)$ that satisfies the condition

$$h(|\Sigma|^{-|w|}, s)\, d(w) = h(|\Sigma|^{-(|w|+1)}, s) \sum_{a\in\Sigma} d(wa)$$

for all $w \in \Sigma^*$.

Notice that our definition of gale (Definition 2.3) corresponds to the scale $h_0(x,s) = x^s$, so an $s^{(h_0)}$-gale is just an s-gale.

Observation 2.10 *Let h, h' be scales, let $s, s' \in [0,\infty)$, and let $d : \Sigma^* \to [0,\infty)$. d is an $s^{(h)}$-gale if and only if*

$$d'(w) = \frac{h(|\Sigma|^{-|w|}, s)}{h'(|\Sigma|^{-|w|}, s')} d(w)$$

is an $s'^{(h')}$-gale.

Success and strong success are defined as follows.

Definition. Let d be an $s^{(h)}$-gale, where h is a scale, let $s \in [0, \infty)$, and let $S \in \Sigma^\infty$. We say that d *succeeds* on S, and we write $S \in S^\infty[d]$, if $\limsup_{t\to\infty} d(S[0..t-1]) = \infty$. We say that d *succeeds strongly* on S, and we write $S \in S^\infty_{str}[d]$, if $\liminf_{t\to\infty} d(S[0..t-1]) = \infty$.

Lutz et al. defined a scaled dimension in Cantor space directly using gales in [27]. Here we introduced a more general concept of scaled dimension for any metric space and now characterize the Cantor space case.

Theorem 2.11 *For all $X \subseteq \Sigma^\infty$,*

$$\dim^{(h)}(X) = \inf \left\{ s \in [0, \infty) \mid \text{there is an } s^{(h)}\text{-gale } d \text{ with } X \subseteq S^\infty[d] \right\}$$

and

$$\mathrm{Dim}^{(h)}(X) = \inf \left\{ s \in [0, \infty) \mid \text{there is an } s^{(h)}\text{-gale } d \text{ with } X \subseteq S^\infty_{str}[d] \right\}.$$

For space reasons, we prefer not to include a full proof of Theorem 2.11 here. The proof can be done by nontrivially adapting the proofs of both parts of Theorem 2.8 that can be found in [45] and [1], respectively.

Our last property identifies the scales for which Cantor space has dimension 1.

Proposition 2.12 *Let h be a scale such that $h(x, s) = \Omega(x)$ for every $s < 1$ and $h(x, s) = O(x)$ for every $s > 1$. Then*

$$\dim^{(h)}(\Sigma^\infty) = \mathrm{Dim}^{(h)}(\Sigma^\infty) = 1.$$

For every $k \in \mathbb{Z}$,

$$\dim^{(k)}(\Sigma^\infty) = \mathrm{Dim}^{(k)}(\Sigma^\infty) = 1.$$

Proof. The property follows from Proposition 2.4.   □

2.4 Effective dimensions

We are mainly interested in subsets of sequences that have some computability or partial computability property, which implies that we will deal with countable sets. Since a countable set of sequences has dimension 0, the classical definitions of (scaled)-Hausdorff and packing dimensions are not useful in this context. The gale characterizations in Theorems 2.11 and 2.8 provide a natural way to generalize them as follows.

Definition. Let Γ be a class of functions. Let $X \subseteq \Sigma^\infty$, the Γ-dimension of X is

$$\dim_\Gamma(X) = \inf \left\{ s \in [0, \infty) \mid \text{there is an } s\text{-gale } d \in \Gamma \text{ with } X \subseteq S^\infty[d] \right\},$$

and the Γ-strong dimension of X is

$$\mathrm{Dim}_\Gamma(X) = \inf\left\{ s \in [0, \infty) \mid \text{there is an } s\text{-gale } d \in \Gamma \text{ with } X \subseteq S_{\mathrm{str}}^\infty[d] \right\}.$$

In the rest of the paper we will use different classes Γ, ranging from constructive to finite state computable functions, and investigate the properties of the corresponding Γ-dimensions inside different sequence sets. The existence of *correspondence principles*, introduced later on, will also imply that the effective dimension coincides with the classical Hausdorff dimension on sufficiently simple sets.

For scaled-dimensions, it is convenient that the scale itself is "computable" inside Γ in order to obtain meaningful results. Given a scale h, we will say that h is Γ-computable if the function $dh : \mathbb{N} \times [0, \infty) \rightarrow [0, \infty)$, $dh(k, s) = h(|\Sigma|^{-k-1}, s)/h(|\Sigma|^{-k}, s)$ is in Γ. The definitions of $\dim_\Gamma^{(h)}$ and $\mathrm{Dim}_\Gamma^{(h)}$ are similar to those of Γ-dimensions, but use $s^{(h)}$-gales in Γ.

3 Finite-state dimension

Our first effectivization of Hausdorff dimension will be the most restrictive of those presented here, and we will go all the way to the level of finite-state computation. In this section we use gales computed by finite-state gamblers to develop the finite-state dimensions of sets of infinite sequences and individual infinite sequences. Finite-state dimension was introduced by Dai et al. in [6], and its dual, strong finite-state dimension is from [1]. The definition has proven to be robust because it has been shown to admit equivalent definitions in terms of information-lossless finite-state compressors [6, 1], finite-state decompression [10], finite-state predictors in the log-loss model [26, 1], and block-entropy rates [2]. In each case, the definitions of $\dim_{\mathrm{FS}}(S)$ and $\mathrm{Dim}_{\mathrm{FS}}(S)$ are exactly dual, differing only in that a limit inferior appears in one definition where a limit superior appears in the other. These two finite-state dimensions are thus, like their counterparts in fractal geometry, robust quantities and not artifacts of a particular definition. In addition, the sequences S satisfying $\dim_{\mathrm{FS}}(S) = 1$ are precisely the normal sequences ([2], also follows from [66]).

In this section we present the finite-state dimension and its characterizations and summarize the main results on Eggleston theorem, existence of low complexity sequences of any dimension, invariance of finite-state dimension under arithmetic operations with rational numbers, and base dependence of the dimensions.

We start by introducing the concept of a finite-state gambler that is used to develop the finite-state dimension. Intuitively, a finite-state gambler is a finite-state device that places a bet on each of the successive symbols of its input sequence. Bets are required to be rational numbers in $\mathbf{B} = \mathbb{Q} \cap [0, 1]$.

Definition. A *finite-state gambler (FSG)* is a 4-tuple $G = (Q, \delta, \beta, q_0)$, where

- Q is a nonempty, finite set of *states*,

- $\delta : Q \times \Sigma \to Q$ is the *transition function*,

- $\beta : Q \times \Sigma \to \mathbf{B}$ is the *betting function*, with $\sum_{a \in \Sigma} \beta(q, a) = 1$ for every $q \in Q$, and

- $q_0 \in Q$ is the *initial state*.

Dai et al. [6] consider an equivalent model, the k-account finite-state gambler, in which the capital is divided into k separate accounts for a fixed k. This model allows simpler descriptions and a smaller number of states in the gambler definitions.

Our model of finite-state gambling has been considered (in essentially equivalent form) by Schnorr and Stimm [66], Feder [18], and others.

Intuitively, if a FSG $G = (Q, \delta, \beta, q_0)$ is in state $q \in Q$ and its current capital is $c \in (\mathbb{Q} \cap [0, \infty))$, then it places the bet $\beta(q, a) \in \mathbf{B}$ on each possible value of the next symbol. If the payoffs are fair, then after this bet, G will be in state $\delta(q, a)$ and it will have capital $|\Sigma| \, c \, \beta(q, a)$.

This suggests the following definition.

Definition. ([6]) Let $G = (Q, \delta, \beta, q_0)$ be a finite-state gambler.

1. The *martingale* of G is the function $d_G : \Sigma^* \to [0, \infty)$ defined by the recursion

$$d_G(\lambda) = 1,$$

$$d_G(wa) = |\Sigma| \, d_{G,i}(w) \, \beta(q, a)$$

for all $w \in \Sigma^*$ and $a \in \Sigma$.

2. For $s \in [0, \infty)$, the *s-gale* of an FSG G is the function $d_G^{(s)} : \Sigma^* \to [0, \infty)$ defined by $d_G^{(s)}(w) = |\Sigma|^{(s-1)|w|} d_G(w)$ for all $w \in \Sigma^*$. In particular, note that $d_G^{(1)} = d_G$.

3. For $s \in [0, \infty)$, a *finite-state s-gale* is an s-gale d for which there exists an FSG G such that $d_G^{(s)} = d$.

We now use finite-state gales to define the finite-state dimension.

Definition. ([6, 1]) Let $X \subseteq \Sigma^\infty$.

1. The *finite-state dimension* of set X is

$$\dim_{\mathrm{FS}}(X) = \inf \left\{ s \in [0, \infty) \,\middle|\, \text{there is a finite-state } s\text{-gale } d \text{ with } X \subseteq S^\infty[d] \right\}.$$

2. The *strong finite-state dimension* of set X is

$$\text{Dim}_{\text{FS}}(X) = \inf\left\{s \in [0, \infty) \mid \text{there is a finite-state } s\text{-gale } d \text{ with } X \subseteq S_{\text{str}}^{\infty}[d]\right\}.$$

3. The *finite-state dimension* and *strong finite-state dimension* of a sequence $S \in \Sigma^{\infty}$ are $\dim_{\text{FS}}(S) = \dim_{\text{FS}}(\{S\})$ and $\text{Dim}_{\text{FS}}(S) = \text{Dim}_{\text{FS}}(\{S\})$.

In general, $\dim_{\text{FS}}(X)$ and $\text{Dim}_{\text{FS}}(X)$ are real numbers satisfying $0 \leq \dim_{\text{H}}(X) \leq \dim_{\text{FS}}(X) \leq \text{Dim}_{\text{FS}}(X) \leq 1$ and $\text{Dim}(X) \leq \text{Dim}_{\text{FS}}(X)$. The finite-state dimension has a finite stability property.

Theorem 3.1 ([6]) *For all $X, Y \subseteq \Sigma^{\infty}$,*

$$\dim_{\text{FS}}(X \cup Y) = \max\left\{\dim_{\text{FS}}(X), \dim_{\text{FS}}(Y)\right\}.$$

The proof of basic properties such as this theorem in [6] benefits greatly from the use of multiple account FSGs, since the equivalence of multiple accounts and our 1-account FSG seems to require an exponential blowup of states.

The main result in this section is that we can characterize the finite-state dimensions of individual sequences in terms of finite-state compressibility. We first recall the definition of an information-lossless finite-state compressor. (This idea is due to Huffman [34]. Further exposition may be found in [35] or [36].)

Definition. A *finite-state transducer* is a 4-tuple $C = (Q, \delta, \nu, q_0)$, where Q is a nonempty, finite set of *states*, $\delta : Q \times \Sigma \rightarrow Q$ is the *transition function*, $\nu : Q \times \Sigma \rightarrow \Sigma^*$ is the *output function*, and $q_0 \in Q$ is the *initial state*.

For $q \in Q$ and $w \in \Sigma^*$, we define the *output from state q on input w* to be the string $\nu(q, w)$ defined by the recursion

$$\nu(q, \lambda) = \lambda,$$
$$\nu(q, wa) = \nu(q, w)\nu(\delta(q, w), a)$$

for all $w \in \Sigma^*$ and $a \in \Sigma$. We then define the *output* of C on input $w \in \Sigma^*$ to be the string $C(w) = \nu(q_0, w)$.

Definition. An *information-lossless finite-state compressor (ILFSC)* is a finite-state transducer $C = (Q, \delta, \nu, q_0)$ such that the function $f : \Sigma^* \rightarrow \Sigma^* \times Q$, $f(w) = (C(w), \delta(w))$ is one-to-one.

That is, an ILFSC is a transducer whose input can be reconstructed from the output and final state reached on that input.

Intuitively, C compresses a string w if $|C(w)|$ is significantly less than $|w|$. Of course, if C is IL, then not all strings can be compressed. Our interest here is in the degree (if any) to which the prefixes of a given sequence $S \in \Sigma^{\infty}$ can be compressed by an ILFSC. We will consider the cases of infinitely often (i.o.) and almost everywhere (a.e.) compression ratio.

Definition.

1. If C is an ILFSC and $S \in \Sigma^\infty$, then the a.e. *compression ratio* of C on S is

$$\rho_C(S) = \liminf_{n \to \infty} \frac{|C(S[0..n-1])|}{n \log |\Sigma|}.$$

2. The *finite-state a.e. compression ratio* of a sequence $S \in \Sigma^\infty$ is

$$\rho_{FS}(S) = \inf \{\rho_C(S) | C \text{ is an ILFSC}\}.$$

3. If C is an ILFSC and $S \in \Sigma^\infty$, then the a.e. *compression ratio* of C on S is

$$R_C(S) = \limsup_{n \to \infty} \frac{|C(S[0..n-1])|}{n \log |\Sigma|}.$$

4. The *finite-state i.o. compression ratio* of a sequence $S \in \Sigma^\infty$ is

$$R_{FS}(S) = \inf \{R_C(S) | C \text{ is an ILFSC}\}.$$

The following theorem says that finite-state dimension and finite-state compressibility are one and the same for individual sequences.

Theorem 3.2 ([6, 1]) *For all* $S \in \Sigma^\infty$,

$$\dim_{FS}(S) = \rho_{FS}(S),$$

and

$$\mathrm{Dim}_{FS}(S) = R_{FS}(S).$$

Doty and Moser [10] remarked that the finite-state dimension can be characterized in terms of decompression by finite-state transducers based on earlier results by Sheinwald et al. [67]. Notice that in this case finite-state machines are not required to be information lossless.

Theorem 3.3 ([10]) *For all* $S \in \Sigma^\infty$,

$$\dim_{FS}(S) = \inf_{\substack{T \text{ finite-state} \\ \text{transducer}}} \liminf_{n \to \infty} \frac{\min_{\pi \in \Sigma^*} \{|\pi| \, | \, T(\pi) = S[0..n-1]\}}{n \log |\Sigma|}$$

and

$$\mathrm{Dim}_{FS}(S) = \inf_{\substack{T \text{ finite-state} \\ \text{transducer}}} \limsup_{n \to \infty} \frac{\min_{\pi \in \Sigma^*} \{|\pi| \, | \, T(\pi) = S[0..n-1]\}}{n \log |\Sigma|}.$$

Theorems 3.2 and 3.3 are instances of the existing relation between dimension and information. It is interesting to view them in comparison with other information characterizations of effective dimension that we will develop in the following sections. In the case of constructive dimension, the characterization is based on general Kolmogorov complexity, which can only be viewed as decompression. For space bounds, dimension can be characterized either by space-bounded compressors or by decompressors, whereas in the case of polynomial-time dimension, the known characterization requires consideration of polynomial-time compressors that are also decompressible in polynomial time. The above results show that the finite-state dimension is similar to space dimension in this matter and apparently simpler than the time-bounded and constructive cases.

We now present a third characterization of the finite-state dimension, this time in terms of block-entropy rates.

Definition. Let $w \in \Sigma^*$, $S \in \Sigma^\infty$.

1. Let $P(w, S[0..k|w| - 1]) = \frac{1}{k}|\{0 \le i < k \mid S[i|w|..(i+1)|w| - 1] = w\}|$.

2. The lth block-entropy rate of S is

$$H_l(S) = \liminf_{k \to \infty} -\frac{1}{l \log |\Sigma|} \sum_{|w|=l} P(w, S[0..kl - 1]) \log(P(w, S[0..kl - 1])).$$

3. The block entropy rate of S is $H(S) = \inf_{l \in \mathbb{N}} H_l(S)$.

4. The lth upper block-entropy rate of S is

$$\overline{H_l}(S) = \limsup_{k \to \infty} -\frac{1}{l \log |\Sigma|} \sum_{|w|=l} P(w, S[0..kl - 1]) \log(P(w, S[0..kl - 1])).$$

5. The upper block-entropy rate of S is $\overline{H}(S) = \inf_{l \in \mathbb{N}} H_l(S)$.

Theorem 3.4 ([2]) *Let $S \in \Sigma^\infty$. $\mathrm{Dim_{FS}}(S) = \overline{H}(S)$, and $\dim_{FS}(S) = H(S)$.*

The first part of Theorem 3.4 follows from [37] and [1].

We can also consider "sliding window" entropy, based on the number of times each string $w \in \Sigma^*$ appears inside an infinite sequence $S \in \Sigma^\infty$ when occurrences can partially overlap.

Definition. Let $w \in \Sigma^*$, $S \in \Sigma^\infty$.

1. Let $P'(w, S[0..n - 1]) = \frac{|w|}{n}|\{0 \le i \le n - |w| \mid S[i..i + |w| - 1] = w\}|$.

2. The lth entropy rate of S is

$$H'_l(S) = \liminf_{n \to \infty} -\frac{1}{l \log |\Sigma|} \sum_{|w|=l} P'(w, S[0..n - 1]) \log(P'(w, S[0..n - 1])).$$

3. The entropy rate of S is $H'(S) = \inf_{l \in \mathbb{N}} H'_l(S)$.

4. The lth upper entropy rate of S is

$$\overline{H'_l}(S) = \limsup_{n \to \infty} -\frac{1}{l \log |\Sigma|} \sum_{|w|=l} P'(w, S[0..n-1]) \log(P'(w, S[0..n-1])).$$

5. The upper entropy rate of S is $\overline{H'}(S) = \inf_{l \in \mathbb{N}} H'_l(S)$.

The following characterization follows from the results in [37] and Theorem 3.4.

Theorem 3.5 *Let $S \in \Sigma^\infty$. $\mathrm{Dim}_{\mathrm{FS}}(S) = \overline{H'}(S)$, and $\dim_{\mathrm{FS}}(S) = H'(S)$.*

Notice that the definitions of entropy consider only frequency properties of the sequence and do not involve finite-state machines; i.e., the finite-state dimension admits a "machine-independent" characterization.

As a consequence of Theorem 3.4 and previous results in [6], the sequences that have finite-state dimension 1 are exactly the (Borel) normal sequences. Therefore, the finite-state dimension is base dependent.

Theorem 3.6 *There exists a real number $\alpha \in [0, 1]$ and $n, m \in \mathbb{N}$ such that the sequences S and S' that represent α in bases n and m, respectively, have different finite-state dimensions.*

The proof of this last theorem is based on the existence of normal sequences that are not absolutely normal, that is, the existence of a real number α and two bases n, m such that the representation of α in base n is a normal sequence, whereas the representation in base m is not normal (proven by Cassels in [5]).

The Hausdorff and packing dimensions are both base independent, and it is known [33] that the polynomial-time dimension is also base independent.

We conclude this section with a summary of the other results on the finite-state dimension.

The theorem of Eggleston [14] (Theorem 2.9) holds for the finite-state dimension.

Theorem 3.7 ([6]) *For all $\alpha \in \mathbb{Q} \cap [0, 1]$,*

$$\dim_{\mathrm{FS}}(\mathrm{FREQ}(\alpha)) = \mathcal{H}(\alpha).$$

The following theorem says that every rational number $r \in [0, 1]$ is the finite-state dimension of a reasonably simple sequence.

Theorem 3.8 ([6]) *For every $r \in \mathbb{Q} \cap [0, 1]$, there exists $S \in \mathrm{AC}_0$ such that $\dim_{\mathrm{FS}}(S) = r$.*

Doty et al. prove that the finite-state dimension is invariant under arithmetical operations with a rational number.

Theorem 3.9 ([9]) *Let $k \in \mathbb{N}$, $q \in \mathbb{Q}$ with $q \neq 0$, $\alpha \in \mathbb{R}$. Then*

$$\dim_{FS}(S_{q+\alpha}) = \dim_{FS}(S_{q\alpha}) = \dim_{FS}(S_\alpha),$$

where S_x is the representation of x in base k. The same result holds for Dim_{FS} in the place of $\dim_{FS}$.

The scaled dimension has not been used in the context of the finite-state dimension. Notice that only scales of the form $x^{f(s)}$ are finite-state-computable.

The finite-state dimension is a real-time effectivization of a powerful tool of fractal geometry. As such it should prove to be a useful tool for improving our understanding of real-time information processing.

4 Constructive dimension

Our next effective version of the Hausdorff dimension is defined by restricting the class of gales to those that are lower semicomputable. We give the definitions of constructive dimension and constructive strong dimension of a set, and also of a sequence, and we relate them and give their main properties, which make it very powerful. We first have absolute stability, which means it can be applied to an arbitrary union of sets. Then there is a precise characterization of the dimension of a sequence in terms of the Kolmogorov complexity of its elements, and finally in many interesting cases, constructive dimension coincides with the classic Hausdorff dimension. We also summarize the known relationships of this concept with Martin–Löf random sequences.

An s-gale d is *constructive* if it is lower semicomputable; that is, its lower graph $\{(w, z) \mid z < d(w)\}$ is c.e. We define constructive dimension as follows.

Definition. ([46, 1]) Let $X \subseteq \Sigma^\infty$.

1. The *constructive* dimension of a set $X \subseteq \Sigma^\infty$ is

   $\mathrm{cdim}(X) = \inf \{s \in [0, \infty) \mid$ there is a constructive s-gale d with $X \subseteq S^\infty[d]\,\}$.

2. The *constructive strong* dimension of a set $X \subseteq \Sigma^\infty$ is

   $\mathrm{cDim}(X) = \inf \{s \in [0, \infty) \mid$ there is a constructive s-gale d with $X \subseteq S^\infty_{\mathrm{str}}[d]\,\}$.

3. The (constructive) dimension and strong dimension of an individual sequence $S \in \Sigma^\infty$ are $\dim(S) = \mathrm{cdim}(\{S\})$ and $\mathrm{Dim}(S) = \mathrm{cDim}(\{S\})$.

By the gale characterizations of the Hausdorff dimension (Theorem 2.8), we conclude that $\mathrm{cdim}(X) \geq \dim_H(X)$ for all $X \subseteq \Sigma^\infty$. But in fact much more is true for certain classes, as Hitchcock shows in [24]. For sets that are low in the arithmetical hierarchy, constructive dimension and Hausdorff dimension coincide.

Theorem 4.1 ([24]) *If $X \subseteq \Sigma^\infty$ is a union of Π_1^0 sets, then $\dim_H(X) = \mathrm{cdim}(X)$.*

Hitchcock also proves that this is an optimal result for the arithmetical hierarchy, since it cannot be extended to sets in Π_2^0. It is open whether such a correspondence principle holds for a strong constructive dimension and a packing dimension.

For the Hausdorff dimension, all singletons have dimension 0 and in fact all countable sets have Hausdorff dimension 0. The situation changes dramatically when we restrict to constructive gales, since a singleton can have a positive constructive dimension and in fact can have any constructive dimension.

Theorem 4.2 ([46]) *For every $\alpha \in [0, 1]$, there is an $S \in \Sigma^\infty$ such that $\dim(S) = \alpha$.*

A sequence is c-regular if its (constructive) dimension and strong dimensions coincide. In fact, these two dimensions can have any arbitrary two values.

Theorem 4.3 ([1]) *For every $\alpha, \beta \in [0, 1]$ with $\alpha \leq \beta$, there is an $S \in \Sigma^\infty$ such that $\dim(S) = \alpha$ and $\mathrm{Dim}(S) = \beta$.*

An interesting example of a c-regular sequence is θ_A^s that generalizes Chaitin's Ω and has been defined by Tadaki [72] and Mayordomo [50]. θ_A^s has dimension and strong dimension s.

The constructive dimension of any set $X \subseteq \Sigma^\infty$ is determined completely by the dimension of the individual sequences in the set.

Theorem 4.4 ([46, 1]) *For all $X \subseteq \Sigma^\infty$,*

$$\mathrm{cdim}(X) = \sup_{x \in X} \dim(x)$$

and

$$\mathrm{cDim}(X) = \sup_{x \in X} \mathrm{Dim}(x).$$

There is no analog of this last theorem for the Hausdorff dimension or for any of the concepts defined in sections 3 and 5. The key ingredient in the proof of Theorem 4.4 is the existence of optimal constructive supergales, that is, constructive supergales that multiplicatively dominate any other constructive supergale. This is analogous to the existence of universal tests of randomness in the theory of random sequences.

Theorem 4.1 together with Theorem 4.4 implies that the classical Hausdorff dimension of every Σ_2^0 set $X \subseteq \Sigma^\infty$ has the pointwise characterization $\dim_H(X) = \sup_{x \in X} \dim(x)$.

Theorem 4.4 immediately implies that constructive and strong constructive dimensions have the *absolute stability* property. Classical Hausdorff and packing dimensions have only countable stability.

Corollary 4.5 ([46, 1]) *For any I*

$$\mathrm{cdim}\left(\bigcup_{i\in I} X_i\right) = \sup_{i\in I} \mathrm{cdim}(X_i),$$

$$\mathrm{cDim}\left(\bigcup_{i\in I} X_i\right) = \sup_{i\in I} \mathrm{cDim}(X_i).$$

The (constructive) dimension of a sequence can be characterized in terms of the Kolmogorov complexities of its prefixes.

Theorem 4.6 ([50]) *For all $A \in \Sigma^\infty$,*

$$\mathrm{dim}(A) = \liminf_{n\to\infty} \frac{\mathrm{K}(A\,[0..n-1])}{n\log|\Sigma|}.$$

This latest theorem justifies the intuition that the constructive dimension of a sequence is a measure of its algorithmic information density. Several authors have studied the close relation of the Hausdorff dimension to measures of information content. Ryabko [57, 58], Staiger [68, 69], and Cai and Hartmanis [3] proved results relating the Hausdorff dimension to Kolmogorov complexity. Ryabko [60] and Staiger [69] studied computable exponents of increase that correspond to a computable dimension [45], defined in terms of computable gales and that is strictly above the constructive dimension. See [46] for a complete chronology.

We note that Theorem 4.6 yields a new proof of Theorem 4.1 above that of Theorem 5 of Staiger [69]. Also, Theorem 4.6 yields a new proof of Theorem 4.2 below that of Lemma 3.4 of Cai and Hartmanis [3].

A dual result holds for the constructive strong dimension as proven in [1]; that is, for any $A \in \Sigma^\infty$,

$$\mathrm{Dim}(A) = \limsup_{n\to\infty} \frac{\mathrm{K}(A\,[0..n-1])}{n\log|\Sigma|}.$$

Alternative characterizations of the constructive dimension in terms of variations of Martin–Löf tests and effectivizations of the Hausdorff measure have been given by Reimann and Stephan [54] and Calude et al. [4]. Doty has considered the Turing reduction compression ratio in [8].

We now briefly state the main results proven so far on constructive dimension, including the existence of sequences of any dimension, the constructive version of Eggleston theorem, and the constructive dimension of sequences that are random relative to a nonuniform distribution.

This is the constructive version of the classical Theorem 2.9 (Eggleston [14]).

Theorem 4.7 ([46]) *If α is Δ_2^0-computable real number in $[0,1]$, then*

$$\mathrm{cdim}(\mathrm{FREQ}(\alpha)) = \mathcal{H}(\alpha).$$

An alternative proof of Theorem 4.7 can be derived from Theorem 4.6 and from earlier results of Eggleston [14] and Kolmogorov [75]. In fact, this approach shows that Theorem 4.7 holds for *arbitrary* $\alpha \in [0, 1]$.

A binary sequence is (Martin–Löf) *random* [48] if it passes every algorithmically implementable test of randomness. This can be reformulated in terms of martingales as follows:

Definition. ([62]) A sequence $A \in \{0, 1\}^\infty$ is (Martin–Löf) *random* if there is no constructive martingale d such that $A \in S^\infty[d]$.

By definition, random sequences have constructive dimension 1. For nonuniform distributions, we have the concept of β-randomness, for β any real number in $(0, 1)$ representing the bias.

Definition. ([62]) Let $\beta \in (0, 1)$.

1. A β-*martingale* is a function $d : \{0, 1\}^* \to [0, \infty)$ that satisfies the condition

$$d(w) = (1 - \beta)\, d(w0) + \beta\, d(w1)$$

   for all $w \in \{0, 1\}^*$.

2. A sequence $A \in \{0, 1\}^\infty$ is (Martin–Löf) *random relative to* β if there is no constructive β-martingale d such that $A \in S^\infty[d]$.

Lutz relates randomness relative to a nonuniform distribution to Shannon information theory.

Theorem 4.8 ([46]) *Let* $\beta \in (0, 1)$ *be a computable real number. Let* $A \in \{0, 1\}^\infty$ *be random relative to* β. *Then* $\dim(A) = \mathcal{H}(\beta)$.

A more general result for randomness relative to sequences of coin-tosses is obtained in [46], and extended in [1], where constructive and constructive strong dimension of such a random sequence are shown to be the lower and upper average entropy of the bias, respectively.

A very recent line of research is the comparison of positive dimension sequences with (Martin–Löf) random sequences (relative to bias 1/2) in terms of their computing power. The main issue is whether positive dimension sequences can substitute random sequences as randomness sources [51]. Doty [7], based on earlier results by Ryabko [57, 58], has proven that a sequence of positive dimension is Turing equivalent to a sequence of strong dimension arbitrarily close to 1. Nies and Reimann [53] and Stephan [70] study the existence of weak-truth-table degrees or lower cones of arbitrary dimension. Gu and Lutz [22] show that positive dimension sequences can substitute randomness in the context of probabilistic polynomial-time computation.

We end this section by going back to scaled dimension. We think that the constructive dimension can benefit particularly from the flexibility provided by using different scales.

Let h be a scale such that (i) $h(x, 1) = \Theta(x)$ and (ii) $dh : \mathbb{N} \times [0, \infty) \rightarrow [0, \infty)$; $dh(k, s) = h(|\Sigma|^{-k-1}, s)/h(|\Sigma|^{-k}, s)$ is a computable function. Given a sequence S we define

$$\dim^{(h)}(S) = \inf \left\{ s \in [0, \infty) \, \middle| \, \text{there is a constructive } s^{(h)}\text{-gale } d \text{ with } S \in S^\infty[d] \right\}.$$

The results in [32] can be extended as follows.

Theorem 4.9 *Let h be a scale as above and such that $h(x, s) \leq (\log(1/x))^{-1-\epsilon}$ for some epsilon (that may depend on s). Then the following are equivalent.*

1. $\dim^{(h)}(S) < s$.

2. $K(S[0..n-1]) < -\log(h(|\Sigma|^{-n}, s))$ for infinitely many n.

There is a strong dimension version of Theorem 4.9 in which the Kolmogorov complexity is bounded for almost every prefix of the sequence. In both cases the upper bound on the scale can be substituted by differentiability of h.

5 Resource-bounded dimension

In this section we briefly review the properties of a resource-bounded dimension more directly related to algorithmic information theory. For a recent summary of dimension in complexity classes, the reader may consult [28].

We will consider polynomial-time and polynomial-space dimensions. We define p to be the class of polynomial-time computable functions, pspace, as the class of polynomial space functions. Let Δ be either p or pspace.

Definition. ([45]) Let $X \subseteq \Sigma^\infty$.

1. The Δ-*dimension* of a set $X \subseteq \Sigma^\infty$ is

$$\dim_\Delta(X) = \inf \{s \in [0, \infty) \mid \text{there is a } s\text{-gale } d \in \Delta \text{ with } X \subseteq S^\infty[d]\}.$$

2. The Δ *strong dimension* of a set $X \subseteq \Sigma^\infty$ is

$$\text{Dim}_\Delta(X) = \inf \{s \in [0, \infty) \mid \text{there is a } s\text{-gale } d \in \Delta \text{ with } X \subseteq S_{\text{str}}^\infty[d]\}.$$

Let us mention that Eggleston theorem also holds for the resource-bounded case [45], for each p-computable (pspace-computable) α, $\dim_p(\text{FREQ}(\alpha)) = \mathcal{H}(\alpha)$ $(\dim_{\text{pspace}}(\text{FREQ}(\alpha)) = \mathcal{H}(\alpha))$, and even for sublinear time-bounds [52].

Hitchcock [25] has characterized the pspace dimension in terms of space-bounded Kolmogorov complexity as follows. Let $\text{KS}^{f(n)}(w)$ be the Kolmogorov complexity of the string w when only space $f(|w|)$ is allowed in the computation of w from its description [40].

Theorem 5.1 ([25]) *For all $X \subseteq \Sigma^\infty$,*

$$\dim_{\text{pspace}}(X) = \inf_c \sup_{A \in X} \liminf_{n \to \infty} \frac{\text{KS}^{n^c}(A[0..n-1])}{n \log |\Sigma|},$$

$$\text{Dim}_{\text{pspace}}(X) = \inf_c \sup_{A \in X} \limsup_{n \to \infty} \frac{\text{KS}^{n^c}(A[0..n-1])}{n \log |\Sigma|}.$$

This result can also be extended to a scaled dimension.

Theorem 5.2 ([32]) *For $z \in \{-1, 0, 1\}$, the following are equivalent.*

1. $\dim_{\text{pspace}}^{(z)}(X) < s.$

2. There exists c such that for every $S \in X$,

$$\text{KS}^{n^c}(S[0..n-1]) < -\log(h_z(|\Sigma|^{-n}, s)) \text{ for infinitely many } n.$$

Theorem 5.2 has a strong dimension version in which the Kolmogorov complexity is bounded for almost every prefix of the sequence.

The case of the polynomial-time dimension seems much harder, since time-bounded Kolmogorov complexity has proven difficult to analyze. After attempts from Hitchcock and Vinodchandran in [29], the right approach seems to be the consideration of polynomial-time compressors that can also be inverted in polynomial time. López-Valdés and Mayordomo [41] prove the following.

Definition. ([41]) Let (C, D) be polynomial-time algorithms such that for every $w \in \Sigma^*$, $D(C(w), |w|) = w$. (C, D) *does not start from scratch* if $\forall \epsilon > 0$ and for almost every $w \in \{0, 1\}^*$, there exists $k = O(\log(|w|))$, $k > 0$, such that

$$\sum_{|u| \leq k} |\Sigma|^{-|C(wu)|} \leq |\Sigma|^{\epsilon k} |\Sigma|^{-|C(w)|}.$$

Let PC be the class of polynomial-time compressors that do not start from scratch.

Theorem 5.3 ([41]) *Let $X \subseteq \Sigma^\infty$,*

$$\dim_{\text{p}}(X) = \inf_{(C,D) \in \text{PC}} \sup_{A \in X} \liminf_{n} \frac{|C(A[0..n-1])|}{n \log |\Sigma|},$$

$$\text{Dim}_{\text{p}}(X) = \inf_{(C,D) \in \text{PC}} \sup_{A \in X} \limsup_{n} \frac{|C(A[0..n-1])|}{n \log |\Sigma|}.$$

Connection of resource-bounded dimension with sequence analysis models from computational learning has proven successful in [31], [26] and [20].

Acknowledgments. I thank an anonymous referee, Philippe Moser, and David Doty for many helpful suggestions.

References

1. K. B. Athreya, J. M. Hitchcock, J. H. Lutz, and E. Mayordomo. Effective strong dimension in algorithmic information and computational complexity. *SIAM Journal on Computing*. To appear.
2. C. Bourke, J. M. Hitchcock, and N. V. Vinodchandran. Entropy rates and finite-state dimension. *Theoretical Computer Science*, 349:392–406, 2005.
3. J. Cai and J. Hartmanis. On Hausdorff and topological dimensions of the Kolmogorov complexity of the real line. *Journal of Computer and Systems Sciences*, 49:605–619, 1994.
4. C. S. Calude, L. Staiger, and S. A. Terwijn. On partial randomness. *Annals of Pure and Applied Logic*, 138:20–30, 2006.
5. J. W. S. Cassels. On a problem of Steinhaus about normal numbers. *Colloquium Mathematicum*, 7:95–101, 1959.
6. J. J. Dai, J. I. Lathrop, J. H. Lutz, and E. Mayordomo. Finite-state dimension. *Theoretical Computer Science*, 310:1–33, 2004.
7. D. Doty. Dimension extractors. Technical Report cs.CC/0606078, Computing Research Repository, 2006.
8. D. Doty. Every sequence is decompressible from a random one. In *Proceedings of Second Conference on Computability in Europe*, Lecture Notes in Computer Science, pages 153–162. Springer-Verlag, 2006.
9. D. Doty, J. H. Lutz, and S. Nandakumar. Finite-state dimension and real arithmetic. In *Proceedings of the 33rd International Colloquium on Automata, Languages, and Programming*, Lecture Notes in Computer Science. Springer-Verlag, 2006.
10. D. Doty and P. Moser. Personal communication, based on [67]. 2006.
11. R. Downey and D. Hirschfeldt. *Algorithmic Randomness and Complexity*. Book Draft, 2006.
12. G. A. Edgar. *Integral, Probability, and Fractal Measures*. Springer-Verlag, 1998.
13. G. A. Edgar. *Measure, Topology, and Fractal Geometry*. Springer-Verlag, 1990.
14. H.G. Eggleston. The fractional dimension of a set defined by decimal properties. *Quarterly Journal of Mathematics*, Oxford Series 20:31–36, 1949.
15. K. Falconer. *The Geometry of Fractal Sets*. Cambridge University Press, 1985.
16. K. Falconer. *Fractal Geometry: Mathematical Foundations and Applications*. John Wiley & Sons, 2003.
17. K. Falconer. *Techniques in Fractal Geometry*. John Wiley & Sons, 2003.
18. M. Feder. Gambling using a finite state machine. *IEEE Transactions on Information Theory*, 37:1459–1461, 1991.
19. H. Federer. *Geometric Measure Theory*. Springer-Verlag, 1969.
20. L. Fortnow and J. H. Lutz. Prediction and dimension. *Journal of Computer and System Sciences*. To appear. Preliminary version appeared in *Proceedings of the 15th Annual Conference on Computational Learning Theory*, LNCS 2375, pages 380–395, 2002.
21. X. Gu, J. H. Lutz, and E. Mayordomo. Points on computable curves. In *Proceedings of the Forty-Seventh Annual IEEE Symposium on Foundations of Computer Science*, 2006. To appear.
22. X. Gu and J. H. Lutz. Dimension characterizations of complexity classes. In *Proceedings of the 31st International Symposium on Mathematical Foundations of Computer Science*, Lecture Notes in Computer Science, pages 471–479. Springer-Verlag, 2006.
23. F. Hausdorff. Dimension und äußeres Maß. *Math. Ann.*, 79:157–179, 1919.

24. J. Hitchcock. Correspondence principles for effective dimensions. In *Proceedings of the 29th Colloquium on Automata, Languages and Programming*. Springer Lecture Notes in Computer Science, 2002. To appear.
25. J. M. Hitchcock. *Effective Fractal Dimension: Foundations and Applications*. PhD thesis, Iowa State University, 2003.
26. J. M. Hitchcock. Fractal dimension and logarithmic loss unpredictability. *Theoretical Computer Science*, 304(1–3):431–441, 2003.
27. J. M. Hitchcock, J. H. Lutz, and E. Mayordomo. Scaled dimension and nonuniform complexity. *Journal of Computer and System Sciences*, 69:97–122, 2004.
28. J. M. Hitchcock, J. H. Lutz, and E. Mayordomo. The fractal geometry of complexity classes. *SIGACT News Complexity Theory Column*, 36:24–38, 2005.
29. J. M. Hitchcock and N. V. Vinodchandran. Dimension, entropy rates, and compression. *Journal of Computer and System Sciences*, 72(4):760–782, 2006.
30. J. M. Hitchcock. Effective fractal dimension bibliography. http://www.cs.uwyo.edu/ jhitchco/bib/dim.shtml.
31. J. M. Hitchcock. Online learning and resource-bounded dimension: Winnow yields new lower bounds for hard sets. *SIAM Journal on Computing*, 2006. To appear.
32. J. M. Hitchcock, M. López-Valdés, and E. Mayordomo. Scaled dimension and the Kolmogorov complexity of Turing-hard sets. In *Proceedings of the 29th International Symposium on Mathematical Foundations of Computer Science*, volume 3153 of *Lecture Notes in Computer Science*, pages 476–487. Springer-Verlag, 2004.
33. J. M. Hitchcock and E. Mayordomo. Base invariance of feasible dimension. Manuscript, 2003.
34. D. A. Huffman. Canonical forms for information-lossless finite-state logical machines. *IRE Trans. Circuit Theory CT-6 (Special Supplement)*, pages 41–59, 1959. Also available in E.F. Moore (ed.), Sequential Machine: Selected Papers, Addison-Wesley, 1964, pages 866-871.
35. Z. Kohavi. *Switching and Finite Automata Theory (Second Edition)*. McGraw-Hill, 1978.
36. A. A. Kurmit. *Information-Lossless Automata of Finite Order*. Wiley, 1974.
37. A. Lempel and J. Ziv. Compression of individual sequences via variable rate coding. *IEEE Transaction on Information Theory*, 24:530–536, 1978.
38. P. Lévy. Propriétés asymptotiques des sommes de variables indépendantes ou enchaînées. *Journal des mathématiques pures et appliquées. Series 9*, 14(4):347–402, 1935.
39. P. Lévy. *Théorie de l'Addition des Variables Aléatoires*. Gauthier-Villars, 1937 (second edition 1954).
40. M. Li and P. M. B. Vitányi. *An Introduction to Kolmogorov Complexity and its Applications* (Second Edition). Springer-Verlag, 1997.
41. M. López-Valdés and E. Mayordomo. Dimension is compression. In *Proceedings of the 30th International Symposium on Mathematical Foundations of Computer Science*, volume 3618 of *Lecture Notes in Computer Science*, pages 676–685. Springer-Verlag, 2005.
42. J. H. Lutz. Almost everywhere high nonuniform complexity. *Journal of Computer and System Sciences*, 44(2):220–258, 1992.
43. J. H. Lutz. The quantitative structure of exponential time. In L. A. Hemaspaandra and A. L. Selman, editors, *Complexity Theory Retrospective II*, pages 225–254. Springer-Verlag, 1997.
44. J. H. Lutz. Resource-bounded measure. In *Proceedings of the 13th IEEE Conference on Computational Complexity*, pages 236–248, 1998.
45. J. H. Lutz. Dimension in complexity classes. *SIAM Journal on Computing*, 32:1236–1259, 2003.

46. J. H. Lutz. The dimensions of individual strings and sequences. *Information and Computation*, 187:49–79, 2003.

47. J. H. Lutz and E. Mayordomo. Dimensions of points in self-similar fractals. Abstract published in the *Proceedings of the Third International Conference on Computability and Complexity in Analysis*.

48. P. Martin-Löf. The definition of random sequences. *Information and Control*, 9:602–619, 1966.

49. P. Matilla. *Geometry of Sets and Measures in Euclidean Spaces: Fractals and Rectifiability*. Cambridge University Press, 1995.

50. E. Mayordomo. A Kolmogorov complexity characterization of constructive Hausdorff dimension. *Information Processing Letters*, 84(1):1–3, 2002.

51. J. S. Miller and A. Nies. Randomness and computability: open questions. *Bulletin of Symbolic Logic*, 12:390–410, 2006.

52. P. Moser. Martingale families and dimension in P. In *Logical Approaches to Computational Barriers, Second Conference on Computability in Europe, CiE 2006*, volume 3988 of *Lecture Notes in Computer Science*, pages 388–397. Springer-Verlag, 2006.

53. A. Nies and J. Reimann. A lower cone in the wtt degrees of non-integral effective dimension. In *Proceedings of IMS Workshop on Computational Prospects of Infinity*, 2006. To appear.

54. J. Reimann and F. Stephan. Effective Hausdorff dimension. In *Logic Colloquium '01*, number 20 in Lecture Notes in Logic, pages 369–385. Association for Symbolic Logic, 2005.

55. J. Reimann and F. Stephan. On hierarchies of randomness tests. In *Proceedings of the 9th Asian Logic Conference 2005*. World Scientific, 2006.

56. C. A. Rogers. *Hausdorff Measures*. Cambridge University Press, 1998. Originally published in 1970.

57. B. Ya. Ryabko. Coding of combinatorial sources and Hausdorff dimension. *Soviets Mathematics Doklady*, 30:219–222, 1984.

58. B. Ya. Ryabko. Noiseless coding of combinatorial sources. *Problems of Information Transmission*, 22:170–179, 1986.

59. B. Ya. Ryabko. Algorithmic approach to the prediction problem. *Problems of Information Transmission*, 29:186–193, 1993.

60. B. Ya. Ryabko. The complexity and effectiveness of prediction problems. *Journal of Complexity*, 10:281–295, 1994.

61. C. P. Schnorr. Klassifikation der Zufallsgesetze nach Komplexität und Ordnung. *Z. Wahrscheinlichkeitstheorie verw. Geb.*, 16:1–21, 1970.

62. C. P. Schnorr. A unified approach to the definition of random sequences. *Mathematical Systems Theory*, 5:246–258, 1971.

63. C. P. Schnorr. Zufälligkeit und Wahrscheinlichkeit. *Lecture Notes in Mathematics*, 218, 1971.

64. C. P. Schnorr. Process complexity and effective random tests. *Journal of Computer and System Sciences*, 7:376–388, 1973.

65. C. P. Schnorr. A survey of the theory of random sequences. In R. E. Butts and J. Hintikka, editors, *Basic Problems in Methodology and Linguistics*, pages 193–210. D. Reidel, 1977.

66. C. P. Schnorr and H. Stimm. Endliche automaten und zufallsfolgen. *Acta Informatica*, 1:345–359, 1972.

67. D. Sheinwald, A. Lempel, and J. Ziv. On compression with two-way head machines. In *Data Compression Conference*, pages 218–227, 1991.

68. L. Staiger. Kolmogorov complexity and Hausdorff dimension. *Information and Computation*, 103:159–94, 1993.

69. L. Staiger. A tight upper bound on Kolmogorov complexity and uniformly optimal prediction. *Theory of Computing Systems*, 31:215–29, 1998.

70. F. Stephan. Hausdorff-dimension and weak truth-table reducibility. Technical Report TR52/05, National University of Singapore, School of Computing, 2005.

71. D. Sullivan. Entropy, Hausdorff measures old and new, and limit sets of geometrically finite Kleinian groups. *Acta Mathematica*, 153:259–277, 1984.

72. K. Tadaki. A generalization of Chaitin's halting probability ω and halting self-similar sets. *Hokkaido Mathematical Journal*, 31:219–253, 2002.

73. C. Tricot. Two definitions of fractional dimension. *Mathematical Proceedings of the Cambridge Philosophical Society*, 91:57–74, 1982.

74. J. Ville. *Étude Critique de la Notion de Collectif.* Gauthier–Villars, 1939.

75. A. K. Zvonkin and L. A. Levin. The complexity of finite objects and the development of the concepts of information and randomness by means of the theory of algorithms. *Russian Mathematical Surveys*, 25:83–124, 1970.

Metamathematical Properties of Intuitionistic Set Theories with Choice Principles

Michael Rathjen*

Department of Pure Mathematics, University of Leeds, Leeds LS2 9JT, United Kingdom, and
Department of Mathematics, Ohio State University, Columbus, OH 43210, U.S.A.
`rathjen@math.ohio-state.edu`

Summary. This paper is concerned with metamathematical properties of intuitionistic set theories with choice principles. It is proved that the *disjunction property*, the *numerical existence property*, *Church's rule*, and several other metamathematical properties hold true for constructive Zermelo–Fraenkel Set Theory and full intuitionistic Zermelo–Fraenkel augmented by any combination of the principles of countable choice, dependent choices, and the presentation axiom. Also Markov's principle may be added. Moreover, these properties hold effectively. For instance from a proof of a statement $\forall n \in \omega \, \exists m \in \omega \, \varphi(n, m)$, one can construct effectively an index e of a recursive function such that $\forall n \in \omega \, \varphi(n, \{e\}(n))$ is provable. Thus we have an explicit method of witness and program extraction from proofs involving choice principles.

As for the proof technique, this paper is a continuation of [29]. In [29], a self-validating semantics for **CZF** is introduced that combines realizability for extensional set theory and truth.

1 Introduction

The objective of this paper is to investigate several metamathematical properties of Constructive Zermelo–Fraenkel Set Theory, **CZF**, and Intuitionistic Zermelo–Fraenkel Set Theory, **IZF**, augmented by choice principles, and to provide an explicit method for extracting computational information from proofs of such theories.

IZF and **CZF** have the same language as **ZF**. Both theories are based on intuitionistic logic. Although **IZF** is squarely built on the idea of basing Zermelo–Fraenkel

---
* This material is based on work supported by the National Science Foundation under Award DMS-0301162.

set theory on intuitionistic logic, **CZF** is a standard reference theory for developing constructive predicative mathematics (cf. [1, 2, 3, 4]).

The axioms of **IZF** comprise Extensionality, Pairing, Union, Infinity, Separation, and Powerset. Instead of Replacement, **IZF** has Collection

$$\forall x \in a \, \exists y \, \varphi(x, y) \rightarrow \exists z \, \forall x \in a \, \exists y \in z \, \varphi(x, y),$$

and rather than Foundation, it has the Set Induction scheme

$$\forall x \, [\forall y \in x \, \psi(y) \rightarrow \psi(x)] \rightarrow \forall x \, \psi(x).$$

The set theoretic axioms of **CZF** are Extensionality, Pairing, Union, Infinity, the Set Induction scheme, and the following:

Bounded Separation scheme $\forall a \, \exists x \, \forall y \, (y \in x \leftrightarrow y \in a \wedge \varphi(y))$, for every *bounded* formula $\varphi(y)$, where a formula $\varphi(x)$ is bounded (or restricted or Δ_0) if all the quantifiers occurring in it are bounded, i.e., of the form $\forall x \in b$ or $\exists x \in b$.

Subset Collection scheme

$$\forall a \, \forall b \, \exists c \, \forall u \, [\forall x \in a \, \exists y \in b \, \psi(x, y, u)$$
$$\rightarrow \exists d \in c \, (\forall x \in a \, \exists y \in d \, \psi(x, y, u) \wedge \forall y \in d \, \exists x \in a \, \psi(x, y, u))].$$

Strong Collection scheme

$$\forall x \in a \, \exists y \, \varphi(x, y) \rightarrow \exists b \, [\forall x \in a \, \exists y \in b \, \varphi(x, y) \wedge \forall y \in b \, \exists x \in a \, \varphi(x, y)]$$

for all formulae $\psi(x, y, u)$ and $\varphi(x, y)$.

There are well-known metamathematical properties such as the disjunction and the numerical existence property that are often considered to be hallmarks of intuitionistic theories. The next definition gives a list of the well-known and some of the lesser-known metamathematical properties that intuitionistic theories may or may not have.

Definition 1.1 Let T be a theory whose language, $L(T)$, encompasses the language of set theory. Moreover, for simplicity, we shall assume that $L(T)$ has a constant ω denoting the set of von Neumann natural numbers and for each n a constant $\bar{n}$ denoting the n-th element of ω.[2]

1. T has the *disjunction property*, **DP**, if whenever $T \vdash \psi \vee \theta$ holds for sentences ψ and θ of T, then $T \vdash \psi$ or $T \vdash \theta$.

2. T has the *numerical existence property*, **NEP**, if whenever $T \vdash (\exists x \in \omega)\phi(x)$ holds for a formula $\phi(x)$ with at most the free variable x, then $T \vdash \phi(\bar{n})$ for some n.

---

[2] The usual language of set theory does not have numerals, strictly speaking. Instead of adding numerals to the language, one could take $\varphi(\bar{n})$ to mean $\exists x \, [\eta_n(x) \wedge \varphi(x)]$, where η_n is a formula defining the natural number n in a canonical way.

3. T has the *existence property*, **EP**, if whenever $T \vdash \exists x \phi(x)$ holds for a formula $\phi(x)$ having at most the free variable x, then there is a formula $\vartheta(x)$ with exactly x free, so that

$$T \vdash \exists! x \, [\vartheta(x) \wedge \phi(x)].$$

4. T has the *weak existence property*, **wEP**, if whenever

$$T \vdash \exists x \phi(x)$$

holds for a formula $\phi(x)$ having at most the free variable x, then there is a formula $\vartheta(x)$ with exactly x free, so that

$$T \vdash \exists! x \, \vartheta(x),$$
$$T \vdash \forall x \, [\vartheta(x) \rightarrow \exists u \, u \in x],$$
$$T \vdash \forall x \, [\vartheta(x) \rightarrow \forall u \in x \, \phi(x)].$$

5. T is closed under *Church's rule*, **CR**, if whenever $T \vdash (\forall x \in \omega)(\exists y \in \omega)\phi(x, y)$ holds for some formula of T with at most the free variables shown, then, for some number e,

$$T \vdash (\forall x \in \omega)\phi(x, \{\bar{e}\}(x)),$$

where $\{e\}(x)$ stands for the result of applying the e-th partial recursive function to x.

6. T is closed under the *Extended Church's rule*, **ECR**, if whenever

$$T \vdash (\forall x \in \omega)[\neg\psi(x) \rightarrow (\exists y \in \omega)\phi(x, y)]$$

holds for formulae of T with at most the free variables shown, then, for some number e,

$$T \vdash (\forall x \in \omega)[\neg\psi(x) \rightarrow \{\bar{e}\}(x) \in \omega \wedge \phi(x, \{\bar{e}\}(x))].$$

Note that $\neg\psi(x)$ could be replaced by any formula that is provably equivalent in T to its double negation. This comprises arithmetic formulae that are both $\vee$-free and $\exists$-free.

7. Let $f : \omega \rightarrow \omega$ convey that f is a function from ω to ω. T is closed under the variant of *Church's rule*, **CR$_1$**, if whenever $T \vdash \exists f \, [f : \omega \rightarrow \omega \wedge \psi(f)]$ (with $\psi(f)$ having no variables but f), then, for some number e, $T \vdash (\forall x \in \omega)(\exists y \in \omega)(\{\bar{e}\}(x) = y) \wedge \psi(\{\bar{e}\}))$.

8. T is closed under the *Unzerlegbarkeits rule*, **UZR**, if whenever $T \vdash \forall x [\psi(x) \vee \neg\psi(x)]$, then

$$T \vdash \forall x \, \psi(x) \quad \text{or} \quad T \vdash \forall x \, \neg\psi(x).$$

9. T is closed under the *Uniformity rule*, **UR**, if whenever $T \vdash \forall x\, (\exists y \in \omega)\psi(x, y)$, then

$$T \vdash (\exists y \in \omega)\, \forall x\, \psi(x, y).$$

Slightly abusing terminology, we shall also say that T enjoys any of these properties if this, strictly speaking, holds only for a definitional extension of T.

Actually, **DP** follows easily from **NEP**, and conversely, **DP** implies **NEP** for systems containing a modicum of arithmetic (see [13]).

Also note that **ECR** entails **CR**, taking $\psi(x)$ to be $x \neq x$.

A detailed historical account of metamathematical properties of intuitionistic set theories can be found in [29]. However, for the reader's convenience, we will quote from the preface to [29]:

"Realizability semantics are of paramount importance in the study of intuitionistic theories. They were first proposed by Kleene [17] in 1945. It appears that the first realizability definition for set theory was given by Tharp [33] who used (indices of) Σ_1 definable partial (class) functions as realizers. This form of realizability is a straightforward extension of Kleene's 1945 realizability for numbers in that a realizer for a universally quantified statement $\forall x\phi(x)$ is an index e of a Σ_1 partial function such that $\{e\}(x)$ is a realizer for $\phi(x)$ for all sets x. In the same vein, e realizes $\exists x\phi(x)$ if e is a pair $\langle a, e' \rangle$ with e' being a realizer for $\phi(a)$. A markedly different strand of realizability originates with Kreisel's and Troelstra's [21] definition of realizability for second-order Heyting arithmetic and the theory of species. Here, the clauses for the realizability relation $\Vdash$ relating to second-order quantifiers are as follows: $e \Vdash \forall X\phi(X) \Leftrightarrow \forall X\, e \Vdash \phi(X)$, $e \Vdash \exists X\phi(X) \Leftrightarrow \exists X\, e \Vdash \phi(X)$. This type of realizability does not seem to give any constructive interpretation to set quantifiers; realizing numbers "pass through" quantifiers. However, one could also say that thereby the collection of sets of natural numbers is conceived generically. On the intuitionistic view, the only way to arrive at the truth of a statement $\forall X\phi(X)$ is a proof. A collection of objects may be called generic if no member of it has an intensional aspect that can make any difference to a proof."

Kreisel–Troelstra realizability was applied to systems of higher order arithmetic and set theory by Friedman [12]. A realizability-notion akin to Kleene's slash [18, 19] was extended to various intuitionistic set theories by Myhill [26, 27]. In [26], it was shown that intuitionistic **ZF** with Replacement was shown instead of Collection (dubbed **IZF**$_R$ henceforth) has the **DP**, **NEP**, and **EP**. In [27], it was proved that the constructive set theory **CST** enjoys the **DP** and the **NEP**, and the theory without the axioms of countable and dependent choice, **CST$^-$**, also has the **EP**. It was left open in [27] whether the full existence property holds in the presence of relativized dependent choice, **RDC**. Friedman and Ščedrov [15] then established that **IZF**$_R$ + **RDC** satisfies the **EP** also. The Myhill–Friedman approach [26, 27] proceeds in two steps. The first, which appears to make the whole procedure noneffective, consists in finding a conservative extension T' of the given theory T, which contains names for all the objects asserted to exist in T. T' is obtained by inductively adding names and defining an increasing sequence of theories T_α through all the countable ordinals $\alpha < \omega_1$ and letting $T' = \bigcup_{\alpha < \omega_1} T_\alpha$.[3] The second step consists in defining a notion of realizability for T' that is a variant of Kleene's "slash."

Several systems of set theory for constructive mathematical practice were propounded by Friedman in [14]. The metamathematical properties of these theories and several others as well were subsequently investigated by Beeson [5, 6]. In particular, Beeson showed that **IZF** has the **DP** and **NEP**. He used a combination of Kreisel–Troelstra realizability and Kleene's [17, 18, 19, 20] q-realizability. However, while Myhill and Friedman developed realizability directly for extensional set theories, Beeson engineered his realizability for nonextensional set theories and obtained results for the extensional set theories of [14] only via an interpretation in their nonextensional counterparts. This detour had the disadvantage that in many cases (where the theory does not have full Separation or Powerset), the **DP** and **NEP** for the corresponding extensional set theory T-ext could only be established for a restricted class of formulas; [5] Theorem 5.2 proves that **NEP** holds for T-ext when T-ext $\vdash (\exists x \in \omega)(x \in Q)$, where Q is a definable set of T. It appears unlikely that the Myhill–Friedman techniques or Beeson's detour through q-realizability for nonextensional set theories can be employed to yield the **DP** and **NEP** for **CZF**. The theories considered by Myhill and Friedman have Replacement instead of Collection, and in all probability, their approach is limited to such theories, whereas Beeson's techniques yield numerical explicit definability, not for all formulae $\varphi(u)$, but only for $\varphi(u)$ of the form $u \in Q$, where Q is a specific definable set. But there was another approach available. McCarty [23, 24] adapted Kreisel–Troelstra realizability directly to extensional set theories. In [23, 24], though, they were concerned with realizability for intuitionistic Zermelo–Fraenkel set theory (having Collection instead of Replacement), **IZF**, and employed transfinite iterations of the powerset operation through all the ordinals in defining a realizability (class) structure. Moreover, in addition to the powerset axiom, this approach also availed itself of unfettered separation axioms. At first blush, this seemed to render the approach unworkable for **CZF** as this theory lacks the powerset axiom and has only bounded separation. Notwithstanding that, it was shown in [30] that these obstacles can be overcome. Indeed, this notion of realizability provides a self-validating semantics for **CZF**, viz. it can be formalized in **CZF**, and demonstrably in **CZF**, it can be verified that every theorem of **CZF** is realized." ([29], pp. 1234–1236.)

The paper [29] introduced a new realizability structure V^*, which arises by amalgamating the realizability structure with the universe of sets in a coherent, albeit rather complicated, way. The main semantical notion presented and used in [29] combines realizability for extensional set theory over V^* with truth in the background universe V. A combination of realizability with truth has previously been considered in the context of realizability notions for first and higher order arithmetic. It was called <u>rnt</u>-

---

[3] This type of construction is due to J.R. Moschovakis [25] §8 & 9.

realizability in [34]. The main metamathematical result obtained via this tool was as follows.

Theorem 1.2 *The* **DP** *and the* **NEP** *hold true for* **CZF** *and* **CZF** + **REA**. *Both theories are closed under* **CR**, **ECR**, **CR**$_1$, **UZR**, *and* **UR**, *too.*

Proof. [29], Theorem 1.2.   □

This paper presents another proof of Beeson's result that **IZF** has the **DP** and the **NEP** and a proof that **IZF** is closed under **CR**, **ECR**, **CR**$_1$, **UZR**, and **UR**. There are a number of further metamathematical results that can be obtained via this technology. For example, it will be shown that Markov's principle can be added to any of the foregoing theories. But the main bulk of this paper is devoted to showing that the technology is particularly suited to the choice principles of countable choice, dependent choices, and the presentation axiom. As a consequence, we can deduce that **CZF** augmented by any combination of these principles also has the properties stated in Theorem 1.2. The same holds for **IZF**.

2 Choice principles

In many a text on constructive mathematics, axioms of countable choice and dependent choices are accepted as constructive principles. This is, for instance, the case in Bishop's constructive mathematics (cf. [8]) as well as in Brouwer's intuitionistic analysis (cf. [35], Chap. 4, Sect. 2). Myhill also incorporated these axioms in his constructive set theory [27].

The weakest constructive choice principle we shall consider is the *Axiom of Countable Choice*, **AC**$_\omega$; i.e., whenever F is a function with domain ω such that $\forall i \in \omega\, \exists y \in F(i)$, then there exists a function f with domain ω such that $\forall i \in \omega\, f(i) \in F(i)$.

Let xRy stand for $\langle x, y \rangle \in R$. A mathematically very useful axiom to have in set theory is the **Dependent Choices Axiom**, **DC**; i.e., for all sets a and (set) relations $R \subseteq a \times a$, whenever

$$(\forall x \in a)\, (\exists y \in a)\, xRy$$

and $b_0 \in a$, then there exists a function $f : \omega \to a$ such that $f(0) = b_0$ and

$$(\forall n \in \omega)\, f(n)Rf(n+1).$$

Even more useful in constructive set theory is the *Relativized Dependent Choices Axiom*, **RDC**. It asserts that for arbitrary formulae ϕ and ψ, whenever

$$\forall x \big[\phi(x) \to \exists y \big(\phi(y) \wedge \psi(x,y)\big)\big]$$

and $\phi(b_0)$, then there exists a function f with domain ω such that $f(0) = b_0$ and

$$(\forall n \in \omega)\big[\phi(f(n)) \wedge \psi(f(n), f(n+1))\big].$$

Let $\mathbf{CZF}^-$ be $\mathbf{CZF}$ without Subset Collection.

Proposition 2.1 *Provably in* $\mathbf{CZF}^-$ *the following hold:*

(i) $\mathbf{DC}$ *implies* $\mathbf{AC}_\omega$.

(ii) $\mathbf{RDC}$ *implies* $\mathbf{DC}$.

Proof. This is a well-known fact. □

The *presentation axiom*, $\mathbf{PAx}$, is an example of a choice principle that is validated on interpretation in type theory. In category theory it is also known as the *existence of enough projective sets*, $\mathbf{EPsets}$ (cf. [7]). In a category $\mathbb{C}$, an object P in $\mathbb{C}$ is *projective* (in $\mathbb{C}$) if for all objects A, B in $\mathbb{C}$, and morphisms $A \xrightarrow{f} B$, $P \xrightarrow{g} B$ with f an epimorphism, there exists a morphism $P \xrightarrow{h} A$ such that the following diagram commutes:

It easily follows that in the category of sets, a set P is projective if for any P-indexed family $(X_a)_{a \in P}$ of inhabited sets X_a, there exists a function f with domain P such that, for all $a \in P$, $f(a) \in X_a$.

$\mathbf{PAx}$ (or $\mathbf{EPsets}$) is the statement that every set is the surjective image of a projective set.

Alternatively, projective sets have also been called *bases*, and we shall follow that usage henceforth. In this terminology, $\mathbf{AC}_\omega$ expresses that ω is a base, whereas $\mathbf{AC}$ amounts to saying that every set is a base.

Proposition 2.2 $(\mathbf{CZF}^-)$ $\mathbf{PAx}$ *implies* $\mathbf{DC}$.

Proof. See [1] or [7], Theorem 6.2. □

The implications of Propositions 2.1 and 2.2 cannot be reversed, not even on the basis of $\mathbf{ZF}$.

Proposition 2.3 $\mathbf{ZF} + \mathbf{DC}$ *does not prove* $\mathbf{PAx}$.

Proof. See [31] Proposition 5.2. □

3 The partial combinatory algebra K1

In order to define a realizability interpretation we must have a notion of realizing functions on hand. A particularly general and elegant approach to realizability builds on structures that have been variably called *partial combinatory algebras*, *applicative structures*, or *Schönfinkel algebras*. These structures are best described as the models of a theory **APP** (cf. [10, 11, 6, 35]). The language of **APP** is a first-order language with a ternary relation symbol App, a unary relation symbol N (for a copy of the natural numbers), and equality, $=$, as primitives. The language has an infinite collection of variables, denoted $x, y, z, \ldots$, and nine distinguished constants: $0, s_N, p_N, k, s, d, p, p_0, p_1$ for, respectively, zero, successor on N, predecessor on N, the two basic combinators, definition by cases, pairing, and the corresponding two projections. No arity is associated with the various constants. The *terms* of **APP** are just the variables and constants. We write $t_1 t_2 \simeq t_3$ for $\text{App}(t_1, t_2, t_3)$.

Formulae are then generated from atomic formulae using the propositional connectives and the quantifiers.

In order to facilitate the formulation of the axioms, the language of **APP** is expanded definitionally with the symbol $\simeq$ and the auxiliary notion of an *application term* is introduced. The set of application terms is given by two clauses:

1. all terms of **APP** are application terms; and

2. if s and t are application terms, then (st) is an application term.

For s and t application terms, we have auxiliary, defined formulae of the form:

$$s \simeq t \; := \; \forall y (s \simeq y \leftrightarrow t \simeq y),$$

if t is not a variable. Here $s \simeq a$ (for a a free variable) is inductively defined by:

$$s \simeq a \; \text{is} \; \begin{cases} s = a, & \text{if } s \text{ is a term of } \mathbf{APP}, \\ \exists x, y [s_1 \simeq x \wedge s_2 \simeq y \wedge \text{App}(x, y, a)] & \text{if } s \text{ is of the form } (s_1 s_2). \end{cases}$$

Some abbreviations are $t_1 t_2 \ldots t_n$ for $((\ldots (t_1 t_2) \ldots) t_n)$; $t \downarrow$ for $\exists y (t \simeq y)$ and $\phi(t)$ for $\exists y (t \simeq y \wedge \phi(y))$.

Some further conventions are useful. Systematic notation for n-*tuples* is introduced as follows: (t) is t, (s, t) is $\text{p} s t$, and $(t_1, \ldots, t_n)$ is defined by $((t_1, \ldots, t_{n-1}), t_n)$. In this paper, the **logic** of **APP** is assumed to be that of intuitionistic predicate logic with identity. **APP**'s **nonlogical axioms** are the following:

Applicative Axioms

1. $\text{App}(a, b, c_1) \wedge \text{App}(a, b, c_2) \rightarrow c_1 = c_2.$

2. $(\mathbf{k}ab) \downarrow \wedge \mathbf{k}ab \simeq a$.

3. $(\mathbf{s}ab) \downarrow \wedge \mathbf{s}abc \simeq ac(bc)$.

4. $(\mathbf{p}a_0a_1) \downarrow \wedge (\mathbf{p}_0a) \downarrow \wedge (\mathbf{p}_1a) \downarrow \wedge \mathbf{p}_i(\mathbf{p}a_0a_1) \simeq a_i$ for $i = 0, 1$.

5. $N(c_1) \wedge N(c_2) \wedge c_1 = c_2 \to \mathbf{d}abc_1c_2 \downarrow \wedge \mathbf{d}abc_1c_2 \simeq a$.

6. $N(c_1) \wedge N(c_2) \wedge c_1 \ne c_2 \to \mathbf{d}abc_1c_2 \downarrow \wedge \mathbf{d}abc_1c_2 \simeq b$.

7. $\forall x \left(N(x) \to \left[\mathbf{s}_N x \downarrow \wedge \mathbf{s}_N x \ne \mathbf{0} \wedge N(\mathbf{s}_N x) \right] \right)$.

8. $N(\mathbf{0}) \wedge \forall x \left(N(x) \wedge x \ne \mathbf{0} \to \left[\mathbf{p}_N x \downarrow \wedge \mathbf{s}_N(\mathbf{p}_N x) = x \right] \right)$.

9. $\forall x \left[N(x) \to \mathbf{p}_N(\mathbf{s}_N x) = x \right]$.

10. $\varphi(\mathbf{0}) \wedge \forall x \left[N(x) \wedge \varphi(x) \to \varphi(\mathbf{s}_N x) \right] \to \forall x \left[N(x) \to \varphi(x) \right]$.

Let $\mathbf{1} := \mathbf{s}_N \mathbf{0}$. The applicative axioms entail that $\mathbf{1}$ is an application term that evaluates to an object falling under N but is distinct from $\mathbf{0}$; i.e., $\mathbf{1} \downarrow$, $N(\mathbf{1})$ and $\mathbf{0} \ne \mathbf{1}$.

Employing the axioms for the combinators $\mathbf{k}$ and $\mathbf{s}$ one can deduce an abstraction lemma yielding λ-terms of one argument. This can be generalized using n–tuples and projections.

Lemma 3.1 ((cf. [10]) (Abstraction Lemma)) *For each application term t there is a new application term t^* such that the parameters of t^* are among the parameters of t minus $x_1, \ldots, x_n$ and such that*

$$\mathbf{APP} \vdash t^* \downarrow \wedge t^* x_1 \ldots x_n \simeq t.$$

$\lambda(x_1, \ldots, x_n).t$ *is written for t^*.*

The most important consequence of the Abstraction Lemma is the Recursion Theorem. It can be derived in the same way as for the λ–calculus (cf. [10], [11], [6], VI.2.7). Actually, one can prove a uniform version of the following in **APP**.

Corollary 3.2 (Recursion Theorem)

$$\forall f \exists g \forall x_1 \ldots \forall x_n \, g(x_1, \ldots, x_n) \simeq f(g, x_1, \ldots, x_n).$$

The "standard" applicative structure is $\mathbf{Kl}$ in which the universe $|\mathbf{Kl}|$ is ω and $\mathrm{App}^{\mathbf{Kl}}(x, y, z)$ is the Turing machine application:

$$\mathrm{App}^{\mathbf{Kl}}(x, y, z) \text{ iff } \{x\}(y) \simeq z.$$

The primitive constants of **APP** are interpreted over $|\mathbf{Kl}|$ in the obvious way. Thus there are nine distinguished elements $\mathbf{0}^{Kl}, \mathbf{s}_N^{Kl}, \mathbf{p}_N^{Kl}, \mathbf{k}^{Kl}, \mathbf{s}^{Kl}, \mathbf{d}^{Kl}, \mathbf{p}^{Kl}, \mathbf{p_0}^{Kl}, \mathbf{p_1}^{Kl}$ of ω pertaining to the axioms of **APP**. For details, see [23], chap. 3, sec. 2, or [6], VI.2.7. In the following we will be solely concerned with the standard applicative structure **Kl**. We will also be assuming that the notion of an applicative structure and in particular the structure **Kl** have been formalized in **CZF**, and that **CZF** proves that **Kl** is a model of **APP**. We will usually drop the superscript "^{Kl}" when referring to any of the special constants of **Kl**.

4 The general realizability structure

If a is an ordered pair, i.e., $a = \langle x, y \rangle$ for some sets x, y, then we use $1^{st}(a)$ and $2^{nd}(a)$ to denote the first and second projection of a, respectively; that is, $1^{st}(a) = x$ and $2^{nd}(a) = y$. For a class X we denote by $\mathcal{P}(X)$ the class of all sets y such that $y \subseteq X$.

Definition 4.1 Ordinals are transitive sets whose elements are transitive also. As usual, we use lowercase Greek letters to range over ordinals.

$$V_\alpha^* = \bigcup_{\beta \in \alpha} \{ \langle a, b \rangle : a \in V_\beta;\ b \subseteq \omega \times V_\beta^*;\ (\forall x \in b)\, 1^{st}(2^{nd}(x)) \in a \}, \quad (1)$$

$$V_\alpha = \bigcup_{\beta \in \alpha} \mathcal{P}(V_\beta),$$

$$V^* = \bigcup_\alpha V_\alpha^*,$$

$$V = \bigcup_\alpha V_\alpha.$$

As the powerset operation is not available in **CZF**, it is not clear whether the classes V and V^* can be formalized in **CZF**. However, employing the fact that **CZF** accommodates inductively defined classes, this can be demonstrated in the same vein as in [30], Lemma 3.4.

The definition of V_α^* in (1) is perhaps a bit involved. Note first that all the elements of V^* are ordered pairs $\langle a, b \rangle$ such that $b \subseteq \omega \times V^*$. For an ordered pair $\langle a, b \rangle$ to enter V_α^* the first conditions to be met are that $a \in V_\beta$ and $b \subseteq \omega \times V_\beta^*$ for some $\beta \in \alpha$. Furthermore, it is required that a contains enough elements from the transitive closure of b in that whenever $\langle e, c \rangle \in b$, then $1^{st}(c) \in a$.

Lemma 4.2 (CZF)

(i) V and V^* *are cumulative: for* $\beta \in \alpha$, $V_\beta \subseteq V_\alpha$ *and* $V_\beta^* \subseteq V_\alpha^*$.

(ii) For all sets a, $a \in V$.

(iii)If a, b are sets, $b \subseteq \omega \times V^*$ and $(\forall x \in b)\, 1^{st}(2^{nd}(x)) \in a$, then $\langle a, b \rangle \in V^*$.

Proof. [29], Lemma 4.2.   □

5 Defining realizability

We now proceed to define a notion of realizability over V^*. We use lowercase gothic letters $\mathfrak{a}, \mathfrak{b}, \mathfrak{c}, \mathfrak{d}, \mathfrak{e}, \mathfrak{f}, \mathfrak{g}, \mathfrak{h}, \mathfrak{n}, \mathfrak{m}, \mathfrak{p}, \mathfrak{q} \dots$ as variables to range over elements of V^*, whereas variables $e, c, d, f, g, \dots$ will be reserved for elements of ω. Each element $\mathfrak{a}$ of V^* is an ordered pair $\langle x, y \rangle$, where $x \in V$ and $y \subseteq \omega \times V^*$; and we define the components of $\mathfrak{a}$ by

$$\mathfrak{a}^\circ := 1^{st}(\mathfrak{a}) = x,$$

$$\mathfrak{a}^* := 2^{nd}(\mathfrak{a}) = y.$$

Lemma 5.1 *For every $\mathfrak{a} \in V^*$, if $\langle e, \mathfrak{c} \rangle \in \mathfrak{a}^*$, then $\mathfrak{c}^\circ \in \mathfrak{a}^\circ$.*

Proof. This is immediate by the definition of V^*. □

If φ is a sentence with parameters in V^*, then φ° denotes the formula obtained from φ by replacing each parameter $\mathfrak{a}$ in φ with $\mathfrak{a}^\circ$.

Definition 5.2 Bounded quantifiers will be treated as quantifiers in their own right; i.e., bounded and unbounded quantifiers are treated as syntactically different kinds of quantifiers.

We define $e \Vdash_{rt} \phi$ for sentences ϕ with parameters in V^*. (The subscript rt is supposed to serve as a reminder of "realizability with truth.")

We shall use the abbreviations (x, y), $(x)_0$, and $(x)_1$ for $\mathbf{p}xy$, $\mathbf{p}_0 x$, and $\mathbf{p}_1 x$, respectively.

$$e \Vdash_{rt} \mathfrak{a} \in \mathfrak{b} \text{ iff } \mathfrak{a}^\circ \in \mathfrak{b}^\circ \ \wedge \ \exists \mathfrak{c} \left[\langle (e)_0, \mathfrak{c} \rangle \in \mathfrak{b}^* \ \wedge \ (e)_1 \Vdash_{rt} \mathfrak{a} = \mathfrak{c} \right]$$

$$e \Vdash_{rt} \mathfrak{a} = \mathfrak{b} \text{ iff } \mathfrak{a}^\circ = \mathfrak{b}^\circ \ \wedge \ \forall f \forall \mathfrak{c} \left[\langle f, \mathfrak{c} \rangle \in \mathfrak{a}^* \ \rightarrow \ (e)_0 f \Vdash_{rt} \mathfrak{c} \in \mathfrak{b} \right]$$

$$\wedge \ \forall f \forall \mathfrak{c} \left[\langle f, \mathfrak{c} \rangle \in \mathfrak{b}^* \ \rightarrow \ (e)_1 f \Vdash_{rt} \mathfrak{c} \in \mathfrak{a} \right]$$

$$e \Vdash_{rt} \phi \wedge \psi \text{ iff } (e)_0 \Vdash_{rt} \phi \ \wedge \ (e)_1 \Vdash_{rt} \psi$$

$$e \Vdash_{rt} \phi \vee \psi \text{ iff } \left[(e)_0 = 0 \ \wedge \ (e)_1 \Vdash_{rt} \phi \right] \ \vee \ \left[(e)_0 \neq 0 \ \wedge \ (e)_1 \Vdash_{rt} \psi \right]$$

$$e \Vdash_{rt} \neg \phi \quad \text{iff } \neg \phi^\circ \ \wedge \ \forall f \ \neg f \Vdash_{rt} \phi$$

$$e \Vdash_{rt} \phi \rightarrow \psi \text{ iff } (\phi^\circ \rightarrow \psi^\circ) \ \wedge \ \forall f \left[f \Vdash_{rt} \phi \ \rightarrow \ ef \Vdash_{rt} \psi \right]$$

$$e \Vdash_{rt} (\forall x \in \mathfrak{a}) \, \phi \text{ iff } (\forall x \in \mathfrak{a}^\circ) \phi^\circ \ \wedge$$

$$\forall f \forall \mathfrak{b} \left(\langle f, \mathfrak{b} \rangle \in \mathfrak{a}^* \ \rightarrow \ ef \Vdash_{rt} \phi[x/\mathfrak{b}] \right)$$

$$e \Vdash_{rt} (\exists x \in \mathfrak{a}) \phi \text{ iff } \exists \mathfrak{b} \left(\langle (e)_0, \mathfrak{b} \rangle \in \mathfrak{a}^* \ \wedge \ (e)_1 \Vdash_{rt} \phi[x/\mathfrak{b}] \right)$$

$$e \Vdash_{rt} \forall x \phi \quad \text{iff } \forall \mathfrak{a} \ e \Vdash_{rt} \phi[x/\mathfrak{a}]$$

$$e \Vdash_{rt} \exists x \phi \quad \text{iff } \exists \mathfrak{a} \ e \Vdash_{rt} \phi[x/\mathfrak{a}].$$

Notice that $e \Vdash_{rt} u \in v$ and $e \Vdash_{rt} u = v$ can be defined for arbitrary sets u, v, viz., not just for $u, v \in \mathrm{V}^*$. The definitions of $e \Vdash_{rt} u \in v$ and $e \Vdash_{rt} u = v$ fall under the scope of definitions by transfinite recursion.

Definition 5.3 By $\in$-recursion we define for every set x a set x^{st} as follows:

$$x^{st} = \langle x, \{\langle 0, u^{st} \rangle \, : \, u \in x\}\rangle. \tag{2}$$

Lemma 5.4 *For all sets x, $x^{st} \in \mathrm{V}^*$ and $(x^{st})^\circ = x$.*

Proof. [29], Lemma 5.4.   □

Lemma 5.5 *If $\psi(\mathfrak{b}^\circ)$ holds for all $\mathfrak{b} \in \mathrm{V}^*$, then $\forall x\, \psi(x)$.*

Proof. [29], Lemma 5.5.   □

Lemma 5.6 *If $\mathfrak{a} \in \mathrm{V}^*$ and $(\forall \mathfrak{b} \in \mathrm{V}^*)[\mathfrak{b}^\circ \in \mathfrak{a}^\circ \to \psi(\mathfrak{b}^\circ)]$, then $(\forall x \in \mathfrak{a}^\circ)\psi(x)$.*

Proof. [29], Lemma 5.6.   □

Lemma 5.7 *If $e \Vdash_{rt} \phi$, then ϕ°.*

Proof. [29], Lemma 5.7.   □

Our hopes for showing **DP** and **NEP** for **CZF** and related systems rest on the following results.

Lemma 5.8 *If $e \Vdash_{rt} (\exists x \in \mathfrak{a})\phi$, then*

$$\exists \mathfrak{b} \left(\langle (e)_0, \mathfrak{b} \rangle \in \mathfrak{a}^* \, \wedge \, \phi^\circ[x/\mathfrak{b}^\circ] \right).$$

Proof. Obvious by 5.7.   □

Lemma 5.9 *If $e \Vdash_{rt} \phi \vee \psi$, then*

$$\left[(e)_0 = 0 \, \wedge \, \phi^\circ \right] \, \vee \, \left[(e)_0 \neq 0 \, \wedge \, \psi^\circ \right].$$

Proof. Obvious by 5.7.   □

Lemma 5.10 *Negated formulae are self-realizing, which is to say, if ψ is a statement with parameters in V^*, then*

$$\neg\psi^\circ \, \to \, 0 \Vdash_{rt} \neg\psi.$$

Proof. Assume $\neg\psi^\circ$. From $f \Vdash_{rt} \psi$ we would get ψ° by Lemma 5.8. But this is absurd. Hence $\forall f \, \neg f \Vdash_{rt} \psi$, and therefore, $0 \Vdash_{rt} \neg\psi$.   □

Definition 5.11 Let t be an application term and ψ be a formula of set theory. Then $t \Vdash_{rt} \psi$ is short for $(\exists e \in \omega)[t \simeq e \, \wedge \, e \Vdash_{rt} \psi]$.

Theorem 5.12 *For every theorem θ of* **CZF**, *there exists a closed application term t such that*

$$\mathbf{CZF} \vdash (t \Vdash_{rt} \theta).$$

Moreover, the proof of this soundness theorem is effective in that the application term t can be effectively constructed from the **CZF** *proof of θ.*

Proof. [29], Theorem 6.1.   □

Remark 5.13 Theorem 5.12 holds also for **CZF** augmented by other large set axioms such as *"Every set is contained in an inaccessible set"* or *"Every set is contained in a Mahlo set."* For definitions of "inaccessible set" and "Mahlo set" see [4, 9]. For example, in the case of the so-called regular extension axiom, this was carried out in [29], Theorem 7.2.

6 Extending the interpretation to IZF

In this section we address several extensions of earlier results. We show that in Theorem 5.12 **CZF** can be replaced by **IZF** and also that Markov's principle may be added.

Theorem 6.1 *For every theorem θ of* **IZF**, *there exists an application term t such that*

$$\mathbf{IZF} \vdash (t \Vdash_{rt} \theta).$$

Moreover, the proof of this soundness theorem is effective in that the application term t can be effectively constructed from the **IZF** *proof of θ.*

Proof. In view of Theorem 5.12 we only need to show that **IZF** proves that the Powerset Axiom and the Full Separation Axiom are realized with respect to $\Vdash_{rt}$.

(Full Separation): Let $\varphi(x)$ be an arbitrary formula with parameters in V^*. We want to find $e, e' \in \omega$ such that for all $\mathfrak{a} \in V^*$, there exists a $\mathfrak{b} \in V^*$ such that

$$\left(e \Vdash_{rt} \forall x \in \mathfrak{b}\,[x \in \mathfrak{a} \wedge \varphi(x)]\right) \wedge \left(e' \Vdash_{rt} \forall x \in \mathfrak{a}[\varphi(x) \to x \in \mathfrak{b}]\right). \tag{3}$$

For $\mathfrak{a} \in V^*$, define

$$\mathrm{Sep}(\mathfrak{a}, \varphi) = \{\langle \mathbf{p}fg, \mathfrak{c}\rangle : f, g \in \omega \wedge \langle g, \mathfrak{c}\rangle \in \mathfrak{a}^* \wedge f \Vdash_{rt} \varphi[x/\mathfrak{c}]\},$$

$$\mathfrak{b} = \langle \{x \in \mathfrak{a}^\circ : \varphi^\circ(x)\}, \mathrm{Sep}(\mathfrak{a}, \varphi)\rangle.$$

$\mathrm{Sep}(\mathfrak{a}, \varphi)$ is a set by full separation, and hence, $\mathfrak{b}$ is a set. To ensure that $\mathfrak{b} \in V^*$, let $\langle h, \mathfrak{c}\rangle \in \mathrm{Sep}(\mathfrak{a}, \varphi)$. Then $\langle g, \mathfrak{c}\rangle \in \mathfrak{a}^*$ and $f \Vdash_{rt} \varphi[x/\mathfrak{c}]$ for some $f, g \in \omega$. Thus $\mathfrak{c}^\circ \in \mathfrak{a}^\circ$, and by Lemma 5.7, $\varphi^\circ[x/\mathfrak{c}^\circ]$, yielding $\mathfrak{c}^\circ \in \{x \in \mathfrak{a}^\circ : \varphi^\circ(x)\}$. Therefore, by Lemma 4.2, we have $\mathfrak{b} \in V^*$.

To verify (3), first assume $\langle h, \mathfrak{c}\rangle \in \mathfrak{b}^*$ and $\mathfrak{c}^\circ \in \mathfrak{b}^\circ$. Then $h = \mathbf{p}fg$ for some $f, g \in \omega$ and $\langle g, \mathfrak{c}\rangle \in \mathfrak{a}^*$ and $f \Vdash_{rt} \varphi[x/\mathfrak{c}]$. Since $\mathfrak{c}^\circ \in \mathfrak{b}^\circ$ holds, it follows that $\mathfrak{c}^\circ \in \mathfrak{a}^\circ$. As a result, $\mathfrak{c}^\circ \in \mathfrak{a}^\circ \wedge \langle g, \mathfrak{c}\rangle \in \mathfrak{a}^* \wedge \mathbf{i}_r \Vdash_{rt} \mathfrak{c} = \mathfrak{c}$, and consequently, we have $\mathbf{p}(h)_1\mathbf{i}_r \Vdash_{rt} \mathfrak{b} \in \mathfrak{a}$ and $(h)_0 \Vdash_{rt} \varphi[x/\mathfrak{c}]$, where $\mathbf{i}_r$ is the realizer of the identity axiom $\forall x \, x = x$ (see [23], Chapter 2, sections 5 and 6). Moreover, we have $(\forall x \in \mathfrak{b}^\circ)(x \in \mathfrak{a}^\circ \wedge \varphi^\circ(x))$. Therefore with $e = \mathbf{p}(\mathbf{p}(\lambda u.(u)_1)\mathbf{i}_r)(\lambda u.(u)_0)$, we get $e \Vdash_{rt} \forall x \in \mathfrak{b} \, [x \in \mathfrak{a} \wedge \varphi(x)]$.

Now assume $\langle g, \mathfrak{c}\rangle \in \mathfrak{a}$, $\mathfrak{c}^\circ \in \mathfrak{a}^\circ$ and $f \Vdash_{rt} \varphi[x/\mathfrak{c}]$. Then $\langle \mathbf{p}fg, \mathfrak{c}\rangle \in \mathfrak{b}^*$ and also $\mathfrak{c}^\circ \in \mathfrak{b}^\circ$ as $\varphi^\circ[x/\mathfrak{c}^\circ]$ is a consequence of $f \Vdash_{rt} \varphi[x/\mathfrak{c}]$ by Lemma 5.7. Therefore $\mathbf{p}(\mathbf{p}fg)\mathbf{i}_r \Vdash_{rt} \mathfrak{c} \in \mathfrak{b}$. Finally, by the very definition of $\mathfrak{b}$, we have $(\forall x \in \mathfrak{a}^\circ)[\varphi^\circ(x) \to x \in \mathfrak{b}^\circ]$, and hence with $e' = \lambda u.\lambda v.\mathbf{p}(\mathbf{p}vu)\mathbf{i}_r$, we get $e' \Vdash_{rt} (\forall x \in \mathfrak{a})[\varphi(x) \to x \in \mathfrak{b}]$.

(Powerset): It suffices to find a realizer for the formula $\forall x \, \exists y \, \forall z \, [z \subseteq x \to z \in y]$ as it implies the Powerset Axiom with the aid of Bounded Separation. Let $\mathfrak{a} \in V^*$. Put $\mathcal{A} = \{\mathfrak{d} : \exists g \, \langle g, \mathfrak{d}\rangle \in \mathfrak{a}^*\}$. For $y \subseteq \omega \times \mathcal{A}$, let

$$\mathfrak{a}_y := \langle \{\mathfrak{c}^\circ : \exists f \, \langle f, \mathfrak{c}\rangle \in y\}, y\rangle.$$

Note that $\mathfrak{a}_y \in V^*$. The role of a set large enough to comprise the powerset of $\mathfrak{a}$ in V^* will be played by the following set:

$$\mathfrak{p} := \langle \mathcal{P}(\mathfrak{a}^\circ), \{\langle 0, \mathfrak{a}_y\rangle : y \subseteq \omega \times \mathcal{A}\}\rangle.$$

$\mathfrak{p}$ is a set in our background theory **IZF**. For $\langle 0, \mathfrak{a}_y\rangle \in \mathfrak{p}^*$, we have $\mathfrak{a}_y^\circ \subseteq \mathfrak{a}^\circ$, and thus $\mathfrak{a}_y^\circ \in \mathcal{P}(\mathfrak{a}^\circ)$, so it follows that $\mathfrak{p} \in V^*$.

Now suppose $e \Vdash_{rt} \mathfrak{b} \subseteq \mathfrak{a}$. Put

$$y_\mathfrak{b} := \{\langle (d, f), \mathfrak{x}\rangle : d, f \in \omega \wedge \langle (df)_0, \mathfrak{x}\rangle \in \mathfrak{a}^* \wedge \exists \mathfrak{c} \, [\langle d, \mathfrak{c}\rangle \in \mathfrak{b}^* \wedge (df)_1 \Vdash_{rt} \mathfrak{x} = \mathfrak{c}]\}. \tag{4}$$

(Recall that (x, y) stands for $\mathbf{p}xy$.) By definition of $y_\mathfrak{b}$, $y_\mathfrak{b} \subseteq \omega \times \mathcal{A}$, and therefore, $\langle 0, \mathfrak{a}_{y_\mathfrak{b}}\rangle \in \mathfrak{p}^*$.

If $\langle f, \mathfrak{c}\rangle \in \mathfrak{b}^*$, it follows that $ef \Vdash_{rt} \mathfrak{c} \in \mathfrak{a}$ since $e \Vdash_{rt} \mathfrak{b} \subseteq \mathfrak{a}$; and hence, there exists $\mathfrak{x}$ such that $\langle (ef)_0, \mathfrak{x}\rangle \in \mathfrak{a}^*$ and $(ef)_1 \Vdash_{rt} \mathfrak{x} = \mathfrak{c}$ when $\langle (e, f), \mathfrak{x}\rangle \in y_\mathfrak{b}$ and therefore $((e, f), (ef)_1) \Vdash_{rt} \mathfrak{c} \in \mathfrak{a}_{y_\mathfrak{b}}$. Thus we can infer that $\lambda f.((e, f), (ef)_1) \Vdash_{rt} \mathfrak{b} \subseteq \mathfrak{a}_{y_\mathfrak{b}}$.

Conversely, if $\langle g, \mathfrak{x}\rangle \in \mathfrak{a}_{y_\mathfrak{b}}^* = y_\mathfrak{b}$, then there exist d, f, and $\mathfrak{c}$ such that $g = (d, f)$, $\langle d, \mathfrak{c}\rangle \in \mathfrak{b}^*$, and $(df)_1 \Vdash_{rt} \mathfrak{c} = \mathfrak{x}$, which entails that $((g)_0, ((g)_0(g)_1)_1) \Vdash_{rt} \mathfrak{x} \in \mathfrak{b}$. As a result, $\eta(e) \Vdash_{rt} \mathfrak{b} = \mathfrak{a}_{y_\mathfrak{b}}$, where $\eta(e) = (\lambda f.((e, f), (ef)_1), \lambda g.((g)_0, ((g)_0(g)_1)_1))$. Hence $(0, \eta(e)) \Vdash_{rt} \mathfrak{b} \in \mathfrak{p}$, so that

$$\lambda e.(0, \eta(e)) \Vdash_{rt} \forall y \, [y \subseteq \mathfrak{a} \to y \in \mathfrak{p}],$$

and therefore, by the genericity of quantifiers,

$$\lambda e.(0, \eta(e)) \Vdash_{rt} \forall x \, \exists y \, \forall z \, [z \subseteq x \to z \in y]. \tag{5}$$

$\square$

Invoking Theorem 7.2 of [29], we may replace **IZF** in the foregoing theorem by **IZF + REA**. Inspection of the preceding proof shows also that Full Separation is self-validating so that the theory **CZF** + Full Separation is self-validating. The same is true of the Powerset Axiom in tandem with Bounded Separation. So we also get the following result:

Corollary 6.2 *Let* T *be any of the theories* **CZF** + Powerset, **CZF** + Full Separation, **CZF + REA** + Powerset, **CZF + REA** + Full Separation, *or* **IZF + REA**. *Then for every theorem* θ *of* T, *there exists an application term* t *such that*

$$T \vdash (t \Vdash_{rt} \theta).$$

More than that, the proof of this soundness theorem is effective in that the application term t *can be effectively constructed from the* T *proof of* θ.

Theorem 6.3 **IZF** *has the* **DP** *and* **NEP**, *and* **IZF** *is closed under* **CR**, **ECR**, **CR$_1$**, **UZR**, *and* **UR**, *too. The same is true of the theories of Corollary 6.2.*

Proof. This follows from Theorem 6.1 and Corollary 6.2 by the proof of [29], Theorem 1.2.   □

Remark 6.4 The preceding Theorem 6.3 and [29], Theorem 1.2, allow for generalizations to extensions of **CZF**, **CZF + REA**, **IZF** (and the theories of Corollary 6.2) via "true" axioms of the form $\neg\psi$. This follows easily from the proofs of these theorems and the fact that negated statements are self-realizing (see Lemma 5.10). As a consequence, we get, for example, that if $\neg\vartheta$ is a true sentence and **CZF** $\vdash$ $\neg\vartheta \rightarrow (\phi \vee \psi)$, then **CZF** $\vdash$ $\neg\vartheta \rightarrow \phi$ or **CZF** $\vdash$ $\neg\vartheta \rightarrow \psi$. Likewise, **CZF** $\vdash \neg\vartheta \rightarrow (\exists x \in \omega)\theta(x)$ implies **CZF** $\vdash (\exists x \in \omega)[\neg\vartheta \rightarrow \theta(x)]$.

The above results can be extended to include a classically valid principle. *Markov's Principle*, **MP**, is closely associated with the work of the school of Russian constructivists. The version of **MP** most appropriate to the set-theoretic context is the schema

$$\forall n \in \omega \left[\varphi(n) \vee \neg\varphi(n)\right] \wedge \neg\neg\exists n \in \omega\, \varphi(n) \rightarrow \exists n \in \omega\varphi(n).$$

The variant

$$\neg\neg\exists n \in \omega\, R(n) \rightarrow \exists n \in \omega R(n),$$

with R being a primitive recursive predicate, will be denoted by **MP$_{PR}$**. Obviously, **MP$_{PR}$** is implied by **MP**.

Theorem 6.5 *Let* T *be any of the theories* **CZF**, **CZF + REA**, **IZF**, *and* **IZF + REA**, *or any of the theories of Corollary 6.2. For every theorem* θ *of* $T +$ **MP**, *there exists an application term* t *such that*

$$T + \mathbf{MP} \vdash (t \Vdash_{rt} \theta).$$

Moreover, the proof of this soundness theorem is effective in that the application term t *can be effectively constructed from the* $T +$ **MP** *proof of* θ.

Proof. Arguing in $T + \mathbf{MP}$, it remains to find realizing terms for $\mathbf{MP}$. We assume that

$$(e)_0 \Vdash_{rt} (\forall x \in \underline{\omega}) \left[\varphi(x) \vee \neg\varphi(x)\right], \tag{6}$$

$$(e)_1 \Vdash_{rt} \neg\neg(\exists x \in \underline{\omega}) \varphi(x). \tag{7}$$

Let $e' = (e)_0$. Unraveling the definition of $\Vdash_{rt}$ for negated formulas, it is a consequence of (7) that $(\forall d \in \omega) \neg (\forall f \in \omega) \ \neg f \Vdash_{rt} (\exists x \in \underline{\omega})\varphi(x)$, and hence that $\neg (\forall f \in \omega) \ \neg f \Vdash_{rt} (\exists x \in \underline{\omega})\varphi(x)$, which implies $\neg\neg(\exists f \in \omega)f \Vdash_{rt} (\exists x \in \underline{\omega})\varphi(x)$ (just using intuitionistic logic), and hence

$$\neg\neg(\exists f \in \omega)(f)_1 \Vdash_{rt} \varphi[x/(\underline{f})_0]. \tag{8}$$

(6) yields that $(\forall n \in \omega)e'n \downarrow$ and

$$(\forall n \in \omega)\left([(e'n)_0 = 0 \wedge (e'n)_1 \Vdash_{rt} \varphi[x/\underline{n}]] \vee [(e'n)_0 \neq 0 \wedge (e'n)_1 \Vdash_{rt} \neg\varphi[x/\underline{n}]]\right).$$

Since $(e'n)_1 \Vdash_{rt} \neg\varphi(\underline{n})$ entails that $\neg(e'n)_1 \Vdash_{rt} \varphi(\underline{n})$, we arrive at

$$(\forall n \in \omega)[\psi(n) \vee \neg\psi(n)], \tag{9}$$

where $\psi(n)$ is the formula $(e'n)_0 = 0 \wedge (e'n)_1 \Vdash_{rt} \varphi[x/\underline{n}]$. Utilizing that $\mathbf{MP}$ holds in the background theory, from (8) and (9), we can deduce that there exists a natural number m such that $\psi(m)$ is true; i.e., $(e'm)_0 = 0$ and $(e'm)_1 \Vdash_{rt} \varphi[x/\underline{m}]$. Then, with $r := \mu n.(e'n)_0 = 0$,

$$(e'r)_1 \Vdash_{rt} \varphi[x/\underline{m}].$$

r can be computed by a partial recursive function ζ from e'. Taking into account that for any instance θ of $\mathbf{MP}$ with parameters in V^*, θ° is an instance of $\mathbf{MP}$, too, the upshot of the foregoing is that $\lambda e.(\zeta((e)_0), ((e)_0\zeta((e)_0))_1)$ is a realizer for $\mathbf{MP}$.   □

Theorem 6.6 *If T is any of the theories $\mathbf{CZF}$, $\mathbf{CZF} + \mathbf{REA}$, $\mathbf{IZF}$, and $\mathbf{IZF} + \mathbf{REA}$, or any of the theories of Corollary 6.2, then $T + \mathbf{MP}$ has the $\mathbf{DP}$ and the $\mathbf{NEP}$, and $T + \mathbf{MP}$ is closed under $\mathbf{CR}$, $\mathbf{ECR}$, $\mathbf{CR}_1$, $\mathbf{UZR}$, and $\mathbf{UR}$.*

Proof. This follows from Theorem 6.5 and the proof of [29], Theorem 1.2.   □

7 Realizability for choice principles

The intent of this section is to show that $\Vdash_{rt}$-realizability can be used to validate the choice principles $\mathbf{AC}_\omega$, $\mathbf{DC}$, $\mathbf{RDC}$, and $\mathbf{PAx}$, providing they hold in the background theory.

7.1 Internal pairing

As choice principles assert the existence of functions, the natural first step in the investigation of choice principles over V^* is the isolation of the V^*-internal versions of pairs and ordered pairs.

If φ is a formula with parameters from V^* we shall frequently write '$V^* \models \varphi$' to convey that there is a closed application term t such that $t \Vdash_{rt} \varphi$. It will be obvious from the context how to construct t.

If $\mathcal{SC}$ is a scheme of formulae we take $V^* \models \mathcal{SC}$ to mean that for every instance φ of $\mathcal{SC}$ there is a closed application term t (uniformly depending on φ) such that $t \Vdash_{rt} \varphi$ holds.

Definition 7.1 *For* $a, b \in V^*$, *set*

$$\overline{\{a, b\}} := \langle \{a^\circ, b^\circ\}, \{\langle 0, a \rangle, \langle 1, b \rangle\} \rangle,$$

$$\overline{\{a\}} := \overline{\{a, a\}},$$

$$\overline{\langle a, b \rangle} := \langle \langle a^\circ, b^\circ \rangle, \{\langle 0, \overline{\{a\}} \rangle, \langle 1, \overline{\{a, b\}} \rangle\} \rangle.$$

Lemma 7.2 *(i)* $\overline{\{a, b\}}^\circ = \{a^\circ, b^\circ\}$.

(ii) $\overline{\langle a, b \rangle}^\circ = \langle a^\circ, b^\circ \rangle$.

(iii) $\overline{\{a, b\}}, \overline{\langle a, b \rangle} \in V^*$.

(iv) $V^* \models c \in \overline{\{a, b\}} \leftrightarrow [c = a \lor c = b]$.

(v) $V^* \models c \in \overline{\langle a, b \rangle} \leftrightarrow [c = \overline{\{a\}} \lor c = \overline{\{a, b\}}]$.

Proof. (i) and (ii) are obvious. To show (iii) we employ Lemma 4.2 (iii). Let $x \in \overline{\{a, b\}}^*$. Then $2^{nd}(x) \in \{a, b\}$ and thus $1^{st}(2^{nd}(x)) \in \overline{\{a, b\}}^\circ$ by (i).

Now let $y \in \overline{\langle a, b \rangle}^*$. Then $2^{nd}(y) \in \{\overline{\{a\}}, \overline{\{a, b\}}\}$, and hence, by (i), $1^{st}(2^{nd}(y)) \in \{\{a^\circ\}, \{a^\circ, b^\circ\}\}$; thus $1^{st}(2^{nd}(y)) \in \overline{\langle a, b \rangle}^\circ$ by (ii).

One easily checks that $(\lambda x.x, \lambda x.dx(1, (x)_1)(x)_0 0)$ provides a realizer for (iv).

In a similar vein one can construct a realizer for (v).   □

7.2 Axioms of choice in V^*

Theorem 7.3 *(i)* $(\mathbf{CZF} + \mathbf{AC}_\omega)$ $V^* \models \mathbf{AC}_\omega$.

(ii) $(\mathbf{CZF} + \mathbf{DC})$ $V^* \models \mathbf{DC}$.

(iii) $(\mathbf{CZF} + \mathbf{RDC})$ $V^* \models \mathbf{RDC}$.

(iv) $(\mathbf{CZF} + \mathbf{PAx})$ $V^* \models \mathbf{PAx}$.

Proof. In the following proof we will frequently use the phrase that "e' *is (effectively) computable from* $e_1, \ldots, e_k$." By this we mean that there exists a closed application term q (which we cannot be bothered to exhibit) such that $q e_1 \ldots e_k \simeq e'$ holds in the partial combinatory algebra **Kl**.

Ad (i): Recall from the proof of [29], Theorem 6.1, that the set ω is represented in V^* by $\underline{\omega}$, which is given via an injection of ω into V^*:

$$\underline{n} = \langle n, \{\langle k, \underline{k} \rangle : k < n\}\rangle, \tag{10}$$

$$\underline{\omega} = \langle \omega, \{\langle n, \underline{n}\rangle : n \in \omega\}\rangle. \tag{11}$$

Now suppose

$$e \Vdash_{rt} \forall x \in \underline{\omega}\, \exists y\, \varphi(x, y).$$

Then $\forall n \in \omega\, [en \downarrow\, \wedge\, en \Vdash_{rt} \exists y\, \varphi(\underline{n}, y)]$, and hence,

$$\forall n \in \omega\, \exists \mathfrak{a}\, [en \downarrow\, \wedge\, en \Vdash_{rt} \varphi(\underline{n}, \mathfrak{a})].$$

Invoking $\mathbf{AC}_\omega$ in the background theory, there exists a function $F : \omega \to V^*$ such that $\forall n \in \omega\, en \Vdash_{rt} \varphi(\underline{n}, F(n))$. Next, we internalize F. Letting $F_0 : \omega \to V$ and $F_1 : \omega \to V^*$ be defined by $F_0(n) := (F(n))^\circ$ and $F_1(n) := \overline{\langle \underline{n}, F(n)\rangle}$, respectively, put

$$\mathfrak{f} = \langle F_0, F_1\rangle.$$

Lemma 7.2 and Lemma 4.2 (iii) entail that $\mathfrak{f} \in V^*$.

First, because of the properties of internal pairing in V^* discerned in Lemma 7.2, it will be shown that, internally in V^*, $\mathfrak{f}$ is a functional relation with domain $\underline{\omega}$ and that this holds with a witness obtainable independently of e. To see that $\mathfrak{f}$ is realizably functional, assume that

$$h \Vdash_{rt} \overline{\langle \mathfrak{a}, \mathfrak{b}\rangle} \in \mathfrak{f} \quad \text{and} \quad j \Vdash_{rt} \overline{\langle \mathfrak{a}, \mathfrak{c}\rangle} \in \mathfrak{f}.$$

Then,

$$h_1 \Vdash_{rt} \overline{\langle \mathfrak{a}, \mathfrak{b}\rangle} = \overline{\langle \underline{h_0}, F(h_0)\rangle} \text{ and } j_1 \Vdash_{rt} \overline{\langle \mathfrak{a}, \mathfrak{c}\rangle} = \overline{\langle \underline{j_0}, F(j_0)\rangle}, \tag{12}$$

where $h_1 = (h)_1$ and $j_1 = (j)_1$. This holds strictly in virtue of the definition of $\mathfrak{f}$ and the conditions on statements of membership. (12) in conjunction with Lemma 7.2 implies that $d \Vdash_{rt} \underline{h_0} = \underline{j_0}$ for some d, and hence, $(\underline{h_0})^\circ = (\underline{j_0})^\circ$ by Lemma 5.7. Thus, in view of the definition of $\underline{n}$, we have $h_0 = j_0$, and consequently, $F(h_0) = F(j_0)$. As a result, $\ell(h, j) \Vdash_{rt} \mathfrak{b} = \mathfrak{c}$, with $\ell(h, j)$ an application term easily constructed from h and j.

Finally, we have to check on the realizability of $\forall x \in \underline{\omega}\, \varphi(x, \mathfrak{f}(x))$. Since $\forall n \in \omega\, en \Vdash_{rt} \varphi(\underline{n}, F(n))$, we deduce by Lemma 5.7 that $\forall n \in \omega\, \varphi^\circ(n, (F(n))^\circ)$ and hence that $\forall n \in \omega\, \varphi^\circ(n, \mathfrak{f}^\circ(n))$ as $\mathfrak{f}^\circ = F_0$. Since $\forall n \in \omega\, en \Vdash_{rt} \varphi(\underline{n}, F(n))$ and

$\mathfrak{f}^* = \{\langle n, \overline{\langle \underline{n}, F(n)\rangle}\rangle : n \in \omega\}$, we can now also construct a $\mathbf{q}$ independent of e such that $\forall n \in \omega \, (\mathbf{q}e)n \Vdash_{rt} \varphi(\underline{n}, \mathfrak{f}(\underline{n}))$. So the upshot of the above is that we can cook up a realizer $\mathbf{r}$ such that

$$\mathbf{r} \Vdash_{rt} \forall x \in \underline{\omega} \, \exists y \, \varphi(x, y) \to \exists f \, [\mathbf{fun}(f) \wedge \mathbf{dom}(f) = \underline{\omega} \wedge \forall x \in \underline{\omega} \, \varphi(x, f(x))].$$

Ad (ii): Suppose

$$e \Vdash \forall x \in \mathfrak{a} \, \exists y \in \mathfrak{a} \, \varphi(x, y) \qquad \text{and} \tag{13}$$

$$d \Vdash \mathfrak{b} \in \mathfrak{a}. \tag{14}$$

Then we have $\mathfrak{b}° \in \mathfrak{a}°$ and there exists $\mathfrak{c}_\mathfrak{b}$ such that

$$\langle (d)_0, \mathfrak{c}_\mathfrak{b}\rangle \in \mathfrak{a}^* \wedge (d)_1 \Vdash_{rt} \mathfrak{b} = \mathfrak{c}_\mathfrak{b}. \tag{15}$$

Moreover, (13) entails that

$$\forall k \, \forall \mathfrak{c} \, (\langle k, \mathfrak{c}\rangle \in \mathfrak{a}^* \to \exists \mathfrak{d} \, [\langle (ek)_0, \mathfrak{d}\rangle \in \mathfrak{a}^* \wedge (ek)_1 \Vdash_{rt} \varphi(\mathfrak{c}, \mathfrak{d})]),$$

and hence,

$$\forall \langle k, \mathfrak{c}\rangle \in \mathfrak{a}^* \, \exists \langle m, \mathfrak{d}\rangle \in \mathfrak{a}^* \, \varphi^{\Vdash}(\langle k, \mathfrak{c}\rangle, \langle m, \mathfrak{d}\rangle), \tag{16}$$

where $\varphi^{\Vdash}(\langle n, \mathfrak{c}\rangle, \langle m, \mathfrak{d}\rangle)$ stands for $en \downarrow \wedge m = (en)_0 \wedge (en)_1 \Vdash_{rt} \varphi(\mathfrak{c}, \mathfrak{d})$.

By $\mathbf{DC}$ in the background theory, there are functions $f : \omega \to \omega$ and $g : \omega \to V^*$ such that $f(0) = (d)_0, g(0) = \mathfrak{c}_\mathfrak{b}, \forall n \in \omega \, \langle f(n), g(n)\rangle \in \mathfrak{a}^*$, and

$$\forall n \in \omega \, \varphi^{\Vdash}(\langle f(n), g(n)\rangle, \langle f(n+1), g(n+1)\rangle). \tag{17}$$

(17) implies that

$$\forall n \in \omega \, [f(n+1) = (e(f(n)))_0 \wedge (e(f(n)))_1 \Vdash_{rt} \varphi(g(n), g(n+1))]. \tag{18}$$

Now put

$$F := \{\langle n, (g(n))° \rangle : n \in \omega\},$$

$$G := \{\langle n, \overline{\langle \underline{n}, g(n)\rangle} \rangle : n \in \omega\},$$

$$\mathfrak{g} := \langle F, G\rangle.$$

Lemma 7.2 and Lemma 4.2 (iii) guarantee that $\mathfrak{g} \in V^*$. First, because of the properties of internal pairing in V^* discerned in Lemma 7.2, it will be shown that, internally in V^*, $\mathfrak{g}$ is a functional relation with domain $\underline{\omega}$ and that this holds with a witness obtainable independently of e and d. To see that $\mathfrak{g}$ is realizably functional, assume that

$$h \Vdash_{rt} \overline{\langle \mathfrak{a}, \mathfrak{b}\rangle} \in \mathfrak{g} \quad \text{and} \quad j \Vdash_{rt} \overline{\langle \mathfrak{a}, \mathfrak{c}\rangle} \in \mathfrak{g}.$$

Then,

$$h_1 \Vdash_{rt} \overline{\langle \mathfrak{a}, \mathfrak{b}\rangle} = \overline{\langle \underline{h_0}, F(h_0)\rangle} \quad \text{and} \quad j_1 \Vdash_{rt} \overline{\langle \mathfrak{a}, \mathfrak{c}\rangle} = \overline{\langle \underline{j_0}, F(j_0)\rangle}, \tag{19}$$

where $h_1 = (h)_1$ and $j_1 = (j)_1$. This holds strictly in virtue of the definition of $\mathfrak{g}$ and the conditions on statements of membership. (12) in conjunction with Lemma 7.2 implies that $d \Vdash_{rt} h_0 = j_0$ for some d, and hence $(h_0)^\circ = (j_0)^\circ$ by Lemma 5.7. Thus, in view of the definition of $\underline{n}$, we have $h_0 = j_0$ and consequently $F(h_0) = F(j_0)$. As a result, $\ell(h, j) \Vdash_{rt} \mathfrak{b} = \mathfrak{c}$, with $\ell(h, j)$ an application term easily constructed from h and j.

Finally, we have to effectively calculate a realizer $\ell(e, d)$ from e and d such that

$$\ell(e, d) \Vdash_{rt} \mathfrak{g}(0) = \mathfrak{b} \;\wedge\; \forall x \in \underline{\omega}\, \varphi(\mathfrak{g}(x), \mathfrak{g}(x + 1)). \tag{20}$$

Since $d \Vdash_{rt} \mathfrak{b} \in \mathfrak{a}$ and $g(0) = \mathfrak{c}_{\mathfrak{b}}$ it follows from (16) that we can construct a realizer $\tilde{d}$ from d such that $\tilde{d} \Vdash_{rt} \mathfrak{g}(0) = \mathfrak{b}$. Moreover, in view of (19) the function f is recursive. Let $\rho(n) := (e(f(n)))_0$. The S-m-n theorem shows how to compute an index of the function ρ from e. Since

$$\mathbf{pni}_r \Vdash_{rt} \overline{\langle \underline{n}, g(n) \rangle} \in \mathfrak{g},$$

$$\rho(n) \Vdash_{rt} \varphi(g(n), g(n + 1)),$$

this shows that we can effectively construct an index $\ell(e, d)$ from e and d such that (20) holds.

Ad (iii): **RDC** implies **DC** (see [28], Lemma 3.4), and on the basis of **CZF** + **DC**, the scheme **RDC** follows from the scheme:

$$\forall x \left(\varphi(x) \;\rightarrow\; \exists y \left[\varphi(y) \wedge \psi(x, y) \right] \right) \;\wedge\; \varphi(\mathfrak{b}) \tag{21}$$
$$\rightarrow \exists z \left(\mathfrak{b} \in z \;\wedge\; \forall x \in z\, \exists y \in z \left[\varphi(y) \wedge \psi(x, y) \right] \right).$$

Thus, in view of part (ii) of this theorem, it suffices to show that, working in **CZF** + **RDC**, V^* validates (21). So suppose $\mathfrak{b} \in V^*$ and

$$e \Vdash \forall x \left(\varphi(x) \;\rightarrow\; \exists y \left[\varphi(y) \wedge \psi(x, y) \right] \right) \qquad \text{and}$$

$$d \Vdash \varphi(\mathfrak{b}).$$

Then, for all $k \in \omega$ and $\mathfrak{a} \in V^*$, we have

$$(k \Vdash \varphi(x)) \;\rightarrow\; \exists \mathfrak{c} \left[(ek)_0 \Vdash_{rt} \varphi(\mathfrak{c}) \;\wedge\; (ef)_1 \Vdash_{rt} \psi(\mathfrak{a}, \mathfrak{c}) \right].$$

By applying **RDC** to the above, we can extract functions $\imath : \omega \to \omega$, $\jmath : \omega \to \omega$, and $\ell : \omega \to V^*$ such that $\imath(0) = d$, $\ell(0) = \mathfrak{b}$, and for all $n \in \omega$:

$$\imath(n) \Vdash_{rt} \varphi(\ell(n)) \quad \text{and} \quad \jmath(n) \Vdash_{rt} \psi(\ell(n), \ell(n + 1)), \tag{22}$$

$$\imath(n + 1) = (e(\imath(n)))_0 \quad \text{and} \quad \jmath(n) = (e(\imath(n)))_1. \tag{23}$$

By the last line, $\imath$ and $\jmath$ are recursive functions whose indices can be effectively computed from e and d. Now set

$$\mathfrak{d} = \langle \{(\ell(n))^\circ : n \in \omega\}, \{\langle n, \ell(n)\rangle : n \in \omega\}\rangle.$$

Obviously, $\mathfrak{d}$ belongs to V^*. We have

$$\mathbf{p0i}_r \Vdash_{rt} \mathfrak{b} \in \mathfrak{d}. \tag{24}$$

(22) entails that

$$\forall n \in \omega \;\; \mathbf{p}(\imath(n+1))(\jmath(n)) \Vdash_{rt} \varphi(\ell(n)) \;\wedge\; \psi(\ell(n), \ell(n+1))$$

and hence that

$$\forall n \in \omega \;\; \mathbf{p}(n+1)\,(\mathbf{p}(\imath(n+1))(\jmath(n))) \Vdash_{rt} \exists y \in \mathfrak{d}\; [\varphi(\ell(n)) \;\wedge\; \psi(\ell(n), y)].$$

Thus choosing an index $\tilde{e}$ such that $\tilde{e}n = \mathbf{p}(n+1)\,(\mathbf{p}(\imath(n+1))(\jmath(n)))$, we arrive at

$$\tilde{e} \Vdash_{rt} \forall x \in \mathfrak{d}\, \exists y \in \mathfrak{d}\; [\varphi(x) \;\wedge\; \psi(x, y)]. \tag{25}$$

Note that $\tilde{e}$ can be effectively calculated from e and d. As a result, (24) and (25) entail that we can construct a realizer $\mathbf{q}$ for (21).

Ad (iv): For the proof of $V^* \models \mathbf{PAx}$, fix an arbitrary $\mathfrak{a}$ in V^*. Since $\mathbf{PAx}$ holds in the background theory we can find bases X and Y and surjections $f : X \to \mathfrak{a}^\circ$ and $g : Y \to \mathfrak{a}^*$. Define

$$\tilde{X} := \{\langle 0, v\rangle : v \in X\}, \tag{26}$$

$$\tilde{Y} := \{\langle g_0(u) + 1, u\rangle : u \in Y\}, \tag{27}$$

where $g_0 : Y \to \omega$ is defined by $g_0(u) := 1^{st}(g(u))$.

As $\tilde{X}$ is in one-to-one correspondence with X and $\tilde{Y}$ is in one-to-one correspondence with Y, $\tilde{X}$ and $\tilde{Y}$ are bases, too. Moreover,

$$B := \tilde{X} \cup \tilde{Y} \tag{28}$$

is a basis as well because $\tilde{X}$ and $\tilde{Y}$ do not have any elements in common, and for an arbitrary $x \in B$, we can decide whether it belongs to $\tilde{X}$ or $\tilde{Y}$ by inspecting $1^{st}(x)$ and determining whether $1^{st}(x) = 0$ or $1^{st}(x) \neq 0$ since $1^{st}(x) \in \omega$. We thus may define a function $\mathcal{F} : B \to \mathfrak{a}^\circ$ by

$$\mathcal{F}(x) = \begin{cases} f(2^{nd}(x)) & \text{if } x \in \tilde{X}, \\ \left(2^{nd}(g(2^{nd}(x)))\right)^\circ & \text{if } x \in \tilde{Y}. \end{cases} \tag{29}$$

Since for $u \in Y$ we have $\left(2^{nd}(g(2^{nd}(\langle g_0(u) + 1, u\rangle)))\right)^\circ = \left(2^{nd}(g(u))\right)^\circ \in \mathfrak{a}^\circ$, $\mathcal{F}$ clearly takes its values in $\mathfrak{a}^\circ$. Moreover, $\mathcal{F}$ is surjective as f is surjective. Now set

$$\wp(u) := \overline{\langle g_0(u) + 1, u^{st}\rangle} \qquad \text{for } u \in Y, \tag{30}$$

$$B^+ := \{\langle\, g_0(u), \wp(u)\rangle : u \in Y\}, \tag{31}$$

$$\mathfrak{b} := \langle B, B^+\rangle. \tag{32}$$

By Lemmata 7.2 and 5.4, and the fact that $(\underline{n})^\circ = n$ (see (10) for the definition of $\underline{n}$), we see that $(\wp(u))^\circ = \left(\overline{\langle g_0(u) + 1, u^{st} \rangle} \right)^\circ = \langle g_0(u) + 1, u \rangle \in B$ for $u \in Y$, it follows that $\mathfrak{b} \in V^*$. The latter also entails that $\wp$ is one-to-one and therefore $u \mapsto \langle g_0(u), \wp(u) \rangle$ is a one-to-one correspondence between Y and B^+, showing that B^+ is a base as well.

We shall verify that, internally in V^*, $\mathfrak{b}$ is a base that can be surjected onto $\mathfrak{a}$. To define this surjection, let

$$\ell(u) := \overline{\langle \wp(u), 2^{nd}(g(u)) \rangle} \qquad \text{for } u \in Y, \tag{33}$$

$$\mathcal{G} := \{ \langle g_0(u), \ell(u) \rangle \; : \; u \in Y \}, \tag{34}$$

$$\mathfrak{h} := \langle \mathcal{F}, \mathcal{G} \rangle. \tag{35}$$

To see that $\mathfrak{h} \in V^*$, let $x \in \mathfrak{h}^*$. Then $x \in \mathcal{G}$, so $x = \langle g_0(u), \ell(u) \rangle$ for some $u \in Y$. Thus $1^{st}(2^{nd}(x)) = (\ell(u))^\circ = \left\langle (\wp(u))^\circ, (2^{nd}(g(u)))^\circ \right\rangle = \left\langle \langle g_0(u) + 1, u \rangle, (2^{nd}(g(u)))^\circ \right\rangle \in \mathcal{F}.$

First, we aim at showing that

$$V^* \models \mathfrak{h} \text{ is a surjection from } \mathfrak{b} \text{ onto } \mathfrak{a}. \tag{36}$$

To verify $V^* \models \mathfrak{h} \subseteq \mathfrak{b} \times \mathfrak{a}$, suppose $e \Vdash_{rt} \overline{\langle \mathfrak{c}, \mathfrak{d} \rangle} \in \mathfrak{h}$. Then there exists $u \in Y$ such that $(e)_0 = g_0(u)$ and $(e)_1 \Vdash_{rt} \overline{\langle \mathfrak{c}, \mathfrak{d} \rangle} = \langle \wp(u), 2^{nd}(g(u)) \rangle$. Hence, because of $\mathbf{p}(g_0(u)) \mathbf{i}_r \Vdash_{rt} 2^{nd}(g(u)) \in \mathfrak{a}$, one can effectively calculate an index e' from e such that $e' \Vdash_{rt} \mathfrak{c} \in \mathfrak{b} \wedge \mathfrak{d} \in \mathfrak{a}$, showing that

$$V^* \models \mathfrak{h} \subseteq \mathfrak{b} \times \mathfrak{a}. \tag{37}$$

To see that $\mathfrak{h}$ is realizably total on $\mathfrak{b}$, assume that $e \Vdash_{rt} \mathfrak{c} \in \mathfrak{b}$. Then there exists $\mathfrak{d}$ such that $\langle (e)_0, \mathfrak{d} \rangle \in \mathfrak{b}^*$ and $(e)_1 \Vdash_{rt} \mathfrak{c} = \mathfrak{d}$. Moreover, by virtue of the definition of $\mathfrak{b}^*$, there exists $u \in Y$ such that $\langle (e)_0, \mathfrak{d} \rangle = \langle g_0(u), \wp(u) \rangle$, and thus, by definition of $\mathfrak{h}$, $(e)_0 \mathbf{i}_r \Vdash_{rt} \overline{\langle \mathfrak{d}, 2^{nd}(g(u)) \rangle} \in \mathfrak{h}$. Therefore an $\tilde{e}$ can be computed from e such that $\tilde{e} \Vdash_{rt} \mathfrak{c}$ is in the domain of $\mathfrak{h}$, so that with (37) we can conclude that for some e^+ effectively obtainable from e, $e^+ \Vdash_{rt} \mathfrak{b}$ is in the domain of $\mathfrak{h}$. As a result, $V^* \models \mathfrak{b} \subseteq \mathbf{dom}(\mathfrak{h})$, so that in view of (37), we have

$$V^* \models \mathbf{dom}(\mathfrak{h}) = \mathfrak{b}. \tag{38}$$

To establish realizable functionality of $\mathfrak{h}$, suppose $e \Vdash_{rt} \overline{\langle \mathfrak{c}, \mathfrak{d} \rangle} \in \mathfrak{h}$ and $d \Vdash_{rt} \overline{\langle \mathfrak{c}, \mathfrak{e} \rangle} \in \mathfrak{h}$. Then there exist $u, v \in Y$ such that $(e)_0 = g_0(u)$, $(d)_0 = g_0(v)$, $(e)_1 \Vdash_{rt} \overline{\langle \mathfrak{c}, \mathfrak{d} \rangle} = \overline{\langle \wp(u), 2^{nd}(g(u)) \rangle}$, and $(d)_1 \Vdash_{rt} \overline{\langle \mathfrak{c}, \mathfrak{e} \rangle} = \langle \wp(v), 2^{nd}(g(v)) \rangle$. Hence $\Vdash_{rt} \wp(u) = \wp(v)$, i.e., $\Vdash_{rt} \overline{\langle g_0(u) + 1, u^{st} \rangle} = \langle g_0(v) + 1, v^{st} \rangle$, and therefore $\Vdash_{rt} u^{st} = v^{st}$, yielding $u = (u^{st})^\circ = (v^{st})^\circ = v$. As a result, $q \Vdash_{rt} \mathfrak{d} = \mathfrak{e}$ for some q effectively computable from e and d. We have thus established that

$$\text{v}^* \models \mathfrak{h} \text{ is a function.} \tag{39}$$

For (36) it remains to be shown that $\mathfrak{h}$ realizably maps onto $\mathfrak{a}$. So let $e \Vdash_{rt} \mathfrak{c} \in \mathfrak{a}$. Then $\langle (e)_0, \mathfrak{d} \rangle \in \mathfrak{a}^*$ and $(e)_1 \Vdash_{rt} \mathfrak{c} = \mathfrak{d}$ for some $\mathfrak{d}$. As g maps Y onto $\mathfrak{a}^*$, there exists $u \in Y$ such that $g(u) = \langle (e)_0, \mathfrak{d} \rangle = \langle g_0(u), 2^{nd}(g(u)) \rangle$. Since $\langle g_0(u), \wp(u) \rangle \in \mathfrak{b}^*$ and $\left\langle g_0(u), \overline{\langle \wp(u), 2^{nd}(g(u)) \rangle} \right\rangle \in \mathfrak{h}^*$, we have $\mathbf{p}(e)_0 \mathbf{i}_r \Vdash_{rt}$ $\wp(u) \in \mathfrak{b}$ and $\mathbf{p}(e)_0 \mathbf{i}_r \Vdash_{rt} \overline{\langle \wp(u), \mathfrak{d} \rangle} \in \mathfrak{h}$. Therefore we can effectively compute an index $\tilde{e}$ from e such that $\tilde{e} \Vdash_{rt} \mathfrak{c}$ is in the range of $\mathfrak{h}$. As a consequence, $\text{v}^* \models \mathfrak{h}$ maps onto $\mathfrak{a}$. The latter in conjunction with (37), (38), and (39) yields (36).

Finally, we have to verify that

$$\text{v}^* \models \mathfrak{b} \text{ is a base.} \tag{40}$$

So assume that

$$e \Vdash_{rt} \forall x \in \mathfrak{b} \, \exists y \, \varphi(x, y) \tag{41}$$

for some formula $\varphi(x, y)$ (parameters from v^* allowed). To ensure (40) we have to describe how to obtain an index e' calculably from e satisfying

$$e' \Vdash_{rt} \exists G \left[\mathbf{fun}(G) \wedge \mathbf{dom}(G) \supseteq \mathfrak{b} \wedge \forall x \in \mathfrak{b} \, \varphi(x, G(x)) \right] . \tag{42}$$

From (41) it follows that $\forall x \in \mathfrak{b}^\circ \, \exists y \, \varphi^\circ(x, y)$, and hence, since $\mathfrak{b}^\circ = B = \tilde{X} \cup \tilde{Y}$,

$$\forall x \in \tilde{X} \, \exists y \, \varphi^\circ(x, y). \tag{43}$$

(41) also implies $\forall \, \langle n, \mathfrak{c} \rangle \in \mathfrak{b}^* \, \exists \mathfrak{d} \, en \Vdash_{rt} \varphi(\mathfrak{c}, \mathfrak{d})$, yielding

$$\forall u \in Y \, \exists \mathfrak{d} \, e(g_0(u)) \Vdash_{rt} \varphi(\wp(u), \mathfrak{d}). \tag{44}$$

$\tilde{X}$ and Y being bases, there exist functions K and L such that $\mathbf{dom}(K) = \tilde{X}$ and $L : Y \to \text{v}^*$ satisfying

$$\forall x \in \tilde{X} \, \varphi^\circ(x, K(x)), \tag{45}$$
$$\forall u \in Y \, e(g_0(u)) \Vdash_{rt} \varphi(\wp(u), L(u)). \tag{46}$$

(46) implies that $\forall u \in Y \, \varphi^\circ (\langle g_0(u) + 1, u \rangle, (L(u))^\circ)$, so that we have $\forall u \in \tilde{Y} \, \varphi^\circ \left(x, (L(2^{nd}(x)))^\circ \right)$. Hence, for the same reasons as in the definition of $\mathcal{F}$, (29) we can define a function M with domain $B = \tilde{X} \cup \tilde{Y}$ by

$$M(x) = \begin{cases} K(x) & \text{if } x \in \tilde{X}, \\ \left(L(2^{nd}(x)) \right)^\circ & \text{if } x \in \tilde{Y}. \end{cases} \tag{47}$$

As a result,

$$\forall x \in \mathfrak{b}^\circ \, \varphi^\circ(x, M(x)). \tag{48}$$

Next, to internalize M in V^* put

$$\mathcal{M} := \{\langle\, g_0(u), \overline{\langle \wp(u), L(u)\rangle}\,\rangle\ :\ u \in Y\}, \tag{49}$$

$$\mathfrak{m} := \langle M, \mathcal{M}\rangle. \tag{50}$$

For $y \in \mathfrak{m}^* = \mathcal{M}$ we have $y = \langle\, g_0(u), \overline{\langle\wp(u), L(u)\rangle}\,\rangle$ for some $u \in Y$, and thus $1^{st}(2^{nd}(y)) = \langle(\wp(u))^\circ, (L(u))^\circ\rangle = \langle\, \langle g_0(u) + 1, u\rangle, (L(u))^\circ\,\rangle$, so that with $x := \langle g_0(u) + 1, u\rangle$, we have $x \in \hat{Y}$ and $(L(u))^\circ = \big(L(2^{nd}(x))\big)^\circ$, showing that $1^{st}(2^{nd}(y)) \in M$. As a consequence, we see that $\mathfrak{m} \in V^*$.

It remains to show that

$$e' \Vdash_{rt} \mathbf{fun}(\mathfrak{m}) \wedge \mathbf{dom}(\mathfrak{m}) \supseteq \mathfrak{b} \wedge \forall x \in \mathfrak{b}\, \varphi(x, \mathfrak{m}(x)) \tag{51}$$

for some index e' that is calculable from e.

To establish the realizable functionality of $\mathfrak{m}$, suppose $a \Vdash_{rt} \overline{\langle \mathfrak{c}, \mathfrak{d}\rangle} \in \mathfrak{m}$ and $b \Vdash_{rt} \overline{\langle \mathfrak{c}, \mathfrak{e}\rangle} \in \mathfrak{m}$. Then there exist $u, v \in Y$ such that $(a)_0 = g_0(u)$, $(b)_0 = g_0(v)$, $(a)_1 \Vdash_{rt} \overline{\langle \mathfrak{c}, \mathfrak{d}\rangle} = \overline{\langle \wp(u), L(u)\rangle}$, and $(b)_1 \Vdash_{rt} \overline{\langle \mathfrak{c}, \mathfrak{e}\rangle} = \overline{\langle \wp(v), L(v)\rangle}$. Hence $\Vdash_{rt} \wp(u) = \wp(v)$, i.e. $\Vdash_{rt} \overline{\langle g_0(u) + 1, u^{st}\rangle} = \overline{\langle g_0(v) + 1, v^{st}\rangle}$, and therefore $\Vdash_{rt} u^{st} = v^{st}$, yielding $u = (u^{st})^\circ = (v^{st})^\circ = v$. As a result, $q \Vdash_{rt} \mathfrak{d} = \mathfrak{e}$ for some q effectively computable from a and b.

Next, we would like to verify that $\mathfrak{m}$ is realizably defined on elements of $\mathfrak{b}$. An element of $\mathfrak{b}^*$ is of the form $\langle g_0(u), \wp(u)\rangle$ for some $u \in Y$. As $\langle\, g_0(u), \overline{\langle\wp(u), L(u)\rangle}\,\rangle \in \mathfrak{m}^*$, it is obvious how to construct $\tilde{q}$ such that $\tilde{q}(g_0(u)) \Vdash_{rt} \overline{\langle g_0(u), \wp(u)\rangle} \in \mathbf{dom}(\mathfrak{m})$, and hence,

$$V^* \models \mathfrak{b} \subseteq \mathbf{dom}(\mathfrak{m}). \tag{52}$$

Finally we have to ensure that

$$\tilde{e} \Vdash_{rt} \forall x \in \mathfrak{b}\, \varphi(x, \mathfrak{m}(x)) \tag{53}$$

for some $\tilde{e}$ computable from e. Now, each element of $\mathfrak{b}^*$ is of the form $\langle g_0(u), \wp(u)\rangle$ for some $u \in Y$. Since $\langle\, g_0(u), \overline{\langle\wp(u), L(u)\rangle}\,\rangle \in \mathfrak{m}^*$ and $e(g_0(u)) \Vdash_{rt} \varphi(\wp(u), L(u))$ holds by (46), we can cook up an index r such that $(re)(g_0(u)) \Vdash_{rt} \varphi(\wp(u), \mathfrak{m}(\wp(u)))$, and therefore, noting that $\forall x \in \mathfrak{b}^\circ\, \varphi^\circ(x, \mathfrak{m}^\circ(x))$ is true, we get $\tilde{e} \Vdash_{rt} \forall x \in \mathfrak{b}\, \varphi(x, \mathfrak{m}(x))$ for an index $\tilde{e}$ effectively computable from e. $\quad\square$

Theorem 7.4 *If T is any of the theories* **CZF**, **CZF + REA**, **IZF**, *or* **IZF + REA** *(or any of the theories of Corollary 6.2), and S is any combination of the axioms and schemes* **MP**, **AC**$_\omega$, **DC**, **RDC**, *and* **PAx**, *then $T + S$ has the* **DP** *and the* **NEP**, *and $T + S$ is closed under* **CR**, **ECR**, **CR**$_1$, **UZR**, *and* **UR**.

Proof. This follows from Theorems 7.3 and 6.5 and the proof of [29], Theorem 1.2.
$\quad\square$

Remark 7.5 Theorem 7.4 can be extended to include large set axioms such as *"Every set is contained in an inaccessible set"* or *"Every set is contained in a Mahlo set."* For definitions of "inaccessible set" and "Mahlo set" see [4, 9]. The proofs are similar to the one for the regular extension axiom, which was carried out in [29], Theorem 7.2.

References

1. P. Aczel: *The type theoretic interpretation of constructive set theory.* In: MacIntyre, A. and Pacholski, L. and Paris, J, editor, *Logic Colloquium '77* (North Holland, Amsterdam 1978) 55–66.

2. P. Aczel: *The type theoretic interpretation of constructive set theory: Choice principles.* In: A.S. Troelstra and D. van Dalen, editors, *The L.E.J. Brouwer Centenary Symposium* (North Holland, Amsterdam 1982) 1–40.

3. P. Aczel: *The type theoretic interpretation of constructive set theory: Inductive definitions.* In: R.B. Marcus, et al. editors, *Logic, Methodology and Philosophy of Science VII* (North Holland, Amsterdam 1986) 17–49.

4. P. Aczel, M. Rathjen: *Notes on constructive set theory,* Technical Report 40, Institut Mittag-Leffler (The Royal Swedish Academy of Sciences, 2001). *http://www.ml.kva.se/preprints/archive2000-2001.php*

5. M. Beeson: *Continuity in intuitionistic set theories.* In: M Boffa, D. van Dalen, K. McAloon, editors, *Logic Colloquium '78* (North-Holland, Amsterdam 1979).

6. M. Beeson: *Foundations of Constructive Mathematics.* (Springer-Verlag, Berlin, Heidelberg, New York, Tokyo 1985).

7. A. Blass: *Injectivity, projectivity, and the axiom of choice.* Transactions of the AMS 255 (1979) 31–59.

8. E. Bishop and D. Bridges: *Constructive Analysis.* (Springer-Verlag, Berlin, Heidelberg, New York, Tokyo 1985).

9. L. Crosilla, M. Rathjen: *Inaccessible set axioms may have little consistency strength.* Annals of Pure and Applied Logic 115 (2002) 33–70.

10. S. Feferman: *A language and axioms for explicit mathematics.* In: J.N. Crossley, editor, *Algebra and Logic*, Lecture Notes in Math. 450 (Springer, Berlin 1975) 87–139.

11. S. Feferman: *Constructive theories of functions and classes.* In: M. Boffa, D. van Dalen, K. McAloon, editors, *Logic Colloquium '78* (North-Holland, Amsterdam 1979) 159–224.

12. H. Friedman: *Some applications of Kleene's method for intuitionistic systems.* In: A. Mathias and H. Rogers, editors, *Cambridge Summer School in Mathematical Logic*, volume 337 of *Lectures Notes in Mathematics* (Springer, Berlin, 1973) 113–170.

13. H. Friedman: *The disjunction property implies the numerical existence property.* Proceedings of the National Academy of Sciences of the United States of America 72 (1975) 2877–2878.

14. H. Friedman: *Set-theoretic foundations for constructive analysis.* Annals of Mathematics 105 (1977) 868–870.

15. H. Friedman, S. Ščedrov: *Set existence property for intuitionistic theories with dependent choice.* Annals of Pure and Applied Logic 25 (1983) 129–140.

16. H. Friedman, S. Ščedrov: *The lack of definable witnesses and provably recursive functions in intuitionistic set theory.* Advances in Mathematics 57 (1985) 1–13.

17. S.C. Kleene: *On the interpretation of intuitionistic number theory.* The Journal of Symbolic Logic 10 (1945) 109–124.

18. S.C. Kleene: *Disjunction and existence under implication in elementary intuitionistic formalisms.* The Journal of Symbolic Logic 27 (1962) 11–18.

19. S.C. Kleene: *An addendum.* Journal of Symbolic Logic 28 (1963) 154–156.

20. S.C. Kleene: *Formalized recursive functionals and formalized realizability.* Memoirs of the AMS 89 (AMS, Providence 1969).

21. G. Kreisel, A.S. Troelstra: *Formal systems for some branches of intuitionistic analysis.* Annals of Mathematical Logic 1 (1970) 229–387.

22. J. Lipton: *Realizability, set theory and term extraction.* In: *The Curry-Howard isomorphism*, Cahiers du Centre de Logique de l'Universite Catholique de Louvain, vol. 8 (1995) 257–364.

23. D.C. McCarty: *Realizability and recursive mathematics*, PhD thesis, Oxford University (1984), 281 pages.

24. D.C. McCarty: *Realizability and recursive set theory*, Annals of Pure and Applied Logic 32 (1986) 153–183.

25. J.R. Moschovakis: *Disjunction and existence in formalized intuitionistic analysis.* In: J.N. Crossley, editor, *Sets, models and recursion theory.* (North-Holland, Amsterdam 1967) 309–331.

26. J. Myhill: *Some properties of Intuitionistic Zermelo-Fraenkel set theory.* In: A. Mathias and H. Rogers (eds.): *Cambridge Summer School in Mathematical Logic*, volume 337 of *Lectures Notes in Mathematics* (Springer, Berlin 1973) 206–231.

27. J. Myhill: *Constructive set theory.* The Journal of Symbolic Logic 40 (1975) 347–382.

28. M. Rathjen: *The anti-foundation axiom in constructive set theories.* In: G. Mints, R. Muskens, editors, *Games, Logic, and Constructive Sets.* (CSLI Publications, Stanford 2003) 87–108.

29. M. Rathjen: *The disjunction and other properties for constructive Zermelo-Fraenkel set theory.* The Journal of Symbolic Logic 70 (2005) 1233–1254.

30. M. Rathjen: *Realizability for constructive Zermelo-Fraenkel set theory.* In: J. Väänänen, V. Stoltenberg-Hansen, editors, *Logic Colloquium '03.* Lecture Notes in Logic 24 (A.K. Peters, 2006) 282–314.

31. M. Rathjen: *Choice principles in constructive and classical set theories.* In: Z. Chatzidakis, P. Koepke, W. Pohlers, editors, *Logic Colloquium '02.* Lecture Notes in Logic 27 (A.K. Peters 2006).

32. M. Rathjen, S. Tupailo: *Characterizing the interpretation of set theory in Martin-Löf type theory.* Annals of Pure and Applied Logic 141 (2006) 442–471.

33. L. Tharp: *A quasi-intuitionistic set theory.* Journal of Symbolic Logic 36 (1971) 456–460.

34. A.S. Troelstra: *Realizability.* In: S.R. Buss, editor, *Handbook of Proof Theory* (Elsevier, Amsterdam 1998) 407–473.

35. A.S. Troelstra, D. van Dalen: *Constructivism in Mathematics, Volumes I, II.* (North Holland, Amsterdam 1988).

New Developments in Proofs and Computations

Helmut Schwichtenberg

Mathematisches Institut der Universität München, D-80333 München, Germany
schwicht@math.lmu.de

It is a tempting idea to use formal existence proofs as a means to precisely and verifiably express algorithmic ideas. This is clearly possible for "constructive" proofs, which are informally understood via the Brouwer–Heyting–Kolmogorov interpretation (BHK-interpretation, for short). This interpretation of intuitionistic (and minimal) logic explains what it means to prove a logically compound statement in terms of what it means to prove its components; the explanations use the notions of *construction* and *constructive proof* as unexplained primitive notions. For prime formulas, the notion of proof is supposed to be given. The clauses of the BHK-interpretation are:

- p proves $A \wedge B$ if and only if p is a pair $\langle p_0, p_1 \rangle$ and p_0 proves A, p_1 proves B;

- p proves $A \to B$ if and only if p is a construction transforming any proof q of A into a proof $p(q)$ of B;

- $\bot$ is a proposition without proof;

- p proves $\forall_{x \in D} A(x)$ if and only if p is a construction such that for all $d \in D$, $p(d)$ proves $A(d)$;

- p proves $\exists_{x \in D} A(x)$ if and only if p is of the form $\langle d, q \rangle$ with d an element of D, and q a proof of $A(d)$.

The problem with the BHK-interpretation is its reliance on the unexplained concepts of construction and constructive proof. Gödel (1958) tried to replace the notion of constructive proof by something more definite, less abstract, his principal candidate being a notion of "computable functional of finite type," which is to be accepted as sufficiently well understood to justify the axioms and rules of his system T, an essentially logic-free theory of functionals of finite type. One only needs to know that certain basic functionals are computable (including primitive recursion operators in finite types), and that the computable functionals are closed under composition.

The general framework for proof interpretations as we understand it is to assign to every formula A a new one $\exists_x A_1(x)$ with $A_1(x)$ $\exists$-free. Then from a derivation $M : A$ we want to extract a "realizing" term r such that $A_1(r)$ can be proved. The intention here is that its meaning should in some sense be related to the meaning of the original formula A. The well-known (modified) realizability interpretation and Gödel's Dialectica interpretation both fall under this scheme (cf. Oliva (2006)). However, Gödel explicitly states in Gödel (1958) that his Dialectica interpretation is *not* the one intended by BHK-interpretation.

One might think that from the informal idea of a particular constructive proof it should be clear what is its algorithmic content. This, however, is not always true. An example is Tait's proof of the existence of normal forms for the simply typed λ-calculus, which uses so-called computability predicates. Somewhat unexpectedly, it turns out that its computational content is the normalization-by-evaluation algorithm. This has first been observed by Berger (1993) and formally treated (including machine extraction of programs) in Berger et al. (2006).

An even greater challenge is the task of finding computational content in proofs of *classical* existence theorems, of the form $\neg\forall_y \neg A_0(y)$ with $A_0(y)$ quantifier-free; we use the shorthand $\tilde{\exists}_y A_0(y)$ for such formulas. It is well known that we need to require that the kernel $A_0(y)$ is quantifier-free. Then the whole proof can be seen as deriving falsity from the (false) assumption $\forall_y \neg A_0(y)$. Now consider the long normal form of this proof. In this long normal form, each instance of the false assumption $\forall_y \neg A_0(y)$ must be applied to a closed term r_i of type $\mathbf{N}$, and for at least one of those r_i, the kernel $\neg A_0(r_i)$ must be false and hence $A_0(r_i)$ is true. This "direct method" has been described in Schwichtenberg (1993); in Berger and Schwichtenberg (1995), it has been shown that it gives the same results as the so-called A-translation of Friedman (1978) (and moreover, that we have the same algorithm in both cases). A refined form of the A-translation has been introduced in Berger et al. (2002) and further studied and applied in Berger et al. (2001) and Seisenberger (2003).

An alternative to extract computational content from proofs of classic existence theorems is Gödel's Dialectica interpretation (1958), which is what we want to concentrate on in the present paper. Gödel assigned to every formula A a new one $\exists_{\vec{x}} \forall_{\vec{y}} A_D(\vec{x}, \vec{y})$ with $A_D(\vec{x}, \vec{y})$ quantifier-free. Here $\vec{x}, \vec{y}$ are lists of variables of finite types; the use of higher types is necessary even when the original formula A was first order. He did this in such a way that whenever a proof of A say in constructive arithmetic was given, one could produce closed terms $\vec{r}$ such that the quantifier-free formula $A_D(\vec{r}, \vec{y})$ is provable in T.

In Gödel (1958), Gödel referred to a Hilbert-style proof calculus. However, since the realizers will be formed in a λ-calculus formulation of system T, Gödel's interpretation becomes a lot more perspicious when it is done for a natural deduction calculus. Such a natural deduction-based treatment of the Dialectica interpretation has been given by Jørgensen (2001) and Hernest (2006). Both authors use a formulation of natural deduction where open assumptions are viewed as *formulas*, and consequently, the necessity of contractions arises when an application of the implication introduction rule $\to^+$ discharges more than one assumption formula. However, it seems to be more in the spirit of the Curry–Howard correspondence (formulas correspond to types, and proofs to terms) to view assumptions as *assumption variables*. This is particularly important when—say in an implementation—one wants to assign object terms ("realizers", in Gödel's T) to proof terms. To see the point, notice that a proof term M may have many occurrences of a free assumption variable u^A. The associated realizer $[\![M]\!]$ then needs to contain an object variable $x_u^{\tau(A)}$ uniquely associated with u^A, again with many occurrences. To organize this in an appropriate way it seems mandatory to be able to refer to an assumption A by means of its "label" u. The present exposition differs from previous ones mainly in this respect.

The rest of the paper is rather technical. We give a detailed natural deduction-based proof of the soundness theorem for the Dialectica interpretation, and extend it to the Dialectica interpretation with majorants (or "monotone" Dialectica interpretation), introduced by Kohlenbach (1992) and Kohlenbach (1996).

The main motivation for this work has been the desire to have a clean and explicit natural deduction-based proof of the soundness theorem, for the exact Dialectica interpretation, as well as for its variant with majorants, in such a way that this proof can be used as a template for an implementation. For the very same reason we have added a simplified and implementation-friendly proof of the fact—first observed by Kohlenbach (1992)—that WKL can be formulated as a $\forall\exists_{\le}\forall$-axiom, and hence is covered by the Dialectica interpretation with majorants. However, it remains to be seen to what extent such an implementation will succeed in producing informative and usable realizers. A promising first step in this direction has been done by Hernest (2006); particularly interesting is his successful integration of the noncomputational ("uniform") quantifiers of Berger (1993), Berger (2005).

We begin in Sec. 1 with a description of the arithmetic HA^ω in finite types that we consider. Sec. 2 contains a proof of the Soundness Theorem for Gödel's Dialectica interpretation, and Sec. 3 gives the majorant-based version of it. The final subsection contains a proof that WKL can be formulated as a $\forall\exists_{\le}\forall$-axiom.

1 Arithmetic in Finite Types

1.1 Types

Our type system is defined by two type-forming operations: arrow types $\rho \to \sigma$ and the formation of inductively generated types $\mu_{\vec{\alpha}}\vec{\kappa}$, where $\vec{\alpha} = (\alpha_j)_{j=1,\dots,N}$ is a list of distinct "type variables", and $\vec{\kappa} = (\kappa_i)_{i=1,\dots,k}$ is a list of "constructor types," whose argument types contain $\alpha_1, \dots, \alpha_N$ in strictly positive positions only.

For instance, $\mu_\alpha(\alpha, \alpha \to \alpha)$ is the type of natural numbers; here the list $(\alpha, \alpha \to \alpha)$ stands for two generation principles: α for "there is a natural number" (the 0), and $\alpha \to \alpha$ for "for every natural number there is a next one" (its successor).

Definition. Let $\vec{\alpha} = (\alpha_j)_{j=1,\ldots,N}$ be a list of distinct type variables. *Types* $\rho, \sigma, \tau, \mu \in \mathrm{Ty}$ and *constructor types* $\kappa \in \mathrm{KT}_{\vec{\alpha}}$ are defined inductively:

$$\frac{\rho, \sigma \in \mathrm{Ty}}{\rho \to \sigma \in \mathrm{Ty}}, \qquad \frac{\vec{\rho}, \vec{\sigma}_1, \ldots, \vec{\sigma}_n \in \mathrm{Ty}}{\vec{\rho} \to (\vec{\sigma}_1 \to \alpha_{j_1}) \to \cdots \to (\vec{\sigma}_n \to \alpha_{j_n}) \to \alpha_j \in \mathrm{KT}_{\vec{\alpha}}} \ (n \geq 0),$$

$$\frac{\vec{\kappa} \in \mathrm{KT}_{\vec{\alpha}}, \forall_{0 < j \leq N} \exists_{j_1, \ldots, j_n < j} \kappa_j = \vec{\rho} \to (\vec{\sigma}_1 \to \alpha_{j_1}) \to \cdots \to (\vec{\sigma}_n \to \alpha_{j_n}) \to \alpha_j}{(\mu_{\vec{\alpha}}(\kappa_1, \ldots, \kappa_k))_j \in \mathrm{Ty}},$$

with $1 \leq N \leq k$; we call $\kappa_1, \ldots, \kappa_N$ *nullary constructor types.* Here $\vec{\rho} \to \sigma$ means $\rho_1 \to \cdots \to \rho_m \to \sigma$, associated with the right. We reserve μ for types of the form $(\mu_{\vec{\alpha}}(\kappa_1, \ldots, \kappa_k))_j$. The *parameter types* of μ are the members of all $\vec{\rho}$ appearing in its constructor types $\kappa_1, \ldots, \kappa_k$.

In the present paper it suffices to only consider the μ-types

$$\mathbf{U} := \mu_\alpha \alpha,$$
$$\mathbf{B} := \mu_\alpha(\alpha, \alpha), \qquad\qquad \mathrm{bin} := \mu_\alpha(\alpha, \alpha \to \alpha, \alpha \to \alpha),$$
$$\mathbf{N} := \mu_\alpha(\alpha, \alpha \to \alpha), \qquad\qquad \rho \wedge \sigma := \mu_\alpha(\rho \to \sigma \to \alpha).$$

A type is *finitary* if it is a μ-type with all its parameter types $\vec{\rho}$ finitary, and all its constructor types are of the form $\vec{\rho} \to \alpha_{j_1} \to \cdots \to \alpha_{j_n} \to \alpha_j$, so the $\vec{\sigma}_1, \ldots, \vec{\sigma}_n$ in the general definition are all empty. For example, $\mathbf{U}, \mathbf{B}, \mathbf{N}$, bin are finitary, and $\rho \wedge \sigma$ is finitary provided its parameter types are.

1.2 Constants

For each of our base types we have *constructors* C_i^μ and *recursion operators* $\mathcal{R}_\mu^\tau$, as follows:

$$\mathbf{t}^{\mathbf{B}} := \mathrm{C}_1^{\mathbf{B}}, \quad \mathbf{ff}^{\mathbf{B}} := \mathrm{C}_2^{\mathbf{B}},$$

$$\mathcal{R}_{\mathbf{B}}^\tau : \mathbf{B} \to \tau \to \tau \to \tau,$$

$$0^{\mathbf{N}} := \mathrm{C}_1^{\mathbf{N}}, \quad \mathrm{S}^{\mathbf{N} \to \mathbf{N}} := \mathrm{C}_2^{\mathbf{N}},$$

$$\mathcal{R}_{\mathbf{N}}^\tau : \mathbf{N} \to \tau \to (\mathbf{N} \to \tau \to \tau) \to \tau,$$

$$1^{\mathrm{bin}} := \mathrm{C}_1^{\mathrm{bin}}, \quad \mathrm{S}_0^{\mathrm{bin} \to \mathrm{bin}} := \mathrm{C}_2^{\mathrm{bin}}, \quad \mathrm{S}_1^{\mathrm{bin} \to \mathrm{bin}} := \mathrm{C}_3^{\mathrm{bin}},$$

$$\mathcal{R}_{\mathrm{bin}}^\tau : \mathrm{bin} \to \tau \to (\mathrm{bin} \to \tau \to \tau) \to (\mathrm{bin} \to \tau \to \tau) \to \tau,$$

$$\left(\wedge_{\rho\sigma}^+\right)^{\rho \to \sigma \to \rho \wedge \sigma} := \mathrm{C}_1^{\rho \wedge \sigma},$$

$$\mathcal{R}_{\rho \wedge \sigma}^\tau : \rho \wedge \sigma \to (\rho \to \sigma \to \tau) \to \tau.$$

1.3 Terms

Terms are inductively defined from typed variables x^ρ and the constants, that is, constructors C_i^μ and recursion operators $\mathcal{R}_\mu^\tau$, by abstraction $(\lambda_{x^\rho} M^\sigma)^{\rho\to\sigma}$ and application $(M^{\rho\to\sigma} N^\rho)^\sigma$. It is well known that every such term has a uniquely determined long normal form w.r.t. β- and $\mathcal{R}$-conversions and η-expansions. We consider two terms to be *definitionally equal* if they have the same long normal form and identify such terms.

Notice that in the more general setting of Schwichtenberg (2006), where we also allow constants defined by computation rules, definitional equality should mean that there is a purely equational proof of their equality based on β- and $\mathcal{R}$-conversions and η-expansions.

Notice also that the boolean "recursion" operator $\mathcal{R}_B^\tau$ does not make any recursive calls. We denote $\mathcal{R}_B^\tau trs$ by $[\text{if } t \text{ then } r \text{ else } s]$ (which also indicates that this term should be evaluated "lazily").

Using the recursion operators we can define boolean-valued functions representing (decidable) equality $=_\mu: \mu \to \mu \to \mathbf{B}$ for finitary base types μ, for instance $\mathbf{N}$:

$$(0 = 0) := \mathbf{tt}, \qquad (S(m) = 0) := \mathbf{ff},$$

$$(0 = S(n)) := \mathbf{ff}, \quad (S(m) = S(n)) := (m = n).$$

The *projections* of a pair to its components can be defined easily:

$$r0 := \mathcal{R}_{\rho\wedge\sigma}^\rho r^{\rho\wedge\sigma}(\lambda_{x^\rho, y^\sigma} x^\rho), \quad r1 := \mathcal{R}_{\rho\wedge\sigma}^\rho r^{\rho\wedge\sigma}(\lambda_{x^\rho, y^\sigma} y^\sigma).$$

We also define the *canonical inhabitant* ε^ρ of a type ρ:

$$\varepsilon^{\mu_j} := C_j^{\vec{\mu}} \varepsilon^{\vec{\rho}}(\lambda_{\vec{x}_1} \varepsilon^{\mu_{j1}}) \cdots (\lambda_{\vec{x}_n} \varepsilon^{\mu_{jn}}), \quad \varepsilon^{\rho\to\sigma} := \lambda_x \varepsilon^\sigma.$$

There are many canonical isomorphisms between types; $(\rho \wedge \sigma \to \tau) \sim (\rho \to \sigma \to \tau)$ is an example. The isomorphism pairs can be constructed explicitly from the functions above.

1.4 Formulas

Atomic formulas are $\text{atom}(r^B)$, indicating that the argument is true. We may also allow other predicate constants, for instance, inductively defined ones, like Leibniz equality.

Notice that there is no need for (logical) falsity $\bot$, since we can take the atomic formula $F := \text{atom}(\mathbf{ff})$ – called *arithmetical falsity*—built from the boolean constant $\mathbf{ff}$ instead.

The *formulas* of HA^ω are built from atomic ones by the connectives $\to$, $\forall$, $\exists$, and $\wedge$. We define *negation* $\neg A$ by $A \to F$.

1.5 Proof terms

We use Gentzen's natural deduction calculus for logical derivations consisting of the well-known rules $\to^+$, $\to^-$, $\forall^+$, and $\forall^-$. It will be convenient to write derivations as terms, where the derived formula is viewed as the type of the term. This representation is known under the name *Curry–Howard correspondence*.

We give an inductive definition of derivation terms in Table 1, where for clarity we have written the corresponding derivations to the left. For the universal quantifier $\forall$,

derivation	term
$u : A$	u^A
$\begin{array}{c} [u:A] \\ \mid M \\ \dfrac{B}{A \to B} \to^+ u \end{array}$	$(\lambda_{u^A} M^B)^{A \to B}$
$\begin{array}{ccc} \mid M & & \mid N \\ \dfrac{A \to B}{} & \dfrac{A}{} & \to^- \\ \multicolumn{3}{c}{B} \end{array}$	$(M^{A \to B} N^A)^B$
$\begin{array}{c} \mid M \\ \dfrac{A}{\forall_x A} \forall^+ x \quad \text{(with var.cond.)} \end{array}$	$(\lambda_x M^A)^{\forall_x A}$ (with var.cond.)
$\begin{array}{c} \mid M \\ \dfrac{\forall_x A(x) \qquad r}{A(r)} \forall^- \end{array}$	$(M^{\forall_x A(x)} r)^{A(r)}$

Table 1. Derivation terms for $\to$, $\forall$

there is an introduction rule $\forall^+ x$ and an elimination rule $\forall^-$, whose right premise is the term r to be substituted. The rule $\forall^+ x$ is subject to the following *(Eigen-) variable condition*: The derivation term M of the premise A should not contain any open assumption with x as a free variable.

1.6 Axioms

The logical axioms are the *truth axiom* Ax_{tt}: $\text{atom}(\text{tt})$, the introduction and elimination axioms $\exists^+$ and $\exists^-$ for existence, and $\wedge^+$, $\wedge^-$ for conjunction:

$$\exists^+ : \forall_z(A \to \exists_z A),$$

$$\exists^- : \exists_z A \to \forall_z(A \to B) \to B \quad (z \notin \text{FV}(B)),$$

$$\wedge^+ : A \to B \to A \wedge B,$$

$$\wedge^- : A \wedge B \to (A \to B \to C) \to C,$$

and the induction axioms

$$\text{Ind}_{p,A} : \forall_p\big(A(\text{tt}) \to A(\text{ff}) \to A(p^{\mathbf{B}})\big),$$

$$\text{Ind}_{n,A} : \forall_m\big(A(0) \to \forall_n(A(n) \to A(\text{S}n)) \to A(m^{\mathbf{N}})\big),$$

$$\text{Ind}_{b,A} : \forall_b\big(A(1) \to \forall_b(A(b) \to A(\text{S}_0 b)) \to \forall_b(A(b) \to A(\text{S}_1 b)) \to A(b^{\text{bin}})\big),$$

$$\text{Ind}_{x,A} : \forall_x\big(\forall_{y^\rho, z^\sigma} A(\langle y, z \rangle) \to A(x^{\rho \wedge \sigma})\big),$$

where $\langle y, z \rangle$ is shorthand for $\wedge^+ yz$. The final axiom expresses that every object of a pair type is a pair; it is sometimes called *pair elimination axiom*.

Using boolean induction $\text{Ind}_{p,A}$, we can derive the arithmetical form of *ex-falso-quodlibet*, that is, $F \to \text{atom}(p^{\mathbf{B}})$ (recall $F := \text{atom}(\text{ff})$), and then $F \to A$ for arbitrary formulas A. Similarly—again using the fact that we only have decidable atoms of the form $\text{atom}(r^{\mathbf{B}})$—we can prove *compatibility*

$$x_1 =_\mu x_2 \to A(x_1) \to A(x_2) \quad (\mu \text{ finitary base type}).$$

Let HA^ω be the theory based on the axioms above including the induction axioms, and ML^ω be the (many-sorted) minimal logic, where the induction axioms are left out.

We define *pointwise equality* $=_\rho$, by induction on the type. $x =_\mu y$ for μ a finitary base type is already defined, and

$$(x =_{\rho \to \sigma} y) := \forall_z(xz =_\sigma yz),$$

$$(x =_{\rho \wedge \sigma} y) := (x0 =_\rho y0) \wedge (x1 =_\sigma y1).$$

The *extensionality axioms* are

$$y_1 =_\rho y_2 \to x^{\rho \to \sigma} y_1 =_\sigma x^{\rho \to \sigma} y_2.$$

We write E-HA^ω when the extensionality axioms are present.

In Troelstra (1973), Howard proved that already the first nontrivial instance of the extensionality scheme

$$y_1 =_1 y_2 \to xy_1 =_N xy_2$$

(with $1 := N \to N$) does not have a Dialectica realizer. In fact, he introduced the majorizing relation as a tool to prove this result. This is in contrast to the realizability interpretation, where extensionality axioms are unproblematic, since they are $\exists$-free.

As a substitute for extensionality one may add the *weak extensionality rule*

$$\frac{A_0 \to r =_\rho s}{A_0 \to t(r) =_\sigma t(s)} \quad (A_0 \text{ quantifier-free})$$

to the formal system considered. This "rule" is special in the sense that its premise must have been derived *without open assumptions*. Since the conclusion is (equivalent to) a purely universal formula, adding the weak extensionality rule does not change the behavior of the formal system w.r.t. the Dialectica interpretation.

We write WE-HA$^\omega$ when the weak extensionality rule is present but not the extensionality axioms.

We will also consider some more axiom schemes. The *axiom of choice* (AC) is the scheme

$$\forall_{x^\rho} \exists_{y^\sigma} A(x, y) \to \exists_{f^{\rho \to \sigma}} \forall_{x^\rho} A(x, f(x)).$$

Independence of premise (IP$_\forall$) is the scheme

$$(A \to \exists_{x^\rho} B) \to \exists_{x^\rho} (A \to B) \quad (x \notin \mathrm{FV}(A))$$

with A of the form $\forall_{y^\sigma} A_0$, A_0 quantifier-free. Moreover, we need the (constructively doubtful) *Markov principle* (MP), for a higher type variable x^ρ and quantifier-free formulas A_0, B_0:

$$(\forall_{x^\rho} A_0 \to B_0) \to \exists_{x^\rho} (A_0 \to B_0) \quad (x^\rho \notin \mathrm{FV}(B_0)).$$

2 Gödel's Dialectica Interpretation

Gödel (1958) assigned to every formula A a new one $\exists_{\vec{x}} \forall_{\vec{y}} A_D(\vec{x}, \vec{y})$ with $A_D(\vec{x}, \vec{y})$ quantifier-free. Here $\vec{x}, \vec{y}$ are lists of variables of finite types; the use of higher types is necessary even when the original formula A was first order. He did this in such a way that whenever a proof of A say in constructive arithmetic was given, one could produce closed terms $\vec{r}$ such that the quantifier-free formula $A_D(\vec{r}, \vec{y})$ is provable in T. Rather than working with tupels of variables and terms, we prefer to work with product types, in order to simplify the implementation. So we assign to every formula A its Gödel translation $\exists_x \forall_y |A|_y^x$, with $|A|_y^x$ quantifier-free.

2.1 Positive and Negative Types

To determine the types of x and y, we assign to every formula A objects $\tau^+(A)$, $\tau^-(A)$ (a type or the "nulltype" symbol ε). $\tau^+(A)$ is intended to be the type of a (Dialectica-)realizer to be extracted from a proof of A, and $\tau^-(A)$ the type of a challenge for the claim that this term realizes A. The definition can be conveniently written if we extend the use of $\rho \to \sigma$ and $\rho \wedge \sigma$ to the nulltype symbol ε:

$$(\rho \to \varepsilon) := \varepsilon, \qquad\qquad (\rho \wedge \varepsilon) := \rho,$$
$$(\varepsilon \to \sigma) := \sigma, \qquad\qquad (\varepsilon \wedge \sigma) := \sigma,$$
$$(\varepsilon \to \varepsilon) := \varepsilon, \qquad\qquad (\varepsilon \wedge \varepsilon) := \varepsilon.$$

With this understanding of $\rho \to \sigma$ and $\rho \wedge \sigma$, we can simply write

$$\tau^+(P(\vec{s})) := \varepsilon, \qquad\qquad \tau^-(P(\vec{s})) := \varepsilon,$$
$$\tau^+(A \wedge B) := \tau^+(A) \wedge \tau^+(B), \qquad \tau^-(A \wedge B) := \tau^-(A) \wedge \tau^-(B),$$
$$\tau^+(\forall_{x^\rho} A) := \rho \to \tau^+(A), \qquad \tau^-(\forall_{x^\rho} A) := \rho \wedge \tau^-(A),$$
$$\tau^+(\exists_{x^\rho} A) := \rho \wedge \tau^+(A), \qquad \tau^-(\exists_{x^\rho} A) := \tau^-(A).$$

and for implication

$$\tau^+(A \to B) := \big(\tau^+(A) \to \tau^+(B)\big) \wedge \big(\tau^+(A) \to \tau^-(B) \to \tau^-(A)\big),$$
$$\tau^-(A \to B) := \tau^+(A) \wedge \tau^-(B).$$

In case $\tau^+(A)$ $(\tau^-(A))$ is $\neq \varepsilon$ we say that A has *positive (negative) computational content*. For formulas without positive or without negative content one can give an easy characterization, involving the well-known notion of positive or negative occurrences of quantifiers in a formula:

$$\tau^+(A) = \varepsilon \leftrightarrow A \text{ has no positive } \exists \text{ and no negative } \forall,$$
$$\tau^-(A) = \varepsilon \leftrightarrow A \text{ has no positive } \forall \text{ and no negative } \exists,$$
$$\tau^+(A) = \tau^-(A) = \varepsilon \leftrightarrow A \text{ is quantifier-free.}$$

Examples.

(a) For quantifier-free A_0, B_0,

$$\tau^+(\forall_{x^\rho} A_0) = \varepsilon, \quad \tau^-(\forall_{x^\rho} A_0) = \rho,$$
$$\tau^+(\exists_{x^\rho} A_0) = \rho, \quad \tau^-(\exists_{x^\rho} A_0) = \varepsilon,$$
$$\tau^+(\forall_{x^\rho} \exists_{y^\sigma} A_0) = (\rho \to \sigma), \quad \tau^-(\forall_{x^\rho} \exists_{y^\sigma} A_0) = \rho.$$

(b) For arbitrary A, B, writing $\tau^{\pm} A$ for $\tau^{\pm}(A)$

$$\tau^{+}(\forall_{z^{\rho}}(A \to B)) = \rho \to (\tau^{+} A \to \tau^{+} B) \wedge (\tau^{+} A \to \tau^{-} B \to \tau^{-} A),$$

$$\tau^{+}(\exists_{z^{\rho}} A \to B) = (\rho \wedge \tau^{+} A \to \tau^{+} B) \wedge (\rho \wedge \tau^{+} A \to \tau^{-} B \to \tau^{-} A),$$

$$\tau^{-}(\forall_{z^{\rho}}(A \to B)) = \rho \wedge (\tau^{+} A \wedge \tau^{-} B),$$

$$\tau^{-}(\exists_{z^{\rho}} A \to B) = (\rho \wedge \tau^{+} A) \wedge \tau^{-} B.$$

It is interesting to note that for an existential formula with a quantifier-free kernel, the positive and negative type is the same, irrespective of the choice of the existential quantifier, constructive or classical.

Lemma. $\tau^{\pm}(\tilde{\exists}_x A_0) = \tau^{\pm}(\exists_x A_0)$ *for A_0 quantifier-free. In more detail,*

(a) $\tau^{+}(\tilde{\exists}_x A) = \tau^{+}(\exists_x A) = \rho \wedge \tau^{+}(A)$ *provided* $\tau^{-}(A) = \varepsilon$,

(b) $\tau^{-}(\tilde{\exists}_x A) = \tau^{-}(\exists_x A) = \tau^{-}(A)$ *provided* $\tau^{+}(A) = \varepsilon$.

Proof. For an arbitrary formula A we have

$$\tau^{+}(\forall_{x^{\rho}}(A \to \bot) \to \bot)$$

$$= \tau^{+}(\forall_{x^{\rho}}(A \to \bot)) \to \tau^{-}(\forall_{x^{\rho}}(A \to \bot))$$

$$= (\rho \to \tau^{+}(A \to \bot)) \to (\rho \wedge \tau^{-}(A \to \bot))$$

$$= (\rho \to \tau^{+}(A) \to \tau^{-}(A)) \to (\rho \wedge \tau^{+}(A)),$$

$$\tau^{+}(\exists_{x^{\rho}} A) = \rho \wedge \tau^{+}(A).$$

Both types are equal if $\tau^{-}(A) = \varepsilon$. Similarly

$$\tau^{-}(\forall_{x^{\rho}}(A \to \bot) \to \bot) = \tau^{+}(\forall_{x^{\rho}}(A \to \bot)) = \tau^{+}(A \to \bot) = \tau^{+}(A) \to \tau^{-}(A),$$

$$\tau^{-}(\exists_{x^{\rho}} A) = \tau^{-}(A).$$

Both types are $= \tau^{-}(A)$ if $\tau^{+}(A) = \varepsilon$. □

2.2 Gödel Translation

For every formula A and terms r of type $\tau^{+}(A)$ and s of type $\tau^{-}(A)$, we define a new quantifier-free formula $|A|_s^r$ by induction on A.

$$|P(\vec{s})|_s^r := P(\vec{s}),$$

$$|\forall_x A(x)|_s^r := |A(s0)|_{s1}^{r(s0)},$$

$$|\exists_x A(x)|_s^r := |A(r0)|_s^{r1},$$

$$|A \wedge B|_s^r := |A|_{s0}^{r0} \wedge |B|_{s1}^{r1},$$

$$|A \to B|_s^r := |A|_{r1(s0)(s1)}^{s0} \to |B|_{s1}^{r0(s0)}.$$

The formula $\exists_x \forall_y |A|^x_y$ is called the *Gödel translation* of A and is often denoted by A^D. Its quantifier-free kernel $|A|^x_y$ is called *Gödel kernel* of A; it is denoted by A_D.

For readability we sometimes write terms of a pair type in pair form:

$$|\forall_z A|^f_{z,y} := |A|^{fz}_y, \qquad |A \wedge B|^{x,z}_{y,u} := |A|^x_y \wedge |B|^z_u,$$

$$|\exists_z A|^{z,x}_y := |A|^x_y, \qquad |A \to B|^{f,g}_{x,u} := |A|^x_{gxu} \to |B|^{fx}_u.$$

Examples.

(a) For quantifier-free formulas A_0, B_0 with $x^\rho \notin \mathrm{FV}(B_0)$

$$\tau^+(\forall_{x^\rho} A_0 \to B_0) = \tau^-(\forall_{x^\rho} A_0) = \rho, \qquad \tau^-(\forall_{x^\rho} A_0 \to B_0) = \varepsilon,$$

$$\tau^+(\exists_{x^\rho}(A_0 \to B_0)) = \rho, \qquad\qquad \tau^-(\exists_{x^\rho}(A_0 \to B_0)) = \varepsilon.$$

Then

$$|\forall_{x^\rho} A_0 \to B_0|^x_\varepsilon \;=\; |\forall_{x^\rho} A_0|^\varepsilon_x \to |B_0|^\varepsilon_\varepsilon = A_0 \to B_0,$$

$$|\exists_{x^\rho}(A_0 \to B_0)|^x_\varepsilon = A_0 \to B_0.$$

(b) For A with $\tau^+(A) = \varepsilon$ and $z \notin \mathrm{FV}(A)$, and arbitrary B

$$\tau^+(A \to \exists_{z^\rho} B) \;= (\rho \wedge \tau^+(B)) \wedge (\tau^+(B) \to \tau^-(A)),$$

$$\tau^+(\exists_{z^\rho}(A \to B)) = \rho \wedge (\tau^+(B) \wedge (\tau^+(B) \to \tau^-(A))),$$

$$\tau^-(A \to \exists_{z^\rho} B) \;= \tau^-(B),$$

$$\tau^-(\exists_{z^\rho}(A \to B)) = \tau^-(B).$$

Then

$$|A \to \exists_{z^\rho} B|^{\langle z,y\rangle,g}_v \;= |A|^\varepsilon_{gv} \to |\exists_{z^\rho} B|^{z,y}_v = |A|^\varepsilon_{gv} \to |B|^y_v,$$

$$|\exists_{z^\rho}(A \to B)|^{z,\langle y,g\rangle}_v = |A \to B|^{y,g}_v = |A|^\varepsilon_{gv} \to |B|^y_v.$$

(c) For arbitrary A, B

$$\tau^+(\forall_{x^\rho} \exists_{y^\sigma} A(x,y)) \qquad= (\rho \to \sigma \wedge \tau^+(A)),$$

$$\tau^+(\exists_{f^{\rho\to\sigma}} \forall_{x^\rho} A(x,fx)) = (\rho \to \sigma) \wedge (\rho \to \tau^+(A)),$$

$$\tau^-(\forall_{x^\rho} \exists_{y^\sigma} A(x,y)) \qquad= \rho \wedge \tau^-(A),$$

$$\tau^-(\exists_{f^{\rho\to\sigma}} \forall_{x^\rho} A(x,fx)) = \rho \wedge \tau^-(A).$$

Then

$$|\forall_{x^\rho}\exists_{y^\sigma} A(x,y)|_{x,u}^{\lambda_x\langle fx,z\rangle} = |\exists_{y^\sigma} A(x,y)|_u^{fx,z} = |A(x,fx)|_u^z,$$

$$|\exists_{f^{\rho\to\sigma}}\forall_{x^\rho} A(x,fx)|_{x,u}^{f,\lambda_x z} = |\forall_{x^\rho} A(x,fx)|_{x,u}^{\lambda_x z} = |A(x,fx)|_u^z.$$

(d) For arbitrary A, writing $\tau^\pm A$ for $\tau^\pm(A)$

$$\tau^+(\forall_{z^\rho}(A \to \exists_{z^\rho} A)) = \rho \to (\tau^+ A \to \rho \wedge \tau^+ A) \wedge (\tau^+ A \to \tau^- A \to \tau^- A),$$

$$\tau^-(\forall_{z^\rho}(A \to \exists_{z^\rho} A)) = \rho \wedge (\tau^+ A \wedge \tau^- A).$$

Then

$$|\forall_{z^\rho}(A \to \exists_{z^\rho} A)|_{z,\langle x,w\rangle}^{\lambda_z\langle\lambda_x\langle z,x\rangle,\lambda_{x,w} w\rangle} = |A \to \exists_{z^\rho} A|_{x,w}^{\lambda_x\langle z,x\rangle,\lambda_{x,w} w}$$

$$= |A|_w^x \to |\exists_{z^\rho} A|_w^{z,x}$$

$$= |A|_w^x \to |A|_w^x.$$

2.3 Characterization

We consider the question when the Gödel translation of a formula A is equivalent to the formula itself.

Theorem. (Characterization)

$$\mathrm{AC} + \mathrm{IP}_\forall + \mathrm{MP} \vdash A \leftrightarrow \exists_x \forall_y |A|_y^x.$$

Proof. Induction on A; we only treat one case.

$$(A \to B) \leftrightarrow (\exists_x \forall_y |A|_y^x \to \exists_v \forall_u |B|_u^v) \quad \text{by IH}$$

$$\leftrightarrow \forall_x (\forall_y |A|_y^x \to \exists_v \forall_u |B|_u^v)$$

$$\leftrightarrow \forall_x \exists_v (\forall_y |A|_y^x \to \forall_u |B|_u^v) \quad \text{by } (\mathrm{IP}_\forall)$$

$$\leftrightarrow \forall_x \exists_v \forall_u (\forall_y |A|_y^x \to |B|_u^v)$$

$$\leftrightarrow \forall_x \exists_v \forall_u \exists_y (|A|_y^x \to |B|_u^v) \quad \text{by } (\mathrm{MP})$$

$$\leftrightarrow \exists_f \forall_x \forall_u \exists_y (|A|_y^x \to |B|_u^{fx}) \quad \text{by } (\mathrm{AC})$$

$$\leftrightarrow \exists_{f,g} \forall_{x,u} (|A|_{gxu}^x \to |B|_u^{fx}) \quad \text{by } (\mathrm{AC})$$

$$\leftrightarrow \exists_{f,g} \forall_{x,u} |A \to B|_{x,u}^{f,g},$$

where the last step is by definition. $\square$

Without the Markov principle one can still prove some relations between A and its Gödel translation. This, however, requires conditions $G^+(A)$, $G^-(A)$ on A, defined inductively by

$$G^{\pm}(P(\vec{s})) := \top,$$

$$G^+(A \to B) := (\tau^-(A) = \varepsilon) \wedge G^-(A) \wedge G^+(B),$$

$$G^-(A \to B) := G^+(A) \wedge G^-(B),$$

$$G^{\pm}(A \wedge B) := G^{\pm}(A) \wedge G^{\pm}(B),$$

$$G^{\pm}(\forall_x A) := G^{\pm}(A), \quad G^{\pm}(\exists_x A) := G^{\pm}(A).$$

Proposition.

$$\mathrm{AC} \vdash \exists_x \forall_y \, |A|_y^x \to A \quad \textit{if } G^-(A), \tag{1}$$

$$\mathrm{AC} \vdash A \to \exists_x \forall_y \, |A|_y^x \quad \textit{if } G^+(A).$$

Proof. Both directions are proved simultaneously, by induction on A. $\qquad\square$

2.4 Soundness

We prove soundness of the Dialectica interpretation, for our natural deduction formulation of the underlying logic.

We first treat some axioms, and show that each of them has a "logical Dialectica realizer," that is, a term t such that $\forall_y |A|_y^t$ can be proved logically.

For $(\exists^+)$ this was proved in Example (d) of 2.2. Conjunction introduction $(\wedge^+)$ and elimination $(\wedge^-)$ have obvious Dialectica realizers.

The axioms $(\exists^-)$, (MP), $(\mathrm{IP}_\forall)$ and (AC) all have the form $C \to D$ where $\tau^+(C) \sim \tau^+(D)$ and $\tau^-(C) \sim \tau^-(D)$, with $\rho \sim \sigma$ indicating that ρ and σ are canonically isomorphic. This has been verfied

- for the existence elimination axiom—written in the equivalent form $\forall_{z^\rho}(A \to B) \to \exists_{z^\rho} A \to B$—in Example (b) of 2.1;

- for (MP), $(\mathrm{IP}_\forall)$ and (AC) in Examples (a)–(c) of 2.2, respectively.

Such canonical isomorphisms can be expressed by λ-terms

$$f^+ : \tau^+(C) \to \tau^+(D), \qquad f^- : \tau^-(C) \to \tau^-(D),$$

$$g^+ : \tau^+(D) \to \tau^+(C), \qquad g^- : \tau^-(D) \to \tau^-(C).$$

(they have been written explicitly in Examples (a)–(c) of 2.2). It is easy to check that the Gödel translations $|C|_{g^-v}^u$ and $|D|_v^{f^+u}$ are equal (modulo β-conversion). But then $\langle f^+, \lambda_u\, g^- \rangle$ is a Dialectica realizer for the axiom $C \to D$, because

$$|C \to D|_{u,v}^{f^+, \lambda_u\, g^-} = |C|_{g^-v}^u \to |D|_v^{f^+u}.$$

Theorem. (Soundness) *Let M be a derivation*

$$\text{WE-HA}^\omega + \text{AC} + \text{IP}_\forall + \text{MP} + \text{Ax}_\forall \vdash A$$

from assumptions $u_i \colon C_i$ ($i = 1, \ldots, n$). Let x_i of type $\tau^+(C_i)$ be variables for realizers of the assumptions, and y be a variable of type $\tau^-(A)$ for a challenge of the goal. Then we can find terms $[\![M]\!]^+ =: t$ of type $\tau^+(A)$ with $y \notin \mathrm{FV}(t)$ and $[\![M]\!]_i^- =: r_i$ of type $\tau^-(C_i)$, and a derivation $\mu(M)$

$$\text{WE-HA}^\omega + \text{Ax}_\forall \vdash |A|_y^t$$

from assumptions $\bar{u}_i \colon |C_i|_{r_i}^{x_i}$.

Proof. Induction on M. We begin with the logical rules and leave treatment of the axioms for the end. The axioms $(\wedge^\pm)$, $(\exists^\pm)$, (MP), (IP$_\forall$), and (AC) have just been dealt with, so we will only need to consider induction, Ax$_\forall$, and the weak extensionality rule.

Case $u \colon A$. Let x of type $\tau^+(A)$ be a variable for a realizer of the assumption u. Define $[\![u]\!]^+ := x$ and $[\![u]\!]^- := y$.

Case $\lambda_{u^A} M^B$. By IH we have a derivation of $|B|_z^t$ from $\bar{u} \colon |A|_r^x$ and $\bar{u}_i \colon |C_i|_{r_i}^{x_i}$, where $\bar{u} \colon |A|_r^x$ may be absent. Substitute $y0$ for x and $y1$ for z. By $(\to^+)$ we obtain $|A|_{r[x,z:=y0,y1]}^{y0} \to |B|_{y1}^{t[x:=y0]}$, which is (up to β-conversion)

$$|A \to B|_y^{\lambda_x t, \lambda_{x,z} r}, \quad \text{from} \quad \bar{u}_i' \colon |C_i|_{r_i[x,z:=y0,y1]}^{x_i}.$$

Here r is the canonical inhabitant of the type $\tau^-(A)$ in case $\bar{u} \colon |A|_r^x$ is absent. Hence we can define the required terms by (assuming that u^A is u_1)

$$[\![\lambda_u M]\!]^+ := (\lambda_x [\![M]\!]^+, \lambda_{x,z} [\![M]\!]_1^-),$$

$$[\![\lambda_u M]\!]_i^- := [\![M]\!]_{i+1}^-[x, z := y0, y1].$$

Case $M^{A \to B} N^A$. By IH we have a derivation of

$$|A \to B|_x^t = |A|_{t1(x0)(x1)}^{x0} \to |B|_{x1}^{t0(x0)} \quad \text{from } |C_i|_{p_i}^{x_i}, |C_k|_{p_k}^{x_k}, \text{ and of}$$

$$|A|_z^s \qquad\qquad\qquad\qquad\qquad \text{from } |C_j|_{q_j}^{x_j}, |C_k|_{q_k}^{x_k}.$$

Substituting $\langle s, y \rangle$ for x in the first derivation and of $t1sy$ for z in the second derivation gives

$$|A|^s_{t1sy} \to |B|^{t0s}_y \quad \text{from } |C_i|^{x_i}_{p'_i}, |C_k|^{x_k}_{p'_k}, \text{ and}$$

$$|A|^s_{t1sy} \quad \text{from } |C_j|^{x_j}_{q'_j}, |C_k|^{x_k}_{q'_k}.$$

Now we contract $|C_k|^{x_k}_{p'_k}$ and $|C_k|^{x_k}_{q'_k}$: since $|C_k|^{x_k}_w$ is quantifier-free, there is a boolean-valued term r_{C_k} such that

$$|C_k|^{x_k}_w \leftrightarrow r_{C_k} w = \mathsf{tt}. \tag{2}$$

Hence with $r_k := [\textbf{if } r_{C_k} p'_k \textbf{ then } q'_k \textbf{ else } p'_k]$ we can derive both $|C_k|^{x_k}_{p'_k}$ and $|C_k|^{x_k}_{q'_k}$ from $|C_k|^{x_k}_{r_k}$. The derivation proceeds by cases on the boolean term $r_{C_k} p'_k$. If it is true, then r_k converts into q'_k, and we only need to derive $|C_k|^{x_k}_{p'_k}$. But this follows by substituting p'_k for w in (2). If $r_{C_k} p'_k$ is false, then r_k converts into p'_k, and we only need to derive $|C_k|^{x_k}_{q'_k}$, from $|C_k|^{x_k}_{p'_k}$. But the latter implies $\mathsf{ff} = \mathsf{tt}$ (substitute again p'_k for w in (2)) and therefore every quantifier-free formula, in particular $|C_k|^{x_k}_{q'_k}$.

Using $(\to^-)$ we obtain

$$|B|^{t0s}_y \quad \text{from } |C_i|^{x_i}_{p'_i}, |C_j|^{x_j}_{q'_j}, |C_k|^{x_k}_{r_k}.$$

Let $[\![MN]\!]^+ := t0s$ and $[\![MN]\!]^-_i := p'_i$, $[\![MN]\!]^-_j := q'_j$, $[\![MN]\!]^-_k := r_k$.

Case $\lambda_x M^{A(x)}$. By IH we have a derivation of $|A(x)|^t_z$ from $\bar{u}_i \colon |C_i|^{x_i}_{r_i}$. Substitute $y0$ for x and $y1$ for z. We obtain $|A(y0)|^{t[x:=y0]}_{y1}$, which is (up to β-conversion)

$$|\forall_x A(x)|^{\lambda_x t}_y, \quad \text{from} \quad \bar{u}'_i \colon |C_i|^{x_i}_{r_i[x,z:=y0,y1]}.$$

Hence we can define the required terms by

$$[\![\lambda_x M]\!]^+ := \lambda_x [\![M]\!]^+,$$

$$[\![\lambda_x M]\!]^-_i := [\![M]\!]^-_i [x, z := y0, y1].$$

Case $M^{\forall_x A(x)} s$. By IH we have a derivation of $|\forall_x A(x)|^t_z = |A(z0)|^{t(z0)}_{z1}$ from $|C_i|^{x_i}_{r_i}$. Substituting $\langle s, y \rangle$ for z gives

$$|A(s)|^{ts}_y \quad \text{from } |C_i|^{x_i}_{r_i[z:=\langle s, y \rangle]}.$$

Let $[\![Ms]\!]^+ := ts$ and $[\![Ms]\!]^-_i := r_i[z := \langle s, y \rangle]$.

We now come to induction, $Ax_\forall$, and the weak extensionality rule. For induction, consider for instance the algebra of natural numbers, given by constructors 0 and S. The induction schema then reads as follows:

$$\forall_n \big(A(0) \to \forall_m (A(m) \to A(m+1)) \to A(n) \big).$$

Let $B(n) := A(0) \to \forall_m(A(m) \to A(m+1)) \to A(n)$. Clearly we can derive $B(0)$ and $B(n) \to B(n+1)$. By those parts of the proof of the Soundness Theorem that we have dealt with already, we obtain realizing terms s and t, r and derivations of $|B(0)|_y^s$ and of $|B(n) \to B(n+1)|_{x,u}^{t,r}$; hence, of

$$|B(n)|_{rxu}^x \to |B(n+1)|_u^{tx},$$

$$\forall_y |B(n)|_y^x \to |B(n+1)|_u^{tx},$$

$$\forall_y |B(n)|_y^x \to \forall_y |B(n+1)|_y^{tx}.$$

So if we define $g(0) := s$ and $g(n+1) := t(g(n))$, then we have proved by induction that $\forall_y |B(n)|_y^{g(n)}$, and hence that $\exists_g \forall_y |\forall_n B(n)|_y^g$.

Now consider a purely universal formula $B = \forall_x A_0$, with A_0 quantifier-free. Then $\tau^+(B) = \varepsilon$, and moreover $|B|_y^\varepsilon = A_0$. Hence such axioms are interpreted by themselves. The weak extensionality rule can be dealt with in the same way.   □

2.5 Practical Aspects of Constructing Dialectica Realizers

In the proof of the Soundness Theorem above, at two points we have made (implicit) use of Dialectica realizers for logically derivable formulas:

- In the treatment of $\exists^-$, the equivalence of $\exists_{z^\rho} A \to \forall_{z^\rho}(A \to B) \to B$ with $\forall_{z^\rho}(A \to B) \to \exists_{z^\rho} A \to B$, and

- for induction, that we can derive $B(0)$ and $B(n) \to B(n+1)$, for $B(n) := A(0) \to \forall_m(A(m) \to A(m+1)) \to A(n)$.

Although these logical derivations are very easy, the fact that the formulas involved contain nested implications makes their Dialectica realizers complex. This shows up drastically in an implementation of the Dialectica interpretation. Two such implementations are presently available, both in the proof assistant and program extraction system Minlog[1]: one by Hernest (2006), and another one by the author, following the present paper.

Much more perspicious Dialectica realizers are obtained if one replaces the existence elimination and induction *axioms* by their equivalent *rule* formulations. Technically in our natural deduction setting with derivation terms, this means that the derivation constants $\exists^-$ and Ind appear with sufficiently many arguments. Clearly this can always be assumed (use η-expansion). Then Dialectica realizers are constructed as follows.

---

[1] See http://www.minlog-system.de

Case $\mathrm{Ind}_{n,A} m M_0^{A(0)} M_1^{\forall_n(A(n)\to A(n+1))}$. By IH we have derivations of

$$|\forall_n(A(n)\to A(n+1))|_{n,f,y}^t$$

$$= |A(n)\to A(n+1)|_{f,y}^{tn}$$

$$= |A(n)|_{tn1fy}^f \to |A(n+1)|_y^{tn0f} \quad \text{from } |C_i|_{r_{i1}(n,f,y)}^{x_i}$$

and of

$$|A(0)|_{x_0}^{t_0} \quad \text{from } |C_i|_{r_{i0}(x_0)}^{x_i}.$$

i ranges over all assumption variables in $\mathrm{Ind}_{n,A} m M_0 M_1$ (if necessary choose canonical terms r_{i0} and r_{i1}). It suffices to construct terms (involving recursion operators) $\tilde{t}, \tilde{r}_i$ with free variables among $\vec{x}$ such that

$$\forall_{m,y}\big((|C_i|_{\tilde{r}_i my}^{x_i})_i \to |A(m)|_y^{\tilde{t}m}\big). \tag{3}$$

For then we can define $[\![\mathrm{Ind}_{n,A} m M_0 M_1]\!]^+ := \tilde{t}m$ and $[\![\mathrm{Ind}_{n,A} m M_0 M_1]\!]_i^- := \tilde{r}_i my$. The recursion equations for $\tilde{t}$ are

$$\tilde{t}0 = t_0, \quad \tilde{t}(n+1) = tn0(\tilde{t}n)$$

and for $\tilde{r}_i$

$$\tilde{r}_i 0y = r_{i0}, \quad \tilde{r}_i(n+1)y = \begin{cases} r_{i1}(n,\tilde{t}n,y) =: s & \text{if } \neg|C_i|_s^{x_i}, \\ \tilde{r}_i n(tn1(\tilde{t}n)y) & \text{otherwise.} \end{cases}$$

$\tilde{t}, \tilde{r}_i$ can be written explicitly with recursion operators:

$$\tilde{t}m = \mathcal{R}mt_0(\lambda_n(tn0)),$$

$$\tilde{r}_i m = \mathcal{R}m(\lambda_y r_{i0})(\lambda_{n,p,y}[\text{if } r_{C_i}s \text{ then } p(tn1(\tilde{t}n)y) \text{ else } s])$$

with s as above. It remains to prove (3). We only consider the successor case. Assume $|C_i|_{\tilde{r}_i(n+1)y}^{x_i}$ for all i. We must show $|A(n+1)|_y^{\tilde{t}(n+1)}$. If $\neg|C_i|_s^{x_i}$ for some i, then by definition $\tilde{r}_i(n+1)y = s$ and we have $|C_i|_s^{x_i}$, a contradiction. Hence $|C_i|_s^{x_i}$ for all i, and therefore $\tilde{r}_i(n+1)y = \tilde{r}_i n(tn1(\tilde{t}n)y)$. The IH (3) with $y := tn1(\tilde{t}n)y$ gives $|A(n)|_{tn1(\tilde{t}n)y}^{\tilde{t}n}$. Recall that the global IH (for the step derivation) gives with $f := \tilde{t}n$

$$(|C_i|_s^{x_i})_i \to |A(n)|_{tn1(\tilde{t}n)y}^{\tilde{t}n} \to |A(n+1)|_y^{tn0(\tilde{t}n)}$$

and we are done.

Case $\exists_{x,A,B}^- M^{\exists_x A} N^{\forall_x(A\to B)}$. We proceed similar to the treatment of $(\to^-)$ above. By IH we have a derivation of

$$|\forall_x (A(x) \to B)|_x^t = |A(x0) \to B|_{x1}^{t(x0)}$$

$$= |A(x0)|_{t(x0)1(x10)(x11)}^{x10} \to |B|_{x11}^{t(x0)0(x10)}$$

from $|C_i|_{p_i}^{x_i}$, $|C_k|_{p_k}^{x_k}$, and of

$$|\exists_x A(x)|_z^s = |A(s0)|_z^{s1} \quad \text{from } |C_j|_{q_j}^{x_j}, |C_k|_{q_k}^{x_k}.$$

Substituting $\langle s0, \langle s1, y \rangle \rangle$ for x in the first derivation and of $t(s0)1(s1)y$ for z in the second derivation gives

$$|A(s0)|_{t(s0)1(s1)y}^{s1} \to |B|_y^{t(s0)0(s1)} \quad \text{from } |C_i|_{p'_i}^{x_i}, |C_k|_{p'_k}^{x_k}, \text{ and}$$

$$|A(s0)|_{t(s0)1(s1)y}^{s1} \qquad\qquad \text{from } |C_j|_{q'_j}^{x_j}, |C_k|_{q'_k}^{x_k}.$$

Now we contract $|C_k|_{p'_k}^{x_k}$ and $|C_k|_{q'_k}^{x_k}$ as in the case $(\to^-)$ above; with

$$r_k := [\textbf{if } r_{C_k} p'_k \textbf{ then } q'_k \textbf{ else } p'_k],$$

we can derive both $|C_k|_{p'_k}^{x_k}$ and $|C_k|_{q'_k}^{x_k}$ from $|C_k|_{r_k}^{x_k}$. Using $(\to^-)$, we obtain

$$|B|_y^{t(s0)0(s1)} \quad \text{from } |C_i|_{p'_i}^{x_i}, |C_j|_{q'_j}^{x_j}, |C_k|_{r_k}^{x_k}.$$

So $[\![\exists^- MN]\!]^+ := t(s0)0(s1)$ and

$$[\![\exists^- MN]\!]_i^- := p'_i, \quad [\![\exists^- MN]\!]_j^- := q'_j, \quad [\![\exists^- MN]\!]_k^- := r_k.$$

2.6 A Unified Treatment of Modified Realizability and the Dialectica Interpretation

Following Oliva (2006), we show that modified realizability can be treated in such a way that similarities with the Dialectica interpretation become visible. To this end, one needs to change the definitions of $\tau^+(A)$ and $\tau^-(A)$ and of the Gödel translation $|A|_y^x$ in the implicational case, as follows.

$$\tau_{\mathrm{mr}}^+(A \to B) := \tau_{\mathrm{mr}}^+(A) \to \tau_{\mathrm{mr}}^+(B),$$

$$\tau_{\mathrm{mr}}^-(A \to B) := \tau_{\mathrm{mr}}^+(A) \wedge \tau_{\mathrm{mr}}^-(B),$$

$$\|A \to B\|_{x,u}^{f} := \forall_y \|A\|_y^x \to \|B\|_u^{fx}.$$

Notice that the (changed) Gödel translation $\|A\|_y^x$ is not quantifier-free any more but only $\exists$-free. Then the standard definition of modified realizability mr (cf. Troelstra (1973)) can be expressed in terms of the (new) $\|A\|_y^x$:

$$\vdash r \text{ mr } A \leftrightarrow \forall_y \|A\|_y^r.$$

This is proved by induction on A. For prime formulas, the claim is obvious. *Case* $A \to B$, with $\tau_{\mathrm{mr}}^+(A) \neq \varepsilon$, $\tau_{\mathrm{mr}}^-(A) \neq \varepsilon$.

$$r \text{ mr } (A \to B) \leftrightarrow \forall_x (x \text{ mr } A \to rx \text{ mr } B) \quad \text{by definition}$$

$$\leftrightarrow \forall_x (\forall_y \|A\|_y^x \to \forall_u \|B\|_u^{rx}) \quad \text{by IH}$$

$$\leftrightarrow \forall_{x,u} (\forall_y \|A\|_y^x \to \|B\|_u^{rx})$$

$$= \forall_{x,u} \|A \to B\|_{x,u}^r \quad \text{by definition.}$$

The other cases are similar (even easier).

2.7 Extraction

As a consequence of the soundness and characterization theorems, we obtain

Theorem. (Extraction) *Assume*

$$\text{WE-HA}^\omega + \text{AC} + \text{IP}_\forall + \text{MP} + \text{Ax}_\forall \vdash \forall_x \exists_y A(x,y)$$

with A arbitrary. Then we can find a closed HA^ω-term t such that

$$\text{WE-HA}^\omega + \text{AC} + \text{IP}_\forall + \text{MP} + \text{Ax}_\forall \vdash \forall_x A(x,tx).$$

Moreover, in case the condition $G^-(A)$ is satisfied, we even have

$$\text{WE-HA}^\omega + \text{AC} + \text{Ax}_\forall \vdash \forall_x A(x,tx).$$

Proof. Recall that

$$|\forall_x \exists_y A(x,y)|_{x,b}^{\lambda_x \langle fx, gx \rangle} = |\exists_y A(x,y)|_b^{fx,gx} = |A(x,fx)|_b^{gx}.$$

By the Soundness Theorem, we obtain closed terms t, s such that

$$\text{WE-HA}^\omega + \text{Ax}_\forall \vdash \forall_{x,b} |A(x,tx)|_b^{sx}$$

and hence,

$$\text{WE-HA}^\omega + \text{Ax}_\forall \vdash \forall_x \exists_a \forall_b |A(x,tx)|_b^a.$$

By the Characterization Theorem, we have

$$\text{AC} + \text{IP}_\forall + \text{MP} \vdash \exists_a \forall_b |A(x,tx)|_b^a \to A(x,tx).$$

By (1), $(\text{IP}_\forall)$ and (MP) are not needed here provided the condition $G^-(A)$ is satisfied. Therefore the claim follows. □

Theorem. (Extraction from classical proofs) *Assume*

$$\text{WE-HA}^\omega + \text{AC} + \text{IP}_\forall + \text{MP} + \text{Ax}_\forall \vdash \forall_x \tilde{\exists}_y A_0(x,y),$$

$A_0(x,y)$ *a quantifier-free formula with at most the displayed variables free. Then we can find a closed HA^ω-term t such that*

$$\text{WE-HA}^\omega + \text{Ax}_\forall \vdash \forall_x A_0(x,tx).$$

Proof. This follws from the Soundness Theorem in 2.4 and

$$|\forall_x \tilde{\exists}_y A_0(x,y)|_x^t = |\tilde{\exists}_y A_0(x,y)|_\varepsilon^{tx} = \neg\neg A_0(x,tx). \qquad \square$$

3 Gödel's Dialectica Interpretation With Majorants

Generally, the Dialectica interpretation has a strong tendency to produce complex extracted terms, as opposed to the realizability interpretation. This is partially due to contraction (necessary in the $\rightarrow^-$-rule). Therefore it is advisable (even more so than for the realizability interpretation) to

- consider derivations from lemmata (whose proofs are not analyzed), and

- try to simplify extracted terms by only aiming at majorants.

This has led Kohlenbach (1992) and Kohlenbach (1996) to develop his "monotone Dialectica interpretation," where one only looks for bounds of realizers rather than exact realizers.

An essential point observed by Kohlenbach (1996) is that when one restricts attention to bounds rather than to exact realizers, then one can conveniently deal with additional assumptions $\text{Ax}_{\forall\exists\leq\forall}$ of the form

$$\forall_{x^\rho} \exists_{y\leq_\sigma rx} \forall_{z^\tau} A_0(x,y,z) \qquad (A_0 \text{ quantifier-free}),$$

with r a closed term of type $\rho \rightarrow \sigma$. We then need to consider strenghtened versions $\text{Ax}'_{\forall\exists\leq\forall}$ of these assumptions as well:

$$\exists_{Y\leq_{\rho\rightarrow\sigma}r} \forall_{x^\rho,z^\tau} A_0(x,Yx,z).$$

Note that with (AC) one can prove the strenghtened version from the original one.

3.1 Majorization

We define *pointwise majorization* $\geq_\rho$, by induction on the type. $x \geq_\mu y$ for μ s finitary base type is already defined, and

$$(x \geq_{\rho\rightarrow\sigma} y) := \forall_z(xz \geq_\sigma yz),$$

$$(x \geq_{\rho\wedge\sigma} y) := (x0 \geq_\rho y0) \wedge (x1 \geq_\rho y1).$$

For simplicity we treat the majorization relation of Howard (1973) just for types built from the base type **N** by $\rho \rightarrow \sigma$. We extend $\geq_{\mathbf{N}}$ to higher types, in a *pointwise* fashion (as we did for $=_\mu$ in 1.6 above)

$$(x_1 \geq_{\rho \to \sigma} x_2) := \forall_y (x_1 y \geq_\sigma x_2 y).$$

Following Howard (1973), we define a relation x^* maj_ρ x (x^* *hereditarily majorizes* x) for $x^*, x \in G_\rho$, by induction on the type ρ:

$$(x^* \ \mathrm{maj}_\mu \ x) \quad := (x^* \geq_\mu x),$$

$$(x^* \ \mathrm{maj}_{\rho \to \sigma} \ x) := \forall_{y^*, y} (y^* \ \mathrm{maj}_\rho \ y \to x^* y^* \ \mathrm{maj}_\sigma \ xy).$$

Lemma.

(a) $\vdash x^* =_\rho \tilde{x}^* \to x =_\rho \tilde{x} \to x^* \ \mathrm{maj}_\rho \ x \to \tilde{x}^* \ \mathrm{maj}_\rho \ \tilde{x}$.

(b) $\vdash x^* \ \mathrm{maj}_\rho \ x \to x \geq_\rho \tilde{x} \to x^* \ \mathrm{maj}_\rho \ \tilde{x}$.

Proof. Induction on ρ. We argue informally and only treat (b). *Case* $\rho \to \sigma$. Assume $y^* \ \mathrm{maj}_\rho \ y$. Then $x^* y^* \ \mathrm{maj}_\sigma \ xy$ and $xy \geq_\sigma \tilde{x}y$, hence by IH $x^* y^* \ \mathrm{maj}_\sigma \ \tilde{x}y$. □

3.2 Majorization of Closed HA$^\omega$-Terms

Let 1 denote the type $\mathbf{N} \to \mathbf{N}$. Clearly, for every monotone function D of type 1, we have $D \ \mathrm{maj} \ D$. Moreover, $\mathcal{R}_\mu^\tau$ is hereditarily majorizable:

Lemma. (Majorization)

(a) *Define* $M : (\mu \to \tau) \to \mu \to \tau$ *with* $\tau = \vec{\rho} \to \mu'$ *by*

$$M f n \vec{x} := \max_{i \leq n} f i \vec{x}.$$

Then HA$^\omega \vdash \forall_n \bar{f} n \ \mathrm{maj} \ fn \to M\bar{f} \ \mathrm{maj} \ f$.

(b) HA$^\omega \vdash f^*, g^* \ \mathrm{maj} \ f, g \to \mathcal{R}_\mu f^* g^* n \ \mathrm{maj} \ \mathcal{R}_\mu fgn$.

(c) *Define* $\mathcal{R}_\mu^* fg := M(\mathcal{R}_\mu fg)$. *Then* HA$^\omega \vdash \mathcal{R}_\mu^* \ \mathrm{maj} \ \mathcal{R}_\mu$.

Proof. We argue informally.

(a) Let $n^* \geq n$ and $\vec{x}^* \ \mathrm{maj} \ \vec{x}$; we must show $M\bar{f}n^*\vec{x}^* \geq fn\vec{x}$.

$$M\bar{f}n^*\vec{x}^* = \max_{i \leq n^*} \bar{f}i\vec{x}^* \geq \bar{f}n\vec{x}^* \geq fn\vec{x}.$$

(b) Induction on n; for simplicity we assume $\mu = \mathbf{N}$. For 0 the claim is obvious, and in the step we have by IH $\mathcal{R}f^*g^*(Sn) := g^*n(\mathcal{R}f^*g^*n) \ \mathrm{maj} \ gn(\mathcal{R}fgn) := \mathcal{R}fg(Sn)$, where $:=$ is definitional equality.

(c) Let $f^*, g^* \ \mathrm{maj} \ f, g$. We must show $M(\mathcal{R}f^*g^*) \ \mathrm{maj} \ \mathcal{R}fg$. By (a) it suffices to prove $\forall_n \mathcal{R}f^*g^*n \ \mathrm{maj} \ \mathcal{R}fgn$. But this holds by (b). □

The following theorem is due to Howard (1973).

Theorem. *Let $r(\vec{x})$ be a HA^ω-term with free variables among $\vec{x}$. Assume that $HA^\omega \vdash c^*$ maj c for all constants c in r. Let r^* be r with all constants c replaced by c^*. Then $HA^\omega \vdash \vec{x}^*$ maj $\vec{x} \to r^*(\vec{x}^*)$ maj $r(\vec{x})$.*

Proof. Induction on r. Case $\lambda_y\, r(y, \vec{x})$. We argue informally. Assume $\vec{x}^*$ maj $\vec{x}$. We must show y^* maj $y \to (\lambda_y\, r^*(y, \vec{x}^*))y^*$ maj $(\lambda_y\, r(y, \vec{x}))y$. So assume y^* maj y. Then by IH $r^*(y^*, \vec{x}^*)$ maj $r(y, \vec{x})$, which is our claim.   □

Hence every closed term r of HA^ω is hereditarily majorizable. In fact, we have constructed a closed term r^* of HA^ω such that r^* maj r.

3.3 Soundness with Majorants

Theorem. (Soundness with majorants) *Let M be a derivation*

$$WE\text{-}HA^\omega + AC + IP_- + MP + Ax_{\forall\exists\leq\forall} \vdash A$$

from assumptions $u_i \colon C_i$ $(i = 1, \ldots, n)$. Let x_i of type $\tau^+(C_i)$ be variables for realizers of the assumptions, and let y of type $\tau^-(A)$ be a variable for a challenge of the goal. Let $\vec{z}$ of type $\vec{\rho}$ be the variables free in M. Then we can find closed terms $[\![\lambda_{\vec{z},\vec{u}} M]\!]^{*+} =: T^*$ of type $\tau^+(C_1) \to \cdots \to \tau^+(C_n) \to \vec{\rho} \to \tau^+(A)$ and $[\![\lambda_{\vec{z},\vec{u}} M]\!]_i^{*-} =: R_i^*$ of type $\tau^+(C_1) \to \cdots \to \tau^+(C_n) \to \vec{\rho} \to \tau^-(A) \to \tau^-(C_i)$, and a derivation $\mu(M)$ in

$$WE\text{-}HA^\omega + Ax'_{\forall\exists\leq\forall}$$

of the formula

$$\exists_{T,R_1,\ldots,R_n}\big(T^* \text{ maj } T \wedge R_1^* \text{ maj } R_1 \wedge \cdots \wedge R_n^* \text{ maj } R_n$$

$$\wedge\, \forall_{\vec{x},\vec{z},y}(|C_1|_{R_1\vec{x}\vec{z}y}^{x_1} \to \cdots \to |C_n|_{R_n\vec{x}\vec{z}y}^{x_n} \to |A|_y^{T\vec{x}\vec{z}})\big).$$

Proof. Induction on M.

Case $u \colon A$. Let x of type $\tau^+(A)$ be a variable for a realizer of the assumption u. We need T^* and R^* such that

$$\exists_{T,R}\big(T^* \text{ maj } T \wedge R^* \text{ maj } R \wedge \forall_{x,y}(|A|_{Rxy}^x \to |A|_y^{Tx})\big).$$

We can take $Tx := x$ and $Rxy := y$, which both majorize themselves.

Case $c \colon A$, c an axiom. Consider an axiom

$$\forall_{x^\rho} \exists_{y \leq_\sigma rx} \forall_{z^\tau} A_0(x, y, z) \qquad (A_0 \text{ quantifier-free}),$$

with r a closed term of type $\rho \to \sigma$. We have to find a majorant of some T such that the following holds:

$$\forall_{x,z} |\forall_{x^\rho} \exists_{y \leq_\sigma rx} \forall_{z^\tau} A_0(x,y,z)|^T_{x,z},$$

$$\forall_{x,z} |\exists_{y \leq_\sigma rx} \forall_{z^\tau} A_0(x,y,z)|^{Tx}_z,$$

$$\forall_{x,z}(Tx \leq rx \wedge |\forall_{z^\tau} A_0(x,Tx,z)|_z),$$

$$\forall_{x,z}(Tx \leq rx \wedge A_0(x,Tx,z)).$$

We now use the corresponding axiom in $\mathrm{Ax}'_{\forall\exists\leq\forall}$:

$$\exists_{Y \leq_{\rho\to\sigma} r} \forall_{x^\rho, z^\tau} A_0(x,Yx,z).$$

Pick this Y as the desired T. Then as a majorant for Y, we can take a closed term r^* majorizing r.

For the other axioms we have already constructed a Dialectica realizer, and we can take an arbitrary majorant of it. However, we can also provide directly a majorant of some Dialectica realizer.

Case $\lambda_{u^A} M^B$. By IH we have a derivation of

$$\exists_{T,R_1,\dots,R_n,R}\big(T^* \text{ maj } T \wedge R_1^* \text{ maj } R_1 \wedge \cdots \wedge R_n^* \text{ maj } R_n \wedge R^* \text{ maj } R$$

$$\wedge \forall_{x_1,\dots,x_n,x,z}(|C_1|^{x_1}_{R_1 x_1 \dots x_n xz} \to \cdots \to |C_n|^{x_n}_{R_n x_1 \dots x_n xz}$$

$$\to |A|^x_{Rx_1 \dots x_n xz} \to |B|^{Tx_1 \dots x_n x}_z)\big).$$

We argue informally. Instantiating x with $y0$ and z with $y1$ gives

$$\forall_{x_1,\dots,x_n,y}(|C_1|^{x_1}_{R_1 x_1 \dots x_n (y0)(y1)} \to \cdots \to |C_n|^{x_n}_{R_n x_1 \dots x_n (y0)(y1)}$$

$$\to |A|^{y0}_{Rx_1 \dots x_n (y0)(y1)} \to |B|^{Tx_1 \dots x_n (y0)}_z),$$

which is

$$\forall_{x_1,\dots,x_n,y}(|C_1|^{x_1}_{R_1 x_1 \dots x_n (y0)(y1)} \to \cdots |C_n|^{x_n}_{R_n x_1 \dots x_n (y0)(y1)}$$

$$\to |A \to B|^{Tx_1 \dots x_n, Rx_1 \dots x_n}_y .$$

Therefore we can define the required $\tilde{T}^*, \tilde{R}_i^*$ by

$$\tilde{T}^* \vec{x} := \langle T^* \vec{x}, R^* \vec{x}\rangle, \quad \tilde{R}_i^* \vec{x} y := R_i^* \vec{x}(y0)(y1).$$

Case $M^{A \to B} N^A$. We argue informally. By IH we have

$$|A{\to}B|_x^{T\vec{x}_i\vec{x}_k} = |A|_{T\vec{x}_i\vec{x}_k1(x0)(x1)}^{x0} \to |B|_{x1}^{T\vec{x}_i\vec{x}_k0(x0)}$$

$$\text{from } |C_i|_{P_i\vec{x}_i\vec{x}_k x}^{x_i}, |C_k|_{P_k\vec{x}_i\vec{x}_k x}^{x_k}$$

$$|A|_z^{S\vec{x}_j\vec{x}_k}\text{from } |C_j|_{Q_j\vec{x}_j\vec{x}_k z}^{x_j}, |C_k|_{Q_k\vec{x}_j\vec{x}_k z}^{x_k}.$$

Instantiating x with $\langle S\vec{x}_j\vec{x}_k, y\rangle$ in the first and z with $T\vec{x}_i\vec{x}_k1(S\vec{x}_j\vec{x}_k)y$ in the second derivation gives

$$|A|_{T\vec{x}_i\vec{x}_k1(S\vec{x}_j\vec{x}_k)y}^{S\vec{x}_j\vec{x}_k} \to |B|_y^{T\vec{x}_i\vec{x}_k0(S\vec{x}_j\vec{x}_k)}\quad\text{from } |C_i|_{p_i'}^{x_i}, |C_k|_{p_k'}^{x_k}, \text{ and}$$

$$|A|_{T\vec{x}_i\vec{x}_k1(S\vec{x}_j\vec{x}_k)y}^{S\vec{x}_j\vec{x}_k}\qquad\qquad\text{from } |C_j|_{q_j'}^{x_j}, |C_k|_{q_k'}^{x_k},$$

with

$$p_i' := P_i\vec{x}_i\vec{x}_k\langle S\vec{x}_j\vec{x}_k, y\rangle,\qquad\quad p_k' := P_k\vec{x}_i\vec{x}_k\langle S\vec{x}_j\vec{x}_k, y\rangle,$$

$$q_j' := Q_j\vec{x}_j\vec{x}_k(T\vec{x}_i\vec{x}_k1(S\vec{x}_j\vec{x}_k)y),\quad q_k' := Q_k\vec{x}_j\vec{x}_k(T\vec{x}_i\vec{x}_k1(S\vec{x}_j\vec{x}_k)y).$$

Hence we can take

$$\tilde{T}^*\vec{x}_i\vec{x}_j\vec{x}_k := T^*\vec{x}_i\vec{x}_k0(S^*\vec{x}_j\vec{x}_k),$$

$$R_i^*\vec{x}_i\vec{x}_j\vec{x}_ky := P_i^*\vec{x}_i\vec{x}_k\langle S^*\vec{x}_j\vec{x}_k, y\rangle,$$

$$R_j^*\vec{x}_i\vec{x}_j\vec{x}_ky := Q_j^*\vec{x}_j\vec{x}_k(T^*\vec{x}_i\vec{x}_k1(S^*\vec{x}_j\vec{x}_k)y),$$

$$R_k^*\vec{x}_i\vec{x}_j\vec{x}_ky := \max(P_k^*\vec{x}_i\vec{x}_k\langle S^*\vec{x}_j\vec{x}_k, y\rangle, Q_k^*\vec{x}_j\vec{x}_k(T^*\vec{x}_i\vec{x}_k1(S^*\vec{x}_j\vec{x}_k)y)).$$

For the verifying derivation we again need to contract $|C_k|_{p_k'}^{x_k}$ and $|C_k|_{q_k'}^{x_k}$: since $|C_k|_w^{x_k}$ is quantifier-free, there is a boolean-valued term r_{C_k} such that

$$|C_k|_w^{x_k} \leftrightarrow r_{C_k}w = \text{tt}.$$

Hence with $r_k := [\textbf{if } r_{C_k}p_k' \textbf{ then } q_k' \textbf{ else } p_k']$, we can derive both $|C_k|_{p_k'}^{x_k}$ and $|C_k|_{q_k'}^{x_k}$ from $|C_k|_{r_k}^{x_k}$. Using $(\to^-)$, we obtain

$$|B|_y^{T\vec{x}_i\vec{x}_k0(S\vec{x}_j\vec{x}_k)}\quad\text{from } |C_i|_{p_i'}^{x_i}, |C_j|_{q_j'}^{x_j}, |C_k|_{r_k}^{x_k}.$$

Case $\lambda_x M^{A(x)}$. By IH we have a derivation of $|A(x)|_z^{Tx_1\ldots x_n x}$ from $|C_i|_{R_ix_1\ldots x_n xz}^{x_i}$. Instantiating x with $y0$ and z with $y1$ gives $|A(y0)|_{y1}^{Tx_1\ldots x_n(y0)}$, which is

$$|\forall_x A(x)|_y^{Tx_1\ldots x_n},\quad\text{from } |C_i|_{R_ix_1\ldots x_n(y0)(y1)}^{x_i}.$$

Hence we can take

$$\tilde{T}^*x_1\ldots x_n := T^*x_1\ldots x_n,$$

$$\tilde{R}_i^*x_1\ldots x_ny := R_i^*x_1\ldots x_n(y0)(y1).$$

Case $M^{\forall_x A(x)}s$. By IH we have in this case a derivation of $|\forall_x A(x)|_z^{Tx_1\cdots x_n}$, which is $|A(z0)|_{z1}^{Tx_1\cdots x_n(z0)}$, from $|C_i|_{R_ix_1\cdots x_n z}^{x_i}$. Instantiating z with $\langle s, y\rangle$ gives

$$|A(s)|_y^{Tx_1\cdots x_n s} \quad \text{from } |C_i|_{R_ix_1\cdots x_n\langle s,y\rangle}^{x_i}.$$

Assume for simplicity that s is closed. Then we can take

$$\tilde{T}^*x_1\ldots x_n := T^*x_1\ldots x_n s^*,$$

$$\tilde{R}_i^*x_1\ldots x_n y := R_i^*x_1\ldots x_n\langle s^*, y\rangle. \qquad \square$$

3.4 The Weak Lemma of König as a $\forall\exists_{\leq}\forall$-Axiom

We show that the "weak" (that is, binary) Lemma of König WKL can be brought into the form of an axiom in $\mathrm{Ax}_{\forall\exists_{\leq}\forall}$. This has been observed by Kohlenbach (1992). Here we give a somewhat simplified proof of this fact; it is based on ideas of Ishihara (2006).

WKL says that every infinite binary tree has an infinite path. When we try to formalize it directly in our (functional) language, it does not quite have the required form, since the assumption that the given tree is infinite needs an additional $\forall$ in the premise. However, one can easily find an equivalent statement of the required form. To this end, we define the "infinite extension" of a given tree, and we let WKL$'$ say that for every t, the infinite extension $I(\hat{t})$ of its "associated tree" $\hat{t}$ has an infinite path. It then is easy to see that WKL and WKL$'$ are equivalent.

Let us first introduce some basic definitions. Let $\mathbf{N}$ be the type of unary and bin the type of binary natural numbers. It is convenient here to view binary numbers as lists of booleans tt, ff, and to write these lists in reverse order, that is, add elements at the end. We fix the types of some variables and state their intended meaning:

a, b, c	of type bin	for nodes,
r, s, t	of type bin $\to \mathbf{B}$	for decidable sets of nodes,
f, g, h	of type $\mathbf{N} \to \mathbf{B}$	for paths,
n, m, k, i, j	of type $\mathbf{N}$	for natural numbers,
p, q	of type $\mathbf{B}$	for booleans.

Let $\mathrm{lh}(a)$ be the *length* of a (viewed as list of booleans). Let $\bar{a}(n)$ denote the initial segment of a of length n, if $n \leq \mathrm{lh}(a)$, and a otherwise. Similarly let $\bar{f}(n)$ denote the initial segment of f of length n, that is, the list $(f(0), f(1), \ldots, f(n-1))$. Let $(a)_n$ denote the n-th element of a, if $n < \mathrm{lh}(a)$, and tt otherwise. f is a *path in* t if all its initial segments $\bar{f}(n)$ are in t. Call t *infinite* if for every n, there is a node of length n in t. Call t a *tree* if it is downwards closed; i.e., $\forall_a\forall_{n\leq\mathrm{lh}(a)}(a \in t \to \bar{a}(n) \in t)$. So WKL says that

$$\forall_t \big(\forall_a \forall_{n \le \mathrm{lh}(a)} (a \in t \to \bar{a}(n) \in t) \to \quad (t \text{ is a tree}),$$

$$\forall_n \exists_{a \in t} \mathrm{lh}(a) = n \to \qquad\qquad (t \text{ is infinite}),$$

$$\exists_f \forall_n \bar{f}(n) \in t) \qquad\qquad\qquad (t \text{ has an infinite path}),$$

which—because of the two premises saying that t is an infinite tree—is not of the required logical form.

To obtain an equivalent formulation in the required form, we introduce some further notions.

$$\hat{t} := \{\, a \mid \forall_{n < \mathrm{lh}(a)} \bar{a}(n) \in t \,\} \quad \text{the associated tree } \hat{t} \text{ for } t,$$

$$b = a * \mathtt{tt}^{\mathrm{lh}(b) - \mathrm{lh}(a)} \qquad\qquad b \text{ is the } \mathtt{tt}\text{-extension of } a,$$

$$\forall_{c; \mathrm{lh}(c) = \mathrm{lh}(b)} c \notin \hat{t} \qquad\qquad b \text{ is } t\text{-big};$$

here $*$ denotes concatenation of lists. Let $\min_{\mathrm{lex}}$ denote the minimum of a set of nodes w.r.t. the lexicographical ordering, and $\mathrm{maxlen}_{<n}(t)$ be the maximal length of all nodes of t of length $< n$. Then $\mathrm{ll}_n(t)$ is the leftmost largest node in t of length $< n$:

$$\mathrm{maxlen}_{<n}(t) := \max\{\, \mathrm{lh}(a) \mid a \in t \wedge \mathrm{lh}(a) < n \,\},$$

$$\mathrm{ll}_n(t) \qquad := \min_{\mathrm{lex}}\{\, c \in t \mid \mathrm{lh}(c) = \mathrm{maxlen}_{<n}(t) \,\}.$$

We can now define the infinite extension $I(t)$ of a tree t:

$$I(t) := \{\, b \mid b \in t \vee (b \text{ is } t\text{-big} \wedge b \text{ is the } \mathtt{tt}\text{-extension of } \mathrm{ll}_{\mathrm{lh}(b)}) \,\}.$$

All these notions are definable in HA^ω. They clearly have the following properties:

$\hat{t}$ is a tree;

if t is a tree, then $\hat{t} = t$;

if t is a tree, then $I(t)$ is an infinite tree extending t;

if t is an infinite tree, then $I(t) = t$.

Then WKL is equivalent (provably in HA^ω) to

$$\mathrm{WKL}' := \forall_t \exists_f \forall_n \bar{f}(n) \in I(\hat{t}).$$

To see this, assume WKL, and let t be arbitrary. Then $I(\hat{t})$ is an infinite tree extending t. By WKL applied to $I(\hat{t})$, $\exists_f \forall_n \bar{f}(n) \in I(\hat{t})$. Conversely, let t be an infinite tree. Then $I(\hat{t}) = t$, and therefore, $\exists_f \forall_n \bar{f}(n) \in t$.

Remark. From the results of Ishihara (1990), it is known that WKL implies Brouwer's fan theorem. Moreover, a direct proof of this implication has been given by Ishihara in 2002 (published in Ishihara (2006)). In Berger and Ishihara (2005), it is shown a weakened form WKL! of WKL, where as an additional hypothesis as it is required that in an effective sense infinite paths are unique, is equivalent to Fan. One direction (WKL! implies Fan) is essentially the proof by Ishihara (2006), enhanced by the additional requirement that the tree extension to be constructed satisfies the effective uniqueness condition (as in Berger and Ishihara (2005)). The main tool of this proof is the construction of $I(\hat{t})$ described above. The other direction (Fan implies WKL!) is far less directly proved in Berger and Ishihara (2005), where the emphasis rather was to provide a fair number of equivalents to Fan, and to do the proof economically by giving a circle of implications. A direct proof of the equivalence of Fan with WKL! is in Schwichtenberg (2005). The latter paper also reports on a formalization in the Minlog proof assistant and gives rather short and perspicious realizing terms (w.r.t. modified realizability) machine-extracted from each of the two directions of this proof.

References

[Berger (1993)] U. Berger. Program extraction from normalization proofs. In M. Bezem and J. Groote, editors, *Typed Lambda Calculi and Applications*, volume 664 of *LNCS*, pages 91–106. Springer Verlag, Berlin, Heidelberg, New York, 1993.

[Berger (2005)] U. Berger. Uniform Heyting arithmetic. *Annals Pure Applied Logic*, 133: 125–148, 2005

[Berger et al. (2001)] U. Berger, H. Schwichtenberg, and M. Seisenberger. The Warshall Algorithm and Dickson's Lemma: Two Examples of Realistic Program Extraction. *Journal of Automated Reasoning*, 26: 205–221, 2001.

[Berger et al. (2002)] U. Berger, W. Buchholz, and H. Schwichtenberg. Refined program extraction from classical proofs. *Annals of Pure and Applied Logic*, 114: 3–25, 2002.

[Berger et al. (2006)] U. Berger, S. Berghofer, P. Letouzey, and H. Schwichtenberg. Program extraction from normalization proofs. *Studia Logica*, 82: 27–51, 2006.

[Berger and Ishihara (2005)] J. Berger and H. Ishihara. Brouwer's fan theorem and unique existence in constructive analysis. *Mathematical Logic Quarterly*, 51 (4): 360–364, 2005.

[Berger and Schwichtenberg (1995)] U. Berger and H. Schwichtenberg. Program development by proof transformation. In H. Schwichtenberg, editor, *Proof and Computation*, volume 139 of *Series F: Computer and Systems Sciences*, pages 1–45. NATO Advanced Study Institute, International Summer School held in Marktoberdorf, Germany, July 20 – August 1, 1993, Springer Verlag, Berlin, Heidelberg, New York, 1995.

[Friedman (1978)] H. Friedman. Classically and intuitionistically provably recursive functions. In D. Scott and G. Müller, editors, *Higher Set Theory*, volume 669 of *Lecture Notes in Mathematics*, pages 21–28. Springer Verlag, Berlin, Heidelberg, New York, 1978.

[Gödel (1958)] K. Gödel. Über eine bisher noch nicht benützte Erweiterung des finiten Standpunkts. *Dialectica*, 12: 280–287, 1958.

[Hernest (2006)] M.-D. Hernest. *Feasible programs from (non-constructive) proofs by the light (monotone) Dialectica interpretation*. PhD thesis, Ecole Polytechnique Paris and LMU München, 2006.

[Howard (1973)] W. A. Howard. Hereditarily majorizable functionals of finite type. In A. Troelstra, editor, *Mathematical Investigation of Intuitionistic Arithmetic and Analysis*, volume 344 of *Lecture Notes in Mathematics*, pages 454–461. Springer Verlag, Berlin, Heidelberg, New York, 1973.

[Ishihara (1990)] H. Ishihara. An omniscience principle, the König lemma and the Hahn-Banach theorem. *Zeitschr. f. math. Logik und Grundlagen d. Math.*, 36: 237–240, 1990.

[Ishihara (2006)] H. Ishihara. Weak König lemma implies Brouwer's fan theorem: a direct proof. *Notre Dame J. Formal Logic*, 47: 249–252, 2006.

[Jørgensen (2001)] K. F. Jørgensen. *Finite type arithmetic*. Master's thesis, University of Roskilde, 2001.

[Kohlenbach (1992)] U. Kohlenbach. Effective bounds from ineffective proofs in analysis: an application of functional interpretation and majorization. *The Journal of Symbolic Logic*, 57(4): 1239–1273, 1992.

[Kohlenbach (1996)] U. Kohlenbach. Analysing proofs in analysis. In W. Hodges, M. Hyland, C. Steinhorn, and J. Truss, editors, *Logic: from Foundations to Applications. European Logic Colloquium (Keele, 1993)*, pages 225–260. Oxford University Press, 1996.

[Oliva (2006)] P. Oliva. Unifying functional interpretations. *Notre Dame J. Formal Logic*, 47: 262–290, 2006.

[Schwichtenberg (1993)] H. Schwichtenberg. Proofs as programs. In P. Aczel, H. Simmons, and S. Wainer, editors, *Proof Theory. A selection of papers from the Leeds Proof Theory Programme 1990*, pages 81–113. Cambridge University Press, 1993.

[Schwichtenberg (2005)] H. Schwichtenberg. A direct proof of the equivalence between Brouwer's fan theorem and König's lemma with a uniqueness hypothesis. *Journal of Universal Computer Science*, 11 (12): 2086–2095, 2005. `http://www.jucs.org/jucs_11_12/a_direct_proof_of`.

[Schwichtenberg (2006)] H. Schwichtenberg. Recursion on the partial continuous functionals. In C. Dimitracopoulos, L. Newelski, D. Normann, and J. Steel, editors, *Logic Colloquium '05*, volume 28 of *Lecture Notes in Logic*, pages 173–201. Association for Symbolic Logic, 2006.

[Seisenberger (2003)] M. Seisenberger. *On the constructive content of proofs*. PhD thesis, Mathematisches Institut der Universität München, 2003.

[Troelstra (1973)] A. S. Troelstra, editor. *Metamathematical Investigation of Intuitionistic Arithmetic and Analysis*, volume 344 of *Lecture Notes in Mathematics*. Springer Verlag, Berlin, Heidelberg, New York, 1973.

Part III

Models of Computation from Nature

From Cells to (Silicon) Computers, and Back

Gheorghe Păun

Institute of Mathematics of the Romanian Academy
014700 Bucureşti, Romania
and
Research Group on Natural Computing
Department of Computer Science and Artificial Intelligence
University of Sevilla
Avda, 41012 Sevilla, Spain
george.paun@imar.ro, gpaun@us.es

Summary. Although the whole history of computer science is marked by events related to and inspired from "computations" taking place in living cells and organisms (human being included), in the last decades, this became a mainstream research direction, with important and well-established areas, such as evolutionary computing and neural computing, and with exciting new areas, such as DNA and membrane (cellular) computing. All these have both consequences on the efficiency of using standard computers, hopefully leading also to new types of hardware, and—maybe more importantly—on the very understanding of the notion of computing and, at the edge of science towards science fiction.

Topics of this kind will be touched in the paper, mainly in relation with DNA and membrane computing.

1 Preliminary warnings

What follows is not a mathematical or computer science paper, is not even a well-structured general essay, but it is mainly a sequence of personal thoughts, shuffled with lecture notes, and based on the personal experience in computer science (over 30 years) and, especially, in bio-inspired (DNA and membrane) natural computing (over 10 years). There are here mainly questions rather than answers, while the paragraphs form only the skeleton on which a much longer construction can/should be built in order to have any chance to cover the subject at hand. What is a computation/computer, does nature compute, are DNA and cellular computing feasible/useful, which are the limits of natural computing, do we dream too much, which can be the consequences of nanocomputers, these and others—more specific or more speculative—questions will be directly or implicitly formulated below, but each of them deserves a separate (chapter in a) book and efforts of a different scale than the present author can afford and the present text can include.

2 What is a computation?

This is already a trap question, context sensitive, and able to push the discussion in many divergent directions. I start from (and adhere to) the standard mathematical definition of what is computable, and the most convincing answer is provided by the *Turing machine*. It was about 70 years ago; there were several attempts to give a mathematical definition to what Hilbert called "mechanically" computable, and the one provided by Turing [62] was the most convincing (Gödel himself was one of the first to acknowledge this). In this framework, a computation is an algorithmic process relating an input to an output, and it is an algorithmic function. Rather reductionistic but accurate and, somewhat surprisingly when materialized in the form of a Turing machine, the most general concept of algorithmic computation, a fact commonly known as the Turing–Church thesis.

It is instructive to note that when defining the "machine" that today bears his name, Turing was explicitly trying to abstract what a clerk from a bank is doing when computing with numbers. What is essential and minimal to this activity? A support of information (the simplest one is a tape with cells that can hold symbols from a finite alphabet), the access to this information (the most reduced one is to see only one cell of the tape, with the possibility to move to left and right along the tape), a way to change the information, locally (the most local action is to change only one symbol at a time), everything under the control of a "state of the mind" (then, let us take states from a finite set given in advance), and according to a precise procedure (the simplest command/instruction, in our framework can be: in a given state, reading a given symbol from a cell of the tape, change that symbol, change the state, and move the control one cell to left, to right, or do not move at all; when changing a symbol or a state, we can also "change" it by itself). If we add the fact that we start in a special initial state, with an input written on the tape (one symbol in a cell), and we consider a computation finished when there is nothing to do (we halt), then we have, in plain words, what is called a Turing machine. It is a toy by all means but still the most general definition of an algorithm, the standard one now.

Actually, what made this "toy" immediately successful was not its definition only, but the fact that it can simulate other computing models (such as Church's lambda calculus) and, mainly, *Turing universality theorem*: there is a universal Turing machine, TU, which can simulate any particular Turing machine TM, in the following sense. If we give as input to TU a code of TM as well as an input x of TM, then TU will provide the same output as TM when starting from x; somewhat formally, $TU(code(TM)x) = TM(x)$. We have skipped several details, e.g., concerning the coding/decoding of inputs/outputs, because the important point is already here: the program of TU is the operating system, $code(TM)$, is the program to run, x is the data handled by this program (note that programs and data are stored in the same place), hence exactly the architecture of a "Turing–von Neumann computer". This is not incidental, as von Neumann has explicitly declared that he was influenced directly by Turing ideas when designing the first computers.

Let us record this observation—from the bank clerk to computers, via Turing machines (also noting that today there is no bank clerk without a computer on the desk!)—and let us return to the idea of computation. Despite its (mathematical) generality, Turing concept is very restrictive. One input, one output, one processor; halting as an essential condition; finitely many states and symbols. It is difficult to see a desk computer from a bank, and still lesser the server of the bank, working day and night, as a materialization of a Turing machine. The same with any parallel machine. Do we overpass in such a case the Turing–Church "barrier" and compute the uncomputable? Not exactly. We just move the discussion in another territory that of *processes that handle information*. This syntagma is so general that in these terms "everything is a computation"; it is a matter of point of view ("for every process there is an observer which can interpret the process as a computation").

3 Does nature compute?

The previous discussion already provides two positive answers to this question: nature computes (at least...) at the level of bank clerks, and depending on the observer, nature computes everywhere.... This explains/illustrates the two opposite positions, the over-orthodox one, which accepts as bio-computations only what *Homo Sapiens* is doing and calling so, and the over-relaxed one, which sees computations everywhere, from the genomic level, to cells and tissues, and to populations of organisms. "Life is computation. Every single living cell reads information from a memory, rewrites it, receives data input (information about the state of its environment), processes the data and acts according to the results of all this computation. Globally, the zillions of cells populating the biosphere certainly perform more computation steps per unit of time than all man made computers put together." We can read this statement at the beginning of [24].

In what follows, I choose a sort of annexionistic (but honest) position, somewhat specific to natural computing: having in mind the mathematical definition of computing, let us look at nature, especially at living nature, in biology, searching ideas, data supports and data structures, operations, architectures, processes, etc. that can inspire abstract computing models and/or new computing devices that can be of interest for humans. Especially by the last words, this looks like a pretty restrictive point of view, ruling out, for instance, "computations" taking place in nature in such a way that we cannot use them at least in theory, but this is not exactly the case. I said "of interest for" not "usable by" humans. Take the case of ciliates. They carry out wonderful list-processing operations (billions of years before McCarthy invented lists...) when unscrambling the micronuclear genes and passing to the macronuclear genes, see [18], with the goal of this process not related to computing; however, we, the humans, can see computations in ciliates, as we can see computations in each cell of a leaf, despite the fact that both the ciliates and the cells of a leaf have a unique purpose to live and nothing to do with (Turing) computations. Still, if we can at least define a (Turing) computing device abstracting from ciliates or leaf cells activity—as this happens already, see [18], [50], etc.—then (the mathematical representation of) the respective processes become computations.

The need to have such a reductionistic/restrictive position appears when we want to rule out "purely" analogical computations (in some sense, most of the computations "carried out" by biological processes are analogical). Just a striking example: when falling down, a drop of liquid instantaneously "solves" difficult differential equations on its surface. Is this a computation? I choose to answer in the negative as long as we cannot learn from here a computing model and/or as long as we cannot put *our* equation in the drop of liquid to have it solved (and the result is "read" somehow after the computation is completed).

A similar position, relating computations to an observer, is expressed in [60]: "We've just seen that it is not useful to call 'computation' just any nontrivial yet somewhat disciplined coupling between state variables. We also want this coupling to have been *intentionally* set up for the purpose of predicting or manipulating—in other words, for *knowing* or *doing* something. This is what shall distinguish bonafide computation from other intriguing function–composition phenomena such as weather patterns or stock-exchange cycles. But now we have new questions, namely, 'Set up by whom or what?', 'What is it good for?', and 'How do we recognize intention?'"

Far from me to want to sneak animistic, spiritualistic, or even simply anthropic considerations into the makeup of computation! *"The concept of computation must emerge as a natural, well-characterized, objective construct, recognizable by and useful to humans, Martians and robots alike"* (my emphasis, Gh.P.).

4 The limits of current computers

It is debatable whether, in newspaper style, "computers are the most important invention of the twentieth century," the one that had—or will have—the most visible impact on humankind (let us hope that this role will not be claimed by the atomic bomb), but what is clear already is that our lives depend today a lot on computers; we live already in a cyberspace that has "confiscated" us step by step without letting us notice it. Despite that—or because of that—the computers as they are present around us have drastic limitations. For good or for bad, this is an issue to discuss (not only the good guys use computers, but also the bad guys—for instance, making insecure the bank transactions). Let us think positively, and note that there still are a lot of things that the computers cannot do and that we would like to have them be able to do. The computers can handle huge amounts of data, at a speed that has continuously increased in the last decades; they are used in so many areas, from technology to medicine and to meteorology, they are every day more reliable and friendly, and they can play chess even better than the world champion. Well, but they are much behind us in speech and text understanding, recognizing faces, writing theorems, and (why not?) in playing GO, the ultimate challenge to AI, as it was written somewhere.

Moving to mathematics ("the numbers are sure things," said Galileo): the Turing–von Neumann computers (uniprocessor, working sequentially) have embarrassing limits: the class of problems tractable by such computers is pretty small; most of the real-life problems are intractable.

A parenthesis is worth opening here. At the beginning it was the question of what it means to compute. After having the Turing machine and many other classes of (mathematical) computing devices—such as Markov algorithms, Post systems, Chomsky grammars—one of the main questions was related to how much we can compute *in principle* (what is computable, what is decidable, how powerful a new computing model is). The discussion was centered on *competence*. Soon, it was realized that this is a beautiful issue from a mathematical and philosophical point of view, but from a practical point of view, it is much more relevant to deal with the *performances* of various computing models, with what can we compute now and here, with the computers we have, and what we can expect to have tomorrow. The first definition of tractability was related to the distinction between polynomial and exponential: a problem solvable in a time that increases at most polynomially in terms of its size was considered tractable. A beautiful theory of computational complexity was developed, aiming to classify the problems according to their time and space complexity; see, e.g., [46]. This beautiful theory has three important "drawbacks": (i) It cannot tell us, yet, whether $\mathbf{P} = \mathbf{NP}$ (with $\mathbf{P}$, the class of polynomially solvable problems being traditionally associated with "tractable," and $\mathbf{NP}$, the class of problems that can be solved in a polynomially time by first guessing a solution and then checking it, a superclass of $\mathbf{P}$ that contains a lot of practically relevant problems about which we do not know whether they are in $\mathbf{P}$ or not, hence whether they are tractable); (ii) it does not take care of "details," such as the coefficients and the degree of polynomials; both a problem that can be solved in $2n^2$ steps and one that needs $k \cdot n^{10000}$, for a large k, are in $\mathbf{P}$, but for a real computation the difference between the two problems is striking; (iii) it deals with the worst cases, i.e., counts the number of steps for solving the most difficult ("pathological") instance of the problem, whereas in reality, the "average" problem is the one of current interest. That is why, the borders of tractability were redefined continuously.

Still, what is clear is that the current computers cannot handle many of the problems that we encounter in reality, and that we would like to have solved. Let us sum–up this in the slogan: *the Turing–von Neumann computers cannot handle problems that request an exponential time to solve.*

One more parenthesis, related to the (in)famous $\mathbf{P} = \mathbf{NP}$ problem—actually, more frequently presented as the $\mathbf{P} \neq \mathbf{NP}$ conjecture. Its importance for computer science, witnessed, for instance, by the fact that it is the first of the seven Millennium Prize Problems of Clay Mathematics Institute (see www.claymath.org), whose solution is rewarded with one million dollars, is difficult to overestimate, but it is also very possible that its practical importance is much overestimated. Discussions about this topic can be found in many places;—see, e.g., [12]. If $\mathbf{P} \neq \mathbf{NP}$, as most computer scientists believe, nothing is changed from a practical point of view (for instance, we can continue to trust cryptographic systems/protocols based on $\mathbf{NP}$-complete problems, and we can also continue to complain that too many problems remain untractable for the Turing–von Neumann computers...). If the equality would be proven in a nonconstructive manner, or in a constructive but intractable manner, then again nothing is changed from a practical point of view (well, almost nothing, because the possibility remains to find easy polynomial solutions to certain problems in a direct way, not based on the proof of $\mathbf{P} = \mathbf{NP}$). If a feasible passage from $\mathbf{NP}$ to $\mathbf{P}$ will be obtained, then, indeed, many things in computer science and computer practice should be reconsidered, but, as said before, the community does not believe in such a possibility (unfortunately, in mathematics the votes do not validate theorems...).

Back to the slogan concerning the current difficulty to solve exponentially hard problems: there are several ways to handle this difficulty, most of them related to natural computing: (i) looking for (massively) parallel computations, (ii) looking for nondeterministic computations, (iii) being satisfied with approximate solutions, and (iv) being satisfied with probabilistic solutions.

5 The promises of natural computing

All four previously mentioned ideas are well investigated in the framework of "standard" computer science, the one based on electronic chips. Multiprocessor computers are already commercially available, but this does not mean massive parallelism. When *many* processors are put together, new difficulties appear, for instance, related to the communication complexity, which, with the increase of the number of processors, starts to be prohibitive (there is already a well-developed theory behind it; see, e.g., [32]). Then, randomized algorithms are provably better than deterministic algorithms—with the problem arising of finding truly random numbers, which is not possible on the usual computer (but this seems to be solved via quantum computing: a quantum device generating random bits and possible to plug in a usual PC was recently launched on the market). Brute force algorithms for addressing hard optimization problems and providing sometimes satisfactory enough solutions were reported since the "old times" of computer science. In turn, the theory of probabilistic algorithms is also well developed.

However, all these should be contrasted with what happens in biology, in a cell, for instance. A huge number of chemicals (ions, simple molecules, macromolecules, DNA molecules, proteins, etc.) evolve together, in a highly parallel manner, with a high degree of nondeterminism, with an intricate coordination, in a robust manner, coping successfully with the environment influences, displaying such attractive features as self-healing, adaptation, and learning. Then, we can discuss other, more technical details, such as reversibility of certain processes, or the energetic efficiency, with the number of operations per Joule much higher than in the case of electronic processing of information. It seems, indeed, that nature has considered the four ideas listed at the end of the previous section and, during billions of years, has polished better and better ways to implement them. Just pointing out that evolution led to the man, a quite parallel "machine," via (nondeterministic) mutations and other evolutionary processes after billions of years, is fully encouraging for computer science, in the attempt to learn from life for the benefit of computability.

And thus, natural computing has appeared: genetic algorithms as a way to drive the search through the space of candidate solutions to optimization problems imitating the Darwinian evolution, and neural networks, trying to imitate the human brain, again leading to approximate solutions, especially to pattern recognition problems. Later, DNA computing, proposing the DNA as a support for computations—hence proposing a massively parallel hardware, of a genuinely new type (bio-chips). Still later, membrane computing, with the aim of abstracting computationally useful (or at least interesting) ideas from the cell structure and functioning. And many others, related, for instance, to the collective behavior of ant colonies or of bacteria populations—not to forget quantum computing, which is also considered part of natural computing, although not biologically inspired.

6 Everything goes back to Turing

What is interesting, and also somewhat confusing, is that in some sense the whole history of computer science is the history of natural computing, of getting inspiration from life. I have mentioned before that Turing, in 1935–1936, when defining his machine, explicitly wanted to abstract and model what a human is doing when computing with numbers. One decade later, McCullock, Pitts, and Kleene founded the finite automata theory starting from modeling the neuron and the neural nets; still later, this led to what is called now neural computing—whose roots can be, however, found in unpublished papers of the same A. Turing.

This is a nice story, about the influence of sociology (of not-over-inspired group leaders) on science (on pure scientists, much more interested in their research than in publicizing it): In 1948, Turing wrote a short paper, "Intelligent machinery," which remained unpublished until 1968, just because his boss at the National Physical Laboratory in London (by the way, it was Sir Charles Darwin, the grandson of the great naturalist with the same name) dismissed the manuscript as a "schoolboy essay." "This paper was the first manifesto of the field of artificial intelligence. In the work (...) the British mathematician not only set out the fundamentals of connectionism but also brilliantly introduced many of the concepts that were later to become central to AI, in some cases after reinvention by others." —citation from [16]. Among others, the paper also introduced two types of randomly connected nets of "neurons," as a step toward creating an intelligent machine; one of the key features of these nets was the possibility of learning, of training them in order to solve problems. Nothing else than neural computing *avant la lettre*, with ideas rediscovered later, without reference to Turing. Details about Turing's "unorganized machines" can be found, e.g., in [58] and [57], whereas from [4] one can learn more about the "official" history of neural computing. At http://www.AlanTuring.net, one can find more about unpublished papers of Turing and recent efforts to reevaluate them.

The same A. Turing, in the same year 1948, proposed "genetical or evolutionary search," probably the first ideas of evolutionary computing. Also this area has had an interesting evolution, with many related branches independently initiated and developed—then merging in what is called today evolutionary computing. The domain now has four main branches: evolutionary programming (Fogel, Owens, Walsh), genetic algorithms (Holland), evolution strategies (Rechenberg, Schwefel), all three initiated in 1960s, and genetic programming (Koza, in 1990s). The first computer experiments on "optimization through evolution and recombination" were carried out by Bremermann, in 1962. For details and bibliographical information, we refer to [19].

It would be nice, and maybe not totally unexpected, to find also DNA and membrane computing ideas in the unpublished manuscripts of Turing....

It is interesting to remark that Turing himself also conceived ways to compute "beyond Turing," looking for devices able to surpass Turing machines. Curiously and sadly, the proposal was exposed in his doctoral thesis at Princeton University, in 1938, hence not in an unknown manuscript, but still the idea has been largely forgotten, as remarked in [16]. The hypermachine imagined by Turing, under the name of O-machine, was a Turing machine with an oracle, an external agent able to solve ("for free" as the computational cost) any decision problem from a given class of problems. This looks a little bit like cheating, because the extra power is explicitly introduced through the oracle, which is a black box whose functioning we do not care about. However, in current computability, the oracles are used in several areas (complexity, learning), without any reference to Turing.

Let us close this paragraph with mentioning that Turing can be seen as a pioneer not only of artificial intelligence, but also of artificial life: in his last years he was interested in morphogenesis, in the simulation of the processes of passing from the genes of a fertilized egg to the structure of the resulting animal.

7 A (simulated) wondering: why are genetic algorithms so good?

In order to see the (sometimes unexpected) benefits we can have when learning from nature how to compute (or to solve problems), it is instructive to examine in a few details the case of genetic algorithms. They are a combination of the slogan "if you do not know the right direction, then walk randomly," with the necessity to define what "randomly" means, and the solution is "like in nature, when evolving species."

More specifically, genetic algorithms try to imitate the bio-evolution in solving optimization problems: the candidate solutions to a problem are encoded as "chromosomes" (strings of abstract symbols, binary numbers, real numbers, representations of permutations, etc.), which are evolved by means of crossover (two chromosomes are cut in several parts and the parts are recombined, thus producing two new chromosomes) and point mutation (local changes, randomly produced) operations and are selected from a generation to the next one by means of a fitness mapping; the trials to improve the fitness mapping continue until either no essential improvement is obtained for a number of steps or until a given number of iterations are performed. The biological metaphors are numerous and obvious. What is not obvious (from a mathematical point of view) is why such a, basically, brute force approach is as successful as it happens to be (with a high probability, the genetic algorithms provide a good enough solution in a large number of applications; in many cases, the genetic algorithms escape from local maxima, converge very rapidly at the beginning of the process, and provide nonintuitive solutions that are hard to imagine otherwise). The most convincing "explanation" is probably "because nature has used the same strategy in improving species." This kind of bio-mystical "explanation" provides a rather optimistic motivation for related research: if genetic algorithms prove to be so successful (despite the lack of a convincing mathematical ground), why not try to bring to computer science other life features; if we are inspired enough (or lucky enough), then we may obtain similarly good ideas for using the existing computers (hence, ideas for new types of algorithms) or, why not, for new types of computers. We have reached in this way one of the most interesting attempts of natural computing: using the DNA molecules as a support for computations.

8 Adleman experiment

Like in most areas of science, there is a history and a pre-history of DNA computing. The history starts with Adleman's experiment [1], of solving a small Instance of the Hamiltonian Path Problem (HPP) in a biochemical laboratory; the pre-history goes back to Feinmann, with his much-quoted phrase "there is plenty of room at the bottom," then to Bennett, Conrad, and others who speculated in the 1970s about the possible use of molecules—in particular, bio-molecules—for computing.

These speculations were confirmed in 1994, when L. Adleman solved HPP in a lab, in linear time (as the number of lab operations), although the problem is known to be **NP**-complete, just by handling DNA by techniques already standard in biochemistry. Now, more than one decade later, the experiment looks strikingly simple (actually, it *is* simple; in one of the many interviews given after the event, Adleman said that after getting the idea, everything was so clear that he was sure that the experiment would succeed, but he effectively carried it out mainly for making the paper publishable...): the nodes of the graph are encoded by single-stranded DNA molecules, of length 20; if two nodes i, j, encoded by x_i, x_j, are linked by an arrow from i to j, then we also construct a single-stranded DNA molecule of length 20, with the first 10 nucleotides being complementary to the last 10 nucleotides from x_i and the last 10 nucleotides being complementary to the first 10 nucleotides from x_j; let us denote by $s_{i,j}$ this molecule. Millions of copies of each molecule of the two types were synthesized and put in a test tube; the temperature was decreased so that the single strands have annealed, producing double-stranded DNA molecules. Thus, the molecules $s_{i,j}$ acted as splints for the molecules x_i, x_j; now, if an arrow exists also from node j to a node k, then a longer molecule is created, and so on, *building in the test tube molecules representing paths in the graph*. It is important to note that everything is done in a massively parallel manner, with *all* paths created at the same time, in one step of the computation.

Of course, there are several details here (the biochemical technicalities are skipped): the paths are created only in the extent of existing enough initial "bricks," copies of molecules of types x_i and $s_{i,j}$; we do not consider here the chronological time of carrying out the experiment, but the computational time, the number of steps as a function of the size of the problem—here the number of nodes in the graph.

Adleman has considered a graph with seven nodes. This means that if a Hamiltonian path exists, then it is represented by a molecule of length 140. Selecting molecules by length is routine in biochemistry: by gel electrophoresis, one can distinguish even molecules differing in a few base pairs. After selecting all molecules of length 140, it was necessary to check whether among them there is one representing a path that visits all nodes (because we know the length, this also implies that each node is visited only once). This is again an easy procedure for biochemists: filtering according to a submolecule (by denaturation, polymerase chain reaction starting with an appropriate primer, gel electrophoresis, repeatedly for each node). In total, a number of steps of the order to the number of nodes: generating all possible paths, selecting the molecules of length 140, filtering once for each node. After a linear number of steps, the answer was found: the considered graph contained a Hamiltonian path.

9 DNA computing, pros and cons

The graph considered by Adleman was small, with the solution visible by a simple inspection (the electronic computers can handle HPP for graphs with 100–200 nodes), and the whole experiment took about one week, but still the enthusiasm was great. In terms of [27], this was a *demo* that we can compute by using DNA. (Hartmanis compares/contrasts computer science, which progresses by demonstrations that something new can be done, with physics, which progresses by means of crucial experiments.) The novelty of the experiment is obvious, making realistic the dream to solve hard problems in a feasible amount of time, by exploiting the massive parallelism possible due to the very compact way of storing information on DNA molecules (bits at the molecular level, with some orders of efficiency over silicon supports). In this way, billions of "computing chips" can be accommodated in a tiny test tube, much more than on electronic supports.

Still, several problems appear. First, we have to observe that we trade here space for time, with the space measured in DNA nucleotides. The same Hartmanis, after expressing his enthusiasm, has also calculated how many molecules we need in order to handle a graph with 100 nodes, and the conclusion was something like ...an Olympic-size swimming pool, [28], which is much more than any biochemist can dream to handle. (In this way, besides the classic measures of computational complexity, time, space, and, more recently, the communication complexity, another one was proposed: the *weight* of the computer! There is no joke here, because in DNA computing the information has a physical representation, each bit of information has a weight...) One may claim that the previous drawback, related to the quantity of necessary DNA, due to the fact that in the first stage one generates all paths in the graph, is specific to Adleman experiment—but this is only partially true. As long as the algorithm is intended to give the exact solution, there is no way to avoid this drawback (this would lead to redefining complexity classes in standard computational complexity theory); if we are satisfied with probabilistic answers to the problem, then we can avoid using an exponential amount of ADN, but then there is no big achievement, probabilistic algorithms are available also on the usual computers.

Actually, the big practical advantage of DNA-based algorithms is in moving forward the feasibility borderlines. We not only handle simultaneously billions of molecules, hence of "processors," but this is done in ways that avoid the difficulties related to using a large number of electronic processors. There is no need for synchronization, controlled communication, and other related difficult issues; everything proceeds in parallel, with a good degree of nondeterminism.

This nondeterminism has good effects and bad effects. Good, when we have "enough" molecules, so that "everything which has a chance to happen, actually happens"; hence we can explore at once a large space of candidate solutions. Bad, because in the case when we get a negative answer, we cannot know whether this is due to the fact that the problem has no solution or the experiment itself failed to find one. Note the fact that Adleman has chosen a graph possessing a Hamiltonian path. This discussion is related to the difference between so-called *false-positive* and *false-negative* solutions: although the first type of solutions can be checked, the second ones cannot. This observation makes DNA computing mainly suitable to address problems where we can check the correctness of the solution, and one good candidate in this respect is cryptography—after getting a key, we check it, and, if it is not correct, we repeat the experiment.

The errors in processing DNA (and any other molecules) are a major issue of research, both from a mathematical point of view (designing the input molecules in such a way to prevent wrong reactions, choosing error resistent problems, and designing error-resistant algorithms) and from a lab point of view (there are several proposals of filtering out the errors).

Anyway, after the initial enthusiasm (with a yearly international conference started in 1995, with books, research groups, projects, PhD theses, a huge bibliography, a lot of reported successful experiments, and surely, many more unsuccessful experiments, never reported...), the overall impression is that from the computer science point of view, the area has reached a sort of deadlock; if no additional breakthrough will appear, then no way is seen to scale up to practically useful computations. On the other hand, the biochemical progresses motivated by DNA computing should not be ignored; see, e.g., the so-called XPCR, reported in [20], or the by-products related to nano-technology (I will come back to this last point).

10 The marvellous DNA molecule

For computer science, DNA computing fuels several hopes, mainly related to the massive parallelism mentioned above; on this basis, we can simulate nondeterminism (which is anyway present in biochemistry), so that one can address in this framework computationally hard problems, with the possibility to push with some steps forward the feasibility barriers—at least for certain problems.

Other good features of DNA are also mentioned as a support for computations (energy efficiency, stability, reversibility of certain processes), but I switch here to a purely theoretical observation, which is simply spectacular from a general computability point of view: in a certain sense, *all Turing computable languages are "hidden" in the DNA molecules, and any particular language can be "read off" from this blueprint of computability by the simplest transducer, the finite state one!*

This informal style statement has a precise mathematical counterpart, which was first mentioned in [55]. Everything starts with an old characterization of recursively enumerable (RE) languages, as the projection of the intersection of a so-called twin-shuffle language with a regular language. However, both the projection and the intersection with a regular language, and the decoding of the symbols of an arbitrary alphabet from codes over a binary alphabet, can be computed by a sequential transducer. Therefore, every RE language is the image through a sequential transducer of the twin-shuffle language over the alphabet with two symbols. Now, a clever observation from [55] relates the twin-shuffle language over two symbols to certain "readings" of DNA molecules (one goes along the two strands of a molecule, step by step but with randomly varying speed, and producing a single string, by interleaving the visited nucleotides; this reading can be done either starting from the same end of a double-stranded molecule, or from opposite ends, for instance, according to the directionality of the two strands). Thus, every RE language can be obtained through a finite state transducer from the pool of readings of DNA molecules! The double-stranded data structure, with the corresponding nucleotides related by the Watson–Crick complementarity relation, is intrinsically universal from a computational point of view.

This observation (mathematical details can be found in [52], together with the following interesting strengthening: it is enough to use three nucleotides, two of them being complementary to each other and one of them its own complementary; from the computability point of view, the DNA is redundant—as it happens in many situations in nature) should bring to theoretical DNA computing a similar degree of optimism as genetic algorithms bring to practical natural computing.

Talking about computations and redundancy, let us also recall that most of the DNA molecule is traditionally considered as "junk DNA", as a way to express the fact that more than 90% of the nucleotides do not encode genes, and it is not clear that is their role; the percentage of "junk DNA" decreased with the advance of genetic knowledge, but still a large part of the DNA molecules seem to be unused. Let us speculate, starting from the observation that in the immunitary activity, the cell has to perform a truly computational task, for instance, when recognizing viruses, which are, basically, sequences of nucleotides; hence they have to be "parsed" as strings in a language. At which level, in the Turing hierarchy of computability, should this "computation" be done? Presumably, at a level as general as possible, maybe at the level of Turing machines themselves, in order to be able to recognize *all* intruders. However, this means that in recognizing a string of a given length, the working space we need can be exponentially large. Maybe the part of the DNA without a genetic function is just the workspace used in parsing aggressing agents—or in other "computations" done by the living cell.

11 Computing by splicing

Let us quit (for a while) this speculative type of discussion and return to DNA computing in a mathematical sense. At a theoretical level, DNA computing started before Adleman, namely in 1987, when T. Head introduced a language-theoretical formalization of what he called *the splicing operation*, a model of the recombinant behavior of DNA molecules under the influence of restriction enzymes and ligase [26]. Somebody said that the restriction enzymes are the most intelligent tools the nature gave us. They are a sort of "context-sensitive scissors": they recognize a short (usually, 6, 8, 10, 12 base pairs) submolecule of a DNA double-stranded molecule, and they cut (in most cases in the interior of this submolecule, but in some cases also outside, at a well-defined distance) the molecule in such a way that (again, in most cases) the two fragments have staggered ends; they are called *sticky ends*, because the single strands with unpaired nucleotides are available to annealing with corresponding complementary single strands. In this way, fragments with identical sticky ends, produced by the same restriction enzyme or by different restriction enzymes that leave identical sticky ends, can be recombined, producing new molecules (the pasting together of fragments also needs the presence of certain enzymes called ligase).

Because this operation is rather interesting, we illustrate it in Figure 1, in a schematic manner. The DNA molecules x and y are cut by two different enzymes, one using the site u_1zu_2 and the second one the site v_1zv_2; important is that the sticky ends are the same, which is indicated here by z. (I have omitted a series of biochemical details, such as the complementarity of the sticky ends, the inverse directionality of the two strand, and the role of the ligase.) The fragments obtained after cutting can either recombine into the starting molecules or into new molecules; if we come back to the molecules x, y, then the enzymes will cut again; hence, after a while "all" molecules x, y will be consumed and only recombined molecules will remain.

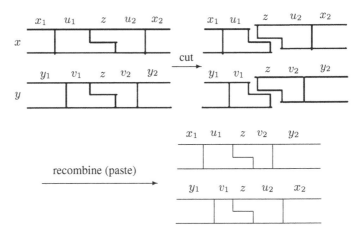

Fig. 1. A schematic representation of the splicing operation

Because of the precise Watson–Crick complementarity and because we can associate the sticky end with one of the contexts of cutting, we can abstract from here, without betraying too much the reality, an operation with strings, the splicing. I present it in the form introduced in [48] (used also in [52]), which is a generalization of the operation from [26].

A splicing rule over an alphabet V is a string $r = u_1\#u_2\$v_1\#v_2$, with u_1, u_2, v_1, v_2 strings over V and $\#, \$$ symbols not in V. For two strings x, y over V, we write

$$(x, y) \vdash_r (w, z) \text{ iff } x = x_1 u_1 u_2 x_2, \ y = y_1 v_1 v_2 y_2,$$

$$w = x_1 u_1 v_2 y_2, \ z = y_1 v_1 u_2 x_2,$$

$$\text{for some } x_1, x_2, y_1, y_2 \in V^*.$$

We say that we have spliced x, y at sites $u_1 u_2, v_1 v_2$, respectively, obtaining the strings w, z.

Now, this operation can be extended in a natural way to sets of strings and to sets of rules, and then it can be iterated. Starting from a finite set of strings and a finite set of splicing rules, we get a language that can be infinite.

It was proved already in 1991, by K. Culik and T. Harju (for the form of splicing rules considered by T. Head), that in this way we obtain at most a regular language. The proof was extended in 1997 by D. Pixton (with a recent easier proof by V. Manca) also to the above type of rules. This is not too much: "computing by splicing" (that is, by using restriction enzymes) is under the competence of finite automata, the simplest class of automata (of restrictions of Turing machines). However, if some simple controls about the use of rules, such as permitting or forbidding conditions (a rule is applied to two string only if they contain—do not contain, respectively—a specified symbol), taking care of the multiplicity of certain molecules, throwing away all molecules that are not used, and continuing only with the results of the splicing, etc., then we can directly jump at the level of Turing machines, and we can generate all languages recognized by Turing machines. Details can be found in [52].

12 What does it mean to compute "in a natural way?"

The previously mentioned results raise a series of interesting questions.

Traditionally, computer science uses in almost all computing models/theories strings as a data structure and the rewriting as the basic operation—where by rewriting we understand a local change in a string. The suggestive example is that of Chomsky phrase structure grammars, with their rules of the form $u \rightarrow v$, where u, v are strings. The same is true for Turing machines, Post systems, Markov algorithms, etc. Nature instead uses mainly recombination, as modeled by splicing, in the process of reproduction, whereas local mutations occur only accidentally, and most of the resulting beings are rejected; only at large intervals of time does the evolution accept a mutated individual. The difference between rewriting and recombination is apparent. Still, both rewriting and recombination (the latter one enhanced with some other context sensing abilities, such as permitting symbols) are computationally universal (hence equal in power). Coming from nature, we may say that the "natural" way to compute is by recombination (cut-and-paste). Why then has computer science not considered recombination as a basic operation? Because the human beings, as macroscopic "computers", are intrinsically closer to the Turing machine than to their internal micro(nano)scopic processes? Because it is easier to process strings locally? Because Hilbert asked what is mechanically computable and related the question (his answer) to formal systems, where substitution is the basic inference operation? It may be a little from all these, but still the question remains of whether it makes sense to reconstruct the computability and the computer science itself in terms of recombination. The results mentioned above show that this is, in principle, possible. Is this also useful?

Then, another observation related to the question in the title of this paragraph is the following. Uncontrolled iterated splicing can characterize the power of finite automata, weakly controlled splicing leads directly to the power of Turing machines, without any simple characterization of intermediate classes, for instance, of classes of languages in Chomsky hierarchy. Does this mean that these intermediate classes are not "natural?" In some sense, this is the case. The definition of context-free languages have a mathematical-linguistic flavor, whereas the context-sensitive languages have a computational complexity definition.

Furthermore, the level of finite automata is not attractive for computability, not only because it is too weak, but also because there is no universality result for finite automata (hence at this level we cannot have programmable computers); going to Turing machines requests an additional control on the splicing operations, of a type that is met but not usual in nature. Otherwise stated, it seems that nature computes by splicing at the level of finite automata only. This raises serious doubts over whether programmable DNA "computers" based on splicing can be ever constructed (of course, such a computer should work autonomously, not having the user as part of the computation—as it is the case, e.g., in Adleman's experiment, where, actually, Adleman has computed, not using paper and pencil, but DNA and enzymes). This observation is worth remembering when discussing the possible intrinsic difficulties/limits of natural computing in general.

13 The marvellous cell

Let us pass now to the smallest living unit, the cell (what is alive/nonalive depends on the definition used, but the cell is unanimously considered alive, whereas, for instance, the viruses are not considered so, because they need other cells to replicate and do not have their own metabolism). The cell is a small but very complex "factory," with a complicate internal structure and activity, robust and sensitive at the same time, facing—alone or in tissues, organs, organisms—the environment aggressions in a wonderful way, handling billions of chemical compounds, from ions to complicatedly folded proteins, as well as information (see again the quote from [24]).

We do not enter here into the fascinating discussion about the role of cells in making possible the life—details can be found in any book about the biology of the cell, such as the comprehensive [2]—but we only cite from [30] the title itself, as a slogan (life means surfaces inside surfaces), then, through [29], a paragraph from [34]: "The secret of life, the wellspring of reproduction, is not to be found in the beauty of Watson–Crick pairing, but in the achievements of collective catalytic closure."

The previous quotes point to one of the essential cell features/ingredients: the membranes. The cell itself exists, can be delimited as an autonomous unit, because it is separated from the environment by its external membrane. Then, inside the cell, several membranes delimit "protected reactors," where a specific biochemical activity takes place. The nucleus, with the hereditary material (in the eucaryotic cells), the complicated Golgy apparatus, mitochondria, vesicles, several membranes arranged in a hierarchical manner, in three or four levels. Moreover, the membranes of a cell do not only delimit compartments where specific reactions take place in solution, hence *inside* the compartments, but many reactions in a cell develop *on the membranes*, catalyzed by the many proteins bound on them. It is said that when a compartment is too large for the local biochemistry to be efficient, life creates membranes, both in order to create smaller "reactors" (small enough that, through the Brownian motion, any two of the enclosed molecules can collide frequently enough), and in order to create further "reaction surfaces."

Rather interesting, all membranes from a cell have in principle the same structure: they are bilayers of phospholipidic molecules, with a hydrophilic head and a hydrophobic tail (consisting of two fatty acids), so that, in the water that both surrounds the cell and fills in the cell these molecules self-arrange to form a bilayer, "hiding" the hydrophobic tails in the middle, and closing a 3D surface—thus spontaneously delimiting an "inside" and an "outside." In between the lipidic molecules there are placed a variety of other molecules, especially proteins, cholesterol, and others, with the proteins either forming channels from a side of the membrane to the other side, or placed on only one side of the membrane. The protein channels have a crucial role in the passage of chemicals across the membrane. Polarized molecules and large molecules cannot pass through a membrane composed only of lipidic molecules, but they can pass through protein channels. This passage is highly selective, with the selection done by size or, the most interesting case, by the very nature of the chemicals. Moreover, the passage of molecules through protein channels is done, by osmosis, from a higher concentration to a lower concentration, hence without a consumption of energy, but proteins also exist that move substances against the gradient—this time with a consumption of metabolic energy. In most cases, the molecules pass one at a time through a protein channel, but there also are cases when two molecules pass together (without being possible for them to pass alone), either in the same direction or in opposite directions. The first situation is called *symport*, and the latter one is called *antiport*. These local phenomena are sometimes rather intricate, with several intermediate steps, or they are coupled in more complex trans-membrane transport processes.

Furthermore, "[m]any proteins in living cells appear to have as their primary function the transfer and processing of information, rather than the chemical transformation of metabolic intermediates or the building of cellular structures. Such proteins are functionally linked through allosteric or other mechanisms into biochemical 'circuits' that perform a variety of simple computational tasks including amplification, integration and information storage." This was the very abstract of [7], whose title is also very suggestive for another direction in which the cell inner activity can be considered: the computational aspects. Actually, as we have seen also from [24], the computational (information processing) aspects of life are a current topic for discussions coming from semiotics (the case of Hoffmeyer), biology (see Bray above, but also [25]—the cytoskeleton itself, biochemically "less active" than the membrane proteins, can be seen as an automaton—while [40] builds a whole theory on the informational aspects of cell life), computer science, artificial life. We close this discussion with a quote from [41], where the issue is synthesized in an equational form with a suggestive computer science flavor: *Life = DNA software + membrane hardware*.

Now, let us remark that only some cells live alone (such as ciliates, bacteria), but in general the cells are organized in tissues, organs, organisms, and communities of organisms. All these suppose a specific organization, starting with the direct communication/cooperation among neighboring cells (adjacent cells can communicate directly, through proteins which are realizing direct channels from a cell to another cell), and ending with the interaction with the environment, at various levels. Also many of these intercellular processes can be considered as computations—not to speak about the fact that the neurons themselves are cells, specialized in information processing.

A short parenthesis here, illustrating once again how intricate and surprising nature can be from a computational point of view. The standard dimensions of computations are time and space. This is true for computer science, not necessarily for the brain, where a great part of computations use time as a sort of resource, as a support of data, in so-called spiking neurons. Such a neuron sends through its axon and dendrites an electrical pulse, which is independent from the inputs to the neuron; however, the time when the impulse is produced depends on the input. This highly differs from the understanding of neurons as captured in artificial neural nets, leading to the "third generation of neural network models"; see [42].

And now the question arises: what can we learn from the cell biology useful to computer science? Membrane computing is an answer to this question.

14 A glimpse to membrane computing

As we have said in the beginning of this discussion, membrane computing is one of the youngest areas of natural computing (the paper [49] was first circulated in

November 1998 on web), and the first one starting explicitly from the cell structure and functioning and systematically investigating computing models inspired both from the life of separate cells and from the functioning of conglomerates of cells, such as tissues or neural nets. The theory is much developed for cell-like membrane systems (P systems), with a considerable recent advance of population P systems (tissue-like P systems), and with a promising beginning of investigations of neural-like P systems, where a lot of work remains to be done.

In what follows, in order to have an idea of what membrane computing means, we only discuss the case of cell-like P systems.

The basic idea is to consider a hierarchical arrangement of membranes, like in a cell, delimiting compartments where various chemicals (we call them *objects*) evolve according to local reaction rules. These objects can also pass through membranes, under the control of specific rules. Because the chemicals in a cell are swimming in an aqueous solution, the data structure we consider is that of a *multiset*—a set with multiplicities associated with its elements. The reaction rules are applied in a parallel manner, with the objects to evolve by them and with the reactions themselves chosen in a nondeterministic manner. In this way, we can define transitions from a configuration to another configuration; hence we can define computations. A computation that halts (reaches a configuration where no rule is applicable) provides a result, for instance, in the form of the number of objects present in the halting configuration in a specified compartment.

There are many variants of this very basic type of a computing device. In all cases, one the fundamental ingredients is that of the *membrane structure*.

The meaning of this notion is illustrated in Figure 2, and this is what we can see when looking (through mathematical glasses, hence abstracting as much as necessary in order to obtain a formal model) to a standard cell.

The typical kind of rule for evolving objects is a multiset-rewriting one, of the form $u \rightarrow v$, where u and v are multisets of objects (this is much like a usual equation describing a chemical reaction), with several variants. A rule of the general form is called cooperative, when u consists of a single objects the rule is called noncooperative (this corresponds to context-free rules in Chomsky grammars), whereas an intermediate (interesting) case is that of catalytic rules, of the form $ca \rightarrow cv$, where c is an object that behaves like a catalyst, it never changes, but only helps object a to get transformed into multiset v.

A very important type of rules are those that correspond to symport and antiport trans-membrane operations: we write (x, in) or (x, out) for a symport moving the objects from multiset x inside, or outside a membrane, respectively, and $(x, out; y, in)$ for an antiport that moves the objects of x outside at the same time with bringing the objects of y inside.

We skip here all technical details, and we refer the interested reader to [50] for a comprehensive presentation (at the level of the spring of 2002), with recent information

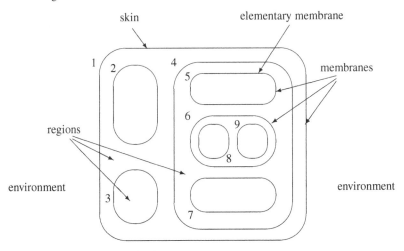

Fig. 2. A membrane structure

available in [65], and we only mention the two main classes of results in membrane computing:

1. Most classes of P systems are computationally universal, equal in power with Turing machines; for instance, this is true for multiset-rewriting systems with catalytic rules (with two catalysts, see [21]), and also for symport/antiport P systems ([47]), even with rules of rather small sizes (see [3]).

2. When systems with an enhanced parallelism are used, e.g., having rules that allow the division of membranes, like in biology; then computationally hard problems (typically, **NP**-complete problems) can be solved in a polynomial (often linear) time in this framework; of course, the trick lies in the time–space trade-off made possible by means of the fact that, by membrane division, an exponential number of membranes (encapsulated "processors") can be created in a linear number of steps.

We can conclude with the observation that the "computing cell" is a very powerful and efficient "computer."

The previous results, both concerning the power (competence) and the efficiency (performance) of membrane computing, are of a theoretical interest, but they are ...theory. However, membrane computing started to become a more and more successful framework for applications. Biology, linguistics, management, and computer science are areas where several applications were reported; see [13] and [65]. The applications to biology take in general processes that develop at the level of the cell, build a model in terms of P systems (in many cases with probabilities/reaction rates associated with the rules), then build a computer program simulating the P system, and experiment with this program, tuning parameters or inputs and following the evolution of the process along a large number of steps. The applications to linguistics and management are mainly using the language of membrane computing, the graphical one, the mathematical formalism, etc. A recent approach to apllications in economics was started in [51], where a genuinely new class of P systems was introduced, with numerical variables placed in membranes, evolving by production–repartition programs. Very promising are the computer science applications, for instance, in computer graphics and in devising approximate algorithms for solving hard optimization problems (this last idea was recently proposed in [44] and was checked with very good results for the traveling salesman problem, the optimization counterpart of HPP).

15 At the edge of science fiction

As mentioned above, there are many convincing achievements of natural computing, many bio-inspired areas of computer science have important practical applications, or/and they are appealing from a theoretical point of view. Sometimes, the usefulness of the bio-inspired models and tools has a somewhat mysterious source/explanation; in other cases, the matter is simpler and more transparent. Anyway, we try here to compose a list of attractive features of this attempt, of learning from the living nature to the benefit of computer science (most of these features can be called "hopes," as not being confirmed yet by current natural computing): in many cases, we look for ideas for improving the use of the existing computers, for new types of algorithms; in other cases, a new hardware is sought for; as new ideas to be found in nature, we can learn new data structures (such as the double strand with complementary pairs of symbols), or new operations (crossover and point mutations, splicing, annealing, and so on and so forth); bio-computing can make available a massive parallelism, reversible computations, nondeterminism, energy efficiency, maybe also evolvable hardware/software, self-healing, robust; new ideas learned from biology can lead to a complete reconstruction of computability theory, on nonstandard bases (e.g., using the splicing operation, quite different from the rewriting operation, which is standard in computability); nature can suggest new computer architectures, new ways to cope with such difficulties of parallel computing as communication, (de)centralization, synchronization, controlling distributed processes, etc.

All these are somewhat standard attempts/hopes, but there are many others that step vigorously in the after tomorrow science, if not directly in science fiction.

Let us return to hypercomputations. Nothing from physics and biology prevents the possibility that devices computing more than Turing machines can exist. In particular, besides the idea of using an oracle, there are several other ideas, in principle, "more realistic," to go 'beyond Turing." Nine other such possibilities are discussed in [45]—among them, accelerated Turing machines. The idea goes back to B. Russell (1936), R. Blake (1926), and H. Weyl (1927), who imagined processes that take one time unit for the first step, a half for the second step, and so on, always halving the time to complete the next step. Imported to computability, this idea leads to Turing machines that finish an infinite computation in two time units, and this makes possible computing what a usual Turing machine cannot compute. Thus, let us recall the observation that nature creates internal membranes in a cell (also) in order to enhance the possibility of chemicals to collide, hence to facilitate reactions. Then, let us assume that the reactions that develop in membranes placed on lower levels in a cell are faster than reactions from the upper levels; let us push this hypothesis to having a continuous acceleration and provide the way of creating inner membranes during a computation. What we obtain is an accelerated P system, and such a device, motivated biologically as we have seen above, can compute more than Turing machines. The proof can be found in [9].

Now, in what concerns the practical consequences of computing beyond Turing, they can be bigger than the consequences of the possible efficient proof of the equality $\mathbf{P} = \mathbf{NP}$—details can be found in [15], [16], [45].

Let us return to the lab: in 2001 an important achievement of DNA computing has been announced—implementing a finite automaton [5]. Not too much as computing power, but very much as strategy: the "machine" was capable of an autonomous computation. Moreover, the degree of parallelism was massive. Here is a paragraph from [5]: "In our implementation 10^{12} automata sharing the same software run independently and in parallel on inputs (which could, in principle, be distinct) in $120 \mu l$ solution at room temperature at a combined rate of 10^9 transitions per second with a transition fidelity greater than 99.8%, consuming less than 10^{-10}W." Fully impressive—with the only "small" detail that the automaton that was implemented had only two states, and as usual in DNA computing, scaling-up at a level of a practical interest is not yet possible.

Still, something important remains: the possibility to do it, for finite automata. Finite automata can parse simple languages, for instance, can identify given patterns in an arbitrary string (this is one of the ways of searching texts, such as the current one, for given words). Now, remember that genes are strings, that many diseases are supposed to be encoded in genes, and that a hope of the future medicine is to cure such diseases by editing the respective parts of genes. Doing it by usually eaten or injected medicines is a waste of medicines, much more efficient looks the idea to send the necessary gene-editing–machinery directly at the place where an intervention is needed. To this aim, we need a vector able to carry the gene editor at the right place and a way to find the right place. The second task is covered by a finite automaton. Nicely enough, also the first task can be, in principle, covered, by building a nano-carrier, able to inspect cells, and inside cells, their genome. Some details of this nano-robot were presented in [6]—not incidentally, the finite automaton was implemented in the same team.

Many things remain to be done, hence the possibility (danger?) to have our body continuously scanned by a gene repairing robot is not at all at horizon. However, DNA motors, moving nano-constructions, periodically opening and closing scissors-like DNA molecules are already realized in several laboratories, as part of the fast growing area of nano-technology (with the mentioning that this means in most cases DNA–based technology). Details and references can be found, e.g., in [54].

Other similar speculations? No problem, because there are no limits of speculating, even starting from very sound scientific grounds. Maybe it deserves to be mentioned here F. Tipler, with his (controversial) eternal life in terms of an informational existence (what else is this than artificial life at an universal scale?), from [59].

16 Do we dream too much?

However, let us come back on the Earth, to natural computing as we have it today and as we hope to have it in the near future, and let us take a more skeptical (realistic?) position, contrasting with the positive attitude of the previous paragraphs and with the enthusiastic or even over-enthusiastic position of many authors.

For promoting a young research area, the enthusiasm is understandable—but natural computing is no longer a young area, by Turing it is as old as computer science. A more lucid position is similarly helpful as a blindly optimistic one, so that we balance here the previous discussion with a list of difficulties of implementing bio-ideas in computer science: nature has (in a certain sense, unlimited) time and resources, nature is cruel, kills what is not fit (all these are difficult to incorporate in computers, let them be based on electronic hardware or on a hypothetic bio-ware); nature has other goals than computing; many biochemical processes have a degree of nondeterminism that we cannot afford/allow in our computations; the life processes are complex, with a high degree of redundancy; biology seems to deal with noncrisp mathematics, with probabilities, with fuzzy estimations, which are not fully manageable in computations. And, last but not least, maybe we dream too much even from a theoretical point of view. First, the space–time trade-off specific to molecular computing, cannot redefine complexity classes, and it is sometimes too costly in space (in the size of used bio-ware). Then, M. Conrad [14] warned us that programmability (universality), efficiency, and evolvability are three contradictory features of any computing model; no computing device can simultaneously have all these three good qualities.... Both observations indicate that there is no free lunch in computer science, even in the bio-inspired one.

Similar to Conrad, impossibility theorems are the result from [22] (see also [39]), about the impossibility of computing beyond Turing as soon as four conditions are to be observed.[1]

---

[1] Specifically, Gandy wanted to have a proof of a theorem of the form "what can be calculated by a machine is computable by a Turing machine," as an anthropic-free version of the Turing–Church thesis, which has, in Turing understanding, the formulation "what can be calculated by an abstract human being working in a routine way is computable by a Turing machine." To this aim, one first introduces four criteria for "being a machine," and then one proves that the behavior of any device, abstract or physical, which satisfies these criteria can

Now, in what concerns the applications of computer science in biology, let us remark that there are serious limits in this respect. Everything starts with the fact that the cell, as small as it is, is a complex system, with a nonlinear behavior, difficult to capture in simple models, difficult also for complex models based on differential equations (with the success from physics, such models were obsessively tried in biology, with not so good results: the models are difficult to understand, difficult to change, difficult to scale up). It was said in several places that, after completing the genome project, the major task of bioinformatics is the simulation of a cell—see the very title of [61].

This last observation has led to a major preoccupation of the last years for the modeling of the cell, in the framework of a new fashion called "systems biology," with several manifestos in well-known journals, such as *Science* [37], *Nature* [38], with many projects in many countries, with serious research efforts put in this attempt, to have an overall model of a cell and to simulate this model on a computer, and then to use this, in relation with data mining, drug and treatment discovery, until transforming biology and medicine into precise engineering (the last words of [38]). The goals are noble (and feasible, at least at a medium run), but the insistence with which the syntagma "systems biology" is repeated made by O. Wolkenhauer to ask himself already from the title of [64] whether this is a genuinely new area of research or is just a "reincarnation of systems theory applied in biology." The paper recalls the efforts made in the 1960s to apply systems theory in biology, with an obvious disappointment at that time, due, among others, to the limits in available data and in the computability power (let us recall that the currently accepted model of the cell membrane, the Singer–Nicolson fluid–mosaic model, dates from 1972 only). But, maybe there is something else that is necessary, coming from the biology itself as a level of maturity (as formal science). This observation is soundly expressed in the last paragraph of [64], where one of the important names of classic systems theory, M. Mesarović, is invoked: "Mihajlo Mesarović wrote in 1968 that 'in spite of the considerable interest and efforts, the application of systems theory in biology has not quite lived up to expectations. (...) One of the main reasons for the existing lag is that systems theory has not been directly concerned with some of the problems of vital importance in biology'. His advice for biologists was that progress could be made by more direct and stronger interactions with systems scientists. 'The real advance in the application of systems theory to biology will come about only when the biologists start *asking questions* which are based on the system–theoretic concepts rather than

---

be simulated by a Turing machine. We have here the starting point of a fascinating debate about issues that we only mention (in a form recently formulated to me by J. Kelemen), and we refer to [31], [35], [36] for well-documented related discussions: what is a computer? how much human features and how much more general features we associate with this concept? Are they living beings with the ability to compute machines? In what sense? It is interesting to note that the answer to these questions has historically evolved, somewhat in parallel with issues related to the nature of robots—from organic, to electro-mechanical, electronic, and, presumably, back to organic; this is well documented, e.g., in [31], and a similar analysis can be carried out also for the history of computing/computers.

using these concepts to represent in still another way the phenomena which are already explained in terms of biophysical or biochemical principles. (...) Then we will not have the applications of engineering principles to biology problems but rather a field of *systems biology* with its own identity and in its own right'." [43].

Transforming biology and medicine into precise engineering can be put in relation with the proposal from [6] and in contrast with the current limits of understanding life, materialized, among others, in the current limits of AI and AL, with many former dreams drastically reshaped. Think, for instance, is still too difficult for any computer Turing test. In general, the computers are perceived as powerful machines, but not at all displaying intelligence (they are already spectacular in IA—intelligence amplification—but not similarly successful in AI—artificial intelligence, [63]). All these suggest, in terms of [8], that, despite the many spectacular achievements, "we might be missing something fundamental and currently unimagined in our models of biology." The computers are good in crunching numbers, but not "at modelling living systems, at small or large scales." The clear intuition is that life is more than biochemistry, but what else should be considered can be something *unimagined*, something "invisible to us right now." "It is not completely impossible that we might discover some new properties of biomolecules or some new ingredient." An example of such "new stuff" can be the quantum effects in the microtubules of nerve cells, which, according to [53], "might be the locus of consciousness at the level of individual cells, which combines in bigger wave functions at the organism level." ([8], page 410).

A similar position is expressed by J. McCarthy in [10]. "Human–level intelligence is a difficult scientific problem and probably needs some new ideas. These are more likely to be invented by a person of genius than as part of a Government or industry project."

And, maybe, we have here one of the sources of introducing difficulty in the stage: on the one hand, we want to make the computers behave like humans, trying to improve them by looking in nature, to the ways life "computes," from cells to the brain, and, on the other hand, we try to understand (model, simulate) the cells themselves, the brain, the life, in computational terms, projecting our computer science experience to biology. Isn't it here a sort of a (vicious) circle?

Anyway, the search for a "new stuff" may start with realizing that we need such a "new stuff," but, as many crucial steps forward in science prove, this does not necessarily make the discovery more plausible or rapid.

17 Closing remarks

The journey from cells (biology) to models (mathematics) and to computers, and then back to cells is fascinating, marked with many achievements and much more unsolved tasks, full of promises and attractive scientifically, with sure payoffs for both partner areas—biology (medicine included) and computer science. The current discussion was only a brief description of this journey, from a subjective point of view, leading to a few general key observations: (i) In all its history, computer science has tried to learn from biology, at various levels; (ii) this is a highly rewarding attempt, at least for computer science; (iii) it starts to be rewarding also for biology; (iv) the future progresses in this area cannot be overestimated; (v) in general we expect too much (and too soon) from the biology–computability symbiosis; (v) we ignore the genuine differences between the two domains, the inherent limits of computability, and the fact that biology is not yet a mathematized science; (vi) it is possible that maybe a new mathematics is not necessary, but something essentially new, related to the difference between living and nonliving; and (vii) independent of any practical usefulness, all this scientific adventure is challenging, attractive, worth carrying on, with possible by-products that cannot be foreseen.

If bio-informatics will not (help to) produce a Golem, it will at least prove why this is not possible... and, to recall the title of [17], this is enough in order to prevent "the end of computing science."

References

1. L.M. Adleman: Molecular computation of solutions to combinatorial problems. *Science*, 226 (Nov. 1994), 1021–1024.
2. B. Alberts, A. Johnson, J. Lewis, M. Raff, K. Roberts, P. Walter: *Molecular Biology of the Cell*, 4th ed. Garland Science, New York, 2002.
3. A. Alhazov, M. Margenstern, V. Rogozhin, Y. Rogozhin, S. Verlan: Communicative P systems with minimal cooperation. In *Membrane Computing. International Workshop WMC5, Milan, Italy, 2004. Revised Papers* (G. Mauri, Gh. Păun, M.J. Pérez–Jiménez, G. Rozenberg, C. Zandron, eds.), *Lecture Notes in Computer Science*, 3365, Springer, Berlin, 2005, 162–178.
4. J.A. Anderson: *An Introduction to Neural Networks*. The MIT Press, Cambridge, MA, 1996.
5. Y. Benenson, T. Paz–Elizur, R. Adar, E. Keinan, Z. Livneh, E. Shapiro: Programmable and autonomous computing machine made of biomolecules. *Nature*, 414 (Nov. 2001), 430–434.
6. Y. Benenson, E. Shapiro, B. Gill, U. Ben–Dor, R. Adar: Molecular computer. A 'smart drug' in a test tube. *Proc. of Tenth DNA Computing Conference, Milano, 2004* (C. Ferretti, G. Mauri, C. Zandron, eds.), Univ. of Milano–Bicocca, 2004, 49 (abstract of invited talk).
7. D. Bray: Protein molecules as computational elements in living cells. *Nature*, 376 (July 1995), 307–312.
8. R. Brooks: The relationship between matter and life. *Nature*, 409 (Jan. 2001), 409–411.
9. C. Calude, Gh. Păun: Bio–steps beyond Turing. *BioSystems*, 77 (2004), 175–194.
10. J. Mc Carthy: Problems and projection in CS for the next 49 years. *Journal of the ACM*, 50, 1 (2003), 73–79.
11. J.L. Casti. Computing the uncomputable, *The New Scientist*, 154/2082, 17 (May 1997), 34.

12. S. Cook: The importance of the P versus NP question. *Journal of the ACM*, 50, 1 (2003), 27–29.
13. G. Ciobanu, Gh. Păun, M.J. Pérez–Jiménez, eds.: *Applications of Membrane Computing*. Springer, Berlin, 2006.
14. M. Conrad: The price of programmability. In *The Universal Turing Machine: A Half–Century Survey* (R. Herken, ed.), Kammerer and Unverzagt, Hamburg, 1988, 285–307.
15. B.J. Copeland: Hypercomputation. *Minds and Machines*, 12, 4 (2002), 461–502.
16. B.J. Copeland, D. Proudfoot: Alan Turing's forgotten ideas in computer science. *Scientific American*, 280 (April 1999), 77–81.
17. E.W. Dijkstra: The end of computer science? *Communications of the ACM*, 44, 3 (2000), 92.
18. A. Ehrenfeucht, T. Harju, I. Petre, D.M. Prescott, G. Rozenberg: *Computation in Living Cells. Gene Assembly in Ciliates*. Springer, Berlin, 2004.
19. A.E. Eiben, J.E. Smith: *Introduction to Evolutionary Computing*. Springer, Berlin, 2003.
20. G. Franco, C. Giabulli, C. Laudana, V. Manca: DNA extraction by cross pairing PCR. *Proc. of Tenth DNA Computing Conference, Milano, 2004* (C. Ferretti, G. Mauri, C. Zandron, eds.), Univ. of Milano–Bicocca, 2004, 193–201.
21. R. Freund, L. Kari, M. Oswald, P. Sosik: Computationally universal P systems without priorities: two catalysts are sufficient. *Theoretical Computer Sci.*, 330, 2 (2005), 251–266.
22. R. Gandy: Church's thesis and principles for mechanisms. In *The Kleene Symposium* (J. Barwise et al., eds.), North–Holland, Amsterdam, 1980, 123–148.
23. M.R. Garey, D.S. Johnson: *Computers and Intractability. A Guide to the Theory of NP–Completeness*. Freeman, San Francisco, CA, 1979.
24. M. Gross: Molecular computation. Chapter 2 of *Non–Standard Computation* (T. Gramss, S. Bornholdt, M. Gross, M. Mitchel, Th. Pellizzari, eds.), Wiley–VCH, Weinheim, 1998.
25. S.R. Hameroff, J.D. Dayhoff, R. Lahoz–Beltra, A.V. Samsonovich, S. Rasmussen: Models for molecular computation: Conformational automata in the cytoskeleton. *Computer*, 25 (Nov. 1992), 30–39.
26. T. Head: Formal language theory and DNA: An analysis of the generative capacity of specific recombinant behaviors. *Bulletin of Mathematical Biology*, 49 (1987), 737–759.
27. J. Hartmanis: About the nature of computer science. *Bulletin of the EATCS*, 53 (June 1994), 170–190.
28. J. Hartmanis: On the weight of computation. *Bulletin of the EATCS*, 55 (Febr. 1995), 136–138.
29. J. Hoffmeyer: Surfaces inside surfaces. On the origin of agency and life. *Cybernetics and Human Knowing*, 5, 1 (1998), 33–42.
30. J. Hoffmeyer: Semiosis and living membranes. *First Seminário Avançado de Comunicação e Semiótica. Biossemiótica e Semiótica Cognitiva*, São Paolo, Brasil, 1998, 9–19.
31. J. Horáková, J. Kelemen: Čapek, Turing, von Neumann, and the 20th century evolution of the concept of machine. In *Proceedings of the International Conference in Memoriam John von Neumann*, Budapest Polytechnic, 2003, 121–135.
32. J. Hromkovic: *Communication Complexity and Parallel Computing*. Springer, Berlin, 1997.
33. S. Ji: The cell as the smallest DNA–based molecular computer. *BioSystems*, 52 (1999), 123–133.
34. S. Kauffman: *At Home in the Universe*. Oxford Univ. Press, New York, 1995.
35. J. Kelemen: *Bodouci Altamira (The New Altamira)*. Votobia, Olomouc, 1996.
36. J. Kelemen: *Kybergolem (Cybergolem)*. Votobia, Olomouc, 2001.
37. H. Kitano: Systems biology: A brief overview. *Science*, 295 (March 2002), 1662–1664.

38. H. Kitano: Computational systems biology. *Nature*, 420 (Nov. 2002), 206–210.
39. V. Kreinovich, L. Longprè: Fast quantum algorithms for handling probabilistic and interval uncertainty. *Math. Logic Quart.*, 50 (2004), 405–416.
40. W.R. Loewenstein: *The Touchstone of Life. Molecular Information, Cell Communication, and the Foundations of Life*. Oxford University Press, New York, Oxford, 1999.
41. S. Marcus: Bridging P systems and genomics: A preliminary approach. In *Membrane Computing. International Workshop, WMC–CdeA 2002, Curtea de Argeş, Romania, Revised Papers* (Gh. Păun, G. Rozenberg, A. Salomaa, C. Zandron, eds.), *Lecture Notes in Computer Science*, 2597, Springer, Berlin, 2003, 371–376.
42. W. Mass: Networks of spiking neurons: The third generation of neural network models. *Neural Networks*, 10, 9 (1997), 1659–1671.
43. M.D. Mesarović: System theory and biology – view of a theoretician. In *System Theory and Biology* (M.D. Mesarović, ed.), Springer, New York, 1968, 59–87.
44. T.Y. Nishida: An application of P system: A new algorithm for NP–complete optimization problems. In *Proceedings of the 8th World Multi–Conference on Systems, Cybernetics and Informatics* (N. Callaos et al, eds.), vol. V, 2004, 109–112.
45. T. Ord: *Hypercomputation: Computing More Than the Turing Machine*. Honours Thesis, Department of Computer Science, University of Melbourne, 2003.
46. C.H. Papadimitriou: *Computational Complexity*. Addison–Wesley, Reading, MA., 1994.
47. A. Păun, Gh. Păun: The power of communication: P systems with symport/antiport. *New Generation Computing*, 20, 3 (2002), 295–306.
48. Gh. Păun: On the splicing operation. *Discrete Appl. Math.*, 70 (1996), 57–79
49. Gh. Păun: Computing with membranes. *Journal of Computer and System Sciences*, 61, 1 (2000), 108–143 (and Turku Center for Computer Science–TUCS Report 208, November 1998, www.tucs.fi).
50. Gh. Păun: *Membrane Computing: An Introduction*. Springer, Berlin, 2002.
51. Gh. Păun, R. Păun: Membrane computing and economics: Numerical P systems. Submitted, 2005 (available at [65]).
52. Gh. Păun, G. Rozenberg, A. Salomaa: *DNA Computing. New Computing Paradigms*. Springer, Berlin, 1998.
53. R. Penrose: *The Emperor's New Mind. Concerning Computers, Minds, and the Laws of Physics*. Oxford University Press, Oxford, 1989.
54. J.H. Reif, T.H. LaBean, S. Sahu, H. Yan, P. Yin: Design, simulation, and experimental demonstration of self–assembled DNA nanostructures and motors. *Proceedings of the Workshop on Unconventional Programming Paradigms, UPP04*, Le Mont Saint–Michel, September 2004, Springer, Berlin, 2005.
55. G. Rozenberg, A. Salomaa: Watson–Crick complementarity, universal computations, and genetic engineering. *Techn. Report* 96–28, Department of Computer Science, Leiden Univ., Oct. 1996.
56. P. Sosik: The computational power of cell division in P systems: Beating down parallel computers? *Natural Computing*, 2, 3 (2003), 287–298.
57. C. Teuscher, ed.: *Alan Turing. Life and Legacy of a Great Thinker*. Springer, Berlin, 2003.
58. C. Teuscher, E. Sanchez: A revival of Turing's forgotten connectionist ideas: exploring unorganized machines. *Proc. Connectionist Models of Learning, Development and Evolution*, Liege, Belgium, 2000 (R.M. French, J.J. Sougne, eds.), Springer-Verlag, London, 2001, 153–162.
59. F. Tipler: *The Physics of Immortality*. Doubleday, New York, 1994.
60. T. Toffoli: Nothing makes sense in computing except in the light of evolution. *Int. J. of Unconventional Computing*, 1 (2005), 3–29.

61. M. Tomita: Whole–cell simulation: A grand challenge of the 21st century. *Trends in Biotechnology*, 19 (2001), 205–210.
62. A.M. Turing: On computable numbers, with an application to the Entscheidungsproblem. *Proceedings of the London Mathematical Society*, Ser. 2, 42 (1936), 230–265; a correction, 43 (1936), 544–546.
63. V. Vinge: Technological singularity. *VISION–21 Symposium*, March 1993 (available at http://www.frc.ri.cmu.edu/ hpm/book98/com.chl/vinge.singularity.html.
64. O. Wolkenhauer: Systems biology: The reincarnation of systems theory applied in biology? *Briefings in Bioinformatics*, 2, 3 (2001), 258–270.
65. The Web Page of Membrane Computing: http://psystems.disco.unimib.it

Computer Science, Informatics, and Natural Computing—Personal Reflections

Grzegorz Rozenberg

Department of Computer Science, University of Colorado at Boulder
Boulder, CO 80309, U.S.A.
and
Leiden Institute of Advanced Computer Science (LIACS), Leiden University,
2300 RA Leiden, the Netherlands
rozenber@liacs.nl

Summary. The paper presents personal reflections on natural computing: its scope, some of its history, and its relationship to (and the influence on) computer science. It also reflects on the nature of computer science, and argues why "informatics" is a better term than "computer science".

The classic notion of computation is firmly rooted in the notion of an algorithm that informally speaking is a set of rules for performing a task. The quest for formalizing the notion of an algorithm so that it could be "mechanised" dates back at least to the work of Gottfried W. Leibnitz. Leibnitz (1646–1716) wanted to formalize human reasoning in such a way that it could be described as a collection of rules, which then could be executed in a mechanistic way.

The motivation for his passionate research was stated by Leibnitz as follows: "[I]t is unworthy of excellent men to lose hours like slaves in the labor of calculation which could safely be relegated to anyone else if the machine was used" (see, e.g., [4]). Thus the motivation was rooted in the specific "negative idea" that the task of performing calculations is a waste of time. This, however, can be turned into a positive motivation, viz., the need to understand which mental processes can be formalized so that they can be automated—this point of view is closer to the current thinking in computer science.

This research into the understanding of the notion of an algorithm was carried on by many outstanding scientists, and it culminated in the works of Kurt Gödel, Alonzo Church, Alan Mathison Turing, and Emil Post approximately in the period 1930–1940.

The formalization of the notion of an algorithm by these four giants, and especially the work by Alan Mathison Turing, led to the construction of the first computers, and to the beginnings of computer science. In formalizing the notion of an algorithm, Turing has focused on what a person performing calculations did when following a set of rules, hence following a given algorithm. Thus the beginnings of computer science were rooted in human-designed computing.

The scope and importance of computer science has grown tremendously since its beginnings. In fact, the spectacular progress in Information and Communication Technology (ICT) is very much supported by the evolution of computer science, which designs and develops the instruments needed for this progress: computers, computer networks, software methodologies, etc. Since ICT has such a tremendous impact on our everyday life, so does computer science.

However, there is much more to computer science than ICT: it is the science of information processing, and as such it is a fundamental science for other scientific disciplines. As a matter of fact, the only common denominator for research done in all the areas so diverse of computer science is thinking about various aspects of information processing. Therefore, the frequently used (mostly in Europe) term "informatics" is much better than "computer science"—the latter stipulates that a specific instrument, viz., computer, is the main research topic of our discipline. On the other hand, one of the important developments of the last century for a number of other scientific disciplines is the adoption of information and information processing as their central notions and thinking habits—biology and physics are beautiful examples here. For these scientific disciplines informatics provides not only instruments but also a way of thinking.

I am convinced that one of the grand challenges of informatics is to understand the world around us in terms of information processing. Each time progress is made in achieving this goal, both the world around us and informatics benefits. Since nature is a dominating part of the world around us, one way to understand this world in terms of information processing is to study computing taking place in nature. Natural computing is concerned with this type of computing as well as with its main benefit for informatics, viz., human-designed computing inspired by nature. Research in natural computing is genuinely interdisciplinary, and therefore, natural computing forms a bridge between informatics and natural sciences. It has already contributed enormously to human-designed computing through the use of paradigms, principles, and mechanisms underlying natural systems. Some disciplines of this type of computing are relatively old (in the young history of computer science) and are well established by now. Well-known examples of such disciplines are evolutionary computing and neural computing. Evolutionary algorithms are based on the concepts of mutation, recombination, and natural selection from the theory of evolution, whereas neural networks are based on concepts originating in the study of the highly interconnected neural structures in the brain and the nervous system. On the other hand, molecular computing and quantum computing are younger disciplines of natural computing: molecular computing is based on paradigms from molecular biology, whereas quantum computing is based on quantum physics and exploits quantum parallelism. Human-designed computing inspired by nature includes also other subdisciplines and paradigms.

Thus research in natural computing had already a big impact on the development of informatics, and in particular, it contributed to our understanding of the nature of computation. Since the understanding of the nature of computation is the main task of theoretical computer science, it is important to point out here that the interaction of theoretical computer science and natural sciences dates back to the very beginnings of computer science and has continued since then. Here are some examples of this interaction.

Some of the most important foundational research in automata theory was inspired by the work of W.S. McCulloch and W. Pitts [13], which considers neurons as binary transmitters of information. The theory of L-systems initiated by A. Lindenmayer (see, e.g., [10] and [14]) was motivated by modeling the development of simple organisms, and it had a fundamental impact on formal language theory, as well as a significant impact on the modeling of plants. The DNA revolution, which in the last 50 years had such tremendous impact on biology and many other areas of science (as well as on our everyday life), had also a big influence on theoretical computer science. For example, the overwhelming success in sequencing of the human and other genomes was to a large extent based on the development of pattern matching and other string processing algorithms. The whole area of design and analysis of pattern matching and editing algorithms benefited enormously from the intense research concerned with sequencing of genomes. It is certainly interesting to mention in this context that A.M. Turing was genuinely interested in natural computing. This is witnessed by his work on morphogenesis [17] and by his work on neural networks architecture (see, e.g., [16]). This interest is very clearly stated in his letter to W. Ross Ashby (see, e.g., [16]): "In working on the ACE [Automatic Computing Engine] I am more interested in the possibility of producing models of the action of the brain than in the practical applications to computing."

As is clear from the above, interdisciplinarity is a key feature of research in natural computing and an important ingredient of evolution of the whole field of informatics.

I consider myself to be an interdisciplinary scientist. The support for this classification is given by my education and my research. My first degree is an engineer of electronics, my second degree is a master of computer science, and my third degree is a Ph.D. in mathematics. Then, for over 30 years, a big part of my research was concerned with the understanding of the principles of biological information processing. A considerable part of this research involved an intense cooperation with biologists. Therefore I would like to say a few words now about interdisciplinary research and, in particular, about research in natural computing. Since my interdisciplinary research involves theoretical computer science on the one hand and biology on the other, my reflections concern this specific research interaction.

Research on formal modeling of biological phenomena requires an a priori realization that the value or utility of the obtained models may be quite temporary. A formal model will at best reflect the biological knowledge at the time of its formulation. Biological knowledge is very dynamic, and new important facts are discovered all the time—some of these findings may change, sometimes dramatically, our understanding of the nature and working of certain biological phenomena. If they involve a phenomenon that we model, then often the model must be adapted, changed, or even totally discarded. Therefore one should strive, whenever possible, that the formal model also be solid and interesting from the formal point of view. In this way, during an often long trajectory of creating, adapting, and modifying a formal model, one may get an interesting and lasting contribution to theoretical computer science.

Another piece of advice based on my experience is that the formulation of the model should begin with an understanding of the biological nature of the problem and then continue with a very critical assessment of the tools that we have available (as specialists in theoretical computer science). Very often such a critical assessment will lead to the conclusion that one needs to formulate a genuinely new model and then to develop tools for its study as one goes along. This is definitely a preferred way of proceeding, rather than to bend (and often distort) biological knowledge in such a way that it fits one's tools!!

Finally in developing a formal model one should not forget that such a model consists of two parts: (1) a mathematical formal construct, and (2) its interpretation in the modeled domain. The second part is too often forgotten, although it is often really crucial in the choice of a "good model": among all models that are equivalent (in some formal sense), the only relevant models are those with good interpretability.

Natural computing is a fast-growing and dynamic research area. When I introduced this name more than 25 years ago, it was considered as a sort of science fiction, but today it is really popular and flourishing. There are institutes, journals, book series, conferences, professorships, ... of/on natural computing (it was interesting for me to learn recently that even Microsoft is establishing a Natural Computing Group).

As I have written above several scientific disciplines, most notably biology and physics, have adopted information and information processing as their central notions and thinking habits. Here are examples of this development.

System biology is a recent (and very fashionable) discipline of biology that attempts to understand biological systems at the system level. The key issue here is the understanding of the dynamics/behavior of the whole system based on the understanding of interactions between (molecular) components (see, e.g., [9]). These interactions are often understood/expressed in computational terms. This is (or certainly should be) especially true in computational system biology, where the interactions of components are considered as a computational process (and often individual components of a system are considered as computational devices); see, e.g., [1].

An interesting example is research concerning the fundamental question, "What is life?" T. Ganti (see, e.g., [8]) proposes that the minimal system of life is chematon, which is a fluid (chemical) automaton consisting of three different units (connected with each other stoichmetrically), each of which is a reproductive fluid automaton. The abstract chematon model, based on chemical reactions, is very much a model of computation in the sense of theoretical computer science.

One of the central goals of synthetic biology is to engineer a synthetic cell (protocell)—it must satisfy 12 requirements for life (see [6]), 3 of which refer explicitly to capturing, using, and mutating information.

Perhaps the most clear and explicit statements concerning the understanding of the nature of life belong to Richard Dawkins ([5]): "If you want to understand life, don't think about vibrant, throbbing gels and oozes, think about information technology."

One of the most compelling examples of the information processing paradigm in biology is a recent, very impressive book *The Regulatory Genome* by E.H. Davidson [3]. Here, gene regulatory networks in development are presented in terms of computational circuits demonstrating in this way (in words of S. Istrail) "the biological computer of cell regulation." This book is a wonderful illustration of the statement by A. Lindenmayer and myself made over 30 years ago ([11]): "The development of an organism... may be considered as the execution of "developmental program" present in the fertilized egg.... A central task of developmental biology is to discover the underlying algorithm from the course of development."

Finally, concerning the role of information processing in physics, the recent book *Programming the Universe* by S. Lloyd ([12]) is a wonderful example of it. Let me quote here the beginning of the introduction to this book: "This book is the story of the universe and the bit. The universe is the biggest thing there is and the bit is the smallest possible chunck of information. The universe is made of bits. Every molecule, atom, and elementary particle registers bits of information. Every interaction between those pieces of the universe processes that information by altering those bits. That is, the universe computes, and because the universe is governed by the laws of quantum mechanics, it computes in an intrinsically quantum-mechanical fashion; its bits are quantum bits. The history of the universe is, in effect, a huge and ongoing quantum computation. The universe is a quantum computer. This begs the question: What does the universe compute? It computes itself. The universe computes its own behavior. As soon as the universe began, it began computing."

I am myself especially fascinated by molecular computing, which is a good example of a research area that has evolved in a very interesting way. It really began as DNA Computing with the initial goal of providing a computing technology that will be a competition to the current silicon technology for computers. However, it has evolved into a science of molecular programming concerned with problems of the following type: "How to design a set of initial molecules so that a certain type of molecular complexes will be formed." In this way a large stream of research in molecular programming became a part of nanoscience and nanoengineering, where, e.g., in human-designed self-assembly one considers the same type of problems. The combination of nanoscale science and engineering with nanoscale computing is certainly an exciting development (see, e.g., [2]), which will have tremendous impact on the science and technology of computing.

Now if we return to the part of natural computing that studies computing taking place in Nature, then the large question is "How does Nature compute?" In order to answer this question, we have to consider and study various processes taking place in Nature as computational processes. But what does "computational" mean here? We will have to redefine the notion of computation, which must be able to accommodate also information processing taking place in nature. This is an exciting adventure that has only just begun. I have no doubts that it will lead to a new science of computation that will provide a broader and deeper understanding of what "computation" is about. As a matter of fact, research in natural computing led already to a reexamination of the axioms/paradigms underlying traditional notions of computation (see, e.g., [15] and [7]).

Let us now conclude by going back to informatics with all its facets. The attractiveness and beauty of informatics as a science is that although it is a fundamental science for a number of scientific disciplines, it is also the main force behind the development of ICT, and through this development, it influences and revolutionizes our everyday life. Natural computing is an important vehicle of progress for both of these facets of informatics. Let us propagate and develop the science of informatics and present it to "the outside world" using this framework. Both informatics, viewed as above, and natural computing have a great future!

Acknowledgments

I am indebted to the editors of this volume for inviting me to write personal reflections on natural computing and computer science (informatics).

References

1. Cardelli, L., Abstract machines of systems biology, *Transactions on Computational Systems Biology*, III: 145–168, 2005.
2. Chen, J., Jonoska, N., and Rozenberg, G. (eds.), *Nanotechnology: Science and Computation*, Springer–Verlag, Berlin, 2006.
3. Davidson, E.H., *The Regulatory Genome: Gene Regulatory Networks in Development and Evolution*, Academic Press/Elsevier, Burlington, 2006.
4. Davis, M., *Engines of Logic*, W.W. Norton and Company, Inc., New York, 2000.
5. Dawkins, R., *The Blind Watchmaker*, Penguin, Harmondsworth, 1986.
6. Deamer, D., A giant step towards artificial life, *Trends in Biotechnology*, 23: 336–338, 2005.
7. Ehrenfeucht, A. and Rozenberg, G., Reaction systems, *Fundamenta Informaticae*, 2006, to appear.
8. Ganti, T., *The Principles of Life*, Oxford University Press, Oxford, 2003.
9. Kitano, H., Systems biology: a brief overview, *Science*, 295: 1662–1664, 2002.
10. Lindenmayer, A., Mathematical models for cellular interaction in development, I and II, *Journal of Theoretical Biology*, 18: 280–315, 1968.
11. Lindenmayer, A. and Rozenberg, G., Introduction, in Lindenmayer, A. and Rozenberg, G. (eds.), *Automata, Languages, Development*, v–vi, North Holland, Amsterdam, 1976.
12. Lloyd, S., *Programming the Universe: A Quantum Computer Scientist Takes on the Cosmos*, Jonathan Cape, London, 2006.
13. McCulloch, W.S. and Pitts, W.H., A logical calculus of the ideas immanent in neural nets, *Bulletin of Mathematical Biophysics*, 5: 115–133, 1943.
14. Rozenberg, G. and Salomaa, A., *The Mathematical Theory of L Systems*, Academic Press, Inc., New York, 1980.
15. Stepney, S., Braunstein, S.L., Clark, J.A., Tyrrell, A., Adamatzky, A., Smith, R.E., Addis, T., Johnson, C., Timmis, J., Welch, P., Milner, R., Partidge, D., Journeys in non-classical computation I: a grand challenge for computing research, *International Journal of Parallel, Emergent and Distributed Systems*, 30(1): 5–19, 2005.
16. Teuscher, C., *Turing's Connectionism: An Investigation of Neural Networks Architectures*, Springer–Verlag, London, 2002.
17. Turing, A.M., The chemical basis of morphogenesis, *Philosophical Transactions of the Royal Society of London*, B 237: 37–72, 1952.

Part IV

Computable Analysis and Real Computation

A Survey on Continuous Time Computations

Olivier Bournez[1,2] and Manuel L. Campagnolo[3,4]

[1] INRIA Lorraine
[2] LORIA (UMR 7503 CNRS-INPL-INRIA-Nancy2-UHP), 54506 Vandœuvre-Lès-Nancy,
France,
Olivier.Bournez@loria.fr
[3] DM/ISA, Technical University of Lisbon, Tapada da Ajuda, 1349-017 Lisboa,
Portugal
[4] SQIG/IT Lisboa
mlc@math.isa.utl.pt

Summary. We provide an overview of theories of continuous time computation. These theories allow us to understand both the hardness of questions related to continuous time dynamical systems and the computational power of continuous time analog models. We survey the existing models, summarizing results, and point to relevant references in the literature.

1 Introduction

Continuous time systems arise as soon as one attempts to model systems that evolve over a continuous space with a continuous time. They can even emerge as natural descriptions of discrete time or space systems. Utilizing continuous time systems is a common approach in fields such as biology, physics or chemistry, when a huge population of agents (molecules, individuals, ...) is abstracted into real quantities such as proportions or thermodynamic data [100], [148].

Several approaches have led to theories on continuous time computations. We will explore in greater depth two primary approaches. One, which we call *inspired by continuous time analog machines*, has its roots in models of natural or artificial analog machinery. The other, which we refer to as *inspired by continuous time system theories*, is broader in scope. It comes from research on continuous time systems theory from a computational perspective. Hybrid systems and automata theory, for example, are two sources.

A wide range of problems related to theories of continuous time computations are encompassed by these two approaches. They originate in fields as diverse as verification (see, e.g., [20]), control theory (see, e.g., [44]), VLSI design (see, e.g., [140],

[141]), neural networks (see, e.g., [160]), and recursion theory on the reals (see, e.g., [145]).

At its beginning, continuous time computation theory was concerned mainly with analog machines. Determining which systems can actually be considered as computational models is a very intriguing question. It relates to the philosophical discussion about what is a programmable machine, which is beyond the scope of this chapter. Nonetheless, some early examples of built analog devices are generally accepted as programmable machines. They include Bush's landmark 1931 Differential Analyzer [50], as well as Bill Phillips's Finance Phalograph, Hermann's 1814 Planimeter, Pascal's 1642 Pascaline, or even the 87 B.C. Antikythera mechanism; see [70]. Continuous time computational models also include neural networks and systems that can be built using electronic analog devices. Since continuous time systems are conducive to modeling huge populations, one might speculate that they will have a prominent role in analyzing massively parallel systems such as the Internet [162].

The first true model of a universal continuous time machine was proposed by Shannon [183], who introduced it as a model of the differential analyzer. During the 1950s and 1960s, an extensive body of literature was published about the programming of such machines.[5] There were also a number of significant publications on how to use analog devices to solve discrete or continuous problems; see, e.g., [200] and the references therein. However, most of this early literature is now only marginally relevant given the ways in which our current understanding of computability and complexity theory have developed.

The research on artificial neural networks, despite the fact that it mainly focused on discrete time analog models, has motivated a change of perspective due to its many shared concepts and goals with today's standard computability and complexity theory [160], [158]. Another line of development of continuous time computation theory has been motivated by hybrid systems, particularly by questions related to the hardness of their verification and control; see, e.g., [44] and [20].

In recent years there has also been a surge of interest in alternatives to classic digital models other than continuous time systems. Those alternatives include discrete-time, analog-space models like artificial neural networks [160], optical models [205], signal machines [76], and the Blum Shub and Smale model [30]. More generally there have also been many recent developments in nonclassical and more-or-less realistic or futuristic models such as exotic cellular automata models [93], molecular or natural computations [96], [3], [122], [163], black hole computations [104], or quantum computations [75], [94], [184], [109]. Some of these contributions are detailed in this volume.

---

[5] See for example the very instructive Doug Coward's web *Analog Computer Museum* [70] and its bibliography. This literature reveals the quite forgotten art of programming continuous time and hybrid (digital–analog) machines, with a level of sophistication that is close to today's engineering programming.

The computational power of discrete time models are fairly well known and understood thanks in large part to the Church–Turing thesis. The Church–Turing thesis states that all reasonable and sufficiently powerful models are equivalent. For continuous time computation, the situation is far from being so clear, and there has not been a significant effort toward unifying concepts. Nonetheless, some recent results establish the equivalence between apparently distinct models [89], [88], [90], and [35], which give us hope that a unified theory of continuous time computation may not be too far in the future.

This text can be considered an up-to-date version of Orponen's 1997 survey [160]. Orponen states at the end of his introduction that the effects of imprecision and noise in analog computations are still far from being understood and that a robust complexity theory of continuous time models has yet to be developed. Although this evaluation remains largely accurate with regard to imprecision and noise, we will see in the current survey that in the intervening decade much progress in understanding the computability and even the complexity of continuous time computations has been made.

This chapter is organized as follows. In Section 2, we review the most relevant continuous time models. In sections 3 and 4, we discuss, respectively, computability and complexity issues in continuous time computations. In these sections we focus mainly on continuous time dynamical systems. In Section 5, we address the effect of imprecision and noise in analog computations. Finally, in Section 6, we conclude with some general insights and directions for further research in the field of continuous time computation.

2 Continuous Time Models

With a historical perspective in mind, we outline in this section several of the major classes of continuous time models that motivated interest in this field. These models also illustrate concepts like continuous dynamics and input/output.

2.1 Models inspired by analog machines

GPAC and other circuit models

Probably, the best known universal continuous time machine is the *Differential Analyzer*, built at MIT under the supervision of Vannevar Bush [50] for the first time in 1931. The idea of assembling integrator devices to solve differential equations dates back to Lord Kelvin in 1876 [195]. Mechanical,[6] and later on electronic, differential analyzers were used to solve various kinds of differential equations primarily related

---

[6] And even *MECANO* machines; see [42].

to problems in the field of engineering; see for, e.g., [42], or more generally [204] for historical accounts. By the 1960s, differential analysers were progressively discarded in favor of digital technology.

The first theoretical study of the computational capabilities of continuous time universal machines was published by Shannon. In [183], he proposed what is now referred to as the *General Purpose Analog Computer (GPAC)* as a theoretical model of Vannevar Bush's differential analyzer. The model, later refined in the series of papers [166], [121], [89], [88], consists of families of circuits built with the basic units presented in Figure 1. There are some restrictions to the kinds of interconnectivity that are allowed to avoid undesirable behavior: e.g., nonunique outputs. For further details and discussions, refer to [89] and [87].

Shannon, in his original paper, already mentions that the GPAC generates polynomials, the exponential function, the usual trigonometric functions, and their inverses (see Figure 2). More generally, he claimed in [183] that a function can be generated by a GPAC if and only if it is differentially algebraic; i.e. it satisfies some algebraic differential equation of the form

$$p\left(t, y, y', ..., y^{(n)}\right) = 0,$$

where p is a nonzero polynomial in all its variables. As a corollary, and noting that the Gamma function $\Gamma(x) = \int_0^\infty t^{x-1}e^{-t}dt$ or the Riemann's Zeta function $\zeta(x) = \sum_{k=0}^\infty \frac{1}{k^x}$ are not d.a. [175], it follows that the Gamma and the Zeta functions are examples of functions that cannot be generated by a GPAC.

However, Shannon's proof relating functions generated by GPACs with differentially algebraic functions was incomplete (as pointed out and partially corrected by [166], [121]). However, for the more robust class of GPACs defined in [89], the following stronger property holds: a scalar function $f : \mathbb{R} \to \mathbb{R}$ is generated by a GPAC if and only if it is a component of the solution of a system $y' = p(t, y)$, where p is a vector of polynomials. A function $f : \mathbb{R} \to \mathbb{R}^k$ is generated by a GPAC if and only if all of its components are also.

The Γ function is indeed GPAC computable, if a notion of computation inspired from recursive analysis is considered [88]. GPAC computable functions in this sense correspond precisely to computable functions over the reals [35].

Rubel proposed [176] an extension of Shannon's original GPAC. In Rubel's model, the Extended Analog Computer (EAC), operations to solve boundary value problems, or to take certain infinite limits were added. We refer to [140] and [141] for descriptions of actual working implementations of Rubel's EAC.

More broadly, a discussion of circuits made of general basic units has been presented recently in [198]. Equational specifications of such circuits, as well as their semantics, are given by fixed points of operators over the space of continuous streams. Under suitable hypotheses, this operator is contracting and an extension of Banach fixed point theorem for metric spaces guarantees existence and unicity of the fixed point. Moreover, that fixed point can also be proved to be continuous and *concretely* computable whenever the basic modules also have those properties.

Hopfield network models

Another well-known continuous time model is the "neural network" model proposed by John Hopfield in 1984 in [105]. These networks can be implemented in electrical [105] or optical hardware [193].

A symmetric Hopfield network is made of a finite number, say n, of simple computational units, or *neurons*. The architecture of the network is given by some (nonoriented) graph whose nodes are the neurons and whose edges are labeled by some weights, the *synaptic weights*. The graph can be assumed to be complete by replacing the absence of a connection between two nodes by an edge whose weight is null.

The state of each neuron i at time t is given by some real value $u_i(t)$. Starting from some given initial state $\vec{u}_0 \in \mathbb{R}^n$, the global dynamic of the network is defined by a system of differential equations

$$C_i u_i'(t) = \sum_j W_{i,j} V_j - u_i/R_i + I_i,$$

where $V_i = \sigma(u_i)$, σ is some saturating function such as $\sigma(u) = \alpha \tan u + \beta$, $W_{i,j} = W_{j,i}$ is the weight of the edge between i, and j, C_i, I_i, R_i are some constants [105].

Hopfield proved in [105], by a Lyapunov-function argument, that such systems are *globally asymptotically stable*; i.e., from any initial state, the system relaxes toward some stable equilibrium state. Indeed, consider for example the energy function [105]

$$E = -\frac{1}{2} \sum_i \sum_j W_{i,j} V_i V_j + \sum_i \frac{1}{R_i} \int_0^{V_i} \sigma^{-1}(V) dV + \sum_i I_i V_i.$$

The function E is bounded, and its derivative is negative. Hence the time evolution of the whole system is a motion in a state space that seeks out (possibly local) minima of E.

This convergence behavior has been used by Hopfield to explore various applications such as associative memory or to solve combinatorial optimization problems [105], [106].

An exponential lower bound on the convergence time of continuous time Hopfield networks has been related to their dimension in [188]. Such continuous time symmetric networks can be proved to simulate any finite, binary-state, discrete-time, recurrent neural network [161], [189].

Networks of spiking neurons

If one classifies, following [129], neural network models according to their activation functions and dynamics, three different generations can be distinguished. The first generation, with discontinuous activation functions, includes multilayer perceptrons, Hopfield networks, and Boltzmann machines (see, for example, [2] for an introduction to all mentioned neural network models). The output of this generation of networks is digital. The second generation of networks uses continuous activation functions instead of step or threshold functions to compute the output signals. These functions include feedforward and recurrent sigmoidal neural network, radial basis functions networks, and continuous time Hopfield networks. Their input and output is analog. The third generation of networks is based on spiking neurons and encodes variables in time differences between pulses. This generation exhibits continuous time dynamics and is the most biologically realistic [133].

There are several mathematical models of spiking neurons of which we will focus on one, whose computational properties have been investigated in depth. The Spiking Neural Network model is represented by a finite directed graph. To each node v (neuron) of the graph is associated a *threshold function* $\theta_v : \mathbb{R}^+ \to \mathbb{R} \cup \{\infty\}$, and to each edge (u, v) (synapse) is associated a *response-function* $\epsilon_{u,v} : \mathbb{R}^+ \to \mathbb{R}$ and a *weight-function* $w_{u,v}$.

For a noninput neuron v, one defines its set F_v of *firing times* recursively. The first element of F_v is $\inf\{t | P_v(t) \geq \theta_v(0)\}$, and for any $s \in F_v$, the next larger element of F_v is $\inf\{t | t > s$ and $P_v(t) \geq \theta_v(t - s)\}$, where

$$P_v(t) = 0 + \sum_u \sum_{s \in F_u, s < t} w_{u,v}(s)\epsilon_{u,v}(t - s).$$

The 0 above can be replaced by some bias function. We use it here to guarantee that P_v is well defined even if $F_u = \emptyset$ for all u with $w_{u,v} \neq 0$. To approximate biological realism, restrictions are placed on the allowed response-functions and bias-functions of these models; see [129], [130], [153], or [131], [132], for discussions on the model. In particular, rapidly fading memory is a biological constraint that prevents chaotic behavior in networks with a continuous time dynamic. Recently, the use of feedback to overcome the limitations of such a constraint was analyzed in [134].

The study of the computational power of several variants of spiking neural networks was initiated in [126]. Noisy extensions of the model have been considered [127], [128], [135]. A survey of complexity results can be found in [190]. Restrictions that are easier to implement in hardware versions have also been investigated in [137].

$\mathbb{R}$-recursive functions

Moore proposed a theory of recursive functions on the reals in [145], which is defined in analogy with classical recursion theory and corresponds to a conceptual analog computer operating in continuous time. As we will see, this continuous time model has in particular the capability of solving differential equations, which similar to an idealized analog integrator of the GPAC. In fact, the theory of $\mathbb{R}$-recursive functions can be seen as an extension of Shannon's theory for the GPAC. A general discussion of the motivations behind $\mathbb{R}$-recursion theory can be found in [150].

A function algebra $[B_1, B_2, ...; O_1, O_2, ...]$ is the smallest set containing basic functions $\{B_1, B_2, ...\}$ and is closed under certain operations $\{O_1, O_2, ...\}$, which take one or more functions in the class and create new ones. Although function algebras have been defined in the context of recursion theory on the integers, and has been widely used to characterize computability and complexity classes [62], they are equally suitable to define classes of real-valued recursive functions.

The $\mathbb{R}$-recursive functions were first defined in [145]. These functions are given by the function algebra $\mathcal{M} = [0, 1, U; \text{comp}, \text{int}, \text{minim}]$,[7] where U is the set of projection functions $U_i(\vec{x}) = x_i$, comp is composition, int is an operation that given f and g returns the solution of the initial value problem $h(\vec{x}, 0) = f(\vec{x})$ and $\partial_y h(\vec{x}, y) = g(\vec{x}, y, h)$, and minim returns the smallest zero $\mu_y f(\vec{x}, y)$ of a given f. Moore also studied the weaker algebra $\mathcal{I} = [0, 1, -1, U; \text{comp}, \text{int}]$ and claimed its equivalence with the class of unary functions generated by the GPAC [145].

Many nonrecursively enumerable sets are $\mathbb{R}$-recursive. Since minim is the operation in $\mathcal{M}$ that gives rise to uncomputable functions, a natural question is to ask whether minim can be replaced by some other operation of mathematical analysis. This was done in [149], where minim is replaced by the operation lim, which returns the infinite limits of the functions in the algebra. These authors stratify $[0, 1, -1, U; \text{comp}, \text{int}, \text{lim}]$ according to the allowed number (η) of nested limits and relate the resulting η-hierarchy with the arithmetical and analytical hierarchies. In [124] it is shown that the η-hierarchy does not collapse (see also [123]), which implies that infinite limits and first-order integration are not interchangeable operations [125].

The algebra $\mathcal{I}$ only contains analytic functions and is not closed under iteration [52]. However, if an arbitrarily smooth extension to the reals θ of the Heaviside function is included in the set of basic functions of $\mathcal{I}$, then $\mathcal{I} + \theta$ contains extensions to the reals of all primitive recursive functions.

The closure of fragments of $\mathcal{I} + \theta = [0, 1, -1, \theta, U; \text{comp}, \text{int}]$ under discrete operations like bounded products, bounded sums, and bounded recursion, has been investigated in the thesis [54] and also in the papers [53], [55], [56].

In particular, several authors studied the function algebra

$$\mathcal{L} = [0, 1, -1, \pi, \theta, U; \text{comp}, \text{LI}],$$

---

[7] We consider that the operator int preserves analyticity (see [52], [55]).

where the LI can only solve *linear* differential equations (i.e., it restricts int to the case $\partial_y h(\vec{x}, y) = g(\vec{x}, y)\, h(\vec{x}, y)$). The class $\mathcal{L}$ contains extensions to the reals of all the elementary functions [53].

Instead of asking which computable functions over $\mathbb{N}$ have extensions to $\mathbb{R}$ in a given function algebra, Bournez and Hainry consider classes of functions over $\mathbb{R}$ computable according to recursive analysis, and they characterize them precisely with function algebras. This was done for the elementarily computable functions [36], characterized as $\mathcal{L}$ closed under a restricted limit schema. This was extended to yield a characterization of the whole class of computable functions over the reals [37], adding a restricted minimisation schema. Those results provide syntactical characterizations of real computable functions in a continuous setting, which is arguably more natural than the higher order Turing machines of recursive analysis.

A more general approach to the structural complexity of real recursive classes, developed in [57], is based on the notion of approximation. This notion was used to lift complexity results from $\mathbb{N}$ to $\mathbb{R}$, and it was applied in particular to characterize $\mathcal{L}$.

Somewhat surprisingly, the results above indicate that two distinct models of computation over the reals (computable analysis and real recursive functions) can be linked in an elegant way.

2.2 Models inspired by continuous time system theories

Hybrid Systems

An increasing number of systems exhibit some interplay between discrete and analog behaviors. The investigation of these systems has led to relevant new results about continuous time computation.

A variety of models have been considered; see, for example, the conference series *Hybrid Systems Computation and Control* or [43]. However, hybrid systems[8] are essentially modeled either as differential equations with discontinuous right-hand sides, as differential equations with continuous and discrete variables, or as hybrid automata. A hybrid automaton is a finite state automaton extended with variables. Its associated dynamics consists of guarded discrete transitions between states of the automaton that can reset some variables. Typical properties of hybrid systems that have been considered are reachability, stability, and controllability.

With respect to the differential equation modeling approach, Branicky proved in [44] that any hybrid system model that can implement a clock and implement general continuous ordinary differential equations can simulate Turing machines. Asarin, Maler,

---

[8] "Hybrid" refers here to the fact that the systems have intermixed discrete and continuous evolutions. This differs from historical literature about analog computations, where "hybrid" often refers to machines with a mixture of analog and digital components.

and Pnueli proved in [20] that piecewise constant differential equations can simulate Turing machines in $\mathbb{R}^3$, whereas the reachability problem for these systems in dimension $d \leq 2$ is decidable [20]. Piecewise constant differential equations, as well as many hybrid systems models, exhibit the so-called *Zeno's phenomenon*: an infinite number of discrete transitions may happen in a finite time. This has been used in [19] to prove that arithmetical sets can be recognized in finite time by these systems. Their exact computational power has been characterized in terms of their dimension in [32] and [33]. The Jordan's theorem-based argument of [20] to get decidability for planar piecewise constant differential equations has been generalized for planar polynomial systems [60] and for planar differential inclusion systems [22].

There is extensive literature on the hybrid automata modeling approach about determining the exact frontier between decidability and nondecidability for reachability properties, according to the type of allowed dynamics, guards, and resets. The reachability property has been proved decidable for timed automata [5]. By reduction to this result, or by a finite bisimulation argument in the same spirit, this has also been generalized to multirate automata [4], to specific classes of updatable timed automata in [38], [39], and to initialized rectangular automata in [98], [171]. There is a multitude of undecidability results, most of which rely on simulations of Minsky two-counter machines. For example, the reachability problem is semi-decidable but nondecidable for linear hybrid automata [4], [156]. The same problem is known to be undecidable for rectangular automata with at least five clocks and one two-slope variable [98], or for timed automata with two skewed clocks [4]. For discussion of these results, see also [21]. Refer to [28] and [66] or to the survey [29] for properties other than reachability (for example, stability and observability).

O-minimal hybrid systems are initialized hybrid systems whose relevant sets and flows are definable in an o-minimal theory. These systems always admit a finite bisimulation [119]. However, their definition can be extended to a more general class of "nondeterministic" o-minimal systems [46], for which the reachability problem is undecidable in the Turing model, as well as in the Blum Shub Smale model of computation [45]. Upper bounds have been obtained on the size of the finite bisimulation for Pfaffian hybrid systems [116] [117] using the word encoding technique introduced in [46].

Automata theory

There have been several attempts to adapt classical discrete automata theory to continuous time; this is sometimes referred to as the general program of Trakhtenbrot [196].

One attempt is related to timed automata, which can be seen as languages recognizers [6]. Many specific decision problems have been considered for timed automata; see survey [7]. Timed regular languages are known to be closed under intersection, union, and renaming, but not under complementation. The membership and empty language problems are decidable, whereas inclusion and universal language problems are undecidable. The closure of timed regular languages under shuffling is investigated in [82]. Several variants of Kleene's theorem are established [15], [12], [16], [40], [41], [18]. There have been some attempts to establish pumping lemmas [23]. A review, with discussions and open problems related to this approach, can be found in [10].

An alternative and independent automata theory over continuous time has been developed in [174], [197], and [173]. Here automata are not considered as language recognizers but as computing operators on signals. A signal is a function from the non-negative real numbers to a finite alphabet (the set of the channel's states). Automata theory is extended to continuous time, and it is argued that the behavior of finite state devices is ruled by so-called finite memory retrospective functions. These are proved to be speed-independent, i.e. independent under "stretchings" of the time axis. Closure properties of operators on signals are established, and the representation of finite memory retrospective functions by finite transition diagrams (transducers) is discussed. See also [84] for a detailed presentation of Trakhtenbrot and Rabinovich's theory and for discussions about the representation of finite memory retrospective operators by circuits.

Finally, another independent approach is considered in [182], where Chomsky-like hierarchies are established for families of sets of piecewise continuous functions. Differential equations, associated with specific memory structures, are used to recognize sets of functions. Ruohonen shows that the resulting hierarchies are not trivial and establishes closure properties and inclusions between classes.

2.3 Other computational models

In addition to the two previously described approaches, several other computational models have led to interesting developments in continuous time computation theory.

The question of whether Einstein's general relativity equations admit space-time solutions that allow an observer to view an eternity in a finite time was investigated and proved possible in [104]. The question of whether this implies that super-tasks can in principle be solved has been investigated in [77], [102], [101], [103], [78], [154], [155], and [203].

Some machine-inspired models are neither clearly digital nor analog. For example, the power of planar mechanisms attracted great interest in England and France in the late 1800s and in the 1940s in Russia. Specifically, these consisted of rigid bars constrained to a plane and joined at either end by rotable rivets. A theorem attributed[9] to Kempe [108] states that they are able to compute all algebraic functions; see for, e.g., [9] or [194].

---

[9] The theorem is very often attributed to Kempe [9], [194], even if he apparently never proved exactly that.

3 ODEs and properties

Most of the continuous time models described above have a continuous dynamics described by differential equations. In Shannon's GPAC and Hopfield networks, the input corresponds to the initial condition, whereas the output is, respectively, the time evolution or the equilibrium state of the system. Other models are language recognizers. The input again corresponds to the initial condition, or some initial control, and the output is determined by some accepting region in the state space of the system. All these systems therefore fall into the framework of dynamical systems.

In this section we will recall some fundamental results about dynamical systems and differential equations and discuss how different models can be compared in this general framework.

3.1 ODEs and dynamical systems

Let us consider that we are working in $\mathbb{R}^n$ (in general, we could consider any vector space with a norm). Let us consider $f : E \to \mathbb{R}^n$, where $E \subset \mathbb{R}^n$ is open. An ODE is given by $y' = f(y)$, and its solution is a differentiable function $y : I \subset \mathbb{R} \to E$ that satisfies the equation.

For any $x \in E$, the fundamental existence-uniqueness theorem (see, e.g., [100]) for differential equations states that if f is Lipschitz on E, i.e., if there exists K such that $||f(y_1) - f(y_2)|| < k||y_1 - y_2||$ for all $y_1, y_2 \in E$, then the solution of

$$y' = f(y), \qquad y(t_0) = x \tag{1}$$

exists and is unique on a certain maximal interval of existence $I \subset \mathbb{R}$. In the terminology of dynamical systems, $y(t)$ is referred to as the *trajectory*, $\mathbb{R}^n$ as the *phase space*, and the function $\phi(t, x)$, which gives the position $y(t)$ of the solution at time t with initial condition x, as the *flow*. The graph of y in $\mathbb{R}^n$ is called the *orbit*.

In particular, if f is continuously differentiable on E, then the existence-uniqueness condition is fulfilled [100]. Most of the mathematical theory has been developed in this case, but it can be extended to weaker conditions. In particular, if f is assumed to be only continuous, then uniqueness is lost, but existence is guaranteed; see, for example, [63]. If f is allowed to be discontinuous, then the definition of the solution needs to be refined. This is explored by Filippov in [81]. Some hybrid system models use distinct and ad hoc notions of solutions. For example, a solution of a piecewise constant differential equation in [20] is a continuous function whose right derivative satisfies the equation.

In general, a dynamical system can be defined as the action of a subgroup $\mathcal{T}$ of $\mathbb{R}$ on a space X, i.e., by a function (a flow) $\phi : \mathcal{T} \times X \to X$ satisfying the following two equations:

$$\phi(0, x) = x, \tag{2}$$

$$\phi(t, \phi(s, x)) = \phi(t + s, x). \tag{3}$$

It is well known that subgroups $\mathcal{T}$ of $\mathbb{R}$ are either dense in $\mathbb{R}$ or isomorphic to the integers. In the first case, the time is called continuous, and in the latter case, discrete.

Since flows obtained by initial value problems (IVP) of the form (1) satisfy equations (2) and (3), they correspond to specific continuous time and space dynamical systems. Although not all continuous time and space dynamical systems can be put in a form of a differential equation, IVPs of the form (1) are sufficiently general to cover a very wide class of such systems. In particular, if ϕ is continuously differentiable, then $y' = f(y)$, with $f(y) = \frac{d}{dt}\phi(t, y)\big|_{t=0}$, describes the dynamical system.

For discrete time systems, we can assume without loss of generality that $\mathcal{T}$ is the integers. The analog of IVP (1) for discrete time systems is a recurrence equation of type

$$y_{t+1} = f(y_t), \qquad y_0 = x. \tag{4}$$

A dynamical system whose space is discrete and that evolves discretely is termed digital; otherwise it is analog. A classification of some computational models according to the nature of their space and time can be found in Figure 3.

3.2 Dissipative and non-dissipative systems

A point x^* of the state space is called an *equilibrium point* if $f(x^*) = 0$. If the system is at x^*, it will remain there. It is said to be *stable* if for every neighborhood U of x^*, there is a neighborhood W of x^* in U such that every solution starting from a point x of W is defined and is in U for all time $t > 0$. The point is *asymptotically stable* if, in addition to the properties above, we have $\lim y(t) = x^*$ [100].

Some local conditions on the differential $Df(x^*)$ of f in x^* have been clearly established. If at an equilibrium point x^* all eigenvalues of $Df(x^*)$ have negative real parts, then x^* is asymptotically stable, and furthermore, nearby solutions approach x^* exponentially. In that case, x^* is called a *sink*. At a stable equilibrium point x^*, no eigenvalue of $Df(x^*)$ can have a positive real part [100].

In practice, Lyapunov's stability theorem applies more broadly (i.e., even if x^* is not a sink). It states that if there exists a continuous function V defined on a neighborhood of x^*, differentiable (except perhaps on x^*) with $V(x^*) = 0$, $V(x) > 0$ for $x \neq x^*$, and $dV(x)/dt \leq 0$ for $x \neq x^*$, then x^* is stable. If, in addition, $dV(x)/dt < 0$ for $x \neq x^*$, then x^* is asymptotically stable; see [100].

If the function V satisfies the previous conditions everywhere, then the system is *globally asymptotically stable*. Whatever the initial point x is, the trajectories will eventually converge to local minima of V. In this context, the Lyapunov function V can be interpreted as an energy, and its minima correspond to attractors of the

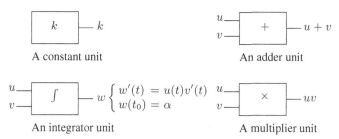

A constant unit

An adder unit

An integrator unit

A multiplier unit

Fig. 1. Different types of units used in a GPAC.

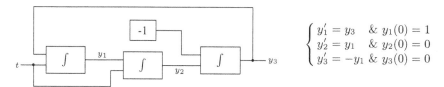

$$\begin{cases} y_1' = y_3 & \& \; y_1(0) = 1 \\ y_2' = y_1 & \& \; y_2(0) = 0 \\ y_3' = -y_1 & \& \; y_3(0) = 0 \end{cases}$$

Fig. 2. Generating cos and sin via a GPAC: circuit version on the left and ODE version on the right. One has $y_1 = \cos$, $y_2 = \sin$, and $y_3 = -\sin$.

Space \ Time	Discrete	Continuous
Discrete	[199] machines [61] lambda calculus [110] recursive functions [164] systems Cellular automata Stack automata Finite state automata ⋮	Discrete time [105] neural networks [186] neural networks [20] PCD systems [31] machines [205] optical machines [76] signal machines [146] dynamical recognizers ⋮
Continuous	[72] BDE models	[183] GPACs Continuous time [105] neural networks [44] hybrid systems [20] PCD systems [5] timed automata [145] $\mathbb{R}$-recursive functions ⋮

Fig. 3. A classification of some computational models, according to their space and time.

dynamical system. These are bounded subsets of the phase space to which regions of initial conditions of nonzero volume converge as time increases.

A dynamical system is called *dissipative* if the volume of a set decreases under the flow for some region of the phase space. Dissipative systems are characterized by the presence of attractors. By opposition, a dynamical system is said to be *volume-preserving* if the volume is conserved. For instance, all Hamiltonian systems are volume-preserving because of Liouville's theorem [8]. Volume-preserving dynamical system cannot be globally asymptotically stable [8].

3.3 Computability of solutions of ODEs

Here we review some results on the computability of solutions of IVPs in the framework of recursive analysis (see, e.g., [201] and the corresponding chapter in this volume).

In general, given a computable function f, one can investigate if the solution of the IVP (1) is also computable in the sense of recursive analysis. If we require that the IVP has a unique solution, then that solution is computable. Formally, if f is computable on $[0,1] \times [-1,1]$ and the IVP $y' = f(t,y)$, $y(0) = 0$ has a unique solution on $[0,b]$, $0 < b \leq 1$, then the solution y is computable on $[0,b]$.

This result also holds for a general n-dimensional IVP if its solution is unique [179]. However, computability of solutions is lost as soon as uniqueness of solutions is relaxed, even in dimension 1. Indeed, the famous result of [167] shows that there exists a polynomial-time computable function $f : [0,1] \times [-1,1] \to \mathbb{R}$, such that the equation $y' = f(t,y)$, with $y(0) = 0$, has nonunique solutions, but none of them is computable on any closed finite time interval.

Similar phenomena hold for other natural equations: the three-dimensional wave equation (which is a partial equation), with computable initial data, can have a unique solution that is nowhere computable[10] [168], [165]. Notice that, even if f is assumed computable and analytic, and the solution unique, it may happen that the maximal interval (α, β) of existence of the solution is noncomputable [92]. This same question is open if f is polynomial. Those authors show, however, that if f and f' are continuous and computable, then the solution of $y' = f(y,t)$, $y(0) = x$, for computable x, is also computable on its maximal interval of existence. Refer also to [169] and [111] for more uncomputability results, and also to [111] and [112] for related complexity issues.

3.4 Static undecidability

As observed in [11] and [181], it is relatively simple but not very informative to get undecidability results with continuous time dynamical systems, if f encodes a

---

[10] However, in all these cases, the problems under study are ill-posed: either the solution is not unique or it is unstable and the addition of some natural regularity conditions to prevent ill-posedness do yield computability [202].

undecidable problem. To illustrate this, we recall the following example in [181]. Ruohonen discusses the event detection problem: given a differential equation $y' = f(t, y)$, with initial value $y(0)$, decide whether a given condition $g_j(t, y(t), y'(t)) = 0, j = 1, \cdots, k$ happens at some time t in a given interval I. Given the Turing machine $\mathcal{M}$, the sequence $f_0, f_1, \cdots$ of rationals defined by

$$f_n = \begin{cases} 2^{-m} & \text{if } \mathcal{M} \text{ stops in } m \text{ steps on input } n \\ 0 & \text{if } \mathcal{M} \text{ does not stop on input } n \end{cases}$$

is not a computable sequence of rationals, but it is a computable sequence of reals, following the nomenclature of [169]. Now, the detection of the event $y(t) = 0$ for the ordinary differential equation $y' = 0$, given n, and the initial value $y(0) = f_n$, is undecidable over any interval containing 0, because $f_n = 0$ is undecidable.

Another modification can be obtained as follows in [181]. He defines the smooth function

$$g(x) = f_{\lfloor x+1/2 \rfloor} e^{-\tan^2 \pi x},$$

which is computable on $[0, \infty)$. The detection of the event $y_1(t) = 0$ for the ODE

$$\begin{cases} y_1' = g(y_2) - 1 \\ y_2' = 0 \end{cases}$$

given an initial value $y_1(0) = 1$, $y_2(0) = n$, where n is a nonnegative integer is then undecidable on $[0, 1]$.

As put forth in [11], undecidability results given by recursive analysis are somehow built similarly.

3.5 Dynamic undecidability

To be able to discuss in more detail computability of differential equations, we will focus on ODEs that encode the transitions of a Turing machine instead of the result of the whole computation simulation.[11] Typically, we start with some (simple) computable injective function that encodes any configuration of a Turing machine M as a point in $\mathbb{R}^n$. Let x be the encoding of the initial configuration of $\mathcal{M}$. Then, we look for a function $f : E \subset \mathbb{R}^{n+1} \to \mathbb{R}^n$ such that the solution of $y'(t) = f(y, t)$, with $y(0) = x$, at time $T \in N$ is the encoding of the configuration of $\mathcal{M}$ after T steps. We will see, in the remainder of this section, that f can be restricted to have low dimension, to be smooth or even analytic, or to be defined on a compact domain.

Instead of stating that the property above is a Turing machine simulation, we can address it as a reachability result. Given the IVP defined by f and x, and any region $A \subset \mathbb{R}^n$, we are interested in deciding if there is a $t \geq 0$ such $y(t) \in A$, i.e., if the flow starting in x crosses A. It is clear that if f simulates a Turing machine in

---

[11] This is called dynamic undecidability in [177].

the previous sense, then reachability for that system is undecidable (just consider A as encoding the halting configurations of $\mathcal{M}$). So, reachability is another way to address the computability of ODEs, and a negative result is often a byproduct of the simulation of Turing machines. Similarly, undecidability of event detection follows from Turing simulation results.

Computability of reachable and invariant sets have been investigated in [64] for continuous time systems and in [65] for hybrid systems.

In general, viewing Turing machines as dynamical systems provides them a physical interpretation that is not provided by the von Neumann picture [54]. This also shows that many qualitative features of (analog or nonanalog) dynamical systems, e.g., questions about basins of attraction, chaotic behavior, or even periodicity, are noncomputable [143]. Conversely, this brings into the realm of Turing machines and computability in general questions traditionally related to dynamical systems. These include in particular the relations between universality and chaos [11], necessary conditions for universality [74], the computability of entropy [113], understanding of edge of chaos [120], and relations with the shadowing property [107].

3.6 Embedding Turing machines in continuous time

The embedding of Turing machines in continuous dynamical systems is often realized in two steps. Turing machines are first embedded into analog space, discrete time systems, and then the obtained systems are in turn embedded into analog space and time systems.

The first step can be realized with low-dimensional systems with simple dynamics: [143], [177], [44], [181] consider general dynamical systems, [114] piecewise affine maps, [187] sigmoidal neural nets, [115] closed form analytic maps, which can be extended to be robust [90], and [118] one-dimensional very restricted piecewise-defined maps.

For the second step, the most common technique is to build a continuous time and space system whose discretization corresponds to the embedded analog space discrete time system.

There are several classical ways to discretize a continuous time and space system; see Figure 4. One way is to use a virtual stroboscope: the flow $x_t = \phi(t, x)$, when t is restricted to integers, defines the trajectories of a discrete time dynamical system. Another possibility is through a Poincaré section: the sequence x_t of the intersections of trajectories with, for example, a hypersurface can provide the flow of a discrete time dynamical system. See [100].

The opposite operation, called *suspension*, is usually achieved by extending and smoothing equations, and it usually requires higher dimensional systems. This explains why Turing machines are simulated by three-dimensional smooth continuous time systems in [143], [144] and [44] or by three-dimensional piecewise constant

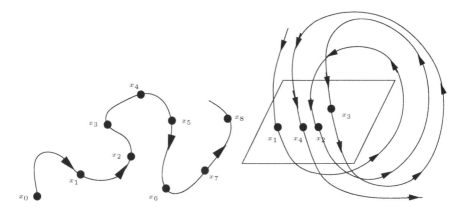

Fig. 4. Stroboscopic map (on left) and Poincaré map (on right) of the dynamic of a continuous time system.

differential equations in [20], while they are known to be simulated in discrete time by only two-dimensional piecewise affine maps in [114]. It is known that two-dimensional piecewise constant differential equations cannot[12] simulate arbitrary Turing machines [20], while the question of whether one-dimensional piecewise affine maps can simulate arbitrary Turing machines is open. Other simulations of Turing machines by continuous time dynamical systems include the robust simulation with polynomial ODEs in [90] and [91]. This result is an improved version of the simulation of Turing machines with real recursive functions in [52], where it is shown that smooth but nonanalytic classes of real recursive functions are closed under iteration. Notice that while the solution of a polynomial ODE is computable on its maximal interval of existence (see Section 3.3), the simulation result shows that the reachability problem is undecidable for polynomial ODEs.

In addition to Turing machines, other discrete models can be simulated by differential equations. Simulating two counter machines can be achieved in two dimensions, or even one dimension, at the cost of a discontinuous ODE [181]. Simulating cellular automata can be done with partial differential equations defined with C^∞ functions [157].

Notice that reversible computations of Turing machines (or counter machines, or register machines) can be simulated by ODEs with backward-unique solutions [177].

Continuous time dynamical systems can in turn be embedded into other continuous time systems. For example, [134] proves that a large class S_n of systems of differential equations are universal for analog computing on time-varying inputs in the following sense: a system of this class can reply to some external input $u(t)$ with the dynamics of any n^{th} order differential equation of the form

---

12 See also already mentioned generalizations of this result in [60] and [22].

$z^{(n)}(t) = G(z(t), z'(t), \cdots, z^{(n-1)}(t)) + u(t)$, if a suitable memoryless feedback and readout functions are added. As the nth order differential equation above can simulate Turing machines, systems from S_n have the power of a universal Turing machine. But since G is arbitrary, systems from S_n can actually simulate any conceivable continuous dynamic response to an input stream. Moreover, this results holds for the case where inputs and outputs are required to be bounded.

3.7 Discussion issues

The key technique in embedding the time evolution of a Turing machine in a flow is to use "continuous clocks" as in [44].[13]

The idea is to start from the function $f : \mathbb{R} \to \mathbb{R}$, preserving the integers, and build the ordinary differential equation over $\mathbb{R}^3$

$$
\begin{aligned}
y_1' &= c(f(r(y_2)) - y_1)^3 \theta(\sin(2\pi y_3)), \\
y_2' &= c(r(y_1) - y_2)^3 \theta(-\sin(2\pi y_3)), \\
y_3' &= 1.
\end{aligned}
$$

Here $r(x)$ is a rounding-like function that has value n whenever $x \in [n - 1/4, n + 1/4]$ for some integer n, and $\theta(x)$ is 0 for $x \le 0$, $\exp(-1/x)$ for $x > 0$, and c is some suitable constant.

The variable $y_3 = t$ is the time variable. Suppose $y_1(0) = y_2(0) = x \in \mathbb{N}$. For $t \in [0, 1/2]$, $y_2' = 0$, and hence y_2 is kept fixed to x. Now, if $f(x) = x$, then y_1 will be kept to x. If $f(x) \ne x$, then $y_1(t)$ will approach $f(x)$ on this time interval, and from the computations in [54], if a large enough number is chosen for c we can be sure that $|y_1(1/2) - f(x)| \le 1/4$. Consequently, we will have $r(y_1(1/2)) = f(x)$. Now, for $t \in [1/2, 1]$, roles are inverted: $y_1' = 0$, and hence y_1 is kept fixed to the value $f(x)$. On that interval, y_2 approaches $f(x)$, and $r(y_2(1)) = f(x)$. The equation has a similar behavior for all subsequent intervals of the form $[n, n + 1/2]$ and $[n+1/2, n+1]$. Hence, at all integer time t, $f^{[t]}(x) = r(y_1(t))$.[14] [124] proposes a similar construction that returns $f^{[\lfloor t \rfloor]}(x)$ for all $t \in \mathbb{R}$.

In other words, the construction above transforms a function over $\mathbb{R}$ into a higher dimensional ordinary differential equation that simulates its iterations. To do so, $\theta(\sin(2\pi y_3))$ is used as a kind of clock. Therefore, the construction is essentially "hybrid" since it combines smooth dynamics with nondifferentiable, or at least non-analytic clocks to simulate the discrete dynamics of a Turing machine. Even if the flow is smooth (i.e., in C^∞) with respect to time, the orbit does not admit a tangent at every point since y_1 and y_2 are alternatively constant. Arguably, one can overcome this limitation by restricting Turing machine simulations to analytic flows and maps.

---

[13] Branicky attributes the idea of a two phase computation to [47] and [48]. A similar trick is actually present in [177]. We will actually not follow [44] but its presentation in [54].

[14] $f^{[t]}(x)$ denotes the tth iteration of f on x.

Although it was shown that analytic maps over unbounded domains are able to simulate the transition function of any Turing machine in [115], only recently it was shown that Turing machines can be simulated with analytic flows over unbounded domains in [90]. It would be desirable to extend the result to compact domains. However, it is conjectured in [147] that this is not possible, i.e., that no analytic map on a compact finite-dimensional space can simulate a Turing machine through a reasonable input and output encoding.

3.8 Time and space contractions

Turing machines can be simulated by ODEs in real time: for example, in the constructions we described above, the state $y(T)$ at time $T \in N$ of the solution of the ordinary differential equation encodes the state after T steps of the Turing machine. However, since continuous time systems might undergo arbitrary space and time contractions, Turing machines, as well as accelerating Turing machines[15] [71], [67], [68] or even oracle Turing machines, can actually be simulated in an arbitrary short time.

In the paragraphs below, we will follow Ruohonen [177] who denotes a continuous time system by the triplet (F, n, A), where F defines the ordinary differential equation $y' = F(y)$ over $\mathbb{R}^n$, with accepting set A: some input x is accepted iff the trajectory starting with initial condition x crosses A.

A machine $\mathcal{M} = (F, n, A)$ can be accelerated: the substitution $t = e^u - 1$ for instance changes $\mathcal{M}$ to $((G, 1), n + 1, A \times \mathbb{R})$, where

$$\frac{dg}{du} = G(g(u), u) = F(g(u))e^u \text{ and } g(u) = y(e^u - 1),$$

yielding an exponential time acceleration. Note that the derivatives of the solution with respect to the new time variable u are exponentially larger. Furthermore, the substitution $t = \tan(\pi u/2)$ gives an infinite time acceleration, i.e., compresses any computation, even an infinite one, into the finite interval $0 \leq u < 1$. Now, the derivatives go to infinity during the course of computation.

Turning to space contraction, replacing the state $y(t)$ of the machine $\mathcal{M} = (F, n, A)$ by $r(t) = y(t)e^{-t}$ gives an exponentially downscaled machine $((H, 1), m + 1, H_1)$ where

$$\frac{dr}{dt} = H(r(t), t) = F(r(t)e^t)e^{-t} - r(t)$$

and

$$H_1 = \{(e^{-t}q, t)|q \in A \text{ and } t \geq 0\}.$$

Obviously, this transformation reduces exponentially the distance between trajectories, which require increased precision to be distinguished.

---

[15] Similar possibilities of simulating accelerating Turing machines through quantum mechanics are discussed in [51].

Hardness results in the levels of the arithmetical or analytical hierarchy for several decision problems about continuous time systems are derived from similar constructions in [177], [178], [145], and [19]. Completeness results, as well as exact characterizations of the recognition power of piecewise constant derivative systems, according to their dimensions have been obtained in [32] and [33]. Notice that such phenomena are instances of the so-called Zeno's phenomena in hybrid systems literature: [5] and [19].

It can be observed that previous constructions yield undecidability results only for functions over infinite or half-open intervals, since positive reals, corresponding to Turing machines integer time, are mapped to intervals of the form $[0, 1)$. An analytical construction is indeed possible over a finite closed domain of the form $[0, 1]$, with a function G that is continuous and bounded on $[0, 1]$, but nondifferentiable. It follows that the event detection problem, for example, is undecidable even with continuous functions over compact intervals [180].

Undecidability is ruled out, however, if the function G is sufficiently smooth (say, in $\mathcal{C}^1$), if both G and the initial value are computable, and if a sufficiently robust acceptance condition is considered. Indeed, problems such as the event detection problem then become decidable, since the system can be simulated effectively [180].

Instead of embedding Turing machines into continuous dynamical systems, it is natural to ask whether there is a better way to think about computation and complexity for the dynamical systems that are commonly used to model the physical world. We address this issue in the next section.

4 Toward a complexity theory

Here we discuss several different views on the complexity of continuous dynamical systems. We consider general systems and question the difficulty of simulating continuous time systems with a digital model. We then focus on dissipative systems, where trajectories converge to attractors. In particular, we discuss the idea that the computation time should be the natural time variable of the ODE. Finally, we review complexity results for more general continuous time systems that correspond to classes of real recursive functions.

4.1 General continuous dynamical systems

In [200] it was asked whether analog computers can be more efficient than digital ones. Vergis et al. also postulated the "Strong Church's Thesis," which states that the time required by a digital computer to simulate any analog computer is bounded by a polynomial function of the resources used by the analog computer. They claim that the Strong Church's Thesis is provably true for continuous time dynamical systems described by any Lipschitzian ODE $y' = f(y)$.

The resources used by an analog computer include the time interval of operation, say $[0, T]$, the size of the system, which can measured by $\max_{t \in [0,T]} \|y(t)\|$, as well as the bound on the derivatives of y. For instance, mass, time of operation, maximum displacement, velocity, acceleration, and applied force are all resources used by a particle described by Newtonian mechanics [200].

The claim above depends on the definition of "simulation." In the article [200] it is considered that the IVP $y' = f(y)$, $y(0) = x$ is simulated if, given T and some precision ε, one can compute an approximation of $y(T)$ with a margin of error of at most ε. Using Euler's method to solve this problem, and considering that the round-off error is less than σ, the total error bound is given by

$$\|y(T) - y_N^*\| \leq \frac{h}{\lambda} \left[\frac{R}{2} + \frac{\sigma}{h2} \right] (e^T \lambda - 1), \tag{5}$$

where y_N^* is the approximation after N steps, h is the step size, λ is the Lipschitz constant for f on $[0, T]$, and $R = \max\{\|y''(t)\|, t \in [0, T]\}$. From the bound in (5), Vergis et al. conclude that the number N of necessary steps in Euler's method is polynomial in R and $\frac{1}{\varepsilon}$. They use this fact to claim that the Strong Church's Thesis is valid for ODEs. However, N is exponential in T, which is the time of operation of this analog computer. This makes the argument in [200] inconclusive, as pointed out in [160].

More recently, Smith discusses in [192] if hypercomputation is possible with respect to the n-body problem in mechanics. In particular, he shows that the exponential dependence in T can be eliminated. As observed in [192], all classical numerical methods of fixed degree for solving differential equations suffer from the same exponential dependence in T. However, by considering a combination of Runge–Kutta methods with degrees varying linearly with T, it is possible to derive a method that only requires N to be polynomial in T, as long as the absolute value of each component of f, y, and the absolute value of each partial derivative of f with respect to any of its arguments, having total differentiation degree k, is in $(kT)^{\mathcal{O}(k)}$ [192]. The implications of these results for Strong Church's Thesis are discussed in [192] and [34].

The same question can be addressed in the framework of recursive analysis. When $f : [0, 1] \times [-1, 1] \to \mathbb{R}$ is polynomial time computable and satisfies a weak form of the Lipschitz condition, the unique solution y on $[0, 1]$ of IVP $y' = f(t, y)$, $y(0) = 0$ is always polynomial space computable [111]. Furthermore, solving in polynomial time a differential equation with this weak Lipschitz condition is essentially as difficult as solving a PSPACE-complete problem, since there exists a polynomial time computable function f as above whose solution y is not polynomial time computable unless $P = PSPACE$ [111], [112].

Ko's results are not directly comparable with the polynomial bound shown in [192]. In recursive analysis, the input's size is the number of bits of precision. If the bound on the error of the approximation of $y(t)$ is measured in bits; i.e., if $\varepsilon = 2^{-d}$, then the required number of steps N in [192] is exponential in d.

If f is analytic, then the solution of $y' = f(y)$ is also analytic. In that case, timestepping methods can be avoided. That is the approach followed in [142], where it is proved using recursive analysis that if f is analytic and polynomial time computable, then the solution is also polynomial time computable.

In short, although Strong Church's Thesis holds for analytic ODEs, it has not yet been fully proved for general systems of Lipschitzian ODEs. Hence, the possibility of super-polynomial computation by differential equations cannot be ruled out, at least in principle. For informal discussions on Strong Church's Thesis, refer to [1] and [139].

Several authors have shown that certain decision or optimization problems (e.g., graph connectivity or linear programming) can be solved by specific continuous dynamical systems. Examples and references can be found in the papers [191], [200], [48], [79], [97] and [27].

4.2 Dissipative systems

We now focus on dissipative systems and review two approaches. The first is about neural network models, such as continuous Hopfield networks, that simulate circuits in a nonuniform manner, and leads to lower bounds on the complexity of such networks. The second deals with convergence to attractors and considers suitable energy functions as ways to measure the complexity of continuous time processes.

When considering dissipative systems, such as Hopfield neural networks, the following approach to a complexity theory is natural. Consider families $(C_n)_{n \in N}$ of continuous time systems, for each input length $n \geq 0$. Given some digital input $w \in \{0, 1\}^*$, the system C_n evolves on input w (or some encoding of w), where n is the length of w. It will eventually reach some stable state, which is considered the result of the computation.

This circuit inspired notion of computability is the most common in the literature about the computational complexity of neural networks models; see survey [190]. With respect to this approach, continuous time symmetric Hopfield networks with a saturated linear activation function have been proved to simulate arbitrary discrete-time binary recurrent neural networks, at the cost of only a linear size overhead [161], [189]. This might be thought counterintuitive, since such symmetric networks, which are constrained by some Liapunov energy function, can only exhibit convergence phenomena, and hence cannot even realize a simple alternating bit. However, the convergence of dissipative systems can be exponentially long in the size of the system [188], and hence the simulation can be accomplished using a subnetwork that provides 2^n clock pulses before converging.

The languages recognized by polynomial size families of discrete-time Hopfield networks have been proved in [159] to correspond to nonuniform complexity class $PSPACE/poly$ for arbitrary interconnection weights, and to $P/poly$ for polynomially bounded weights. Therefore, families of continuous time symmetric Hopfield networks have the same lower bounds. However, these lower bounds may be not tight, since upper bounds for continuous time dynamics are not known [189], [190].

Let us now turn our attention to dissipative systems with a Lyapunov function E.

Gori and Meer [86] consider a computational model that has the capability of finding the minimizers (i.e., the points of local or global minimum) of E. To prevent the complexity of a problem from being hidden in the description of E, this function must be easy to compute. In that setting, a problem Π is considered easy if there exists a unimodal function E (i.e., all local minimizers of E are global minimizers) such that the solution of Π can be obtained from the global minimum of E.

More precisely, Gori and Meer investigate in [86] a model where a problem Π over the reals is considered to be solved if there exists a family $(E_n)_n : \mathbb{R}^n \times \mathbb{R}^{q(n)} \to \mathbb{R}$ of energy functions, given by a uniform family of straight line programs (q is some fixed polynomial), and another family $(N_n)_n$ of straight line programs, such that for all input d, a solution $\Pi(d)$ of the problem can be computed using $N_{q(n)}(w^*)$, from a global minimizer w^* of $w \to E_n(d, w)$.

Gori and Meer define classes U and NU in analogy with P and NP in classical complexity. U corresponds to the above-mentioned case where for all d, $w \to E_n(d, w)$ is unimodal, in opposition to NU where it needs not be unimodal. Notions of reductions are introduced, and it is proved that the natural optimization problem "find the minimum of some linear objective function over a set defined by quadratic multivariate polynomial constraints" is NU-hard. They show that there exist (artificial) NU complete problems. These ideas are generalized to obtain a polynomial hierarchy, with complete problems [86].

Actually, Gori and Meer's proposed framework is rather abstract, avoiding several problems connected to what one might expect of a true complexity theory for continuous time computations. Nonetheless, it has the great advantage of not relying on any particular complexity measure for the computation of trajectories. See the interesting discussion in [86].

However, one would like to understand the complexity of approaching the minima of energy functions, which correspond to the equilibria of dynamical systems. First steps toward this end have been investigated in [27], where dissipative systems with exponential convergence are explored. Recall that if x^* is a sink, then the rate of convergence toward x^* satisfies

$$|x(t) - x^*| \equiv e^{-\lambda t},$$

where $-\lambda$ is the largest real part of an eigenvalue of $Df(x^*)$. This means that $\tau = 1/\lambda$ is a natural characteristic time of the attractor: every $\tau \log 2$ time units, a new bit of the attractor is computed.

For the systems considered in [27], each sink has an attracting region, where the trajectories are trapped. One can define the computation time t_c of a dissipative continuous time dynamical system as $t_c = \max(t_c(\epsilon), t_c(U))$, where $t_c(\epsilon)$ is the time required to reach some ϵ vicinity of some attractor, and $t_c(U)$ is the time required to reach its attracting region. Then, $T = \frac{t_c}{\tau}$ is a dimensionless complexity measure, invariant under any linear time contraction.

Two continuous time algorithms, MAX to compute the maximum of n numbers, and $FLOW$ to compute the maximum flow problem have been studied in this framework in [27]. MAX has been shown to belong to proposed complexity class $CLOG$ (continuous log time) and $FLOW$ to CP (continuous polynomial time). The authors conjecture that CP corresponds to classical polynomial time [27]. Both MAX and $FLOW$ algorithms are special cases of a flow proposed in [80] to solve linear programming problems, which are further investigated in [24] and [25]. Variations on definitions of complexity classes, as well as ways to introduce nondeterministic classes in relation to computations by chaotic attractors, have also been discussed in [185].

4.3 Complexity and real recursive functions

Real recursive functions are a convenient way to analyze the computational power of certain operations over real functions. Additionally, given a continous time model, if its equivalence with a function algebra of real recursive functions can be established, then some properties of the model can be proved inductively over the function algebra.

Since many time and space complexity classes have recursive characterizations over $\mathbb{N}$ [62], structural complexity results about discrete operations may imply lower and upper bounds on the computational complexity of real recursive functions. This approach was followed in [53] to show that $\mathcal{L}$ contains extensions of the elementary functions, and it was further developed in [56] to obtain weaker classes that relate to the exponential space hierarchy. This tells us something about the computational complexity of certain dynamical systems. For instance, $\mathcal{L}$ corresponds to cascades of finite depth, each level of which depends linearly on its own variables and the output of the level before it.

Results about the idea of lifting computability questions over $\mathbb{N}$ to $\mathbb{R}$ have been discussed before. Concerning complexity, the question $P = NP$ in classical complexity has been investigated using real recursive functions by Costa and Mycka. In particular, they propose two classes of real recursive functions such that their inequality would imply $P \neq NP$ in [69] and [151]. More generally a part of Costa and Mycka's program, which is explicitly stated in [150] and [152], uses recursion theory on the reals to establish a bridge between computability and complexity theory and mathematical analysis.

5 Noise and Robustness

Up to this point we have considered continuous time computations in idealized, noise-free spaces. As was also the case in the survey by Orponen [160], most of the results we discussed disregard the impact of noise and imprecision in continuous time systems. This is a recurrently criticized weakness of the research in the field. Although there have not been major breakthroughs with regard to these problems as they relate specifically to continuous time computations, some interesting developments concerning noise and imprecision have come about in discrete time analog computation studies. In this section we will broaden our scope to discuss a number of discrete time results. We believe that some of these studies and results might be generalized to, or at least provide some insight into, the effects of noise and imprecision on continuous time systems, although this work has yet to be done.

We first focus on systems with a bounded state space about which a folklore conjecture claims that robustness implies decidability. We review some results that support this conjecture as well as others that challenge it. At the end of this section, we discuss continuous time systems with unbounded state spaces.

Common techniques to simulate Turing machines by dynamical systems in bounded state spaces require the encoding of the configuration of the Turing machine into real numbers. Since Turing machines have unbounded tapes (otherwise they would degenerate into finite automata), these simulations are destroyed if the real numbers or the functions involved are not represented with infinite precision. This leads to the folklore conjecture, popular in particular in the verification community, which states that undecidability do not hold for "realistic," "unprecise," "noisy," "fuzzy," or "robust" systems. See, for example, [85] and [83] for various statements of this conjecture and [13] for discussions on other arguments that lead to this conjecture.

There is no consensus on what is a realistic noise model. A discussion of this subject would require to question what are good models of the physical world. In the absence of a generally accepted noise model, one can however consider various models for noise, imprecision, or smoothness conditions, and one can investigate the properties of the resulting systems.

In particular, there have been several attempts to show that noisy analog systems are at best equivalent to finite automata. Brockett proved that continuous time dynamical systems can simulate arbitrary finite automata in [47]. Using topological arguments based on homotopy equivalence relations and associated Deck transformations, he showed in [49] that some automata can be associated with dissipative continuous time systems.

Maass and Orponen proved that the presence of bounded noise reduces the power of a large set of discrete time analog models to that of finite automata in [136]. This extends a previous result established in [58] and [59] for the case where the output is assumed to be perfectly reliable (i.e., $\rho = 1/2$ in what follows).

Maass and Orponen's idea is to replace a perfect discrete time dynamic of type $x_{i+1} = f(x_i, a_i)$, where a_i is the symbol input at time i, over a compact domain, by a probabilistic dynamic

$$\text{Probability } (x_{i+1} \in B) = \int_{q \in B} z(f(x_i, a_i), q) d\mu, \qquad (6)$$

where B is any Borel set. Here, z is a density kernel reflecting arbitrary noise, which is assumed to be piecewise equicontinuous. This means that, for all ϵ, there exists δ such that for all r, p, q, $\|p - q\| \leq \delta$ implies $|z(r, p) - z(r, q)| \leq \epsilon$. They denote by $\pi_x(q)$ the distribution of states after string x is processed from some fixed initial state q, and they consider the following robust acceptance condition: a language L is recognized, if there exists $\rho > 0$ such that $x \in L$ iff $\int_F \pi_{xu}(q) d\mu \geq 1/2 + \rho$ for some $u \in \{U\}^*$, and $x \notin L$ iff $\int_F \pi_{xu}(q) d\mu \leq 1/2 - \rho$ for all $u \in \{U\}^*$, where U is the blank symbol, and F is the set of accepting states. Then, they show that the space of functions $\pi_x(.)$ can be partitioned into finitely many classes C such that two functions $\pi_x(.)$ and $\pi_y(.)$ in the same class satisfy $\int_r |\pi_x(r) - \pi_y(r)| d\mu \leq \rho$. Therefore, two words x, y in the same class satisfy $xw \in L$ iff $yw \in L$ for all words w.

In fact, for any common noise, such as Gaussian noise, which is nonzero on a sufficiently large part of the state space, systems described by (6) are unable to recognize even arbitrary regular languages [138]. They recognize precisely the definite languages introduced by [172], as shown in [138] and [26]. If the noise level is bounded, then all regular languages can be recognized [136]. Feedback continuous time circuits in [134] have the same computational power when subject to bounded noise.

As an alternative to the probabilistic approach of Maass and Orponen, noise can be modeled through nondeterminism. One can associate with deterministic noise-free discrete time dynamical system S defined by $x_{i+1} = f(x_i)$, the nondeterministic ϵ-perturbated system S_ϵ whose trajectories are sequences $(x_n)_n$ with $\|x_{i+1} - f(x_i)\| \leq \epsilon$. For a dynamical system S, it is natural to consider the predicate $\text{Reach}[S](x, y)$ (respectively, $\text{Reach}_n[S](x, y)$), which is true if there exists a trajectory of S from x to y (resp. in $i \leq n$ steps). Then, algorithmic verification of safety of state properties is closely related to the problem of computing reachable states. Given S, and a subset of initial states S_0, let $\text{Reach}[S]$ denote the set of y's such that $\text{Reach}[S](x, y)$ for some $x \in S_0$. Given a state property p (i.e., a property that is either true or false in a state s), let $[[\neg p]]$ denote the subset of states s where p is false. Then S is safe (p is an invariant) iff $\text{Reach}[S] \cap [[\neg p]] = \emptyset$ (see, for example, [4] and [156]).

If the class of systems under consideration is such that relation $\text{Reach}_n[S](x, y)$ is recursive[16] (assuming that S_0 recursively enumerable), then $\text{Reach}[S]$ is recursively enumerable because $\text{Reach}[S] = \bigcup_n \text{Reach}_n[S]$. Several papers have been devoted to prove that $\text{Reach}[S]$ is indeed recursive for classes of dynamical systems under different notions of robustness. We now review several of them.

Fränzle observes in [85] that the computation of $\text{Reach}[S_\epsilon]$ by $\text{Reach}[S_\epsilon] = \bigcup_n \text{Reach}_n[S_\epsilon]$ must always terminate if $\text{Reach}[S_\epsilon]$ has a strongly finite diameter. This means that there exists an infinite number of points in $\text{Reach}[S_\epsilon]$ at a mutual distance of at least ϵ, which is not possible over a bounded domain. It follows that if we call robust a system that is either unsafe or whose ϵ-perturbated system is safe for some ϵ, then safety is decidable for robust systems over compact domains [85].

Consider as in [170] the relation $\text{Reach}_\omega[S] = \bigcap_{\epsilon > 0} \text{Reach}[S_\epsilon]$, corresponding to states that stay reachable when noise converges to 0. Asarin and Bouajjani prove in [14] that for large classes of discrete and continuous time dynamical systems (Turing machines, piecewise affine maps, piecewise constant differential equations), $\text{Reach}_\omega[S]$ is co-recursively enumerable. Furthermore, any co-recursively enumerable relation is of form $\text{Reach}_\omega[S]$ for some S for the classes that Asarin and Bouajjani consider. Therefore, if we call robust a system such that $\text{Reach}[S] = \text{Reach}_\omega[S]$, then computing $\text{Reach}[S]$ is decidable for robust systems.

Asarin and Collins considered in [17] a model of Turing machines exposed to a small stochastic noise, whose computational power have been characterized to correspond to Π_2^0. It is interesting to compare this result with previous results where a small nondeterministic noise lead to Π_1^0 (co-recursively enumerable sets) computational power only.

We now turn our attention to results that challenge the conjecture that robustness implies decidability. A first example is that the safety of a system is still undecidable if the transition relation of the system is open, as proved in [99], and [13]. However, the question for the restriction to a uniform nondeterministic noise bounded from below is open [13].

Noise can also be modeled by perturbating trajectories. Gupta, Henzinger, and Jagadeesan consider in [95] a metric over trajectories of timed automata, and assume that if a system accepts a trajectory, then it must accept neighboring trajectories also. They prove that this notion of robustness is not sufficient to avoid undecidability of complementation for Timed automata. Henzinger and Raskin prove in [99] that major undecidability results about verification of hybrid systems are still undecidable for robust systems in that sense.

Finally, we review a recent robustness result for continuous time dynamical systems with unbounded state space. Graça, Campagnolo, and Buescu prove in [91] that polynomial differential equations can simulate robustly Turing machines in real time. More precisely, let us consider that $\theta : \mathbb{N}^3 \to \mathbb{N}^3$ is the transition function

---

[16] Recursive in $x, y,$ and n.

some Turing machine M whose configuration is encoded on $\mathbb{N}^3$. Then, there is a $\epsilon > 0$, a solution f of a polynomial ODE, and an initial condition $f(0)$ such that the solution of $y' = f(t, y)$ encodes the state of M after t steps with error at most ϵ. Moreover, this holds for a neighborhood of any integer t even if f and the initial condition $f(0)$ are perturbed. Obviously, this kind of simulation requires the system to have an unbounded state space.

6 Conclusion

Having surveyed the field of continuous time computation theory, we see that it provides insights into many diverse areas such as verification, control theory, VLSI design, neural networks, analog machines, recursion theory, theory of differential equations, and computational complexity.

We have attempted to give a systematic overview of the most relevant models and results on continuous time computations. In the last decade many new results have been obtained, indicating that this is an active field of research. We reviewed recent developments of the theory of continuous time computation with respect to computability, complexity, and robustness to noise, and we identified several open problems. To conclude, we will discuss some directions for future research related to these areas.

Computability

It is not clear whether a unifying concept similar to the Church–Turing thesis exists for continuous time computation. Although it has been shown that some continuous time models exhibit super Turing power, these results rely on the use of an infinite amount of resources such as time, space, precision, or energy. In general, it is believed that "reasonable" continuous time models cannot compute beyond Turing machines. This raises the question if physically realistic continuous time computation can be as powerful as digital computation. We saw that if we restrict continuous time systems to evolve in a bounded state space and to be subjected to noise, then they become comparable with finite automata. However, with a bounded state space, Turing machines also degenerate into finite automata. Since analytic and robust continuous time systems can simulate Turing machines in an unbounded state space, we believe that digital computation and analog continuous time computation are equally powerful from the computability point of view. Moreover, as we saw, several recent results establish the equivalence between functions computable by polynomial ODEs, GPAC-computable functions and real computable functions in the framework of recursive analysis. These kind of results reinforce the idea that there could be an unified framework for continuous time computations, analogous to what occurs in classical computation theory.

We feel that a general paradigm of realistic continuous time computations ideally should only involve analytic functions, since these are often considered as the most acceptable from a physical point of view. Continuous dynamical systems are a natural form of representing continuous time processes. Classic systems like the van der Pol equation, the Lotka–Volterra system, or the Lorenz equations are described with differential equations with an analytic, even polynomial, right-hand side. These physics-related arguments combined with the computability properties of systems of polynomial differential equations lead us to suggest that this continuous time model is a possible candidate for a general paradigm of continuous time computation. We believe that this idea deserves further investigation.

Complexity

We saw that a complexity theory for continuous time computation is still under way and that there has not been an agreement between authors on basic definitions such as computation time or input size. The results described in Section 4 are either derived from concepts that are intrinsic to the continuous time systems under study or related to classical complexity theory. As computable analysis is a well-established and understood framework for the study of computational complexity of continuous time systems, we believe that understanding relations between different approaches and computable analysis from a complexity point of view is of first importance. There are still many open questions about upper bounds for continuous time models. For example, upper bounds are not known for Hopfield networks and general systems of Lipschitizian ODEs, which compromises the validity of the Strong Turing thesis. We saw that this thesis might hold for systems of analytic ODEs. This leads us to ask whether a continuous time computation theory based on polynomial ODEs could be naturally extended to a complexity theory.

Computable analysis also permits study of the complexity of real recursive functions. One of the most intriguing areas of research in continuous time computation tries to explore the link between real recursive functions and computational complexity to establish a translation of open problems of classic complexity into analysis.

Robustness

We saw that very little research has been done with respect to the robustness and tolerance to noise of continuous time systems. One might ask how the power of analog computations increases with their precision. This question was raised and formalized for discrete time analog systems, in particular for dynamical recognizers, in [147], but most of the research in that direction has yet to be done. Many interesting open questions arise if one asks whether undecidability results for continuous time systems still hold for robust systems. This is of first importance for example for the verification of hybrid systems, since this question is closely related to the question of termination of automatic verification procedures. A better understanding of the hypotheses under which noise yield decidability or undecidability is required. For example, nondeterministic noise on open systems does not rule out undecidability, but the question is unanswered for a uniform noise bounded from below [13].

Acknowledgments. We would like to thank all our colleagues in a wide sense, since this survey benefited from recent and old discussions with a long list of people. Many of them have their work cited in the text. We would also like to deeply thank Kathleen Merrill for her careful reading of the text and for her suggestions to improve its clarity and an anonymous referee for his/her helpful advice. This work was partially supported by EGIDE and GRICES under the Program *Pessoa* through the project *Calculabilité et complexité des modèles de calculs à temps continu*, by *Fundação para a Ciência e a Tecnologia* and FEDER via the Center for Logic and Computation - CLC and the project ConTComp POCTI/MAT/45978/2002.

References

1. Aaronson, S. (2005). NP-complete problems and physical reality. *ACM SIGACT News*, 36(1):30–52.
2. Abdi, H. (1994). A neural network primer. *Journal of Biological Systems*, 2:247–281.
3. Adleman, L. M. (1994). Molecular computation of solutions to combinatorial problems. *Science*, 266:1021–1024.
4. Alur, R., Courcoubetis, C., Halbwachs, N., Henzinger, T. A., Ho, P. H., Nicollin, X., Olivero, A., Sifakis, J., and Yovine, S. (1995). The algorithmic analysis of hybrid systems. *Theoretical Computer Science*, 138(1):3–34.
5. Alur, R. and Dill, D. L. (1990). Automata for modeling real-time systems. In Paterson, M., editor, *Automata, Languages and Programming, 17th International Colloquium, ICALP90, Warwick University, England, July 16-20, 1990, Proceedings*, volume 443 of *Lecture Notes in Computer Science*, pages 322–335. Springer.
6. Alur, R. and Dill, D. L. (1994). A theory of timed automata. *Theoretical Computer Science*, 126(2):183–235.
7. Alur, R. and Madhusudan, P. (2004). Decision problems for timed automata: A survey. In Bernardo, M. and Corradini, F., editors, *Formal Methods for the Design of Real-Time Systems, International School on Formal Methods for the Design of Computer, Communication and Software Systems, SFM-RT 2004, Bertinoro, Italy, September 13-18, 2004, Revised Lectures*, volume 3185 of *Lecture Notes in Computer Science*, pages 1–24. Springer.
8. Arnold, V. I. (1989). *Mathematical methods of classical mechanics*, volume 60 of *Graduate Texts in Mathematics*. Springer, second edition.
9. Artobolevskii, I. (1964). *Mechanisms for the Generation of Plane Curves*. Macmillan, New York. Translated by R.D. Wills and W. Johnson.
10. Asarin (2004). Challenges in timed languages: From applied theory to basic theory. *Bulletin of the European Association for Theoretical Computer Science*, 83:106–120.
11. Asarin, E. (1995). Chaos and undecidabilty (draft). Avalaible in `http://www.liafa.jussieu.fr/$\tilde{\}$asarin/`.

12. Asarin, E. (1998). Equations on timed languages. In Henzinger, T. A. and Sastry, S., editors, *Hybrid Systems: Computation and Control, First International Workshop, HSCC'98, Berkeley, CA, April 13-15, 1998, Proceedings*, volume 1386 of *Lecture Notes in Computer Science*, pages 1–12. Springer.
13. Asarin, E. (2006). Noise and decidability. Continuous Dynamics and Computability Colloquium. Video and sound available trough "Diffusion des savoirs de l'Ecole Normale Supérieure," on http://www.diffusion.ens.fr/en/index.php?res=conf\&idconf=1226.
14. Asarin, E. and Bouajjani, A. (2001). Perturbed Turing machines and hybrid systems. In *Proceedings of the 16th Annual IEEE Symposium on Logic in Computer Science*, pages 269–278, Los Alamitos, CA. IEEE Computer Society Press.
15. Asarin, E., Caspi, P., and Maler, O. (1997). A Kleene theorem for timed automata. In *Proceedings, 12th Annual IEEE Symposium on Logic in Computer Science*, pages 160–171, Warsaw, Poland. IEEE Computer Society Press.
16. Asarin, E., Caspi, P., and Maler, O. (2002). Timed regular expressions. *Journal of the ACM*, 49(2):172–206.
17. Asarin, E. and Collins, P. (2005). Noisy Turing machines. In Caires, L., Italiano, G. F., Monteiro, L., Palamidessi, C., and Yung, M., editors, *Automata, Languages and Programming, 32nd International Colloquium, ICALP 2005, Lisbon, Portugal, July 11-15, 2005, Proceedings*, volume 3580 of *Lecture Notes in Computer Science*, pages 1031–1042. Springer.
18. Asarin, E. and Dima, C. (2002). Balanced timed regular expressions. *Electronic Notes in Theoretical Computer Science*, 68(5).
19. Asarin, E. and Maler, O. (1998). Achilles and the tortoise climbing up the arithmetical hierarchy. *Journal of Computer and System Sciences*, 57(3):389–398.
20. Asarin, E., Maler, O., and Pnueli, A. (1995). Reachability analysis of dynamical systems having piecewise-constant derivatives. *Theoretical Computer Science*, 138(1):35–65.
21. Asarin, E. and Schneider, G. (2002). Widening the boundary between decidable and undecidable hybrid systems. In Brim, L., Jancar, P., Kretínský, M., and Kucera, A., editors, *CONCUR 2002 - Concurrency Theory, 13th International Conference, Brno, Czech Republic, August 20-23, 2002, Proceedings*, volume 2421 of *Lecture Notes in Computer Science*, pages 193–208. Springer.
22. Asarin, E., Schneider, G., and Yovine, S. (2001). On the decidability of the reachability problem for planar differential inclusions. In Benedetto, M. D. D. and Sangiovanni-Vincentelli, A. L., editors, *Hybrid Systems: Computation and Control, 4th International Workshop, HSCC 2001, Rome, Italy, March 28-30, 2001, Proceedings*, volume 2034 of *Lecture Notes in Computer Science*, pages 89–104. Springer.
23. Beauquier, D. (1998). Pumping lemmas for timed automata. In Nivat, M., editor, *Foundations of Software Science and Computation Structure, First International Conference, FoSSaCS'98, Held as Part of the European Joint Conferences on the Theory and Practice of Software, ETAPS'98, Lisbon, Portugal, March 28 - April 4, 1998, Proceedings*, volume 1378 of *Lecture Notes in Computer Science*, pages 81–94. Springer.
24. Ben-Hur, A., Feinberg, J., Fishman, S., and Siegelmann, H. T. (2003). Probabilistic analysis of a differential equation for linear programming. *Journal of Complexity*, 19(4):474–510.
25. Ben-Hur, A., Feinberg, J., Fishman, S., and Siegelmann, H. T. (2004a). Random matrix theory for the analysis of the performance of an analog computer: a scaling theory. *Physics Letters A*, 323(3–4):204–209.
26. Ben-Hur, A., Roitershtein, A., and Siegelmann, H. T. (2004b). On probabilistic analog automata. *Theoretical Computer Science*, 320(2–3):449–464.

27. Ben-Hur, A., Siegelmann, H. T., and Fishman, S. (2002). A theory of complexity for continuous time systems. *Journal of Complexity*, 18(1):51–86.
28. Blondel, V. D. and Tsitsiklis, J. N. (1999). Complexity of stability and controllability of elementary hybrid systems. *Automatica*, 35(3):479–489.
29. Blondel, V. D. and Tsitsiklis, J. N. (2000). A survey of computational complexity results in systems and control. *Automatica*, 36(9):1249–1274.
30. Blum, L., Cucker, F., Shub, M., and Smale, S. (1998). *Complexity and Real Computation*. Springer.
31. Blum, L., Shub, M., and Smale, S. (1989). On a theory of computation and complexity over the real numbers; NP completeness, recursive functions and universal machines. *Bulletin of the American Mathematical Society*, 21(1):1–46.
32. Bournez, O. (1999a). Achilles and the Tortoise climbing up the hyper-arithmetical hierarchy. *Theoretical Computer Science*, 210(1):21–71.
33. Bournez, O. (1999b). *Complexité Algorithmique des Systèmes Dynamiques Continus et Hybrides*. PhD thesis, Ecole Normale Supérieure de Lyon.
34. Bournez, O. (2006). How much can analog and hybrid systems be proved (super-)Turing. *Applied Mathematics and Computation*, 178(1):58–71.
35. Bournez, O., Campagnolo, M. L., Graça, D. S., and Hainry, E. (2007). Polynomial differential equations compute all real computable functions on computable compact intervals. *Journal of Complexity*. To appear.
36. Bournez, O. and Hainry, E. (2005). Elementarily computable functions over the real numbers and $\mathbb{R}$-sub-recursive functions. *Theoretical Computer Science*, 348(2–3): 130–147.
37. Bournez, O. and Hainry, E. (2006). Recursive analysis characterized as a class of real recursive functions. *Fundamenta Informaticae*, 74(4):409–433.
38. Bouyer, P., Dufourd, C., Fleury, E., and Petit, A. (2000a). Are timed automata updatable? In Emerson, E. A. and Sistla, A. P., editors, *Computer Aided Verification, 12th International Conference, CAV 2000, Chicago, IL, July 15-19, 2000, Proceedings*, volume 1855 of *Lecture Notes in Computer Science*, pages 464–479. Springer.
39. Bouyer, P., Dufourd, C., Fleury, E., and Petit, A. (2000b). Expressiveness of updatable timed automata. In Nielsen, M. and Rovan, B., editors, *Mathematical Foundations of Computer Science 2000, 25th International Symposium, MFCS 2000, Bratislava, Slovakia, August 28 - September 1, 2000, Proceedings*, volume 1893 of *Lecture Notes in Computer Science*, pages 232–242. Springer.
40. Bouyer, P. and Petit, A. (1999). Decomposition and composition of timed automata. In Wiedermann, J., van Emde Boas, P., and Nielsen, M., editors, *Automata, Languages and Programming, 26th International Colloquium, ICALP'99, Prague, Czech Republic, July 11-15, 1999, Proceedings*, volume 1644 of *Lecture Notes in Computer Science*, pages 210–219. Springer.
41. Bouyer, P. and Petit, A. (2002). A Kleene/Büchi-like theorem for clock languages. *Journal of Automata, Languages and Combinatorics*, 7(2):167–186.
42. Bowles, M. D. (1996). U.S. technological enthusiasm and British technological skepticism in the age of the analog brain. *IEEE Annals of the History of Computing*, 18(4): 5–15.
43. Branicky, M. S. (1995a). *Studies in Hybrid Systems: Modeling, Analysis, and Control*. PhD thesis, Laboratory for Information and Decision Systems, Massachusetts Institute of Technology, Cambridge, MA.
44. Branicky, M. S. (1995b). Universal computation and other capabilities of hybrid and continuous dynamical systems. *Theoretical Computer Science*, 138(1):67–100.

45. Brihaye, T. (2006). A note on the undecidability of the reachability problem for o-minimal dynamical systems. *Math. Log. Q*, 52(2):165–170.
46. Brihaye, Th. and Michaux, Ch. (2005). On the expressiveness and decidability of o-minimal hybrid systems. *Journal of Complexity*, 21(4):447–478.
47. Brockett, R. W. (1989). Smooth dynamical systems which realize arithmetical and logical operations. In Nijmeijer, H. and Schumacher, J. M., editors, *Three Decades of Mathematical Systems Theory*, volume 135 of *Lecture Notes in Computer Science*, pages 19–30. Springer.
48. Brockett, R. W. (1991). Dynamical systems that sort lists, diagonalize matrices, and solve linear programming problems. *Linear Algebra and its Applications*, 146:79–91.
49. Brockett, R. W. (1994). Dynamical systems and their associated automata. In U. Helmke, R. M. and Saurer, J., editors, *Systems and Networks: Mathematical Theory and Applications*, volume 77, pages 49–69. Akademi-Verlag, Berlin.
50. Bush, V. (1931). The differential analyser. *Journal of the Franklin Institute*, 212(4): 447–488.
51. Calude, C. S. and Pavlov, B. (2002). Coins, Quantum measurements, and Turing's barrier. *Quantum Information Processing*, 1(1-2):107–127.
52. Campagnolo, M., Moore, C., and Costa, J. F. (2000). Iteration, inequalities, and differentiability in analog computers. *Journal of Complexity*, 16(4):642–660.
53. Campagnolo, M., Moore, C., and Costa, J. F. (2002). An analog characterization of the Grzegorczyk hierarchy. *Journal of Complexity*, 18(4):977–1000.
54. Campagnolo, M. L. (2001). *Computational complexity of real valued recursive functions and analog circuits*. PhD thesis, IST, Universidade Técnica de Lisboa.
55. Campagnolo, M. L. (2002). The complexity of real recursive functions. In Calude, C., Dinneen, M., and Peper, F., editors, *Unconventional Models of Computation, UMC'02*, Volume 2509 in *Lecture Notes in Computer Science*, pages 1–14. Springer.
56. Campagnolo, M. L. (2004). Continuous time computation with restricted integration capabilities. *Theoretical Computer Science*, 317(4):147–165.
57. Campagnolo, M. L. and Ojakian, K. (2007). The elementary computable functions over the real numbers: applying two new techniques. *Archive for Mathematical Logic*. To appear.
58. Casey, M. (1996). The dynamics of discrete-time computation, with application to recurrent neural networks and finite state machine extraction. *Neural Computation*, 8: 1135–1178.
59. Casey, M. (1998). Correction to proof that recurrent neural networks can robustly recognize only regular languages. *Neural Computation*, 10:1067–1069.
60. Ceraens, K. and Viksna, J. (1996). Deciding reachability for planar multi-polynomial systems. In *Hybrid Systems III*, volume 1066 of *Lecture Notes in Computer Science*, page 389. Springer-Verlag.
61. Church, A. (1936). An unsolvable problem of elementary number theory. *American Journal of Mathematics*, 58:345–363. Reprinted in [73].
62. Clote, P. (1998). Computational models and function algebras. In Griffor, E. R., editor, *Handbook of Computability Theory*, pages 589–681. North-Holland, Amsterdam.
63. Coddington, E. A. and Levinson, N. (1972). *Theory of Ordinary Differentiel Equations*. McGraw-Hill.
64. Collins, P. (2005). Continuity and computability on reachable sets. *Theoretical Computer Science*, 341:162–195.
65. Collins, P. and Lygeros, J. (2005). Computability of finite-time reachable sets for hybrid systems. In *Proceedings of the 44th IEEE Conference on Decision and Control and the European Control Conference*, pages 4688–4693. IEEE Computer Society Press.

66. Collins, P. and van Schuppen, J. H. (2004). Observability of piecewise-affine hybrid systems. In Alur, R. and Pappas, G. J., editors, *Hybrid Systems: Computation and Control, 7th International Workshop, HSCC 2004, Philadelphia, PA, March 25-27, 2004, Proceedings*, volume 2993 of *Lecture Notes in Computer Science*, pages 265–279. Springer.

67. Copeland, B. J. (1998). Even Turing machines can compute uncomputable functions. In Calude, C., Casti, J., and Dinneen, M., editors, *Unconventional Models of Computations*. Springer.

68. Copeland, B. J. (2002). Accelerating Turing machines. *Minds and Machines*, 12: 281–301.

69. Costa, J. F. and Mycka, J. (2006). The conjecture $P \neq NP$ given by some analytic condition. In Bekmann, A., Berger, U., Löwe, B., and Tucker, J., editors, *Logical Approaches to Computational Barriers, Second conference on Computability in Europe, CiE 2006*, pages 47–57, Swansea, UK. Report CSR 7-26, Report Series, University of Wales Swansea Press, 2006.

70. Coward, D. (2006). Doug Coward's Analog Computer Museum. http://dcoward. best.vwh.net/analog/.

71. Davies, E. B. (2001). Building infinite machines. *The British Journal for the Philosophy of Science*, 52:671–682.

72. Dee, D. and Ghil, M. (1984). Boolean difference equations, I: Formulation and dynamic behavior. *SIAM Journal on Applied Mathematics*, 44(1):111–126.

73. Davis M. (ed.) (1965) *The Undecidable: Basic Papers on Undecidable Propositions, Unsolvable Problems and Computable Functions*, Raven, NY.

74. Delvenne, J.-C., Kurka, P., and Blondel, V. D. (2004). Computational universality in symbolic dynamical systems. In Margenstern, M., editor, *MCU: International Conference on Machines, Computations, and Universality*, volume 3354 of *Lecture Notes in Computer Science*, pages 104–115. Springer.

75. Deutsch, D. (1985). Quantum theory, the Church-Turing principle and the universal quantum computer. *Proceedings of the Royal Society (London), Series A*, 400:97–117.

76. Durand-Lose, J. (2005). Abstract geometrical computation: Turing-computing ability and undecidability. In Cooper, S. B., Löwe, B., and Torenvliet, L., editors, *New Computational Paradigms, First Conference on Computability in Europe, CiE 2005, Amsterdam, The Netherlands, June 8-12, 2005, Proceedings*, volume 3526 of *Lecture Notes in Computer Science*, pages 106–116. Springer.

77. Earman, J. and Norton, J. D. (1993). Forever is a day: Supertasks in Pitowksy and Malament-Hogarth spacetimes. *Philosophy of Science*, 60(1):22–42.

78. Etesi, G. and Németi, I. (2002). Non-Turing computations via Malament-Hogarth spacetimes. *International Journal Theoretical Physics*, 41:341–370.

79. Faybusovich, L. (1991a). Dynamical systems which solve optimization problems with linear constraints. *IMA Journal of Mathematical Control and Information*, 8:135–149.

80. Faybusovich, L. (1991b). Hamiltonian structure of dynamical systems which solve linear programming problems. *Physics*, D53:217–232.

81. Filippov, A. (1988). *Differential equations with discontinuous right-hand sides*. Kluwer Academic Publishers.

82. Finkel, O. (2006). On the shuffle of regular timed languages. *Bulletin of the European Association for Theoretical Computer Science*, 88:182–184. Technical Contributions.

83. Foy, J. (2004). A dynamical system which must be stable whose stability cannot be proved. *Theoretical Computer Science*, 328(3):355–361.

84. Francisco, A. P. L. (2002). Finite automata over continuous time. Diploma Thesis. Universidade Técnica de Lisboa, Instituto Superior Técnico.

85. Fränzle, M. (1999). Analysis of hybrid systems: An ounce of realism can save an infinity of states. In Flum, J. and Rodríguez-Artalejo, M., editors, *Computer Science Logic (CSL'99)*, volume 1683 of *Lecture Notes in Computer Science*, pages 126–140. Springer Verlag.

86. Gori, M. and Meer, K. (2002). A step towards a complexity theory for analog systems. *Mathematical Logic Quarterly*, 48(Suppl. 1):45–58.

87. Graça, D. (2002). The general purpose analog computer and recursive functions over the reals. Master's thesis, IST, Universidade Técnica de Lisboa.

88. Graça, D. S. (2004). Some recent developments on Shannon's general purpose analog computer. *Mathematical Logic Quarterly*, 50(4–5):473–485.

89. Graça, D. S. and Costa, J. F. (2003). Analog computers and recursive functions over the reals. *Journal of Complexity*, 19(5):644–664.

90. Graça, D., Campagnolo, M., and Buescu, J. (2005). Robust simulations of Turing machines with analytic maps and flows. In Cooper, B., Loewe, B., and Torenvliet, L., editors, *Proceedings of CiE'05, New Computational Paradigms*, volume 3526 of *Lecture Notes in Computer Science*, pages 169–179. Springer.

91. Graça, D. S., Campagnolo, M. L., and Buescu, J. (2007). Computability with polynomial differential equations. *Advances in Applied Mathematics*. To appear.

92. Graça, D. S., Zhong, N., and Buescu, J. (2006). Computability, noncomputability and undecidability of maximal intervals of IVPs. *Transactions of the American Mathematical Society*. To appear.

93. Grigorieff, S. and Margenstern, M. (2004). Register cellular automata in the hyperbolic plane. *Fundamenta Informaticae*, 1(61):19–27.

94. Gruska, J. (1997). *Foundations of Computing*. International Thomson Publishing.

95. Gupta, V., A., T., and Jagadeesan, R. (1997). Robust timed automata. In Maler, O., editor, *Hybrid and Real-Time Systems, International Workshop. HART'97, Grenoble, France, March 26-28, 1997, Proceedings*, volume 1201 of *Lecture Notes in Computer Science*, pages 331–345. Springer.

96. Head, T. (1987). Formal language theory and DNA: An analysis of the generative capacity of specific recombinant behaviors. *Bulletin of Mathematical Biology*, 49:737–759.

97. Helmke, U. and Moore, J. (1994). *Optimization and Dynamical Systems*. Communications and Control Engineering Series. Springer Verlag, London.

98. Henzinger, T. A., Kopke, P. W., Puri, A., and Varaiya, P. (1998). What's decidable about hybrid automata? *Journal of Computer and System Sciences*, 57(1):94–124.

99. Henzinger, T. A. and Raskin, J.-F. (2000). Robust undecidability of timed and hybrid systems. In Lynch, N. A. and Krogh, B. H., editors, *Hybrid Systems: Computation and Control, Third International Workshop, HSCC 2000, Pittsburgh, PA, March 23-25, 2000, Proceedings*, volume 1790 of *Lecture Notes in Computer Science*, pages 145–159. Springer.

100. Hirsch, M. W., Smale, S., and Devaney, R. (2003). *Differential Equations, Dynamical Systems, and an Introduction to Chaos*. Elsevier Academic Press.

101. Hogarth, M. (1994). Non-Turing computers and non-Turing computability. In *Proceedings of the Philosophy of Science Association (PSA'94)*, volume 1, pages 126–138.

102. Hogarth, M. (1996). *Predictability, Computability and Spacetime*. PhD thesis, Sidney Sussex College, Cambridge.

103. Hogarth, M. (2006). Non-Turing computers are the new non-Euclidean geometries. In *Future Trends in Hypercomputation*. Sheffield, 11–13 September 2006. Available for download on www.hypercomputation.net.

104. Hogarth, M. L. (1992). Does general relativity allow an observer to view an eternity in a finite time? *Foundations of Physics Letters*, 5:173–181.

105. Hopfield, J. J. (1984). Neural networks with graded responses have collective computational properties like those of two-state neurons. *Proceedings of the National Academy of Sciences of the United States of America*, 81:3088–3092.
106. Hopfield, J. J. and Tank, D. W. (1985). 'Neural' computation of decisions in optimization problems. *Biological Cybernetics*, 52:141–152.
107. Hoyrup, M. (2006). Dynamical systems: stability and simulability. Technical report, Département d'Informatique, ENS Paris.
108. Kempe, A. (1876). On a general method of describing plane curves of the n–th degree by linkwork. *Proceedings of the London Mathematical Society*, 7:213–216.
109. Kieu, T. D. (2004). Hypercomputation with quantum adiabatic processes. *Theoretical Computer Science*, 317(1-3):93–104.
110. Kleene, S. C. (1936). General recursive functions of natural numbers. *Mathematical Annals*, 112:727–742. Reprinted in [73].
111. Ko, K.-I. (1983). On the computational complexity of ordinary differential equations. *Information and Control*, 58(1-3):157–194.
112. Ko, K.-I. (1991). *Complexity Theory of Real Functions*. Progress in Theoretical Computer Science. Birkhäuser, Boston.
113. Koiran, P. (2001). The topological entropy of iterated piecewise affine maps is uncomputable. *Discrete Mathematics & Theoretical Computer Science*, 4(2):351–356.
114. Koiran, P., Cosnard, M., and Garzon, M. (1994). Computability with low-dimensional dynamical systems. *Theoretical Computer Science*, 132(1-2):113–128.
115. Koiran, P. and Moore, C. (1999). Closed-form analytic maps in one and two dimensions can simulate universal Turing machines. *Theoretical Computer Science*, 210(1):217–223.
116. Korovina, M. V. and Vorobjov, N. (2004). Pfaffian hybrid systems. In Marcinkowski, J. and Tarlecki, A., editors, *Computer Science Logic, 18th International Workshop, CSL 2004, 13th Annual Conference of the EACSL, Karpacz, Poland, September 20-24, 2004, Proceedings*, volume 3210 of *Lecture Notes in Computer Science*, pages 430–441. Springer.
117. Korovina, M. V. and Vorobjov, N. (2006). Upper and lower bounds on sizes of finite bisimulations of Pfaffian hybrid systems. In Beckmann, A., Berger, U., Löwe, B., and Tucker, J. V., editors, *Logical Approaches to Computational Barriers, Second Conference on Computability in Europe, CiE 2006, Swansea, UK, June 30-July 5, 2006, Proceedings*, volume 3988 of *Lecture Notes in Computer Science*, pages 267–276. Springer.
118. Kurganskyy, O. and Potapov, I. (2005). Computation in one-dimensional piecewise maps and planar pseudo-billiard systems. In Calude, C., Dinneen, M. J., Paun, G., Pérez-Jiménez, M. J., and Rozenberg, G., editors, *Unconventional Computation, 4th International Conference, UC 2005, Sevilla, Spain, October 3-7, 2005, Proceedings*, volume 3699 of *Lecture Notes in Computer Science*, pages 169–175. Springer.
119. Lafferriere, G. and Pappas, G. J. (2000). O-minimal hybrid systems. *Mathematics of Control, Signals, and Systems*, 13:1–21.
120. Legenstein, R. and Maass, W. (2007). What makes a dynamical system computationally powerful? In Haykin, S., Principe, J. C., Sejnowski, T., and McWhirter, J., editors, *New Directions in Statistical Signal Processing: From Systems to Brain*, pages 127–154. MIT Press, Cambridge, MA.
121. Lipshitz, L. and Rubel, L. A. (1987). A differentially algebraic replacement theorem, and analog computability. *Proceedings of the American Mathematical Society*, 99(2): 367–372.
122. Lipton, R. J. (1995). DNA solution of hard computational problems. *Science*, 268: 542–545.

123. Loff, B. (2007). A functional characterisation of the analytical hierarchy. In *Computability in Europe 2007: Computation and Logic in the Real World.*

124. Loff, B., Costa, J. F., and Mycka, J. (2007a). Computability on reals, infinite limits and differential equations. *Applied Mathematics and Computation.* To appear.

125. Loff, B., Costa, J. F., and Mycka, J. (2007b). The new promise of analog computation. In *Computability in Europe 2007: Computation and Logic in the Real World.*

126. Maass, W. (1996a). Lower bounds for the computational power of networks of spiking neurons. *Neural Computation*, 8(1):1–40.

127. Maass, W. (1996b). On the computational power of noisy spiking neurons. In Touretzky, D., Mozer, M. C., and Hasselmo, M. E., editors, *Advances in Neural Information Processing Systems*, volume 8, pages 211–217. MIT Press, Cambridge, MA.

128. Maass, W. (1997a). A model for fast analog computations with noisy spiking neurons. In Bower, J., editor, *Computational Neuroscience: Trends in research*, pages 123–127.

129. Maass, W. (1997b). Networks of spiking neurons: the third generation of neural network models. *Neural Networks*, 10:1659–1671.

130. Maass, W. (1999). Computing with spiking neurons. In Maass, W. and Bishop, C. M., editors, *Pulsed Neural Networks*, pages 55–85. MIT Press, Cambridge, MA.

131. Maass, W. (2002). Computing with spikes. *Special Issue on Foundations of Information Processing of TELEMATIK*, 8(1):32–36.

132. Maass, W. (2003). Computation with spiking neurons. In Arbib, M. A., editor, *The Handbook of Brain Theory and Neural Networks*, pages 1080–1083. MIT Press, Cambridge, MA. 2nd edition.

133. Maass, W. and Bishop, C. (1998). *Pulsed Neural Networks*. MIT Press, Cambridge, MA.

134. Maass, W., Joshi, P., and Sontag, E. D. (2007). Computational aspects of feedback in neural circuits. *Public Library of Science Computational Biology*, 3(1):1–20. e165.

135. Maass, W. and Natschläger, T. (2000). A model for fast analog computation based on unreliable synapses. *Neural Computation*, 12(7):1679–1704.

136. Maass, W. and Orponen, P. (1998). On the effect of analog noise in discrete-time analog computations. *Neural Computation*, 10(5):1071–1095.

137. Maass, W. and Ruf, B. (1999). On computation with pulses. *Information and Computation*, 148(2):202–218.

138. Maass, W. and Sontag, E. (1999). Analog neural nets with gaussian or other common noise distributions cannot recognize arbitrary regular languages. *Neural Computation*, 11(3):771–782.

139. MacLennan, B. J. (2001). Can differential equations compute? `citeseer.ist.psu.edu/maclennan01can.html`.

140. Mills, J. (1995). Programmable VLSI extended analog computer for cyclotron beam control. Technical Report 441, Indiana University Computer Science.

141. Mills, J. W., Himebaugh, B., Allred, A., Bulwinkle, D., Deckard, N., Gopalakrishnan, N., Miller, J., Miller, T., Nagai, K., Nakamura, J., Ololoweye, B., Vlas, R., Whitener, P., Ye, M., , and Zhang, C. (2005). Extended analog computers: A unifying paradigm for VLSI, plastic and colloidal computing systems. In *Workshop on Unique Chips and Systems (UCAS-1). Held in conjunction with IEEE International Symposium on Performance Analysis of Systems and Software (ISPASS05)*, Austin, Texas.

142. Müller, N. and Moiske, B. (1993). Solving initial value problems in polynomial time. In *Proc. 22 JAIIO - PANEL '93, Part 2*, pages 283–293.

143. Moore, C. (1990). Unpredictability and undecidability in dynamical systems. *Physical Review Letters*, 64(20):2354–2357.

144. Moore, C. (1991). Generalized shifts: unpredictability and undecidability in dynamical systems. *Nonlinearity*, 4(3):199–230.
145. Moore, C. (1996). Recursion theory on the reals and continuous-time computation. *Theoretical Computer Science*, 162(1):23–44.
146. Moore, C. (1998a). Dynamical recognizers: real-time language recognition by analog computers. *Theoretical Computer Science*, 201(1–2):99–136.
147. Moore, C. (1998b). Finite-dimensional analog computers: Flows, maps, and recurrent neural networks. In Calude, C. S., Casti, J. L., and Dinneen, M. J., editors, *Unconventional Models of Computation (UMC'98)*. Springer.
148. Murray, J. D. (2002). *Mathematical Biology. I: An Introduction*. Springer, third edition.
149. Mycka, J. and Costa, J. F. (2004). Real recursive functions and their hierarchy. *Journal of Complexity*, 20(6):835–857.
150. Mycka, J. and Costa, J. F. (2005). What lies beyond the mountains? Computational systems beyond the Turing limit. *European Association for Theoretical Computer Science Bulletin*, 85:181–189.
151. Mycka, J. and Costa, J. F. (2006). The $P \neq NP$ conjecture in the context of real and complex analysis. *Journal of Complexity*, 22(2):287–303.
152. Mycka, J. and Costa, J. F. (2007). A new conceptual framework for analog computation. *Theoretical Computer Science*, 374:277–290.
153. Natschläger, T. and Maass, W. (2002). Spiking neurons and the induction of finite state machines. *Theoretical Computer Science: Special Issue on Natural Computing*, 287(1):251–265.
154. Németi, I. and Andréka, H. (2006). New physics and hypercomputation. In Wiedermann, J., Tel, G., Pokorný, J., Bieliková, M., and Stuller, J., editors, *SOFSEM 2006: Theory and Practice of Computer Science, 32nd Conference on Current Trends in Theory and Practice of Computer Science, Merín, Czech Republic, January 21-27, 2006, Proceedings*, volume 3831 of *Lecture Notes in Computer Science*, page 63. Springer.
155. Németi, I. and Dávid, G. (2006). Relativistic computers and the Turing barrier. *Applied Mathematics and Computation*, 178:118–142.
156. Nicollin, X., Olivero, A., Sifakis, J., and Yovine, S. (1993). An approach to the description and analysis of hybrid systems. In Grossman, R. L., Nerode, A., Ravn, A. P., and Rischel, H., editors, *Hybrid Systems*, volume 736 of *Lecture Notes in Computer Science*, pages 149–178. Springer.
157. Omohundro, S. (1984). Modelling cellular automata with partial differential equations. *Physica D*, 10D(1–2):128–134.
158. Orponen, P. (1994). Computational complexity of neural networks: a survey. *Nordic Journal of Computing*, 1(1):94–110.
159. Orponen, P. (1996). The computational power of discrete Hopfield nets with hidden units. *Neural Computation*, 8(2):403–415.
160. Orponen, P. (1997). A survey of continuous-time computation theory. In Du, D.-Z. and Ko, K.-I., editors, *Advances in Algorithms, Languages, and Complexity*, pages 209–224. Kluwer Academic Publishers.
161. Orponen, P. and Šíma, J. (2000). A continuous-time Hopfield net simulation of discrete neural networks. In *Proceedings of the 2nd International ICSC Symposium on Neural Computations (NC'2000)*, pages 36–42, Berlin, Germany. ICSC Academic Press, Wetaskiwin (Canada)
162. Papadimitriou, C. (2001). Algorithms, games, and the Internet. In *Proceedings of the 33rd Annual ACM Symposium on Theory of Computing: Hersonissos, Crete, Greece, July 6–8, 2001*, pages 749–753, New York, NY. ACM Press.

163. Păun, G. (2002). *Membrane Computing. An Introduction.* Springer-Verlag, Berlin.

164. Post, E. (1946). A variant of a recursively unsolvable problem. *Bulletin of the American Math. Soc.*, 52:264–268.

165. Pour-El, M. and Zhong, N. (1997).   The wave equation with computable initial data whose unique solution is nowhere computable. *Mathematical Logic Quarterly*, 43(4):499–509.

166. Pour-El, M. B. (1974).   Abstract computability and its relation to the general purpose analog computer (some connections between logic, differential equations and analog computers). *Transactions of the American Mathematical Society*, 199:1–28.

167. Pour-El, M. B. and Richards, J. I. (1979). A computable ordinary differential equation which possesses no computable solution. *Annals of Mathematical Logic*, 17:61–90.

168. Pour-El, M. B. and Richards, J. I. (1981).   The wave equation with computable initial data such that its unique solution is not computable. *Advances in Mathematics*, 39: 215–239.

169. Pour-El, M. B. and Richards, J. I. (1989). *Computability in Analysis and Physics.* Springer.

170. Puri, A. (1998). Dynamical properties of timed automata. In Ravn, A. P. and Rischel, H., editors, *Formal Techniques in Real-Time and Fault-Tolerant Systems, 5th International Symposium, FTRTFT'98, Lyngby, Denmark, September 14-18, 1998, Proceedings*, volume 1486 of *Lecture Notes in Computer Science*, pages 210–227. Springer.

171. Puri, A. and Varaiya, P. (1994). Decidability of hybrid systems with rectangular differential inclusion. In Dill, D. L., editor, *Computer Aided Verification, 6th International Conference, CAV '94, Stanford, CA. June 21-23, 1994, Proceedings*, volume 818 of *Lecture Notes in Computer Science*, pages 95–104. Springer.

172. Rabin, M. O. (1963). Probabilistic automata. *Information and Control*, 6(3):230–245.

173. Rabinovich, A. (2003). Automata over continuous time. *Theoretical Computer Science*, 300(1–3):331–363.

174. Rabinovich, A. M. and Trakhtenbrot, B. A. (1997). From finite automata toward hybrid systems (extended abstract). In Chlebus, B. S. and Czaja, L., editors, *Fundamentals of Computation Theory, 11th International Symposium, FCT '97, Kraków, Poland, September 1-3, 1997, Proceedings*, volume 1279 of *Lecture Notes in Computer Science*, pages 411–422. Springer.

175. Rubel, L. A. (1989). A survey of transcendentally transcendental functions. *American Mathematical Monthly*, 96(9):777–788.

176. Rubel, L. A. (1993). The extended analog computer. *Advances in Applied Mathematics*, 14:39–50.

177. Ruohonen, K. (1993). Undecidability of event detection for ODEs. *Journal of Information Processing and Cybernetics*, 29:101–113.

178. Ruohonen, K. (1994).   Event detection for ODEs and nonrecursive hierarchies.   In Karhumäki, J. and Maurer, H., editors, *Proceedings of the Colloquium in Honor of Arto Salomaa. Results and Trends in Theoretical Computer Science (Graz, Austria, June 10-11, 1994)*, volume 812 of *Lecture Notes in Computer Science*, pages 358–371. Springer, Berlin.

179. Ruohonen, K. (1996). An effective Cauchy-Peano existence theorem for unique solutions. *International Journal of Foundations of Computer Science*, 7(2):151–160.

180. Ruohonen, K. (1997a).   Decidability and complexity of event detection problems for ODEs. *Complexity*, 2(6):41–53.

181. Ruohonen, K. (1997b). Undecidable event detection problems for ODEs of dimension one and two. *Theoretical Informatics and Applications*, 31(1):67–79.

182. Ruohonen, K. (2004). Chomskian hierarchies of families of sets of piecewise continuous functions. *Theory of Computing Systems*, 37(5):609–638.
183. Shannon, C. E. (1941). Mathematical theory of the differential analyser. *Journal of Mathematics and Physics MIT*, 20:337–354.
184. Shor, P. W. (1994). Algorithms for quantum computation: Discrete logarithms and factoring. In Goldwasser, S., editor, *Proceedings of the 35th Annual Symposium on Foundations of Computer Science*, pages 124–134, Los Alamitos, CA. IEEE Computer Society Press.
185. Siegelmann, H. T. and Fishman, S. (1998). Analog computation with dynamical systems. *Physica D*, 120:214–235.
186. Siegelmann, H. T. and Sontag, E. D. (1994). Analog computation via neural networks. *Theoretical Computer Science*, 131(2):331–360.
187. Siegelmann, H. T. and Sontag, E. D. (1995). On the computational power of neural nets. *Journal of Computer and System Sciences*, 50(1):132–150.
188. Šíma and Orponen (2003a). Exponential transients in continuous-time Liapunov systems. *Theoretical Computer Science*, 306(1–3):353–372.
189. Šíma, J. and Orponen, P. (2003b). Continuous-time symmetric Hopfield nets are computationally universal. *Neural Computation*, 15(3):693–733.
190. Šíma, J. and Orponen, P. (2003c). General-purpose computation with neural networks: A survey of complexity theoretic results. *Neural Computation*, 15(12):2727–2778.
191. Smith, W. D. (1998). Plane mechanisms and the downhill principle. `http://citeseer.ist.psu.edu/475350.html`.
192. Smith, W. D. (2006). Church's thesis meets the N-body problem. *Applied Mathematics and Computation*, 178(1):154–183.
193. Stoll, H. M. and Lee, L. S. (1988). A continuous-time optical neural network. In *IEEE Second International Conference on Neural Networks (2nd ICNN'88)*, volume II, pages 373–384, San Diego, CA. IEEE Society Press.
194. Svoboda, A. (1948). *Computing Mechanisms and Linkages*. McGraw Hill. Reprinted by Dover Publications in 1965.
195. Thomson, W. (1876). On an instrument for calculating the integral of the product of two given functions. In *Proceedings of the Royal Society of London*, volume 24, pages 266–276.
196. Trakhtenbrot, B. (1995). Origins and metamorphoses of the trinity: Logic, nets, automata. In Kozen, D., editor, *Proceedings of the 10th Annual IEEE Symposium on Logic in Computer Science San Diego, CA, June 26-29, 1995*, pages 506–507. IEEE Computer Society, Press.
197. Trakhtenbrot, B. A. (1999). Automata and their interaction: Definitional suggestions. In Ciobanu, G. and Paun, G., editors, *Fundamentals of Computation Theory, 12th International Symposium, FCT '99, Iasi, Romania, August 30 - September 3, 1999, Proceedings*, volume 1684 of *Lecture Notes in Computer Science*, pages 54–89. Springer.
198. Tucker, J. V. and Zucker, J. I. (2007). Computability of analog networks. *Theoretical Computer Science*, 371(1-2):115–146.
199. Turing, A. (1936). On computable numbers, with an application to the Entscheidungsproblem. *Proceedings of the London Mathematical Society*, 42(2):230–265. Reprinted in [73].
200. Vergis, A., Steiglitz, K., and Dickinson, B. (1986). The complexity of analog computation. *Mathematics and Computers in Simulation*, 28(2):91–113.
201. Weihrauch, K. (2000). *Computable Analysis*. Springer.

202. Weihrauch, K. and Zhong, N. (2002). Is wave propagation computable or can wave computers beat the Turing machine? *Proceedings of the London Mathematical Society*, 85(3):312–332.
203. Welch, P. D. (2006). The extent of computation in Malament-Hogarth spacetimes. http://www.citebase.org/abstract?id=oai:arXiv.org:gr-qc/0609035.
204. Williams, M. R. (1996). About this issue. *IEEE Annals of the History of Computing*, 18(4).
205. Woods, D. and Naughton, T. J. (2005). An optical model of computation. *Theoretical Computer Science*, 334(1-3):227–258.

A Tutorial on Computable Analysis

Vasco Brattka[1], Peter Hertling[2], and Klaus Weihrauch[3]

[1] Laboratory of Foundational Aspects of Computer Science, Department of Mathematics and Applied Mathematics, University of Cape Town, Rondebosch 7701, South Africa
Vasco.Brattka@uct.ac.za

[2] Institut für Theoretische Informatik und Mathematik, Fakultät für Informatik, Universität der Bundeswehr München, 85577 Neubiberg, Germany
Peter.Hertling@unibw.de

[3] Faculty of Mathematics and Computer Science, University of Hagen, 58084 Hagen, Germany
Klaus.Weihrauch@fernuni-hagen.de

Summary. This tutorial gives a brief introduction to computable analysis. The objective of this theory is to study algorithmic aspects of real numbers, real number functions, subsets of real numbers, and higher type operators over the real numbers. In this theory, the classical computability notions and complexity notions based on the Turing machine model and studied in computability theory and computational complexity theory are applied to computational problems involving real numbers.

1 Introduction

This tutorial gives a brief introduction to computable analysis. In computable analysis, computational problems over the real numbers are studied from the point of view of computability theory and computational complexity theory.

A large part of today's computing power all over the world is spent on computational problems that can best be modeled as computational problems involving real numbers. These are mostly numerical problems but also geometric problems and problems studied in other fields, e.g., in the theory of neural networks. Therefore, a theory is needed that tells which computational problems over the real numbers can be solved on a digital computer and how much time or how much computer memory this will take. Computable analysis studies these questions on the basis of the notions studied in computability theory and computational complexity theory that are defined via the Turing machine model.

Computable analysis is based on the one hand on analysis and numerical analysis and on the other hand on computability theory and computational complexity theory. Other mathematical or computer science disciplines that are closely related to computable analysis are as follows:

- Constructive analysis. There are several schools and varieties of constructive analysis. We give the following references: Bishop and Bridges [8], Bridges and Richman [27], Troelstra and van Dalen [100, 101], Šanin [85], and Kušner [59].

- Domain theory. Domains are used for defining semantics of programming languages; see, e.g., Stoltenberg-Hansen et al. [98]. In our context this is especially interesting for programming languages over the real numbers. One can also define computability notions over the real numbers via domains; see, for instance, Edalat, Escardo, and Blanck [38, 37, 9].

- Mathematical logic. In fact, computability theory, constructive analysis, and domain theory can all be considered as subdisciplines of logic.

- Interval analysis; see, e.g., Moore [69]. Note that intervals of real numbers form a domain.

- Algebraic complexity theory over the real numbers, as studied by Blum et al. [10].

- Information-based complexity theory, a theory that studies mainly the computational complexity of problems involving function spaces over the real numbers in an algebraic computation model; see Traub et al. [99].

Since real numbers and many other objects studied in analysis are "infinite" objects containing an "infinite amount of information," one has to approximate them by "finite" objects containing only a "finite amount of information" and to perform the actual computations on these finite objects. This leads to the following three levels of study of computational problems in analysis:

- Topology: how can one approximate infinite objects?

- Computability theory: how can one compute with infinite objects?

- Computational complexity theory: how can one compute efficiently with infinite objects?

In this tutorial we will introduce the basic notions of computable analysis and present selected results that are supposed to illustrate important aspects and ideas of computable analysis and to give an overview of current areas of research in computable analysis. The main focus of this article is on approximation and computability. Because of limited space, many important or interesting results, for example, from complexity theory, are not mentioned, and the bibliography is far from being complete. We apologize to all whose work is insufficiently, or not, mentioned. A bibliography on computable analysis can be found through the web pages of the CCA (= Computability and Complexity in Analysis) network: http://cca-net.de. We will mostly use the approach via representations, which was developed by Hauck (see, e.g., [42, 43]), and by Kreitz and Weihrauch [58]. A more detailed presentation of computable analysis based on representations is the textbook by the third author [109]. Another approach to computable analysis can be found in the textbook by Pour-El and Richards [80]: via sequential computability and effective uniform continuity of functions and via computability structures. This approach has been generalized from normed spaces to metric spaces by Yasugi et al. [114]. The textbook by Ko [55] covers a large part of the complexity theoretic results in computable analysis and is based on the notion of an oracle Turing machine; see the last two sections. The paper [25] by Braverman and Cook is a short introduction into important basic notions and ideas from computable analysis. This tutorial is based mostly on the slides of the tutorial on computable analysis given by the first author at the conference "Computability in Europe 2005" in June/July 2005 in Amsterdam, the Netherlands. It should be readable by anyone with a basic knowledge in computability theory and analysis and related fields. Here is the table of contents of the paper:

1. Introduction

2. Preliminaries

3. Computable Real Numbers

4. Computable Functions

5. Computability Notions for Subsets of Euclidean Space

6. Representations and Topological Considerations

7. Solvability of Some Problems Involving Sets and Functions

8. Computability of Linear Operators

9. Degrees of Unsolvability

10. Computational Complexity of Real Numbers and Real Number Functions

11. Computational Complexity of Sets and Operators over the Real Numbers

2 Preliminaries

First we introduce some notation. By $\mathbb{N}$ we denote the set of natural numbers, i.e., $\mathbb{N} = \{0, 1, 2, 3, \ldots\}$, by $\mathbb{Q}$ the set of rational numbers, by $\mathbb{R}$ the set of real numbers, and by $\mathbb{C}$ the set of complex numbers. By $d : \mathbb{R}^n \times \mathbb{R}^n \to \mathbb{R}$ we denote the Euclidean distance on $\mathbb{R}^n$. Any finite set containing at least two elements is called an *alphabet*. Usually we denote alphabets by uppercase Greek letters, e.g., Σ, Γ. Furthermore, unless stated otherwise, we assume that 0 and 1 are elements of Σ. If Σ is an alphabet, then Σ^* denotes the set of all finite strings over Σ, $\Sigma^\omega = \{p \mid p : \mathbb{N} \to \Sigma\}$ the set of all one-way infinite sequences over Σ, and λ denotes the empty word. The i-th component of a one-way infinite sequence p is written $p(i)$ or p_i. If X and Y are sets, then $f :\subseteq X \to Y$ denotes a function whose domain $\mathrm{dom}(f)$ is contained in X and whose range $\mathrm{range}(f)$ is contained in Y. If $\mathrm{dom}(f) = X$, then we call f *total* and may write $f : X \to Y$. A *sequence* in X is simply a total function $x : \mathbb{N} \to X$ and is often written as $(x_n)_{n \in \mathbb{N}}$.

We assume that the reader is familiar with the Turing machine model. Via the Turing machine model one defines computability of functions on Σ^*. We also consider computability notions on $\mathbb{N}^n$ and $\mathbb{Q}^m$ for $n, m \in \mathbb{N} \setminus \{0\}$ as basic. For completeness sake and in order to avoid misunderstandings, we formally introduce computability of functions on or between these sets via the Turing machine model. In order to do that we have to represent elements of $\mathbb{N}^n$ and of $\mathbb{Q}^m$ by strings. Let $\nu_{\mathbb{N}^1} := \nu_{\mathbb{N}} :\subseteq \Sigma^* \to \mathbb{N}$ be the usual binary notation of natural numbers. Using the standard bijection $\langle \cdot, \cdot \rangle : \mathbb{N}^2 \to \mathbb{N}$ defined by $\langle x, y \rangle := \frac{(x+y)\cdot(x+y+1)}{2} + y$ and the derived bijections $\langle \ldots \rangle : \mathbb{N}^{k+1} \to \mathbb{N}$ defined by $\langle x_1, \ldots, x_k, x_{k+1} \rangle := \langle \langle x_1, \ldots, x_k \rangle, x_{k+1} \rangle$ for $k \geq 2$, we define notations $\nu_{\mathbb{N}^n} :\subseteq \Sigma^* \to \mathbb{N}^n$ for $n \geq 2$ by

$$\nu_{\mathbb{N}^n}(w) := (x_1, \ldots, x_n) : \iff \nu_{\mathbb{N}}(w) = \langle x_1, \ldots, x_n \rangle.$$

Using the surjection $\nu_Q : \mathbb{N} \to \mathbb{Q}$ defined by $\nu_Q(\langle i, j, k \rangle) := \frac{i-j}{k+1}$, we define notations $\nu_{\mathbb{Q}^n} :\subseteq \Sigma^* \to \mathbb{Q}^n$ for $n \geq 1$ (and use $\nu_{\mathbb{Q}} := \nu_{\mathbb{Q}^1}$) by

$$\nu_{\mathbb{Q}^n}(w) := (q_1, \ldots, q_n) : \iff q_i = \nu_Q(\nu_{\mathbb{N}^n}(w)_i) \text{ for } i = 1, \ldots, n.$$

Finally, for any alphabet Σ^* we define $\nu_{\Sigma^*} := \mathrm{id}_{\Sigma^*}$. Let now X and Y by any of the spaces Σ^*, $\mathbb{N}^n$, and $\mathbb{Q}^m$ for some $n, m \geq 1$. A function $f :\subseteq X \to Y$ is called *computable* if there is a Turing machine that, on input $v \in \Sigma^*$, never stops if $v \notin \mathrm{dom}(f\nu_X)$, and that stops after finitely many steps with some output w satisfying $\nu_Y(w) = f\nu_X(v)$ if $v \in \mathrm{dom}(f)$. We will also use the notions "decidable" and "computably enumerable" for subsets of Σ^*, $\mathbb{N}^n$, and $\mathbb{Q}^m$ in the usual sense; see Section 5 for definitions.

3 Computable Real Numbers

For a long time, mathematics has been concerned with computation problems that would nowadays be considered as computation problems over the real numbers. But only in the twentieth century, a mathematical theory of algorithms, of computability, and of complexity over the real numbers has been developed. Quite early in the twentieth century, constructive mathematics was born. Results in constructive mathematics, e.g., by Brouwer [28, 29] and by Bishop and Bridges [8], certainly have algorithmic content. But these constructive mathematical theories are based on certain logical principles, not on a formal computation model. The so-called "Russian school of constructive analysis," founded by Markov (see Šanin [85] or Kušner [59]) is closer to our approach. We will come back to it in Section 6. Perhaps the real starting point of computable analysis was the landmark paper "On computable numbers, with an application to the Entscheidungsproblem" by Turing [102]. In this paper, Turing asked which real numbers should be considered as computable and, in order to answer this question, developed the theoretical computer model that subsequently was called "Turing machine model" and has become the standard computation model in computability theory and computational complexity theory. Turing defined the computable real numbers as those real numbers that have a binary expansion that can be computed by a Turing machine. We will formulate several equivalent conditions.

Definition 3.1. A real number x is *computable* if it satisfies one (and then all) of the conditions in the following theorem.

Theorem 3.2. *For $x \in \mathbb{R}$, the following conditions are equivalent:*

1. *There exists a Turing machine that outputs a binary expansion of x (without input and without ever stopping).*

2. *There exists a computable sequence of rational numbers $(q_n)_{n \in \mathbb{N}}$ that converges rapidly to x; i.e., $|x - q_i| < 2^{-i}$ for all i.*

3. *There exists a computable sequence of rational shrinking intervals enclosing only x; i.e., there exist two computable sequences $(a_n)_{n \in \mathbb{N}}$ and $(b_n)_{n \in \mathbb{N}}$ of rational numbers with $a_0 < \cdots < a_n < a_{n+1} < \cdots x \cdots < b_{n+1} < b_n < \cdots < b_0$ and with $\lim_{n \to \infty} a_n = x = \lim_{n \to \infty} b_n$.*

4. *$\{q \in \mathbb{Q} \mid q < x\}$ is a decidable set of rational numbers.*

5. *There exist an integer $z \in \mathbb{Z}$ and a decidable set $A \subseteq \mathbb{N}$ such that $x = z + x_A$, where $x_A := \sum_{i \in A} 2^{-i-1}$.*

6. *x is rational (then it has a finite continued fraction expansion) or x has a computable infinite continued fraction expansion, i.e., there exist an integer $z \in \mathbb{Z}$ and a computable function $f : \mathbb{N} \to \mathbb{N} \setminus \{0\}$ such that*

$$x = z + \cfrac{1}{f(0) + \cfrac{1}{f(1) + \cfrac{1}{f(2) + \cdots}}}.$$

Proof. We prove only that the first two notions are equivalent.

"1 $\Rightarrow$ 2": Let us assume that x has a binary expansion that can be computed by a Turing machine. Then let q_n be the rational number that one obtains when in this binary expansion one replaces all digits from the $(n+2)$-th digit after the binary point on by zeros. The sequence $(q_n)_{n \in \mathbb{N}}$ is a computable sequence of rational numbers and converges rapidly to x.

"2 $\Rightarrow$ 1": If x happens to be of the form: integer divided by a power of two, then it has a binary expansion in which only finitely many digits are different from zero. Obviously, such a binary expansion can be computed by a Turing machine. Now, let us assume that x is not of this form. Then x has a uniquely determined binary expansion. Let $(q_n)_{n \in \mathbb{N}}$ be a computable sequence of rational numbers converging rapidly to x. We wish to show that one can compute the binary expansion of x. First, let us determine the part of the binary expansion of x in front of the binary point. Since x is not an integer, there is an i such that the interval $[q_i - 2^{-i}, q_i + 2^{-i}]$ does not contain an integer. By computing sufficiently many q_i for $i = 0, 1, 2, \ldots$ we will find such an i. Then the part of the binary expansion of q_i in front of the binary point is equal to the part of the binary expansion of x in front of the binary point. Now, let us assume that we have computed the part of the binary expansion of x in front of the binary point and, for some $n \in \mathbb{N}$, also the first n digits after the binary point in the binary expansion of x. We can determine whether the next digit is a 0 or a 1 in a similar way as we determined the part of the binary expansion of x in front of the binary point: since x is not of the form: integer divided by 2^{n+1}, there is an i such that the interval $[q_i - 2^{-i}, q_i + 2^{-i}]$ does not contain an integer divided by 2^{n+1}. Then the binary expansions of x and of q_i are identical up to the $(n+1)$-th digit after the binary point. $\square$

The argument just given works as well for the decimal representation or the representation with base $b \geq 2$ instead of the binary representation. Thus we have for any base $b \geq 2$: a real number is computable if, and only if, its has a computable base b expansion.

Remark 3.3. Note that it is easy to go from a computable binary expansion of a real number x to a computable sequence of rational numbers converging rapidly to x. But in the proof of the other direction we made a case distinction that is not a result of a computation. Furthermore, the algorithm in the second case of this case distinction might ask for a very good approximation (with precision much higher than 2^{-i}) of x in order to compute the i-th digit of the binary expansion of x. This shows that the two underlying representations of real numbers used in the two conditions above are quite different. Indeed, Turing himself noticed in a correction [103] to the paper cited above [102] that the binary representation is fine for defining the class of computable real numbers but unsuitable for performing computations on real numbers. We will come back to this. In contrast, the representation of real numbers by rapidly converging sequences of rational numbers is also suitable for performing computations on real numbers, as we will see soon.

Theorem 3.4 (Rice [83]). *The set of computable real numbers $\mathbb{R}_c$ forms a real algebraically closed field.*

That means it is a subfield of the field $\mathbb{R}$ of real numbers, and it contains the real zeros of any polynomial $a_n x^n + \cdots + a_1 x + a_0$ whose coefficients $a_n, \ldots, a_1, a_0$ are computable real numbers.

Examples 3.5.   1. All rational numbers are computable real numbers. Even more, all real algebraic numbers are computable.

2. Also π and e are computable real numbers.

Remark 3.6. In the Russian school of constructive analysis, computability is defined for functions mapping computable real numbers to computable real numbers; see Šanin [85] or Kušner [59]. We shall mainly consider a different notion of computability for functions mapping real numbers to real numbers. In Section 4 we will explain the connection between these two notions and other notions of computability for real number functions.

We list a few more properties of the set of computable real numbers.

Proposition 3.7. *The set of computable real numbers is countably infinite.*

Proof. It is clear that it is infinite since it contains all rational numbers. And it is countable because for every computable real number, there exists a Turing machine computing its binary expansion, and there are only countably many Turing machines. □

Now it is natural to ask whether one can effectively list all computable real numbers. This turns out to be impossible. In order to make this statement precise we need the notion of a computable sequence of real numbers.

Definition 3.8. A sequence $(x_n)_{n \in \mathbb{N}}$ of real numbers is called *computable* if there exists a computable sequence $(q_k)_{k \in \mathbb{N}}$ of rational numbers such that $|x_n - q_{\langle n,i \rangle}| < 2^{-i}$ for all $n, i \in \mathbb{N}$.

Obviously, any member of a computable sequence of real numbers is itself a computable real number.

Proposition 3.9. *No computable sequence of real numbers contains all computable real numbers.*

The proof goes by diagonalization; see, e.g., Weihrauch [106, p. 486].

Since there are uncountably many real numbers, by Proposition 3.7, there are uncountably many noncomputable real numbers.

Proposition 3.10. *The set of computable real numbers is not complete.*

Proof. Any real number, also a noncomputable one, is the limit of a sequence of rational numbers, hence, the limit of a sequence of computable real numbers. □

Many noncomputable real numbers are an even limit of a computable sequence of reals. Nevertheless, the set of computable real numbers is "computably complete" in a certain sense, as we will see now.

Definition 3.11. 1. If $(r_n)_{n \in \mathbb{N}}$ is a convergent sequence of real numbers with limit x, and $m : \mathbb{N} \rightarrow \mathbb{N}$ is a function such that, for all $i, n \in \mathbb{N}$, if $i \geq m(n)$ then $|x - r_i| < 2^{-n}$, then we call m a *modulus of convergence* of the sequence $(r_n)_{n \in \mathbb{N}}$.

2. A sequence $(r_n)_{n \in \mathbb{N}}$ of real numbers *converges computably* if it has a computable modulus of convergence.

Proposition 3.12. *The limit of any computable sequence of real numbers that converges computably is again a computable real number.*

Proof. Let $(r_n)_{n \in \mathbb{N}}$ be a computable sequence of real numbers that converges computably. Let m be a computable modulus of convergence, and let $(q_k)_{k \in \mathbb{N}}$ be a computable sequence of rational numbers proving the computability of $(r_n)_{n \in \mathbb{N}}$ in the sense of Definition 3.8. Then the sequence $(p_n)_{n \in \mathbb{N}}$ defined by $p_n := q_{\langle m(n+1), n+1 \rangle}$ is a computable sequence of rational numbers, and it converges rapidly to the limit of $(r_n)_{n \in \mathbb{N}}$. ◻

We end this section by an excursion into the class of noncomputable real numbers. Quite a lot of research has been going and is still going into various types of effectivity properties for real numbers that are more general than computability. We describe some of them.

Definition 3.13. A real number x is *left-computable* (often called *c.e.*) if it satisfies one and then all of the conditions in the following proposition. A real number x is *right-computable* if $-x$ is left-computable.

Proposition 3.14. *For a real number x, the following conditions are equivalent.*

1. *There exists a computable, strictly increasing sequence of rational numbers with limit x.*

2. *There exists a computable, nondecreasing sequence of rational numbers with limit x.*

3. *The set $\{q \in \mathbb{Q} \mid q < x\}$ is a c.e. subset of $\mathbb{Q}$.*

Further characterizations of left-computable real numbers have been given by Calude et al. [30, Theorem 4.1]. Right-computable real numbers can be characterized in a similar way. The following lemma is obvious.

Lemma 3.15. *A real number is computable if, and only if, it is left-computable and right-computable.*

Examples of left-computable but not computable real numbers can be constructed easily as follows.

Examples 3.16.   1. For $A \subseteq \mathbb{N}$, let

$$x_A := \sum_{i \in A} 2^{-i-1}.$$

This is a number in the interval $[0, 1]$. The number x_A is computable if, and only if, A is decidable. If A is c.e., then x_A is left-computable. Hence, if A is c.e. but not decidable, then x_A is left-computable but not computable (Specker [96]).

2. It is interesting that there are left-computable real numbers in $[0, 1]$ that are not of the form x_A with c.e. $A \subseteq \mathbb{N}$. Indeed, if $B \subseteq \mathbb{N}$ is c.e. but not decidable, then $B \oplus B^c := \{2n \mid n \in B\} \cup \{2n + 1 \mid n \notin B\}$ is not c.e., but $x_{B \oplus B^c}$ is nevertheless left-computable (Jockusch 1969, unpublished; see Soare [95]).

Besides the left-computable real numbers and among many others, the following classes of real numbers, defined by effectivity properties, have been studied:

- The *strongly c.e.* reals. These are the reals x_A with c.e. $A \subseteq \mathbb{N}$.

- The *weakly computable* reals; see Ambos-Spies et al. [2] and Zheng and Rettinger [116]. These are the differences of left-computable reals. These real numbers form a field, which is the arithmetical closure of the left-computable reals.

- The *computably approximable* reals; see, e.g., Ho [52] and Barmpalias [5]. These are the limits of computable and converging (but not necessarily computably converging) sequences of rational numbers. These real numbers form a field as well. This field is closed under application of computable real number functions (computable real number functions will be defined soon).

- The *Ω numbers* defined by Chaitin [33] as the halting probabilities of universal self-delimiting Turing machines. These real numbers can also be characterized as those real numbers in $[0, 1]$ that are left-computable and random, and also as those left-computable real numbers in $[0, 1]$ such that any computable, strictly increasing sequence of rational numbers converging to such a number converges as slowly as it is possible for a computable, strictly increasing sequence of rational numbers; see Calude et al. [30] and Kučera and Slaman [60].

- The *trivial* reals; see Downey et al. [36]. A real in $[0, 1]$ is *trivial* if the Kolmogorov complexity of the prefix of length l of its binary expansion differs at most by a constant from the Kolmogorov complexity of a string of l zeros (or, which is equivalent, from the Kolmogorov complexity of the prefix of length l of the binary expansion of some computable real number).

One can also define a hierarchy of real numbers corresponding to the arithmetical hierarchy of sets of natural numbers; see Zheng and Weihrauch [117]. For an overview of many classes of real numbers, defined by effectivity properties, the reader is referred to Zheng [115]. Effective randomness notions for real numbers or infinite binary sequences constitute an area of research of its own; the paper [66] by Miller and Nies contains a list of open problems in that area and many references.

Finally, for $n \geq 1$, a point $x \in \mathbb{R}^n$ is *computable*, if each of its components is a computable real number. Obviously, a point $x \in \mathbb{R}^n$ is computable if, and only if, there is a computable sequence $(q_i)_{i \in \mathbb{N}}$ of rational points $q_i \in \mathbb{Q}^n$ rapidly converging to x. Also Definitions 3.8 and 3.11 and Propositions 3.7, 3.9, 3.10, and 3.12 can be generalized directly to points in $\mathbb{R}^n$.

4 Computable Functions

Computable real numbers and the various kinds of weaker effectivity properties for real numbers are certainly interesting and worth studying. But in most computational problems over the real numbers, the task is not to compute some specific real number. Instead, there is some input, often one or several real numbers, and one wishes to compute some real number depending on this input. That is, given a vector x of real numbers, one wishes to compute the value $f(x)$ of some real number function f (of course, there are also more complicated types of computational problems; we will discuss them later on).

How can one compute a real number function $f :\subseteq \mathbb{R}^n \to \mathbb{R}$? We want to define a precise computability notion that describes exactly the real number functions that can be computed by a digital computer, except that at this moment we do not want to worry about time or space constraints (later we shall consider also the question of computational complexity). We use the Turing machine model as a model for digital computers.

What does it mean to compute a real number function $f :\subseteq \mathbb{R}^n \to \mathbb{R}$ in practice? Since a real number contains an infinite amount of information, it is unrealistic to expect that, given x, one can compute $f(x)$ exactly in finitely many steps. Rather, in numerical computations one aims at computing a "good" rational approximation of $f(x)$. We replace "good" by "arbitrarily good." Furthermore, the computer should not need x with infinite precision, but a "good" rational approximation should be sufficient. Again, we will allow the computer to ask for "arbitrarily good" rational approximations to x, i.e., approximations as good as it likes. Here is our first definition of computable real number functions.

Definition 4.1. A function $f :\subseteq \mathbb{R}^n \to \mathbb{R}$ is *computable* if there is an oracle Turing machine that, given any $k \in \mathbb{N}$, may ask for arbitrarily good rational approximations of the input $x \in \mathrm{dom}(f)$; i.e., it may ask finitely many questions of the kind "Give me a vector $p \in \mathbb{Q}^n$ of rational numbers with $d(x, p) < 1/2^i$," where the exponent i may depend on the answers to the previous questions, and after finitely many steps, it writes a rational number q on the output tape with $|f(x) - q| < 2^{-k}$.

Here, the precision i of a request as above has to be written in binary form on a special tape, the *oracle tape*, and the machine has to enter a special state, the *query state*. Then the answer will be provided in one step on the oracle tape, with the tape head on the field to the left of the answer.

That means the input $x \in \mathbb{R}^n$ is given to the machine only through rational approximations that the machine has to request. One can obviously modify this idea slightly without changing the computability notion by assuming that via some input tape(s) the machine has access to a rapidly converging rational sequence with limit x.

Once one has done that, one might as well change the model even more by not asking the machine to produce a rational approximation of precision 2^{-k} to $f(x)$, for any given $k \in \mathbb{N}$, but rather asking the machine to produce such approximations directly for all $k \in \mathbb{N}$, i.e., to ask the machine to produce a rapidly converging rational sequence with limit $f(x)$. This is slightly more elegant because in this notion the additional input k does not appear anymore, and input and output are of the same type (both are rapidly converging rational sequences).

Now we want to make this idea precise. We will directly formulate the basic definitions more generally since they will be useful for defining computability for many other kinds of objects than just real number functions.

Definition 4.2. A *representation* of a set X is a surjective function $\delta :\subseteq \Sigma^\omega \rightarrow X$, where Σ is some alphabet. Then for any $x \in X$ and any $p \in \Sigma^\omega$ with $\delta(p) = x$, the sequence p is called a δ-*name* of x.

Example 4.3. The usual decimal representation can be defined precisely so that it is a representation $\rho_{10} :\subseteq \Sigma^\omega \rightarrow \mathbb{R}$ in this sense, i.e., such that $\rho_{10}(1.414\ldots) = \sqrt{2}$, $\rho_{10}(3.141\ldots) = \pi$, $\rho_{10}(-0.999\ldots) = -1$. Similarly, one can define representations to any other base $b \in \mathbb{N}$ with $b \geq 2$.

Definition 4.4. The *Cauchy representation* $\rho :\subseteq \Sigma^\omega \rightarrow \mathbb{R}$ of the real numbers is a representation where a real number x is represented by a one-way infinite stream of symbols if this one-way infinite stream encodes a sequence of rational numbers converging rapidly to x:

$$\rho(w_0 \# w_1 \# w_2 \# \ldots) = x : \Longleftrightarrow |x - \nu_\mathbb{Q}(w_i)| < 2^{-i}, \text{ for all } i \in \mathbb{N}.$$

Here, the alphabet Σ contains at least the symbols $0, 1, \#$.

Often, it is useful to consider names of objects that enumerate basic information about the object.

Definition 4.5. We say that a $p \in \Sigma^\omega$ *enumerates* a set $A \subseteq \mathbb{N}$ if it satisfies the following two conditions:

1. if $\# w \#$ is a substring of p, then w is either empty or the binary name of an element of A,

2. for every element $n \in A$, the string $\# \nu_\mathbb{N}^{-1}(n) \#$ is a substring of p.

By using notations as in Section 2, one can generalize this notion straightforwardly to subsets A of $\mathbb{N}^n \times \mathbb{Q}^m$.

We use this idea in order to define two more representations of real numbers that we will need later.

Definition 4.6. The representation $\rho_< :\subseteq \Sigma^\omega \to \mathbb{R}$ is defined by

$$\rho_<(p) = x : \iff \quad p \text{ enumerates all } q \in \mathbb{Q} \text{ with } q < x.$$

Analogously, the representation $\rho_>$ is defined.

How can one perform computations with one-way infinite streams of symbols? One can feed the input stream(s) symbol by symbol into a Turing machine and demand that the Turing machine produces the output stream symbol by symbol.

Definition 4.7. Let Σ be an alphabet. A function $F :\subseteq \Sigma^\omega \to \Sigma^\omega$ is *computable* if there exists a Turing machine that, given a $p \in \mathrm{dom}(F)$ as a stream on an input tape, writes the output stream $F(p)$ symbol by symbol, without ever stopping, on a one-way output tape, and that, given a $p \in \Sigma^\omega \setminus \mathrm{dom}(F)$ does not write infinitely many symbols on the output tape.

Remark 4.8. Note that we demand that the output tape is one way. This ensures that output symbols that have already been written will never be erased or changed again. With a two-way output tape it would be possible to erase or change already written output symbols later again. But then any output symbol written after finitely many steps might be false and would therefore be useless.

Definition 4.9. Let $f :\subseteq X \to Y$ be a function and $\delta_X :\subseteq \Sigma^\omega \to X$ and $\delta_Y :\subseteq \Sigma^\omega \to Y$ be representations. A function $F :\subseteq \Sigma^\omega \to \Sigma^\omega$ is called a (δ_X, δ_Y)-*realizer* of f if
$$\delta_Y F(p) = f \delta_X(p) \quad \text{for all } p \in \mathrm{dom}(f\delta_X),$$
i.e., such that for any δ_X-name of some $x \in \mathrm{dom}(f)$, the value $F(p)$ is a δ_Y-name of $f(x)$. The function f is (δ_X, δ_Y)-*computable* if a computable (δ_X, δ_Y)-realizer of f exists.

Remark 4.10. Note that we do not impose any condition on how the realizer F should behave for input $p \notin \mathrm{dom}(f\delta_X)$. The motivation for this is that for any f that one wishes to compute, one is usually interested more in computing $f(x)$ for all valid input values x than in characterizing the domain of definition of f. That is a different problem, and it seems convenient to treat it separately. We have taken care of the domain only in our definition of computable F on Σ^ω, in Definition 4.7. Note that according to our definition, any restriction of a (δ_X, δ_Y)-computable function is (δ_X, δ_Y)-computable as well.

Since many real number functions that one wishes to compute expect as input not one real number but two or more, we need a representation of $\mathbb{R}^k$ for $k \geq 2$.

Definition 4.11.   1. Let $\delta :\subseteq \Sigma^\omega \to X$, $\delta' :\subseteq \Sigma^\omega \to X'$ be representations of sets X and X'. Then the representation $[\delta, \delta'] :\subseteq \Sigma^\omega \to X \times Y$ is defined by

$$[\delta, \delta']\langle p, q \rangle = (x, y) : \iff \delta(p) = x \text{ and } \delta'(q) = y,$$

where for $p, q \in \Sigma^\omega$ we define $\langle p, q \rangle := p(0)q(0)p(1)q(1)p(2)q(2)\ldots \in \Sigma^\omega$.

2. For $k \geq 1$, the representation δ^k of X^k is defined recursively by $\delta^1 := \delta$, $\delta^{k+1} := [\delta^k, \delta]$.

3. Sometimes we will also combine a notation $\nu :\subseteq \Sigma^* \to X$ and a representation $\delta :\subseteq \Sigma^\omega \to X'$ to a representation $[\nu, \delta] :\subseteq \Sigma^\omega \to X \times X'$, defined by

$$[\nu, \delta](0a_10a_20\ldots a_{n-1}0a_n1p) = (x, x') : \iff \nu(a_1a_2\ldots a_{n-1}a_n) = x$$
$$\text{and } \delta(p) = x'.$$

The discussion after Definition 4.1 shows that one can characterize the computable real number functions via the Cauchy representation ρ.

Lemma 4.12. *A real number function* $f :\subseteq \mathbb{R}^k \to \mathbb{R}$ *is computable in the sense of Definition 4.1 if, and only if, it is* (ρ^k, ρ)*-computable.*

Given the prominent role played by the decimal representation in daily life, it is natural to look at real number functions that are computable with respect to the decimal representation. But something goes wrong with the decimal representation, as the following observation shows.

Proposition 4.13. *The function* $f : \mathbb{R} \to \mathbb{R}$, $f(x) := 3 \cdot x$ *is not* (ρ_{10}, ρ_{10})*-computable.*

Proof. For the sake of a contradiction, assume that there is a Turing machine that, given a ρ_{10}-name of an arbitrary real number x, produces a ρ_{10}-name of $3 \cdot x$. Let us feed $0.3333\ldots$ into the machine. Then the machine has to produce either $0.99999\ldots$ or $1.00000\ldots$. Let us first assume that it produces $0.99999\ldots$. The machine will write the first symbol (a zero) of this output after finitely many steps, thus, after reading only finitely many symbols of the input, say, after reading at most the prefix 0.3^n, for some $n \in \mathbb{N}$. But then the first output symbol of the machine on input $0.3^n4444\ldots$ must be a zero as well although the numerical value of $3 \cdot \rho_{10}(0.3^n4444\ldots)$ is greater than 1, which means that it does not have ρ_{10}-name starting with a zero. Contradiction! In the other case, when on input $0.3333\ldots$ the machine produces $1.00000\ldots$, one arrives at a similar contradiction. We can summarize the argument as follows: it is impossible to determine even the first symbol of a decimal name for the real number $1 = 3 \cdot \rho_{10}(0.33333\ldots)$ correctly after reading only finitely many symbols of $0.33333\ldots$. □

Thus, not even multiplication with 3 is computable with respect to the decimal representation. Similarly one shows that also addition is not computable with respect to the decimal representation. This shows that the decimal representation is not suitable for computing real number functions. In fact, these negative statements are all due to topological reasons. This will be discussed in Section 6. There we will also discuss other representations of real numbers than the decimal representation and the Cauchy representation. Right now we stick to the computability notion for functions introduced above that can be described as computability with respect to the Cauchy representation (Lemma 4.12). This notion captures our intuition well that computations over the reals are approximative in nature and actually performed on rational numbers.

Our computability notion has properties that one expects from a computability notion for real number functions.

Theorem 4.14. *The following functions are computable:*

1. *The arithmetical operations* $+, -, \cdot, / :\subseteq \mathbb{R} \times \mathbb{R} \to \mathbb{R}$.

2. *The absolute value function* $\mathrm{abs} : \mathbb{R} \to \mathbb{R}, x \mapsto |x|$.

3. *The functions* $\min, \max : \mathbb{R} \times \mathbb{R} \to \mathbb{R}$.

4. *The constant functions* $\mathbb{R} \to \mathbb{R}, x \mapsto c$ *with computable value* $c \in \mathbb{R}$.

5. *The projections* $\mathrm{pr}_i : \mathbb{R}^n \to \mathbb{R}, (x_1, \ldots, x_n) \mapsto x_i$.

6. *All polynomials* $p : \mathbb{R}^n \to \mathbb{R}$ *with computable coefficients.*

7. *The exponential function and the trigonometric functions* $\exp, \sin, \cos : \mathbb{R} \to \mathbb{R}$.

8. *The square root function* $\sqrt{}\{x \in \mathbb{R} \mid x \geq 0\} \subseteq \mathbb{R} \to \mathbb{R}$ *and the logarithm function* $\log : \{x \in \mathbb{R} \mid x > 0\} \subseteq \mathbb{R} \to \mathbb{R}$.

Proof. We sketch the proof that addition $f : \mathbb{R} \times \mathbb{R} \to \mathbb{R}$ is computable. Given two sequences $(q_n)_{n \in \mathbb{N}}$ and $(r_n)_{n \in \mathbb{N}}$ of rational numbers that rapidly converge to x and y, respectively, we can compute the sequence $(p_n)_{n \in \mathbb{N}}$ of rational numbers defined by $p_n := q_{n+1} + r_{n+1}$. This sequence converges rapidly to $x + y$:

$$|x + y - p_n| \leq |x - q_{n+1}| + |y - r_{n+1}| < 2^{-n-1} + 2^{-n-1} = 2^{-n}.$$

Since addition on rational number can be computed by Turing machines, it follows that f is computable as well. □

Remark 4.15. Addition requires only a uniform lookahead of one step that does not depend on the input. For functions that are not uniformly continuous, such as multiplication, the "modulus of continuity" and thus the lookahead depend on the input.

Also more complicated functions such as the Gamma function or Riemann's zeta function are computable.

Now we list some basic properties of computable real number functions:

- they map computable real numbers to computable real numbers,

- they also map computable sequences of real numbers to computable sequences of real numbers,

- and they are closed under composition.

We can easily state these results more generally with respect to arbitrary representations of arbitrary representable sets.

Definition 4.16. 1. A sequence $p \in \Sigma^\omega$ is *computable* if there is a Turing machine that, given $k \in \mathbb{N}$ in binary form, produces the k-th symbol of p. It is equivalent to demand that there is a Turing machine with a one-way output tape that, without input and without ever stopping, writes p symbol by symbol on the output tape.

2. Let $\delta_X :\subseteq \Sigma^\omega \to X$ be a representation of a set X. An element $x \in X$ is δ_X-*computable* if it possesses a computable δ_X-name.

Example 4.17. By Theorem 3.2, the following conditions are equivalent for a real number x:

1. x is a computable real number,

2. x is a ρ_{10}-computable real number,

3. x is a ρ-computable real number.

Proposition 4.18. *If* $F :\subseteq \Sigma^\omega \to \Sigma^\omega$ *is a computable function and* $p \in \Sigma^\omega$ *is a computable element in the domain of* F, *then* $F(p)$ *is computable as well.*

Proof. One simply has to combine a Turing machine computing p with a Turing machine computing F in order to obtain a Turing machine computing $F(p)$. □

Corollary 4.19. *Let* X *and* Y *be sets with representations* δ_X *and* δ_Y. *If* $f :\subseteq X \to Y$ *is* (δ_X, δ_Y)-*computable and* $x \in X$ *is a* δ_X-*computable element in the domain of* f, *then* $f(x)$ *is* δ_Y-*computable.*

In particular, a computable real number function maps any computable real number in its domain to a computable real number.

Everything that we just said about computable elements is also true for computable sequences of elements. Computable functions preserve computability of sequences as well.

Definition 4.20. 1. A sequence $(s^{(n)})_{n \in \mathbb{N}}$ of elements $s^{(n)} \in \Sigma^\omega$ is *computable* if a Turing machine, given $n, k \in \mathbb{N}$ in binary form, produces the k-th symbol of

$s^{(n)}$. It is equivalent to demand that there is a Turing machine with a one-way output tape that, given $n \in \mathbb{N}$ in binary form, writes $s^{(n)}$ symbol by symbol on the output tape.

2. Let $\delta_X :\subseteq \Sigma^\omega \to X$ be a representation of a set X. A sequence $(x_n)_{n \in \mathbb{N}}$ of elements of X is δ_X-*computable* if there is a computable sequence of δ_X-names for the x_n.

Example 4.21. A sequence of real numbers is computable if, and only if, it is ρ-computable.

Proposition 4.22. *If $F :\subseteq \Sigma^\omega \to \Sigma^\omega$ is a computable function and $(s^{(n)})_{n \in \mathbb{N}} \in (\Sigma^\omega)^\omega$ is a computable sequence of elements in the domain of F, then the sequence $(F(s^{(n)}))_{n \in \mathbb{N}}$ is computable as well.*

Proof. Similar to the proof of Proposition 4.18.   □

Corollary 4.23. *Let X and Y be sets with representations δ_X and δ_Y. If $f :\subseteq X \to Y$ is (δ_X, δ_Y)-computable and $(x_n)_{n \in \mathbb{N}} \in X$ is a δ_X-computable sequence of elements in the domain of f, then $(f(x_n))_{n \in \mathbb{N}}$ is a δ_Y-computable sequence.*

In particular, a computable real number function maps any computable sequence of real numbers in the domain of the function to a computable sequence of real numbers.

Finally, the composition of computable functions is computable again.

Proposition 4.24. *If $F, G :\subseteq \Sigma^\omega \to \Sigma^\omega$ are computable, then their composition $G \circ F :\subseteq \Sigma^\omega \to \Sigma^\omega$ is computable as well.*

Proof. We combine Turing machines T_G for G and T_F for F in the following way. T_G uses the output tape of T_F as input tape and T_F receives p on its input tape. We start T_G. Whenever T_G needs to read a symbol of $F(p)$, we check whether T_F has already written this symbol. If not, we let T_G pause and start T_F and let it run until it has written this symbol. Then we let it pause again and take up the computation of T_G again. And so on. Does this composition of machines compute the composition $G \circ F$ of the functions F and G? Almost. In fact, this composition of the two machines might compute an extension of the composition $G \circ F$. On input $p \in \text{dom}(G \circ F)$, it correctly computes the value $G \circ F(p)$. But, when there is some input $p \in \Sigma^\omega \setminus \text{dom}(F)$, where T_F produces only finitely many output symbols, and T_G happens to need no more than these for producing some infinite output, then this composition of machines produces some infinite output on input p although $G \circ F$ is not defined at p. Therefore, we introduce the following modification of the composition of the two machines described above: whenever G is about to write the n-th output symbol, we check whether F has already produced at least n symbols. If not, we let F run until it has (if it never does, G will never write an n-th output symbol). This modified machine indeed computes the function $G \circ F$.   □

Corollary 4.25. *Let X, Y, Z be sets with representations $\delta_X, \delta_Y, \delta_Z$, respectively. If $f :\subseteq X \to Y$ is (δ_X, δ_Y)-computable and $g :\subseteq Y \to Z$ is (δ_Y, δ_Z)-computable, then $g \circ f$ is (δ_X, δ_Z)-computable.*

In particular, the composition of computable real number functions is computable as well.

Now we come to a very important property of computable real number functions: they are continuous. In fact, they are exactly the real number functions that are effectively continuous in the following sense. Here $B(x, \varepsilon) := \{y \in \mathbb{R}^n \mid d(x, y) < \varepsilon\}$ is the open ball in $\mathbb{R}^n$ with midpoint x and radius ε, for $x \in \mathbb{R}^n$ and $\varepsilon > 0$. We define a total numbering B^n of all open rational balls in $\mathbb{R}^n$ by

$$B^n(\langle i_1, \ldots, i_n, j, k \rangle) := B\left((\nu_Q(i_1), \ldots, \nu_Q(i_n)), \frac{j+1}{k+1} \right).$$

Definition 4.26. A function $f :\subseteq \mathbb{R}^n \to \mathbb{R}$ is *effectively continuous* if there is a c.e. subset $S \subseteq \mathbb{N}$ with the following two properties:

1. For any $\langle i, j \rangle \in S$, $f(B^n(i)) \subseteq B^1(j)$.

2. For any $x \in \mathrm{dom}(f)$ and any $\varepsilon > 0$, there is some $\langle i, j \rangle \in S$ such that $x \in B^n(i)$ and such that the radius of $B^1(j)$ is at most as large as ε.

Theorem 4.27. *A function $f :\subseteq \mathbb{R}^n \to \mathbb{R}$ is computable if, and only if, it is effectively continuous.*

Proof. For simplicity, we consider only the case $n = 1$. First, assume that f is effectively continuous. Then, given a ρ-name of some point $x \in \mathrm{dom}(f)$, one can, using an enumeration of S, compute $f(x)$ with arbitrary precision. Thus, f is computable.

Now, assume that f is computable. Then a Turing machine, given any k and given any name p of any point $x \in \mathrm{dom}(f)$, computes a rational 2^{-k}-approximation q of $f(x)$, i.e., a rational number q with $f(x) \in B(q, 2^{-k})$. The machine does so after reading at most a finite prefix of p having the form $w_0 \# w_1 \# w_2 \# \ldots w_l \#$. Then $U := \bigcap_{i=0}^{l} B(\nu_Q(w_i), 2^{-i})$ is an open neighborhood of x, and any point in this open neighborhood has a ρ-name starting with this prefix. Thus, given a name starting with $w_0 \# w_1 \# w_2 \# \ldots w_l \#$ of any point in this neighborhood, the Turing machine will produce the same output q. This implies $f(U) \subseteq B(q, 2^{-k})$, and that means that f is continuous. By systematically enumerating all prefixes of ρ-names of real numbers and testing the Turing machine on them, one can construct a c.e. set S that shows that f is even effectively continuous. $\qquad \square$

The fact that even a function as simple as the sign function

$$\mathrm{sign} : \mathbb{R} \to \mathbb{R}, \quad \mathrm{sign}(x) := \begin{cases} 0 & \text{if } x < 0, \\ 1 & \text{if } x \geq 0 \end{cases}$$

is not computable may seem counter-intuitive at first. But, if the input real number x is given only through approximations (even arbitrarily good ones) and if x happens to be equal to a point of discontinuity of the function, then the value of the function can simply not be computed from the input. Of course, for example, the sign function is computable in various weaker senses (e.g., lower semi-computable, a notion we will define soon), which are also useful, and it is also a computable point in a certain function space (this will be defined soon as well). Thus, it is simple in certain senses. But in the numerical sense considered by us, it is definitely not computable.

It is interesting that continuity is a feature not only of this computability notion but also of two other natural computability notions for real number functions that can be defined via the Turing machine model. Both of them apply to functions that are defined only on computable real numbers and that map computable real numbers to computable real numbers. In the following we list four computability notions on real numbers or on computable real numbers and explain the relations between them.

1. Computability for functions $f :\subseteq \mathbb{R} \to \mathbb{R}$, i.e., the computability notion for real number functions that we have considered so far. It goes back to Grzegorczyk [41] and Lacombe [62].

2. Markov computability for functions $f :\subseteq \mathbb{R}_c \to \mathbb{R}_c$. If one calls a standard description of a Turing machine a *program*, and if the Turing machine computes a ρ-name of a real number, then this description can be called a *program* for this computable real number. A function $f :\subseteq \mathbb{R}_c \to \mathbb{R}_c$ is *Markov computable* if there exists a computable function mapping finite strings to finite strings that maps any program for any $x \in \mathrm{dom}(f)$ to a program for $f(x)$. This computability notion has been considered in the Russian school of constructive analysis, e.g., by Ceĭtin [31], Šanin [85], and Kušner [59]. For a purely computability theoretic presentation of many results in this direction, the reader is referred to Aberth [1].

3. Sequential computability for functions $f :\subseteq \mathbb{R}_c \to \mathbb{R}_c$. A function $f :\subseteq \mathbb{R}_c \to \mathbb{R}_c$ is *sequentially computable* if it maps any computable sequence of real numbers in $\mathrm{dom}(f)$ to a computable sequence of real numbers. This notion has been studied by Mazur [63].

4. Computable invariance for functions $f :\subseteq \mathbb{R} \to \mathbb{R}$. A function $f :\subseteq \mathbb{R} \to \mathbb{R}$ is *computably invariant* if it maps any computable real number in its domain to a computable real number.

Now we explain the connections between these notions.

- First we observe that one can apply the computability notion for functions $f :\subseteq \mathbb{R} \to \mathbb{R}$, of course, also to functions $f :\subseteq \mathbb{R}_c \to \mathbb{R}_c$ and that this gives rise to a fifth class of functions. We have already seen that any computable function $f :\subseteq \mathbb{R} \to \mathbb{R}$ is computably invariant. And it is clear that any restriction of a computable real number function is a computable real number function as well.

Thus, if $f : \subseteq \mathbb{R} \to \mathbb{R}$ is computable, then its restriction $f|_{\mathbb{R}_c} : \subseteq \mathbb{R}_c \to \mathbb{R}_c$ is computable again. But, there exists a computable function $g : \mathbb{R}_c \to \mathbb{R}_c$ defined on all computable real numbers which cannot be extended to a continuous function defined on all real numbers; see Aberth [1].

- Now we compare computable functions $f : \subseteq \mathbb{R}_c \to \mathbb{R}_c$ and Markov computable functions $f : \subseteq \mathbb{R}_c \to \mathbb{R}_c$. It is easy to show that any computable function $f : \subseteq \mathbb{R}_c \to \mathbb{R}_c$ is Markov computable. The converse is not true in general; see Slisenko [93] or Weihrauch [109, Example 9.6.5]. But it is an important fact due to Ceĭtin [31] that the converse is true for Markov computable functions $f : \subseteq \mathbb{R}_c \to \mathbb{R}_c$ with *computably separable* domain, i.e., a domain $\mathrm{dom}(f)$ such that there exists a computable sequence $(x_n)_{n \in \mathbb{N}}$ of real numbers such that $\{x_n \mid n \in \mathbb{N}\}$ is a dense subset of $\mathrm{dom}(f)$. Related results were obtained by Kreisel et al. [57] and by Moschovakis [70]. See Kušner [59] for a careful presentation and discussion of Ceĭtin's result and related results. In Theorem 4.27, we observed that computability is equivalent to effective continuity. So, Ceĭtin's result says that any Markov computable function $f : \subseteq \mathbb{R}_c \to \mathbb{R}_c$ with a computably separable domain is effectively continuous.

- Next, we compare Markov computable functions $f : \subseteq \mathbb{R}_c \to \mathbb{R}_c$ and sequentially computable functions $f : \subseteq \mathbb{R}_c \to \mathbb{R}_c$. It is clear that any Markov computable function $f : \subseteq \mathbb{R}_c \to \mathbb{R}_c$ is sequentially computable. The converse is not true: there exists even a sequentially computable function defined on all computable real numbers that is not Markov computable; see Hertling [49]. Nevertheless, surprisingly, any sequentially computable function with computably separable domain is continuous. This has already been observed by Mazur [63] (for functions defined on an interval).

- Finally, we compare sequentially computable functions $f : \subseteq \mathbb{R}_c \to \mathbb{R}_c$ and computably invariant functions $f : \subseteq \mathbb{R} \to \mathbb{R}$. Trivially, any real number function that maps any computable sequence of real numbers in its domain to a computable sequence of real numbers is computably invariant. The converse is not true. For example, the restriction of the sign function to the computable real numbers is a total function from $\mathbb{R}_c$ to $\mathbb{R}_c$ and computably invariant but not continuous, hence, not sequentially computably.

Functions considered in analysis that are not computable are often not computable simply because they are discontinuous. But they may still be computably invariant, like the sign function. It is often, not only for real number functions, but for many other kinds of functions, an interesting task to show that some noncomputable function is not even computably invariant, i.e., that there exists a computable input element such that the output element is not computable. We will see some examples of this later on.

5 Computability Notions for Subsets of Euclidean Space

Over the natural numbers, not only computable functions are of fundamental importance, but also effectivity notions for sets are important. The two most important classes of subsets of $\mathbb{N}^n$ defined by computability conditions are certainly

1. the computable (or decidable or recursive) subsets,

2. and the computably enumerable (or recursively enumerable) subsets.

We have already learned about computable functions over the real numbers. In this section we will see that there are natural computability notions for sets of real numbers that correspond to computable or to computably enumerable sets of natural numbers. Furthermore, these notions have a natural computational meaning, and they generalize the notions for subsets of $\mathbb{N}$.

Let us first consider the computable subsets. A subset $A \subseteq \mathbb{N}$ is called *computable* or *decidable* or *recursive* if its characteristic function $\chi_A : \mathbb{N} \to \mathbb{N}$, defined by

$$\chi_A(n) := \begin{cases} 0 & \text{if } n \in A, \\ 1 & \text{if } n \notin A, \end{cases}$$

is computable. That means, a set A is decidable if a Turing machine, given a natural number n, will after finitely many steps give the answer yes or no, saying whether n is an element of A. Let us first try to translate this idea to subsets of $\mathbb{R}^n$.

Definition 5.1. Let X be a set with a representation δ_X. Let us call a subset $A \subseteq X$ δ_X-*decidable* if the characteristic function $\chi_A : X \to \mathbb{R}$, defined by

$$\chi_A(x) := \begin{cases} 0 & \text{if } x \in A, \\ 1 & \text{if } x \notin A, \end{cases}$$

is (δ_X, ρ)-computable. The (ρ^n, ρ)-decidable subsets of $\mathbb{R}^n$ are called *decidable*.

Instead of considering the output values 0 and 1 as elements of $\mathbb{R}$, one might as well consider them as elements of $\mathbb{N}$. That means, a set $A \subseteq \mathbb{R}^n$ is decidable if, and only if, a Turing machine, given a ρ^n-name of a point $x \in \mathbb{R}^n$, will after finitely many steps give the answer yes or no, saying whether x is an element of A. Unfortunately, only two trivial subsets of $\mathbb{R}^n$ are decidable in this sense.

Lemma 5.2. *The only decidable subsets of $\mathbb{R}^n$ are $\emptyset$ and $\mathbb{R}^n$.*

Proof. The function $\chi_\emptyset$ and $\chi_{\mathbb{R}^n}$ are constant functions with value 1 and value 0, respectively, and obviously computable. The characteristic function of any other subset of $\mathbb{R}^n$ is discontinuous and therefore not computable, according to Theorem 4.27. $\qquad\square$

Thus, this notion of decidability is not very interesting. What is wrong here? Is in the end our definition of computability of real number functions unsuitable? We saw that it can be characterized as computability with respect to the representation ρ. Is there perhaps a representation of the real numbers that is better suited for real number computations than ρ, perhaps a representation that would make basic nontrivial subsets of $\mathbb{R}^n$ decidable, e.g., which would allow comparison of real numbers? This is not the case.

Proposition 5.3. *There is no representation $\delta :\subseteq \Sigma^\omega \to \mathbb{R}$ of the real numbers such that any of the tests $=, <, \leq$ is decidable with respect to δ, i.e., such that any of the three following subsets of $\mathbb{R}^2$ is δ^2-decidable:*

$$\{(x,y) \in \mathbb{R}^2 \mid x = y\}, \quad \{(x,y) \in \mathbb{R}^2 \mid x < y\}, \quad \{(x,y) \in \mathbb{R}^2 \mid x \leq y\}.$$

Proof. We give the proof for the equality test. Let us assume that there is a representation δ such that the equality of two real numbers is decidable with respect to δ. Now consider an arbitrary real number x and an arbitrary δ-name p for x. On input $\langle p, p \rangle$, after finitely many steps, the Turing machine will have written a prefix of a ρ-name of 0 that cannot be a prefix of a ρ-name of 1. That means, the machine has come to the conclusion "equal." During these finitely many steps it can have read only a finite prefix of its input. Let n be a number such that it has read at most the first $2n$ symbols of its input $\langle p, p \rangle$. Then, for any input $\langle p, q \rangle$ such that q starts with $u := p(0)p(1) \ldots p(n-1)$, the machine would also come to the conclusion "equal." This implies that $\delta(q) = \delta(p) = x$ for every δ-name q starting with the string u. Since there are only countably many strings in Σ^* and $\delta :\subseteq \Sigma^\omega \to \mathbb{R}$ is surjective, this implies that $\mathbb{R}$ is countable. Contradiction. $\qquad\square$

Thus, trying to change to another representation does not help. Therefore, we will stick to our original notion of computability of real number functions and accept that one cannot effectively make a discontinuous decision. This observation leads to the question of whether, maybe, there is a useful "smooth" replacement for discontinuous decisions? Let us consider a nonempty closed subset $A \subseteq \mathbb{R}^n$. The distance function $d_A : \mathbb{R}^n \to \mathbb{R}$, defined by

$$d_A(x) := \inf_{y \in A} d(x, y),$$

can be considered as a "smooth" version of the characteristic function of A.

Definition 5.4. We call a closed subset $A \subseteq \mathbb{R}^n$ *computable* or *recursive* if either it is empty or d_A is computable. We call an open subset $U \subseteq \mathbb{R}^n$ *computable* or *recursive* if its complement (which is a closed set) is computable.

First, we notice that this notion is a generalization of the decidability notion for subsets of $\mathbb{N}^n$.

Proposition 5.5. *A subset $A \subseteq \mathbb{N}^n$ is decidable in the classical sense (as a subset of $\mathbb{N}^n$) if, and only if, it is computable when considered as a closed subset of $\mathbb{R}^n$ (under the natural embedding $\mathbb{N}^n \to \mathbb{R}^n$).*

The proof is straightforward.

Examples 5.6.   1. The empty set $\emptyset$ and the full set $\mathbb{R}^n$ are computable closed sets and computable open sets.

2. For any $x \in \mathbb{R}^n$, the set $\{x\}$ is computable if, and only if, x is a computable point.

3. The equality relation and the $\leq$-Relation are computable; i.e., the closed subsets $\{(x,y) \in \mathbb{R}^2 \mid x = y\}$ and $\{(x,y) \in \mathbb{R}^2 \mid x \leq y\}$ of $\mathbb{R}^2$ are computable.

4. The open ball $B(x, \varepsilon) = \{y \in \mathbb{R}^n \mid d(x,y) < \varepsilon\}$ and the closed ball $\overline{B}(x, \varepsilon) = \{y \in \mathbb{R}^n \mid d(x,y) \leq \varepsilon\}$ are computable if $x \in \mathbb{R}^n$ is a computable point and $\varepsilon > 0$ is a computable real number.

5. For $a, b \in \mathbb{R}$ with $a \leq b$, the closed interval $[a, b]$ is computable if, and only if, a and b are computable real numbers. The same is true for the open interval (a, b).

6. For any total function $f : \mathbb{R}^n \to \mathbb{R}$, the graph $\mathrm{graph}(f) := \{(x_1, \ldots, x_n, y) \mid x_1, \ldots, x_n, y \in \mathbb{R}, \ y = f(x_1, \ldots, x_n)\}$ is a computable closed set if, and only if, f is a computable function.

The computability notion for closed sets has a very intuitive meaning: the following proposition can be interpreted as saying that a closed subset $A \subseteq \mathbb{R}^n$ is computable if, and only if, one can plot pixel images of it with any desired precision. This condition, i.e., the second condition in the following proposition, will be used later for defining the computational complexity of a closed set. Here, computability on $\mathbb{Z}$ can be reduced to computability on $\mathbb{N}$ via the bijection $\nu : \mathbb{N} \to \mathbb{Z}, \nu(2n) := n, \nu(2n + 1) := -n - 1$.

Proposition 5.7. *For a closed set $A \subseteq \mathbb{R}^k$, the following two conditions are equivalent.*

1. A is computable.

2. There exists a computable function $f : \mathbb{N} \times \mathbb{Z}^k \to \mathbb{N}$ with $\mathrm{range}(f) \subseteq \{0, 1\}$ and such that for all $n \in \mathbb{N}$ and $z \in \mathbb{Z}^k$

$$f(n, z) = \begin{cases} 0 & \text{if } d_A(\frac{z}{2^n}) < 2^{-n}, \\ 1 & \text{if } d_A(\frac{z}{2^n}) > 2 \cdot 2^{-n}, \\ 0 \text{ or } 1 & \text{otherwise.} \end{cases}$$

Now let us see whether we can find also a natural generalization of the notion of a computably enumerable subset of $\mathbb{N}^n$ to subsets of $\mathbb{R}^n$. We will see that there are even two natural generalizations, one for open subsets and one for closed subsets. As is well known, a set $A \subseteq \mathbb{N}$ is called *computably enumerable* (short: *c.e.*) or *recursively enumerable* if it satisfies one (and then both) of the following two equivalent conditions:

1. there is a computable function $f :\subseteq \mathbb{N} \to \mathbb{N}$ with $\mathrm{dom}(f) = A$; i.e., a Turing machine, given (a binary name of) some $n \in \mathbb{N}$, stops after finitely many steps if, and only if, $n \in A$,

2. A is empty or there is a total computable function $f : \mathbb{N} \to \mathbb{N}$ whose range is equal to A.

A set $A \subseteq \mathbb{N}$ is called *co-c.e.* if its complement is c.e. Using the pairing function $\langle \cdot, \cdot \rangle$, one can generalize these notions to subsets of $\mathbb{N}^k$ for $k \geq 2$. And using $\nu_\mathbb{Q}$ from Section 2, one can translate it to subsets of $\mathbb{Q}^m$: we call a subset $S \subseteq \mathbb{Q}^m$ *c.e.* if the set $\{(n_1, \ldots, n_m) \in \mathbb{N}^m \mid (\nu_\mathbb{Q}(n_1), \ldots, \nu_\mathbb{Q}(n_m)) \in S\}$ is a c.e. subset of $\mathbb{N}^m$. The idea in the first characterization can be transferred directly to a represented set.

Definition 5.8. Let X be a set with a representation $\delta :\subseteq \Sigma^\omega \to X$. A set $U \subseteq X$ is called *δ-c.e.* if a Turing machine, given an arbitrary δ-name p of an arbitrary element $x \in X$, halts after finitely many steps if, and only if, $x \in U$.

Applying the idea in the first characterization to real numbers leads to the first condition in the following proposition, whereas the second condition above corresponds to the second and third conditions in the proposition.

Proposition 5.9. *Let $U \subseteq \mathbb{R}^n$ be open and $A := \mathbb{R}^n \setminus U$ be its (closed) complement. The following conditions are equivalent.*

1. U is ρ^n-c.e.

2. $U = \bigcup_{(q,\varepsilon) \in S} B(q, \varepsilon)$ for some c.e. set $S \subseteq \mathbb{Q}^n \times \mathbb{Q}_+$ (here $\mathbb{Q}_+ := \{q \in \mathbb{Q} \mid q > 0\}$).

3. The set $\{(q, \varepsilon) \in \mathbb{Q}^n \times \mathbb{Q}_+ \mid \overline{B}(q, \varepsilon) \subseteq U\}$ is computably enumerable.

4. $A = f^{-1}(\{0\})$ for some total computable function $f : \mathbb{R}^n \to \mathbb{R}$.

5. The function $\chi_A : \mathbb{R}^n \to \mathbb{R}$ is lower semicomputable; i.e., it is $(\rho^n, \rho_<)$-computable.

6. Either $A = \emptyset$ or the function $d_A : \mathbb{R}^n \to \mathbb{R}$ is lower semicomputable.

Definition 5.10. Let $U \subseteq \mathbb{R}^n$ be open and $A := \mathbb{R}^n \setminus U$ be its (closed) complement. If one (and then all) of the conditions in the previous proposition are satisfied, then U is called *c.e. open* and A is called *co-c.e. closed*.

The third condition above can be understood as saying that we can enumerate "negative" information about the closed set A, namely all closed rational balls contained in the complement of A. This corresponds to the following well-known topology on the space of all closed subsets (see Beer [6] for hyperspace topologies).

Definition 5.11. The *upper Fell topology* on the space of all closed subsets of $\mathbb{R}^n$ is the topology generated by the subbase consisting of all sets

$$\{A \subseteq \mathbb{R}^n \mid A \text{ is closed and } A \cap K = \emptyset\},$$

for compact sets $K \subseteq \mathbb{R}^n$.

There is another natural generalization of computable enumerability to subsets of $\mathbb{R}^n$.

Proposition 5.12. *Let $A \subseteq \mathbb{R}^n$ be closed. The following are equivalent.*

1. *The set $\{(q, \varepsilon) \in \mathbb{Q}^n \times \mathbb{Q}_+ \mid \mathrm{B}(q, \varepsilon) \cap A \neq \emptyset\}$ is computably enumerable.*

2. *Either A is empty or there is a computable sequence $(x_i)_{i \in \mathbb{N}}$ of points $x_i \in \mathbb{R}^n$ such that A is the closure of the set $\{x_i \mid i \in \mathbb{N}\}$.*

3. *Either A is empty or the function $d_A : \mathbb{R}^n \to \mathbb{R}$ is upper semicomputable; i.e., it is $(\rho^n, \rho_>)$-computable.*

Definition 5.13. Let $A \subseteq \mathbb{R}$ be closed. If one (and then all) of the conditions in the previous proposition are satisfied, then A is called *c.e. closed*, and its (open) complement is called *co-c.e. open*.

The first condition in the proposition can be understood as saying that we can enumerate "positive" information about the closed set A, namely all rational balls having nonempty intersection with A. As above, this corresponds also to a well-known topology on the space of all closed subsets.

Definition 5.14. The *lower Fell topology* on the space of all closed subsets of $\mathbb{R}^n$ is the topology generated by the subbase consisting of all sets

$$\{A \subseteq \mathbb{R}^n \mid A \text{ is closed and } A \cap U \neq \emptyset\},$$

for open sets $U \subseteq \mathbb{R}^n$.

Both of these two notions of computable enumerability for open or closed subsets of $\mathbb{R}^n$ are generalizations of the corresponding notions for subsets of $\mathbb{N}^n$.

Proposition 5.15. *1. A subset $A \subseteq \mathbb{N}^n$ is c.e. in the classical sense (as a subset of $\mathbb{N}^n$) if, and only if, it is c.e. closed when considered as a closed subset of $\mathbb{R}^n$ (under the natural embedding $\mathbb{N}^n \to \mathbb{R}^n$).*

2. *A subset $A \subseteq \mathbb{N}^n$ is co-c.e. in the classical sense (as a subset of $\mathbb{N}^n$) if, and only if, it is co-c.e. closed when considered as a closed subset of $\mathbb{R}^n$ (under the natural embedding $\mathbb{N}^n \to \mathbb{R}^n$).*

A subset of $\mathbb{N}^n$ is decidable if, and only if, it is c.e. and its complement is c.e. as well. This generalizes to subsets of $\mathbb{R}^n$ as well.

Proposition 5.16. *An open or closed subset of $\mathbb{R}^n$ is computable if, and only if, it is c.e. and its complement is c.e. as well.*

Proof. It suffices to prove this for closed sets. It follows from the fact that a real number function is computable if, and only if, it is upper semicomputable and lower semicomputable. Apply this to the function d_A for some nonempty closed set $A \subseteq \mathbb{R}^n$. □

For any class of subsets of $\mathbb{R}^n$, it is natural to ask whether it is closed under basic set theoretic operations.

Proposition 5.17. *Let A and B be closed subsets of $\mathbb{R}^n$.*

1. *If A and B are co-c.e. closed, so are $A \cup B$ and $A \cap B$.*

2. *If A and B are c.e. closed, so is $A \cup B$.*

3. *If A and B are computable, so is $A \cup B$.*

Proof.   1. Let $f, g : \mathbb{R}^n \to \mathbb{R}$ be computable functions with $A = f^{-1}(\{0\})$ and $B = g^{-1}(\{0\})$. Then $A \cup B = (f \cdot g)^{-1}(\{0\})$ and $A \cap B = (|f| + |g|)^{-1}(\{0\})$, and $f \cdot g$ and $|f| + |g|$ are also computable functions.

2. Remember that a closed set is c.e. closed if, and only if, the set of all open rational balls intersecting the set can be enumerated effectively. Now note that any open (rational) ball intersects $A \cup B$ if, and only if, it intersects at least one of the two sets A and B.

3. This follows from the first two statements.

□

We illustrate a typical technique that is used in computable analysis. We encode a c.e. but nondecidable set $K \subseteq \mathbb{N}$ into a subset of the reals in order to obtain a counterexample.

Example 5.18. There are two computable closed sets $A, B \subseteq \mathbb{R}$ such that $A \cap B$ is not even c.e. closed. Indeed, let $h : \mathbb{N} \to \mathbb{N}$ be a total computable function with nondecidable range $K := \text{range}(h)$. We define $A := \bigcup_{n=0}^{\infty} A_n$ with

$$A_n := \begin{cases} [n - \frac{1}{2}, n] & \text{if } n \notin K, \\ [n - \frac{1}{2}, n - 2^{-k-2}] & \text{if } n \in K \text{ and } k = \min\{i \mid h(i) = n\}. \end{cases}$$

It is not hard to check that A is computable. Furthermore, we define $B := \mathbb{N} \subseteq \mathbb{R}$. Then $A \cap B = \mathbb{N} \setminus K$. This is a closed, but not a c.e. closed, subset of $\mathbb{R}$.

We conclude this section with an open problem. The famous Mandelbrot set M is a subset of the complex plane $\mathbb{C}$ that can be defined as follows:

$$M = \{c \in \mathbb{C} \mid \text{all numbers in the sequence } (z_i)_{i \in \mathbb{N}} \text{ of complex numbers}$$
$$\text{defined by } z_0 := 0 \text{ and } z_{i+1} := z_i^2 + c \text{ satisfy } |z_i| \leq 2\}.$$

The Mandelbrot set is a closed subset of the complex plane contained in the circle with radius 2 around the origin. In the following we identify $\mathbb{C}$ with $\mathbb{R}^2$. It is known that the complement of the Mandelbrot set is c.e. open (Weihrauch [109, Exercise 5.1.32]) and that the boundary of the Mandelbrot set is c.e. closed (Hertling [50]).

Problem 5.19.   1. Is the Mandelbrot set computable?

2. Is the interior of the Mandelbrot set c.e. open?

The first question is equivalent to the question of whether the Mandelbrot set is c.e. closed: compare Proposition 5.16. The second question is stronger than the first because of the following simple lemma.

Lemma 5.20. *The closure of a c.e. open subset $U \subseteq \mathbb{R}^n$ is c.e. closed.*

The famous *hyperbolicity conjecture* says that a certain subset of the interior of the Mandelbrot set, the union of its so-called "hyperbolic components," is actually equal to the interior of the Mandelbrot set; compare, e.g., the introductory text [13] by Branner. Since the union of the hyperbolic components is c.e. open (Hertling [50]), one arrives at the following conclusion.

Proposition 5.21 (Hertling [50]). *If the hyperbolicity conjecture is true, then the answer to both questions in Problem 5.19 is yes.*

6 Representations and Topological Considerations

We have introduced computability notions for real numbers, for real number functions, and for sets of real numbers. We have seen that all these notions can be characterized using Turing machines and the Cauchy representation. We have also seen that the decimal representation is unsuitable for real number computations. But one can define many other representations of real numbers. Maybe some other representation is even more suitable for real number computations than the Cauchy representation? In this section we will analyze representations more systematically. We will compare various real number representations, and we will make some general observations about representations that will be useful when in later sections we consider computability on spaces that are "more complicated" than the space of real numbers.

We will see soon that a good deal of the difficulties arising in the context of representations is of a topological nature. Indeed, already our first definition of computable real number functions contained the idea of approximating the real value $f(x)$ with arbitrary precision. Approximation is a notion from topology. So, here topology enters the stage, and we will see that it plays a very important role for computability notions over the real numbers or other nondiscrete sets.

First, the set Σ^ω, on which our Turing machines carry out the actual computations, carries a natural topology, namely the usual product topology of the discrete topology on Σ. Endowed with this topology Σ^ω is called *Cantor space*. A base of the topology is the set of all sets $w\Sigma^\omega := \{p \in \Sigma^\omega \mid p \text{ starts with } w\}$ with $w \in \Sigma^*$. This topology is also induced by the metric d on the Cantor space defined by

$$d(p, q) := \begin{cases} 0 & \text{if } p = q, \\ 2^{-\min\{n \mid p(n) \neq q(n)\}} & \text{otherwise.} \end{cases}$$

Thus, Σ^ω is a separable metric space. In addition, it is complete and (if Σ is finite, which we always assume) compact. A function $F :\subseteq \Sigma^\omega \to \Sigma^\omega$ is continuous if for any finite prefix w of $F(p)$, there exists a finite prefix v of p such that $F(v\Sigma^\omega \cap \mathrm{dom}(F)) \subseteq w\Sigma^\omega$, i.e., such that w is a prefix of $F(q)$ for any $q \in v\Sigma^\omega \cap \mathrm{dom}(F)$. That means, any symbol of $F(p)$ depends only on finitely many symbols in p. An important observation is that any computable function $F :\subseteq \Sigma^\omega \to \Sigma^\omega$ is continuous.

Proposition 6.1. *Any computable function $F :\subseteq \Sigma^\omega \to \Sigma^\omega$ is continuous.*

Proof. A Turing machine that computes F has to compute any finite prefix w of $F(p)$ for any $p \in \mathrm{dom}(F)$ within finitely many steps. But until then it can have read only a finite prefix v of p. Then for any $q \in \mathrm{dom}(F)$ that starts with v as well, the Turing machine will also produce w within the same number of steps. Since the Turing machine has a one-way output tape, this implies that $F(q)$ starts with w as well. Hence, $F(v\Sigma^\omega \cap \mathrm{dom}(F)) \subseteq w\Sigma^\omega$. □

Similarly as in Theorem 4.27 for computable real number functions, the computable functions $F :\subseteq \Sigma^\omega \to \Sigma^\omega$ can be characterized as exactly the effectively continuous functions $F :\subseteq \Sigma^\omega \to \Sigma^\omega$.

This proposition allows us to define the following purely topological weakening of our relative computability notion defined in Definition 4.9.

Definition 6.2. Let X and Y be sets with representations δ_X and δ_Y. A function $f :\subseteq X \to Y$ is called (δ_X, δ_Y)-*continuous* if a continuous (δ_X, δ_Y)-realizer F of f exists.

Often, a function that is not computable with respect to some representations is so for purely topological reasons; that is, it is not continuous with respect to the representations. This is for example the case for some real number functions that one certainly wants to be able to compute but that are not computable with respect to the decimal representation.

Example 6.3. We have seen that multiplication of real numbers by 3 is not (ρ_{10}, ρ_{10})-computable. A short look at the proof shows that this operation is not even (ρ_{10}, ρ_{10})-continuous.

By a similar argument one shows that also addition is not even continuous with respect to the decimal representation. As mentioned, already Turing noticed in a correction [103] to his paper [102] that the decimal (or binary) representation is unsuitable for defining computability of real number functions. Turing suggested to use a different representation instead. We have done that as well; namely, we have used and will be using in the future the Cauchy representation, which is "equivalent" to the representation suggested by Turing. Later we will consider other useful "equivalent" representations of the set of real numbers. What does "equivalent" mean here? As with relative computability and relative continuity, we define a computability-theoretic and a topological version.

Definition 6.4. Let $\delta, \delta' :\subseteq \Sigma^\omega \to X$ be representations of some set X.

1. δ is *reducible* to δ', written $\delta \leq \delta'$, if there is a computable function $F :\subseteq \Sigma^\omega \to \Sigma^\omega$ with $\delta(p) = \delta' F(p)$ for all $p \in \text{dom}(\delta)$.

2. δ is *equivalent* to δ', written $\delta \equiv \delta'$, if $\delta \leq \delta'$ and $\delta' \leq \delta$.

If one asks F only to be continuous instead of computable, one obtains topological variants of these notions: *continuous reducibility*, written $\leq_t$, and *continuous equivalence*, written $\equiv_t$.

Example 6.5. We can define a variant ρ' of the Cauchy representation ρ of the real numbers as follows:

$$\rho'(w_0 \# w_1 \# w_2 \# \ldots) = x :\Longleftrightarrow \lim_{i \to \infty} \nu_\mathbb{Q}(w_i) = x$$
$$\text{and } (\forall i < j) \, |\nu_\mathbb{Q}(w_i) - \nu_\mathbb{Q}(w_j)| \leq 2^{-i}.$$

If $w_0 \# w_1 \# w_2 \# \ldots$ is a ρ'-name of a real number x, then $w_1 \# w_2 \# \ldots$ is a ρ-name of the same number, and vice versa. Thus, ρ and ρ' are equivalent.

The following lemma is easy to prove.

Lemma 6.6. *1. δ is reducible (continuously reducible) to δ' if, and only if, the identity id $\to X$ is (δ, δ')-computable $((\delta, \delta')$-continuous).*

2. (Continuous) reducibility is a reflexive and transitive relation on the representations of a fixed set X.

We have already met several representations of the real numbers: the decimal representation ρ_{10} as well as the analogously defined representations ρ_b to any base $b \geq 2$, the Cauchy representation ρ, and the two representations $\rho_<$ and $\rho_>$. Here is another one.

Example 6.7. The *naive Cauchy representation* ρ_{nC} is defined like the Cauchy representation with the difference that the sequence of rational numbers listed by a ρ_{nC}-name of some real number x has to converge to x, but it does not need to converge rapidly, not even computably.

We have also learned how to combine representations of spaces into a representation of the product space. One can also combine different representations of a space into one representation as follows.

Definition 6.8. Let $\delta, \delta' :\subseteq \Sigma^\omega \to X$ be representations of a set X. Then the representation $\delta \wedge \delta' :\subseteq \Sigma^\omega \to X$ is defined by

$$(\delta \wedge \delta')\langle p, q \rangle = x : \iff \delta(p) = x \text{ and } \delta'(q) = x.$$

That means, a $\delta \wedge \delta'$-name of an element x contains the information about x contained in a δ-name for x and the information about x contained in a δ'-name for x.

In Figure 1 we describe the relationships between some real number representations. We observe that the differences between the representations in this diagram are really of a topological nature: whenever a representation in the diagram is not reducible to another one, it is not even continuously reducible to it.

Of course, equivalent representations induce the same computability notions.

Lemma 6.9. *Let $\gamma, \gamma' :\subseteq \Sigma^\omega \to X$ be representations with $\gamma \leq \gamma'$.*

1. Any γ-computable element of X is also γ'-computable.

2. Any γ-computable sequence in X is also γ'-computable.

3. Any γ'-c.e. subset $U \subseteq X$ is also γ-c.e.

4. If, additionally, $\delta, \delta' :\subseteq \Sigma^\omega \to Y$ are representations with $\delta \leq \delta'$, then any (γ', δ)-computable function $f :\subseteq X \to Y$ is also (γ, δ')-computable. The same holds true with "continuous" instead of "computable."

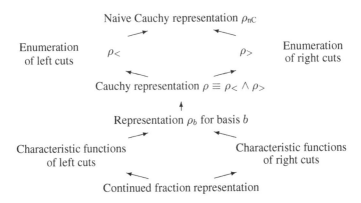

Fig. 1. Each arrow means $\leq$ and $\not\leq_t$.

Proof. By Lemma 6.6 and by Corollary 4.19 for the first statement, by Corollary 4.23 for the second statement, and by Corollary 4.25 for the fourth statement. For the third statement one combines a Turing machine computing a reduction from γ to γ' with a Turing machine proving that U is γ'-c.e. For the topological version of the fourth statement, use that the composition of continuous functions is continuous as well. $\square$

Corollary 6.10. *If* $\gamma, \gamma' :\subseteq \Sigma^\omega \to X$ *are equivalent representations, then*

1. *an element* $x \in X$ *is* γ-*computable if, and only if, it is* γ'-*computable,*

2. *a sequence of elements of* X *is* γ-*computable if, and only if, it is* γ'-*computable,*

3. *a subset* $U \subseteq X$ *is* γ-*c.e. if, and only if, it is* γ'-*c.e.*

If additionally $\delta, \delta' :\subseteq \Sigma^\omega \to Y$ *are equivalent representations, then*

4. *a function* $f :\subseteq X \to Y$ *is* (γ, δ)-*computable if, and only if, it is* (γ', δ')-*computable.*

Also the topological version of the last statement holds true.

It is interesting that representations that are not equivalent may nevertheless define the same class of computable elements: although the four lower representations in Figure 1 are not equivalent to the Cauchy representation ρ, the real numbers that are computable with respect to these representation are exactly the computable real numbers, i.e., the ρ-computable real numbers. The $\rho_<$-computable real numbers are exactly the left-computable real numbers, the $\rho_>$-computable real numbers are exactly the right-computable real numbers, and the ρ_{nC}-computable real numbers are exactly the computably approximable real numbers.

In contrast, already for sequences the notion of computability with respect to a representation is very sensitive with respect to the representation (Mostowski [72]), which is similar to functions (Turing [103]).

That the equivalence class of the Cauchy representation is really the right choice for defining computability over the real numbers can also be justified by the observation that it is the only equivalence class of representations with respect to which certain elementary operations are computable that correspond to the structure of the real numbers; see Hertling [48]. Very soon we will also characterize the continuous equivalence class of the Cauchy representation in a natural way.

One may now ask whether this equivalence class contains some representations with special properties, e.g., an injective or a total representation. But the different nature of the spaces Σ^ω and $\mathbb{R}$ implies that this is not the case.

Proposition 6.11.   *1. No injective representation $\delta :\subseteq \Sigma^\omega \to \mathbb{R}$ is continuously equivalent to ρ.*

  2. No total representation $\delta : \Sigma^\omega \to \mathbb{R}$ is continuously equivalent to ρ (given that Σ is finite).

In the following sections, we will consider computability on spaces "more complicated" than the space of real numbers, e.g., on function spaces or spaces of subsets of Euclidean space. We have already developed a machinery that allows us to introduce approximative notions of computability on such spaces in a natural and simple way: the machinery of representations and computability with respect to representations. But for any given space and any computational task involving that space the question arises of which representation to use. A natural requirement is that the representation should fit with the topology on the space with respect to which we wish to perform approximate computations. A very useful definition in this context is the following one.

Definition 6.12. A representation δ of a topological space X is called *admissible* if δ is continuous and $\delta' \leq_t \delta$ holds true for any continuous representation of X.

Examples 6.13.   1. The Cauchy representation ρ of $\mathbb{R}$ is admissible with respect to the usual Euclidean topology on $\mathbb{R}$.

  2. The representations $\rho_<$ and $\rho_>$ are admissible with respect to the lower and upper Euclidean topology, respectively (these topologies are generated by the intervals (x, ∞) and $(-\infty, x)$, for $x \in \mathbb{R}$, respectively).

We have already observed that any computable real number function is continuous. In fact, one can show that a function $f :\subseteq \mathbb{R} \to \mathbb{R}$ is continuous if, and only if, it is (ρ, ρ)-continuous. This statement carries over to arbitrary admissible representations.

Theorem 6.14 (Kreitz and Weihrauch [58] and Schröder [91]). *Let (X, δ_X) and (Y, δ_Y) be admissibly represented topological spaces. Then a function $f :\subseteq X \to Y$ is (δ_X, δ_Y)-continuous if, and only if, it is sequentially continuous.*

A function $f :\subseteq X \to Y$ between topological spaces is *sequentially continuous* if it preserves convergence of sequences (in the domain of f). For T_0-spaces with countable base, e.g., $\mathbb{R}^n$ with the Euclidean topology, sequential continuity is equivalent to ordinary continuity. But in general it is a slightly weaker notion. The theorem has been generalized by Schröder to so-called weak limit spaces [89]. Schröder's results allow the definition of useful representations even for spaces that are not countably based. This has turned out to be very useful, e.g., for defining computability on spaces of generalized functions; see Zhong and Weihrauch [119]. Another interesting generalization by Schröder are multi-representations; see [90].

We conclude this section by mentioning that there are very useful standard ways for constructing new representations from given ones. We have seen this already for the product of represented sets in Definition 4.11. If (X, δ_X) and (Y, δ_Y) are spaces with representations, then obviously the projection $X \times Y \to X$ is $([\delta_X, \delta_Y], \delta_X)$-computable, and the projection $X \times Y \to Y$ is $([\delta_X, \delta_Y], \delta_Y)$-computable. But one can construct also a representation with natural properties for, e.g., the space of all relatively continuous functions.

Proposition 6.15. *Let (X, δ_X) and (Y, δ_Y) be spaces with representations. There is a representation $[\delta_X \to \delta_Y]$ of the set of all total (δ_X, δ_Y)-continuous functions $f : X \to Y$ such that the following statements are true.*

1. The function

$$(f, x) \mapsto f(x)$$

is $([[\delta_X \to \delta_Y], \delta_X], \delta_Y)$-computable (evaluation)*.*

2. For any represented space (Z, δ_Z), any function $f : Z \times X \to Y$ is $([\delta_Z, \delta_X], \delta_Y)$-computable if, and only if, the function

$$z \mapsto (x \mapsto f(z, x))$$

is $(\delta_Z, [\delta_X \to \delta_Y])$-computable (type conversion)*.*

A suitable representation $[\delta_X \to \delta_Y]$ can be defined via a standard encoding of the continuous functions $h :\subseteq \Sigma^\omega \to \Sigma^\omega$ whose domains are effective G_δ-sets.

Lemma 6.16. *The representation $[\rho^n \to \rho]$ is admissible with respect to the compact-open topology.*

The following result indicates that the class of so-called sequential T_0-spaces with admissible representations form a natural class for computability considerations.

Theorem 6.17 (Schröder [91]). *The category consisting of admissibly represented sequential T_0-spaces as objects and total continuous functions between them as morphisms is cartesian closed.*

7 Solvability of Some Problems Involving Sets and Functions

For many numerical problems the input is not a vector of real numbers, but rather a real number function, e.g., for integration and zero finding. In order to compute with functions as input, we have to feed some descriptions of them into the computer. Therefore, we need a representation. Many function spaces and other spaces on which one wishes to perform computations turn out to be computable metric spaces.

Definition 7.1. A triple (X, d, α) is called a *computable metric space*, if

1. $d : X \times X \to \mathbb{R}$ is a metric on X,

2. $\alpha : \mathbb{N} \to X$ is a sequence such that the set $\{\alpha(n) \mid n \in \mathbb{N}\}$ is dense in X,

3. $d \circ (\alpha \times \alpha) : \mathbb{N}^2 \to \mathbb{R}$ is a computable (double) sequence in $\mathbb{R}$.

Definition 7.2. Let (X, d, α) be a computable metric space. Then we define the *Cauchy representation* $\delta_X :\subseteq \Sigma^\omega \to X$ by

$$\delta_X(01^{n_0+1}01^{n_1+1}01^{n_2+1}...) := \lim_{i \to \infty} \alpha(n_i)$$

for any sequence $(n_i)_{i \in \mathbb{N}}$ such that $(\alpha(n_i))_{i \in \mathbb{N}}$ converges rapidly (and $\delta_X(p)$ is undefined for all other input sequences $p \in \Sigma^\omega$). We say that a sequence $(x_i)_{i \in \mathbb{N}}$ *converges rapidly*, if $d(x_i, \lim_{n \to \infty} x_n) < 2^{-i}$ for all i.

Examples 7.3. 1. $(\mathbb{R}^n, d, \alpha_{\mathbb{R}^n})$ with the *Euclidean metric*

$$d(x, y) := \sqrt{\sum_{i=1}^{n} |x_i - y_i|^2}$$

and the standard numbering $\alpha_{\mathbb{R}^n}$ of $\mathbb{Q}^n$ defined by

$$\alpha_{\mathbb{R}^n}(\langle i_1, \ldots, i_n \rangle) := (\nu_Q(i_1), \ldots, \nu_Q(i_n))$$

is a computable metric space.

2. $(\mathcal{C}[0, 1], d_{\mathcal{C}}, \alpha_{\mathcal{C}})$ with the *supremum metric*

$$d_{\mathcal{C}}(f, g) := ||f - g|| := \sup_{x \in [0,1]} |f(x) - g(x)|$$

and some standard numbering α_C of $\mathbb{Q}[x]$, for instance, $\alpha_C(\langle k, \langle i_0, \ldots, i_k \rangle \rangle) := \sum_{j=0}^{k} \nu_\mathbb{Q}(i_j) \cdot x^j$, is a computable metric space. Note that the Cauchy representation $\delta_{C[0,1]}$ of C allows effective evaluation: the function $(f, x) \mapsto f(x)$ is $([\delta_{C[0,1]}, \rho^{[0,1]}], \rho)$-computable (where $\rho^{[0,1]}$ is the restriction of ρ to names of real numbers in $[0, 1]$). Even more, $\delta_{C[0,1]}$ is equivalent to the representation $[\rho^{[0,1]} \to \rho]$. This implies that the computable points in this space are exactly the computable functions $f : [0, 1] \to \mathbb{R}$. This is in fact an effective version of the Weierstrass Approximation Theorem; compare Pour-El and Richards [80].

3. $(\mathcal{K}(X), d_\mathcal{K}, \alpha_\mathcal{K})$ with the set $\mathcal{K}(X)$ of nonempty compact subsets of a computable metric space (X, d, α), with the *Hausdorff metric*

$$d_\mathcal{K}(A, B) := \max \left\{ \sup_{a \in A} \inf_{b \in B} d(a, b), \sup_{b \in B} \inf_{a \in A} d(a, b) \right\},$$

and with some standard numbering $\alpha_\mathcal{K}$ of the nonempty finite subsets of range(α); e.g., $\alpha_\mathcal{K}(-1 + \sum_{i \in E} 2^i) = \{\alpha(i) \mid i \in E\}$, for any finite nonempty $E \subseteq \mathbb{N}$, is a computable metric space. The computable points are exactly the nonempty computable compact subsets $A \subseteq X$.

Proposition 7.4. *Let X and Y be computable metric spaces with Cauchy representations δ_X and δ_Y, respectively. Then a function $f : X \to Y$ is (δ_X, δ_Y)-continuous if, and only if, it is continuous in the ordinary sense.*

Proof. It is clear that any computable metric space is a countably based T_0-space. We also note that the Cauchy representation of a computable metric space is admissible with respect to the topology induced by the metric. The assertion follows now from Theorem 6.14.   ☐

To give an example, we consider several versions of the problem to find a zero of a continuous function. We remind the reader of the classic Intermediate Value Theorem.

Proposition 7.5. *For every continuous function $f : [0, 1] \to \mathbb{R}$ with $f(0) \cdot f(1) < 0$, there exists some $x \in [0, 1]$ with $f(x) = 0$.*

A number x with $f(x) = 0$ is called *a zero* of f. Is there an effective version of this statement? Under what circumstances can one find a zero of a continuous function $f : [0, 1] \to \mathbb{R}$ with $f(0) \cdot f(1) < 0$? First we are modest and restrict ourselves to functions that have exactly one zero. Then we can compute it.

Theorem 7.6. *The partial function on $C[0, 1]$ that maps any $f \in C[0, 1]$ with $f(0) \cdot f(1) < 0$ and exactly one zero to its zero is $(\delta_{C[0,1]}, \rho)$-computable.*

Proof. Note that in the classical bisection method functions, values are compared with 0. We have seen that such a comparison cannot be performed reliably in finite time. Therefore, we use the *trisection* method, which is a stable variant of the classic bisection method. Assume that a $\delta_{C[0,1]}$-name of a function as above is given. Remember that using this $\delta_{C[0,1]}$-name and given any ρ-name of a real number in $x \in [0, 1]$, we can compute $f(x)$.

We start with $a_0 := 0$ and $d_0 := 1$. Now assume that we have computed two numbers $a_i, d_i \in [0, 1]$ with $d_i - a_i = (2/3)^i$ such that $f(a_i) \cdot f(d_i) < 0$ (this is true for $i = 0$). In parallel we compute the two values

$$f(a_i) \cdot f\left(a_i + \frac{2}{3}(d_i - a_i)\right) \quad \text{and} \quad f\left(a_i + \frac{1}{3}(d_i - a_i)\right) \cdot f(d_i)$$

with higher and higher precision until one of them turns out to be smaller than zero. Note that at least one of them must be smaller than zero! If the first one for which we verify this is the left one, we set $a_{i+1} := a_i$, $d_{i+1} := a_i + \frac{2}{3}(d_i - a_i)$, and if it is the right one, we set $a_{i+1} := a_i + \frac{1}{3}(d_i - a_i)$, $d_{i+1} := d_i$. In this way we obtain a sequence of intervals $[a_i, d_i]$ of length $(2/3)^i$ converging to the unique zero of f. This can easily be turned into a ρ-name of the zero. $\qquad \square$

For functions with more than one zero, things are not that easy.

Theorem 7.7. *There is no continuous (and, hence, no $(\delta_{C[0,1]}, \rho)$-computable) function $Z :\subseteq C[0,1] \to \mathbb{R}$ with $f(Z(f)) = 0$ for all $f \in \{f \in C[0,1] \mid f(0) \cdot f(1) < 0,$ and f has at most three zeros$\}$.*

The proof is quite simple and quite similar to the even slightly easier proof of Theorem 7.10, given below. We omit it.

But for functions with not too many zeros, we can still compute some zero.

Theorem 7.8. *There is a Turing machine which, given a $\delta_{C[0,1]}$-name of some function $f \in C$ with $f(0) \cdot f(1) < 0$ and such that $f^{-1}(\{0\})$ does not contain an interval, computes a (ρ-name of a) zero of f.*

Proof. First, we note that if f is such a function and $a < d$ are numbers in $[0, 1]$ with $f(a) \cdot f(d) < 0$, then for any $n \in \mathbb{N}$ there are rational numbers b and c with $a \leq b < c \leq d$, with $c - b \leq 2^{-n}$, and with $f(b) \cdot f(c) < 0$.

Let us assume that a $\delta_{C[0,1]}$-name of such a function f is given. The algorithm works similarly to the trisection algorithm: it starts with $a_0 := 0$, $d_0 := 1$, and once two rational numbers a_i and d_i with $a_i < d_i$, with $d_i - a_i \leq 2^{-i}$, and with $f(a_i) \cdot f(d_i) < 0$ are found, it searches for two rational numbers a_{i+1}, d_{i+1} with $a_i \leq a_{i+1} < d_{i+1} \leq d_i$, with $d_{i+1} - a_{i+1} \leq 2^{-i-1}$, and with $f(a_{i+1}) \cdot f(d_{i+1}) < 0$. $\qquad \square$

Note that, due to Theorem 7.7, the zero found by a Turing machine as in Theorem 7.8 cannot depend only on the input function f. But it depends continuously on the actual $\delta_{C[0,1]}$-name given to the Turing machine. Hence, for some function f, the algorithm will compute different zeros of f depending on the $\delta_{C[0,1]}$-name of f given to the algorithm. This can be expressed in the terminology of representation-based computability using multi-valued functions. In the following, we call a relation $f \subseteq X \times Y$ a *multi-valued function* and write it as $f :\subseteq X \rightrightarrows Y$.

Definition 7.9. A multi-valued function is called (δ_X, δ_Y)-*computable*, if there exists a computable function $F :\subseteq \Sigma^\omega \to \Sigma^\omega$ such that

$$\delta_Y F(p) \in f\delta_X(p)$$

for all $p \in \mathrm{dom}(f\delta)$. Analogously, f is called (δ_X, δ_Y)-*continuous*, if there exists a continuous F with this property.

For example, a multi-valued function $Z :\subseteq C[0, 1] \rightrightarrows \mathbb{R}$ can be defined by

$$\mathrm{dom}(Z) := \{f \in C[0, 1] \mid f(0) \cdot f(1) < 0\},$$

$$Z(f) := f^{-1}(\{0\}).$$

In this terminology, Theorem 7.8 can be expressed as follows: the restriction of Z to the set

$$\{f \in C[0, 1] \mid f(0) \cdot f(1) < 0 \text{ and } f^{-1}(\{0\}) \text{ does not contain an interval}\}$$

is $(\delta_{C[0,1]}, \rho)$-computable. In the general unrestricted case, one cannot even determine a zero in this weaker multi-valued sense.

Theorem 7.10. *The multi-valued function Z defined above is not $(\delta_{C[0,1]}, \rho)$-continuous.*

Proof. For the sake of a contradiction, let us assume that Z is $(\delta_{C[0,1]}, \rho)$-continuous. Consider the continuous, piecewise linear function $h_y : [0, 1] \to \mathbb{R}$ with breakpoints $(0, -1), (1/3, y), (2/3, y), (1, 1)$, for arbitrary $y \in \mathbb{R}$. Given a ρ-name of y, we can compute a $\delta_{C[0,1]}$-name of h_y. Thus, if Z were $(\delta_{C[0,1]}, \rho)$-continuous, then there would also be a continuous function mapping any ρ-name of any y to a ρ-name of a zero of h_y. But in any neighborhood of a ρ-name of $y = 0$, there are names of numbers $y < 0$ and of numbers $y > 0$. The unique zero of h_y for $y < 0$ is greater than $2/3$, whereas the unique zero of h_y for $y > 0$ is smaller than $1/3$. Hence, given a name of $y = 0$, a continuous function producing a name of a zero of h_y would have to produce a name of a number $\geq 2/3$ and at the same time a name of a number $\leq 1/3$. Impossible! $\square$

All the results so far are *uniform* results in the sense that we asked what one can compute if (a $\delta_{C[0,1]}$-name of) a continuous function is given. The following statement is a *nonuniform* result in the sense that we only ask whether a computable zero exists, not whether we can compute it from some input data.

Theorem 7.11. *Every computable function* $f : [0, 1] \to \mathbb{R}$ *with* $f(0) \cdot f(1) < 0$ *has a computable zero.*

Proof. We distinguish two cases:

1. The set $f^{-1}(\{0\})$ does not contain an interval. Then, given a computable name of f, an algorithm as in Theorem 7.8 will produce a computable name of a zero of f. Hence, this zero is computable.

2. The set $f^{-1}(\{0\})$ contains an interval. Then it contains also a rational, hence, computable number.

$\square$

In general, by Corollary 4.19, if one has a positive uniform result, also the non-uniform positive version holds true. But the previous two results show that the nonuniform version may be true even if the corresponding uniform version is false.

So far we looked at zero-finding for arbitrary continuous functions with a sign change on the unit interval. Now let us look at the more specialized problem to compute the zeros of a polynomial. By the Fundamental Theorem of Algebra, any polynomial

$$p(z) = a_n z^n + a_{n-1} z^{n-1} + \cdots + a_1 z + a_0 \in \mathbb{C}[z]$$

of degree n, i.e., with $a_n \neq 0$, has exactly n complex zeros, when counted with multiplicity. Can one compute them? In the following, for computation purposes, we will always identify $\mathbb{C}$ with $\mathbb{R}^2$ and $\mathbb{C}^n$ with $\mathbb{R}^{2n}$ and we define a representation of $\mathbb{C}$ by $\rho_{\mathbb{C}} := \rho^2$. On first sight, the following well-known statement seems to say no. For $n \geq 1$, set

$$P_n := \{(a_n, \ldots, a_0) \in \mathbb{R}^{n+1} \mid a_n \neq 0 \text{ and } p(z) := a_n z^n + \cdots + a_0 \text{ has}$$

$$\text{exactly } n \text{ pairwise different zeros}\}.$$

Lemma 7.12. *Fix some* $n \geq 2$. *There is no continuous function* $Z : P_n \to \mathbb{C}^n$ *such that* $Z(p)$ *contains all zeros of* p, *for any* $p \in P_n$.

What does it mean? Let us fix some $n \geq 2$. It means that even when restricted to polynomials that have exactly n pairwise different zeros, the problem to find a vector of all zeros contains a discontinuity. By Proposition 7.4 this implies that "there is no Turing machine which, given a $p \in P_n$, would produce a vector depending only on p and containing the zeros of p." But the statement in quotation marks should be read very carefully: in the statement and in Lemma 7.12, we are saying that for each polynomial $p \in P_n$, we wish to obtain a vector of zeros where the vector is determined by the polynomial. But a vector(!) of zeros contains actually more information than what we are interested in: in addition to the information that complex numbers are the zeros, in the vector the zeros are ordered! It turns out that this ordering is the only problem and the reason for the discontinuity and, hence, unsolvability of this version of the problem. If we forget about the ordering, then the problem can be solved, even for arbitrary polynomials.

Theorem 7.13 (Weyl [112]). *There is a Turing machine which, given some* $n \geq 1$ *and a* $\rho_{\mathbb{C}}^{n+1}$*-name of the coefficients of a polynomial*

$$p(z) = a_n z^n + a_{n-1} z^{n-1} + \cdots + a_1 z + a_0 \in \mathbb{C}[z] \text{ with } a_n \neq 0,$$

computes a $\rho_{\mathbb{C}}^n$*-name of a vector in* $\mathbb{C}^n$ *containing all zeros of p with correct multiplicity.*

For any fixed $n \geq 1$, this can be expressed also as follows: the multi-valued function that associates with any polynomial of degree n all permutations of its n zeros (counted with multiplicity) is computable. One could formulate this also without using the notion of a multi-valued function if instead of $\mathbb{C}^n$ as the output space one uses the computable metric space $\mathbb{C}^n/\mathcal{S}_n$, where $\mathcal{S}_n$ is the group of permutations of n elements; see Specker [97]. Note the difference to the negative statement above: the algorithm described in Theorem 7.13 still computes a vector of zeros when given a (standard description of a) polynomial, but the order in which the zeros appear in the vector may depend on the description of the polynomial given to the algorithm. It does not need to be the same for all descriptions of the polynomial (as it would have to be for an algorithm computing a Z as in Lemma 7.12).

By Proposition 4.18, Theorem 7.13 implies that the zeros of any polynomial with computable coefficients are computable again; compare Theorem 3.4. We conclude this discussion with the remark that it is still an open problem to determine the precise "degree of discontinuity" or "topological complexity" of functions Z as in Lemma 7.12; compare Smale [94], and Vassiliev [104]. More on "degrees of discontinuity" can be found in Hertling [44, 45, 46].

Now we wish to consider computational problems that receive not only numbers and functions as input but also subsets of $\mathbb{R}^n$. Therefore, we need representations of subsets.

Definition 7.14. Let (X, d, α) be a computable metric space.

1. We define a representation ϑ of all open subsets of X by

$$\vartheta(p) = U : \Longleftrightarrow p \text{ enumerates a set } A \subseteq \mathbb{N} \times \mathbb{Q}_+ \text{ such that}$$

$$U = \bigcup_{(n,\varepsilon) \in A} B(\alpha(n), \varepsilon).$$

2. We define three representations of the set $\mathcal{A}(X) := \{A \subseteq X \mid A \text{ is closed}\}$ by

   a) $\psi_>(p) = A : \Longleftrightarrow \vartheta(p) = X \setminus A,$

b) $\psi_<(p) = A : \iff p$ enumerates all pairs $(n, \varepsilon) \in \mathbb{N} \times \mathbb{Q}_+$ with
$A \cap B(\alpha(n), \varepsilon) \neq \emptyset$,

c) $\psi_= := \psi_< \wedge \psi_>$.

Note that the ϑ-computable sets are exactly the c.e. open sets, that the $\psi_>$-computable elements of $\mathcal{A}(\mathbb{R}^n)$ are exactly the co-c.e. closed subsets of $\mathbb{R}^n$, that the $\psi_<$-computable elements of $\mathcal{A}(\mathbb{R}^n)$ are exactly the c.e. closed subsets of $\mathbb{R}^n$, and that the $\psi_=$-computable elements of $\mathcal{A}(\mathbb{R}^n)$ are exactly the computable closed subsets of $\mathbb{R}^n$. The characterizations of Propositions 5.9 and 5.12 can be generalized to the case of (certain) computable metric spaces, and they even hold uniformly (Brattka and Presser [21]). Proposition 5.17 can be formulated uniformly, which gives the first two of the statements in the following lemma. The counterexample in Example 5.18 implies the noncomputability part of the third statement.

Lemma 7.15.   *1. The union operator* $\cup : \mathcal{A}(X) \times \mathcal{A}(X) \rightarrow \mathcal{A}(X)$ *is computable with respect to* $\psi_<, \psi_>,$ *and* $\psi_=$*; i.e., for any* $\psi \in \{\psi_<, \psi_>, \psi_=\}$*, it is* $([\psi, \psi], \psi)$*-computable.*

2. *The intersection operator* $\cap : \mathcal{A}(X) \times \mathcal{A}(X) \rightarrow \mathcal{A}(X)$ *is computable with respect to* $\psi_>$.

3. *But the intersection operator* $\cap$ *is not computable with respect to* $\psi_<$ *or* $\psi_=$. *It is not even* $([\psi_=, \psi_=], \psi_<)$*-continuous.*

The representations $\psi_>, \psi_<, \psi_=$ are admissible with respect to natural topologies on $\mathcal{A}(X)$.

Lemma 7.16. *The representation* $\psi_<$ *is admissible with respect to the lower Fell topology, the representation* $\psi_>$ *is admissible with respect to the upper Fell topology, and the representation* $\psi_=$ *is admissible with respect to the Fell topology (which is the topology generated by the lower and the upper Fell topology).*

We use these representations in order to formulate an effective version of the Riemann Mapping Theorem. For completeness sake, we remind the reader of its statement. It says that for a subset U of the complex plane $\mathbb{C}$, the following two conditions are equivalent:

1. U is a nonempty, proper, open, connected, and simply connected subset of $\mathbb{C}$.

2. There exists a conformal function $f : D \rightarrow \mathbb{C}$ with $f(D) = U$.

Here $D := \{z \in \mathbb{C} : ||z|| < 1\}$ is the open unit disk in the complex plane. We identify $\mathbb{C}$ with $\mathbb{R}^2$ again, and we use the representation $\rho_{\mathbb{C}} = \rho^2$ for $\mathbb{C}$.

Theorem 7.17 (Hertling [47]).

1. Given a $[\rho_{\mathbb{C}} \rightarrow \rho_{\mathbb{C}}]$*-name of a conformal mapping defined on* D*, one can compute a* ϑ*-name of* $U := f(D)$ *and a* $\psi_<$*-name of the boundary of* U.

2. *Given a ϑ-name of some nonempty, proper, open, connected, and simply connected subset U of $\mathbb{C}$, a $\psi_<$-name of its boundary and a $\rho_{\mathbb{C}}$-name of a point z_0 in U, one can compute a $[\rho_{\mathbb{C}} \to \rho_{\mathbb{C}}]$-name of the uniquely determined conformal mapping $f : D \to \mathbb{C}$ with $f(D) = U$, with $f(0) = z_0$, and with $f'(0) > 0$.*

By looking at computable objects (see Corollary 4.19), one arrives at the following nonuniform version.

Corollary 7.18 (Hertling [47]). *For $U \subseteq \mathbb{C}$ the following are equivalent:*

1. *There exists a computable holomorphic bijection $f : D \to U$.*

2. *U is a nonempty, proper, open, connected, simply connected subset of $\mathbb{C}$ that is c.e. open and has a c.e. closed boundary ∂U.*

Next, we formulate an effective version of the Baire Category Theorem. The classical result can be stated as follows: if X is a complete metric space and $(A_n)_{n\in\mathbb{N}}$ a sequence of closed nowhere dense subsets of X, then $X \setminus \bigcup_{n\in\mathbb{N}} A_n$ is nonempty; i.e., it contains at least one point. In order to formulate the uniform version, we need a representation of a sequence of closed sets.

Definition 7.19. Let (X, δ) be a represented set. Then by

$$\delta^\omega(p) = (x_n)_{n\in\mathbb{N}} : \iff \quad \text{for each } n \in \mathbb{N}, \text{ the sequence } (i \mapsto p(\langle n, i \rangle))$$
$$\text{is a } \delta\text{-name of } x_n,$$

one defines a natural representation of the space of all sequences $(x_n)_{n\in\mathbb{N}}$ of elements in X.

Theorem 7.20 (Brattka [15]). *Let X be a complete computable metric space. The multi-valued function that associates with any sequence $(A_n)_{n\in\mathbb{N}}$ of closed nowhere dense subsets $A_n \subseteq X$ all sequences $(x_i)_{i\in\mathbb{N}}$ that are dense in $X \setminus \bigcup_{n=0}^\infty A_n$ is $(\psi_>^\omega, \delta_X^\omega)$-computable.*

Corollary 7.21 (Yasugi et al. [114]). *Let X be a complete computable metric space. For any computable sequence $(A_n)_{n\in\mathbb{N}}$ of co-c.e. closed nowhere dense subsets A_n of X, there exists a computable sequence $(x_i)_{i\in\mathbb{N}}$ that is dense in $X \setminus \bigcup_{n=0}^\infty A_n$.*

If X has no isolated points, then all singletons $\{x\}$ are nowhere dense and we obtain the following corollary, which is a generalization of Proposition 3.9.

Corollary 7.22 (Brattka [15]). *If X is a complete computable metric space without isolated points, then there is no computable sequence $(x_n)_{n\in\mathbb{N}}$ such that the set $\{x_n \mid n \in \mathbb{N}\}$ contains all computable points of X.*

The following statement can be proved directly, but it can also be deduced from Corollary 7.21.

Corollary 7.23 (Brattka [15]). *There exists a computable but nowhere differentiable function* $f : [0, 1] \to \mathbb{R}$.

With the Intermediate Value Theorem we have seen a result that has a nonuniform computable version but no general uniform computable version. The Riemann Mapping Theorem and the Baire Category Theorem both admit fully uniform computable versions. The Brouwer Fixed Point Theorem is an example of a theorem that does not even admit a nonuniform computable version. It states that any continuous function $f : [0, 1]^2 \to [0, 1]^2$ possesses a fixed point. This statement does not hold true computationally: there is the following surprising negative result.

Theorem 7.24 (Orevkov [78], Baigger [4]). *There exists a computable function* $f : [0, 1]^2 \to [0, 1]^2$ *without a computable fixed point.*

We conclude this section with another interesting result in the context of the Fixed Point Theorem. Let us call a subset $A \subseteq [0, 1]^n$ *fixable* if there exists a computable function $f : [0, 1]^n \to [0, 1]^n$ such that $f(A) = A$.

Theorem 7.25 (Miller [67]). *Let* $A \subseteq [0, 1]^n$ *be a co-c.e. closed set. Then the following are equivalent:*

1. *A is fixable,*

2. *A contains a nonempty co-c.e. closed connected component,*

3. *A contains a nonempty co-c.e. closed connected subset,*

4. $f(A)$ *contains a computable real for all computable* $f : [0, 1]^n \to \mathbb{R}$.

8 Computability of Linear Operators

Many problems in numerical analysis can be expressed by linear operators over normed spaces or even Banach spaces.

We start our selection of observations concerning computability of linear operators with several simple facts about linear (or affine) mappings on Euclidean spaces. The first lemma is pretty obvious.

Lemma 8.1. *Consider a matrix* $A \in \mathbb{R}^{m \times n}$. *The linear function mapping any vector* $x \in \mathbb{R}^n$ *to* $A \cdot x \in \mathbb{R}^m$ *is computable if, and only if, A is computable, i.e., if, and only if, all coefficients of A are computable real numbers.*

If A is invertible, then the range of this linear function is equal to $\mathbb{R}^m$. Otherwise, the problem to decide whether a given vector $b \in \mathbb{R}^m$ is in the range of this function is discontinuous and therefore unsolvable. But even if one knows in advance that b is in the range, computing a preimage is impossible in general, due to a well-known discontinuity.

Proposition 8.2. *The multi-valued function that associates with a matrix $A \in \mathbb{R}^{m \times n}$ and a vector $b \in \text{range}(x \mapsto Ax) \subseteq \mathbb{R}^m$ all solutions $x \in \mathbb{R}^n$ of the equation $A \cdot x = b$ is not $(\rho^{m(n+1)}, \rho^n)$-continuous. This implies that no Turing machine, given (a $\rho^{m \cdot n}$-name of) A and (a ρ^m-name of) b, would compute (a ρ^n-name of) a solution x to $A \cdot x = b$.*

This changes if one knows the dimension of the solution space.

Theorem 8.3 (Ziegler and Brattka [120]). *Let $A \in \mathbb{R}^{m \times n}$ be a matrix and $b \in \mathbb{R}^m$ a vector, and let $L := \{x \in \mathbb{R}^n \mid A \cdot x = b\}$ be the affine solution space of the corresponding linear equation. If A and b are computable, then L is a computable closed set. Given names with respect to ρ of some A and b such that L is not empty and given the dimension d of L, we can even compute a ψ-name of L. This implies that we can computably find a solution $x \in L$ and in case $d > 0$ also a basis of the homogeneous solution space $\{x \in \mathbb{R}^n \mid A \cdot x = 0\}$.*

In fact, instead of a $\nu_\mathbb{N}$-name of the dimension, a $\rho_>$-name of the dimension is sufficient.

Now let us turn to linear operators over normed spaces. Many normed spaces used in the practice of numerical computation are computable according to the following definitions. In the following, $\mathbb{F}$ is either $\mathbb{R}$ or $\mathbb{C}$, and for computation purposes, we identify $\mathbb{C}$ with $\mathbb{R}^2$, and use $\alpha_\mathbb{C} := \alpha_{\mathbb{R}^2}$.

Definition 8.4. $(X, \| \ \|, e)$ is called a *computable normed space*, if

1. X is a linear space,

2. $\| \ \| \to \mathbb{R}$ is a norm on X,

3. the linear span of the sequence $e : \mathbb{N} \to X$ is dense in X,

4. (X, d, α_e) with $d(x, y) := \|x - y\|$ and $\alpha_e \langle k, \langle n_0, \ldots, n_k \rangle \rangle := \sum_{i=0}^{k} \alpha_\mathbb{F}(n_i) e_i$ is a computable metric space.

Note that in any computable normed space, the point 0 is a computable point and that vector space addition and scalar multiplication are automatically computable operations.

In the following, whenever we say "computable" in connection with a computable normed space X, we always mean computable with respect to the Cauchy representation δ_X of X.

Theorem 8.5. *Let X, Y be computable normed spaces, let $T : X \to Y$ be a linear operator, and let $(s_n)_{n \in \mathbb{N}}$ be a computable sequence in X whose linear span is dense in X. Then the following are equivalent:*

1. $T : X \to Y$ is computable,

2. $(T(s_n))_{n\in\mathbb{N}}$ *is a computable sequence and* T *is bounded,*

3. T *maps computable sequences to computable sequences and is bounded,*

4. $\mathrm{graph}(T)$ *is a computable closed subset of* $X \times Y$ *and* T *is bounded,*

5. $\mathrm{graph}(T)$ *is an c.e. closed subset of* $X \times Y$ *and* T *is bounded.*

In case X *and* Y *are even Banach spaces, one can omit boundedness in the last two conditions, because it follows from the Closed Graph Theorem.*

Examples 8.6. We use the norm $||f|| := \sup_{x\in[0,1]} |f(x)|$ on $C[0,1]$ and the norm $||f||_1 := ||f|| + ||f'||$ on $C^1[0,1]$.

1. The integration operator

$$I : C[0,1] \to C[0,1], f \mapsto \left(x \mapsto \int_0^x f(t)\,dt \right)$$

is a linear computable operator.

2. The differentiation operator

$$d :\subseteq C[0,1] \to C[0,1], f \mapsto f',$$

defined for continuously differentiable functions on $[0,1]$, is a linear unbounded operator with a closed graph.

3. The differentiation operator

$$D : C^1[0,1] \to C[0,1], f \mapsto f'$$

is a computable linear operator. This is not surprising because a Cauchy name of a function $f \in C^1[0,1]$ contains already the information needed for computing f'.

4. The differentiation operator

$$D :\subseteq C^2[0,1] \to C[0,1], f \mapsto f'$$

is computable if for the elements of $C^2[0,1]$ one uses the representation δ defined by

$$\delta\langle p, q\rangle = f : \iff \delta_{C[0,1]}(p) = f \text{ and } \rho(q) > ||f''||.$$

Corollary 8.7. *If a computable function* $f \in C[0,1]$ *has a continuous second derivative, then* f' *is computable as well.*

We had remarked at the end of Section 4 that a noncomputable function may nevertheless be computably invariant; i.e., map computable elements to computable elements, and if it is not computably invariant, it can be a challenge to construct a computable element of the domain that is mapped to a noncomputable element. Pour-El and Richards [80] have shown a general result that shows that for linear operators the situation is simpler.

Theorem 8.8 (Pour-El and Richards [80]). *Let X, Y be computable normed spaces, let $T : X \to Y$ be a linear operator with closed graph, and let $(s_n)_{n \in \mathbb{N}}$ be a computable sequence in X whose linear span is dense in X and such that $(T(s_n))_{n \in \mathbb{N}}$ is a computable sequence in Y. Then T is unbounded if, and only if, there exists a computable $x \in X$ such that $T(x)$ is noncomputable.*

We have seen that there exists a computable nowhere differentiable function. The result by Pour-El and Richards, applied to the differentiation operator in Example 8.6.2, gives the following observation.

Corollary 8.9 (Myhill [76]). *There exists a computable and continuously differentiable function $f : [0,1] \to \mathbb{R}$ whose derivative $f' : [0,1] \to \mathbb{R}$ is not computable.*

A striking application of the theorem by Pour-El and Richards is the following statement concerning the three-dimensional wave equation.

Theorem 8.10 (Pour-El and Richards [79, 80]). *There exists a computable function $f : \mathbb{R}^3 \to \mathbb{R}$ such that the unique solution u of the three-dimensional wave equation*

$$\begin{cases} u_{tt} = \Delta u \\ u(0, x) = f, \ u_t(0, x) = 0, \ t \in \mathbb{R}, x \in \mathbb{R}^3 \end{cases}$$

has the following properties: the wave $x \mapsto u(0, x) = f(x)$ at time 0 is computable, but the wave $x \mapsto u(1, x)$ at time 1 is not computable.

This result has led to a number of misinterpretations because it appears as if the wave equation produces some noncomputability that emerges out of nothing. However, a careful analysis of the result shows that the noncomputability in the counterexample is already hidden in the derivative of the initial condition, and hence, the example can be considered as yet another instance of Corollary 8.9. The solvability of the wave equation was analyzed further by Washihara [105] and by Weihrauch and Zhong [111].

Theorem 8.11 (Weihrauch and Zhong [111]). *The solution operator of the wave equation $S : C^k(\mathbb{R}^3) \to C^{k-1}(\mathbb{R}^4), f \mapsto u$ is computable.*

Remark 8.12. Weihrauch and Zhong [111] proved also that the operator of wave propagation is computable without any loss of the degree of differentiability if physically appropriate Sobolev spaces are used.

The next result deals with the Hahn–Banach Theorem.

Theorem 8.13 (Metakides and Nerode [64]). *Let X be a finite-dimensional computable Banach space with some c.e. closed linear subspace $Y \subseteq X$. For any computable linear functional $f : Y \to \mathbb{R}$ with computable norm $||f||$, there exists a computable linear extension $g : X \to \mathbb{R}$ with $||g|| = ||f||$.*

It follows from results of Metakides et al. [65] that this result cannot be extended to infinite-dimensional spaces. However, it is worth mentioning that for normed spaces X with strictly convex dual X', the norm-preserving extension of functionals is always unique, and in the unique case, the extension can always be computed (even uniformly, that is given f one can compute a suitable g); see Brattka [18]. This class of spaces with strictly convex dual includes, for instance, all Hilbert spaces.

Many results from functional analysis can be analyzed with respect to their computational content; see, e.g., Pour-El and Richards [80], Brattka and Dillhage [19], and Brattka and Yoshikawa [22].

Often, a computational problem can be formulated as the problem to compute the inverse of a given linear operator. This poses the question under which circumstances the inverse of a linear operator can be computed. Let X, Y be Banach spaces, and let $T : X \to Y$ be a linear operator. By Banach's Inverse Mapping Theorem, any linear, bijective, and bounded T has a bounded inverse T^{-1}. If we can compute T, can we compute T^{-1} as well? Interestingly, here we have a positive nonuniform answer but a negative uniform answer.

Theorem 8.14 (Brattka [16]).

1. *Let X, Y be Banach spaces, and let $T : X \to Y$ be a linear operator. If T is bijective and computable, then T^{-1} is computable as well.*

2. *There exist Banach spaces X and Y such that the function that maps any bijective linear operator $T : X \to Y$ to its inverse $T^{-1} : Y \to X$ is not $([\delta_X \to \delta_Y], [\delta_Y \to \delta_X])$-continuous.*

We conclude this section with an interesting application of the first, positive, statement of the previous theorem to the theory of differential equations. Although the first statement of the previous theorem is nonuniform, the following application is a uniform statement. Here we represent $C^n[0, 1]$ by the Cauchy representation for the norm $||f||_n := \sum_{i=0}^n ||f^{(i)}||$ that includes information on all the derivatives.

Theorem 8.15 (Brattka [16]). *Let $n \geq 1$, and let $f_0, \ldots, f_n : [0, 1] \to \mathbb{R}$ be computable with $f_n \neq 0$. The solution operator $L : C[0, 1] \times \mathbb{R}^n \to C^n[0, 1]$ that maps each $(y, a_0, \ldots, a_{n-1}) \in C[0, 1] \times \mathbb{R}^n$ to the unique $x \in C^n[0, 1]$ such that*

$$\sum_{i=0}^n f_i(t)x^{(i)}(t) = y(t) \text{ with } x^{(j)}(0) = a_j \text{ for } j = 0, \ldots, n-1,$$

is computable.

Proof. The following operator is linear and computable:

$$L^{-1} : C^n[0, 1] \to C[0, 1] \times \mathbb{R}^n, x \mapsto \left(\sum_{i=0}^n f_i x^{(i)}, x^{(0)}(0), \ldots, x^{(n-1)}(0)\right).$$

By the computable Inverse Mapping Theorem it follows that L is computable too.

$\square$

9 Degrees of Unsolvability

In this section we describe several approaches for classifying unsolvable computational problems over the real numbers. First, remember that computability of real number functions can be expressed as effective continuity. We generalize this notion using the Borel hierarchy. In the second approach one looks only at computable input and uses the arithmetical hierarchy for the classification. Finally, we ask about Turing degrees of real number functions and look at a reducibility relation on functions on computable metric spaces, which is introduced in a natural way using Cauchy representations.

In Example 8.6 we have seen that the differentiation operator

$$d :\subseteq \mathcal{C}[0,1] \to \mathcal{C}[0,1], f \mapsto f',$$

defined for continuously differentiable functions on $[0,1]$, is a linear unbounded operator and therefore not computable. Can one characterize the degree of noncomputability of differentiation more precisely? In order to do this we remind the reader of the (finite beginning of the) Borel hierarchy. In the following, X and Y are computable metric spaces unless stated otherwise. Let $\Sigma_1^0(X)$ be the set of all open subsets of X, and for $k \geq 1$,

$\Pi_k^0(X)$ be the set of all complements of sets in $\Sigma_k^0(X)$,

$\Sigma_{k+1}^0(X)$ be the set of all countable unions of sets in $\Pi_k^0(X)$.

This hierarchy can be extended to transfinite levels. Here, we will consider only finite levels. One can define representations of the finite levels of the Borel hierarchy in a straightforward manner.

Definition 9.1. Let (X, d, α) be a computable metric space. For $k \geq 1$, we define representations $\delta_{\Sigma_k^0(X)}$ of $\Sigma_k^0(X)$ and $\delta_{\Pi_k^0(X)}$ of $\Pi_k^0(X)$ as follows:

$$\delta_{\Sigma_1^0(X)}(p) := \bigcup_{(i,\varepsilon)\in\mathbb{N}\times\mathbb{Q}_+ \text{ enumerated by } p} B(\alpha(i), \varepsilon),$$

$$\delta_{\Sigma_{k+1}^0(X)}(p) := \bigcup_{i=0}^{\infty} \delta_{\Pi_k^0(X)}^{\omega}(p)(i),$$

$$\delta_{\Pi_k^0(X)}(p) := X \setminus \delta_{\Sigma_k^0(X)}(p).$$

The representation $\delta_{\Sigma_1^0(X)}$ is identical to the representation ϑ of open subsets as defined earlier.

Definition 9.2. Let X, Y be computable metric spaces. A function $f :\subseteq X \to Y$ is Σ_k^0-*measurable* if for any $U \in \Sigma_1^0(Y)$ there is a set $V \in \Sigma_k^0(X)$ with $f^{-1}(U) = V \cap \mathrm{dom}(f)$. It is called *effectively* Σ_k^0-*measurable* or Σ_k^0-*computable*, if a Turing machine, given a $\delta_{\Sigma_1^0(X)}$-name of such a U, computes a $\delta_{\Sigma_k^0(X)}$-name of such a V.

Remark 9.3. 1. A function $f :\subseteq X \to Y$ is Σ_1^0-measurable if, and only if, it is continuous.

2. In Theorem 4.27 we have seen that the computable real number functions are exactly the real number functions that are effectively continuous in a certain sense. Using this, one can easily show that they are exactly the Σ_1^0-computable real number functions. And this can easily be generalized to functions between computable metric spaces: a function $f :\subseteq X \to Y$ is computable (with respect to the Cauchy representations) if, and only if, it is Σ_1^0-computable.

3. The total Σ_k^0-computable functions $f : X \to Y$ are in fact the Σ_k^0-recursive functions in the sense of Moschovakis [71]; see Brattka [17, page 24].

The following result is an extension of Theorem 6.14.

Theorem 9.4 (Brattka [17]). *Let X, Y be computable metric spaces and k be a positive integer. Then a total function $f : X \to Y$ is Σ_k^0-measurable (Σ_k^0-computable) if, and only if, there is a Σ_k^0-measurable (Σ_k^0-computable) function $F :\subseteq \Sigma^\omega \to \Sigma^\omega$ with $f\delta_X(p) = \delta_Y F(p)$ for all $p \in \mathrm{dom}(f\delta_X)$.*

Because of this result, it is reasonable to extend the definition of Σ_k^0-computability to arbitrary represented spaces as follows.

Definition 9.5. Let X, Y be sets with representations δ_X, δ_Y, respectively. Then a function $f :\subseteq X \to Y$ is called Σ_k^0-computable (with respect to (δ_X, δ_Y)) if, and only if, it has a Σ_k^0-computable realizer.

Due to the previous theorem, this is a conservative extension of the concept of Σ_k^0-computability. With respect to a suitable reducibility relation, for each $k \geq 1$, there are complete problems in the class of Σ_k^0-computable functions.

Definition 9.6. Consider two functions $f :\subseteq \Sigma^\omega \to \Sigma^\omega$ and $g :\subseteq \Sigma^\omega \to \Sigma^\omega$. We say that f is *computably reducible* to g, written $f \leq_c g$, if there are computable functions $A :\subseteq \Sigma^\omega \times \Sigma^\omega \to \Sigma^\omega$ and $B :\subseteq \Sigma^\omega \to \Sigma^\omega$ such that

$$f(p) = A(p, g(B(p)))$$

for all $p \in \mathrm{dom}(f)$.

A topological version of this reducibility relation, also on more general topological spaces, has been studied by Hirsch [51], Weihrauch [108], and Hertling [44, 46].

Proposition 9.7 (Brattka [17]). *For $k \geq 0$, there exist functions $C_k :\subseteq \Sigma^\omega \to \Sigma^\omega$ such that any $F :\subseteq \Sigma^\omega \to \Sigma^\omega$ is Σ_{k+1}^0-computable if, and only if, F is computably reducible to C_k.*

We wish to compare not only functions on Σ^ω but on more general spaces. Transferring Definition 9.6 in a straightforward manner to functions $f :\subseteq X \to Y$ and $g :\subseteq U \to V$ does not look very promising because then it becomes important whether the spaces are connected. For example, in case $X = \mathbb{R}$, and $U = \Sigma^\omega$, there is no nontrivial continuous function $B : X \to U$. Instead, we remember that computations are performed on names.

Definition 9.8. Let X, Y, U, V be sets with representations $\delta_X, \delta_Y, \delta_U, \delta_V$, respectively, and let $f :\subseteq X \to Y$ and $g :\subseteq U \to V$ be functions. We say that f is *computably realizer reducible* to g, written $f \leq_c g$, if there are computable functions $A :\subseteq \Sigma^\omega \times \Sigma^\omega \to \Sigma^\omega$ and $B :\subseteq \Sigma^\omega \to \Sigma^\omega$ such that for any (δ_U, δ_V)-realizer G of g, the function

$$p \mapsto A(p, G(B(p)))$$

is a (δ_X, δ_Y)-realizer of f.

Definition 9.9. Let X, Y be sets with representations δ_X, δ_Y, respectively. We say that a function $f :\subseteq X \to Y$ is Σ_k^0-*complete* (with respect to (δ_X, δ_Y)), if it is Σ_k^0-computable with respect to (δ_X, δ_Y) and any other Σ_k^0-computable function is computably realizer reducible to it.

Note that the functions C_k whose existence is stated in Proposition 9.7 are Σ_{k+1}^0-complete. A linear operator satisfying the assumptions in Theorem 8.8 by Pour-El and Richards is at least as difficult to compute as C_1.

Theorem 9.10 (Brattka [14]). *Let X, Y be computable Banach spaces, and let $f :\subseteq X \to Y$ be a linear and unbounded operator with closed graph. Let $(s_n)_{n \in \mathbb{N}}$ be a computable sequence in $\mathrm{dom}(f)$ whose linear span is dense in X (note that this makes $\mathrm{dom}(f)$ a computable metric space), and let $(f(s_n))_{n \in \mathbb{N}}$ be a computable sequence in Y. Then $C_1 \leq_c f$.*

This can be used to prove that certain operators such as differentiation are Σ_2^0-complete. For nonlinear maps, other techniques are required (see for instance Brattka [14] and Gherardi [40]).

Example 9.11. The following functions are examples of complete functions:

1. The differentiation $d :\subseteq \mathcal{C}[0, 1] \to \mathcal{C}[0, 1], f \mapsto f'$ is Σ_2^0-complete.

2. The limit map $\lim :\subseteq \mathbb{R}^{\mathbb{N}} \to \mathbb{R}, (x_n)_{n \in \mathbb{N}} \mapsto \lim_{n \to \infty} x_n$ is Σ_2^0-complete.

3. The boundary operation $\partial : \mathcal{A}(\mathbb{R}^n) \to \mathcal{A}(\mathbb{R}^n), A \mapsto \partial A$ is Σ_2^0-complete.

4. The derived set operator $D : \mathcal{A}(\mathbb{R}^n) \to \mathcal{A}(\mathbb{R}^n), A \mapsto A'$ is Σ_3^0-complete.

Here we assume that $\mathcal{A}(\mathbb{R}^n)$ is represented by ψ.

It is worth mentioning that any Σ^0_{k+1}-complete map has the property that it maps some computable input to some Δ^0_{k+1}-computable output that is not Δ^0_k-computable (Brattka [17]). Here the light face classes Δ^0_{k+1} refer to the arithmetical hierarchy. Applied to the differentiation operator d, it gives us yet another proof of Corollary 8.9. Applied to the limit map, it yields a computably approximable real number that is not computable and applied to the boundary operator it yields a computable closed set whose boundary is not a computable closed set.

This brings us to the second approach for classifying unsolvable problems that we wish to describe. It uses the arithmetical hierarchy. Cenzer and Remmel [32] consider operators whose input are real number functions. They use a Gödel numbering of computable real number functions and characterize the set of Gödel numbers of computable input functions that are mapped by the operator to computable output. If the operator is computable, this is of course a trivial question because we know that computable operators are computably invariant; i.e., they map every computable input to computable output. The question becomes interesting for noncomputable operators. We define a Gödel numbering of the total computable functions $f : \mathbb{R}^n \to \mathbb{R}$, for each $n \geq 1$. In the following definition, φ denotes a total standard numbering of all computable functions $F :\subseteq \mathbb{N} \to \mathbb{N}$; i.e., $(\varphi_i)_{i \in \mathbb{N}}$ is a list containing exactly all computable functions $F :\subseteq \mathbb{N} \to \mathbb{N}$ such that φ satisfies the utm property and the smn property; compare any book on computability theory, e.g., Rogers [84] and Weihrauch [106].

Definition 9.12. Fix some $n \geq 1$. We write $F^n_e = f$ if the c.e. set $S := \mathrm{dom}(\varphi_e)$ describes a total effectively continuous function $f : \mathbb{R}^n \to \mathbb{R}$ as in Definition 4.26. Furthermore, $I^n := \mathrm{dom}(F^n)$.

By Theorem 4.27, F^n is a numbering of all total computable functions $f : \mathbb{R}^n \to \mathbb{R}$. Actually, Cenzer and Remmel [32] used a slightly different numbering. But the two numberings are equivalent, with the consequence that the following theorem holds also for our numbering F^n. The following theorem lists several results by Cenzer and Remmel [32] related to computational problems that we considered earlier.

Theorem 9.13 (Cenzer and Remmel [32]). *The following sets are Σ^0_3-complete.*

1. *The set of all $e \in I^1$ such that F^1_e is differentiable and the derivative is computable.*

2. *The set of all $e \in I^2$ such that the differential equation $y'(x) = F^2_e(x, y(x))$ with initial condition $y(0) = 0$ has a computable solution on the interval $[-\delta, \delta]$, for some $\delta > 0$.*

3. *The set of all $e \in I^3$ such that the wave equation*

$$u_{xx} + u_{yy} + u_{zz} - u_{tt} = 0$$

with initial conditions $u_t(x, y, z, 0) = 0$ and $u(x, y, z, 0) = F^3_i(x, y, z)$ has a computable solution.

We note that Cenzer and Remmel also found natural Π_3^0-complete problems and Σ_4^0-complete problems over the real numbers [32].

Finally, we describe work by Miller [68] concerning degrees of uncomputability of real number functions and functions over computable metric spaces. A large part of classic computability theory deals with Turing reducibility and Turing degrees. Thus, in the context of computable analysis, it is natural to try to characterize the degree of uncomputability of a real number function by asking for its Turing degree. Let X be a computable metric space with Cauchy representation δ_X. It is natural to define the *Turing degree* $\deg_T(x)$ of an element $x \in X$ by

$$\deg_T(x) := \min\{\deg_T(p) \mid \delta_X(p) = x\}.$$

Indeed, for $X := \mathbb{R}$ and any real number $x \in \mathbb{R}$, the Turing degree $\deg_T(x)$ is well defined and coincides with the Turing degree of the fractional part of the binary expansion of x. But this approach does not work well anymore for real number functions.

Theorem 9.14 (Miller [68]). *Consider $X := \mathcal{C}[0, 1]$. There exists some $f \in X$ such that $\deg_T(f)$ is not well defined; i.e., there is no Cauchy name for f with minimal Turing degree.*

So, another approach is needed. Turing reducibility is defined for subsets of $\mathbb{N}$ or for their characteristic sequences: a sequence $p \in \{0, 1\}^\omega$ is Turing reducible to a sequence $q \in \{0, 1\}^\omega$ if there is a computable function $F :\subseteq \Sigma^\omega \to \Sigma^\omega$ with $F(q) = p$. By using Cauchy representations, one can translate this reducibility to elements of computable metric spaces.

Definition 9.15. Let X and Y be computable metric spaces. An element $x \in X$ is *representation reducible* to an element $y \in Y$ if there is a computable (i.e., (δ_X, δ_Y)-computable) function $f :\subseteq Y \to X$ with $f(y) = x$.

This relation is clearly reflexive and transitive. Let us call the equivalence classes of elements of computable metric spaces under this reducibility relation *continuous degrees*.

Proposition 9.16 (Miller [68]). *Every continuous degree contains a real-analytic function.*

Theorem 9.17 (Miller [68]). *There is a natural, nontrivial embedding from the Turing degrees into the continuous degrees and a natural, nontrivial embedding from the continuous degrees into the enumeration degrees.*

10 Computational Complexity of Real Numbers and Real Number Functions

In the previous section we discussed unsolvable computational problems over the real numbers. In this and the following section, we turn in the other direction: we analyze the computational complexity of solvable problems. The two most important complexity measures in classical complexity theory are time complexity and space complexity. We will restrict ourselves to time complexity. We wish to classify computational problems over the real numbers according to the time one needs in order to solve them. We measure the time by counting the steps of a Turing machine. In classic complexity theory, discrete problems are considered, and the time complexity is usually analyzed in dependence of the input size of the concrete instance of the computational problem. Often, this input size corresponds quite well to the input dimension of the concrete instance of the problem. But for continuous problems whose output is real-valued, another parameter turns out to be at least as important in the practice of numerical analysis: the desired precision of the output value. In fact, for a single real number, this is the only parameter on which the time can depend. And for any fixed real number function, the dimension is fixed as well. This has the consequence that the time complexity of computable real numbers and computable real number functions is usually analyzed as a function of the desired output precision. In the following definition, we use a standard binary notation $\nu_{\mathbb{D}}$ of the set

$$\mathbb{D} := \{z/2^n \mid z \in \mathbb{Z}, \, n \in \mathbb{N}\}$$

of dyadic rational numbers, defined by

$$\nu_{\mathbb{D}}(a_{-k} \ldots a_0 \cdot a_1 a_2 \ldots a_n) := \sum_{i=-k}^{n} a_i \cdot 2^{-i} \quad \text{and}$$

$$\nu_{\mathbb{D}}(-a_{-k} \ldots a_0 \cdot a_1 a_2 \ldots a_n) := -\sum_{i=-k}^{n} a_i \cdot 2^{-i},$$

where $k, n \in \mathbb{N}$, $a_i \in \{0, 1\}$, and $a_{-k} = 0 \Rightarrow k = 0$. Such a string $a_{-k} \ldots a_0 \cdot a_1 a_2 \ldots a_n$ or $-a_{-k} \ldots a_0 \cdot a_1 a_2 \ldots a_n$ with n digits after the binary point will be called a *binary name of precision* n of the corresponding dyadic rational number. Our first definition of the time complexity of computable real numbers and computable real number functions is in analogy to our first definition of computable real number functions (Definition 4.1).

Definition 10.1. Let $t : \mathbb{N} \to \mathbb{N}$ be a total function.

1. A real number x is *precision-computable in time* $\mathcal{O}(t)$, if there are a constant c and a Turing machine that, given any $n \in \mathbb{N}$ in binary notation, computes within $c \cdot t(n) + c$ steps a binary name of precision n of a dyadic rational number q with $|x - q| < 2^{-n}$.

2. Consider a compact set $K \subseteq \mathbb{R}^k$. A function $f :\subseteq \mathbb{R}^k \to \mathbb{R}$ with $K \subseteq \mathrm{dom}(f)$ is *precision-computable in time* $\mathcal{O}(t)$ *on* K if there are a constant c and an oracle Turing machine that, given any $n \in \mathbb{N}$ in binary notation, may ask for arbitrarily good dyadic approximations of the input $x \in K$; i.e., it may ask finitely many questions of the kind "Give me a k-vector of binary names with precision i of the components of a dyadic vector $r \in \mathbb{D}^k$ with $\max_{l=1,\dots,k} |x_l - r_l| < 1/2^i$," where the exponent i may depend on the answers to the previous questions, and after at most $c \cdot t(n) + c$ steps, it stops, having written onto the output tape a binary name with precision n of a dyadic rational number q with $|f(x) - q| < 2^{-k}$.

Here, the precision i of a request as above has to be written in binary form on a special tape, the *oracle tape*, and the machine has to enter a special state, the *query state*. Then the answer will be provided in one step on the oracle tape, with the tape head on the field to the left of the answer. For reading the answer, the Turing machine will need time as for any usual reading operation on any other tape.

We consider here only functions on compact sets because for arbitrarily large input numbers, one may need arbitrarily much time just to read the digits in front of the binary point. Then in general, one will be unable to give any bound on the time needed even for computing the result with precision 1.

We had characterized computability of real number functions not only by an oracle approach as in Definition 4.1 but also via the Cauchy representation and equivalent representations. And we had seen that both approaches were equivalent. Can one say the same also for the time complexity of real number functions and of real numbers? One would say that a real number x is computable in time $t : \mathbb{N} \to \mathbb{N}$ with respect to a representation δ of $\mathbb{R}$ if a Turing machine computes a δ-name p for x in time $t(n)$; i.e., without ever stopping it writes p onto the output tape and it needs at most $t(n)$ steps for writing the first n symbols of p. Using the Cauchy representation is not a good idea because for any rational number one can easily construct arbitrarily long $\nu_\mathbb{Q}$-names. This has the consequence that for any computable real number x, there is a Cauchy name for x computable in linear time. Thus, a definition of time complexity of real numbers via the Cauchy representation does not make any sense. One could "repair" the Cauchy representation ρ by using names of the form $w_0 \# w_1 \# w_2 \# \cdots$, where each w_i is not a $\nu_\mathbb{Q}$-name of a rational number but a binary name of precision i of a dyadic rational number. But such a name would be quite wasteful because information that is contained already in w_i for some i will be repeated in all w_j with $j > i$. The following representation is a much more elegant way of representing real numbers. It is defined just like the usual binary representation except that besides the digits 0 and 1 also a digit $\bar{1}$, standing for -1, can be used. Avizienis [3] studied it in the context of computer arithmetic. This representation as well as similar "redundant number representations" have turned out to be very important in computer arithmetic; see, e.g., Muller [73].

Definition 10.2. The *signed digit representation* $\rho_{\mathrm{sd}} :\subseteq \Sigma^\omega \to \mathbb{R}$ is defined by

$$\rho_{\mathrm{sd}}(a_n a_{n-1} \ldots a_0 \bullet a_{-1} a_{-2} \ldots) := \sum_{i=n}^{-\infty} a_i \cdot 2^i$$

for all sequences with $a_i \in \{\overline{1}, 0, 1\}$ and $n \geq -1$, and with the additional properties that $a_n \neq 0$, if $n \geq 0$ and $a_n a_{n-1} \notin \{1\overline{1}, \overline{1}1\}$, if $n \geq 1$, where we interpret $\overline{1}$ as -1.

As in the binary representation, the number of digits after the binary point of any prefix of a name of a real number corresponds directly to the precision by which this prefix already describes the real number. But, in contrast to the binary representation, the signed digit representation is redundant in a symmetric way. If w is a prefix of a ρ_{sd}-name, then $\rho_{\mathrm{sd}}(w\Sigma^\omega)$ is a closed interval, $\rho_{\mathrm{sd}}(w\overline{1}\Sigma^\omega)$ is the left half of $\rho_{\mathrm{sd}}(w\Sigma^\omega)$, $\rho_{\mathrm{sd}}(w0\Sigma^\omega)$ covers the two middle quarters of $\rho_{\mathrm{sd}}(w\Sigma^\omega)$, and $\rho_{\mathrm{sd}}(w1\Sigma^\omega)$ is the right half of $\rho_{\mathrm{sd}}(w\Sigma^\omega)$. If additionally w contains the binary point and n digits after the binary point, then the interval $\rho_{\mathrm{sd}}(w\Sigma^\omega)$ has length $2 \cdot 2^{-n}$.

Proposition 10.3. *The signed digit representation ρ_{sd} is equivalent to the Cauchy representation of the real numbers.*

Definition 10.4. Let $t : \mathbb{N} \to \mathbb{N}$ be a total function. A real number x is *computable in time* $\mathcal{O}(t)$ if there are a constant c and a Turing machine that, without ever stopping, produces a ρ_{sd}-name of x and, after $c \cdot t(n) + c$ steps, has written at least the prefix containing n digits after the binary point.

Before we compare this definition with the previous definition of "precision-computability in time $\mathcal{O}(t)$," we explain how one can define the time complexity of real number functions via the signed digit representation. As we will see soon, it is useful not only to count the steps needed by a Turing machine but also to count how many digits of the input the machine needs in order to compute the first n output symbols.

Definition 10.5. Let $t : \mathbb{N} \to \mathbb{N}$ and $l : \mathbb{N} \to \mathbb{N}$ be total functions. Consider a compact set $K \subseteq \mathbb{R}^k$. A function $f :\subseteq \mathbb{R}^k \to \mathbb{R}$ with $K \subseteq \mathrm{dom}(f)$ is *computable in time* $\mathcal{O}(t)$ *with input lookahead* l *on* K if a Turing machine with k input tapes, given ρ_{sd}-names $p_1, \ldots, p_k$ of the k components of any $x \in K$, does the following:

1. without ever stopping it computes a ρ_{sd}-name q of $F(x)$,

2. after $c \cdot t(n) + c$ steps it has written at least the prefix of q containing n symbols after the binary point,

3. until it has written this prefix of q, for any $i \in \{1, \ldots, k\}$, it has read at most the first $l(n)$ symbols after the binary point of p_i.

Lemma 10.6. *For any co-c.e. closed, compact $K \subseteq \mathbb{R}^k$ and any computable f : $K \to \mathbb{R}$ there exist computable $t, l : \mathbb{N} \to \mathbb{N}$ such that f is computable in time $\mathcal{O}(t)$ and with input lookahead l.*

Remark 10.7. In order to be able to speak about a uniform time bound for the function, we consider the function only on a compact set. For a function with, e.g., domain $\mathbb{R} = \bigcup_{i=1}^{\infty}[-i, i]$, one can consider the time complexity of f on $[-i, i]$ for each positive i and then use i as an additional parameter that corresponds to the "size" of the input.

Now let us compare the two definitions of time complexity of real numbers and real number functions that we have formulated. A real number x or a real number function $f :\subseteq \mathbb{R}^k \to \mathbb{R}$ on a compact set $K \subseteq \operatorname{dom}(f)$ is precision-computable in time $\mathcal{O}(t)$ (see Definition 10.1) if it is computable in time $\mathcal{O}(t)$ (see Definition 10.4 and Definition 10.5). In general, the converse does not need to be true. But it is true for so-called regular time bounds.

Definition 10.8. A total function $t : \mathbb{N} \to \mathbb{N}$ is *regular* if it is nondecreasing and there exist constants $N, c > 0$ such that $t(N) > 0$ and

$$2 \cdot t(n) \leq t(2n) \leq c \cdot t(n)$$

for all $n > N$.

A regular function grows polynomially and at least linearly. Any function obtained by multiplying n^k for some $k \geq 1$ by a power of logarithms or multiple logarithms is regular. For the practice of computing, the regular functions are presumably the most important time bounds.

Proposition 10.9. *Let $t : \mathbb{N} \to \mathbb{N}$ be regular. A real number x or a real number function $f : K \to \mathbb{R}$ for some compact set $K \subseteq \mathbb{R}^k$ is precision-computable in time $\mathcal{O}(t)$ if, and only if, it is computable in time $\mathcal{O}(t)$.*

Proof. We sketch the proof for the case of real numbers.

First, let us assume that x is a real number computable in time $\mathcal{O}(t)$. Then a Turing machine M_{sd} produces a signed-digit name of x in time $\mathcal{O}(t)$. Given some n in binary notation, one can simulate M_{sd} until it has computed a signed digit name of x up to $n + 2$ digits after the binary point. This prefix describes a dyadic number q_1 with denominator 2^{-n-2}. Then one computes a dyadic number q_2 with denominator 2^{-n} as close as possible to q_1 and writes its binary name with precision n onto the output tape. All this takes time $\mathcal{O}(t(n + 2) + n) \subseteq \mathcal{O}(t(n))$. Thus, x is precision computable in time $\mathcal{O}(t)$.

Now, let us assume that x is precision computable in time $\mathcal{O}(t)$. Let M_{prec} be a Turing machine which, given an n in binary form, computes a dyadic rational q with precision n and with $|x - q| < 2^{-n}$. In order to produce a signed digit name of x, one can proceed as follows. One can simulate M_{prec} repeatedly with input 2^i for $i = 0, 1, 2, 3, \ldots$ and, after each simulation, use the new dyadic approximation of x in order to improve the previously computed prefix of a signed digit name of x. In fact, the number of computed digits after the decimal point doubles after each simulation. In order to compute the first n digits after the binary point, the new machine will need about $\mathcal{O}(\sum_{i=0}^{\lceil \log_2(n) \rceil} t(2^i))$ time. It is easy to verify that regularity of t implies that $\mathcal{O}(t)$ is an upper bound for this term. $\square$

The following two statements are in analogy to Theorem 3.4 for computable real numbers. Let $\mathcal{M}(n) := n \cdot \log_2(n) \cdot \log_2(\log_2(n))$ be the time bound for the Schönhage–Strassen algorithm for multiplying two natural numbers given by binary words of length n.

Theorem 10.10 (Müller [74]). *Let t be a regular function with $\mathcal{M} \in \mathcal{O}(t)$. The set of real numbers computable in time $\mathcal{O}(t)$ forms a real algebraically closed field.*

Corollary 10.11 (Ko and Friedman [56, 55]). *The set of polynomial time computable real numbers forms a real algebraically closed field.*

In particular, the zeros of a polynomial whose coefficients can be computed fast, e.g., in polynomial time, can be computed fast as well, e.g., in polynomial time. The problem to find fast algorithms for computing the zeros of polynomials has received a lot of attention; see Neff and Reif [77]. Also numbers like π and e are polynomial time computable. The time complexity of π has been analyzed in great detail. Finding even faster algorithms for computing π as well as actually computing π with precision as high as possible is still a challenge; see, e.g., Borwein and Borwein [12].

Remark 10.12. We have seen in Section 3 that for defining the set of computable real numbers, the decimal (or binary) representation is as good as the Cauchy representation and therefore as good as the signed digit representation. But the set of real numbers with polynomial time computable decimal name (or with polynomial time computable binary name) is a proper subset of the set of polynomial time computable real numbers, and it is not even closed under addition; see Ko [55].

Now we turn to functions.

Theorem 10.13 (Brent [26], Schröder [87, 88]). *For each of the following functions $f :\subseteq \mathbb{R}^m \to \mathbb{R}$ and domains $K \subseteq \mathbb{R}^m$, there exists a Turing machine M which, computes f in time $\mathcal{O}(t)$ with input lookahead l:*

function	domain	time	lookahead
$f(x_1, \ldots, x_m)$	K	$\mathcal{O}(t(n))$	$l(n)$
$-x_1$	$\mathbb{R}$	n	n
$x_1 + x_2$	$[-1,1] \times [-1,1]$	n	$n + c$
$x_1 \cdot x_2$	$[-1,1] \times [-1,1]$	$\mathcal{M}(n)$	$2n + c$
$x_1 \cdot x_2$	$[-1,1] \times [-1,1]$	$\mathcal{M}(n) \cdot \log_2(n)$	$n + c$
$1/x_1$	$[7/8, 2]$	$\mathcal{M}(n)$	$2n + c$
$1/x_1$	$[7/8, 2]$	$\mathcal{M}(n) \cdot \log_2(n)$	$n + c$
$\exp, \sin, \cos$	$[-1,1]$	$\mathcal{M}(n) \cdot \log_2(n)$	$n + c$

A machine computing a real function f may read more digits from the input ρ_{sd}-name p than necessary to define the result with precision 2^{-n}. Let $\mathrm{Dep}(p)(n)$ be the smallest number m such that m digits after the binary point suffice to define n digits after the binary point of a ρ_{sd}-name of $f\rho_{\mathrm{sd}}(p)$. Call a machine k-*input-optimal*, if for any input name p it reads at most the first $\mathrm{Dep}(p)(n) + k$ digits after the point for writing the first n digits after the point of the result. For some functions there is a trade-off between input information and computation time.

Theorem 10.14 (Weihrauch [107]). *There exists a real number function $f : [0, 1] \to \mathbb{R}$ with the following properties:*

1. *The function f can be computed in time $\mathcal{O}(n)$ with input lookahead $4n$.*

2. *For any k, the function f can be computed by a k-input-optimal machine in polynomial time if, and only if, $\mathrm{P} = \mathrm{NP}$.*

If the input is itself the result of some (expensive) computation, it is desirable to use as few input digits as possible. However, as we have just seen, this might increase the computation time substantially. In case of a composition of functions, for optimizing the total computation time, one has to balance input lookahead and time complexity of the second function in an appropriate way.

We have seen that if $f : [0, 1] \to \mathbb{R}$ is computable and has a continuous second derivative, then f' is computable as well (Corollary 8.7). In fact, the condition that the function has a continous second derivative can be replaced by the weaker condition that the derivative is Lipschitz continuous; see Pour-El and Richards [80].

Theorem 10.15 (Müller [75]). *Let t be a regular function with $\mathcal{M} \in \mathcal{O}(t)$. If $f : [0, 1] \to \mathbb{R}$ is a function that is computable in time $\mathcal{O}(t)$ and that has a Lipschitz continuous derivative, then the derivative f' is also computable in time $\mathcal{O}(t)$.*

We conclude this section with a comment on the complexity of real number functions as it is defined in the real random access machine model, most notably by Blum et al. [11]. In this model, one assumes that a computer can process real numbers with infinite precision, and that each comparison and arithmetic operation takes one step. This is certainly not realistic. On the other hand, the idea to consider real numbers as entities on which one can operate directly without having to refer to names or representations looks attractive. Brattka and Hertling [20] have shown that one can express the computability and polynomial time complexity notion for real number functions as defined in this section also via a real random access machine model. This model is a natural generalization of the random access machine model for computations over the natural numbers with logarithmic cost measure.

11 Computational Complexity of Sets and Operators over the Real Numbers

So far we have discussed the time complexity of real numbers and of real number functions. The next step would be to define the time complexity of subsets of $\mathbb{R}^k$. Before we do that, let us look from a more general perspective at the problem to define time complexity for arbitrary mathematical objects that one might wish to compute in computations over the real numbers. We had observed that the Cauchy representation was unsuitable even for defining the time complexity of real numbers. Why did we come to that conclusion? Well, for defining the time complexity of a real number one would certainly take a name with "minimal time complexity." The problem with the Cauchy representation is that for any real number there are too many Cauchy names and such a minimum does not exist. A sufficient condition on a representation δ of a set X would be that for each $x \in X$, the fiber $\delta^{-1}(x)$ should be compact. Since we wish to have uniform time bounds even for all objects in a compact set, e.g., for all real numbers in a closed, finite interval, it makes sense to demand more: that for any compact set $K \subseteq X$ the set $\delta^{-1}(K)$ is compact as well. Let us call a representation δ of a topological space X *proper* if for each compact $K \subseteq X$ the set $\delta^{-1}(K)$ is compact.

Proposition 11.1. *The signed-digit representation is a proper representation of the real numbers equivalent to the Cauchy representation ρ. In particular, it is admissible.*

Schröder [86, 92] has characterized the spaces that possess admissible representations that have compact fibers or are even proper. Weihrauch [110] has continued this study and has shown how one can introduce in a natural way computational complexity on computable metric spaces. Labhalla, Lombardi, and Moutai [61] suggested other approaches for defining computational complexity on computable metric spaces.

Now let us introduce time complexity of compact subsets of Euclidean spaces. The metric space of compact subsets of $\mathbb{R}$ (we consider the Hausdorff metric) possesses proper admissible representations. A name in such a representation should in addition have the property that prefixes of increasing length should describe the set with increasing precision, e.g., by describing the set with a sequence of "pixels." A problem with such a name is that in order to describe the set with precision 2^{-n} one will in general need an exponential number of "pixels." Thus, simply writing down such a prefix requires exponential time. This does not look very natural. The problem seems to be that a name of a set describes the set as a whole. Often one may be happy to be able to describe some details of the set with high precision. This idea is realized in the following definition of "local time complexity" of a compact set, as defined by Rettinger and Weihrauch [82].

Definition 11.2. A compact set $K \subseteq [0,1]^2$ is called *(locally) computable in time* $\mathcal{O}(t)$, *for some total function* $t : \mathbb{N} \to \mathbb{N}$ if there exists a function $f : (\{0,1\}^*)^3 \to \{0,1\}^*$ computable in time $\mathcal{O}(t)$ and defined on all triples $(0^n, \nu_{\mathbb{N}}^{-1}(i), \nu_{\mathbb{N}}^{-1}(j))$ with $n \in \mathbb{N}$, $i, j \in \{0, \ldots, 2^n\}$ and with

$$
f(0^n, \nu_{\mathbb{N}}^{-1}(i), \nu_{\mathbb{N}}^{-1}(j)) = \begin{cases} 1 & \text{if } d_K(\frac{i}{2^n}, \frac{j}{2^n}) < 2^{-n}, \\ 0 & \text{if } d_K(\frac{i}{2^n}, \frac{j}{2^n}) > 2 \cdot 2^{-n}, \\ 0 \text{ or } 1 & \text{otherwise.} \end{cases}
$$

This definition realizes the idea that being able to compute a compact subset of $\mathbb{R}^2$ means being able to plot it with any desired precision, as in Proposition 5.7. Given a screen of print area of fixed resolution, say with k pixels, it requires $\mathcal{O}(k \cdot t(n))$ time to print a set $K \subseteq \mathbb{R}^2$ with (local) time complexity t and a zoom factor 2^n. This definition can be generalized straightforwardly to compact subsets of $\mathbb{R}^m$ for any $m \geq 1$.

Remark 11.3. Chou and Ko [34] (see also [54, 55]) have defined and analyzed several other notions of computational complexity of two-dimensional regions. For example, they call a subset $K \subseteq \mathbb{R}^2$ *polynomial-time approximable*, resp. *polynomial-time recognizable*, if a Turing machine, given a point $x \in \mathbb{R}^2$, decides whether x is in K within polynomial time and makes errors only on a subset of $\mathbb{R}^2$ of measure at most 2^{-n}, resp. makes errors only for points of distance at most 2^{-n} from the boundary of K. If a Turing machine of this kind never errs for points in K, then they call the set K *strongly polynomial-time approximable*, resp. *strongly polynomial-time recognizable*. If a set is locally computable in polynomial time, then it is strongly polynomial-time recognizable, but the converse is true if, and only if, P = NP (Braverman [23]).

We had characterized the nonempty, closed, computable subsets of $\mathbb{R}^k$ also as those nonempty closed subsets that have a computable distance function. It looks natural to ask for the time complexity of the distance function of a compact subset of $\mathbb{R}^k$. The following result clarifies the relation between the notion in Definition 11.2 and the time complexity of the distance function.

Theorem 11.4 (Braverman [23]). *Let* $K \subseteq \mathbb{R}^n$ *be a nonempty compact subset.*

1. *If the distance function* $d_K : \mathbb{R}^n \to \mathbb{R}$ *of* K *is polynomial-time computable, then* K *is polynomial-time computable.*

2. If $n = 1$, then the converse holds.

3. If $n > 1$, then the converse holds if, and only if, P $=$ NP.

Remark 11.5. A similar result for strongly polynomial-time recognizable sets with strongly polynomial-time recognizable complement and for the distance to the boundary of the set has been shown by Chou and Ko [35].

In view of the intuitively appealing meaning of the time complexity of compact sets as in Definition 11.2, we will stick to that definition.

Very interesting compact subsets of Euclidean spaces are fractal sets, in particular the Julia sets. For computability and complexity considerations, we will identify $\mathbb{C}$ with $\mathbb{R}^2$.

Definition 11.6. Let $f : \mathbb{C} \to \mathbb{C}$ be a polynomial function of degree ≥ 2.

- A point $z \in \mathbb{C}$ is called a *periodic point* of f, if there exists a $p \in \mathbb{N}$ such that $f^p(z) = z$.

- A periodic point $z \in \mathbb{C}$ of f is called *repelling*, if $|(f^p)'(z)| > 1$.

- The Julia set $J(f)$ of f is the closure of the set of repelling periodic points of f.

- A point $z \in \mathbb{C}$ is called *critical*, if $f'(z) = 0$.

- The function f is called *hyperbolic*, if $J(f)$ is disjoint from the closure of the orbits $\bigcup_z$ critical $\bigcup_{n=0}^{\infty} f^n(z)$.

In 1998 Zhong [118] showed that a Turing machine, given the coefficients of a hyperbolic polynomial f, computes the Julia set $J(f)$ of f; i.e., given in addition a point $z \in \mathbb{C}$ and a number $k \in \mathbb{N}$, it computes the distance of z from $J(f)$ with precision 2^{-k}, or it computes a function for $K := J(f)$ as described in Definition 11.2. Rettinger and Weihrauch [82] looked at the time complexity of Julia sets and showed the following first result.

Theorem 11.7 (Rettinger and Weihrauch [82]). *There is a Turing machine which, given a signed-digit name of a $c \in \mathbb{C}$ with $|c| < \frac{1}{4}$, computes in time $\mathcal{O}(n^2 \cdot \mathcal{M}(n)) \subseteq \mathcal{O}(n^3 \cdot \log n \cdot \log \log n)$ a function f as in Definition 11.2 for the Julia set of the quadratic polynomial $z \mapsto z^2 + c$.*

This result was generalized and strengthened in 2004 independently by Braverman [24] and Rettinger [81] to hyperbolic polynomial and hyperbolic rational functions, respectively, with the time bound $\mathcal{O}(n \cdot \mathcal{M}(n))$. Since then, Binder, Braverman, and Yampolsky have obtained a series of positive and negative results concerning computability and polynomial time computability of Julia sets; see [7] and the references therein.

Now we come to the time complexity of numerical operators, e.g., of type

$$F : \mathcal{C}[0, 1] \to \mathcal{C}[0, 1]$$

such as integration or differentiation. It is difficult to define a uniform notion of complexity for such operators because $\mathcal{C}[0, 1]$ is not locally compact, and therefore, there is no obvious way to define a uniform notion of complexity (not even parameterized over $\mathbb{N}$ as we had suggested to do it for functions on $\mathbb{R}^k$; see Remark 10.7). But one can study the complexity of such operators restricted to compact subspaces $K \subseteq \mathcal{C}[0, 1]$. In fact, this is often done in numerical analysis, although often only in an algebraic computation model; see Traub et al. [99]. It is likely that for many numerically stable problems over function spaces, the time complexity in the Turing machine model will not be very different from the "arithmetic" complexity in an algebraic computation model; see Woźniakowski [113]. But a thorough study of the time complexity in the sense of Weihrauch [110] or Labhalla et al. [61] of higher type numerical operators in the Turing machine model seems to be missing.

There is a different approach, due to Ko and Friedman, see e.g. [56], and covered by Ko's book [55]: if F is a numerical operator receiving real number functions as input, one can study the time complexity of $F(f)$ for polynomial time computable input f. This approach leads to the insight that the complexity of many numerical problems can be characterized by discrete complexity classes or, the other way around, to the insight that many famous open problems concerning discrete complexity classes can be found to be hidden in numerical problems. We formulate only three results and refer the reader to Ko [55] for many more results and many open problems in this context.

Theorem 11.8 (Friedman [39]). *The following two statements are equivalent:*

1. P = NP.

2. For each polynomial-time computable $f : [0, 1] \to \mathbb{R}$, the maximum function $g : [0, 1] \to \mathbb{R}$, defined by

$$g(x) := \max\{f(y) : 0 \le y \le x\}$$

for all $x \in [0, 1]$, is polynomial-time computable.

Proof (Sketch of Proof). The direction "1.$\Longrightarrow$2." of the proof is based on the idea that

$$z \le g(x) \iff (\exists y \in [0, x]) \, z \le f(y).$$

Using the Polynomial-Time Projection Theorem, this is (approximately) decidable in polynomial time if P = NP. By a binary search over z, one can determine $g(x)$ in polynomial time.   □

Theorem 11.9 (Friedman [39]). *The following two statements are equivalent:*

1. FP = #P.

2. For each polynomial-time computable $f : [0,1] \rightarrow \mathbb{R}$, *the integral function* $g : [0,1] \rightarrow \mathbb{R}$, *defined by*

$$g(x) := \int_0^x f(t)\, dt$$

for all $x \in [0,1]$, *is polynomial-time computable.*

Here #P denotes the class of functions that count the number of accepting computations of a nondeterministic polynomial-time Turing machine.

Proof (Sketch of Proof). For the proof of "1.$\Longrightarrow$2." one can guess a number of points (t, y) with $0 \le t \le x$ and then count those with $y \le f(t)$ to get an approximation for the integral $g(x)$ (in case f is positive). □

Theorem 11.10 (Ko [53]). *For the following three statements, the implications* $1 \Rightarrow 2$ *and* $2 \Rightarrow 3$ *hold true.*

1. P = PSPACE.

2. For each polynomial-time computable $f : [0,1] \times [-1,1] \rightarrow \mathbb{R}$ *that satisfies the Lipschitz condition*

$$|f(x, z_1) - f(x, z_2)| \le L \cdot |z_1 - z_2|$$

for some $L > 0$, *the unique solution* $y : [0,1] \rightarrow \mathbb{R}$ *of the differential equation*

$$y'(x) = f(x, y(x)), \qquad y(0) = 0$$

is polynomial-time computable.

3. FP = #P.

Acknowledgments

The first author has been supported by the National Research Foundation of South Africa. The second author has been supported by the German Research Council (DFG).

References

1. O. Aberth. *Computable Analysis*. McGraw-Hill, New York, 1980.

2. K. Ambos-Spies, K. Weihrauch, and X. Zheng. Weakly computable real numbers. *Journal of Complexity*, 16(4):676–690, 2000.

3. A. Avizienis. Signed-digit number representations for fast parallel arithmetic. *IRE Transactions on Electronic Computers*, 10:389–400, 1961.

4. G. Baigger. Die Nichtkonstruktivität des Brouwerschen Fixpunktsatzes. *Arch. Math. Logik Grundlag.*, 25:183–188, 1985.

5. G. Barmpalias. The approximation structure of a computably approximable real. *The Journal of Symbolic Logic*, 68(3):885–922, 2003.

6. G. Beer. *Topologies on Closed and Closed Convex Sets*, volume 268 of *Mathematics and Its Applications*. Kluwer Academic, Dordrecht, 1993.

7. I. Binder, M. Braverman, and M. Yampolsky. On computational complexity of Siegel Julia sets. *Comm. Math. Phys.*, 264(2):317–334, 2006.

8. E. Bishop and D. S. Bridges. *Constructive Analysis*, volume 279 of *Grundlehren der Mathematischen Wissenschaften*. Springer, Berlin, 1985.

9. J. Blanck. Domain representations of topological spaces. *Theoretical Computer Science*, 247:229–255, 2000.

10. L. Blum, F. Cucker, M. Shub, and S. Smale. *Complexity and Real Computation*. Springer, New York, 1998.

11. L. Blum, M. Shub, and S. Smale. On a theory of computation and complexity over the real numbers: NP-completeness, recursive functions and universal machines. *Bulletin of the American Mathematical Society*, 21(1):1–46, 1989.

12. J. M. Borwein and P. B. Borwein. *Pi and the AGM*. John Wiley & Sons, New York, 1987.

13. B. Branner. The Mandelbrot set. In R. L. Devaney and L. Keen, editors, *Chaos and Fractals. The Mathematics Behind the Computer Graphics*, volume 39 of *Proceedings of Symposia in Applied Mathematics*, pages 75–105, Providence, Rhode Island, 1989. American Mathematical Society, 1989.

14. V. Brattka. Computable invariance. *Theoretical Computer Science*, 210:3–20, 1999.

15. V. Brattka. Computable versions of Baire's category theorem. In J. Sgall, A. Pultr, and P. Kolman, editors, *Mathematical Foundations of Computer Science 2001*, volume 2136 of *Lecture Notes in Computer Science*, pages 224–235, Berlin, 2001. Springer. 26th International Symposium, MFCS 2001, Mariánské Lázně, Czech Republic, August 27–31, 2001.

16. V. Brattka. The inversion problem for computable linear operators. In H. Alt and M. Habib, editors, *STACS 2003*, volume 2607 of *Lecture Notes in Computer Science*, pages 391–402, Berlin, 2003. Springer. 20th Annual Symposium on Theoretical Aspects of Computer Science, Berlin, Germany, February 27–March 1, 2003.

17. V. Brattka. Effective Borel measurability and reducibility of functions. *Mathematical Logic Quarterly*, 51(1):19–44, 2005.

18. V. Brattka. On the Borel complexity of Hahn-Banach extensions. In V. Brattka, L. Staiger, and K. Weihrauch, editors, *Proceedings of the 6th Workshop on Computability and Complexity in Analysis*, volume 120 of *Electronic Notes in Theoretical Computer Science*, pages 3–16, Amsterdam, 2005. Elsevier. 6th International Workshop, CCA 2004, Wittenberg, Germany, August 16–20, 2004.

19. V. Brattka and R. Dillhage. On computable compact operators on Banach spaces. In D. Cenzer, R. Dillhage, T. Grubba, and K. Weihrauch, editors, *Proceedings of the Third International Conference on Computability and Complexity in Analysis*, volume 167 of *Electronic Notes in Theoretical Computer Science*, Amsterdam, 2007. Elsevier. CCA 2006, Gainesville, Florida, November 1–5, 2006.

20. V. Brattka and P. Hertling. Feasible real random access machines. *Journal of Complexity*, 14(4):490–526, 1998.

21. V. Brattka and G. Presser. Computability on subsets of metric spaces. *Theoretical Computer Science*, 305:43–76, 2003.

22. V. Brattka and A. Yoshikawa. Towards computability of elliptic boundary value problems in variational formulation. *Journal of Complexity*, 22(6):858–880, 2006.

23. M. Braverman. *Computational Complexity of Euclidean Sets: Hyperbolic Julia Sets are Poly-Time Computable*. Master thesis, Department of Computer Science, University of Toronto, 2004.

24. M. Braverman. Hyperbolic Julia sets are poly-time computable. In V. Brattka, L. Staiger, and K. Weihrauch, editors, *Proceedings of the 6th Workshop on Computability and Complexity in Analysis*, volume 120 of *Electronic Notes in Theoretical Computer Science*, pages 17–30, Amsterdam, 2005. Elsevier. 6th International Workshop, CCA 2004, Wittenberg, Germany, August 16–20, 2004.

25. M. Braverman and S. Cook. Computing over the reals: Foundations for scientific computing. *Notices of the AMS*, 53(3):318–329, 2006.

26. R. Brent. Fast multiple-precision evaluation of elementary functions. *Journal of the Association for Computing Machinery*, 23(2):242–251, 1976.

27. D. Bridges and F. Richman. *Varieties of Constructive Mathematics*, volume 97 of *London Mathematical Society Lecture Note Series*. Cambridge University Press, Cambridge, 1987.

28. L. Brouwer. *Collected Works, Vol. 1, Philosophy and Foundations of Mathematics*. North-Holland, Amsterdam, 1975. Heyting, A. (ed).

29. L. Brouwer. *Collected Works, Vol. 2, Geometry, Analysis, Topology and Mechanics*. North-Holland, Amsterdam, 1976. Freudenthal, H. (ed).

30. C. S. Calude, P. H. Hertling, B. Khoussainov, and Y. Wang. Recursively enumerable reals and Chaitin ω numbers. *Theoretical Computer Science*, 255:125–149, 2001.

31. G. Ceĭtin. Algorithmic operators in constructive metric spaces. *Tr. Mat. Inst. Steklov*, 67:295–361, 1962. (in Russian, English trans. in AMS Trans. 64, 1967).

32. D. Cenzer and J. B. Remmel. Index sets for computable differential equations. *Mathematical Logic Quarterly*, 50(4,5):329–344, 2004.

33. G. J. Chaitin. A theory of program size formally identical to information theory. *Journal of the Association for Computing Machinery*, 22:329–340, 1975.

34. A. Chou and K.-I. Ko. Computational complexity of two-dimensional regions. *SIAM Journal on Computing*, 24:923–947, 1995.

35. A. W. Chou and K.-I. Ko. The computational complexity of distance functions of two-dimensional domains. *Theoretical Computer Science*, 337:360–369, 2005.

36. R. G. Downey, D. R. Hirschfeldt, A. Nies, and F. Stephan. Trivial reals. In R. Downey, D. Decheng, T. S. Ping, Q. Y. Hui, and M. Yasugi, editors, *Proceedings of the 7th and 8th Asian Logic Conferences*, pages 63–102, Singapore, 2003. World Scientific. 7th Conference: Hsi-Tou, Taiwan, June 6–10, 1999; 8th Conference: Chongqing, China, August 29–September 2, 2002.

37. A. Edalat. Domains for computation in mathematics, physics and exact real arithmetic. *Bulletin of Symbolic Logic*, 3(4):401–452, 1997.

38. M. H. Escardó. PCF extended with real numbers. In K.-I. Ko and K. Weihrauch, editors, *Computability and Complexity in Analysis*, volume 190 of *Informatik Berichte*, pages 11–24. FernUniversität Hagen, Sept. 1995. CCA Workshop, Hagen, August 19–20, 1995.

39. H. Friedman. On the computational complexity of maximization and integration. *Advances in Mathematics*, 53:80–98, 1984.

40. G. Gherardi. Effective Borel degrees of some topological functions. *Mathematical Logic Quarterly*, 52(6):625–642, 2006.
41. A. Grzegorczyk. On the definitions of computable real continuous functions. *Fundamenta Mathematicae*, 44:61–71, 1957.
42. J. Hauck. Berechenbare reelle Funktionen. *Zeitschrift für Mathematische Logik und Grundlagen der Mathematik*, 19:121–140, 1973.
43. J. Hauck. Konstruktive Darstellungen reeller Zahlen und Folgen. *Zeitschrift für Mathematische Logik und Grundlagen der Mathematik*, 24:365–374, 1978.
44. P. Hertling. A topological complexity hierarchy of functions with finite range. Technical Report 223, Centre de recerca matematica, Institut d'estudis catalans, Barcelona, Barcelona, Oct. 1993. Workshop on Continuous Algorithms and Complexity, Barcelona, October, 1993.
45. P. Hertling. Topological complexity with continuous operations. *Journal of Complexity*, 12:315–338, 1996.
46. P. Hertling. *Unstetigkeitsgrade von Funktionen in der effektiven Analysis*. PhD thesis, Fachbereich Informatik, FernUniversität Hagen, 1996.
47. P. Hertling. An effective Riemann Mapping Theorem. *Theoretical Computer Science*, 219:225–265, 1999.
48. P. Hertling. A real number structure that is effectively categorical. *Mathematical Logic Quarterly*, 45(2):147–182, 1999.
49. P. Hertling. A Banach-Mazur computable but not Markov computable function on the computable real numbers. *Annals of Pure and Applied Logic*, 132(2-3):227–246, 2005.
50. P. Hertling. Is the Mandelbrot set computable? *Mathematical Logic Quarterly*, 51(1):5–18, 2005.
51. M. D. Hirsch. Applications of topology to lower bound estimates in computer science. In *From Topology to Computation: Proceedings of the Smalefest, Berkeley, CA, 1990*, pages 395–418, New York, 1993. Springer.
52. C.-K. Ho. Relatively recursive reals and real functions. *Theoretical Computer Science*, 210(1):99–120, 1999.
53. K.-I. Ko. On the computational complexity of ordinary differential equations. *Inform. Contr.*, 58:157–194, 1983.
54. K.-I. Ko. Approximation to measurable functions and its relation to probabilistic computation. *Annals of Pure and Applied Logic*, 30:173–200, 1986.
55. K.-I. Ko. *Complexity Theory of Real Functions*. Progress in Theoretical Computer Science. Birkhäuser, Boston, 1991.
56. K.-I. Ko and H. Friedman. Computational complexity of real functions. *Theoretical Computer Science*, 20:323–352, 1982.
57. G. Kreisel, D. Lacombe, and J. Shoenfield. Partial recursive functionals and effective operations. In A. Heyting, editor, *Constructivity in Mathematics*, Studies in Logic and the Foundations of Mathematics, pages 290–297, Amsterdam, 1959. North-Holland. Proc. Colloq., Amsterdam, Aug. 26–31, 1957.
58. C. Kreitz and K. Weihrauch. Theory of representations. *Theoretical Computer Science*, 38:35–53, 1985.
59. B. A. Kušner. *Lectures on Constructive Mathematical Analysis*, volume 60 of *Translations of Mathematical Monographs*. American Mathematical Society, Providence, Rhode Island, 1984.
60. A. Kučera and T. A. Slaman. Randomness and recursive enumerability. *SIAM J. Comput.*, 31(1):199–211, 2001.

61. S. Labhalla, H. Lombardi, and E. Moutai. Espaces métriques rationnellement présentés et complexité, le cas de l'espace des fonctions réelles uniformément continues sur un intervalle compact. *Theoretical Computer Science*, 250:265–332, 2001.

62. D. Lacombe. Extension de la notion de fonction récursive aux fonctions d'une ou plusieurs variables réelles III. *Comptes Rendus Académie des Sciences Paris*, 241:151–153, 1955. Théorie des fonctions.

63. S. Mazur. *Computable Analysis*, volume 33. Razprawy Matematyczne, Warsaw, 1963.

64. G. Metakides and A. Nerode. The introduction of non-recursive methods into mathematics. In A. Troelstra and D. v. Dalen, editors, *The L.E.J. Brouwer Centenary Symposium*, volume 110 of *Studies in Logic and the foundations of mathematics*, pages 319–335, Amsterdam, 1982. North-Holland. Proceedings of the conference held in Noordwijkerhout, June 8–13, 1981.

65. G. Metakides, A. Nerode, and R. Shore. Recursive limits on the Hahn-Banach theorem. In M. Rosenblatt, editor, *Errett Bishop: Reflections on Him and His Research*, volume 39 of *Contemporary Mathematics*, pages 85–91, Providence, Rhode Island, 1985. American Mathematical Society. Proceedings of the memorial meeting for Errett Bishop, University of California, San Diego, September 24, 1983.

66. J. Miller and A. Nies. Randomness and computability: Open questions. *Bull. Symb. Logic*, 12(3):390–410, 2006.

67. J. S. Miller. *Pi-0-1 Classes in Computable Analysis and Topology*. PhD thesis, Cornell University, Ithaca, New York, 2002.

68. J. S. Miller. Degrees of unsolvability of continuous functions. *The Journal of Symbolic Logic*, 69(2):555–584, 2004.

69. R. E. Moore. *Interval Analysis*. Prentice Hall, Englewood Cliffs, New Jersey, 1966.

70. Y. N. Moschovakis. Recursive metric spaces. *Fundamenta Mathematicae*, 55:215–238, 1964.

71. Y. N. Moschovakis. *Descriptive Set Theory*, volume 100 of *Studies in Logic and the Foundations of Mathematics*. North-Holland, Amsterdam, 1980.

72. A. Mostowski. On computable sequences. *Fundamenta Mathematicae*, 44:37–51, 1957.

73. J. M. Muller. *Elementary Functions*. Birkhäuser, Boston, 2nd edition, 2006.

74. N. T. Müller. Subpolynomial complexity classes of real functions and real numbers. In L. Kott, editor, *Proceedings of the 13th International Colloquium on Automata, Languages, and Programming*, volume 226 of *Lecture Notes in Computer Science*, pages 284–293, Berlin, 1986. Springer.

75. N. T. Müller. Polynomial time computation of Taylor series. In *Proceedings of the 22th JAIIO - Panel'93, Part 2*, pages 259–281, 1993. Buenos Aires, 1993.

76. J. Myhill. A recursive function defined on a compact interval and having a continuous derivative that is not recursive. *Michigan Math. J.*, 18:97–98, 1971.

77. C. A. Neff and J. H. Reif. An efficient algorithm for the complex roots problem. *Journal of Complexity*, 12:81–115, 1996.

78. V. Orevkov. A constructive mappping of the square onto itself displacing every constructive point (Russian). *Doklady Akademii Nauk*, 152:55–58, 1963. Translated in: Soviet Math. - Dokl., 4 (1963) 1253–1256.

79. M. B. Pour-El and J. I. Richards. The wave equation with computable inital data such that its unique solution is not computable. *Advances in Math.*, 39:215–239, 1981.

80. M. B. Pour-El and J. I. Richards. *Computability in Analysis and Physics*. Perspectives in Mathematical Logic. Springer, Berlin, 1989.

81. R. Rettinger. A fast algorithm for Julia sets of hyperbolic rational functions. In V. Brattka, L. Staiger, and K. Weihrauch, editors, *Proceedings of the 6th Workshop on Computability and Complexity in Analysis*, volume 120 of *Electronic Notes in Theoretical*

Computer Science, pages 145–157, Amsterdam, 2005. Elsevier. 6th International Workshop, CCA 2004, Wittenberg, Germany, August 16–20, 2004.

82. R. Rettinger and K. Weihrauch. The computational complexity of some Julia sets. In M. X. Goemans, editor, *Proceedings of the 35th Annual ACM Symposium on Theory of Computing*, pages 177–185, New York, 2003. ACM Press. San Diego, California, June 9–11, 2003.

83. H. Rice. Recursive real numbers. *Proc. Amer. Math. Soc.*, 5:784–791, 1954.

84. H. Rogers. *Theory of Recursive Functions and Effective Computability*. McGraw-Hill, New York, 1967.

85. N. Šanin. *Constructive Real Numbers and Constructive Function Spaces*, volume 21 of *Translations of Mathematical Monographs*. American Mathematical Society, Providence, 1968.

86. M. Schröder. Topological spaces allowing type 2 complexity theory. In K.-I. Ko and K. Weihrauch, editors, *Computability and Complexity in Analysis*, volume 190 of *Informatik Berichte*, pages 41–53. FernUniversität Hagen, 1995. CCA Workshop, Hagen, August 19–20, 1995.

87. M. Schröder. Fast online multiplication of real numbers. In R. Reischuk and M. Morvan, editors, *STACS 97*, volume 1200 of *Lecture Notes in Computer Science*, pages 81–92, Berlin, 1997. Springer. 14th Annual Symposium on Theoretical Aspects of Computer Science, Lübeck, Germany, February 27–March 1, 1997.

88. M. Schröder. Online computations of differentiable functions. *Theoretical Computer Science*, 219:331–345, 1999.

89. M. Schröder. Admissible representations of limit spaces. In J. Blanck, V. Brattka, and P. Hertling, editors, *Computability and Complexity in Analysis*, volume 2064 of *Lecture Notes in Computer Science*, pages 273–295, Berlin, 2001. Springer. 4th International Workshop, CCA 2000, Swansea, UK, September 2000.

90. M. Schröder. Effectivity in spaces with admissible multirepresentations. *Mathematical Logic Quarterly*, 48(Suppl. 1):78–90, 2002.

91. M. Schröder. Extended admissibility. *Theoretical Computer Science*, 284(2):519–538, 2002.

92. M. Schröder. Spaces allowing type-2 complexity theory revisited. *Mathematical Logic Quarterly*, 50(4,5):443–459, 2004.

93. A. Slisenko. Examples of a nondiscontinuous but not continuous constructive operator in a metric space. *Trudy Mat. Inst. Steklov*, 72:524–532, 1964. (in Russian, English trans. in AMS Trans. 100, 1972).

94. S. Smale. On the topology of algorithms, I. *Journal of Complexity*, 3:81–89, 1987.

95. R. Soare. Cohesive sets and recursively enumerable Dedekind cuts. *Pacific J. Math.*, 31:215–231, 1969.

96. E. Specker. Nicht konstruktiv beweisbare Sätze der Analysis. *The Journal of Symbolic Logic*, 14(3):145–158, 1949.

97. E. Specker. The fundamental theorem of algebra in recursive analysis. In B. Dejon and P. Henrici, editors, *Constructive Aspects of the Fundamental Theorem of Algebra*, pages 321–329, London, 1969. Wiley-Interscience.

98. V. Stoltenberg-Hansen, I. Lindström, and E. Griffor. *Mathematical Theory of Domains*, volume 22 of *Cambrige Tracts in Theoretical Computer Science*. Cambridge University Press, Cambridge, 1994.

99. J. F. Traub, G. Wasilkowski, and H. Woźniakowski. *Information-Based Complexity*. Computer Science and Scientific Computing. Academic Press, New York, 1988.

100. A. Troelstra and D. v. Dalen. *Constructivism in Mathematics, Volume 1*, volume 121 of *Studies in Logic and the Foundations of Mathematics*. North-Holland, Amsterdam, 1988.

101. A. Troelstra and D. v. Dalen. *Constructivism in Mathematics, Volume 2*, volume 123 of *Studies in Logic and the Foundations of Mathematics*. North-Holland, Amsterdam, 1988.

102. A. M. Turing. On computable numbers, with an application to the "Entscheidungsproblem". *Proceedings of the London Mathematical Society*, 42(2):230–265, 1936.

103. A. M. Turing. On computable numbers, with an application to the "Entscheidungsproblem". A correction. *Proceedings of the London Mathematical Society*, 43(2):544–546, 1937.

104. V. Vassiliev. Cohomology of braid groups and the complexity of algorithms. *Funktsional. Anal. i Prilozhen.*, 22(3):15 – 24, 1989. Englische Übers. in *Functional. Anal. Appl.*, 22:182–190, 1989.

105. M. Washihara. Computability and Fréchet spaces. *Mathematica Japonica*, 42(1):1–13, 1995.

106. K. Weihrauch. *Computability*, volume 9 of *EATCS Monographs on Theoretical Computer Science*. Springer, Berlin, 1987.

107. K. Weihrauch. On the complexity of online computations of real functions. *Journal of Complexity*, 7:380–394, 1991.

108. K. Weihrauch. The TTE-interpretation of three hierarchies of omniscience principles. Informatik Berichte 130, FernUniversität Hagen, Hagen, Sept. 1992.

109. K. Weihrauch. *Computable Analysis*. Springer, Berlin, 2000.

110. K. Weihrauch. Computational complexity on computable metric spaces. *Mathematical Logic Quarterly*, 49(1):3–21, 2003.

111. K. Weihrauch and N. Zhong. Is wave propagation computable or can wave computers beat the Turing machine? *Proceedings of the London Mathematical Society*, 85(2):312–332, 2002.

112. H. Weyl. Randbemerkungen zu Hauptproblemen der Mathematik. *Math. Zeitschrift*, 20:131–150, 1924.

113. H. Woźniakowski. Why does information-based complexity use the real number model? *Theoretical Computer Science*, 219:451–465, 1999.

114. M. Yasugi, T. Mori, and Y. Tsujii. Effective properties of sets and functions in metric spaces with computability structure. *Theoretical Computer Science*, 219:467–486, 1999.

115. X. Zheng. Recursive approximability of real numbers. *Mathematical Logic Quarterly*, 48(Suppl. 1):131–156, 2002.

116. X. Zheng and R. Rettinger. Weak computability and representation of reals. *Mathematical Logic Quarterly*, 50(4,5):431–442, 2004.

117. X. Zheng and K. Weihrauch. The arithmetical hierarchy of real numbers. *Mathematical Logic Quarterly*, 47(1):51–65, 2001.

118. N. Zhong. Recursively enumerable subsets of R^q in two computing models: Blum-Shub-Smale machine and Turing machine. *Theoretical Computer Science*, 197:79–94, 1998.

119. N. Zhong and K. Weihrauch. Computability theory of generalized functions. *Journal of the Association for Computing Machinery*, 50(4):469–505, 2003.

120. M. Ziegler and V. Brattka. Computability in linear algebra. *Theoretical Computer Science*, 326(1–3):187–211, 2004.

A Continuous Derivative for Real-Valued Functions

Abbas Edalat

Department of Computing, Imperial College London, London, UK
ae@doc.ic.ac.uk

Summary. We develop a notion of derivative of a real-valued function on a Banach space, called the L-derivative, which is constructed by introducing a generalization of Lipschitz constant of a map. As with the Clarke gradient, the values of the L-derivative of a function are nonempty, weak,* compact, and convex subsets of the dual of the Banach space. The L-derivative, however, is shown to be upper semicontinuous, a result that is not known to hold for the Clarke gradient. We also formulate the notion of primitive maps dual to the L-derivative, an extension of Fundamental Theorem of Calculus for the L-derivative, and a domain for computation of real-valued functions on a Banach space with a corresponding notion of effectivity. For real-valued functions on finite-dimensional Euclidean spaces, the L-derivative can be obtained within an effectively given continuous domain. We also show that in finite dimensions the L-derivative and the Clarke gradient coincide, thus providing a computable representation for the latter in this case.

This paper is dedicated to the historical memory of Sharaf al-din Tusi (d. 1213), the Iranian mathematician who was the first to use the derivative systematically to solve for roots of cubic polynomials and find their maxima.

1 Introduction

The notion of derivative of functions has been the key fundamental concept in the advent and development of differential calculus and is at the basis of some of the most crucial branches of mathematics, including ordinary and partial differential equations, dynamical systems, mathematical physics, differential geometry, and differential topology. These comprise what is often referred to as continuous mathematics, one of the two main branches of mathematics, with discrete mathematics as the other distinguished branch.

The first systematic use of the derivative of functions was undertaken by the Iranian mathematician Sharaf al-din Tusi (d. 1213) who introduced a technique, which is algebraically equivalent to what we now call the Ruffini–Horner method, for finding the roots of cubic polynomials by an iterative process using the derivative of the polynomial [14]. Although he never put a name to it in Arabic, which like Latin later on in Europe was the language of scholarship in the Muslim world, he also used the derivative to find the maxima of polynomials, which until recently historians of mathematics had attributed to the 16th century French mathematician *François Viète* [15, 19]. Sharaf al-din Tusi who died only six years before the cataclysmic Mongol invasion of Iran in 1219 is now considered to be the forerunner of algebraic geometry.

Nearly two centuries after the ground-breaking work of Newton and Leibniz on the foundation of differential calculus in the 17th century, modern mathematical analysis was born with the introduction of the mathematical limit by Cauchy in the nineteenth century, which provided a precise notion for the existence of the derivative of a function at a point. This led to new and surprising results about the derivative. In 1872, based on what was by then a rigorous mathematical framework, Weierstrass constructed a continuous function that was nowhere differentiable.

In the early 20th century, the French mathematicians *Gâteaux* and *Fréchet* extended the notion of derivative in two distinct ways to functions of infinite-dimensional Banach spaces. These higher dimensional derivatives have now applications in quantum field theory, but like the classical derivative of a real-valued function of a single variable, they may not exist and when they do exist they may not give rise to continuous functions. For a comprehensive modern account of the various notions of derivative in topological linear spaces see [27].

In the 1980s, Frank Clarke, motivated by problems in non-smooth analysis and control theory, introduced the notion of a generalized gradient of a function, which is now named after him [4]. Clarke's gradient of a locally Lipschitz real-valued function on a Banach space always exists and is a set-valued function: on finite-dimensional Euclidean spaces, it takes nonempty compact and convex subsets of the Euclidean space as its values and the gradient is upper semicontinuous. On an infinite-dimensional Banach space, the Clarke gradient is a nonempty, weak* compact, and convex subset of the dual of the Banach space. It is however not known if Clarke's gradient is also upper semicontinuous on infinite-dimensional Banach spaces [3].

A few decades earlier, following the seminal work of Alan Turing [24, 25] and the advent of computer science in the 1930s, computable analysis took shape in the 1950s with the work of Grzegorczyk [17, 18]. A fundamental thesis established in the subject is that a computable function is necessarily a continuous function [21, 26]. Indeed, if a function is to be computed at a real number, which is given as the limit of a sequence of rational numbers, then the continuity of the function is required to be able to compute the value of the function as the limit of its values at the elements of the sequence.

Since the derivative of functions plays a fundamental role in mathematics, one would expect a real interest in a notion of a derivative that is always continuous in computability theory. However, surprisingly, no attempt was made to develop a continuous derivative for functions and the work of Clarke went unnoticed by researchers in computable analysis, who have only worked with the classical derivative of functions.

A new approach to differential calculus based on mathematical structures in computer science, called domains [6, 2, 16], was introduced in [9, 10] first for real-valued functions of a real variable and then for multivariable functions. The motivation here has arisen from computer science and computable analysis to formulate and use, in particular, a notion of continuous derivative for functions.

In the domain-theoretic framework, a continuous derivative for functions, a corresponding notion of primitive maps, an extension of fundamental theorem of calculus, and a domain for differentiable functions have been developed. These have led to data types for presenting differentiable functions and solving ordinary differential equations [7, 11], a constructive version of the inverse and implicit function theorems [12], and a denotational semantics for hybrid systems [13].

The concept of a derivative of a real-valued function that was developed in [10] depends, somewhat unsatisfactorily, on the choice of the coordinate system used. In fact, the value of the derivative of a locally Lipschitz real valued function on a finite Euclidean space turns out to be the smallest hyperrectangle, with edges parallel to the given coordinate axes, containing Clarke's gradient.

In this paper, inspired by the above domain-theoretic framework, we introduce a coordinate free approach to develop the notion of the *L-derivative* of a real-valued function on a Banach space; it is constructed by formulating a generalized Lipschitz property of functions. The local generalized Lipschitz properties of the function, which provide finitary information about the rate of growth of the function in local neighbourhoods, are used to define the L-derivative of the function globally. Like the Clarke gradient, the values of the L-derivative are nonempty, weak* compact and convex subsets of the dual of the Banach space.

The L-derivative, developed here from the local to the global and from the discrete to the continuum, is shown to be upper semicontinuous for real-valued locally Lipschitz functions on any Banach space, a result that is not known for the Clarke gradient as we have already mentioned above.

For a C^1 function, i.e., one with a continuous Fréchet derivative, the L-derivative and the Fréchet derivative coincide. More generally, when the function fails to be C^1, the L-derivative contains the Clarke gradient, and also the Gâteaux and the Fréchet derivatives, whenever the latter two exist.

The L-derivative gives rise to an extension of the Fundamental Theorem of Calculus. The class of functions from the Banach space into the collection of nonempty, weak* compact and convex subsets of the dual of the Banach space, which are generated by step functions, is dual via the L-derivative to families of real-valued, locally Lipschitz functions on the Banach space. The L-derivative is also employed to construct a domain of computation for real-valued functions on Banach spaces that carries an effective structure when the space is separable. These results extend those for finite dimensions in [9, 10].

For functions on finite Euclidean spaces, the L-derivative is an element of a countably based continuous domain that can be given an effective structure that characterizes computable functions with computable L-derivatives. Any continuous function and its L-derivative can be obtained as the supremum of an increasing sequence of pairs of finitary and consistent information about the function and its L-derivative.

Although they are defined using very different techniques, we show here that in finite dimensions the Clarke gradient and the L-derivative coincide. Thus, in finite dimensions, the construction of the L-derivative provides a new computable representation for the Clarke gradient.

1.1 Background definitions

For the remainder of this section we will present the basic background definitions of the various notions of derivative that we will need in this paper.

Let X and Y be Banach spaces, and let $U \subset X$ be an open subset. We recall that the (one sided) directional derivative of $f : U \to Y$ at $x \in U$ in the direction $v \in X$ is

$$F'(x; v) = \lim_{t \downarrow 0} \frac{f(x + tv) - f(x)}{t},$$

if the limit exists. If the above directional derivative exists for all $v \in X$, then $D(f)(x) \to Y$ with $D(f)(x)(v) := F'(x; v)$ is the *Gâteaux* derivative of f at x if $D(f)(x)$ is a bounded linear map [27, 20].

The *Fréchet* derivative [27] of a map $f : U \to Y$ at $x \in U$, when it exists, is a bounded linear map $T : X \to Y$ with

$$\lim_{\|x-y\| \to 0} \frac{\|f(x) - f(y) - T(x - y)\|}{\|x - y\|} = 0.$$

The linear map T is denoted by $f'(x)$. When the Fréchet derivative exists at x, so does the Gâteaux derivative and they are equal. However, the *Fréchet* derivative at x can fail to exist even if the Gâteaux derivative exists at x and is a bounded linear map.

From now on we will assume that $Y = \mathbb{R}$. We next aim to define the generalized (Clarke) gradient of a function [4, Chapter two] and explain its properties. Let $f : U \to \mathbb{R}$ be Lipschitz near $x \in U$ and $v \in X$. The *generalized directional derivative* of f at x in the direction of v is

$$f^\circ(x; v) = \limsup_{\substack{y \to x \\ t \downarrow 0}} \frac{f(y + tv) - f(y)}{t}.$$

Let us denote by X^* the dual of X, i.e., the set of real-valued continuous linear functions on X. Unless otherwise stated, we will consider X^* with its weak* topology. Recall that the weak* topology is the weakest topology on X^* in which for any $x \in X$ the map $f \mapsto f(x)^* \to \mathbb{R}$ is continuous.

The *generalized gradient* of f at x, denoted by $\partial f(x)$, is the subset of X^* given by

$$\{A \in X^* : f^\circ(x; v) \geq A(v) \text{ for all } v \in X\}.$$

It is shown in [4, page 27] that

- $\partial f(x)$ is a nonempty, convex, weak* compact subset of X^*.

- For $v \in X$, we have

$$f^\circ(x; v) = \max\{A(v) : A \in \partial f(x)\}.$$

There is an alternative characterization of the generalized gradient when X is finite dimensional, say $X = \mathbb{R}^n$. In this case, by Rademacher's theorem [5, p 148], a locally Lipschitz map $f : U \to \mathbb{R}$ is Fréchet differentiable almost everywhere with respect to the Lebesgue measure. If Ω_f is the nullset where f fails to be differentiable, then

$$\partial f(x) = \mathrm{Co}\{\lim f'(x_i) : x_i \to x, \ x_i \notin \Omega_f\}, \tag{1}$$

where $\mathrm{Co}(S)$ is the convex hull of a subset $S \subset \mathbb{R}^n$ [4, page 63]. The above expression is interpreted as follows. Consider all sequences $(x_i)_{i \geq 0}$, with $x_i \notin \Omega_f$, for $i \geq 0$, which converge to x such that the limit $f'(x_i)$ exists. Then the generalized gradient is the convex hull of all such limits. Note that, in the above definition, since f is locally Lipschitz at x, it is differentiable almost everywhere in a neighbourhood of x and thus there are plenty of sequences $(x_m)_{m \geq 0}$ such that $\lim_{m \to \infty} x_m = x$ and $\lim_{m \to \infty} f'(x_m)$ exist.

Recall that for a Hausdorff space Z, we can define three topologies on the set of nonempty compact subsets of Z as follows. The *upper topology* has as a base the collection of subsets of the form $\Box O = \{C : C \subseteq O\}$, whereas the *lower topology* has as a subbase the collection of subsets of the form $\Diamond O = \{C : C \cap O \neq \emptyset\}$, where $O \subset Z$ is an open subset. The *Vietoris topology* is the refinement of the upper and lower topologies and is Hausdorff, and when Z is a metric space, it is equivalent to the topology induced by the *Hausdorff metric* d_h defined by $d_h(A, B) = \max(d(A, B), d(B, A))$, where for compact sets C and D, $d(C, D)$ is the infimum of positive numbers δ such that C is contained in the δ-*parallel body* of D defined as $D_\delta = \{x \in Z \mid \exists y \in D. d(x, y) \leq \delta\}$; see [23, page 737]. We write $\mathbf{U}(Z)$, $\mathbf{L}(Z)$, and $\mathbf{V}(Z)$, respectively, for the three topological spaces or *hyperspaces*, called respectively the *upper space*, the *lower space*, and the *Vietoris space* of Z, obtained by considering, respectively, the upper topology, the lower topology, and the Vietoris topology on the set of nonempty compact subsets of Z. The upper space and the lower space are non-Hausdorff.

In finite dimensions, the Clarke gradient is *upper semicontinuous*; i.e., it is continuous with respect to the *upper topology* on the space of the nonempty compact subsets of $\mathbb{R}^n$. It is not known if a similar result holds in infinite dimensions [3], i.e., if the Clarke gradient is continuous with respect to the upper topology on the space of non-empty weak* compact subsets of X^*.

For $X = \mathbb{R}^n$, we let ∇f denote the classical gradient of f, when it exists; i.e.,

$$(\nabla f)_i(x) = \frac{\partial f}{\partial x_i}$$

$$= \lim_{x_i' \to x_i} \frac{f(x_1, \ldots, x_i, \ldots, x_n) - f(x_1, \ldots, x_i', \ldots, x_n)}{x_i - x_i'},$$

for $1 \leq i \leq n$. Recall that, in finite dimensions, if the (Fréchet) derivative exists at a point, then the gradient also exists at that point and is the same linear map.

We also recall that for a function $f : U \to \mathbb{R}$, where U is an open subset of $\mathbb{R}^n$, Dini's lower and upper partial derivatives, for $1 \leq i \leq n$, are defined, respectively, as

$$(\nabla f)_i^l(x) = \liminf_{x_i' \to x_i} \frac{f(x_1, \ldots, x_i, \ldots, x_n) - f(x_1, \ldots, x_i', \ldots, x_n)}{x_i - x_i'}, \qquad (2)$$

$$(\nabla f)_i^u(x) = \limsup_{x_i' \to x_i} \frac{f(x_1, \ldots, x_i, \ldots, x_n) - f(x_1, \ldots, x_i', \ldots, x_n)}{x_i - x_i'}. \qquad (3)$$

Note that the Dini's lower and upper partial derivatives always exist as extended real numbers.

2 Some properties related to the dual of a Banach space

Let X be a Banach space. For an open subset $U \subset X$, let $U \to \mathbb{R}$ be the set of all continuous functions of type $U \to \mathbb{R}$ with respect to the norm topology on X. For $x \in X$ and $f \in X^*$, we write $f(x)$ for the real number obtained by the action of f on x and $x(f)$ for the same real number when x is considered as a linear functional on X^*.

The operator norm on X^* extends pointwise to an interval valued map on the weak* compact subsets of X^*. Note that a weak* compact subset is bounded with respect to the operator norm. If b is a nonempty, weak* compact and convex subset of X^*, then $\|b\| = \{\|\lambda\| : \lambda \in b\}$ with $\|b\| = [\|b\|^-, \|b\|^+]$ is a compact real interval. In particular, for $X = \mathbb{R}$, if $b \subset \mathbb{R}$ is a compact interval, then so is $|b| = \{|r| : r \in b\}$.

We will consider the extension of the action of bounded linear operators on X (i.e., the mapping $X^* \times X \to \mathbb{R}$ given by $(f, x) \mapsto f(x)$) to the three hyperspaces to obtain three maps E_U : $\mathbf{U}(X^*) \times X \to \mathbf{U}(\mathbb{R})$, E_L : $\mathbf{L}(X^*) \times X \to \mathbf{L}(\mathbb{R})$, and E_V : $\mathbf{V}(X^*) \times X \to \mathbf{V}(\mathbb{R})$, which are defined with respect to the different three topologies but have the same action given by $(b, x) \mapsto \{f(x) : f \in b\}$. We write $b(x) = \{f(x) : f \in b\}$, which is a compact subset as b is weak* compact.

Proposition 2.1 *The three maps E_U, E_L, and E_V are each continuous separately in their two arguments with respect to the norm topology on X.*

Proof. To prove the continuity of the three maps when the second argument is fixed, let $x \in X$. It is sufficient to show that, for any open set $I \subset \mathbb{R}$, the preimage of $\square I$ is open in $\mathbf{U}(X^*)$ and the preimage of $\lozenge I$ is open in $\mathbf{L}(X^*)$. First we prove that the preimage of $\square I$ is open in $\mathbf{U}(X^*)$. Let b be a nonempty, weak* compact subset of X^* with $b(x) \subset I$. Let $\epsilon > 0$ be such that $(b(x))_\epsilon \subset I$. For any $f \in b$, the weak* open set $O(x, f, \epsilon) = \{g \in X^* : |f(x) - g(x)| < \epsilon\}$ gives an open neighbourhood of f. Put $O(x, \epsilon) = \bigcup_{f \in b} O(x, f, \epsilon)$. Then, we have $b \subset O(x, \epsilon)$. If c is a weak* compact set with $c \subset O(x, \epsilon)$, then by compactness, there exists a finite number of functions $f_i \in b$ $(i = 1, \ldots, n)$ such that $c \subset \bigcup_{1 \le i \le n} O(x, f_i, \epsilon)$. Then for any $g \in c$, there exists $i \in \{1, \ldots, n\}$ such that $f_i \in b$ with $|f_i(x) - g(x)| < \epsilon$ and thus $g(x) \in I$. It follows that $c(x) \subset I$, and therefore, the preimage of $\square I$ is open in $\mathbf{U}(X^*)$. Next we consider the preimage of $\lozenge I$. Let b be a nonempty, weak* compact subset of X^* with $b(x) \cap I \ne \emptyset$. Let $f \in b$ with $f(x) \in I$ take any $\epsilon > 0$ such that the ϵ open ball centred at $f(x)$ is contained in I. Then, $b \cap O(x, f, \epsilon) \ne \emptyset$, and for any weak* compact subset c of X^* with $c \cap O(x, f, \epsilon) \ne \emptyset$, we have $c(x) \cap I \ne \emptyset$, which shows that the preimage of $\lozenge I$ is open in $\mathbf{L}(X^*)$. Finally, we prove the continuity of the three maps when the first argument b, say, is fixed. Let $b(x) \in O$ where O is either $\square I$ and $\lozenge I$ for any open subset $I \subset \mathbb{R}$. Since any weak* compact subset of X^* is bounded with respect the operator norm, $\|b\| \le K$ for some $K > 0$. Hence, for any given $\epsilon > 0$ and any $f \in b$, the relation $\|x - y\| < \epsilon/K$ implies $|f(x) - f(y)| \le \|f\|\|x - y\| \le K\epsilon/K = \epsilon$. From this property, the result follows easily. $\square$

As usual, we consider the upper space $\mathbf{U}(Z)$ of any Hausdorff space Z partially ordered with reverse inclusion so that $\mathbf{U}(Z)$ becomes a dcpo; we also include in this dcpo a least element represented by Z. Thus, the map $x \mapsto \{x\} : Z \to \mathbf{U}(Z)$ is a topological embedding onto the set of maximal elements of $\mathbf{U}(Z)$. We identify the input and output of this embedding and write $\{x\}$ simply as x.

We also recall that the *Scott* topology on any dcpo has as open sets those sets O which are upper sets (that is $x \in O$ and $x \sqsubseteq y$ implies $y \in O$) and that are inaccessible by directed sets; i.e., if $\sup_{i \in I} a_i \in O$ for a directed set $(a_i)_{i \in I}$, then there exists $i \in I$ such that $a_i \in O$ [22]. A function $f : D \to E$ of dcpo's D and E is continuous with respect to the Scott topologies on D and E iff it is monotone ($x \sqsubseteq y$ implies $f(x) \sqsubseteq f(y)$) and preserves the lubs of directed subsets; i.e., for any directed set $(a_i)_{i \in I}$ in D, we have $\sup_{i \in I} f(a_i) = f(\sup_{i \in I} a_i)$.

We then have the following:

Proposition 2.2 [6, Propositions 3.1(iii) and 3.3]

(i) *For any Hausdorff space Z, the Scott topology on $\mathbf{U}(Z)$ refines the upper topology.*

(ii) *If Z is locally compact, then $\mathbf{U}(Z)$ is a continuous dcpo, on which the Scott topology and the upper topology coincide.* $\qquad\square$

Consider the poset, denoted by $\mathbf{C}(X^*)$, consisting of X^* and its nonempty, weak* compact and convex subsets partially ordered by reverse inclusion so that it has the least element $\bot = X^*$. Note that $\mathbf{C}(X^*)$ is a bounded complete dcpo and a sub-dcpo of the upper space $\mathbf{U}(X^*)$ of X^*.

When $X = \mathbb{R}$, we consider the sub-dcpo of $\mathbf{U}(\mathbb{R})$ denoted by $\mathbb{IR}$ of all nonempty compact intervals of $\mathbb{R}$ ordered by reverse inclusion; it is a countably based bounded complete continuous domain. We now restrict the first component of G in Proposition 2.1 to convex subsets so that the range of G will become nonempty compact intervals. Restricting to continuity with respect to the Scott topology, Proposition 2.1 reduces to:

Corollary 2.3 *The map $G : \mathbf{C}(X^*) \times X \to \mathbb{IR}$ is continuous separately in its two arguments with respect to the Scott topology on $\mathbf{C}(X^*)$ and $\mathbb{IR}$ and the norm topology on X. In particular, for any directed set $b_i \in \mathbf{C}(X^*)$, $i \in I$, and $v \in X$, we have $\bigcap_{i \in I}(b_i(v)) = (\bigcap_{i \in I} b_i)(v)$.* $\qquad\square$

The following result plays a crucial role in the construction of the L-derivative.

Theorem 2.4 *Let S and T be disjoint nonempty convex subsets of the dual X^* of a Banach space X such that, with respect to the weak* topology, S is closed and T is compact. Then there exists a hyperplane in X^* induced by an element of X which separates S and T; i.e., there exist $x \in X$ and $c \in \mathbb{R}$ such that $x(f) < c$ for $f \in S$ and $x(f) > c$ for $f \in T$.*

Proof. Since S and T are disjoint closed sets, for each $\lambda \in T$, there exist $\epsilon > 0$ and a finite number of elements $x_1, \ldots x_n \in X$ such that the open neighbourhood of λ defined by $\{\alpha : |x_j(\alpha) - x_j(\lambda)| < \epsilon$ for $1 \leq j \leq n\}$ is disjoint from S. By compactness of T we can find a finite number of functionals, say, $\lambda_i \in X^*$ $(1 \leq i \leq m)$ and elements $x_{i1}, x_{i2}, \ldots x_{in_i} \in X$ and $\epsilon_i > 0$ for $1 \leq i \leq m$ such that the open subsets

$$N_i = \{\alpha : |x_{ij}(\alpha) - x_{ij}(\lambda_i)| < \epsilon_i, \text{ for } 1 \leq j \leq n_i\},$$

for $1 \leq i \leq m$, are disjoint from S and cover T. Define $\Phi : X^* \to \mathbb{R}^{n_1 + n_2 + \cdots + n_m}$ by

$$\alpha \mapsto$$

$$(x_{11}(\alpha), \ldots, x_{1n_1}(\alpha); \ldots; x_{i1}(\alpha), \ldots, x_{in_i}(\alpha); \ldots; x_{m1}(\alpha), \ldots, x_{mn_m}(\alpha)).$$

Then $\Phi(S)$ and $\Phi(T)$ are convex subsets of $\mathbb{R}^{n_1+n_2+\cdots,+n_m}$ and $\Phi(T)$ is compact. Let $P_i : \mathbb{R}^{n_1} \times \cdots \times \mathbb{R}^{n_i} \times \cdots \times \mathbb{R}^{n_m} \to \mathbb{R}^{n_i}$ be the projection onto $\mathbb{R}^{n_i}$. Note that $\Phi(T)$ is contained in the open subset of $\mathbb{R}^{n_1+\cdots+n_m}$ given by the intersection of the m infinite open strips $\mathbb{R}^{n_1} \times \mathbb{R}^{n_2} \times \cdots \times \mathbb{R}^{n_{i-1}} \times R_i \times \mathbb{R}^{n_{i+1}} \times \cdots \times \mathbb{R}^{n_m}$, where

$$R_i = \{z : \|z - P_i(\Phi(x_i))\| < \epsilon_i\},$$

with $\|v\|$ being the max norm of $v \in \mathbb{R}^{n_i}$. Therefore, $\Phi(T)$ is disjoint from the closure of $\Phi(S)$. Note that for any two disjoint convex subsets of $\mathbb{R}^k$, with one closed and the other compact, a hyperplane in $\mathbb{R}^k$ separates them (the normal to such a hyperplane is given by the line through two boundary points of the sets, which give the closest distance of the two sets). Thus, is a hyperplane $\sum_{1 \leq i \leq m} \sum_{1 \leq j \leq n_i} c_{ij} z_{ij} = c$ in $\mathbb{R}^{n_1+n_2+\cdots+n_m}$, for some real numbers $c, c_{ij} \in \mathbb{R}$ ($1 \leq i \leq m$ and $1 \leq j \leq n_i$), separates $\Phi(T)$ from the closure of $\Phi(S)$. Let $x = \sum_{1 \leq i \leq m} \sum_{1 \leq j \leq n_i} c_{ij} x_{ij} \in X$. It follows that the hyperplane $x(\alpha) = c$, where $\alpha \in X^*$, separates S and T in X^*.   □

3 Ties of functions

The local differential property of a function is formalized in our framework by the notion of an interval Lipschitz constant. Assume $U \subset X$ is an open subset of a Banach space X.

Definition 3.1 Let f be a real-valued function with domain $\mathrm{dom}(f) \subset U$. We say that $f : \mathrm{dom}(f) \to \mathbb{R}$ has *an interval Lipschitz constant* $b \in \mathbf{C}(X^*)$ in a convex open subset $a \subset \mathrm{dom}(f)$ if for all $x, y \in a$, we have $b(x - y) \sqsubseteq f(x) - f(y)$. The *single tie* $\delta(a, b)$ of a with b is the collection of all real-valued partial functions f on U with $a \subset \mathrm{dom}(f) \subset U$, which have an interval Lipschitz constant b in a. We call a the *domain* of the single tie.

Since a single tie provides a local Lipschitz property for a family of functions, it is sufficient in Definition 3.1 to restrict the domain of a single tie to a convex open subset. As an example, if $X = \mathbb{R}^2$ and b is the compact rectangle $b_1 \times b_2$ (with compact intervals $b_1, b_2 \subset \mathbb{R}$), the information relation above reduces to

$$b_1(x_1 - y_1) + b_2(x_2 - y_2) \sqsubseteq f(x) - f(y).$$

Lemma 3.2 *For $b \in \mathbf{C}(X^*)$ and $z \in X$, we have $|b(z)|^+ \leq \|b\|^+ \|z\|$.*

Proof. We have $|b(z)|^+ = |\{f(z) : f \in b\}|^+ = \{|f(z)| : f \in b\}^+ \leq \{\|f\| : f \in b\}^+ \|z\| = \|b\|^+ \|z\|$.   □

Proposition 3.3 *If $f \in \delta(a, b)$ for $a \neq \emptyset$ and $b \neq \perp$, then $f : a \to \mathbb{R}$ is Lipschitz: for all $x, y \in a$ we have $|f(x) - f(y| \leq \|b\|^+ \|x - y\|$.*

Proof. Suppose $f \in \delta(a, b)$ and $x, y \in a$. It follows from $f(x) - f(y) \sqsupseteq b(x - y)$ that $|f(x) - f(y)| \leq \|b\|^+ \|x - y\|$.   □

For any topological space Z and any bounded complete dcpo D with bottom $\perp$, let $Z \to D$ be the bounded complete dcpo of Scott continuous functions from Z to D. The domain of $f : Z \to D$ is defined as $\mathrm{dom}(f) = \{x : f(x) \neq \perp\}$. In particular, for any open subset $a \subset Z$ and any non-bottom $b \in D$, the *single step function* $a \searrow b : Z \to D$, with $(a \searrow b)(x) = b$ if $x \in a$ and $(a \searrow b)(x) = \perp$ if $x \notin a$, is Scott continuous and has domain a. A *step function* is then the supremum of any finite set of consistent single step functions. In the sequel, we consider the dcpo $U \to \mathbf{C}(X^*)$ of Scott continuous functions with $U \subset X$ equipped with its the norm topology.

The following proposition justifies our definition of the interval Lipschitz constant. Let a be a convex open subset of X.

Proposition 3.4 *If $f : a \to \mathbb{R}$ is $C^1(a)$, i.e., f is Fréchet differentiable and $f' : a \to X^*$ is continuous, then the following three conditions are equivalent: (i) $f \in \delta(a, b)$, (ii) $\forall z \in a.\ f'(z) \in b$, and (iii) $a \searrow b \sqsubseteq f'$.*

Proof. (i) $\Rightarrow$ (ii). Suppose, for the sake of a contradiction, that for some $z \in a$, we have $L := f'(z) \notin b$. By Theorem 2.4, there exists a unit vector $s \in X$ and $c \in \mathbb{R}$ such that $(s(L))^+ < c$ and $(s(b))^- > c$. From $f \in \delta(a, b)$, we obtain for sufficiently small h that $\frac{f(z+hs)-f(z)}{h} \in s(b)$. But by Fréchet differentiability at z, we have

$$\lim_{h \to 0} \left| \frac{f(z + hs) - f(z)}{h} - s(L) \right| = 0,$$

which is a contradiction.

(ii) $\Rightarrow$ (i). Assume $x, y \in a$. Then, since the convex set a contains the straight line from x to y, by the mean value theorem, there exists $z \in a$ such that $f(x) - f(y) = f'(z)(x - y) \in b(x - y)$.

(iii) $\Longleftrightarrow$ (ii). Obvious. $\qquad\square$

Note that the convexity of the domain of a single tie is crucial in establishing the equivalence in Proposition 3.4.

We will now show that ties have a dual property in relation to step functions of type $U \to \mathbf{C}(X^*)$.

Proposition 3.5 *Suppose $a \neq \emptyset$ and $b \neq \perp$. We have $\delta(a, b) \supseteq \delta(c, d)$ iff $c \supseteq a$ and $b \sqsubseteq d$.*

Proof. The "if" part follows easily from the definition of $\delta(a, b)$. To show the "only if" part, we take any $f \in \delta(c, d)$ such that $\mathrm{dom}(f) = c$. Then, since $f \in \delta(a, b)$, we have $a \subset \mathrm{dom}(f) = c$. On the other hand if $b \sqsubseteq d$ does not hold, take $\gamma \in d \setminus b$ and consider the function $f : c \to \mathbb{R}$ with $f(x) = \gamma(x)$. Then, $f \in \delta(c, d) \setminus \delta(a, b)$. $\quad\square$

Corollary 3.6 *Suppose $a, c \neq \emptyset$ and $b, d \neq \perp$. We have $\delta(a, b) = \delta(c, d) \iff a = c$ & $b = d$. Furthermore $\delta(a, b) \supseteq \delta(c, d)$ iff $a \searrow b \sqsubseteq c \searrow d$.* $\quad\square$

For the rest of this section, we assume we are in an infinite-dimensional Banach space or in the finite-dimensional space $\mathbb{R}^n$ with $n \geq 2$. The case $n = 1$ is completely covered in [9].

Definition 3.7 A *tie* of partial real-valued functions on U is any intersection $\Delta = \bigcap_{i \in I} \delta(a_i, b_i)$, for an arbitrary indexing set I. The *domain* of a nonempty tie Δ is defined as $\text{dom}(\Delta) = \bigcup_{i \in I}\{a_i \mid b_i \neq \bot\}$.

If a nonempty tie is given by the intersection of a finite number of single ties, then it gives us a family of functions with a *finite* set of consistent differential properties. Generally, a nonempty tie gives a family of functions with a consistent set of differential properties.

Similar to Proposition 3.3, we have the following result. Recall that a function $f : U \to \mathbb{R}$ defined on the open set $U \subseteq X$ is *locally Lipschitz* if it is Lipschitz in a neighbourhood of any point in U.

Proposition 3.8 *If Δ is a tie and $f \in \Delta$, then f is locally Lipschitz on $\text{dom}(\Delta)$.*

Proof. Let $x \in \text{dom}(\Delta)$. Then there exists a tie $\delta(a, b)$ with $x \in a$ and $f \in \Delta \subseteq \delta(a, b)$, and the result follows from Proposition 3.3. $\square$

We now collect some fundamental properties of ties, which we will use later. The next proposition, whose proof uses Theorem 2.4, is the key technical result for the development of our theory.

Proposition 3.9 *For any indexing set I, the family of step functions $(a_i \searrow b_i)_{i \in I}$ is consistent if $\bigcap_{i \in I} \delta(a_i, b_i) \neq \emptyset$.*

Proof. Suppose $f \in \bigcap_{i \in I} \delta(a_i, b_i)$. We will show that every finite subfamily of $(a_i \searrow b_i)_{i \in I}$ is consistent, from which the result follows as $\mathbf{C}(X^*)$ is bounded complete. It suffices to prove that for any finite subset $J \subseteq I$, we have $\bigcap_{j \in J} b_j \neq \emptyset$ if $\bigcap_{j \in J} a_j \neq \emptyset$. This we will show by induction on the cardinality $|J|$ of J. For $|J| = 1$, there is nothing to prove. Suppose now $|J| > 1$ and $\bigcap_{j \in J} a_j \neq \emptyset$. Let $k \in J$. Then by the inductive hypothesis $\bigcap_{j \in J \setminus \{k\}} b_j \neq \emptyset$. If $\bigcap_{j \in J} b_j = \emptyset$, then by Theorem 2.4 there exists a vector $z \in X$ and $c \in \mathbb{R}$ such that the disjoint, nonempty compact convex sets $\bigcap_{j \in J \setminus \{k\}} b_j$ and b_k are on the opposite sides of the affine space $\{\lambda : z(\lambda) = c\}$. Take elements $x, y \in \bigcap_{j \in J} a_j$ such that $x - y = lz$ for some $l > 0$. It follows that the two intervals $(\bigcap_{j \in J \setminus \{k\}} b_j)(x - y)$ and $b_k(x - y)$ are disjoint. But by our assumption that $f \in \bigcap_{i \in I} \delta(a_i, b_i) \subseteq \bigcap_{j \in J} \delta(a_j, b_j)$, we have $b_k(x - y) \sqsubseteq f(x) - f(y)$ and $b_j(x - y) \sqsubseteq f(x) - f(y)$ for $j \neq k$, which implies $(\bigcap_{j \in J \setminus \{k\}} b_j)(x - y) \sqsubseteq f(x) - f(y)$, a contradiction. $\square$

Corollary 3.10 *The family $(a_i \searrow b_i)_{i \in I}$ is consistent if for any finite subfamily $J \subseteq I$, we have $\bigcap_{i \in J} \delta(a_i, b_i) \neq \emptyset$.* $\square$

Proposition 3.11 *If $a \searrow b \sqsubseteq \sup_{i \in I} a_i \searrow b_i$, then $\delta(a, b) \supseteq \bigcap_{i \in I} \delta(a_i, b_i)$.*

Proof. Let $b \sqsubseteq \bigcap_{a_i \supseteq a} b_i$, and assume $f \in \bigcap_{i \in I} \delta(a_i, b_i)$. Let $x, y \in a$. For each $i \in I$ with $a_i \supseteq a$, we have

$$b_i(x - y) \sqsubseteq f(x) - f(y).$$

Therefore, we get

$$b(x - y) \sqsubseteq \bigcap_{a_i \supseteq a} b_i(x - y) \sqsubseteq f(x) - f(y),$$

as required. $\qquad\qquad\square$

Corollary 3.12 *If* $\sup_{i \in I} a_i \searrow b_i \sqsubseteq \sup_{i \in J} a_i \searrow b_i$, *then*

$$\bigcap_{i \in I} \delta(a_i, b_i) \supseteq \bigcap_{i \in J} \delta(a_i, b_i). \qquad\qquad\square$$

Let $(\mathbf{T}(U), \supseteq)$ be the partial order of ties of $U \to X$ ordered by reverse inclusion.

Proposition 3.13 $(\mathbf{T}(U) \setminus \{\emptyset\}, \supseteq)$ *is a dcpo.*

Proof. Suppose $(\Delta_j)_{j \in J}$ is a directed set in $(\mathbf{T}(U) \setminus \{\emptyset\}$ with respect to the partial order $\supseteq$; i.e., $\Delta_{j_1} \cap \Delta_{j_2} \neq \emptyset$ for $j_1, j_2 \in J$. Let $\Delta_j = \bigcap_{i \in I_j} \delta(a_i, b_i)$, where we assume $I_{j_1} \cap I_{j_2} = \emptyset$ for $j_1 \neq j_2$. Consider the collection $(\delta(a_i, b_i))_{i \in \bigcup_{j \in J} I_j}$. By Corollary 3.10, it suffices to show that any finite subfamily of this collection has a nonempty intersection. Suppose $i_t \in \bigcup_{j \in J} I_j$ for $1 \leq t \leq n$. Then $\delta(a_{i_t}, b_{i_t}) \in \Delta_{j_t}$ for some $j_t \in J$ $(1 \leq t \leq n)$. By assumption $\bigcap_{1 \leq t \leq n} \Delta_{j_t} \neq \emptyset$. Hence, $\bigcap_{1 \leq t \leq n} \delta(a_{i_t}, b_{i_t}) \supseteq \bigcap_{1 \leq t \leq n} \Delta_{j_t} \neq \emptyset$. $\qquad\square$

For any topological space Z and any bounded complete dcpo D, let $Z \to_s D$ be the subset of $Z \to D$ consisting of Scott continuous functions that are supremums of step functions; i.e., $f = \sup_{i \in I} a_i \searrow b_i$ for a family $(a_i \searrow b_i)_{i \in I}$ of step functions with a_i an open subset of Z and $b_i \in D$. We note that $Z \to D$ is bounded, complete, continuous dcpo iff the lattice of open subsets of Z is continuous [16]. Thus, $Z \to_s D$ is the whole function space $Z \to D$ iff the lattice of open subsets of Z is continuous.

Consider $U \to_s \mathbf{C}(X^*)$. Since any open set $a \subset X$ is the union of open balls, we can assume without loss of generality that the open subsets a_i $(i \in I)$ in the expression for f above are convex. It is easy to check that $U \to_s \mathbf{C}(X^*)$ is a dcpo.

We now show that, for any Banach space X, the set of maximal elements of $U \to_s \mathbf{C}(X^*)$ contains the set of functions of type $U \to X^*$, which are continuous with respect to the norm topology on U and X^*. Recall that a metric space is separable if it has a countable dense subset.

Proposition 3.14 (i) *If $f : U \to X^*$ is continuous with respect to the norm topologies on U and X^*, then $f \in U \to_s \mathbf{C}(X^*)$. Moreover, if X is separable with a countable dense subset $P \subset X$, then f is the lub of single step functions of the form $a \searrow b$, where a is an open ball centred at a point of P with rational radius, whereas b is a closed ball centred at a point of P with a rational radius.*

(ii) *If $f : U \to \mathbb{R}$ is continuous with respect to the norm topology on U, then $f \in U \to_s \mathbb{IR}$. Moreover, if X is separable with a countable dense subset $P \subset X$, then f is the lub of single step functions of the form $a \searrow b$, where a is an open ball centred at a point of P with rational radius, whereas b is a rational compact interval.*

Proof. (i) By continuity of f, for $x \in U$ and an open ball $B_\epsilon(f(x))$ of radius ϵ around $f(x)$, there exists an open neighbourhood a of x such that $f[a] \subset B_\epsilon(f(x))$. Since the closed ball $\overline{B_\epsilon(f(x))}$ is weak* compact by Alaoglu's theorem, it follows that $a \searrow \overline{B_\epsilon(f(x))} \sqsubseteq f$. Since $\epsilon > 0$ is arbitrary and $\bigcap_{\epsilon > 0} \overline{B_\epsilon(f(x))} = f(x)$, we conclude that f is the supremum of step functions below it. It is easy to check that the second statement also holds.

(ii) This is proved similar to (i). □

We are finally in a position to define the L-primitives of a Scott continuous function; in fact now we can do more and define

Definition 3.15 The *L-primitive map* $\int : (U \to_s \mathbf{C}(X^*)) \to \mathbf{T}(U)$ is defined by

$$\int f = \bigcap_{a \searrow b \sqsubseteq f} \delta(a, b).$$

We call $\int f$ the *L-primitives* of f. The following result depends crucially on the fact that the domain of the L-primitive map is defined to be $U \to_s \mathbf{C}(X^*)$ rather than the bigger function space $U \to \mathbf{C}(X^*)$.

Proposition 3.16 *If $f = \sup_{i \in I} a_i \searrow b_i$, then $\int f = \bigcap \delta(a_i, b_i)$.*

Proof. This follows easily from Corollary 3.12.

The above property leads us to believe that $U \to_s \mathbf{C}(X^*)$, respectively, $Z \to D$, may have wider applications in Banach space theory, respectively, abstract domain theory, beyond this paper.

Proposition 3.17 *The L-primitive map is continuous and onto the set of nonempty tie.*

Proof. Clearly the primitive map is monotone. Let $(g_i)_{i \in I}$ be a directed set in $U \to_s \mathbf{C}(X^*)$ with $g_i = \sup_{j \in I_i} a_j \searrow b_j$. Then,

$$\int \sup_{i \in I} g_i = \int \sup_{i \in I} \sup_{j \in I_i} a_j \searrow b_j$$

$$= \bigcap_{i \in I} \bigcap_{j \in I_i} \delta(a_j, b_j) = \bigcap_{i \in I} \int g_i = \sup_{i \in I} \int g_i.$$

By Proposition 3.9, any nonempty tie is the L-primitive of some element.   □

If $X = \mathbb{R}^n$, for $n \geq 2$ or if X is infinite dimensional, the L-primitive map will have the empty tie in its range, a situation that does not occur for $n = 1$. This is similar to the situation in classical analysis in which a continuous vector field in $\mathbb{R}^n$ for $n > 1$ may not be an exact differential.

Example 3.18 Let $g \in \mathbb{R}^2 \to \mathbf{C}(R^2)$ be the maximal function given by $g(x, y) = (g_1(x, y), g_2(x, y))$ with $g_1(x, y) = 1$ and $g_2(x, y) = x$. Then $\frac{\partial g_1}{\partial y} = 0 \neq 1 = \frac{\partial g_2}{\partial x}$, and it will follow as in classical analysis that $\int g = \emptyset$.

4 The L-derivative

Given a Scott continuous function $f : U \to \mathbb{R}$, the relation $f \in \delta(a, b)$ provides, as we have seen, finitary information about the local interval Lipschitz properties of f. By collecting all such local information, we obtain the complete differential properties of f, namely its L-derivative.

Definition 4.1 The *L-derivative* of a continuous function $f : U \to \mathbb{R}$ is the map

$$\mathcal{L}f : U \to \mathbf{C}(X^*),$$

given by

$$\mathcal{L}f = \sup_{f \in \delta(a,b)} a \searrow b.$$

Theorem 4.2 (i) *The L-derivative is well defined and Scott continuous.*

(ii) *If $f \in C^1(U)$, then $\mathcal{L}f = f'$.*

(iii) $f \in \delta(a, b)$ *iff* $a \searrow b \sqsubseteq \mathcal{L}f$.

Proof. (i) Let the indexing set I be defined by $i \in I \iff f \in \delta(a_i, b_i)$. Then $\bigcap_{i \in I} \delta(a_i, b_i) \neq \emptyset$. Hence, $(a_i \searrow b_i)_{i \in I}$ is consistent by Proposition 3.9. Therefore, $\mathcal{L}f = \sup_{i \in I} a_i \searrow b_i$ is well defined and is Scott continuous.

(ii) By Proposition 3.4, $f \in \delta(a, b) \iff a \searrow b \sqsubseteq f'$. Hence,

$$f' \sqsupseteq \sup_{f \in \delta(a,b)} a \searrow b.$$

To show equality, let $z \in U$ and put $L := f'(z)$. By the continuity of the Fréchet derivative $f' : U \to X^*$ at z, for each integer $n > 0$, there exists an open ball $a \subset U$

with $z \in a$ such that $f'(x) \in B_{1/n}(L)$ for $x \in a$, where $B_r(L)$ is the open ball of radius r and centre $L \in X^*$. By Proposition 3.4, we have $f' \sqsupseteq \mathcal{L}f \sqsupseteq a \searrow \overline{B_{1/n}(L)}$, where $\overline{B_r(L)}$ is the closed ball centred at L with radius r, which is convex and weak* compact by Alaoglu's Theorem. Since $\bigcap_{n \geq 0} \overline{B_{1/n}(L)} = f'(z)$, we conclude that $f' = \mathcal{L}f$.

(iii) Obvious.   □

Since the Scott topology refines the upper topology on $\mathbf{C}(X^*)$, we also obtain

Corollary 4.3 *The L-derivative of any continuous function $X \to \mathbb{R}$ is upper semicontinuous.*   □

We now obtain the generalization of Theorem 4.2(iii) to ties, which provides a duality between the L-derivative and the L-primitives and can be considered as a general version of the Fundamental Theorem of Calculus.

Theorem 4.4 (Fundamental Theorem of Calculus) *For any $g \in U \to_s \mathbf{C}(X^*)$,*

$$f \in \int g \iff g \sqsubseteq \mathcal{L}f.$$

Proof. Let $g \in U \to_s \mathbf{C}(X^*)$. Then by Theorem 4.2(iii):

$$f \in \int g \iff f \in \bigcap_{a \searrow b \sqsubseteq g} \delta(a, b)$$

$$\iff a \searrow b \sqsubseteq \mathcal{L}f \text{ if } a \searrow b \sqsubseteq g \iff g \sqsubseteq \mathcal{L}f.$$ □

We will now show that the Gâteaux derivative, if it exists, is always in the L-derivative.

Lemma 4.5 *For any locally Lipschitz map $f : U \to \mathbb{R}$ and any $x, v \in X$, we have*

$$(\mathcal{L}f(x))(v) = \bigcap \{b(v) : f \in \delta(a, b), x \in a\}.$$

Proof. This follows immediately from Corollary 2.3.   □

Lemma 4.6 *Let $U \subset X$, $x \in U$, and $f : U \to \mathbb{R}$ be locally Lipschitz. Then, for any $v \in X$.*

$$\limsup_{y \to x \ t \downarrow 0} \frac{f(y + tv) - f(y)}{t} \leq (\mathcal{L}f(v))^+,$$

$$\liminf_{y \to x \ t \downarrow 0} \frac{f(y + tv) - f(y)}{t} \geq (\mathcal{L}f(v))^-.$$

Proof. If $f \in \delta(a, b)$ with $x \in a$, then for y sufficiently close to x and $t > 0$ sufficiently small, we have $f(y + tv) - f(y) \in tb(v)$ and thus

$$\limsup_{y \to x \; t \downarrow 0} \frac{f(y + tv) - f(y)}{t} \leq (b(v))^+,$$

which implies

$$\limsup_{y \to x \; t \downarrow 0} \frac{f(y + tv) - f(y)}{t} \leq \inf\{(b(v))^+ : f \in \delta(a, b), x \in a\}.$$

Since $\mathcal{L}f(x) = \bigcap\{b : f \in \delta(a, b), x \in a\}$, the proof of the first inequality follows from Lemma 4.5. The second inequality is proved in a similar way. □

Corollary 4.7 *The Gâteaux derivative of f at x, when it exists, belongs to the L-derivative. Similarly for the Fréchet derivative.* □

In order to obtain the next corollary we first need the following characterization of the generalized gradient.

Lemma 4.8 *For any locally Lipschitz function f, we have $A \in \partial f(x)$ iff for all $v \in X$,*

$$\liminf_{y \to x \; t \downarrow 0} \frac{f(y + tv) - f(y)}{t} \leq A(v) \leq \limsup_{y \to x \; t \downarrow 0} \frac{f(y + tv) - f(y)}{t}.$$

Proof. The "if" part follows by definition. For the "only if" part, the second inequality is just the definition of the generalized gradient. For the first inequality, assume $A \in \partial f(x)$ and $v \in X$. Then, by the definition of the generalized gradient, with v replaced by $-v$, we have

$$-A(v) \leq \limsup_{y \to x \; t \downarrow 0} \frac{f(y - tv) - f(y)}{t}$$

or

$$A(v) \geq \liminf_{y \to x \; t \downarrow 0} \frac{-f(y - tv) + f(y)}{t}.$$

Setting $z = y - tv$, the latter inequality reduces to

$$A(v) \geq \liminf_{z \to x \; t \downarrow 0} \frac{f(z + tv) - f(z)}{t},$$

as required. □

Corollary 4.9 *The generalized (Clarke) gradient is contained in the L-derivative.*

Proof. This follows from Lemma 4.6 and Lemma 4.8. □

We do not know if the L-derivative and the Clarke gradient coincide on an infinite-dimensional Banach space. We do know, however, that in finite dimensions, they are the same, as we will show in Section 8.

5 Domain for Lipschitz functions

We will construct a domain for locally Lipschitz functions and for $C^1(U)$. The idea is to use step functions in $U \to_s \mathbb{IR}$ to represent the function and step functions in $U \to \mathbf{C}(X^*)$ to represent the differential properties of the function. Note that a continuous partial function f of type $U \to \mathbb{R}$, as we have considered in defining ties of functions in Section 3, can be regarded as an element $\hat{f}$ of $U \to_s \mathbb{IR}$ with $\hat{f}(x) = f(x)$ if $f(x)$ is defined and $\hat{f}(x) = \bot = \mathbb{R}$; otherwise, we always identify f and $\hat{f}$. Furthermore, a function $f \in U \to \mathbb{IR}$ is given by a pair of, respectively, lower and upper semicontinuous functions $f^-, f^+ : U \to \mathbb{R}$ with $f(x) = [f^-(x), f^+(x)]$.

Consider the *consistency* relation

$$\mathsf{Cons} \subset (U \to_s \mathbb{IR}) \times (U \to_s \mathbf{C}(X^*)),$$

defined by $(f, g) \in \mathsf{Cons}$ if $\uparrow\!f \cap \int g \neq \emptyset$. For a consistent (f, g), we think of f as the *function part* or the *function approximation* and g as the *derivative part* or the *derivative approximation*. We will show that the consistency relation is Scott closed. The proofs of the rest of results in this section are essentially as in [9] for the case of $X = \mathbb{R}$. We will present them here for a general Banach space X for the sake of completeness.

Proposition 5.1 *Let $g \in U \to_s \mathbf{C}(X^*)$ and $(f_i)_{i \in I}$ be a nonempty family of functions $f_i : dom(g) \to \mathbb{R}$ with $f_i \in \int g$ for all $i \in I$. If $h_1 = \inf_{i \in I} f_i$ is real-valued, then $h_1 \in \int g$. Similarly, if $h_2 = \sup_{i \in I} f_i$ is real-valued, then $h_2 \in \int g$.*

Proof. Suppose h_1 is real-valued. Let $a \searrow b \sqsubseteq g$. We have $f_i(x) - f_i(y) \in b(x-y)$ for all $i \in I$. Thus, $(b(x-y))^- \leq f_i(x) - f_i(y) \leq (b(x-y))^+$. Thus, $\inf_{i \in I} f_i(x) \leq f_i(y) + (b(x-y))^+$. Taking infimum again, we obtain $\inf_{i \in I} f_i(x) \leq \inf_{i \in I} f_i(y) + (b(x-y))^+$, and hence, $h_1(x) - h_1(y) \leq (b(x-y))^+$. Similarly, $(b(x-y))^- \leq h_1(x) - h_1(y)$, and the result follows. The case of h_2 is similar. $\square$

Let $R[0, 1]$ be the set of partial maps of $[0, 1]$ into the extended real line. Consider the two dcpo's $(R[0,1], \leq)$ and $(R[0,1], \geq)$. Define the maps $s : (U \to_s \mathbb{IR}) \times (U \to_s \mathbf{C}(X^*)) \to (R, \leq)$ and $t : (U \to_s \mathbb{IR}) \times (U \to_s \mathbf{C}(X^*)) \to (R, \geq)$ by

$$s : (f, g) \mapsto \inf\{h : dom(g) \to \mathbb{R} \mid h \in \int g \ \& \ h \geq f^-\},$$

$$t : (f, g) \mapsto \sup\{h : dom(g) \to \mathbb{R} \mid h \in \int g \ \& \ h \leq f^+\}.$$

We use the convention that the infimum and the supremum of the empty set are ∞ and $-\infty$, respectively. Note that given a connected component A of $dom(g)$ with $A \cap dom(f) = \emptyset$; then $s(f, g)(x) = -\infty$ and $t(s, f)(x) = \infty$ for $x \in A$. In words, $s(f, g)$ is the least primitive map of g that is greater than the lower part of f, whereas $t(f, g)$ is greatest primitive map of g less than the upper part of f.

Proposition 5.2 *The following are equivalent:*

(i) $(f, g) \in$ Cons.

(ii) $s(f, g) \leq t(f, g)$.

(iii) *There exists a locally Lipschitz function* $h : dom(g) \to \mathbb{R}$ *with* $g \sqsubseteq \mathcal{L}h$ *and* $f \sqsubseteq h$ *on* $dom(g)$.

Proof. If $dom(f) \cap dom(g) = \emptyset$, then the three statements hold trivially. So assume in the following proof that $dom(f) \cap dom(g) \neq \emptyset$.

(ii) $\Rightarrow$ (i). Suppose $s(f, g) \leq t(f, g)$. Then, $s(f, g) \in \uparrow\!f \cap \int g$ and hence $(f, g) \in$ Cons.

(i) $\Rightarrow$ (ii). Suppose $(f, g) \in$ Cons. Assume $h \in \uparrow\!f \cap \int g$. Then, the induced map $h : dom(g) \to \mathbb{R}$ satisfies $h \in \int g$. Hence, $f^- \leq h \leq f^+$ and thus $s(f, g) \leq t(f, g)$.

(ii) $\Rightarrow$ (iii). Suppose $s(f, g) \leq t(f, g)$. Put $h = s(f, g)$.

(iii) $\Rightarrow$ (ii). We have $s(f, g) \leq h \leq t(f, g)$. $\qquad\square$

Moreover, s and t are well behaved:

Proposition 5.3 *The maps* s *and* t *are Scott continuous.*

Proof. Consider the map s. If $f_1 \sqsubseteq f_2$ and $g_1 \sqsubseteq g_2$, then we have $\int g_1 \supseteq \int g_2$ and $f_1^- \leq f_2^-$ and it follows that $s(f_1, g_1) \leq s(f_2, g_2)$. Let $\{(f_i, g_i)\}_{i \in I}$ be a directed set, and put $f = \sup_{i \in I} f_i$ and $g = \sup_{i \in I} g_i$. To show the continuity of s, we need to show that $\sup_{i \in I} s(f_i, g_i) \geq s(f, g)$ on any connected component of $dom(g) = \bigcup_{i \in I} dom(g_i)$. Take any such connected component $A \subseteq dom(g)$. If $A \cap dom(f) = \emptyset$, then $s(f, g) = -\infty$ on A and the result follows. Assume that $A \cap dom(f) \neq \emptyset$; i.e., $dom f_{i_0} \cap dom g_{i_0} \neq \emptyset$ for some $i_0 \in I$. If $s(f_i, g_i) = \infty$ on $A \cap dom(g_i)$ for some $i \geq i_0$, then $\sup_{i \in I} s(f_i, g_i) = \infty$ on A and the result follows again. Otherwise, assume without loss of generality that $-\infty < s(f_i, g_i) < \infty$ on $A \cap dom(g_i)$ for all $i \in I$. Then from $(s(f_i, g_i))\!\restriction_A \in \int g_i$, it follows that $\forall i \geq j$. $(s(f_i, g_i))\!\restriction_A \in \int g_j$, and hence, by Proposition 5.1, $(\sup_{i \in I}(s(f_i, g_i)\!\restriction_A)) \in \int g_j$. Thus $(\sup_{i \in I} s(f_i, g_i))\!\restriction_A \in \sup_j \int g_j = \int \sup g_j$. On the other hand, $s(f_i, g_i) \geq f_i^-$ on A implies $\sup_{i \in I} s(f_i, g_i) \geq f_i^-$ on A and hence $\sup_{i \in I} s(f_i, g_i) \geq f^-$ on A. This shows that s is continuous. Similarly t is continuous. $\qquad\square$

This enables us to deduce:

Corollary 5.4 *The relation* Cons *is Scott closed.*

Proof. Let $(f_i, g_i)_{i \in I} \subset (U \to_s \mathbb{IR}) \times (U \to_s C\mathbb{R}^n)$ be a directed set with $(f_i, g_i) \in$ Cons for all $i \in I$. Then, by Proposition 5.2, $s(f_i, g_i) \leq t(f_i, g_i)$ for all $i \in I$. Hence, $s(f, g) = \sup_{i \in I} s(f_i, g_i) \leq \inf_{i \in I} t(f_i, g_i) = t(f, g)$. $\qquad\square$

We can now sum up the situation for a consistent pair of function and derivative information.

Corollary 5.5 *Let $(f, g) \in$ Cons. Then in each connected component A of the domain of definition of g that intersects the domain of definition of f, there exist two locally Lipschitz functions $s : A \to \mathbb{R}$ and $t : A \to \mathbb{R}$ such that $s, t \in \uparrow f \cap \int g$, and for each $u \in \uparrow f \cap \int g$, we have with $s(x) \leq u(x) \leq t(x)$ for all $x \in A$.*

We now can define a basic construct of this paper:

Definition 5.6 Define

$$D^1(U) = \{(f, g) \in (U \to_s \mathbb{IR}) \times (U \to_s \mathbf{C}(X^*)) : (f, g) \in \text{Cons}\}.$$

From Corollary 5.4, we obtain

Corollary 5.7 *The poset $D^1(U)$ is a bounded complete dcpo.*

Proposition 5.8 *For any $f \in (U \to \mathbb{R})$, the element $(f, \mathcal{L}f)$ is a maximal element of $D^1(U)$.*

Proof. By Corollary 4.4, we have $f \in \int \mathcal{L}f$, and thus, $(f, \mathcal{L}f) \in D^1 U$. We now show that $(f, \mathcal{L}f)$ is maximal. If $\mathcal{L}f \sqsubseteq g$ and $(f, g) \in D^1 U$, then we have $f \in \int g$, which implies $g \sqsubseteq \mathcal{L}f$; i.e., $g = \mathcal{L}f$ and $(f, \mathcal{L}f)$ is maximal. $\square$

For a locally Lipschitz function $f : U \to \mathbb{R}$, the L-derivative satisfies $\mathcal{L}f(x) \neq \perp$ for all $x \in U$, whereas for a piecewise C^1 function f, we further have the property that $\mathcal{L}f(x)$ is maximal except for a finite set of points.

6 L-derivative in finite dimensions

Assume $X = \mathbb{R}^n$ and $U \subset \mathbb{R}^n$ is an open subset. Then we can identify $X^* = X = \mathbb{R}^n$. Moreover $\mathbf{C}(\mathbb{R}^n)$ and $U \to \mathbf{C}(\mathbb{R}^n)$ are both countably based bounded complete continuous dcpo's with $U \to_s \mathbf{C}(\mathbb{R}^n) = U \to \mathbf{C}(\mathbb{R}^n)$.

In the finite-dimensional case, we can deduce the following proposition, which relates the L-derivative to its classical counterpart. For any compact subset $c \subset \mathbb{R}^n$, we denote its diameter by $w(c)$. For a nonempty compact interval $c = [c^-, c^+] \subset \mathbb{R}$, we thus have $w(c) = c^+ - c^-$. The following result generalizes Theorem 4.2(i) in finite dimensions; we do not know whether it can be extended to infinite-dimensional Banach spaces.

Proposition 6.1 *If $\mathcal{L}f(y) \in \mathbf{C}(\mathbb{R}^n)$ is maximal for some $y \in U$, then the Fréchet derivative of f exists at y and $f'(y) = \mathcal{L}f(y)$.*

Proof. Put $c := \mathcal{L}f(y) = \bigcap\{b \mid y \in a \text{ \& } f \in \delta(a,b)\}$. Let $\epsilon > 0$ be given. Take $a \searrow b \sqsubseteq \mathcal{L}f$ with $y \in a$ and $w(b) < \epsilon$. Note that $b \sqsubseteq c$ and there exists $\delta > 0$ such that $\|x - y\| < \delta$ implies $x \in a$. We have $b(x - y) \sqsubseteq f(x) - f(y)$ for $x \in a$, and

$$w(b(x - y)) \leq w(b)|x - y| \leq \epsilon |x - y|.$$

Since $b(x - y) \sqsubseteq c(x - y)$, we obtain $|f(x) - f(y) - c(x - y)| \leq \epsilon |x - y|$ and the result follows by the definition of Fréchet derivative. $\qquad\square$

We can now obtain the following result in finite dimensions, which is simply the classical version of the Fundamental Theorem of Calculus.

Corollary 6.2 *Suppose $g : U \to \mathbb{R}$ is a continuous function. Then $f \in \int g$ implies that f' exists in U and we have $\mathcal{L}f = f' = g$.*

Proof. By Theorem 4.4, $g \sqsubseteq \mathcal{L}f$, and thus, $\mathcal{L}f = g$ since g is maximal. By Proposition 6.1, we also obtain $\mathcal{L}f = f'$. $\qquad\square$

We now consider a given Cartesian coordinate system denoted say by e with basis $(e_1, \ldots, e_n)$. Let $\mathbf{Q}_e(\mathbb{R}^n)$ be the collection of all n-dimensional compact hyper-rectangles with edges parallel with e_i's and define the *rectangular L-derivative with respect to e* as

$$(\mathcal{L}f)^e = \sup\{a \searrow b : f \in \delta(a,b) \text{ \& } b \in \mathbf{Q}_e(\mathbb{R}^n)\}.$$

It immediately follows from the definition that $(\mathcal{L}f)^e(y)$, at each point $y \in U$, is an n-dimensional compact hyper-rectangle with edges parallel to the basis vectors e_i. Moreover, if E denotes the collection of all Cartesian coordinate systems in $\mathbb{R}^n$, we have

Proposition 6.3 *For each point $y \in U$, we have*

$$\mathcal{L}f(y) = \bigcap_{e \in E} (\mathcal{L}f)^e(y). \qquad\square$$

For $b \in \mathbf{Q}_e(\mathbb{R}^n)$, i.e., $b = b_1 \times \ldots \times b_n$, the relation $b(x - y) \sqsubseteq f(x) - f(y)$, which defines the single tie $\delta(a,b)$, can be computed in the coordinate system e simply as $\sum_{j=1}^n b_j(x_j - y_j) \sqsubseteq f(x) - f(y)$. Furthermore, if $x_j = y_j$ for all $j \neq i$, the relation reduces to $b_i(x_i - y_i) \sqsubseteq f(x) - f(y)$. This suggests a characterization of $(\mathcal{L}f)^e(y)$ in terms of Dini's derivatives; in fact, we can deduce the following result.

Proposition 6.4 *The components of the rectangular L-derivative with respect to the basis e are given by*

$$(\mathcal{L}f)_i^e(x) = \left[\liminf_{y \to x} (\nabla f)_i^l(y), \limsup_{y \to x} (\nabla f)_i^u(y) \right],$$

when the two limits are finite for all $i = 1, \ldots, n$ and $\mathcal{L}f(x) = \bot$ otherwise.

Proof. If $f \in \delta(a,b)$ for some $a \subseteq U$ with $x \in a$ and $b = [b_1^-, b_1^+] \times \cdots \times [b_n^-, b_n^+]$, then $b_i^- \leq (\nabla f)_i^l(y) \leq (\nabla f)_i^u(y) \leq b_i^+$ for $y \in a$, and thus, $b_i^- \leq \liminf_{y \to x}(\nabla f)_i^l(y) \leq \limsup_{y \to x}(\nabla f)_i^u(y) \leq b_i^+$. It follows that

$$\left[\liminf_{y \to x}(\nabla f)_i^l(y), \limsup_{y \to x}(\nabla f)_i^u(y)\right] \subseteq (\mathcal{L}f)_i^e(x).$$

On the other hand, if $\liminf_{y \to x}(\nabla f)_i^l(y)$ and $\limsup_{y \to x}(\nabla f)_i^u(y)$ are finite for all $i = 1, \ldots, n$, then for any $\epsilon > 0$, there exists an open $a \subseteq U$ containing x such that, for all $y \in a$ and all $i = 1, \ldots, n$,

$$K_i(x, \epsilon) := \liminf_{y \to x}(\nabla f)_i^l(y) - \epsilon < (\nabla f)_i^l(y),$$

$$L_i(x, \epsilon) := \limsup_{y \to x}(\nabla f)_i^u(y) + \epsilon > (\nabla f)_i^u(y).$$

Let c be the interior of a hypercube containing x with $\bar{c} \subset a$, and fix i with $1 \leq i \leq n$. By the first inequality above, we can cover $\bar{c}$ with a finite number of open hyper-rectangles such that for any pair of points $y \geq z$ in each hyper-rectangle with $y_j = z_j$ for $j \neq i$ and $y_i \geq z_i$, we have $K_i(x, \epsilon)(y_i - z_i) \leq f(y) - f(z)$. It thus follows, by adding a finite number of inequalities one for each open hyper-rectangle, that for all $y, z \in c$ with $y_j = z_j$ for $j \neq i$ and $y_i \geq z_i$, we have $K_i(x, \epsilon)(y_i - z_i) \leq f(y) - f(z)$, and similarly, by using the second inequality above, $f(y) - f(z) \leq L_i(x, \epsilon)(y_i - z_i)$. Thus, for all $y, z \in c$ with $y_j = z_j$ for $j \neq i$, we have $b_i(y_i - z_i) \sqsubseteq f(y) - f(z)$, where $b_i = [K_i(x, \epsilon), L_i(x, \epsilon)]$. For any pair $y, z \in c$, consider the $n + 1$ points $y = p^0, p^1, p^2, \ldots p^{n-1}, p^n = z$, such that $p_i^i = z_i$ and $p_j^i = y_j$ for $j \neq i$. Therefore,

$$f(y) - f(z)$$
$$= (f(p^0) - f(p^1)) + (f(p^1) - f(p^2)) + \ldots + (f(p^i) + f(p^{i+1}))$$
$$+ \cdots (f(p^n) - f(p^{n+1})) \sqsupseteq \sum_{i=1}^n b_i(y_i - z_i).$$

It follows that $f \in \delta(c, b)$. Since $\epsilon > 0$ is arbitrary, we conclude that

$$\left[\liminf_{y \to x}(\nabla f)_i^l(y), \limsup_{y \to x}(\nabla f)_i^u(y)\right] \supseteq (\mathcal{L}f)_i^e(x). \qquad \square$$

The domain-theoretic derivative developed in [10] is indeed $(\mathcal{L}f)^e$, the rectangular L-derivative with respect to a given coordinate axis e. We do not know whether there is an analogue of the above Proposition for infinite-dimensional separable Hilbert spaces.

7 Computability

Let Z be a topological space with a countable basis M of its open subsets and D a bounded complete dcpo with a countable subset $E \subset D$. Let $(f_i)_{i \geq 0}$ be an effective enumeration of the class of step functions of $Z \to D$ made from single step functions $a \searrow b$, where $a \in M$ and $b \in E$. We say $f \in U \to_s D$ is *computable* with respect to this enumeration if there exists a total recursive function $\phi : \mathbb{N} \to \mathbb{N}$ such that $(f_{\phi(n)})_{n \geq 0}$ is an increasing sequence with $f = \sup_{n \geq 0} f_{\phi(n)}$.

When, in addition, Z is locally compact and D is a countably based continuous dcpo, then $Z \to D$ is a countably based bounded complete continuous dcpo, which can be given an effective structure. In this case, we obtain the same class of computable elements with any effective change of a countable basis of D. In general, however, the computable elements will depend on the enumeration of the countable subset E.

Suppose now that X is a separable Banach space, with a countable dense set $P \subset X$. Then the collection of open balls centred at points of P with rational radii provides a countable basis of the norm topology on X. We use the rational compact intervals as a countable basis of $\mathbb{IR}$ and the collection of closed balls of X^* with centres at points P with rational radii as a countable subset of $\mathbf{C}(X^*)$ to generate two countable sets, S_1 and S_2 say, of step functions for the two dcpo's $U \to_s \mathbb{IR}$ and $U \to_s \mathbf{C}(X^*)$. We then obtain an enumeration $(f_i)_{i \geq 0}$ of S_1 and an enumeration $(g_i)_{i \geq 0}$ of S_2.

By Proposition 3.14, we know that any continuous function $f : U \to \mathbb{R}$ and any function $g : U \to \mathbf{C}(X^*)$ continuous with respect to the norm topology on X and X^*, is the supremum of step functions in S_1 and S_2, respectively. We say that f is *computable* with respect to the enumeration $(f_i)_{i \geq 0}$, respectively, g is *computable* with respect to $(g_i)_{i \geq 0}$, if f considered as an element of $U \to \mathbb{IR}$, respectively, g considered as an element of $U \to \mathbf{C}(X^*)$, is computable with respect to the enumeration.

We then use an oracle to decide whether $(f_i, g_j) \in \mathsf{Cons}$ for $i, j \geq 0$, which enables us to construct an enumeration $(h_i)_{i \geq 0}$ of a countable set, S_3 say, of step functions of $D^1(U)$, where $h_i = (f_{p(i)}, g_{q(i)})$ for $i \geq 0$ with $p, q : \mathbb{N} \to \mathbb{N}$ total recursive functions. By Proposition 3.14, we know that if $f : U \to \mathbb{R}$ is Fréchet differentiable, then (f, f') is the lub of step functions in S_3. We thus say that f and its Fréchet derivative f' are computable with respect to $(h_i)_{i \geq 0}$ if (f, f') considered as a maximal element of $D^1(U)$ is computable with respect to this enumeration.

As we will see in the next section, when X is finite dimensional, $D^1(U)$ can be given an effective structure with respect to which Cons is decidable, obviating the need for an oracle.

7.1 An effectively given domain for Lipschitz functions

In the finite-dimensional case, $X = \mathbb{R}^n$, the countably based bounded complete continuous dcpo's $U \to \mathbb{IR}$ and $U \to \mathbf{C}(\mathbb{R}^n)$ each have a canonical basis, respectively, made from single step functions $a \searrow b$, where a is an open ball with a rational radius centred at a point in U with rational coordinates and b is a rational compact interval, respectively, a convex compact polyhedra in $\mathbb{R}^n$ with vertices having rational coordinates.

In [10], it is shown that, when the rectangular L-derivative $(\mathcal{L}f)^e$ with respect to a given coordinate axis e is used, the corresponding consistency predicate Cons^e, defined on $(U \to \mathbb{IR}) \times (U \to \mathbb{IR}^n)$ by $(f, g) \in \mathrm{Cons}^e$ if there exists $h : U \to \mathbb{R}$ such that $f \sqsubseteq h$ and $g \sqsubseteq (\mathcal{L}h)^e$ is decidable on the basis elements. The proof of decidability is fairly simple for $n = 1$ with an algorithm to test consistency, which is linear in the total number of single step functions in the function and derivative parts [7]. In higher dimensions, the existing proof of decidability in [10] is long and the algorithm to test consistency is super-exponential. First, one checks, by a generalization of Green's theorem, if $\int^e g \neq \emptyset$ where $\int^e$ is the primitive map dual to the rectangular L-derivative with respect to e; i.e., $f \in \int^e g$ if $g \sqsubseteq (\mathcal{L}f)^e$. If the test for integrability of g is positive, then one checks if $s^e(f, g) \leq t^e(f, g)$ where s^e and t^e are defined as s and t in Section 5 except that $\int^e$ is used in their definitions.

The technique for proving the decidability of Cons^e on basis elements can be extended to prove that Cons is also decidable on basis elements of $D^1(U)$. Since the proof and the corresponding algorithm to test consistency is very long, they will be presented elsewhere.

Using the decidability of Cons on basis elements, we can provide an effective structure for $D^1(U)$. In particular this will characterize real-valued functions on $U \subset \mathbb{R}^n$ that are computable and have a computable L-derivative as pairs $(f, \mathcal{L}f)$ for which there exists a total recursive function $\phi : \mathbb{N} \to \mathbb{N}$ with $(f, \mathcal{L}f) = \sup_{i \geq 0}(f_{p(\phi(i))}, g_{q(\phi(i))})$ in the notation of Section 7.

If $f : U \to R$ is C^{m-1} for some open subset $U \subset \mathbb{R}^n$, i.e., if it has continuous Fréchet derivatives $f^{(d)}$ of order d with $1 \leq d \leq m - 1$, then the L-derivative of components of $f^{(m-1)}$ exists. One can extend the construction of $D^1(U)$ to higher derivatives and build a domain $D^m(U)$ for representing and approximating a function together with its $m - 1$ Fréchet derivatives and its mth L-derivative $\mathcal{L}f^{(m-1)}$. The basis of this domain will consist of $m + 1$ step functions representing approximations to the function, its first $m - 1$ Fréchet derivatives and its mth L-derivative. We will discuss the question of decidability of consistency for basis elements of this domain in the final section.

8 Relation with generalized gradient

Recall that by Rademacher's theorem [5, p 148], a function $f : U \to \mathbb{R}$ that is Lipschitz in an open neighbourhood of $U \subset \mathbb{R}^n$ is differentiable almost everywhere with respect to the n-dimensional Lebesgue measure in that neighbourhood. Let $\Omega_f \subseteq U$ denote the set of points, where f is not differentiable.

We now establish the equality of the L-derivative and the generalized gradient in finite dimensions..

Theorem 8.1 *For any function $f : U \rightarrow \mathbb{R}$, the rectangular L-derivative with respect to a given Cartesian coordinate system, at a point where the function is locally Lipschitz, is the smallest hyper-rectangle with sides parallel to the coordinate planes that contains the generalized gradient at that point.*

Proof. Fix a Cartesian coordinate system e. By Corollary 4.9 and Proposition 6.3, we already know that

$$\partial f(x) \subseteq (\mathcal{L}f(x))^e. \tag{4}$$

We show that $(\mathcal{L}f(x))^e$ is the smallest hyper-rectangle with sides parallel to the coordinate planes, which contains $\partial f(x)$. Assume $\epsilon > 0$ is given, $1 \leq i \leq n$, and let $B \subset \mathbb{R}^n$ be the unit closed ball centred at the origin. For $1 \leq i \leq n$, let $\pi_i : \mathbb{R}^n \rightarrow \mathbb{R}$ be the projection to the i coordinate axis and consider the pointwise extension of π_i to compact subsets of $\mathbb{R}^n$. From Equation 1, we have

$$(\pi_i(\partial f(x)))^+ = \limsup\{(\nabla f)_i(y) : y \rightarrow x, \, y \notin \Omega_f\}.$$

Thus, there exists $\delta > 0$ such that for all $y \in x + \delta B$, we have $(\nabla f)_i(y) \leq (\pi_i(\partial f(x)))^+ + \epsilon$. Consider the line segment $L_y = \{y + te_i : 0 < t < \delta/2\}$, where e_i is the unit vector in the direction of the ith coordinate axis. Since Ω_f has a zero n-dimensional Lebesgue measure in $x + \delta B$, it follows from Fubini's theorem that for almost all $y \in x + \frac{\delta}{2}B$, the line segment L_y meets Ω_f in a set of zero one-dimensional Lebesgue measure. If y is such a point and $0 < t < \delta/2$, we obtain

$$f(y + te_i) - f(y) = \int_0^t (\nabla f)_i(y + se_i) \, ds,$$

since, by Rademacher's theorem [5, p 148], f' exists almost everywhere on L_y. On the other hand, $(\nabla f)_i(y + se_i) \leq (\pi_i(\partial f(x)))^+ + \epsilon$, since $\|y + se_i - x\| < \delta$ for $0 < s < t$. Thus,

$$f(y + te_i) - f(y) \leq t(\pi_i(\partial f(x)))^+ + \epsilon). \tag{5}$$

Equation (5) holds for almost all y within $\delta/2$ of x and for all $t \in (0, \delta/2)$. Since f, being Lipschitz, is continuous, it follows that Equation (5) holds for all y within $\delta/2$ of x and for all $t \in (0, \delta/2)$. Thus, $(\nabla f)_i^u(y) \leq (\pi_i(\partial f(x)))^+ + \epsilon$ for all y within $\delta/2$ of x, and using Proposition 6.4, we conclude that

$$((\mathcal{L}f(x))_i^e)^+ = \limsup_{y \rightarrow x}(\nabla f)_i^u(y) \leq (\pi_i(\partial f(x)))^+. \tag{6}$$

Similarly,

$$((\mathcal{L}f(x))_i^e)^- = \liminf_{y \rightarrow x}(\nabla f)_i^l(y)$$

$$\geq \liminf\{(\nabla f)_i(y) : y \rightarrow x, \, y \notin \Omega_f\} = (\pi_i(\partial f(x)))^-. \tag{7}$$

Comparing Equations (6) and (7) with Equation (4), it follows that $\partial f(x)$ touches all the $2n$ sides of the hyper-rectangle $(\mathcal{L}f(x))^e$ and the proof is complete.   □

Corollary 8.2 *For any locally Lipschitz map* $f : U \to \mathbb{R}$, *the L-derivative and the Clarke gradient are equal:* $\mathcal{L}f = \partial f$.   □

Thus, in finite dimensions, the L-derivative gives a new representation for the Clarke gradient and the construction of an effectively given domain for locally Lipschitz functions provides a new computational framework for its applications. We note that the proof of Theorem 8.1 uses Proposition 6.4, for which we do not know any infinite-dimensional analogue.

9 Further work and open problems

As pointed out, it remains an open question whether the L-derivative coincides with the Clarke gradient on infinite-dimensional Banach spaces. It is also unknown whether the Clarke gradient is upper semicontinuous in infinite dimensions, a property that holds for the L-derivative as we have shown in this paper. On the other hand, it will be interesting to see whether the L-derivative can be extended to functions from a Banach space to a finite-dimensional Banach space, for example, to the complex plane, a case which has applications in quantum field theory.

There are quite a few unsolved problems in finite dimensions. For $n = 1$, the algorithm for testing consistency of basis elements in $D^1(U)$ is linear as already mentioned. For $D^2(U)$, consistency on basis elements is decidable, but the present algorithm to test it is super-exponential in the total number of single step functions for the three approximations of the function part, the derivative part and the second derivative part [1]. Decidability of consistency for $D^m(U)$ when $m > 2$ is unknown. For $n = 2$, consistency on basis elements for $D^1(U)$ is decidable, but the algorithm to test it in [10] is super-exponential. The complexity of the consistency test in this case is unknown as is the question of decidability of consistency of basis elements for $D^m(U)$ when $m > 1$.

Based on the domain-theoretic framework for differential calculus, one can embark on the task of constructing a domain for orientable Euclidean manifolds, which would extend the set-theoretic model for computational geometry and solid modelling presented in [8] to the piecewise smooth setting.

Acknowledgement

I would like to thank André Lieutier and Dirk Pattinson for reading and checking various parts of this work.

References

1. S. Abolfathbeigi and M. Mahmoudi. Consistency for approximating twice different functions, 2003. Manuscript in Persian, Department of Mathematical Sciences, Sharif University of Techonology, Tehran, Iran.
2. S. Abramsky and A. Jung. Domain theory. In S. Abramsky, D. M. Gabbay, and T. S. E. Maibaum, editors, *Handbook of Logic in Computer Science*, volume 3. Clarendon Press, 1994.
3. F. H. Clarke. Private communications. Summer 2005.
4. F. H. Clarke. *Optimization and Nonsmooth Analysis*. Wiley, 1983.
5. F. H. Clarke, Yu. S. Ledyaev, R. J. Stern, and P. R. Wolenski. *Nonsmooth Analysis and Control Theory*. Springer, 1998.
6. A. Edalat. Dynamical systems, measures and fractals via domain theory. *Information and Computation*, 120(1):32–48, 1995.
7. A. Edalat, M. Krznarić, and A. Lieutier. Domain-theoretic solution of differential equations (scalar fields). In *Proceedings of MFPS XIX*, volume 83 of *Electronic Notes in Theoretical Computer Science*, 2003. Full paper in www.doc.ic.ac.uk/~ae/papers/scalar.ps.
8. A. Edalat and A. Lieutier. Foundation of a computable solid modelling. *Theoretical Computer Science*, 284(2):319–345, 2002.
9. A. Edalat and A. Lieutier. Domain theory and differential calculus (Functions of one variable). *Mathematical Structures in Computer Science*, 14(6):771–802, 2004.
10. A. Edalat, A. Lieutier, and D. Pattinson. A computational model for multi-variable differential calculus. In V. Sassone, editor, *Proc. FoSSaCS 2005*, volume 3441, pages 505–519, 2005.
11. A. Edalat and D. Pattinson. A domain-theoretic account of {P}icard's theorem. *LMS Journal of Computation and Mathematics*, 10:83–118, 2007.
12. A. Edalat and D. Pattinson. Inverse and implicit functions in domain theory. In P. Panangaden, editor, *Proc. 20th IEEE Symposium on Logic in Computer Science (LICS 2005)*, pages 417–426, 2005.
13. A. Edalat and D. Pattinson. Denotational semantics of hybrid automata. In L. Aceto and A. Ingofsdottir, editors, *Proc. FoSSaCS 2006*, volume 3921, pages 231–245, 2006.
14. N. Fars. Aspects analytiques dans la mathematique de shraf al-din al-tusi. *Historia Sc.*, 5(1), 1995.
15. N. Fars. Le calcul du maximum et la 'derive' selon shraf al-din al-tusi. *Arabic Sci. Philos.*, 5(2):219–237, 1995.
16. G. Gierz, K. H. Hofmann, K. Keimel, J. D. Lawson, M. Mislove, and D. S. Scott. *Continuous Lattices and Domains*. Cambridge University Press, 2003.
17. A. Grzegorczyk. Computable functionals. *Fund. Math.*, 42:168–202, 1955.
18. A. Grzegorczyk. On the definition of computable real continuous functions. *Fund. Math.*, 44:61–71, 1957.
19. J. P. Hogendijk. *Shraf al-din al-Tusi* on the number of positive roots of cubic equations. *Fund. Math.*, 16(1):69–85, 1989.
20. K. Lau and C. Weil. Differentiability via directional derivatives. *Proceedings of American Mathematical Society*, 70(1):11–17, 1978.
21. M. B. Pour-El and J. I. Richards. *Computability in Analysis and Physics*. Springer-Verlag, 1988.
22. D. S. Scott. Outline of a mathematical theory of computation. In *4th Annual Princeton Conference on Information Sciences and Systems*, pages 169–176, 1970.

23. M. B. Smyth. Topology. In S. Abramsky, D. Gabbay, and T. S. E. Maibaum, editors, *Handbook of Logic in Computer Science*, chapter 5. Oxford University Press, 1992.
24. A. Turing. On computable numbers with an application to the Entscheidungsproblem. *Proc. London Mathematical Soc.*, 42:230–265, 1936.
25. A. Turing. On computable numbers with an application to the Entscheidungsproblem. *Proc. London Mathematical Soc.*, 43:544–546, 1937.
26. K. Weihrauch. *Computable Analysis (An Introduction)*. Springer, 2000.
27. S. Yamamuro. *Differential calculus in Topological Linear Spaces*, volume 374 of *Lecture Notes in Mathematics*. Springer-Verlag, 1970.

Infinite Time Computable Model Theory

Joel David Hamkins[*1], Russell Miller[2], Daniel Seabold[3], and Steve Warner[4]

[1] The College of Staten Island of The City University of New York, Mathematics, Staten Island, NY 10314, U.S.A.
and
The Graduate Center of The City University of New York, Ph.D. Program in Mathematics, New York, NY 10016, U.S.A.
jhamkins@gc.cuny.edu, http://jdh.hamkins.org
[2] Queens College of The City University of New York, Mathematics, Flushing, New York 11367, U.S.A.
and
The Graduate Center of The City University of New York, Ph.D. Program in Computer Science, New York, NY 10016, U.S.A.
Russell.Miller@qc.cuny.edu
[3] Department of Mathematics, Hofstra University, Hempstead, NY 11549-1030, U.S.A.
matdes@hofstra.edu
[4] Department of Mathematics, Hofstra University, Hempstead, NY 11549-1030, U.S.A.
matsjw@hofstra.edu

Summary. We introduce infinite time computable model theory, the computable model theory arising with infinite time Turing machines, which provide infinitary notions of computability for structures built on the reals $\mathbb{R}$. Much of the finite time theory generalizes to the infinite time context, but several fundamental questions, including the infinite time computable analog of the Completeness Theorem, turn out to be independent of ZFC.

1 Introduction

Computable model theory is model theory with a view to the computability of the structures and theories that arise (for a standard reference, see [2]). Infinite time computable model theory, which we introduce here, carries out this program with the infinitary notions of computability provided by infinite time Turing machines.

[*]MSC: 03D60; 03D45; 03C57; 03E15. Keywords: infinite time Turing machines, computable model theory. The research of the first two authors has been supported in part by grants from the Research Foundation of CUNY, and the first author is additionally thankful to the Institute for Logic, Language and Computation and the NWO (Bezoekersbeurs B62-612) for supporting his summer 2005 stay at Universiteit van Amsterdam.

The motivation for a broader context is that, although finite time computable model theory is necessarily limited to countable models and theories, the infinitary context naturally allows for uncountable models and theories, while retaining the computational nature of the undertaking. Many constructions generalize from finite time computable model theory, with structures built on $\mathbb{N}$, to the infinitary theory, with structures built on $\mathbb{R}$. In this article, we introduce the basic theory and consider the infinitary analogs of the completeness theorem, the Löwenheim–Skolem Theorem, Myhill's theorem, and others. It turns out that, when stated in their fully general infinitary forms, several of these fundamental questions are independent of ZFC. The analysis makes use of techniques from both computability theory and set theory. This article follows up [4].

1.1 Infinite time Turing machines

The definitive introduction to infinite time Turing machines appears in [5], but let us quickly describe how they work. The

start

input:	1	1	1	0	1	0	$\cdots$
scratch:	0	0	0	0	0	0	$\cdots$
output:	0	0	0	0	0	0	$\cdots$

hardware of an infinite time Turing machine is identical to a classical (three tape) Turing machine, with a head reading and writing 0s and 1s on the one-way infinite tapes, following the instructions of a finite program with finitely many states. Computation begins with the input on the *input* tape and the head on the left-most cell in the *start* state. Successor steps of computation are determined by the program in exactly the classic manner. At any limit ordinal stage, as a matter of definition, the machine resets the head to the left-most cell, assumes the *limit* state, and updates the tape so that every cell exhibits the lim sup of the previous values displayed in that cell. This is equivalent to using the limit value, if the value displayed by the cell has stabilized, and otherwise 1. Computation ceases only when the *halt* state is explicitly obtained, and in this case, the output is whatever is written on the output tape. (If the head falls off the tape, no output is given.) If p is a program, it computes a function φ_p, defined by $\varphi_p(x) = y$ if and only if on input x the computation determined by p leads to output y. The natural context here for input and output is the Cantor space $^{\omega}2$ of all infinite binary sequences, which we will denote by $\mathbb{R}$ and refer to as the set of reals. A (partial) function $f : \mathbb{R} \rightarrow \mathbb{R}$ is infinite time *computable* if it is φ_p for some program p. Binary and n-ary functions can be equivalently modeled either by adding additional input tapes or by viewing a single real as the interleaving of the digits of n many reals. A set $A \subseteq \mathbb{R}$ is infinite time *decidable* if its characteristic function is infinite time computable. The set A is infinite time *semi-decidable* if the function $1 \upharpoonright A$ with domain A and constant value 1 is computable. In this article, we will freely use the terms *computable* and *decidable* to mean infinite time computable and infinite time decidable, although we will sometimes specify "infinite time" for clarity. When referring to the classical notions of computability, we will always say "finite time computable" and "finite time decidable." We regard the natural numbers $\mathbb{N}$ as coded in $\mathbb{R}$ by identifying n with the binary sequence consisting of n ones followed by zeros. A real is *writable* if it is $\varphi_p(0)$ for some program p. A real is *accidentally* writable if it appears on one of the tapes during any computation $\varphi_p(0)$. A real is *eventually* writable if it appears on the output tape of a (not necessarily halting) computation $\varphi_p(0)$, and from some point on in that computation, it is never changed. An ordinal α is *clockable* if there is a computation $\varphi_p(0)$ moving to the *halt* state exactly on the αth computational step.

The growing body of literature on infinite time Turing machines includes [5], [13], [12], [7], [11], [6], [3], [8], [1], [4], [14], and [10].

1.2 Basic definitions

The main idea will be that a computable model is one whose underlying set is decidable and whose functions and relations are uniformly computable. In order to make this precise, let us first be more specific about our syntax and how it is represented. A language consists of a collection of function, relation, and constant symbols, with each function and relation symbol assigned a finite arity. In addition, every language has the logical connective symbols $\wedge$, $\vee$, $\neg$, $\rightarrow$, $\leftrightarrow$, parentheses, the equality symbol $=$, quantifiers $\forall$, $\exists$, variable symbols v_0, v_1, and so on. In finite time computable model theory, in order to bring these syntactic objects into the realm of computability, one views each symbol in the (countable) language as being represented by a particular natural number, its Gödel code, so that the various syntactic objects—such as terms, formulas, and sentences—are simply finite sequences of these codes, which can in turn be coded with a single natural number.

Infinite time computable model theory, however, offers the possibility of *uncountable* computable models. And because we will want to consider the elementary or atomic diagrams of such models, the possibility of uncountable languages is unavoidable. Clearly, we cannot expect to code such languages using Gödel codes only in $\mathbb{N}$. Therefore, we work in a more general context, where the symbols of a language are represented with Gödel codes in $\mathbb{R}$, rather than $\mathbb{N}$. This conforms with the philosophy of infinite time computability, where the fundamental inputs and outputs of computations are real numbers. A *computable presentation* of a language $\mathcal{L}$ is the assignment of a Gödel code $\ulcorner s \urcorner$ to every function, relation, and constant symbol s in the language, in such a way that the set of such codes for symbols in $\mathcal{L}$ is decidable, and there are computable functions telling us, given any $\ulcorner s \urcorner$, what kind of symbol s is and, when it is a function or relation symbol, what arity it has. We assume that the basic logical symbols (logical connectives, $=$ symbol, parentheses, variable symbols, quantifiers) have simple Gödel codes in $\mathbb{N}$.

Given the Gödel codes of the underlying symbols, one develops the Gödel coding of all the usual syntactic notions. For example, a term τ is a particular kind of finite sequence of function, constant, and variable symbols, and we may assign the Gödel code $\ulcorner \tau \urcorner$ via the usual manner of coding finite sequences of reals with reals. Similarly, any formula φ in the language is a finite sequence of symbols from the language, and we can assign it a natural Gödel code. We assume that the Gödel coding of the language is undertaken in such a way that we can unambiguously determine whether a given Gödel code is the code of a formula or an individual symbol, and what kind; that from the Gödel code of a formula or term we can compute the Gödel codes of the subformulas and subterms; and that the Gödel codes are uniquely readable. For any computable presentation $\mathcal{L}$, it follows that all the elementary syntactic notions are computable from the Gödel codes, such as finding the inductive construction history of a formula or term or determining whether a given occurrence of a variable in a formula is free.

Definition 1 In the infinite time context, a *computable model* is a structure $\mathcal{A} = \langle A, f^{\mathcal{A}}, R^{\mathcal{A}}, c^{\mathcal{A}} \rangle_{f,R,c \in \mathcal{L}}$ in a language $\mathcal{L}$, with a fixed computable presentation of $\mathcal{L}$, such that the underlying set $A \subseteq \mathbb{R}$ of the model is decidable and the functions, relations, and constants of $\mathcal{A}$ are uniformly computable from their input and the Gödel codes of their symbols. A structure has a *computable presentation* if it is isomorphic to a computable model.

A simple recursive argument shows that the value of any term $\tau(\vec{a})$ is uniformly computable from its Gödel code $\ulcorner \tau \urcorner$ and the input $\vec{a}$. It follows that one can compute the truth in $\mathcal{A}$ of any given atomic formula. Specifically, the atomic diagram of $\mathcal{A}$ is the set $\Delta_0(\mathcal{A}) = \{ \varphi[\vec{a}] \mid \varphi \text{ atomic}, \vec{a} \in A^{<\omega}, \mathcal{A} \models \varphi[\vec{a}] \}$, and if $\mathcal{A}$ is a computable model, then we can decide, on input $\ulcorner \varphi \urcorner$ and $\vec{a}$, whether $\varphi[\vec{a}] \in \Delta_0(\mathcal{A})$. More generally, we define:

Definition 2 A model $\mathcal{A}$ is *(infinite time) decidable* if the full elementary diagram of the structure $\Delta(\mathcal{A}) = \{ \varphi[\vec{a}] \mid \mathcal{A} \models \varphi[\vec{a}] \}$ is infinite time decidable.

We caution the reader that in the infinite time context, a decidable model might not be computable (see Corollary 10). This is a consequence of the phenomenon in infinite time computability that a function can have a decidable graph without being a computable function. The classical algorithm to compute a function from its graph relies on having an effective enumeration of the possible values of the function, but in the infinite time context, we have no effective method to enumerate $\mathbb{R}$. For a purely relational model, with no function or constant symbols in the language, however, this phenomenon is avoided and the model is computable if and only if its atomic diagram is decidable.

Another departure from the classic theory is that every computable model $\mathcal{A}$ with underlying set contained in $\mathbb{N}$ is decidable. The point is that the infinite time algorithm can systematically check the truth of any first-order statement φ in $\mathcal{A}$, given the Gödel code $\ulcorner \varphi \urcorner$, by inductively applying the Tarski definition of truth. If φ has the form $\exists x \, \psi(x)$, then the algorithm simply checks the truth of all $\psi(n)$ for $n \in A$.

More generally, if an infinite time Turing machine has the capacity for a complete search through the domain of a structure—for example if the domain consisted of a writable set of writable reals—then we will be able effectively to carry out the Tarski definition of truth. So one might want to regard such a situation as a special or trivial case in infinite time computable model theory. We refer to such a structure as a *writable structure*; a formal definition appears on page 533.

A theory (meaning any set of sentences in a fixed language) is *computably axiomatizable* if there is a theory T_0, having the same consequences as T, such that the set of Gödel codes $\{ \ulcorner \varphi \urcorner \mid \varphi \in T_0 \}$ is decidable. A theory T is *decidable* if the set of Gödel codes of its consequences $\{ \ulcorner \varphi \urcorner \mid T \vdash \varphi \}$ is decidable. If the underlying language is coded in $\mathbb{N}$, then every computably axiomatizable theory is decidable, because an infinite time algorithm is easily able to search through all proofs. More generally, if an algorithm can write a real listing all the Gödel codes of symbols in the language, then it can systematically generate the Gödel codes of all sentences in that language, determine which are axioms in T_0, and then generate a list of all possible proofs. This shows that any theory with a writable set of axioms has a writable set of theorems.

1.3 Coding with reals

We would like to view our algorithms as engaging with arbitrary countable objects, such as countable ordinals or theories, even though formally the machines treat only infinite binary sequences. So let us introduce a method of coding. We regard any real $x \in \mathbb{R}$ as coding a relation $\lhd$ on $\mathbb{N}$ by $i \lhd j$ if and only if the $\langle i, j \rangle^{\text{th}}$ bit of x is 1, using a bijective pairing function $\langle \cdot, \cdot \rangle$ on $\mathbb{N}$. For every countable ordinal α, there is such a relation $\lhd$ on $\mathbb{N}$ with $\langle \alpha, < \rangle \cong \langle A, \lhd \rangle$, where A is the field of $\lhd$. The set WO consists of the reals x coding such well-ordered relations $\lhd$, and we refer to these as the reals coding ordinals. This is well known to be a complete Π^1_1 set of reals. One of the early results of [5] showing the power of infinite time Turing machines is that this set is decidable. We sketch the proof because the method will be useful for other purposes here.

Theorem 3 ([5, Theorem 2.2]) WO *is infinite time decidable.*

Proof. Given a real x, we first check whether x codes a linear order $\lhd$, by systematically checking all instances of transitivity, reflexivity, trichotomy, and anti-symmetry, in ω many steps of computation. Assuming $\lhd$ is a linear order, we next attempt to find the $\lhd$-least element in the field of the relation. This can be done by placing a current guess for the least element on the scratch tape and by searching for a $\lhd$-smaller element. When such a better guess is found, the algorithm overwrites it on the scratch tape and flashes a special flag on and then off. At the next limit stage, if the flag is on, then the guess was changed infinitely many times, and so the real is rejected, because it does not code a well order. If the flag is off at a limit, then the guesses stabilized on the current $\lhd$-least element, which now appears on the scratch tape.

Next, the algorithm erases all mention of this element from the field of the relation coded on the input tape and then continues to find (and subsequently erase) the next least element, and so on. The algorithm should detect limits of limit stages, so that the scratch tape and the flag can be accordingly reset. Eventually, the well-ordered initial segment of $\lhd$ is erased from the field of the relation coded on the input tape. By detecting when the tape is empty, the algorithm can know whether the original real coded a well order. If not, the algorithm will detect the ill-founded part of it and will reject it at that stage. $\qquad\square$

Since WO is a complete Π^1_1 set, any Π^1_1 question reduces to a question about WO, and so we obtain:

Corollary 4 *Any Π^1_1 set is infinite time decidable. Hence, any Σ^1_1 set is also decidable.*

Any real x can be viewed as the code of an ω-sequence of reals $\langle (x)_n \mid n < \omega \rangle$ by $(x)_n(m) = x(\langle n, m \rangle)$. Thus, if we are also given a real z coding a relation $\lhd$ on $\mathbb{N}$ of order type α, then any $\beta < \alpha$ is represented by some n with respect to $\lhd$, and we may view x as coding via z an α-sequence $\langle x_\beta \mid \beta < \alpha \rangle$ of reals. The real x_β is $(x)_n$, where n is the βth element with respect to $\lhd$.

More generally, any hereditarily countable set a can be coded with a real as follows. Suppose b is any countable transitive set containing a as an element, such as the transitive closure $\mathrm{TC}(\{a\})$, and let E be a relation on a subset $A \subseteq \mathbb{N}$ such that there is an isomorphism $\pi : \langle A, E \rangle \cong \langle b, \in \rangle$. Since this isomorphism π must be the Mostowski collapse of E, the set a is determined by E and the natural number n such that $\pi(n) = a$. We view the pair $\langle n, E \rangle$, coded by a real, as representing the set a. Of course, a set a generally has many different codes. In analogy with WO, let us define HC to be the set of such reals coding hereditarily countable sets in this way. Given two such codes x and y, define $x \equiv y$ if x and y are codes for the same set, and $x \in^* y$ if the set coded by x is an element of the set coded by y.

Theorem 5 *The structure $\langle \mathrm{HC}, \in^*, \equiv \rangle$ is infinite time computable but not infinite time decidable.*

Proof. The elements of HC are precisely the reals coding pairs $\langle n, E \rangle$, where E is a well-founded relation on some $A \subseteq \mathbb{N}$, where A is the field of E, the natural number n is in A, and the structure $\langle A, E \rangle$ satisfies extensionality. Thus, the set HC is Π^1_1 definable and, hence, decidable. The relation $x \equiv y$ is satisfied, where $x = \langle n, E \rangle$ and $y = \langle n', E' \rangle$ if and only if there is an isomorphism from the part of the field of E below n to the field of E' below n'. This is a Σ^1_1 property in the codes and, hence, decidable. Similarly, the relation $x \in^* y$ simply asserts that there is some m in the field of E' such that $\langle n, E \rangle \equiv \langle m, E' \rangle$, which is also Σ^1_1 and, hence, decidable. The structure is not infinite time decidable since the halting problem $0^\triangledown$ is expressible in this language. $\qquad\square$

The quotient structure HC $/\equiv$, under the induced relation $\in^*$, is of course isomorphic to the transitive collection H_{ω_1} of hereditarily countable sets.

Theorem 6 *The satisfaction relation for hereditarily countable sets* $\langle b, \in \rangle \models \varphi[\vec{a}]$ *is infinite time decidable, given any code* $\langle n, E \rangle \in$ HC *for b, the Gödel code* $\ulcorner\varphi\urcorner$*, and the code* $\vec{n}$ *of* $\vec{a}$ *with respect to* E.

Proof. This is simply an instance of the earlier remark we made, that when an algorithm has access to the entire domain of a structure, it can carry out the Tarski definition of truth. In this case, the code for b effectively provides the structure $\langle b, \in \rangle$ as a subset of $\mathbb{N}$. Alternatively, one could simply observe that the satisfaction relation has complexity Δ_1^1 and is therefore decidable.   □

The constructible hierarchy of Gödel is the transfinite hierarchy of sets L_α, defined by $L_0 = \emptyset$; $L_{\alpha+1}$ is the collection of definable subsets of L_α; for limit ordinals, $L_\eta = \bigcup_{\alpha < \eta} L_\alpha$. The constructible universe L is the proper class $\bigcup_\alpha L_\alpha$, and Gödel proved that $\langle L, \in \rangle$ is a (class) model of ZFC + GCH and much more.

Theorem 7

1. *There is an infinite time algorithm such that on input, a code for* L_α *for some countable ordinal* α *writes a code of* $L_{\alpha+1}$.

2. *There is an infinite time algorithm such that on input, a code of a countable ordinal* α *writes a code of* L_α.

Proof. Given a code of L_α, one divides the tape into infinitely many copies of $\mathbb{N}$, systematically considering each definition and each parameter, and by repeated applications of Theorem 6, one can write down codes for each of the definable subsets. This produces a code for $L_{\alpha+1}$. Given a code for α, one views $\mathbb{N}$ as an α-sequence of copies of $\mathbb{N}$. On each copy of $\mathbb{N}$, the algorithm may iteratively apply the previous method to produce codes for the successive new elements of L_β for each $\beta \leq \alpha$.   □

The next theorem asserts that there is a real c such that an infinite time Turing machine can recognize whether a given real is c, but no algorithm can produce c on its own. This is like a person who is able to recognize a particular song, a lost melody, when someone else sings it, but who is unable to sing it on his or her own. The idea of the proof leads to the concept of L-codes for sets and ordinals, of which we will make extensive use later.

Lost Melody Theorem 8 ([5]) *There is a real c such that* $\{c\}$ *is infinite time decidable but c is not writable.*

Proof. We sketch the proof from [5]. Results there show that every infinite time Turing machine computation either halts or repeats by some countable stage. Let β be the supremum of the stages by which all computations of the form $\varphi_p(0)$ have either halted or repeated. (Welch proved in [13] that $\beta = \Sigma$, the supremum of the accidentally writable ordinals.) The structure L_β is able to carry out all the computations $\varphi_p(0)$ for any length up to β, and so the defining property of β is expressible in L_β. One can use the defining property of β to show that there is a map from ω unbounded in β that is a definable subset of L_β. This map is therefore an element of $L_{\beta+1}$, and consequently, β is countable in $L_{\beta+1}$. So there is some L-least real $c \in L_{\beta+1}$ coding a relation of order type β. This is the real we seek.

Notice that $\{c\}$ is decidable, because if we are given any candidate real c', we can check that it codes an ordinal β', and if so, we can write down a code for $L_{\beta'+1}$, and check whether $L_{\beta'+1}$ satisfies that β' is the supremum of the repeat points for all computations $\varphi_p(0)$. This will be true if and only if $\beta' = \beta$. Next, we check that c' is the least real in $L_{\beta'+1} = L_{\beta+1}$ coding $\beta' = \beta$. This will be true if and only if $c' = c$. So we can decide whether any given real is c.

Finally, c is not writable, because β is necessarily larger than every clockable ordinal and, hence, larger than every writable ordinal. So β is not coded by any writable real. $\square$

Corollary 9 *There is a function f that is not infinite time computable, but whose graph is infinite time decidable.*

Proof. Let $f(x) = c$ be the constant function with value c, the lost melody real. Since $\{c\}$ is decidable, we can decide the graph of f, which consists of all pairs (x, y) for which $y = c$. But f is not computable, since $c \neq \varphi_p(0)$ for every program p. $\square$

Corollary 10 *There is an infinite time decidable model that is not infinite time computable.*

Proof. Let $\mathcal{A} = \langle \mathbb{R}, f \rangle$, where $f(x) = c$ is the constant function with value c, given by the Lost Melody Theorem, and $\ulcorner f \urcorner \in \mathbb{N}$. This is not a computable model, because the function f is not computable. Nevertheless, we will show that the elementary diagram of $\mathcal{A}$ is decidable. First, we consider the atomic diagram. We can use $f(f(x)) = f(x)$ to reduce the complexity of terms and then observe that $f(x) = f(y)$ is always true and $f(x) = y$ amounts to $y = c$, which is decidable. So any atomic assertion is decidable. To decide the full elementary diagram, we observe that it admits the effective elimination of quantifiers down to Boolean combinations of assertions of the form $x = c$ and $x = y$ (plus true and false). The quantifier case essentially amounts to observing that $\exists x\, (x = c\ \&\ x \neq y)$ is equivalent to $y \neq c$ and $\exists x\, (x \neq c\ \&\ x \neq y)$ is simply true. So $\mathcal{A}$ is decidable, but not computable, concluding the proof. $\square$

This Corollary can also be proved by using a language with a single constant symbol 0, with $\ulcorner 0 \urcorner \in \mathbb{N}$. The structure $\mathcal{B} = \langle \mathbb{R}, c \rangle$, interpreting 0 as the Lost Melody real c, is not a computable model because the value of the constant is not computable from its Gödel code. But the structure $\mathcal{B}$ is simply an infinite model with a distinguished constant, which admits the elimination of quantifiers, and since one can decide all statements of the form $x = c$, it follows that $\mathcal{B}$ has a decidable theory.   □

The idea of the Lost Melody Theorem provides a method of coding countable ordinals in L with unique codes. Specifically, for any $\alpha < \omega_1^L$, let β be least above α such that β is countable in $L_{\beta+1}$, and let c be the L-least real in $L_{\beta+1}$ coding a relation $\lhd$ on $\mathbb{N}$ with order type β. The ordinal α is represented by some natural number n with respect to $\lhd$, and so we will define $\langle n, c \rangle$ to be the L-code of α. Note that every ordinal α that is countable in L has exactly one L-code, since α determines β, which determines c, which determines $\lhd$, which determines n. Since the L-code of α is also a code of α in the sense of HC, we can computably determine by Theorem 5 whether $\alpha < \beta$, given L-codes for α and β. And just as with HC in this case, we can computably construct the isomorphism from the field of the relation coding α to the appropriate initial segment of the field of the relation coding β, and find the particular natural number representing α with respect to the code for β.

Lemma 11 *The set of L-codes for countable ordinals is infinite time decidable.*

Proof. Given a real coding a pair $\langle n, c \rangle$, we can determine whether c is the code of a relation $\lhd$ on $\mathbb{N}$ that is a well order of some order type β. If so, we can construct a code for $L_{\beta+1}$ and check that $L_{\beta+1}$ satisfies that β is countable and that the L-least real coding a relation of order type β is c. Finally, we can check that $L_{\beta+1}$ thinks that β is least such that it satisfies that α, the ordinal coded by n with respect to $\lhd$, is countable. If all these tests are passed, then the pair $\langle n, c \rangle$ is the L-code of α.   □

More generally, we have L-codes for any set that is hereditarily countable in L. Specifically, suppose that a is any set that is hereditarily countable in L. Let β be least such that $a \in L_\beta$ and β is countable in $L_{\beta+1}$. It follows that L_β is countable in $L_{\beta+1}$, so there is some L-least real c coding a relation E such that $\langle \mathbb{N}, E \rangle \cong \langle L_\beta, \in \rangle$. The set a is represented by some natural number n with respect to E, and the L-*code* of a is the pair $\langle n, c \rangle$. Let LC be the set of such L-codes for hereditarily countable sets in L. Since these are also codes for sets in the sense of HC, it follows by Theorem 5 that we may computably decide the relation $\in^*$ on the codes induced by the $\in$ relation on the sets coded.

Theorem 12 *The structure $\langle L_{\omega_1^L}, \in \rangle$ has an infinite time computable presentation as $\langle \mathrm{LC}, \in^* \rangle$.*

Proof. The set $L_{\omega_1^L}$ is precisely HC^L, the sets that are hereditarily countable in L, and this is isomorphic to $\langle \mathrm{LC}, \in^* \rangle$ via the L-codes. The $\in^*$ relation is decidable on the L-codes, just as in Theorem 5. And the set of L-codes LC is decidable just as in Lemma 11.   □

Similarly, using the L-codes for ordinals, we see that the structure $\langle \omega_1^L, < \rangle$ has an infinite time computable presentation.

2 Arithmetic on the real line

As a straightforward example of an infinite time computable structure, we consider the most prominent uncountable structure in mathematics, the real line under arithmetic.

Lemma 13 *The standard structure $\mathcal{R}$ of the real line under addition, multiplication, subtraction, division, and the order relation $<$ is infinite time computably presentable.*

We use "the real line" to describe this structure and refer to its elements as "points," because elsewhere in this paper we use the term "real number" to refer to elements of 2^ω. Also, since division by zero is usually undefined, let us regard it as a function on the computable domain $\mathcal{R} \times (\mathcal{R} - \{0\})$.

Proof. It is straightforward to identify points x on the real line uniquely with binary sequences $C \in 2^\omega$ such that $C(2n) = 0$ for infinitely many n and $C(2n+1) = 0$ for all but finitely many n and $C \neq \langle 1000 \cdots \rangle$. The element C corresponds to the real point

$$(-1)^{C(0)} \left(\sum_{n=0}^{\infty} 2^n \cdot C(2n+1) + \sum_{n=1}^{\infty} \frac{C(2n)}{2^n} \right),$$

and C is called the *presentation* of this real point. (The condition $C \neq \langle 1000 \cdots \rangle$ rules out the second presentation of the point 0 as -0.) The domain of our structure $\mathcal{R}$ is the set of all presentations of real points and is decidable in infinite time, since each of the conditions can be checked in ω many steps. Also, it is easy to give a process for deciding (in infinite time) whether two given domain elements are equal, and if not, which is larger under $<$.

Of course, all of the usual arithmetic operations on these representations, such as sum, difference, product, and quotient, have complexity (much less than) Δ_1^1 in the input, and therefore, by Corollary 4, these are all infinite time computable operations. Nevertheless, for illustration let us show in moderate detail how to compute the sum C'' of two presentations C and C' of positive real points. First, in ω many steps, we find the greatest $k > 0$ such that $C(2k) = C'(2k)$, or else we establish that there are infinitely many such k. Then we have two cases. If there is a greatest k, then beyond the k-th bit, C and C' complement each other perfectly, and there are only finitely many bits left to add. Otherwise, there are infinitely many k with $C(2k) = C'(2k)$, and we build the sum from the inside, by always searching for the next greater bit k with $C(2k) = C'(2k)$ and computing C'' up to that bit. (The point is that when $C(2k) = C'(2k)$, we know right away whether we need to "carry" a 1 from $C''(2k)$ when calculating $C''(2k-2)$, even without knowing $C''(2k)$ itself yet.) This defines the entire sequence C''. Note that if the representation of the sum happens to have $C''(2k) = 1$ for a tail segment, then one must switch to the preferred representation by changing these bits to 0 and performing an additional carry.

Notice that each of the two cases could be carried out in finite time computability, producing each bit $C''(n)$ in finitely many steps, assuming that one was given oracles presenting C and C'. Infinite time is required only to decide which of the two cases to use and (in the first case) to find the greatest k.

Addition of two negative real points can be defined using the above algorithm conjugated by the negation map $x \mapsto -x$, which is immediately seen to be computable. To get addition of a positive to a negative, we define subtraction of a positive real C' from another one $C > C'$, by taking finite approximations of the difference, adding them to C', and checking whether each finite approximation yields a sum $> C$ or $\leq C$.

It is tempting to bypass the discussion for subtraction by saying that the difference $C - C'$ should be that domain element D such that $C' + D = C$, since we have already given a method of computing the sum of positive domain elements. However, this does not suffice to prove computability, and indeed it illustrates a fundamental difference between the contexts of finite and infinite time: in infinite time computability, we may no longer have such effective search procedures. Without an infinite-time-computable enumeration of the domain of $\mathcal{R}$, there is no guarantee that we would ever find the element D described above, even though it must lie somewhere in the domain of $\mathcal{R}$. Therefore, it is necessary to compute D directly in infinite time, rather than searching for a D that satisfies $C' + D = C$. Decidability of subtraction as a ternary relation (that is, decidability of the statement $C - C' = D$) does follow from decidability of the addition relation, which follows from computability of addition as a function, but computability of subtraction is stronger.

For multiplication of positive domain elements C and C', we simply multiply C by each individual bit of C' (for instance, if $C'(2n) = 1$, then the product of C with that bit maps each bit of C n places to the right) and add the results together, one by one, in ω^2 many steps. Clearly each bit on the output tape does converge to a limit, since $C(2n+1) = 0$ for all but finitely many n, and the final output is the product of C and C'. This extends easily to the case of non-positive domain elements, so multiplication is computable. Finally, for division, we can check whether the divisor is the real point 0, and if not, we define it using the multiplication function, just as subtraction was defined using addition. Thus division is indeed a computable function on the domain $\mathcal{R} \times (\mathcal{R} - \{0\})$. □

One can expand the real field $\mathcal{R}$ to include all the usual functions of analysis: e^x, $\sqrt{x}$, $\ln x$, $\sin x$, and so on. Since (the bit values of) these functions have complexity below Δ_1^1, they are all infinite time computable by Corollary 4.

Let us turn now to the subfield $\mathcal{R}_w$, consisting of those real points having a writable presentation. It is clear from the algorithms given in the proof of Lemma 13 that $\mathcal{R}_w$ is a substructure of the real line $\mathcal{R}$. Moreover, we have the following lemma.

Lemma 14 *In the infinite time context, the ordered field $\mathcal{R}_w$ is computably presentable, and more generally, the ordered field $\mathcal{R}_w^X$ of those real points that have* *presentations writable using any oracle $X \subseteq \mathbb{R}$ is X-computably presentable. In each presentation, there is a computable (resp. X-computable) function from domain elements to the binary expansions of the real points they represent.*

Proof. The main difficulty is in getting the domain of our presentation of $\mathcal{R}_w$ to be decidable. For our domain S, we take the set of pairs $\langle e, c \rangle \in \omega \times \{c\}$, where c is the Lost Melody real of Theorem 8, the eth infinite time program outputs a presentation of a real point, and no $e' < e$ is the index of a program outputting the same point. This is indeed a decidable domain: given any pair, we first check whether the second element is c (since the set $\{c\}$ is decidable), and, if so, use c to check the remaining conditions, which we can now do because c codes an ordinal α so large that every program that halts at all must halt by stage α, as seen in the proof of Theorem 8.

Given any two elements $\langle e, c \rangle$ and $\langle e', c \rangle$ of S, we need to compute their sum, product, difference, and quotient and to compute the relation $<$. For each of the four operations, the proof of Lemma 13 gives a program P_{e_0} that writes a presentation of the resulting real point, with e_0 being infinite time computable uniformly in e and e'. So the result of the operation is the element $\langle e_1, c \rangle$, where e_1 is the least index of a program that outputs a presentation of the same real point as e_0. We were given c itself, of course, as part of the points $\langle e, c \rangle$ and $\langle e', c \rangle$, and with it it is simple to find the least such e_1. Thus each operation is computable on the domain S. The final claim is clear, since an element of S contains an algorithm for writing out a presentation of the corresponding real point, which in turn quickly yields every digit of the binary expansion of that point. From this, the relation $<$ on S is easily computed.

For $\mathcal{R}_w^X$, one simply relativizes the entire proof (including the choice of the Lost Melody real) to the oracle X.   □

Lemma 14 shows how the infinite time computable model theory differs from its finite time analog. Although we have proved that the ordered field $\mathcal{R}_w$ of infinite time computable reals (i.e., the writable reals) has an infinite time computable presentation, the corresponding fact in finite time is not true, for the finite time computable reals have no finite time computable presentation.

Proposition 15 *Let $\mathcal{R}_w$ be the ordered field of infinite time computable real points, and let $\mathcal{R}_f$ be the ordered subfield of finite time computable real points. Then neither $\mathcal{R}_w$ nor $\mathcal{R}_f$ is finite time computably presentable (in domain $\mathbb{N}$), but both are computably presentable in infinite time.*

Since the rational ordered field $\mathbb{Q}$ embeds uniquely and densely into $\mathcal{R}$, it follows that every ordered subfield $\mathbb{F}$ of $\mathcal{R}$, such as $\mathcal{R}_w$ or $\mathcal{R}_f$, embeds uniquely into our presentation of $\mathcal{R}$. We show next that this unique embedding is computable.

Lemma 16 *If $\mathbb{F}$ is any computable presentation of an ordered subfield of $\mathcal{R}$, then the unique embedding of $\mathbb{F}$ into $\mathcal{R}$ is computable.*

Proof. Given any $x \in \mathbb{F}$, we may use the computable functions of $\mathbb{F}$ to systematically compute the $\mathbb{F}$-representations of the rationals $\frac{m}{2^n}$, and we make comparisons of these rationals with x using the order of $\mathbb{F}$. This allows us to know the binary representation of x and, therefore, the representation of x in our presentation of $\mathcal{R}$. Thus, we have computed the unique embedding of $\mathbb{F}$ into our presentation of $\mathcal{R}$.   □

Proof of Proposition 15. An infinite time computable presentation of $\mathcal{R}_w$ was shown above to exist, and $\mathcal{R}_f$ is an infinite time computable subset of the domain, since infinite time Turing machines can easily simulate finite time ones.

If $\mathbb{F}$ were a finite time computable presentation of $\mathcal{R}_f$, then given any element $x \in \mathbb{F}$, we could compute the n-th digit of the binary expansion of the real point corresponding to $\mathbb{F}$, in finite time and uniformly in x and n. If $\mathbb{F} \cong \mathcal{R}_f$, this would give a simultaneous uniform finite time computation of all finite time computable sets, which of course is impossible. If $\mathbb{F} \cong \mathcal{R}_w$, then it would give a simultaneous uniform finite time computation of all infinite time writable reals, which again is easy to diagonalize against. This completes the proof of Proposition 15.   □

We note that the same diagonalization against finite time computable presentations of all finite time computable sets can be used to show that there is no infinite time writable presentation of all infinite time writable reals. Therefore we ask how it is that $\mathcal{R}_w$ is infinite time computably presentable. The answer is that although the domain of the presentation of $\mathcal{R}_w$ is a countable decidable set, it is not the image of ω under any infinite time computable function. The use of the Lost Melody real c in the domain of $\mathcal{R}_w$ makes this clear, and indeed, without using c or a similar element, we could not decide in infinite time which programs output infinite time computable reals.

A concise statement of the foregoing argument is to say that there is no writable presentation of $\mathcal{R}_w$, even though there is a computable presentation. A *writable structure* is an infinite time computable structure $\mathcal{A}$ such that there exists a single writable real $r \in 2^\omega$ whose first row $r^{[0]}$ codes the entire atomic diagram of $\mathcal{A}$ and whose remaining rows name all elements of the domain of $\mathcal{A}$. That is,

$$r^{[0]} = \{\ulcorner \varphi \urcorner \ : \ \varphi \in D_a(\mathcal{A})\},$$

$$\mathrm{dom}(\mathcal{A}) = \{r^{[n]} \ : \ n \in \omega - \{0\}\}.$$

(An equivalent definition requires that $r^{[n]} \neq r^{[m]}$ whenever $0 < n < m < 1 + |\mathcal{A}|$, and $r^{[m]} = 0$ if $m > |\mathcal{A}|$.) We assume for these purposes that the language of $\mathcal{A}_A$ is also coded into ω, with $2n - 1$ coding the constant symbol for the element named as $r^{[n]}$. Thus we have a computable enumeration of the elements of $\mathcal{A}$, from which it is immediate that the complete diagram of $\mathcal{A}$ is infinite time decidable. Since they allow computable searches of the entire domain, writable structures behave something like an analog to the finite structures in the classical theory.

Let us conclude this section with a brief generalization. Let $\mathcal{R}_a$ be the structure of the real points having an accidentally writable presentation, and similarly, let $\mathcal{R}_e$ consist of those having an eventually writable presentation.

Theorem 17 $\mathcal{R}_f \prec \mathcal{R}_w \prec \mathcal{R}_e \prec \mathcal{R}_a \prec \mathcal{R}$.

Proof. The point is that each of these structures is a real closed ordered field. Before explaining this, let us first iron out a wrinkle with $\mathcal{R}_a$. In order to see even that this structure is closed under addition, it is useful to know that the set of accidentally writable reals is closed under pairing. To see this, consider the algorithm that simulates all programs on input 0, and for each accidentally writable real x observed during this master simulation, the algorithm starts another master simulation that produces all accidentally writable reals y that appear before the first appearance of x. Then, for each such y, our main algorithm writes a real coding the pair $\langle x, y \rangle$ on the scratch tape. This algorithm shows that if x and y are accidentally writable, then the pair $\langle x, y \rangle$ is also accidentally writable. Using this and the observations of Lemma 13, it now follows that $\mathcal{R}_a$ is a field.

Each of the fields is closed under square roots for its positive elements, since the digits of the square root can be systematically computed. Also, for any odd degree polynomial, one can use successive approximations (for example, by Newton's method) to find a computable root. Since the theory of real closed fields is model complete, the theorem now follows.   □

One can naturally extend this theorem by oracles and have a rich lattice of relatively computable subfields of $\mathcal{R}$. Each of the extensions in the theorem is strict, by [5, Theorem 6.15], and it follows that each is a transcendental extension of the previous. Finally, we observe that $\mathcal{R}_w$ can have no writable transcendence basis over $\mathbb{Q}$ or $\mathcal{R}_f$, since then we would be able to produce a writable list of all writable reals, which we have observed is impossible by a simple diagonalization. Similarly, $\mathcal{R}_e$ has no eventually writable transcendence basis over $\mathcal{R}_w$ and $\mathcal{R}_a$ has no accidentally writable transcendence basis over $\mathcal{R}_e$.

3 The infinite time computable Completeness Theorem

The Completeness Theorem asserts that every consistent theory has a model. The finite time effective version of this asserts that any finite time decidable theory has a finite time decidable model. And in the infinite time context, at least for languages coded in $\mathbb{N}$, this proof goes through without any hitch. In fact, the infinitary context gives a slightly stronger result:

Theorem 18 *In the infinite time context, if T is a consistent theory in a computable language coded in $\mathbb{N}$ and T has a computable axiomatization, then T has a decidable computable model. In fact, such a theory has a model coded by a writable real.*

Proof. The point is that the classical Henkin construction is effective for infinite time Turing machines. Note that if T has a computable axiomatization in a language coded in $\mathbb{N}$, then it is actually decidable, since the infinite time Turing machines can search through all proofs in ω steps. We may assume that there is an infinite supply of new constant symbols, by temporarily rearranging the Gödel codes of the symbols in the original language if necessary. Enumerate the sentences in the expanded language as $\langle \sigma_n \mid n \in \mathbb{N} \rangle$, and build a complete, consistent Henkin theory in the usual manner: at stage n, we add σ_n, if this is consistent with what we have already added to T, or else $\neg \sigma_n$, if it is not. Since T is decidable, this is computable. In addition, if σ_n has the form $\exists x\, \varphi(x)$ and we added it to the theory, then we also add $\varphi(c)$ for the first new constant symbol c that has not yet been considered. The result of this construction is a complete consistent Henkin theory $\bar{T}$ extending T. The theory $\bar{T}$ is decidable, because for any σ, the infinite time algorithm can run the construction until σ is considered, and answer accordingly as it was added to $\bar{T}$ or not. As usual, we may use the Henkin constants to build a model of T. Specifically, let $c \equiv d$ if $\bar{T} \vdash c = d$, and define the $R([\vec{c}]) \iff \bar{T} \vdash R(\vec{c})$ and $f([\vec{c}]) = [d] \iff \bar{T} \vdash f(\vec{c}) = d$. The classical induction shows that the resulting structure $M_{\bar{T}}$ of equivalence classes satisfies $\varphi([\vec{c}])$ if and only if $\bar{T} \vdash \varphi(\vec{c})$, so this is a model of T. Finally, for any constant symbol d, one may compute the (numerically) least element of $[d]$ by simply testing each of the smaller constants c to determine whether $c \equiv d$. Thus, by replacing each equivalence class with its least member, we construct a computable presentation of $M_{\bar{T}}$. Since the underlying set of this model is contained in $\mathbb{N}$, an algorithm can write down the entire structure as a writable real. □

Many theories, including some very powerful theories, have infinite time computable axiomatizations, and so this result provides numerous interesting decidable models. For example, the theory of *true arithmetic* $\mathrm{TA} = \mathrm{Th}(\langle \mathbb{N}, +, \cdot, 0, 1, < \rangle)$ is infinite time decidable, because arithmetic truth is infinite time decidable, and so the theory $\mathrm{TA} + \{ n < c \mid n \in \mathbb{N} \}$ is a computable axiomatization of the theory of the nonstandard models of true arithmetic. Similar observations establish:

Corollary 19 *There are infinite time, decidable, computable, nonstandard models of the theories PA, TA, ZFC, ZFC + large cardinals, and so on, provided that these theories are consistent.*

The infinite time realm, therefore, lies considerably beyond the computable models of the finite time theory. What is more, as we have emphasized, the infinite time context allows for uncountable computable models and uncountable languages, which cannot be coded in $\mathbb{N}$. So Theorem 18 does not tell the full story. In the general context, where languages are coded in the reals, we ask whether the full infinite time analog of the Completeness Theorem holds:

Question 20 *Does every consistent infinite time decidable theory have an infinite time decidable model? Does every such theory have an infinite time computable model?*

One of the convenient features of the classical theory, when working with a language coded in $\mathbb{N}$, is that one can enumerate the function, relation, and constant symbols of the language $s_0, s_1, \ldots$ in such a way that from any symbol s_n, one can reconstruct the list $\langle s_m \mid m \leq n \rangle$ of prior symbols. This is a triviality in the context of computable languages coded in $\mathbb{N}$, because we simply enumerate the symbols in the order of their Gödel codes. Given any such code, one simply tests all the smaller natural numbers in turn to discover the list of prior codes for symbols. But in the uncountable context, a computable representation of a language may not have this feature. Let us therefore define that a computable representation $\mathcal{L}$ of a language is computably *well presented* if there is an enumeration $\langle s_\alpha \mid \alpha < \delta \rangle$ of all of the function, relation, and constant symbols of the language, for some $\delta \leq \omega_1$, such that from any $\ulcorner s_\alpha \urcorner$, we can (uniformly, in infinite time) compute a code for the sequence $\langle \ulcorner s_\beta \urcorner \mid \beta \leq \alpha \rangle$ of prior symbols. In this case, we can prove the infinite time computable analog of the Completeness Theorem.

Theorem 21 *Every consistent infinite time decidable theory in a computably well-presented language has an infinite time decidable model in this language.*

We begin with a few preliminary lemmas. Let us say that a computable presentation $\mathcal{L}$ of a language admits a *computably stratified enumeration* of formulas if there is an enumeration of all $\mathcal{L}$-formulas $\langle \varphi_\alpha \mid \alpha \leq \delta \rangle$, for some $\delta \leq \omega_1$, such that from the Gödel code $\ulcorner \varphi_\alpha \urcorner$, one can (uniformly in infinite time) compute a real coding the sequence $\langle \ulcorner \varphi_\beta \urcorner \mid \beta \leq \alpha \rangle$ of Gödel codes of the prior formulas.

Lemma 21.1 *If a language $\mathcal{L}$ is computably well presented, then it admits a computably stratified enumeration of formulas.*

Proof. Suppose that a language $\mathcal{L}$ is computably well presented by the enumeration $\langle s_\alpha \mid \alpha \leq \delta \rangle$. Given a well-ordered list of function, relation, and constant symbols, one can systematically produce a list of all formulas in that language, as follows. The first ω many formulas are those not using any of the symbols; the next ω many formulas are those using the first symbol only; the next ω many formulas use the second symbol and possibly the first. There is a (finite time) computable list of countably many first-order formula templates, with holes for the function, constant, and relation symbols, and the actual formulas are obtained by plugging codes for actual function, relation, and constant symbols (of the appropriate arity) into those holes. From the presentation of the symbols, we systematically generate a list of all finite sequences of the symbols, and from these and the templates, one can generate the list of all formulas. We therefore generate the formulas in blocks of length ω, and all formulas in the αth block are required to use the symbol s_α and may use earlier symbols. This defines the enumeration of the formulas $\langle \varphi_\alpha \mid \alpha \leq \gamma \rangle$.

Given any formula $\ulcorner \varphi \urcorner$, we can inspect it for the symbols s that appear in it, and from each $\ulcorner s \urcorner$, we can generate the corresponding list of prior symbols $\langle \ulcorner s_\beta \urcorner \mid \beta \leq \alpha \rangle$, where $s = s_\alpha$. By comparing the lengths of these sequences, we can tell which symbol was the last to appear in the enumeration of $\mathcal{L}$. For this maximal α, we know that φ appears in the αth block of formulas. From the list of symbols $\langle \ulcorner s_\beta \urcorner \mid \beta \leq \alpha \rangle$, we can regenerate the list of formulas up to and including the αth block of formulas, thereby producing the prior list of formulas $\langle \ulcorner \varphi_\xi \urcorner \mid \xi \leq \eta \rangle$, where $\varphi = \varphi_\eta$. $\qquad \square$

A fundamental construction of first-order logic is to expand a language by adding infinitely many new constant symbols. In the context of computable model theory, whether finite or infinite time, if the presentation of a language $\mathcal{L}$ already uses all the available Gödel codes, then one is forced to consider translations of the language in order to free up space in the Gödel codes to represent the expanded language. For example, even in the finite time context, if one has a model in a language with infinitely many constant symbols, and the Gödel codes of the symbols already use up all of $\mathbb{N}$, then in order to add constants to the language, one seems forced to use a translation of the language. A given language can have many different computable presentations, and in general these may not be computably equivalent. For two presentations of the language, there may be no computable method of translating symbols or formulas from one representation to the other. (And this phenomenon occurs already in the finite time context.) In the infinite time context, where we represent symbols with real numbers, this phenomenon can occur even in finite languages, since the Gödel codes for a symbol may be reals that are incomparable in the infinite time Turing degrees. If we have two computable presentations $\mathcal{L}$ and $\mathcal{L}'$ of a language, and it happens that there is a computable function mapping every $\mathcal{L}'$ code for a symbol to the $\mathcal{L}$ code for the same symbol, then we will say that $\mathcal{L}'$ is a *computable translation* of $\mathcal{L}$. In such a case, syntactic questions about $\mathcal{L}'$ can be reduced computably to syntactic questions about $\mathcal{L}$. This relation is not necessarily symmetric (because in the infinite time context, a function can be computable without its inverse being computable). If both languages are computable translations of each other, we say that the languages are computably isomorphic translations.

Lemma 21.2 *If a language $\mathcal{L}$ is well-presented computably, then there is a computably isomorphic translation of it to a well-presented language $\mathcal{L}_0$, preserving the order of the enumeration of symbols, and a well-presented expansion $\mathcal{L}_1$ of $\mathcal{L}_0$ containing ω many new constant symbols c_s^n for every symbol s of $\mathcal{L}$, such that from $\ulcorner s \urcorner$ and n, one can uniformly compute $\ulcorner c_s^n \urcorner$ and conversely.*

Proof. For each symbol s of $\mathcal{L}$, let its code in $\mathcal{L}_0$ be obtained by simply adding a 0 to the front of $\ulcorner s \urcorner$ in $\mathcal{L}$. For $\mathcal{L}_1$, the code of the constant symbol c_s^n is obtained by adding $n+1$ many 1s plus 0 to the front of $\ulcorner s \urcorner$ in $\mathcal{L}$. Thus, from $\ulcorner s \urcorner$ in $\mathcal{L}$, we can easily compute every $\ulcorner c_s^n \urcorner$ and $\ulcorner s \urcorner$ in $\mathcal{L}_1$ and vice versa. So it is clear that $\mathcal{L}_0$ is a computably isomorphic translation of the language $\mathcal{L}$. The enumeration of the symbols of $\mathcal{L}_1$ simply replaces each symbol s of $\mathcal{L}$ with the block of symbols s, c_s^0, c_s^1, and so on. From any of these symbols, we can reconstruct the prior list of symbols in $\mathcal{L}$, and from those symbols, we can reconstruct the corresponding constant symbols, so as to generate the prior list of symbols in $\mathcal{L}_1$. □

Proof of Theorem 21. We carry out the proof of Theorem 18 in this more general context. Suppose that T is a computably axiomatized consistent theory in the well-presented language $\mathcal{L}$. Let $\mathcal{L}'$ be the well-presented language of Lemma 21.2, with infinitely many new constant symbols for each symbol of $\mathcal{L}$. Because it is well presented, this expanded language has a computably stratified enumeration

$\langle \ulcorner \varphi_\alpha \urcorner \mid \alpha < \delta \rangle$ of formulas. We assume that this language is enumerated in the manner of Lemma 21.1, in blocks of length ω containing all formulas with a given symbol and earlier symbols. Because we arranged that every symbol s of $\mathcal{L}$ gives rise to an infinite list of new constant symbols c_s^n, we may arrange that from any $\ulcorner \varphi_\alpha \urcorner$, we may compute uniformly the code of a distinct new constant symbol c not appearing in any earlier φ_β.

We now recursively build the theory $\bar{T}$ in stages: at stage α, if φ_α is a sentence, then we add it to $\bar{T}$ if this remains consistent; otherwise we add $\neg \varphi_\alpha$. In addition, if φ_α is a sentence of the form $\exists x \, \psi(x)$ and we had added it to $\bar{T}$, then we also add a sentence of the form $\psi(c)$, where c is the distinct new constant symbol that has not yet appeared in any earlier formula. The usual model theoretic arguments show that $\bar{T}$ is a complete, consistent Henkin theory extending T.

We argue that $\bar{T}$ is decidable. Given any $\mathcal{L}'$ formula φ_α, we may use the computable stratification to write down a code of $\langle \ulcorner \varphi_\beta \urcorner \mid \beta \leq \alpha \rangle$. From this, we may computably reconstruct $\bar{T}$ up to stage α. The question of whether to add φ_β or $\neg \varphi_\beta$ at stage β reduces to a question about whether the theory constructed up to stage β proves $\neg \varphi_\beta$. But since the algorithm has a real coding the theory constructed up to stage β, it can enumerate computably all finite combinations of the formulas it is committed to adding to T, and check whether T proves that any of those finite combinations of formulas proves $\neg \varphi_\beta$. This is a decidable question, since T is decidable and we may translate computably between the languages $\mathcal{L}$ and $\mathcal{L}'$. Thus, $\bar{T}$ is computable.

Next, we build a decidable model of $\bar{T}$. Define the equivalence relation $c \equiv d \iff \bar{T} \vdash c = d$, and from each equivalence class $[c]$, select the constant $c_{s_\alpha}^n$ such that the pair $\langle \alpha, n \rangle$ is lexicographically least, where s_α is the αth symbol in the original presentation of $\mathcal{L}$. The set of such least constants is decidable, because from any constant $c_{s_\alpha}^n$ we may construct the list of prior symbols, and therefore the ω-blocks of the symbols in $\mathcal{L}'$, and therefore all the corresponding formulas φ_β containing only those symbols. By reconstructing the theory $\bar{T}$ up to that point, we can tell whether $\bar{T}$ proves $c_{s_\alpha}^n = c_{s_\xi}^m$, for any $\xi < \alpha$. So the set of such least representatives is decidable. We may now impose the usual structure on these representatives, to get a decidable model of $\bar{T}$. Since we have a computable isomorphism of $\mathcal{L}'$ with $\mathcal{L}$, it is no problem to translate between the two languages, and so we may use the original language presentation $\mathcal{L}$ when imposing this structure, resulting in a decidable model of T in the original language $\mathcal{L}$, as desired. $\qquad \square$

Theorem 22 *If $V = L$, then every consistent infinite time decidable theory has an infinite time decidable model, in a computable translation of the language.*

Proof. The first step is to translate to a computably well-presented language.

Lemma 22.1 *If $V = L$, then every computably presented language has a computable translation to a computably well-presented language.*

Proof. Assume $V = L$, and suppose that $\mathcal{L}$ is a computably presented language. Let $S \subseteq \mathbb{R}$ be the corresponding computable set of Gödel codes for the function, relation, and constant symbols of $\mathcal{L}$. Let $\langle s_\alpha \mid \alpha < \delta \rangle$ be the enumeration of the elements of S in order type $\delta \leq \omega_1$, using the canonical L-ordering of $\mathbb{R}^L$. For each $\alpha < \delta$, let γ_α be the smallest countable ordinal above α such that L_{γ_α} satisfies "ω_1 exists" and s_β exists for every $\beta \leq \alpha$. By this latter assertion, we mean that for every $\beta \leq \alpha$, the structure L_{γ_α} computes that S has at least β many elements in the L-order. Notice that because it satisfies "ω_1 exists," this structure correctly computes all infinite time computations for input reals that it has. Therefore, it correctly computes $S \cap L_{\gamma_\alpha}$, which has $\langle s_\beta \mid \beta \leq \alpha \rangle$ as an initial segment in the L order. In particular, $\langle s_\beta \mid \beta \leq \alpha \rangle \in L_{\gamma_\alpha}$. Let t_α be the L-code of the pair $\langle \alpha, \gamma_\alpha \rangle$. We will use t_α to represent the symbol coded by s_α in $\mathcal{L}$. Denote this new translation of the language by $\mathcal{L}'$.

First, we observe that the set $\{\, t_\alpha \mid \alpha \leq \delta \,\}$ is decidable. Given any real t, we can check if it is the L-code of a pair of ordinals $\langle \alpha, \gamma \rangle$ and, if so, whether γ is least such that L_γ satisfies "ω_1 exists" and s_β exists for every $\beta \leq \alpha$. If so, then we accept t. Necessarily, in this case $t = t_\alpha$. These questions are all decidable, because we know how to recognize an L-code for a pair of ordinals, and given the code of an ordinal γ, we can construct a code of L_γ and then check the truth of any statement in that structure by Theorem 6.

What is more, from t_α we can construct all earlier t_β for $\beta \leq \alpha$, because with an L-code for γ, we can look for the least $\gamma' \leq \gamma$ such that $L_{\gamma'}$ satisfies "ω_1 exists" and s_ξ exists for all $\xi \leq \beta$. Thus, our new language is computably well presented via $\mathcal{L}'$. Finally, $\mathcal{L}'$ is a computable translation of $\mathcal{L}$ because from t_α we can compute s_α. □

We remark that the translation from $\mathcal{L}$ to $\mathcal{L}'$, although perhaps not a computably isomorphic translation, is nevertheless relatively mild. Specifically, from s_α and any code for a sufficiently large ordinal, one can compute t_α. In this sense, the two representations of the language are close.

We now complete the proof of Theorem 22. Assume $V = L$, and suppose that T is a consistent decidable theory in a language $\mathcal{L}$. (By testing whether certain tautologies are well formed, it follows that the language itself is computable.) By Lemma 22.1, there is a computable translation of $\mathcal{L}$ to a well-presented language $\mathcal{L}'$. Let T' be the corresponding translation of T into this translated language. Note that T' remains decidable in $\mathcal{L}'$, because the question $T' \vdash \sigma'$ computably reduces to a question of the form $T \vdash \sigma$, which is decidable. By Theorem 21, the theory T' has a decidable model, as desired. □

So it is at least consistent with ZFC that the infinite time computable Completeness Theorem holds, if one allows computable translations of the language, and in this sense, one may consistently hold a positive answer to Question 20. Does this settle the matter? No, for we will now turn to negative instances of the Completeness

Theorem. The fact is that in some models of set theory, there are consistent decidable theories having no decidable model, and so the infinitary computable Completeness Theorem is actually independent of ZFC.

Theorem 23 *It is relatively consistent with* ZFC *that there is an infinite time decidable theory, in a computably presented language, having no infinite time computable or decidable model in any translation of the language (computable or not).*

This theorem relies on the following fact from descriptive set theory. For a proof, see [9, Theorem 25.23].

Lemma 23.1 (Mansfield–Solovay) *If $A \subseteq \mathbb{R}$ is Σ_2^1 and $A \nsubseteq L$, then A contains a perfect subset.*

The crucial consequence for us will be:

Lemma 23.2 *If ω_1^L is countable and the* CH *fails, then there are no Σ_2^1 sets of size ω_1. Hence, under these hypotheses, there are also no decidable sets or semi-decidable sets of size ω_1.*

Proof. Every decidable or semi-decidable set $A \subseteq \mathbb{R}$ is Δ_2^1 and, hence, Σ_2^1. If ω_1^L is countable and $A \subseteq L$, then A is countable. If $A \nsubseteq L$, then by Lemma 23.1, it contains a perfect subset and, hence, has cardinality 2^ω. Under $\neg$CH, this excludes the possibility that A has cardinality ω_1. $\qquad\square$

Proof of Theorem 23. Suppose that ω_1^L is countable and the CH fails. An elementary forcing argument shows that this hypothesis is relatively consistent with ZFC. Lemma 23.2 now shows that there are no Σ_2^1 sets of size ω_1. Consider the following theory, in the language with a constant c_x for every $x \in$ WO (for simplicity, let $\ulcorner c_x \urcorner = x$), a binary relation $\equiv$ and a function symbol f. The theory T is the atomic diagram of the structure $\langle$WO$, \equiv\rangle$, where $\equiv$ is the relation of coding the same ordinal, together with the axiom asserting that f is a choice function on the equivalence classes. That is, T contains all the atomic facts that are true about the constants c_x for $x \in$ WO, plus the assertions "$x \equiv f(x)$" and "$x \equiv y \implies f(x) = f(y)$." This theory is computably axiomatizable, because $\equiv$ is a decidable relation on WO. So as a set of sentences, the axioms of T are decidable.

But actually, the theory T is fully decidable. First, we observe that it admits elimination of quantifiers. The point is that T is essentially similar to the theory of an equivalence relation with infinitely many equivalence classes, all infinite. Note that the theory T implies $f(f(x)) = f(x)$, and $x \equiv f(y)$ is the same as $x \equiv y$. Also, $x = f(y)$ is equivalent to $x = f(x)$ & $x \equiv y$. By combining these reductions with the usual induction, it suffices to eliminate quantifiers from assertions of the form $\exists x\, x \equiv y$ & $x \neq z$ & $x = f(x)$ and $\exists x\, x \equiv y$ & $x \neq z$ & $x \neq f(x)$. But these are both equivalent to $y \neq z$, since in the former case one may use $x = f(y)$, and in the latter case some x equivalent to y, other than $f(y)$. This inductive reduction provides a computable method of finding, for any given formula, a quantifier-free formula that is equivalent to it under T. The point now is that any quantifier-free sentence is a Boolean combination of assertions about the constants c_x of the form $c_x \equiv c_y$, $c_x = c_z$, and $f(c_x) = c_y$. The first two of these are computable, since they are equivalent to $x \equiv y$ and $x = z$, respectively. The assertion $f(c_x) = c_y$ is false if $x \not\equiv y$, which is computable, and otherwise it is not settled by T, since there are models of T where $f(c_x)$ is any desired c_y with $y \equiv x$. For any finite list of constants c_y, it is consistent that $f(c_x)$ is equal to any of them (at most one of them), provided $x \equiv y$, or none of them. Because of this, we can decide computably whether T proves any given quantifier-free assertion in the language of T. So T is decidable.

Finally, suppose toward contradiction that T has a computable or decidable model $M = \langle A, \equiv^M, f^M, c_x^M \rangle_{x \in \mathrm{WO}}$. In this case, both the graph of f and the relation $z = c_x^M$ are decidable, and so the set $\{ f(c_x^M) \mid x \in \mathrm{WO} \}$ has complexity Σ_2^1. But this set also has cardinality ω_1, contradicting Lemma 23.2. So T can have no computable or decidable model under these set theoretic hypotheses. Since the set theoretic hypotheses are relatively consistent with ZFC, it is relatively consistent with ZFC that there is an infinite time decidable theory with no computable or decidable model.   □

With Theorems 22 and 23, we have now established the following:

Theorem 24 *The infinite time computable Completeness Theorem is independent of ZFC.*

For this theorem, we take the infinite time computable Completeness Theorem to be the following assertion: every consistent decidable theory in a computably presented language has a decidable model in a computable translation of the language.

4 The infinite time computable Löwenheim–Skolem Theorem

The classic Löwenheim–Skolem Theorem has two parts: the upward theorem asserts that every infinite model has arbitrarily large elementary extensions, in every cardinality at least as large as the original model and the language; the downward theorem asserts that every infinite model has elementary substructures of every smaller infinite cardinality at least as large as the language. Here, of course, we are interested in the infinite time computable analogs of these assertions, which concern computable or decidable models.

Question 25 *Does every infinite time decidable model have an infinite time decidable elementary extension of size continuum?*

Question 26 *Does every infinite time decidable infinite model (in a language coded in $\mathbb{N}$, say) have a countable infinite time decidable elementary substructure?*

These questions have many close variants, depending, for example, on whether the models are decidable or computable, and on whether the languages or models are well presented. One could ask in Question 25 merely for a proper elementary extension, or for an uncountable extension, rather than for one of size continuum, and in Question 26, merely for a proper elementary substructure rather than for a countable one (when the original model is uncountable). We regard all such variations as infinite time computable analogs of the Löwenheim–Skolem Theorem.

If the Continuum Hypothesis fails badly, then it is too much to ask for computable models of every cardinality between ω and 2^ω. To be sure, this is clearly impossible if the continuum is too large (if $2^\omega \geq \aleph_{\omega_1}$), for in this case there would be uncountably many such intermediate cardinalities but only countably many decidable models. More importantly, however, Lemma 23.1 shows that there can be no decidable sets of cardinality strictly between ω_1^L and 2^ω. Thus, the possible cardinalities of decidable sets of reals are finite, countable, ω_1^L, and 2^ω.

We do not know the full answers to either of the questions above, although we do know the answers to some of the variants. For the upward version, if a model is well presented, then we can find an infinite time decidable proper elementary extension (see Theorem 27); if $V = L$, then we can arrange this extension to be uncountable (see Theorem 28). So it is consistent that the upward Löwenheim–Skolem Theorem holds. For the downward version, if an uncountable decidable model is well presented, then we can always find a countable decidable elementary substructure (see Theorem 29); but if one broadens Question 26 to the case of computable models, rather than decidable models, then we have a strong negative answer, for there is a computable structure on $\mathbb{R}$ having no computable proper elementary substructures (see Theorem 30).

In analogy with well-presented languages, let us define that an infinite time computable model $\mathcal{A} = \langle A, \cdots \rangle$ is *well presented* if the language of its elementary diagram is well presented. This means that there is an enumeration $\langle s_\alpha \mid \alpha < \delta \rangle$, for some $\delta \leq \omega_1$, including every Gödel code for a symbol in the language and every element of A, such that from s_α one can compute a code for $\langle s_\beta \mid \beta \leq \alpha \rangle$. The models produced in the computable Completeness Theorem 21, for example, have this property.

Theorem 27 *If $\mathcal{A}$ is a well-presented infinite time decidable infinite model, then $\mathcal{A}$ has a proper elementary extension with an infinite time decidable presentation.*

Proof. Let T be the elementary diagram of $\mathcal{A}$, in a well-presented language. Let $\mathcal{L}'$ be the language of T together with new constants, as in Lemma 21.2. Let T' be the theory T together with the assertion that these new constants are not equal to each other or to the original constants. Since T is decidable, it is easy to see that T' is decidable, since any question about whether T' proves an assertion about the new constants can be decided by replacing them with variables and the assumption that those variables are not equal. Thus, by Theorem 21, there is an infinite time decidable model of T'. Such a model provides a decidable presentation of a proper elementary extension of $\mathcal{A}$. □

Theorem 28 *If* $V = L$, *then every infinite time decidable infinite model* $\mathcal{A}$ *elementarily embeds into an infinite time decidable model of size the continuum, in a computable translation of the language.*

Proof. Assume $V = L$, and suppose that $\mathcal{A}$ is an infinite time decidable infinite model. We may assume, by taking a computably isomorphic copy of the language that all the Gödel codes of symbols and elements in $\mathcal{A}$ begin with the digit 0. So there are continuum many additional codes, beginning with 1, that we use as the Gödel codes of new constant symbols. If T is the elementary diagram of $\mathcal{A}$, then let T' be T together with the assertion that these new constants are not equal. The theory T' is decidable, because any question about whether T' proves an assertion reduces to a question about whether T proves an assertion about some new arbitrary but unequal elements. This can be decided by replacing those new constant symbols with variable symbols plus the assertion that they are distinct. Thus, by Theorem 22, there is a decidable model $\mathcal{A}' \models T'$. The model $\mathcal{A}'$ has size continuum because of the continuum many new constants we added, and $\mathcal{A}$ embeds elementarily into $\mathcal{A}'$ because $\mathcal{A}'$ satisfies the elementary diagram of $\mathcal{A}$.   □

We note that the graph of the elementary embedding of $\mathcal{A}$ into $\mathcal{A}'$ is infinite time decidable, because from the code of a symbol in the expanded language, one can compute the code of the corresponding symbol in the original language. There seems little reason to expect in general that this embedding should be a computable function, and it cannot be if the original presentation was not well presented.

Let us turn now to the infinite time computable analogs of the downward Löwenheim–Skolem Theorem.

Theorem 29 *If* $\mathcal{A}$ *is an uncountable, well-presented infinite time decidable model in a language coded by a writable real, then there is an infinite time decidable, countable elementary substructure* $\mathcal{B} \prec \mathcal{A}$.

Proof. The idea is to effectively verify the Tarski–Vaught criterion on the shortest initial elementary cut of the well-presented enumeration of $\mathcal{A}$. So, suppose that $\langle a_\alpha \mid \alpha < \omega_1 \rangle$ is the well-presented enumeration of the underlying set of $\mathcal{A}$. By classic methods, there is a closed unbounded set of countable initial segments of this enumeration that form elementary substructures of $\mathcal{A}$. Let β be least such that $B = \{ a_\alpha \mid \alpha < \beta \}$ forms an elementary substructure $\mathcal{B} \prec \mathcal{A}$. Thus, β is least such that the set $\{ a_\alpha \mid \alpha < \beta \}$ satisfies the Tarski–Vaught criterion in $\mathcal{A}$. We will argue that B is infinite time decidable as a set. Given any a_ξ, we can generate the sequence $\langle a_\alpha \mid \alpha < \xi \rangle$ and for each $\xi' \leq \xi$ we can check whether $\{ a_\alpha \mid \alpha < \xi' \}$ satisfies the Tarski–Vaught criterion in $\mathcal{A}$. To check this, we use the writable real coding the language to generate a list of all formulas φ in the language. For every such formula φ and every finite sequence $a_{\alpha_0}, \ldots, a_{\alpha_n}$ with each $\alpha_i < \xi'$, we use the decidability of $\mathcal{A}$ to inquire whether $\exists x \, \varphi(x, a_{\alpha_0}, \ldots, a_{\alpha_n})$ is true in $\mathcal{A}$. If so, then we check that there is some $\alpha < \xi'$ with $\varphi(a_\alpha, a_{\alpha_0}, \ldots, a_{\alpha_n})$ true in $\mathcal{A}$. These checks will all be satisfied if and only if $\{\, a_\alpha \mid \alpha < \xi' \,\}$ satisfies the Tarski–Vaught criterion. Consequently, if such a ξ' exists with $\xi' \leq \xi$, then by the minimality of β, it must be that $\beta \leq \xi'$, and so a_ξ is not in B. If no such ξ' exists up to ξ, then $\xi < \beta$ and so $a_\xi \in B$. Therefore, as a set, B is decidable. The corresponding model $\mathcal{B}$ is therefore a decidable model and a countable elementary substructure of $\mathcal{A}$, as desired.   □

Finally, we have a strong violation to the infinite time computable downward Löwenheim–Skolem Theorem, when it comes to computable models. For infinite time Turing machines, a *computation snapshot* is a real coding the complete description of a machine configuration, namely, the program that the machine is running, the head position, the state, and the contents of the cells.

Theorem 30 *There is an infinite time computable structure with underlying set $\mathbb{R}$ having no infinite time computable proper elementary substructure.*

Proof. Define the relation $U_p(x, y)$ if y codes the computation sequence of program p on input x showing it to have been accepted. That is, y codes a well-ordered sequence of computation snapshots $\langle y_\alpha \mid \alpha \leq \beta \rangle$, such that (i) the first snapshot y_0 is the starting configuration of the computation of program p on input x; (ii) successor snapshots $y_{\alpha+1}$ are updated correctly from the prior snapshot y_α and the operation of p; (iii) limit snapshots y_ξ correctly show the head on the left-most cell in the *limit* state, with the tape updated correctly from the prior tape values in $\langle y_\alpha \mid \alpha < \xi \rangle$; and lastly, (iv) the final snapshot y_β shows that the computation halted and accepted the input. This is a computable property of $\langle p, x, y \rangle$, since one can computably verify that y codes such a well-ordered sequence of snapshots by counting through the underlying order of y and systematically checking each of the requirements. So the structure $\mathcal{R} = \langle \mathbb{R}, U_p \rangle_{p \in \mathbb{N}}$ is a computable structure. (One could reduce this to a finite language with a trinary predicate $U(p, x, y)$, by regarding programs as reals and ensuring that the programs are necessarily in any elementary substructure.)

Suppose that there is a computable proper elementary substructure $\mathcal{A} \prec \mathcal{R}$. Let p_0 be a program deciding the underlying set A of $\mathcal{A}$. Since every real $a \in A$ is accepted by p_0, there will be a real y in $\mathbb{R}$ coding the computation sequence and witnessing $U_{p_0}(a, y)$. Thus, $\mathcal{A} \models \forall a \, \exists y \, U_{p_0}(a, y)$. By elementarity $\mathcal{A} \prec \mathcal{R}$, we conclude that $\mathcal{R}$ also satisfies this assertion. So every real is accepted by p_0. Thus, $A = \mathbb{R}$ and the substructure is not a proper substructure after all.   □

Since this model is only infinite time computable and not infinite time decidable (the halting problem $0^\triangledown$ is expressible in the Σ_1 diagram), the following question remains open:

Question 31 *Is there an infinite time decidable model with underlying set $\mathbb{R}$ having no proper infinite time computable elementary substructure?*

Such a model would be a very strong counterexample to the infinite time computable downward Löwenheim–Skolem Theorem.

5 Computable quotient presentations

Recall from Definition 1 that a structure has an infinite time *computable presentation* if it is isomorphic to an infinite time computable structure. Weakening this concept slightly, let us define that a structure $\mathcal{A}$ has an infinite time computable *quotient presentation* if there is an infinite time computable structure $\mathcal{B} = \langle B, \ldots \rangle$ and an infinite time computable equivalence relation $\equiv$ on B such that $\mathcal{A}$ is isomorphic to the quotient structure $\mathcal{B}/\equiv$. In particular, $\equiv$ should be a *congruence relation* on $\mathcal{B}$, meaning that the functions and relations of $\mathcal{B}$ are well defined on the $\equiv$-equivalence classes $[b]_\equiv$ for $b \in B$, while the quotient structure $\mathcal{B}/\equiv$ consists precisely of these equivalence classes with the induced functions and relations.

Every computable structure, of course, has a computable quotient presentation, using the equivalence relation of identity. Other more elaborate and interesting quotient presentations involve nontrivial equivalence relations. The difference between the two kinds of presentation has to do with the two possibilities in first-order logic of treating $=$ as a logical symbol, insisting that it be interpreted as identity in a model, or treating it axiomatically, so that it can be interpreted merely as an equivalence relation. The natural question here, of course, is whether the two notions coincide.

Question 32 *Does every structure with an infinite time computable quotient presentation have an infinite time computable presentation?*

This is certainly true in the context of finite time computability, because one can build a computable presentation by using the least element of each equivalence class. More generally, for the same reason, it is true in the infinite time context for structures having a quotient presentation whose underlying set is contained in the natural numbers. Specifically, if $\mathcal{A} = \langle A, \ldots, \equiv \rangle$ is computable and $A \subseteq \mathbb{N}$, where $\equiv$ is a congruence on $\mathcal{A}$, then $\mathcal{A}/\equiv$ has a computable presentation. This is because the function s, mapping every $n \in A$ to the least element $s(n)$ in the equivalence class of n, is computable. To compute $s(n)$, one may simply try out all the smaller values in turn to discover the least representative. It follows that the set $B = \{ s(n) \mid n \in A \}$ is a computable choice set for the collection of equivalence classes. For any relation symbol R in the language of $\mathcal{A}$, we may now naturally define $R^{\mathcal{B}}(\vec{n}) \iff R^{\mathcal{A}}(\vec{n})$; and for any function symbol f we define $f^{\mathcal{B}}(\vec{n}) = s(f^{\mathcal{A}}(\vec{n}))$. These are clearly computable functions and relations, and since $\equiv$ is a congruence, it follows that $\mathcal{A}/\equiv$ is isomorphic to $\mathcal{B}$, as desired. This argument shows more generally that if a structure has a computable quotient presentation $\langle A, \ldots, \equiv \rangle$, and there is a computable function s mapping every element to a representative for its equivalence class, then the quotient structure $\mathcal{A}/\equiv$ has a computable presentation. (Note: the range of such a computable choice function will be decidable, because it is precisely the collection of x in the original structure for which $s(x) = x$.) Such a function s is like a computable choice function on the equivalence classes.

In the general infinite time context, of course, one does not expect necessarily to be able to effectively compute representatives from each equivalence class. In fact, we will show that the answer to Question 32 is independent of ZFC. In order to illustrate the ideas, let us begin with the simple example of the uncountable well order $\langle \omega_1, < \rangle$.

Theorem 33

1. *The uncountable well-ordered structure $\langle \omega_1, < \rangle$ has an infinite time computable quotient presentation.*

2. *It is relatively consistent with ZFC that $\langle \omega_1, < \rangle$ has no infinite time computable presentation.*

Proof. For the first claim, observe that the structure $\langle \mathrm{WO}, <, \equiv \rangle$ is an infinite time computable quotient presentation of $\langle \omega_1, < \rangle$. For any $x \in \mathrm{WO}$, the equivalence class $[x]_\equiv$ is exactly the set of reals coding the same ordinal as x, and so $\langle \mathrm{WO}, < \rangle / \equiv$ is isomorphic to $\langle \omega_1, < \rangle$, as desired.

For the second claim, observe that by forcing, one may easily collapse ω_1^L and add sufficient Cohen generic reals, so that in the forcing extension $V[G]$ we have that ω_1^L is countable and the CH fails. By Lemma 23.2, therefore, the model $V[G]$ has no computable structures of size ω_1. In particular, in $V[G]$ the structure $\langle \omega_1, < \rangle$ has no computable presentation, as desired. $\square$

Thus, it is consistent that the answer to Question 32 is negative. We turn now to the possibility of a positive answer. Let us begin with a positive answer for the specific structure $\langle \omega_1, < \rangle$.

Theorem 34 *If $\omega_1 = \omega_1^L$ (a consequence of $V = L$), then the structure $\langle \omega_1, < \rangle$ has an infinite time computable presentation.*

Proof. We already observed after Theorem 12 that $\langle \omega_1^L, < \rangle$ has a computable presentation using the L-codes for ordinals. $\square$

It seems likely that one does not really need the failure of CH in the proof of Theorem 33, and we suspect that the particular structure $\langle \omega_1, < \rangle$ has a computable presentation if and only if $\omega_1 = \omega_1^L$. That is, we suspect that the converse of Theorem 34 also holds.

Corollary 35 *The question of whether the structure $\langle \omega_1, < \rangle$ has an infinite time computable presentation is independent of ZFC.*

Proof. On the one hand, by Theorem 33 it is relatively consistent that $\langle \omega_1, < \rangle$ has no computable presentation. On the other hand, if $V = L$ or merely $\omega_1^L = \omega_1$, then $\langle \omega_1, < \rangle$ has a computable presentation. $\square$

Rather than studying just one structure, however, let us now turn to the possibility of a full positive solution to Question 32. Under $V = L$, one has a full affirmative answer.

Theorem 36 *If $V = L$, then every structure with an infinite time computable quotient presentation has an infinite time computable presentation.*

Proof. Assume $V = L$, and suppose that $\mathcal{A} = \langle A, \ldots, \equiv \rangle$ is a computable structure, where $\equiv$ is a congruence with respect to the rest of the structure. We would like to show that $\mathcal{A}/\equiv$ has a computable presentation. Our argument will be guided by the idea of building a computable presentation of $\mathcal{A}/\equiv$ by selecting the L-least representatives of each equivalence class. We will not, however, be able to do exactly this, because we may not be able to recognize that a given real is the L-least representative of its equivalence class. Instead, we will attach an escort y to every such L-least representative x of an equivalence class $[x]$, where y codes an ordinal sufficiently large to allow us computably to verify that x is the L-least representative of its equivalence class. We will then build the computable presentation out of these escorted pairs $\langle x, y \rangle$.

First, for simplicity, consider the case that $\mathcal{A}$ is a relational structure. Let B be the set of pairs $\langle x, y \rangle$ such that y is an L-code for the least ordinal α such that x is an element of L_α and L_α satisfies that x is in A, that "ω_1 exists," and that x is the L-least real that is equivalent to x. The assertions about membership in A or equivalence can be expressed in L_α using the programs that compute these relations. Note that because $L_\alpha \models$ "ω_1 exists," all the computations for reals in L_α either halt or repeat before α, and so L_α has access to the full, correct computations for the reals in L_α.

We claim that B is decidable. First, the set of L-codes is decidable. Next, given that y is the L-code of an ordinal α, we can by Theorem 7 compute a code for the whole structure L_α, and so questions of satisfaction in this structure will be decidable. Next, we can check that x is an element of L_α, and that L_α satisfies all those other properties, as desired. Checking that α is least with those properties amounts to checking that L_α thinks there is no β having an L-code that works.

Next, observe that if $\langle x, y \rangle \in B$, then x really is the L-least representative of $[x]$ in $\mathcal{A}$. The reason is that if $z \equiv x$ and z precedes x in the L order, then z would be in L_α also, where y codes α, and so L_α would know that z precedes x. And it is correct about whether $z \equiv x$, since it has the computation checking this. The point is that L_α can see x and all its L predecessors, and it knows whether they are equivalent. So L_α will be correct about whether x is the L-least representative of $[x]$.

Finally, we put a structure on B as follows. For a relation symbol R, let $R^{\mathcal{B}}(\langle x_0, y_0 \rangle, \ldots, \langle x_n, y_n \rangle)$ hold if and only if $R^{\mathcal{A}}(x_0, \ldots, x_n)$, which is computable. For each $a \in A$, there is an L-least representative x in $[a]$, and a least ordinal α large enough so that x is in L_α and L_α satisfies all those tests. If y is the L-code of α, then $\langle x, y \rangle$ will be in B. By mapping $[a]$ to $\langle x, y \rangle$, it is clear that $\mathcal{A}/\equiv$ is isomorphic to $\mathcal{B}$, providing a computable presentation.

When the language has function symbols, we define $f^{\mathcal{B}}(\langle x_0, y_0 \rangle, \ldots, \langle x_n, y_n \rangle) = \langle x, y \rangle$, where x is the L-least member of $f^{\mathcal{A}}(x_0, \ldots, x_n)$ and y is the L-code for which $\langle x, y \rangle \in B$. The point now is that since $f^{\mathcal{A}}(x_0, \ldots, x_n)$ is the result of a computation in L_α, where α is the largest of the ordinals arising from $y_0, \ldots, y_n$, with the structure L_α, we will be able to find the L-least member x of the corresponding equivalence class and the L-code y putting $\langle x, y \rangle$ into B. Thus, we will be able to compute this information from $\langle x_0, y_0 \rangle, \ldots, \langle x_n, y_n \rangle$, and so f^B is a computable function. Once again $\mathcal{A}/\equiv$ is isomorphic to B, as desired.   □

The argument does not fully use the hypothesis that $V = L$, but rather only that $A \subseteq L$, since in this case we might as well live inside L. In particular, any structure that has a computable quotient presentation using only writable reals or even accidentally writable reals has a computable presentation.

Corollary 37 *The answer to Question 32 is independent of* ZFC.

Proof. By Theorem 33, it is relatively consistent that there is a structure with a computable quotient presentation but no computable presentation. On the other hand, by Theorem 36, it is also relatively consistent that every structure with a computable quotient presentation has a computable presentation.   □

Another way to express what the argument shows is the following. Let us say that a function $f : \mathbb{R} \to \mathbb{R}$ is *semi-computable* if its graph is semi-decidable.

Theorem 38 *If $V = L$ and $\equiv$ is an infinite time computable equivalence relation on a decidable set, then there is a semi-computable function f such that $x \equiv y$ if and only if $f(x) = f(y)$. Succinctly, every computable equivalence relation on a decidable set reduces to equality via a semi-computable function.*

Proof. Suppose $\equiv$ is an infinite time decidable equivalence relation on $\mathbb{R}^L$. Let $f(u) = \langle x, y \rangle$ where x is the L-least member of the equivalence class $[u]_\equiv$ and y is the L-code of the least α such that $x \in L_\alpha \models$ "ω_1 exists." The relation $f(u) = \langle x, y \rangle$ is decidable, since given u and $\langle x, y \rangle$, we can computably verify that $u \equiv x$ and that y is the L-code of an ordinal α; if so, we can compute a code for L_α, and from this code we can check whether α is least such that $x \in L_\alpha \models$ "ω_1 exists" and x is the L-least member of its equivalence class. The structure L_α is correct about this because it has all the earlier reals in the L-order and it has the full computations determining whether they are equivalent to x. So f is semi-computable. Finally, notice that $u \equiv v$ if and only if $f(u) = f(v)$, since the value of f depended only on the equivalence classes $[u] = [v]$.   □

This observation opens up a number of natural questions for further analysis. One naturally wants to consider computable reductions, for example, rather than semi-computable reductions. What is the nature of the resulting reducibility hierarchy? To what extent does it share the features of the hierarchy of Borel equivalence relations under Borel reducibility? For starters, can one show that there is no computable reduction of the relation E_0 (eventual equality of two binary strings) to equality?

On a different topic, Theorem 36 will allow us to show that a positive answer to the following question is consistent with ZFC.

Question 39 *Does every infinite time decidable structure have an infinite time computable presentation?*

Although this question remains open, we offer two partial solutions. First, we show in Theorem 40 that when the language is particularly simple, the answer is affirmative. Second, we show in Theorem 41 that a fully general affirmative answer, for all languages, is consistent with ZFC. We do not know whether a negative answer is consistent with ZFC.

Theorem 40 *In a purely relational language, or in a language with only relation symbols plus one unary function symbol, every infinite time decidable model has an infinite time computable presentation.*

Proof. In a purely relational language, every decidable structure is already computable. So let us suppose that $\mathcal{A}$ is an infinite time decidable structure in a language with relation symbols plus one unary function symbol f. We assume that the language is computably presented, so that $\{\ulcorner f \urcorner\}$ is decidable. For each $a \in \mathcal{A}$, let a^* be the real coding the list $\langle a, \ulcorner f \urcorner, f(a), f^2(a), f^3(a), \ldots \rangle$. Let $\mathcal{A}^*$ be the set of all such a^*. This is an infinite time decidable set, because if we are given a real x coding $\langle x_0, x_1, x_2, \ldots \rangle$, we can check whether $x_0 \in \mathcal{A}$ using the fact that the underlying set of $\mathcal{A}$ is decidable; we can check whether $x_1 = \ulcorner f \urcorner$ using the decision algorithm for the language, and after this, we can check whether $x_2 = f(x_0)$, $x_3 = f(x_2)$, and so on, using the decidability of $\mathcal{A}$. So we can check whether $x = a^*$ for some a. Next, we put a structure on $\mathcal{A}^*$. For each relation symbol U of $\mathcal{A}$, define U on $\mathcal{A}^*$ by $U(a_1^*, \ldots, a_n^*)$ if and only if $U(a_1, \ldots, a_n)$. This is computable because a is computable from a^*. Next, define $f^{\mathcal{A}^*}(a^*) = (f(a))^* = \langle f(a), \ulcorner f \urcorner, f^2(a), f^3(a), f^4(a), \ldots \rangle$. The point is that this is computable from a^*, since a^* lists all this information directly. So the structure $\mathcal{A}^*$ is computable (and decidable). Since $a \mapsto a^*$ is clearly an isomorphism, this proves the theorem. $\quad\square$

If the language involves countably many unary function symbols and there is a writable real listing the Gödel codes of these function symbols, then a similar construction, using $a^* = \oplus\{\tau(a) \mid \tau \text{ is a term}\}$, would provide a computable presentation. This idea, however, does not seem to work with binary function symbols.

Theorem 41 *It is relatively consistent with ZFC that all infinite time decidable structures are infinite time computably presentable. Thus, it is consistent with ZFC that the answer to Question 39 is yes.*

Proof. Suppose that $\mathcal{A}$ is an infinite time decidable structure. Augment the language by adding a constant symbol for every element of $\mathcal{A}$, and let $\mathcal{A}^*$ be the set of all terms in this expanded language. The function symbols have their obvious interpretations and are computable; the relations have their natural interpretations and are decidable (since $\mathcal{A}$ is decidable). Define $t_1 \equiv t_2$ if $\mathcal{A} \models t_1 = t_2$. This is a computable equivalence relation, because $\mathcal{A}$ is decidable. Since $\mathcal{A}^*/\equiv$ is isomorphic to $\mathcal{A}$, we have provided an infinite time computable quotient presentation for $\mathcal{A}$. By Theorem 36, it is relatively consistent with ZFC that all such structures have a computable presentation. □

We note that in Theorem 41, the computable presentation may involve a computable translation of the language.

6 The infinite time analog of Schröder–Cantor–Bernstein–Myhill

In this section, we prove the infinite time computable analogs of the Schröder–Cantor–Bernstein Theorem and the Myhill's Theorem. With the appropriate hypotheses, as in Theorems 46 and 47, the proofs go through with a classic argument. But let us first discuss the need for careful hypotheses. The usual proofs of the Myhills Theorem and the Cantor–Schröder–Bernstein Theorem involve iteratively applying the functions in a zigzag pattern between the two sets. And one of the useful properties of computable injective functions in the classic finite time context is that their corresponding inverse functions are also automatically computable: to compute $f^{-1}(b)$, one simply searches the domain for an a such that $f(a) = b$. Unfortunately, this method does not work in the infinite time context, where we generally have no ability to enumerate effectively the domain, and indeed, there are infinite time *one-way* computable functions f, meaning that f is computable but f^{-1} is not computable. An easy example of such a function is provided by the Lost Melody Theorem 8, where we have a real c such that $\{c\}$ is decidable, but c is not writable. It follows that the function $c \mapsto 1$ on the singleton domain $\{c\}$ is computable, but its inverse is not. Building on this, we can provide a decidable counterexample to a direct infinitary computable analog of the Myhills Theorem.

Theorem 42 *In the infinite time context, there are decidable sets A and B with computable total injections $f : \mathbb{R} \to \mathbb{R}$ and $g : \mathbb{R} \to \mathbb{R}$ such that $x \in B \iff f(x) \in A$ and $x \in A \iff g(x) \in B$, but there is no computable bijection $h : A \to B$.*

Proof. Let $A = \mathbb{N}$ and $B = \mathbb{N} \cup \{c\}$, where c is the real of the Lost Melody Theorem. Define $f(c) = 0$, $f(n) = n + 1$ for $n \in \mathbb{N}$ and otherwise $f(x) = x$. Clearly, f is a computable total injection and $x \in B \iff f(x) \in A$. To help define g, for any real x (infinite binary sequence), let x^* be the real obtained by omitting the first digit, and let $x^{*(n)}$ be the real obtained by omitting the first n digits. Now let $g(c) = c^*$ and more generally $g(c^{*(n)}) = c^{*(n+1)}$, and otherwise $g(x) = x$. This function g is clearly total and injective, and it is computable because given any x, we can by adding various finite binary strings to the front of x determine whether $x = c^{*(n)}$ for some n and thereby compute $g(x)$. Since c is not periodic, we have $c \notin \mathrm{ran}(g)$ and

$x \in A \iff g(x) \in B$. Finally, there can be no computable onto map from A to B, since c is not the output of any computable function with natural number input. $\quad\square$

In this example, the function $f \restriction B$ is actually a computable bijection in the converse direction, from B to A, but this does not contradict the theorem because f^{-1} is not computable from A to B, since it maps 0 to c. What we really want in the infinitary context is not merely a computable bijection from A to B, but rather a computable bijection whose inverse is also computable, so that the relation is symmetric. The next example shows that we cannot achieve this even when we have computable bijections in both directions.

Theorem 43 *In the infinite time context, there are decidable sets A and B with computable bijections $f : A \longrightarrow B$ and $g : B \longrightarrow A$, for which there is no computable bijection $h : A \longrightarrow B$ whose inverse h^{-1} is also computable.*

To construct A and B, we first generalize the Lost Melody Theorem by recursively building a sequence of reals that can each be recognized, but not written, by an infinite time Turing machine using the preceding reals of the sequence.

Lemma 43.1 *There exists a sequence $\langle d_k \mid k \in \omega \rangle$ of reals such that*

1. *for each k, the real d_k is not writable from $\langle d_i \mid i < k \rangle$ and*

2. *there is an infinite time program that, for any z and any k, can decide on input $\langle d_0, d_1, \ldots, d_{k-1}, z \rangle$ whether $z = d_k$.*

Proof. The repeat-point of a computation is the least ordinal stage by which the computation either halts or enters a repeating loop from which it never emerges. For each $k \geq 0$, let δ_k be the supremum of the repeat-points of all computations of the form $\varphi_p(\langle d_i \mid i < k \rangle)$. Note that δ_k is countable in L. Let β_k be the smallest ordinal greater than δ_k such that $L_{\beta_k + 1} \models$ "β_k is countable." Finally, let d_k be the L-least real coding β_k. The real d_k is not writable on input $\langle d_i \mid i < k \rangle$, for if it were, then we could solve the halting problem relative to $\langle d_i \mid i < k \rangle$ by writing d_k and using it to check whether any given program halts within β_k steps on input $\langle d_i \mid i < k \rangle$. Next, on input $\langle d_0, d_1, \ldots, d_{k-1}, z \rangle$, let us explain how to determine whether $z = d_k$. We first check whether z codes an ordinal α, and if so, we simulate every computation $\varphi_p(\langle d_i \mid i < k \rangle)$ for α many steps. By inspecting these computations, we can verify that they all halt or repeat by stage α and thereby verify that $\alpha \geq \delta_k$. By Theorem 7, we can now write down a real coding $L_{\alpha+1}$ and verify that z is the L-least code for α in $L_{\alpha+1}$. If all these tests are passed, then $z = d_k$. $\quad\square$

Proof of Theorem 43. We use the sequence $\langle d_k \mid k \in \omega \rangle$ to construct a bi-infinite sequence $\langle c_k \mid k \in \mathbb{Z} \rangle$ as follows: for $k > 0$, let c_k be a real coding $\langle d_i \mid i < k \rangle$ in the usual manner, and for $k \leq 0$, let $c_k = k$. Let $A = \{c_{2k} \mid k \in \mathbb{Z}\}$ and $B = \{c_{2k+1} \mid k \in \mathbb{Z}\}$, and define bijections $f : A \to B$ by $f : c_{2k} \mapsto c_{2k-1}$ and $g : B \to A$ by $f : c_{2k+1} \mapsto c_{2k}$. It follows immediately from the definition of c_k that f and g are computable.

We next show that A is decidable. Given a real z, we first verify that either z is an even integer less than or equal to zero, in which case we accept it immediately, or else it codes a sequence $\langle z_0, \ldots z_{n-1} \rangle$ of even length, in which case we use the lemma iteratively to verify that $z_i = d_i$ for each $i < n$. Since the real z is an element of A if and only if it passes this test, A is decidable. Similarly, B is decidable.

We conclude by showing that if $h : A \longrightarrow B$ is a bijection, then h and h^{-1} cannot both be computable. From clause (1) of the lemma and the definition of c_n, it follows that for positive n, c_n cannot be written by any machine on input c_k if $k < n$. Thus, if h is computable, then $h(c_2)$ must equal c_k for some $k < 2$. But then $h^{-1}(c_k) = c_2$ so h^{-1} is not computable. $\qquad\Box$

Corollary 44 *In the infinite time context, there are decidable sets A and B and a computable permutation $\pi : \mathbb{R} \to \mathbb{R}$ such that π " $A = B$ and π " $B = A$, but there is no computable bijection $h : A \to B$ for which h^{-1} is also computable.*

Proof. Let A and B be as in the proof of Theorem 43. Since A and B are disjoint, the function $\pi = f \cup g \cup \mathrm{id}$, where we use the identity function outside $A \cup B$, is a permutation of $\mathbb{R}$. Since f and g are computable and A and B are decidable, it follows that π is computable. Since π " $A = f$ " $A = B$ and π " $B = g$ " $B = A$, the proof is completed by mentioning that Theorem 43 shows that there is no computable bijection from A to B whose inverse is also computable. $\qquad\Box$

If one assumes merely that the inverses of the injections are computable, then this is insufficient to get a computable bijection:

Theorem 45 *In the infinite time context, there are semi-decidable sets A and B with computable injections $f : A \to B$ and $g : B \to A$ whose inverses are also computable, such that there is no computable bijection $h : A \to B$.*

Proof. In fact, there will be no computable surjection from A to B. Let $A = \mathbb{N}$ be the set of all natural numbers, and let $B = 0^\triangledown = \{ p \mid \varphi_p(0) \downarrow \}$ be the infinite time halting problem. Define an injective function $f : A \to B$ by setting $f(n)$ to be the nth program on a decidable list of obviously halting programs (such as the program with n states and all transitions leading immediately to the *halt* state). The function f is clearly computable, and by design its inverse is also computable and $\mathrm{ran}(f)$ is decidable. Conversely, construing programs as natural numbers, the inclusion map $g : B \to A$ is a computable injection whose inverse is also computable, since $\mathrm{dom}(g^{-1}) = 0^\triangledown$ is semi-decidable. So we have defined the required computable injections. Suppose now that $h : A \to B$ is a computable surjection of $\mathbb{N}$ to $0^\triangledown$. In this case, an infinite time computable function could systematically compute all the values $h(0)$, $h(1)$, and so on, and thereby write $0^\triangledown$ on the tape. This contradicts the fact that $0^\triangledown$ is not a writable real. So there can be no such computable bijection from A to B. $\qquad\Box$

In the classical finite time context, of course, there is a computable bijection between $\mathbb{N}$ and $0'$ (or any infinite c.e. set), mapping each n to the nth element appearing in the canonical enumeration of it. This idea does not work in the infinitary context, however, because the infinitary halting problem 0^∇ is not computably enumerated in order type ω, but rather in the order type λ of the clockable ordinals. And λ is not a writable ordinal, so there is no way to effectively produce a real coding it.

Finally, with the right hypotheses, we prove the positive results, starting with the effective content of the Cantor–Schröder–Bernstein Theorem.

Theorem 46 *In the infinite time context, suppose that A and B are semi-decidable sets, with computable injections $f : A \to B$ and $g : B \to A$, whose inverses are computable and whose ranges are decidable. Then there is a computable bijection $h : A \to B$ whose inverse is computable.*

Proof. Let A_0 be the set of a such that there is some finite zigzag pre-image $(g^{-1}f^{-1})^k g^{-1}(a) \notin \mathrm{ran}(f)$ for $k \in \mathbb{N}$. Our hypotheses ensure that this set is infinite time decidable, since we can systematically check all the corresponding pre-images to see that when and if they stop it was because they landed outside $\mathrm{ran}(f)$ in B. The usual proof of the Cantor–Schröder–Bernstein Theorem now shows that the function $h = (g^{-1} \restriction A_0) \cup (f \restriction A \setminus A_0)$ is a bijection between A and B. Note that h is computable because g^{-1} and f are each computable, A_0 is decidable, and A is semi-decidable. To see that h^{-1} is computable, let $B_0 = g^{-1}A_0$ and observe that $h^{-1} = (g \restriction B_0) \cup (f^{-1} \restriction B \setminus B_0)$. Since these components are each computable, h^{-1} is computable and the proof is complete.   □

We may drop the assumption that A and B are semi-decidable if we make the move to total functions, as in the classical Myhill Theorem. Define that a set of reals A is *reducible* to another set B by the function $f : \mathbb{R} \to \mathbb{R}$ if $x \in A \iff f(x) \in B$.

Theorem 47 *In the infinite time context, suppose that A and B are reducible to each other by computable one-to-one total functions f and g, whose inverses are computable and whose ranges are decidable. Then there is a computable permutation $\pi : \mathbb{R} \to \mathbb{R}$ with π^{-1} also computable and $\pi \; '' A = B$.*

Proof. As in Theorem 46, let A_0 be the set of a such that some finite zigzag pre-image $(g^{-1}f^{-1})^k g^{-1}(a) \notin \mathrm{ran}(f)$ for $k \in \mathbb{N}$, and again this is decidable. Let $\pi = (g^{-1} \restriction A_0) \cup (f \restriction \mathbb{R} \setminus A_0)$. The usual Cantor–Schröder–Bernstein argument shows that this is a permutation of $\mathbb{R}$. As above, both π and π^{-1} are computable. Finally, we have both $x \in A \iff \pi(x) \in B$ and $x \in B \iff \pi(x) \in A$, since $\pi(x)$ is either $f(x)$ or $g^{-1}(x)$, both of which have the desired properties. It follows that $\pi \; '' A = B$ and the proof is complete.   □

7 Some infinite time computable transitive models of set theory

Because the power of the machines are connected intimately with well-orders and countable ordinals, it is not surprising that there are many interesting models of a set theoretic nature. We have already seen that the hereditarily countable sets have an infinite time computable quotient presentation $\langle \mathrm{HC}, \in, \equiv \rangle$. In addition, we have provided infinite time computable presentations of the model $\langle L_{\omega_1^L}, \in \rangle$ and of $\langle L_\alpha, \in \rangle$, given a real coding α. In this section we will show, however, that depending on the set theoretic background, one can transcend these, by actually producing infinite time decidable presentations of *transitive* models of ZFC, or even ZFC plus large cardinals. Each of these presentations, however, will involve a somewhat strange manner of coding information into the individual elements of the model or into the language, even while the model and language technically remains computable. We begin by proving that the task is impossible without such subterfuge.

Theorem 48 *There is no infinite time computable presentation of a transitive model of* ZFC *with underlying set* $\mathbb{N}$ *and Gödel codes of the language entirely in* $\mathbb{N}$.

Proof. The operation of an infinite time Turing machine is absolute to any transitive model of ZFC containing the input. Thus, all transitive models of ZFC agree on the elements of the halting problem $0^\triangledown$. If $\mathcal{M} = \langle \mathbb{N}, E \rangle$ is a computable presentation of such a model, then there is some natural number k representing $0^\triangledown$ in $\mathcal{M}$. Assuming that $\ulcorner \in \urcorner$ is writable, then we can computably determine for each natural number p the element k_p representing it in $\mathcal{M}$. In this case, we could compute $0^\triangledown = \{ p \mid k_p \, E \, k \}$, contradicting the fact that $0^\triangledown$ is not computable. $\square$

Of course, this argument uses much less than ZFC. It shows that there can be no computable presentation, using underlying set $\mathbb{N}$ and writable presentation of the language, of a transitive model computing $0^\triangledown$ correctly. For example, it would be enough if the model satisfied "ω_1 exists," or even less, that every infinite time Turing computation either halted or reached its repeat point.

Despite Theorem 48, however, computable models and languages are not in general limited to the domain $\mathbb{N}$, and in this general setting we can actually find computable presentations of transitive, well-founded models of ZFC.

Theorem 49 *If there is a transitive model of* ZFC, *then the smallest transitive model of* ZFC *has an infinite time decidable computable presentation.*

Proof. If there is a transitive model of ZFC, then there is one satisfying $V = L$. A Löwenheim–Skolem argument, followed by the Mostowski collapse, shows that there must be a countable such model, and any such model will be L_α for some countable ordinal α. By minimizing α, we see in this case that there is a smallest transitive model $L_\alpha \models$ ZFC. Let c be the L-code for this minimal α. Note that $\{ c \}$ is decidable, since on input x, we can check whether it is an L-code for an ordinal ξ such that $L_\xi \models$ ZFC and, if so, whether ξ is the smallest such ordinal.

If so, it must be that $\xi = \alpha$ and $x = c$. From c, we may compute a relation E on $\mathbb{N}$ such that $\langle L_\alpha, \in \rangle \cong \langle \mathbb{N}, E \rangle$. Let M be the collection of pairs $\langle c, n \rangle$, where $n \in \mathbb{N}$. This is a decidable set, because $\{c\}$ is decidable. The idea is that $\langle c, n \rangle$ represents the set coded by n with respect to E. Define $\langle c, n \rangle$ $\bar{E}$ $\langle c, m \rangle$ if n E m. This is a computable relation, because E is c-computable and $\{c\}$ is decidable. Clearly, $\langle M, \bar{E} \rangle$ is isomorphic to $\langle \mathbb{N}, E \rangle$, which is isomorphic to $\langle L_\alpha, \in \rangle$. So we have a computable presentation of $\langle L_\alpha, \in \rangle$.

Let us point out that this presentation is nearly decidable, in that we can decide $\mathcal{M} \models \varphi[x_1, \ldots, x_n]$ on input $\ulcorner \varphi \urcorner, x_1, \ldots, x_n$, provided $n \geq 1$. Specifically, from x_1, we can compute the real c, and from c, we can enumerate effectively the whole structure $\langle M, E \rangle$. Having done so, we can compute whether $\mathcal{M} \models \varphi[x_1, \ldots, x_n]$ according to the Tarskian definition of truth. This method makes fundamental use of the information c that is present in any of the parameters, so it does not help us to decide whether a given sentence holds in $\mathcal{M}$, if we are not given such a parameter.

To make the model fully decidable, therefore, we assume $\ulcorner \in \urcorner = c$. Since $\{c\}$ is decidable, this language remains decidable (although no longer enumerable in any nice sense). The point now is that if we are given a sentence σ, and the symbol $\in$ appears in it, then we can compute the real c from $\ulcorner \in \urcorner$ and thereby once again enumerate effectively the whole structure $\mathcal{M}$, allowing us to compute whether σ holds. If $\in$ does not occur in σ, then σ is an assertion in the language of equality, which either holds or fails in all infinite models, and we can determine computably this in $\omega + 1$ many steps.   □

If one allows $\ulcorner \in \urcorner = c$, then one can actually take the underlying set of $\mathcal{M}$ to be $\mathbb{N}$, since if one has already coded c into the language, there is no additional need to code c into the individual elements of $\mathcal{M}$. In this case, one has a decidable presentation of the form $\langle \mathbb{N}, E \rangle$. We caution in this case that the relation E is not computable but only computable relative to c. This does not prevent the model from being a computable model, however, since in order to be a computable model, the relations need only be computable from their Gödel codes. This may be considered to be a quirk in the definition of computable model, but in order to allow for uncountable languages, we cannot insist that the relations of a computable model are individually computable but rather only computable from their Gödel codes.

Similar arguments establish:

Theorem 50 *If there is a transitive model of ZFC with an inaccessible cardinal (or a Mahlo cardinal or ω^2 many weakly compact cardinals, etc.), then the smallest such model has an infinite time decidable presentation.*

Proof. If there is a transitive model of ZFC plus any of these large cardinal hypotheses, then there is one satisfying $V = L$. Hence, as argued in Theorem 48, the theory holds in some countable L_α. By using the L-code c of the minimal such model, we can build a decidable presentation as above.   □

If one wants to consider set theoretic theories inconsistent with $V = L$, then a bit more care is needed.

Theorem 51 *If there is a transitive model of* ZFC, *then there is a transitive model of* ZFC $+ \neg$CH *with an infinite time decidable computable presentation.*

Proof. Let L_α be the minimal transitive model of ZFC. This is a countable transitive model, and so there is an L-least set G in L such that G is L_α-generic for the forcing $\text{Add}(\omega, \omega_2)^{L_\alpha}$. Thus, $L_\alpha[G] \models \text{ZFC} + \neg$CH. The set G appears in some countable L_β, where $\alpha < \beta < \omega_1$. Let d be the L-code of the pair $\langle \alpha, \beta \rangle$. Thus, $\{ d \}$ is decidable, because given any real z, we can check whether z is an L-code for a pair $\langle \alpha', \beta' \rangle$ such that $L_{\alpha'}$ is the smallest model of ZFC and β' is smallest such that $L_{\beta'}$ has an $L_{\alpha'}$-generic filter G for $\text{Add}(\omega, \omega_2)^{L_{\alpha'}}$. Using the real d, we can compute a relation E on $\mathbb{N}$ such that $\langle L_\alpha[G], \in \rangle \cong \langle \mathbb{N}, E \rangle$. Let N be the set of pairs $\langle d, n \rangle$ where $n \in \mathbb{N}$, and define $\langle d, n \rangle \ \bar{E} \ \langle d, m \rangle$ if $n \ E \ m$. Again, this structure is computable, and it is isomorphic to $\langle L_\alpha[G], \in \rangle$, as desired. By taking $\ulcorner \in \urcorner = d$, the model is decidable as in Theorem 49. $\qquad\square$

Clearly this method is very flexible; it provides decidable presentations of transitive models of any theory having a transitive model in L.

8 Future directions

We close this paper by mentioning a number of topics for future research.

Infinitary languages $L_{\omega_1, \omega}$. In the context of infinite time computable model theory, it is very natural to consider infinitary languages, which are still easily coded into the reals. With any writable structure or for a structure whose domain we can search, one can still compute the Tarskian satisfaction relation. What other examples and phenomenon exist here?

Infinite time computable equivalence relation theory. The idea is to investigate the analog of the theory of Borel equivalence relations under Borel reducibility. Here, one wants to consider infinite time computable reductions. Some of these issues are present already in our analysis of the computable quotient presentation problem in Section 5 and particularly Theorem 38. How much of the structure of Borel equivalence relations translates to the infinite time computable context?

Infinite time computable cardinalities. The computable cardinalities are the equivalence classes of the decidable sets by the computable equinumerousity relation. What is the structure of the computable cardinalities?

Infinite time computable Löwenheim–Skolem theorems. Although Theorem 28 shows that the infinite time computable upward Löwenheim–Skolem Theorem holds in L, our analysis leaves open the question of whether it is consistent with ZFC that there could be a decidable countable model having no size continuum decidable elementary extension. If so, the infinite time, computable, upward Löwenheim–Skolem Theorem will be independent of ZFC. In addition, our analysis does not fully settle the infinite time, computable, downward Löwenheim–Skolem Theorem.

References

1. Vinay Deolalikar, Joel David Hamkins, and Ralf-Dieter Schindler. $P \neq NP \cap \text{co-}NP$ for infinite time turing machines. *Journal of Logic and Computation*, 15(5):577–592, 2005.
2. Yuri L. Ershov, Sergey S. Goncharov, Anil Nerode, and Jeffrey B. Remmel, editors. *Handbook of Recursive Mathematics, Volume 1: Recursive Model Theory*, volume 138 of *Studies in Logic and the Foundations of Mathematics*. Elsevier, 1998.
3. Joel David Hamkins. Infinite time turing machines. *Minds and Machines*, 12(4):521–539, 2002. (special issue devoted to hypercomputation).
4. Joel David Hamkins. Infinitary computability with infinite time Turing machines. In Barry S. Cooper and Benedikt Löwe, editors, *New Computational Paradigms*, volume 3526 of *LNCS*, Amsterdam, June 8-12 2005. CiE, Springer-Verlag.
5. Joel David Hamkins and Andy Lewis. Infinite time Turing machines. *J. Symbolic Logic*, 65(2):567–604, 2000.
6. Joel David Hamkins and Andy Lewis. Post's problem for supertasks has both positive and negative solutions. *Archive for Mathematical Logic*, 41(6):507–523, 2002.
7. Joel David Hamkins and Daniel Seabold. Infinite time Turing machines with only one tape. *Mathematical Logic Quarterly*, 47(2):271–287, 2001.
8. Joel David Hamkins and Philip Welch. $P^f \neq NP^f$ for almost all f. *Mathematical Logic Quarterly*, 49(5):536–540, 2003.
9. Thomas Jech. *Set Theory*. Springer Monographs in Mathematics, 3rd edition, 2003.
10. Peter Koepke. Turing computations on ordinals. *Bulletin of Symbolic Logic*, 11(3):377–397, 2005.
11. Benedikt Löwe. Revision sequences and computers with an infinite amount of time. *Logic Comput.*, 11(1):25–40, 2001.
12. Philip Welch. Eventually infinite time Turing machine degrees: Infinite time decidable reals. *Journal of Symbolic Logic*, 65(3):1193–1203, 2000.
13. Philip Welch. The lengths of infinite time Turing machine computations. *Bulletin of the London Mathematical Society*, 32(2):129–136, 2000.
14. Philip Welch. The transfinite action of 1 tape Turing machines. In Barry S. Cooper and Benedikt Löwe, editors, *New Computational Paradigms*, volume 3526 of *LNCS*, Amsterdam, June 8-12 2005. CiE, Springer-Verlag.

Index